Speech Processing

Signal Processing and Communications

1. Digital Signal Processing for Multimedia Systems, *edited by Keshab K. Parhi and Takao Nishitani*
2. Multimedia Systems, Standards, and Networks, *edited by Atul Puri and Tsuhan Chen*
3. Embedded Multiprocessors: Scheduling and Synchronization, *Sundararajan Sriram and Shuvra S. Bhattacharyya*
4. Signal Processing for Intelligent Sensor Systems, *David C. Swanson*
5. Compressed Video over Networks, *edited by Ming-Ting Sun and Amy R. Reibman*
6. Modulated Coding for Intersymbol Interference Channels, *Xiang-Gen Xia*
7. Digital Speech Processing, Synthesis, and Recognition: Second Edition, Revised and Expanded, *Sadaoki Furui*
8. Modern Digital Halftoning, *Daniel L. Lau and Gonzalo R. Arce*
9. Blind Equalization and Identification, *Zhi Ding and Ye (Geoffrey) Li*
10. Video Coding for Wireless Communication Systems, *King N. Ngan, Chi W. Yap, and Keng T. Tan*
11. Adaptive Digital Filters: Second Edition, Revised and Expanded, *Maurice G. Bellanger*
12. Design of Digital Video Coding Systems, *Jie Chen, Ut-Va Koc, and K. J. Ray Liu*

13. Programmable Digital Signal Processors: Architecture, Programming, and Applications, *edited by Yu Hen Hu*
14. Pattern Recognition and Image Preprocessing: Second Edition, Revised and Expanded, *Sing-Tze Bow*
15. Signal Processing for Magnetic Resonance Imaging and Spectroscopy, *edited by Hong Yan*
16. Satellite Communication Engineering, *Michael O. Kolawole*
17. Speech Processing: A Dynamic and Optimization-Oriented Approach, *Li Deng and Douglas O'Shaughnessy*

Additional Volumes in Preparation

Multidimensional Discrete Unitary Transforms: Representation, Partitioning, and Algorithms, *Artyom M. Grigoryan*

Domain-Specific Embedded Multiprocessors: Systems, Architectures, Modeling, and Simulation, *Shuvra Bhattacharyya, Ed Deprettere, and Jürgen Teich*

High-Resolution and Robust Signal Processing, *Yingbo Hua, Alex Gershman, and Qi Cheng*

Speech Processing

A Dynamic and Optimization-Oriented Approach

LI DENG
Microsoft Research
Redmond, Washington, U.S.A.

DOUGLAS O'SHAUGHNESSY
INRS Energy, Materials, and Telecommunications
University of Quebec
Montreal, Quebec, Canada

MARCEL DEKKER, INC. NEW YORK • BASEL

Library of Congress Cataloging-in-Publication Data
A catalog record for this book is available from the Library of Congress.

ISBN: 0-8247-4040-8

This book is printed on acid-free paper.

Headquarters
Marcel Dekker, Inc., 270 Madison Avenue, New York, NY 10016, U.S.A.
tel: 212-696-9000; fax: 212-685-4540

Distribution and Customer Service
Marcel Dekker, Inc., Cimarron Road, Monticello, New York 12701, U.S.A.
tel: 800-228-1160; fax: 845-796-1772

Eastern Hemisphere Distribution
Marcel Dekker AG, Hutgasse 4, Postfach 812, CH-4001 Basel, Switzerland
tel: 41-61-260-6300; fax: 41-61-260-6333

World Wide Web
http://www.dekker.com

The publisher offers discounts on this book when ordered in bulk quantities. For more information, write to Special Sales/Professional Marketing at the headquarters address above.

Current printing (last digit):

10 9 8 7 6 5 4 3 2 1

PRINTED IN THE UNITED STATES OF AMERICA

To our families,
students, and colleagues

SERIES INTRODUCTION

Over the past 50 years, digital signal processing has evolved as a major engineering discipline. The fields of signal processing have grown from the origin of fast Fourier transform and digital filter design to statistical spectral analysis and array processing, image, audio, and multimedia processing, and shaped developments in high-performance VLSI signal processor design. Indeed, there are few fields that enjoy so many applications—signal processing is everywhere in our lives.

When one uses a cellular phone, the voice is compressed, coded, and modulated using signal processing techniques. As a cruise missile winds along hillsides searching for the target, the signal processor is busy processing the images taken along the way. When we are watching a movie in HDTV, millions of audio and video data are being sent to our homes and received with unbelievable fidelity. When scientists compare DNA samples, fast pattern recognition techniques are being used. On and on, one can see the impact of signal processing in almost every engineering and scientific discipline.

Because of the immense importance of signal processing and the fast-growing demands of business and industry, this series on signal processing serves to report up-to-date developments and advances in the field. The topics of interest include but are not limited to the following:

- Signal theory and analysis
- Statistical signal processing
- Speech and audio processing
- Image and video processing
- Multimedia signal processing and technology
- Signal processing for communications
- Signal processing architectures and VLSI design

We hope this series will provide the interested audience with high-quality, state-of-the-art signal processing literature through research monographs, edited books, and rigorously written textbooks by experts in their fields.

PREFACE

Aims and intended audience

This book was written to serve four main purposes, categorized according to the potential readers of the book. First and principally, the book has been designed and organized as a text for a graduate level (or senior undergraduate level) course for students in Electrical and Computer Engineering, Systems Design, Computer Science, Applied Mathematics or Physics, Cognitive Science, or Linguistics and Phonetics (computation oriented). The materials contained in this book have grown out of our lecture notes prepared and used since 1990, with continual updating, in teaching the courses entitled "Digital Speech Processing" and "Speech Communication," offered formally to the graduate and senior undergraduate students at the University of Waterloo and at McGill University, respectively. Most of the students specialized in Communication, Signal Processing, and Information Systems in the respective Departments of Electrical and Computer Engineering. While writing this book, serious effort was made to extract and reorganize our earlier lecture notes, with the aim of providing earnest students with a systematic textbook tool to acquire comprehensive knowledge and to build a solid intellectual foundation in both speech science and speech technology. In particular, the materials included in the book have been chosen and organized to underscore the importance of analytical skills in student training that we believe are essential for them to meet future challenges in scientific and technological advances in the field upon graduation.

Second, this book is intended also for professionals and engineering practitioners who are nevertheless newcomers to some specific fields of speech technology and who desire to acquire necessary background knowledge. Research and technology in speech processing have advanced significantly over the last decade and are continuing to make remarkable progress. The trend is expected to continue for many years to come as computing technology continues to advance. This trend should be particularly strong since more and more applications, such as those in telecommunications and in multimedia information systems, enabled by voice technology, are fast emerging from laboratories to the real world. Because many researchers and engineers are likely to find themselves involved in the area of speech processing for the first time, we have structured the book to contain sufficient background information related to speech but also to provide pointers to skipping certain mathematical materials for those who are already equipped with mathematical backgrounds based on training in engineering disciplines other than speech technology.

Third, this book can also serve as a reference book for seasoned researchers and engineers in speech processing. In this book, we have emphasized the notions of mathematical abstraction and of engineering optimization that are used to unify many seemingly distinct approaches to problem solving in speech processing. The generality of the many common methods described in this book will enable readers to advance their understanding of speech processing theory. Researchers with limited experience may use this book as an introduction to the field. Those with more experience in speech processing may find in this book useful techniques and solutions to specific problems of interest. More advanced readers may explore further research topics based on the platform provided in this book.

Finally, the multidisciplinary nature of speech research provides an incentive for a book that, in an integrated view and in a balanced fashion, is capable of covering several rather distinct disciplinary areas that jointly underpin the field of speech processing.

During our years of teaching and research, we have found that students or researchers specializing in one or more of these multidisciplinary areas often have a very strong desire to know other less familiar subareas in speech processing based on the expertise of their own specialized subarea(s). For example, electrical engineers or computer scientists who are doing advanced research on speech recognition often are interested in the true linguistic and phonetic nature of the speech signal they are dealing with everyday, and are interested in how humans perform speech recognition and why their performance is so much better than that of current computers. On the other hand, phoneticians, phonologists, speech production and perception researchers (many of them engineers by training), and linguistics-oriented cognitive scientists often wonder whether their observations and abstract theories can be put to the test and can gain support by comprehensive computer models, systems, and practical performances related to automatic, large-scale speech processing. One important objective that we kept in mind while writing this book was to contribute to bridging the gap among the various traditionally distinct disciplines, all of which nevertheless are dealing with the identical physical entity of human speech yet with drastically different disciplinary traditions in conjunction with drastically different habitual approaches. Moreover, since in this book many problems of probabilistic inference and statistical learning (two types of estimation issues) are treated in a unified view in the context of speech modeling, researchers from statistics (e.g., time series analysis, data analysis, and computer-intensive methods, etc.), neural networks, pattern recognition, machine learning, and artificial intelligence can also benefit from reading this book.

Scope of the book

The main body of this book contains four parts, which have been organized into a total of 14 chapters. A summary of the book's subjects is provided below:

- **PART I**

  Chapters 1-6 cover mathematical background and signal processing techniques that are necessary for speech analysis at the sampled-waveform level and for the remaining materials in the book focusing on statistical modeling at higher levels of the speech process. Throughout this book, we deal almost exclusively with discrete-time, as opposed to continuous-time, speech signals. This decision was made based on the computation-oriented nature of the book (and on the belief that speech technology advancement would lose ground without close ties to digital-computing technology). Conversion from continuous-time signals to the discrete-time counterpart is covered in the very beginning of the book, where rigorous treatment is made on the popular sampling theorem. Other essential concepts and techniques in digital signal processing (DSP), including signal transformations, digital filter characterization and analysis, frequency responses of digital filters, and stationary and nonstationary frequency-domain signal analysis, are covered in the remainder of Chapter 1. The material in Chapter 2 encompasses major topics in speech analysis, a conventional subarea of speech processing that exemplifies much of the natural applications of the DSP techniques covered in Chapter 1.

  While the first two chapters deal with analysis and processing of deterministic signals, the remaining chapters in Part I cover statistical aspects of mathematical background. Basics of probability theory are covered in Chapter 3, which are used subsequently in Chapter 4 to describe the canonical forms of two very widely used signal models in statistical signal processing: the static linear model and the dynamic system model. This is followed by Chapter 5, which covers major results in estimation theory and associated optimization methods. Estimation theory is one core topic in this book since the techniques reviewed here permeate through most areas of speech technology presented in Part IV. In particular, the training and decoding problems in speech recognition have been cast as two aspects of the estimation problem, parameter estimation and state estimation. The last chapter of Part I introduces concepts, main results, and techniques in statistical pattern recognition. Results described in this chapter are used in formulating decision rules for speech recognition.

- **PART II**

The two chapters contained in this part provide fundamentals of speech science, focusing on scientific/linguistic findings and principles governing the nature of human speech communication and processing. Chapters 7 and 8 are devoted, respectively, to two rather distinct disciplines in speech science by tradition — those pertinent to the phonetic and phonological processes of human speech. The phonetic process pertains to the low-level, physical aspects of speech with continuous, numeric attributes. Three main subfields of phonetics are covered in Chapter 7: articulatory, acoustic, and auditory phonetics, together with major theories and models in phonetics. (In Part II, we have used the word "model" in a rather narrow sense to mean a mental or computational abstraction of the speech process for the purpose of scientific understanding of human speech; this contrasts against the "computational model" used in most places in Parts I, III and IV.) The phonological process pertains to high-level, structural aspects and sound patterns of speech and language. It is symbolic in nature and is often referred to as the organizational units of speech. A range of such units, from the atomic units such as features to the large, external organizational units such as syllable and other prosodic units, are discussed in Chapter 8.

Both the phonetic and the phonological processes, as well as their interface, embody speech production and perception as an integrated, closed-loop system. The dynamic nature of such a system receives special attention descriptively in the chapters constituting Part II.

- **PART III**

The three chapters in this part represent a significant transition from speech science (Part II) to speech technology (Part IV); that is, a transition from human processing of speech to computer processing of speech. A key step for a successful transition of such a kind is to firmly establish computational models of the speech process. By **computational model**, we mean mathematical abstraction of the true physical process, with necessary simplification for mathematical tractability, so as to make such abstraction amenable to computer implementation for useful purposes. We have tried to use this definition as consistently as possible throughout the entire book. More specifically, computational models of the speech process are the mathematical abstraction of the phonological (symbolic) and phonetic (numeric) aspects of human speech, which can be subject to discrete-time computation for purposes of computer speech recognition and understanding, synthesis, analysis, enhancement, coding, speaker recognition, and so on.

Chapter 9 is devoted to computational models for the symbolic, phonological process, while Chapters 10 and 11 treat two major aspects of computational phonetics.

Special treatment is made of a hierarchy of statistical models of speech dynamics, including the hidden Markov model, explicitly defined trajectory model, and recursively defined trajectory model (state-space model). These models are placed under the heading of computational models for speech production, because all of them can be classified into one of the three levels of human speech production for which the dynamic properties of the speech process are directly represented in the models. The three levels include the "task" level where the goal of speech production is defined, the articulatory level which constitutes the implementation system for speech production, and the acoustic level which gives the final outcome of speech production. The statistical formulation of these computational models allows them to be easily subject to computational optimization (such as training and decoding guided by Estimation Theory). As a result, these computational models have found useful applications in various areas of speech technology. In presenting these models in Chapter 10, which forms another core topic of this book, emphasis is placed on the tradeoff between model complexity and the degree to which the model is accurately (or inaccurately) approximating realistic dynamic properties of the speech process (its phonetic aspects in particular).

- **PART IV**

The final three chapters constitute selected areas in speech technology, which bear fruits from the materials of all previous chapters building up the mathematical, scientific, and computational foundations. The emphasis is on technology manifestation of speech science and of the associated computational frameworks. Several case studies are included in the chapters of Part IV to detail the speech technology applications.

Chapter 12 is dedicated to speech recognition, the most important area of speech technology and the final core topic of this book. This chapter is focused on model optimization and system design aspects of speech recognition, while placing the signal processing or preprocessing portion more appropriately, in our judgment, in Chapter 2 on speech analysis. The roles of phonetics and phonology in speech recognition are elaborated on carefully, and are put into as balanced a perspective as possible. The specific topics covered in this chapter include the mathematical formulation and general statistical paradigm of speech recognition, various kinds of statistical models for acoustic modeling of speech, Bayesian-oriented robust techniques for acoustic modeling and recognizer design, and statistical language modeling, which provides prior constraints for speech recognition.

Chapter 13 covers speech enhancement, or noise reduction, another important area in speech technology. Basic principles are described, with an emphasis on the key roles that dynamic modeling and optimization play in the design of the enhancement algorithms and systems. Major approaches in speech enhancement are presented, including spectral subtraction, Wiener filtering, and use of statistical models as prior information for speech enhancement. Important applications of speech enhancement for robust speech recognition are also discussed in this chapter. The techniques presented have been characterized by active use of speech dynamic information and by use of various kinds of optimization criteria. At the heart of these techniques is statistical optimization, which is made possible because of the statistical formulation of the speech enhancement problem, based on various statistical models established to describe speech dynamics.

Contents

Part I

ANALYTICAL BACKGROUND AND TECHNIQUES

Chapter 1

Discrete-Time Signals, Systems, and Transforms

In this chapter, the basic principles for analyzing general discrete-time signals and systems will be reviewed. These principles underlie most of the speech analysis techniques to be discussed in Chapter 2, and form a basis for most speech technology applications. Although physical signals such as speech are continuous in time, i.e., they vary at an uncountably infinite number of times $t \in \mathcal{R}$, we in this book largely restrict ourselves to discrete-time speech signals. This is because most applications of speech processing involve use of digital computers, whose operation requires that continuous signals be converted to discrete-number sequences. The computation-oriented nature of speech processing and the close ties of speech processing techniques to digital-computing technology warrant an adequate coverage of the basics of **digital signal processing** (DSP), to which this chapter is devoted.

As a short introduction, DSP concerns filtering, transforming, modifying, or processing of signals based on a set of techniques and algorithms carried out by a digital computer or processor. Comprehensive coverage of the entire DSP field can be found in numerous textbooks [Opp 99, Porat 97, Mitra 98, Lim 88]. Briefly, a real-time DSP system, including a digital speech processing system, typically contains the following components: input low-pass filter (anti-aliasing filter), analogue-to-digital converter, digital computer or digital signal processor, digital-to-analogue converter, and output low-pass filter (anti-imaging filter). An understanding of such a system and its components firstly requires adequate knowledge of conversion from continuous-time signals to the discrete-time counterpart, which is covered at the beginning of this chapter. Other common techniques in DSP useful in speech processing, including discrete-time signal transformations, filterings, linear filter characterization and analysis, and time-varying spectral analysis, are covered in the remainder of this chapter.

1.1 Signal Sampling

Signals encountered in real-world applications such as speech, audio, image, video, etc., are usually in continuous time. The process of converting a continuous-time signal to a sequence of numbers is called **sampling**. Signal sampling essentially is a selection of a finite number of data points at a finite time interval to represent the infinite amount of data points that are contained in the continuous-time signal within the same interval. A very familiar example of signal sampling is a motion picture where a continuously varying scene is converted by the camera to a sequence of "discrete-time" frames. Sampling is a fundamental concept and operation in DSP.

1.1.1 Sampling basics

Let a deterministic signal $x(t)$ be a continuous-time function, where time t is real, or $t \in \mathcal{R}$. Sampling of $x(t)$ is a process of picking the values of $x(t)$ at certain time points. In particular, when the sampled time points are $nT_0, n \in \mathcal{N}$ (integer set), we have **uniform sampling**, and T_0 is called the **sampling interval**. The inverse of the sampling interval is called the **sampling frequency** or sampling rate $f_0 = \frac{1}{T_0}$. The **angular sampling frequency** is $\omega_0 = 2\pi f_0 = \frac{2\pi}{T_0}$. The sampled continuous-time signal is called the **discrete-time signal**, or **sequence**:

$$x[n] = x(nT_0), \quad n \in \mathcal{N}.$$

An alternative view of the above *point sampling* operation is to express the sampled signal as a new continuous-time signal, $x_p(t)$, constructed by *impulse sampling* according to

$$x_p(t) = x(t)[\sum_{n=-\infty}^{\infty} \delta(t - nT_0)] = \sum_{n=-\infty}^{\infty} x(t)\delta(t - nT_0) = \sum_{n=-\infty}^{\infty} x(nT_0)\delta(t - nT_0).$$

The above impulse sampling essentially amounts to multiplying the continuous-time signal by a pulse (δ function) train with the period of T_0: $\sum_{n=-\infty}^{\infty} \delta(t - nT_0)$.

1.1.2 Sampling theorem

Let the continuous-time signal $x(t)$ have a (continuous-time) **Fourier transform** defined by

$$X(\omega) = \int_{-\infty}^{\infty} x(t) \exp(-j\omega t).$$

The inverse Fourier transform is given by

$$x(t) = \frac{1}{2\pi} \int_{-\infty}^{\infty} X(\omega) \exp(j\omega t) d\omega, \quad t \in \mathcal{R}.$$

Likewise, the Fourier transform of the impulse-sampled signal $x_p(t)$ is

$$X_p(\omega) = \int_{-\infty}^{\infty} x_p(t) \exp(-j\omega t) dt, \quad \omega \in \mathcal{R}. \tag{1.1}$$

We now define the **discrete-time Fourier transform** for the point-sampled, discrete-time signal $x[n]$ to be

$$X^{dt}(\theta) = \sum_{n=-\infty}^{\infty} x[n] \exp(-j\theta n), \quad \theta \in \mathcal{R}. \tag{1.2}$$

The discrete-time inverse Fourier transform, which will recover the time-domain signal from its Fourier transform, can be shown to be

$$x[n] = \frac{1}{2\pi}\int_{-\pi}^{\pi} X^{dt}(\theta)\exp(j\theta n)d\theta, \quad n \in \mathcal{N}. \tag{1.3}$$

The relationship among the three separate Fourier transforms in Eqs. 1.1.2, 1.1, and 1.2 is addressed by the following celebrated **sampling theorem** credited to Nyquist.

Sampling Theorem: The discrete-time Fourier transform of a sampled signal is a summation of infinite replicas of the Fourier transform of the original continuous-time signal according to

$$X^{dt}(\theta) = \frac{1}{T_0}\sum_{k=-\infty}^{\infty} X\left(\frac{\theta - 2\pi k}{T_0}\right), \quad -\infty < \theta < \infty \tag{1.4}$$

where each replica is shifted horizontally by an integer multiple of the angular sampling frequency $\omega_0 = \frac{2\pi}{T_0}$.

Proof: Using the inverse Fourier transform Eq. 1.1.2 at a sampling time $t = nT_0$, we have

$$x(nT_0) = \frac{1}{2\pi}\int_{-\infty}^{\infty} X(\omega)\exp(j\omega nT_0)d\omega.$$

This integral can be decomposed into an infinite sum, of $2\pi/T_0$ length each, to give

$$x(nT_0) = \frac{1}{2\pi}\sum_{k=-\infty}^{\infty}\int_{-(2k-1)\pi/T_0}^{-(2k+1)\pi/T_0} \exp(j\omega nT_0)d\omega.$$

Using variable substitution $\omega = \omega' - \frac{2\pi k}{T_0}$, this becomes

$$\begin{aligned} x(nT_0) &= \frac{1}{2\pi}\sum_{k=-\infty}^{\infty}\int_{-\pi/T_0}^{\pi/T_0} X\left(\omega' - \frac{2\pi k}{T_0}\right)\exp\left[j\left(\omega' - \frac{2\pi k}{T_0}\right)nT_0\right]d\omega' \\ &= \frac{1}{2\pi}\sum_{k=-\infty}^{\infty}\int_{-\pi/T_0}^{\pi/T_0} X(\omega' - \frac{2\pi k}{T_0})\exp(j\omega' nT_0)d\omega', \end{aligned}$$

where we used the property

$$\exp\left[j(\omega' - \frac{2\pi k}{T_0})nT_0\right] = \exp\left[j\omega' nT_0\right]\ \exp\left[j\frac{2\pi k}{T_0}nT_0\right] = \exp[j\omega' nT_0]\ \exp[j2\pi kn] = \exp[j\omega' nT_0].$$

Using another variable substitution $\omega' = \frac{\theta}{T_0}$ and exchanging summation and integration, we obtain

$$x(nT_0) = \frac{1}{2\pi}\int_{-\pi}^{\pi}\left[\frac{1}{T_0}\sum_{k=-\infty}^{\infty} X\left(\frac{\theta - 2\pi k}{T_0}\right)\right]\exp(j\theta n)d\theta.$$

Comparing the above with the discrete-time inverse Fourier transform Eq. 1.3, we see that the quantity inside the brackets is the discrete-time Fourier transform of $x(nT_0)$ (i.e., $x[n]$); that is,

$$\left[\frac{1}{T_0}\sum_{k=-\infty}^{\infty} X\left(\frac{\theta - 2\pi k}{T_0}\right)\right] = X^{dt}(\theta).$$

The sampling theorem, Eq. 1.4, provides a practical way of computing the discrete-time Fourier transform of a sampled signal, $X^{dt}(\theta)$, directly from the Fourier transform of the original continuous-time signal, $X(\omega)$. This involves three steps: 1) linear scaling of the frequency axis according to $\theta = \omega T_0$; 2) summation of an infinite number of frequency-shifted (continuous-time) Fourier transforms; and 3) scaling of the amplitude axis by $1/T_0$.

Due to the infinite summation in the sampling theorem, the discrete-time inverse Fourier transform, $X^{dt}(\theta)$, according to Eq. 1.3 will be periodic in θ with period 2π. This is the fundamental reason for the folklore statement in DSP — sampling in the time domain produces periodicity in the frequency domain.

While the sampling theorem of Eq. 1.3 is expressed in terms of the relationship between the discrete-time Fourier transform of a sampled signal and the Fourier transform of the original continuous-time signal, it can also be equivalently expressed in terms of the relationship to the Fourier transform of the impulse sampled signal. This relationship can be easily proved to be

$$X^{dt}(\theta) = X_p\left(\frac{\theta}{T_0}\right);$$

that is, the two Fourier transforms are the same except for the linear scaling in the frequency axis. This result is not surprising because the two alternative views of sampling, point sampling and impulse sampling, are equivalent. Hence, their spectra determined by the respective Fourier transforms ought to be the same as well.

The sampling theorem is extremely important in DSP because it guides us in understanding the consequences of sampling, and in assessing the nature of possible distortions introduced by sampling. More importantly, the sampling theorem also allows us to determine the precise condition under which no sampling distortion would occur. This becomes the guidance for designing anti-aliasing filters. We will now discuss these practical issues guided by the sampling theorem.

1.1.3 Practical cases of sampling

The practical cases of sampling discussed now are classified according to the relationship between the sampling rate and a quantity called **bandwidth** of the signal. A continuous-time signal $x(t)$ is called **band-limited** if its Fourier transform satisfies that

$$X(\omega) = 0, \qquad \text{for } |\omega| \geq \omega_b \tag{1.5}$$

where ω_b is called the bandwidth of signal $x(t)$.

A continuous-time signal $x(t)$ is called **band-pass** if its Fourier transform satisfies that

$$X(\omega) = 0, \qquad \text{for } |\omega| \geq \omega_h \quad \text{and} \quad |\omega| \leq \omega_l. \tag{1.6}$$

Strictly speaking, the bandwidth of a band-pass signal is ω_h. However, it is common to consider the bandwidth of a band-pass signal as being $\omega_h - \omega_l$. This has been due to an important sampling property we will discuss soon.

Given a band-limited signal with its bandwidth $\omega_b = 2\pi f_b$, we define a critically important quantity, called the **Nyquist rate** or Nyquist frequency, to be $2f_b$. The Nyquist rate plays a critical role in understanding practical signal sampling discussed below.

Signal sampling above or at the Nyquist rate

Let us sample a band-limited signal $x(t)$ where we choose the sampling rate to be above the Nyquist rate, or $f_0 = \frac{1}{T_0} \geq 2f_b$. We now want to analyze the relationship between the spectra of the sampled signal $x[n]$ and of the original signal $x(t)$. Such a relationship is given by the sampling theorem Eq. 1.4:

$$\begin{aligned} X^{dt}(\theta) &= \frac{1}{T_0}\sum_{k=-\infty}^{\infty} X\left(\frac{\theta - 2\pi k}{T_0}\right), \qquad -\infty < \theta < \infty \\ &= \frac{1}{T_0} X\left(\frac{\theta}{T_0}\right), \qquad -\pi \leq \theta \leq \pi. \end{aligned} \tag{1.7}$$

The second line above is due to the property of the band-limited signal described by Eq. 1.5.

What Eq. 1.7 says is that the replicas of the spectrum that appear in the sampling theorem Eq. 1.4 do not overlap along the frequency axis due to the property of the band-limited signal and due to sufficient separation of these replicas; that is, the discrete-time Fourier transform (spectrum) of the sampled signal in the principal frequency range $[-\pi, \pi]$ faithfully preserves the shape of the spectrum of the original signal. This situation is shown in Fig. 1.1 where the Fourier transform $X(\omega)$ of signal $x(t)$ and the discrete-time Fourier transform $X^{dt}(\theta)$ of the sampled signal are aligned according to $\theta = \omega T_0$, and plotted.

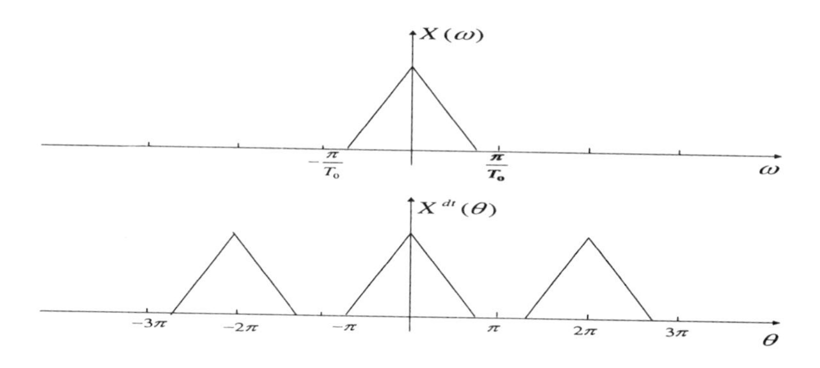

Figure 1.1: Signal sampling above the Nyquist rate. Upper: Fourier transform of the continuous-time, band-limited signal; Lower: Discrete-time Fourier transform of the digitally sampled signal.

Since the shape of the (true) spectrum is preserved over a frequency region when sampling a band-limited signal above or at the Nyquist rate, the spectrum can be recovered faithfully from the sampled signal. This recovery process is called **reconstruction**, based on a low-pass filtering operation. In time domain, reconstruction can be accomplished by the following Shannon's reconstruction theorem:

Reconstruction Theorem: A band-limited signal $x(t)$ can be exactly reconstructed from its samples $x(nT_0)$ according to

$$x(t) = \sum_{n=-\infty}^{\infty} x(nT_0)\text{sinc}\left(\frac{t - nT_0}{T_0}\right),$$

if the sampling rate $\frac{1}{T_0}$ is at or above the Nyquist rate. In the above, the *sinc* function is defined by

$$\text{sinc}(t) = \begin{cases} \frac{\sin(\pi t)}{\pi t}, & \text{if } t \neq 0 \\ 0, & \text{if } t = 0. \end{cases}$$

Signal sampling below the Nyquist rate

If we perform signal sampling below, rather than above or at, the Nyquist rate, i.e., $f_0 = \frac{1}{T_0} \leq 2f_b$, then we will have a very different situation from the earlier case. Fig. 1.2 shows an example for the new case where the sampling rate is set at $f_0 = 1.5f_b$, 25% lower than the Nyquist rate. Now, the second line of Eq. 1.7 no longer holds because the adjacent replicas of the spectrum that appear in the sampling theorem partially overlap along the frequency axis. The overlapped portion distorts the shape of the spectrum $X(\omega)$, making it difficult to recover the (true) spectrum $X(\omega)$ from that of the sampled signal.

In analyzing the nature of this distortion, we see that the high-frequency contents of the continuous-time signal have been "folded" back into a low-frequency range, and have been added to the original contents. This phenomenon is called **aliasing** . Aliasing results from overlapping of the replicas in the sampling theorem Eq. 1.4, and would generally occur when signal sampling rate is below the Nyquist rate.

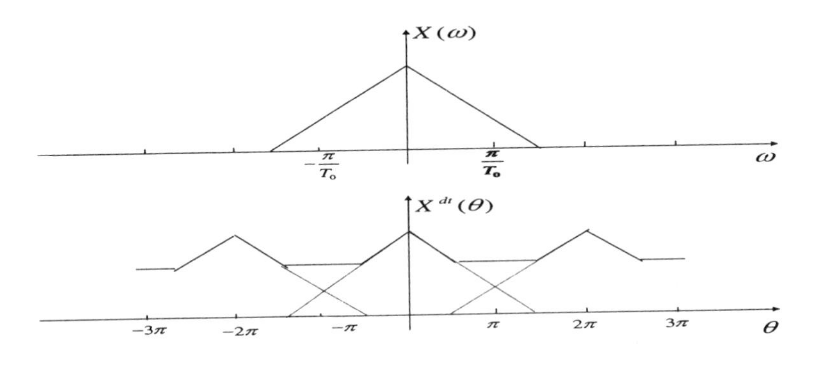

Figure 1.2: Signal sampling below the Nyquist rate. Upper: Fourier transform of the continuous-time, band-limited signal; Lower: Discrete-time Fourier transform of the digitally sampled signal.

Sampling signals with infinite bandwidth

In this case, the replicas of the spectrum which appear in the sampling theorem will always overlap along the frequency axis. Viewing it another way: since the Nyquist rate, which is twice the signal bandwidth, is infinite, sampling rate has to be infinite as well to avoid aliasing or spectral distortion.

Real-world signals always have (theoretically) infinite bandwidth, and hence will be aliased subject to sampling. However, most real-world signals have their bandwidths which are *practically* finite; that is, the energy outside a certain frequency range is negligible from a practical standpoint. Therefore, it is possible to sample a real-world signal at twice of its practical bandwidth. In this case, the aliasing effect, due to the overlapping of negligible portions of the signal energy, will be negligibly small as well.

As an example, the practical bandwidth of a (wide-band) speech signal is about 8 kHz, meaning that the percentage of the speech signal energy beyond 8 kHz is negligibly small. According to the previous discussion, if we sample this speech signal at a rate of 16 kHz, or 16,000 samples per second, the aliasing effect will be very small.

As another example, since the human auditory system is more sensitive to musical distortion than speech distortion, the practical bandwidth of a musical audio signal is about 20 kHz, significantly higher than that of speech. Standard compact-disc digital audio uses a sampling rate of 44.1 kHz, higher than the practical Nyquist rate of 40 kHz, which makes the aliasing effect negligibly small.

Such small aliasing effects can be eliminated using a common technique of **anti-aliasing filtering**. The continuous-time signal is low-pass filtered before passing it to the sampler. This low-pass filter is an analog one, with its cutoff frequency lower than half of the sampling rate.

Sampling of band-pass signals

Since a band-pass signal has its highest frequency of $\frac{\omega_h}{2\pi}$, the previous analysis concludes that it can be sampled at or above the Nyquist rate of $2 \times \frac{\omega_h}{2\pi} = \frac{\omega_h}{\pi}$ without suffering from aliasing. However, such a condition for setting the sampling rate is only a sufficient condition to avoid aliasing, not a necessary condition. *The necessary and sufficient condition to avoid aliasing is that no overlapping occur among the replicas of the true spectrum demanded by the sampling theorem described by Eq. 1.4.* Indeed, the replicas in the sampling theorem provide the fundamental cause for the possibility of aliasing. Let us return to this basic understanding and examine whether we can reduce the Nyquist sampling rate of $\frac{\omega_h}{\pi}$ for a band-pass signal without subjecting it to aliasing. Note that with the same amount of signal distortion, a lower sampling rate is better than a higher one in any practical DSP application.

We will show below that a band-pass signal can be sampled at a rate of $\frac{\omega_h - \omega_l}{\pi}$, which is often significantly smaller than $\frac{\omega_h}{\pi}$, while remaining aliasing-free. Let us assume that ω_h is an integer (M) multiple of the "bandwidth" $\omega_h - \omega_l$:

$$\omega_h = M(\omega_h - \omega_l),$$

or

$$\omega_l = (M - 1)(\omega_h - \omega_l).$$

In this case, the sampling rate can be reduced by a factor of M without suffering from aliasing as shown below.

Consider using this reduced sampling rate

$$f_0 = \frac{\omega_h - \omega_l}{\pi},$$

or the "enlarged" sampling interval

$$T_0 = \frac{\pi}{\omega_h - \omega_l} = \frac{\pi M}{\omega_h}.$$

Using the sampling theorem expressed by the impulse-sampled signal, $x_p(t)$, we can get its Fourier transform as

$$X_p(\omega) = \frac{1}{T_0} \sum_{k=-\infty}^{\infty} X\left(\omega - \frac{2\pi k}{T_0}\right) = \frac{1}{T_0} \sum_{k=-\infty}^{\infty} X\left(\omega - 2\pi k(\omega_h - \omega_l)\right).$$

Using the definition of the band-pass signal, Eq. 1.6, which specifies only a narrow frequency range of $\omega_h - \omega_l$ over which the spectrum is non-zero, we see that each term in the above, $X\left(\omega - 2\pi k(\omega_h - \omega_l)\right)$, is non-zero only in the range of

$$\omega_l \leq |\omega - 2\pi k(\omega_h - \omega_l)| \leq \omega_h.$$

This is equivalent to

$$\omega_l \leq \omega - 2\pi k(\omega_h - \omega_l) \leq \omega_h,$$

or

$$\omega_l \geq -[\omega - 2\pi k(\omega_h - \omega_l)] \geq \omega_h.$$

Using $\omega_l = (M-1)(\omega_h - \omega_l)$ and $\omega_h = M(\omega_h - \omega_l)$, we rewrite the above inequalities for non-zero regions as

$$(2k - M)(\omega_h - \omega_l) \leq \omega \leq (2k - M + 1)(\omega_h - \omega_l),$$

or

$$(2k + M - 1)(\omega_h - \omega_l) \leq \omega \leq (2k + M)(\omega_h - \omega_l).$$

This shows that the different replicas (corresponding to different values of k) do not overlap. The situation for $M = 3$ is illustrated in Fig. 1.3 to demonstrate the non-overlapping nature of the spectral replicas. Hence, the sampled signal using the "reduced" sampling rate will not be aliased.

The same conclusion also holds for the general case where ω_h may not be an integer multiple of the bandwidth. We will omit such analysis, as it is only slightly more involved

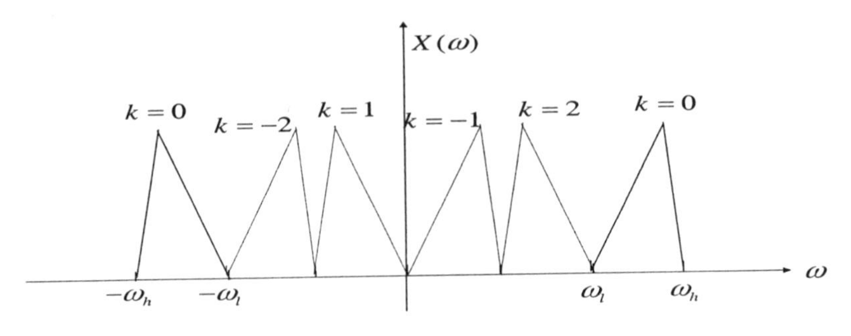

Figure 1.3: Sampling of a band-pass signal when $M = 3$.

1.2 Discrete-Time Systems and z-Transforms

A **discrete-time system**, also called a **digital filter**, is defined mathematically as a transformation mapping from an input sequence, $x[n]$, to an output sequence, $y[n]$, where $n \in \mathcal{N}$ is the integer time index. Therefore, system operation, transformation, filtering, modification, and mapping are all the same concept based on the following general mathematical operation on a sequence that produces a new sequence:

$$y[n] = \mathcal{T}\{x[1], x[2], \cdots, x[n], x[n+1], \cdots, \}, \qquad n \in \mathcal{N}$$

The output of a system is often called the system response. In general, a system transforms the input sequence by taking into account not only the current value but a combination of current, past, and possibly future values of the input in producing the current value in the system response.

1.2.1 Classification of systems

Classification of systems is generally based on the constraints and the nature of the generic transformation $\mathcal{T}\{\cdot\}$. Some commonly used classification schemes for discrete-time systems are described below. These different ways of classifying systems are largely separate from each other.

Linear versus nonlinear systems

The principle of **superposition** defines the class of **linear systems**. Superposition includes both **additive** and **scaling** properties; that is, the transformation $\mathcal{T}\{\cdot\}$ as the system operator satisfies the following superposition property:

$$\mathcal{T}\{a_1 x_1[n] + a_2 x_2[n]\} = a_1 \mathcal{T}\{x_1[n]\} + a_2 \mathcal{T}\{x_2[n]\}, \quad n \in \mathcal{N}$$

for arbitrary constants a_1 and a_2. This property can be easily generalized to more than two input sequences.

The acoustic system of speech production can be closely approximated by a linear system, which will be covered extensively in a later chapter. The use of linear system for speech production also underlies important speech technology applications.

Example: The accumulator system defined by the input-output transformation

$$y[n] = \sum_{l=-\infty}^{n} x[l]$$

is a linear system. It can be easily verified that the principle of superposition holds for this system.

A system which does not obey the principle of superposition is called a **nonlinear system**. A nonlinear system is much more difficult to analyze than a linear system. Many problems in speech processing, however, require use of nonlinear system analysis, as will be presented amply in later chapters of the book. For example, the input-output relationship between articulatory positions and the resulting speech acoustics is nonlinear, and so is the input-output relationship between motor commands and the resulting articulatory movements.

Example: The system defined by the input-output transformation of

$$y[n] = \sum_{l=-\infty}^{n} x[l]x[l+1]$$

is a nonlinear system. (This transformation is called lag-one autocorrelation function). It can be easily verified that the principle of superposition does not hold for this system.

Time-invariant versus time-varying systems

In a **time-invariant** or shift-invariant system, a time shift or delay of the input sequence causes the same time shift in the output as the system response; that is, if a time-invariant system maps the input sequence $x[n]$ into the output sequence $y[n]$, then the same system must also map the input sequence $x[n-n_0]$ into the output sequence $y[n-n_0]$ for an arbitrary integer $n_0 \in \mathcal{N}$. A system which does not have the above property is called a time-varying system.

Example: The accumulator system above is a time-invariant system, since a shifted input produces the same shift in the output as required of a time-invariant system.

Example: A "delta-parameter" system defined by

$$y[n] = x[n] - x[n-1], \quad n \in \mathcal{N}$$

is a time-invariant system. (It is also a linear system.) Such a system has found wide use in speech recognition technology.

Example: The compressor system defined by the input-output transformation of

$$y[n] = x[Ln], \quad n \in \mathcal{N}$$

is a time-varying system. In this system, the time-shifted input, $x[n-n_0]$, produces the output $x[Ln-n_0]$ according to the system definition. This is different from what is required of a time-invariant system:

$$y[n-n_0] = x[L(n-n_0)] = x[Ln - Ln_0] \neq x[Ln - n_0].$$

Example: The system defined by

$$y[n] = nx[n], \quad n \in \mathcal{N}$$

is a linear, time-varying system.

In the last example above, we see that a linear system whose coefficient(s) or parameter(s) in the linear term(s) are a function of time is necessarily a time-varying system. (This can be easily verified by showing contradiction to the definition of a time-invariant system.) Systems of such a type are typical of speech production models where a linear, acoustic model for speech wave generation is modulated by time-varying parameters attributed to articulatory motions resulting in changes of the vocal-tract area function over time.

Causal versus noncausal systems

In a **causal** or a nonanticipative system, the system response at any particular time $n = n_0$ is a function of only the input sequence before or at that time: $n \leq n_0$. A system whose output depends on the future values of the input sequence is called a non-causal system.

Example: The ideal delay system defined by

$$y[n] = x[n - n_d], \quad n, \ n_d \in \mathcal{N}$$

is a causal system if $n_d \geq 0$, and is a non-causal system if $n_d < 0$.

All physical systems, including the human speech production and perception (e.g., cochlea) systems are causal systems. However, man-made systems, such as mathematical transformations implemented in a computer for data processing, can be either causal or noncausal. Allowing for noncausality, i.e., providing the possibility of using anticipated future data as the system input, gives more flexibility and greater power for achieving desired data processing results. Such noncausal systems, however, cannot be implemented in physical devices or hardwares.

Stable versus unstable systems

In a **stable** or bounded-input, bounded-output system, every bounded (i.e., amplitude limited) input sequence necessarily produces a bounded output sequence; that is, when the input sequence is such that there exists a fixed positive finite value B_1 for which

$$|x[n]| \leq B_1 < \infty, \quad n \in \mathcal{N}$$

then there also exists a fixed positive finite value B_2 for which

$$|y[n]| \leq B_2 < \infty, \quad n \in \mathcal{N}.$$

Memoryless systems versus systems with memory

In a **memoryless** or static system, the output $y[n]$ at every time index n is a function of solely the input $x[n]$ at the same time index n; that is, the system does not memorize the past history of the input sequence, and neither does the system anticipate the future values of the input sequence. A memoryless system is necessarily a causal system.

A system for which the output $y[n]$ at time index n depends on the input at time indices before and/or after n is called a system with memory.

Example: A rectification system defined by

$$y[n] = |x[n]|, \quad n \in \mathcal{N}$$

is a memoryless system. It is also a nonlinear, causal, time-invariant, and stable system.

Example: A system defined by

$$y[n] = n|x[n]|, \quad n \in \mathcal{N}$$

is also a memoryless system, but this is a time-varying system.

For a time-invariant, memoryless system, we can often drop the time indices n in describing the system, since the input-output relationship does not depend on the time index n any more. Therefore, for the rectification system above, we can simply describe the system by

$$y = |x|.$$

In general, one can describe a time-invariant, memoryless, and nonlinear system by

$$y = g(x).$$

A special case is a time-invariant, memoryless, and linear (or affine) system which can be described by

$$y = Ax + b.$$

This is also called a linear regression system or model, widely used in many areas of engineering including speech processing.

1.2.2 Fundamentals of linear time-invariant systems

We have discussed in the above a number of different ways of classifying a general system. Among them, the classification schemes based on linearity and time-invariance are the most common and important. The simplest yet most frequently used system in these two classification schemes is the **linear time-invariant** system. A system with both linearity and time-invariance properties in combination leads to a particularly convenient representation for such a system. This representation is the **impulse response** of the linear system. We now discuss the fundamental theory of linear time-invariant systems based on this representation.

The impulse response of a linear time-invariant system, denoted by $h[n]$, is defined as the output of the system to the input sequence of a delta function $\delta[n]$; that is,

$$h[n] \doteq \mathcal{T}\{\delta[n]\}.$$

We now show, by construction, that this impulse response completely characterizes the system behavior. According to the property of time invariance, the system output will be $h[n-k]$ if the input is $\delta[n-k]$ or when the input impulse occurs at $n = k$; that is,

$$h[n-k] = \mathcal{T}\{\delta[n-k]\}.$$

Using the generic expression for a sequence,

$$x[n] = \sum_{k=-\infty}^{\infty} x[k]\delta[n-k],$$

we first write the system output to $x[n]$ as

$$y[n] = \mathcal{T}\{x[n]\} = \mathcal{T}\left[\sum_{k=-\infty}^{\infty} x[k]\delta[n-k]\right].$$

The superposition property of a linear system simplifies the above to the following fundamental equation of the linear system:

$$y[n] = \sum_{k=-\infty}^{\infty} x[k]\mathcal{T}\{\delta[n-k]\} = \sum_{k=-\infty}^{\infty} x[k]h[n-k]. \tag{1.8}$$

The implication of Eq. 1.8 is profound: The impulse response $h[n]$ of a linear, time invariant system $\mathcal{T}\{\cdot\}$ completely characterizes the system in the sense that it allows computation of the system output sequence $y[n]$ for an arbitrary input sequence $x[n]$ according to Eq. 1.8. The computation at the right hand side of Eq. 1.8 is commonly called (linear) **convolution**, which is also denoted by

$$y[n] = x[n] * h[n] \doteq \sum_{k=-\infty}^{\infty} x[k]h[n-k].$$

Convolution is a commutative operation, and hence we also have the system response of

$$x[n] * h[n] = h[n] * x[n] = \sum_{k=-\infty}^{\infty} h[k]x[n-k].$$

Most books on DSP and on discrete-time system theory give ample examples of convolution computation and interpretation, e.g., [Opp 99, Porat 97, Mitra 98], to which we refer the readers of this book.

1.2.3 z-transforms

The **z-transform** for discrete-time signals is the counterpart of the Laplace transform for continuous-time signals. They each have a similar relationship to the corresponding Fourier transform. The z-transform is a generalization of the discrete-time Fourier transform. The generalization is motivated by the convenience for time-domain interpretation of linear time-invariant system responses, and by the possibility of analyzing systems that do not possess valid Fourier transforms. Use of the z-transform allows block-diagram manipulation of systems consisting of subsystems connected in various ways (e.g., parallel, cascade, and feedback). It also allows decomposition of systems into constituent building blocks. In short, the z-transform is a powerful tool for the analysis of discrete-time signals and systems.

The z-transform of a sequence $x[n]$ is defined as the complex-valued function

$$X(z) = \sum_{k=-\infty}^{\infty} x[k]z^{-n}, \tag{1.9}$$

where z belongs to a subset of the complex numbers: $z \in \mathcal{C}$. For all z values within this subset, the series of Eq. 1.9 is convergent, or

$$\sum_{k=-\infty}^{\infty} |x[k]z^{-n}| < \infty.$$

This subset constitutes the **region of convergence** (ROC) of the z-transform. For the z values outside the ROC, the z-transform defined by Eq. 1.9 is divergent and is hence not valid. It can be easily proved that for any sequence $x[n]$, the shape of the ROC includes the annulus:

$$R_i < |z| < R_o$$

in a complex plane.

The **inverse z-transform** can be shown [Porat 97] to be

$$x[n] = \frac{1}{2\pi j} \oint X(z) z^{n-1} dz.$$

This allows recovery of the sequence $x[n]$ from its z-transform $X(z)$ in a principled way, although in practice more convenient methods are used for the same purpose, without the need to carry out this complex contour integration.

1.3 Characterizations of Digital Filters

In this section, we will use the powerful tool of the z-transform to perform basic analysis of discrete-time systems or digital filters.

1.3.1 Filter transfer functions

We noted in Section 1.2.2 that a linear, time-invariant digital filter is fully characterized by its impulse response sequence $h[n]$; that is, $h[n]$ contains all there is to know about the filter behavior (in the temporal domain). The z-transform $H(z)$ of the impulse response sequence $h[n]$, when having a non-empty ROC, is called the **transfer function** of the filter; that is,

$$H(z) \doteq \sum_{k=-\infty}^{\infty} h[k] z^{-n}, \qquad z \in \mathcal{C} \text{ (ROC)},$$

where $z \in \mathcal{C}$ (ROC), which guarantees

$$\sum_{k=-\infty}^{\infty} |h[k] z^{-n}| < \infty. \tag{1.10}$$

We now illustrate how the transfer function is useful in understanding the basic properties of a filter. The stability property of a filter can be identified by the (temporal) property of the impulse response according to (proof omitted):

$$\sum_{k=-\infty}^{\infty} |h[k]| < \infty. \tag{1.11}$$

Alternatively, the stability property can also be identified by the transfer function $H(z)$ of the filter: the ROC of $H(z)$ contains the unit circle $|z| = 1$. To see this, simply substitute $|z| = 1$ into Eq. 1.10 to obtain the same result as Eq. 1.11. We have thus proved that *a linear filter is stable if and only if the ROC of its transfer function includes the unit circle.*

The causality of a filter can also be identified by its transfer function in a simple way. It is easy to prove that *a linear filter is causal if and only if the ROC of its transfer function has the outward shape: $R_i < |z| < \infty$* . Combining the above two results, we have the following result based on the transfer function properties of a filter:

For a linear filter to be both stable and causal, its ROC of the transfer function must include the unit circle and all points outside the unit circle.

This result illustrates the importance of the ROC in defining the transfer function of a linear filter. It is possible that two different linear filters have an identical rational function in the transfer function, with different ROCs. The two different ROCs in the otherwise identical transfer function(s) give different impulse responses as inverse z-transforms of the transfer function(s). These impulse responses correspond to different properties concerning stability and causality of the linear filters. In most cases, given the form of the rational function in the transfer function, uniqueness of the filter can be determined by the stability and causality properties. These properties are equivalent to the specification of the ROC in the transfer function, and are essential in uniquely determining the impulse response of the filter from the rational form in the transfer function.

1.3.2 Filters described by difference equations

A digital filter, with its stability property specified, can be described by a difference equation. Linear difference equations with constant coefficients are of special importance in the theory of digital filters or discrete-time systems. The general form of a linear difference equation is

$$y[n] + a_1 y[n-1] + a_2 y[n-2] + ... + a_p y[n-p] = b_0 x[n] + b_1 x[n-1] + ... + b_q x[n-q],$$

or

$$y[n] = -\sum_{i=1}^{p} a_i y[n-i] + \sum_{i=0}^{q} b_i x[n-i]. \tag{1.12}$$

A system or filter defined by the difference equation Eq. 1.12 is linear, causal, and time-invariant. It may or may not be stable.

Eq. 1.12 directly specifies the input-output relation in the time domain. To specify the relation in the transformed domain, we take the z-transform on both sides of Eq. 1.12 to obtain

$$Y(z) = -\sum_{i=1}^{p} a_i z^{-i} Y(z) + \sum_{i=0}^{q} b_i z^{-i} X(z).$$

From this, we obtain the rational form of the transfer function:

$$H(z) = \frac{Y(z)}{X(z)} = \frac{\sum_{i=0}^{q} b_i z^{-i}}{1 + \sum_{i=1}^{p} a_i z^{-i}}. \tag{1.13}$$

The impulse response $h[n]$ of the filter can be found by performing an inverse z-transform on the transfer function in Eq. 1.13. This should be exactly the same as the time-domain solution of $y[n]$ for Eq. 1.12 when the input is the delta function $x[n] = \delta[n]$. The time-domain solution can be obtained by direct recursion using Eq. 1.12.

1.3.3 Poles and zeros in a digital filter

Definitions

The rational form, expressed by Eq. 1.13, of the transfer function of a digital filter is a ratio of two polynomials in the variable z^{-1}. Eq. 1.13 can be rewritten in the variable z.

For the case $p \geq q$ and $b_0 \neq 0$, we multiply $z^p z^q$ on both the denominator and numerator to obtain

$$H(z) = z^{p-q} \frac{\sum_{i=0}^{q} b_i z^{q-i}}{z^p + \sum_{i=1}^{p} a_i z^{p-i}} = b_0 z^{p-q} \frac{z^q + \sum_{i=1}^{q}(\frac{b_i}{b_0}) z^{q-i}}{z^p + \sum_{i=1}^{p} a_i z^{p-i}}. \tag{1.14}$$

Since any polynomial of order n can be written as a product of n factors

$$\prod_{i=1}^{n}(z - \lambda_i),$$

we can decompose Eq. 1.14 into the following form,

$$H(z) = b_0 z^{p-q} \frac{\prod_{i=1}^{q}(z - \beta_i)}{\prod_{i=1}^{p}(z - \alpha_i)}. \tag{1.15}$$

The values α_i are called the **poles** of the filter, and the values β_i are called the **zeros** of the filter. (Here we ignore the poles and zeros that occur at $z = 0$ since they are not of any practical importance.) Complex poles or zeros, if they exist, always come in conjugate pairs when the coefficients of the difference equation or the equivalent polynomial coefficients in the transfer function are real.

A filter that does not contain any zero is called an **all-pole** filter. An all-pole filter has the transfer function of

$$H(z) = \frac{1}{1 + \sum_{i=1}^{p} a_i z^{-i}}.$$

It plays a crucial role in speech analysis (linear predictive analysis, to be covered in Chapter 2), and is a good functional model for the vocal-tract acoustic system responsible for production of vocalic sounds in speech.

A filter that does not contain any pole is called an **all-zero** filter. An all-zero filter has the transfer function of

$$H(z) = \sum_{i=0}^{q} b_i z^{-i}.$$

FIR and IIR filters as related to poles and zeros

It can be easily proved that a filter containing one or more poles has its impulse response lasting for $n \to \infty$. For this reason, we also call this type of filter an **infinite impulse response (IIR)** filter. An IIR filter may have any number of zeros as long as the zeros do not coincide with *all* poles. (Coincidence of a pair of a pole and a zero is called, often found in speech production literature, pole-zero cancellation.) Returning to the difference-equation representation of a filter, we see that an IIR filter corresponds to output $y[n]$ recursion, or filter feedback, in the time domain.

It can also be easily proved that an all-zero filter has its impulse response lasting for only a finite time span; that is, $h[n] = 0$ for $n > q$. For this reason, we also call an all-zero filter a **finite impulse response (FIR)** filter. In the difference-equation representation, an FIR filter contains input $x[n]$ recursion, but no output $y[n]$ recursion or filter feedback.

Impulse responses as related to pole locations

The impulse response of a filter is the inverse z-transform of its transfer function and is fully characterized by its pole and zero locations. However, the general shape of the impulse response is mainly determined by the pole location(s), and is less sensitive to the zero location(s). In general, the determining factors are whether the pole is located inside, on, or outside the unit circle, and whether the pole is real or complex.

Example: Consider a causal IIR filter with a single real pole a. The transfer function of this filter, including its ROC, is

$$H(z) = \frac{1}{1 - az^{-1}}, \quad |z| > |a|.$$

Its impulse response can be found by inverse z-transform to be

$$h[n] = a^n u[n] = a^n, \quad 0 < n < \infty.$$

From this, we see that this geometric (exponential) impulse response is

- divergent to ∞ for $|a| > 1$ (pole outside the unit circle, giving rise to an unstable system);
- with constant amplitude for $|a| > 1$ (pole on the unit circle);
- convergent to zero for $|a| < 1$ (pole inside the unit circle);
- oscillatory for $a < 0$, either divergent, convergent, or with a constant amplitude.

Example: Consider a causal IIR filter with a pair of complex poles at

$$\alpha_1 = A\exp(j\theta), \qquad \alpha_2 = A\exp(-j\theta),$$

and with a single zero at

$$\beta = A\cos\theta.$$

The corresponding transfer function is

$$H(z) = \frac{1 - A\cos\theta z^{-1}}{(1 - A\exp(j\theta)z^{-1})(1 - A\exp(-j\theta)z^{-1})} = \frac{1 - A\cos\theta z^{-1}}{1 - 2A\cos\theta z^{-1} + A^2 z^{-2}}, \quad |z| > |A|.$$

Its impulse response can be found by inverse z-transforming the transfer function:

$$h[n] = A^n \cos(\theta n)u[n] = A^n \cos(\theta n), \quad 0 < n < \infty.$$

This sinusoidally modulated exponential shape of the impulse response is related to the pole locations in the following way:

- always oscillatory due to the cosine term in $h[n]$;
- divergent to ∞ for $|A| > 1$;
- convergent to zero for $|A| < 1$
- neither divergent to ∞ nor convergent to zero for $|A| = 1$, and the shape is rather irregular.

1.4 Frequency Responses of Digital Filters

Let a stable linear time-invariant digital filter have a rational transfer function; that is, its input-output relation satisfies the difference equation of Eq. 1.12. Then, its **frequency response**, $H^{dt}(\theta)$, defined as the discrete-time Fourier transform of the impulse response $h[n]$, can be obtained directly from the transfer function of the z-transform according to

$$H^{dt}(\theta) = H(z)\mid_{z=e^{j\theta}} = \frac{\sum_{i=0}^{q} b_i e^{-j\theta i}}{1+\sum_{i=1}^{p} a_i e^{-j\theta i}} = b_0 e^{j\theta(p-q)} \frac{\prod_{i=1}^{q}(e^{j\theta}-\beta_i)}{\prod_{i=1}^{p}(e^{j\theta}-\alpha_i)}.$$

The magnitude of the frequency response is thus

$$|H^{dt}(\theta)| = b_0 \frac{\prod_{i=1}^{q}|e^{j\theta}-\beta_i|}{\prod_{i=1}^{p}|e^{j\theta}-\alpha_i|}, \tag{1.16}$$

and the phase of the frequency response is

$$\angle H^{dt}(\theta) = \theta(p-q) + \sum_{i=1}^{q} \angle(e^{j\theta}-\beta_i) - \sum_{i=1}^{p} \angle(e^{j\theta}-\alpha_i). \tag{1.17}$$

Using the frequency response, the input-output relation, $y[n]$ and $x[n]$, of the filter, expressed in the frequency domain, will satisfy the same product relation as in the z-transform domain:

$$Y^{dt}(\theta) = H^{dt}(\theta) X^{dt}(\theta).$$

1.4.1 Frequency response as related to pole and zero locations

Given the pole and zero locations in the filter transfer function, there exist convenient graphical procedures [Opp 99, Porat 97] to determine the magnitude and phase of the frequency response based on Eqs. 1.16 and 1.17. Such graphical procedures allow coarse but intuitive estimation of the general shape of the frequency response, especially the magnitude of the response. Useful insight can also be gained on the fundamental nature of the frequency response.

The following general observations can be made from the graphical procedures:

- A real zero near $z = 1$ leads to a low magnitude at low frequency ($\theta = 0$);
- A real zero near $z = -1$ leads to a low magnitude at high frequency ($\theta = \pi$);
- A real pole near $z = 1$ leads to a high magnitude at low frequency;
- A real pole near $z = -1$ leads to a high magnitude at high frequency;
- A pair of complex zeros near the unit circle leads to a low magnitude near the frequencies corresponding to the phase angles of the zeros;
- A pair of complex poles near the unit circle leads to a high magnitude near the frequencies corresponding to the phase angles of the poles;

1.4.2 Digital resonator

Consider a causal, second-order, all-pole filter with a pair of complex poles located at

$$\alpha_1 = A\exp(j\theta_0), \qquad \alpha_2 = A\exp(-j\theta_0).$$

Transfer function

The transfer function of the above filter, according to Eq. 1.15, is

$$H(z) = \frac{b_0}{(1 - A\ \exp(j\theta_0)z^{-1})(1 - A\ \exp(-j\theta_0)z^{-1})} = \frac{b_0}{1 - 2A\ \cos\theta_0 z^{-1} + A^2 z^{-2}},$$

where $|z| > |A|$, and the input gain is b_0.

Difference equation

The difference equation representing this filter, determined directly from the above z-transform, is

$$y[n] - 2A\ \cos\theta_0 y[n-1] + A^2 y[n-2] = b_0 x[n].$$

From this, we see that this filter is a recursive one, with feedback from the two previous samples of the output.

Impulse response

Using the inverse z-transform of the transfer function, we obtain the impulse response of the filter:

$$h[n] = A^n \frac{\sin[\theta_0\ (n+1)]}{\sin\ \theta_0} u[n],$$

where $\theta_0 \neq 0$. This is a sinusoidally modulated (note that $sin\ \theta_0$ is a non-zero constant) exponential sequence. It will be damped to zero as n increases if $A < 1$ (stable system), or will diverge to infinity if $A > 1$ (unstable system). This behavior is reminiscent of an over-damped or an under-damped second-order continuous-time resonant system. If $\theta_0 = 0$, we have a critically damped system where the impulse response becomes

$$h[n] = (n+1)A^n u[n];$$

this is analogous to a critically damped, second-order, continuous-time, resonant system.

Frequency response magnitude

Based on Eq. 1.16, the magnitude of the frequency response of the above filter is

$$|H^{dt}(\theta)| = \frac{b_0}{[1 + A^2 - 2A\cos(\theta - \theta_0)]^{1/2}[1 + A^2 - 2A\cos(\theta + \theta_0)]^{1/2}}. \tag{1.18}$$

This expression has a unique maximum (which occurs at $\theta = \theta_0$) over the meaningful range of frequency $0 \leq \theta < \pi$; that is, this digital filter is a **resonant system** with a single resonance occurring at the frequency corresponding to the phase angle (θ_0) of the pole (within $0 \leq \theta < \pi$). If $A < 1$, the maximal value is finite. If $A = 1$, the maximal value becomes infinitely large.

Actually, without going through the above analysis involving Eq. 1.18, the resonance behavior associated with this filter can also be quickly established from the general observations using the graphical procedure discussed in the preceding section.

The resonance frequency is sometimes called a **formant** frequency when applied to the speech signal. The preceding analysis thus gives a way of estimating the (single) formant based on the pole's phase angle in the transfer function. If the sampling rate is T_0, then the formant frequency in the units of radians is θ/T_0, or in the units of Hz, is

$$F = \frac{\theta_0}{2\pi T_0}. \tag{1.19}$$

Using Eq. 1.18, we can further determine that the 3-dB bandwidth of the resonance (or formant bandwidth for speech) is solely a function of the magnitude A of the pole according to

$$B_w = -\frac{\log A}{\pi T_0} \tag{1.20}$$

in the units of Hz.

Klatt formant synthesizer

It is interesting to point out that the digital resonator analyzed above is essentially the same one used in the Klatt formant synthesizer embedded in the MITalk text-to-speech systems [Klatt 87b]. In [Klatt 87b], the resonator is described by the second-order filter

$$y[n] - By[n-1] - Cy[n-2] = b_0 x[n].$$

The coefficients of the filter are determined by the formant frequency F and bandwidth B_w, supplied from a higher-level component of the synthesizer, according to

$$C = -e^{2\pi B_w T_0}, \quad B = 2e^{\pi B_w T_0}\cos(2\pi F T_0). \tag{1.21}$$

The input gain is set to be $b_0 = 1 - B - C$. It can be easily verified that with a change of variables:

$$B = 2A \,\cos\,\theta_0, \qquad C = -A^2,$$

Eq. 1.21 is equivalent to Eqs. 1.19 and 1.20.

1.4.3 All-pass filter

Another example which illustrates the use of frequency responses is the all-pass filter. The concept of the filter's frequency response is fundamental to understanding a class of digital filters of such a type. Consider a filter with the transfer function

$$H(z) = \frac{z^{-1} - a^*}{1 - az^{-1}},$$

with a pole a and a zero $(a^*)^{-1}$. To find the magnitude frequency response of this filter, we substitute $e^{j\theta}$ for z to obtain the discrete-time Fourier transform:

$$|H^{dt}(\theta)| = \frac{e^{-j\theta} - a^*}{1 - ae^{-j\theta}} = e^{-j\theta}\frac{1 - a^* e^{j\theta}}{1 - ae^{-j\theta}}.$$

Since the denominator and numerator are complex conjugates of each other, they have the same magnitude. Hence,

$$|H^{dt}(\theta)| = 1.$$

This filter is called an **all-pass filter** since it passes all of the frequency components of its input with constant magnitude gain.

In the above example filter, if a is real, then the filter has an identical pole and zero. In this case, we say there is a perfect pole-zero cancellation. Thus, while a pole typically represents a resonance or a magnitude peak in the frequency response, under the condition of pole-zero cancellation the peak can be completely "masked" by the companion zero. This results in a flat magnitude frequency response. Disappearance of some formants, typically the low-frequency ones, in many consonants of speech is attributed to pole-zero cancellation.

It can be easily shown that the most general form of an all-pass filter has the transfer function of

$$H(z) = \prod_{i=1}^{R} \frac{z^{-1} - d_i}{1 - d_i z^{-1}} \prod_{i=1}^{C} \frac{(z^{-1} - e_i^*)(z^{-1} - e_i)}{(1 - e_i z^{-1})(1 - e_i^* z^{-1})},$$

where d_i's are the real poles, and e_i's are the complex poles. All zeros are inverses of corresponding poles.

1.5 Discrete Fourier Transform

The Discrete Fourier Transform, **DFT** for short, is related to, but different from, the discrete-time Fourier transform discussed earlier. Why need yet another transform? This is because the discrete-time Fourier transform requires knowledge of the transform over a continuum of frequencies: $-\pi \leq \theta \leq \pi$ in order to uniquely recover the original signal sequence (via the inverse transform). A finite subset of frequencies, required by a practical computer system or digital processor, is generally insufficient for the unique recovery.

However, if the signal sequence has a finite duration, say of length N, then it is possible to sample the discrete-time Fourier transform at N points, which allow faithful reconstruction of the signal sequence. This type of sampling on the discrete-time Fourier transform for a finite-length sequence is called the DFT. The DFT is a very useful and important tool in DSP.

1.5.1 Definition of DFT

Let us first examine the familiar discrete-time Fourier transform for a finite-length signal sequence $x[n], 0 \leq n \leq N-1$. The definition in Eq. 1.2 becomes

$$X^{dt}(\theta) = \sum_{n=0}^{N-1} x[n] \exp(-j\theta n), \quad \theta \in \mathcal{R}. \tag{1.22}$$

Uniform sampling of the frequency axis θ over $[0, 2\pi]$ in Eq. 1.22 gives the sampling points of

$$\theta[k] = \frac{2\pi k}{N}, \quad 0 \leq k \leq N-1.$$

This turns Eq. 1.22 to the definition of the DFT:

$$X^d[k] = \sum_{n=0}^{N-1} x[n] \exp\left(-\frac{j2\pi kn}{N}\right) = \sum_{n=0}^{N-1} x[n] W^{-kn}, \quad 0 \le k \le N-1, \tag{1.23}$$

where $W \doteq \exp\left(\frac{j2\pi}{N}\right)$.

The inverse DFT allows unique recovery of the finite-length signal sequence from its DFT points according to

$$x[n] = \frac{1}{N} \sum_{k=0}^{N-1} X^d[k] W^{kn}, \quad 0 \le n \le N-1, \tag{1.24}$$

The correctness of the above inverse DFT formula can be easily verified as follows:

$$\begin{aligned}
\frac{1}{N} \sum_{k=0}^{N-1} X^d[k] W^{kn} &= \frac{1}{N} \sum_{k=0}^{N-1} \left[\sum_{l=0}^{N-1} x[l] W^{-kl} \right] W^{kn} \\
&= \frac{1}{N} \sum_{l=0}^{N-1} x[l] \left[\sum_{k=0}^{N-1} W^{(n-l)k} \right] \\
&= \frac{1}{N} \sum_{l=0}^{N-1} x[l] \; N\delta[(n-l) \bmod N] = x[n].
\end{aligned}$$

1.5.2 Frequency range and frequency resolution of the DFT

According to the frequency sampling relationship:

$$\theta[k] = \frac{2\pi k}{N}, \quad 0 \le k \le N-1,$$

the frequency range $0 \le k \le N/2$ (assuming N even) of the DFT corresponds to frequencies $0 \le \theta \le \pi$ of the discrete-time Fourier transform, and the frequency range $N/2 \le k \le N-1$ of the DFT corresponds to frequencies $\pi \le \theta \le 2\pi$ or equivalently $-\pi \le \theta \le 0$. Therefore, the point $k = 0$ in the DFT is associated with zero frequency $\theta = 0$ in the unsampled discrete-time Fourier transform, and the point $k = N/2$ is associated with the high frequency $\theta = \pi$.

Frequency resolution of the DFT is an important concept for understanding many practical issues for DFT applications. It refers to the "frequency bin" of the DFT defined as the spacing between two adjacent frequencies in the unsampled discrete-time Fourier transform:

$$\Delta\theta = \frac{2\pi}{N}.$$

Based on the earlier discussion of signal sampling, when the sampling interval is T_0, the frequency resolution of the DFT is related to the physical frequency resolution, in the units of radians, according to

$$\Delta\omega = \frac{\Delta\theta}{T_0} = \frac{2\pi}{NT_0}.$$

In the units of Hz, this corresponds to

$$\Delta f = \frac{\Delta\omega}{2\pi} = \frac{1}{NT_0}.$$

Since NT_0 is the total length of the continuous-time signal before the sampling, we conclude from the above analysis that *the best frequency separation or resolution in the physical signal provided by the DFT analysis is the exact inverse of the length of the continuous-time signal.* In other words, given the fixed duration of the signal, it is not possible to arbitrarily improve the frequency resolution (the smaller, the better) by solely increasing the number of signal sampling points N and to perform the DFT using these additional points.

1.5.3 Zero-padding technique

Because the frequency resolution is limited by the fixed duration of the signal used in the DFT analysis, there will be many frequency "holes" in the results of DFT. This is especially serious when the signal duration is short, rendering poor frequency resolution. The **zero-padding technique** commonly used in DFT applications can "fill in" these frequency "holes" and thus improve the visual continuity of the DFT analysis output plots.

The zero-padding technique involves the addition of zeros to the tail of the initial relatively *short* sampled signal sequence $x[n]$ according to

$$x_{zp} = \begin{cases} x[n] & 0 \leq n \leq N-1, \\ 0 & N \leq n \leq M-1, \end{cases}$$

where $M > N$. Applying an M-point DFT to the new sequence x_{zp}, we have

$$X^d_{zp}[k] = \sum_{n=0}^{M-1} x[n] \exp\left(-\frac{j2\pi kn}{M}\right) = \sum_{n=0}^{N-1} x[n] \exp\left(-\frac{j2\pi kn}{M}\right).$$

From this, we see that the new DFT $X^d_{zp}[k]$ is simply a new frequency sampling of the original discrete-time Fourier transform $X^{dt}(\theta)$. It samples at M equally spaced frequency points:

$$\theta[k] = \frac{2\pi k}{M}, \quad 0 \leq k \leq M-1,$$

and since $M > N$, it provides more detailed sampling points on the identical function $X^{dt}(\theta)$ than the DFT $X^d[k]$ which would be computed from the original non-zero-padded sequence $x[n]$.

1.6 Short-Time Fourier transform

One very interesting and useful extension of the discrete-time Fourier transform and DFT analysis techniques discussed earlier in this chapter is the **short-time Fourier transform (STFT)**. It plays a significant role in DSP, notably in speech analysis where the conventional time-domain and frequency-domain signal processing techniques are combined into a single, consistent, and integrated framework. What motivated the use of STFT in speech analysis is the time-varying properties of the speech signal's frequency-spectral contents. These time-varying properties root in the time-varying properties of the speech production system due to the continual movement of speech articulators under their higher-level, linguistically motivated control. For example, a natural speech signal is typically characterized by the movement of frequency peaks (formants) as a function of time. The conventional discrete-time Fourier transform and DFT analysis described earlier are inadequate for such characterization since only one transform is defined for the entire signal. It is thus impossible to characterize the movement of formants over time. Extension of the conventional Fourier transform technique is needed to overcome such a limitation.

The STFT technique is devised just for such a purpose. The idea behind the STFT is very simple and yet powerful. It consists of a separate Fourier transform for each sample (or for each block of samples) in the time-domain signal, using the signal samples in the neighborhood of that sample. We confine all discussions of STFT in this book to the version of the STFT applied only to discrete-time, rather than continuous-time, signals.

1.6.1 Definition of STFT: two alternative views

Fourier transform view

An extension of the discrete-time Fourier transform (Eq. 1.3)

$$X^{dt}(\theta) = \sum_{m=-\infty}^{\infty} x[m] \exp(-j\theta m), \quad \theta \in \mathcal{R}$$

gives the following definition of the **discrete-time STFT**:

$$X(n,\theta) = \sum_{m=-\infty}^{\infty} x[m]w[n-m] \exp(-j\theta m), \quad \theta \in \mathcal{R}, \quad n \in \mathcal{N} \tag{1.25}$$

where $w(n-m)$ is called the "windowing" function centered at $m = n$. We thus see that the discrete-time STFT is a sequence of discrete-time Fourier transforms for a new, "short-time" section of the signal

$$x_n[m] \doteq x[m]w[n-m].$$

The window function is often chosen to be symmetric at $m = n$, and to satisfy the temporal localization property:

$$x_n[m] = \begin{cases} x[m], & \text{for } m \text{ near } n \\ 0, & \text{for } m \text{ far away from } n \end{cases}$$

An extension of the DFT (Eq. 1.23)

$$X^d[k] = \sum_{m=0}^{N-1} x[m]W^{-km}, \quad 0 \le k \le N-1,$$

gives the following definition of the *discrete* STFT:

$$X[n,k] = \sum_{m=0}^{N-1} x[m]w[n-m]W^{-km}, \quad 0 \le k, m \le N-1. \tag{1.26}$$

Again, the discrete STFT is a sequence of DFT's for the "short-time" section of the signal. Just as the DFT is the frequency-sampled discrete-time Fourier transform, the discrete STFT is the frequency-sampled discrete-time STFT. In the remainder of Chapter

1, we will focus on the discrete-time STFT with all principles and results equally applied to the discrete STFT.

Filtering view

The above definition of the STFT has been based on the Fourier transform of the short-time section of the original time sequence. An identical result can be obtained using an alternative view based on linear digital filtering, discussed earlier in this chapter. To see this, we first rewrite Eq. 1.25 as the convolution

$$X(n,\theta) = \sum_{m=-\infty}^{\infty} \{x[m]\exp(-j\theta m)\}w[n-m] = \{x[n]\exp(-j\theta n)\} * w[n].$$

According to the fundamental equation of the linear system expressed by Eq. 1.8, the STFT of $X(n,\theta)$ can be interpreted as the linear filter output for the input sequence

$$x[n]\exp(-j\theta n),$$

with the impulse response of the filter being $w[n]$. The input sequence can be interpreted as the modulation of $x[n]$ to the fixed frequency θ, or amplitude modulation by the sinusoid $\exp(-j\theta n)$.

1.6.2 STFT magnitude (spectrogram)

The magnitude of the STFT just defined, from either the Fourier transform view or the filtering view, is called a spectrogram. A spectrogram is a function of both time index n and frequency θ (or sampled frequency index k, as in the discrete STFT). It can be displayed as a plot on a paper: with the time index as the x-axis, frequency as the y-axis, and the STFT magnitude as the darkness of the point located at the corresponding time and frequency. The spectrogram plays a critical role in speech analysis and applications. Articulatory movement underlying the speech signal can be partially traced from the corresponding spectrogram. Visual cues in the spectrogram are related to aspects of human speech perception and computer speech recognition. The human auditory system extracts some perceptually relevant information directly from a spectrogram-like representation of speech.

Narrowband and wideband spectrogram

Depending on the nature of the time windowing function $w[n]$ used in constructing a STFT, a spectrogram can be classified to be either a narrowband or a wideband one. If the time window $w[n]$ is of a short duration, then its Fourier transform will be of a wide bandwidth. In this case, the spectrogram is called a **wideband spectrogram**. The large bandwidth in the wideband spectrogram gives poor frequency localization. This is because the fine detail of the spectrum, such as individual spectral harmonics of speech, is smeared by the short-windowing, multiplication operation in the time domain. This operation is equivalent to the convolution operation in the frequency domain, and in the case of a large bandwidth window, the convolution necessarily creates strong smearing effects in the frequency domain.

The smearing effects in the frequency domain can be reduced by using a long-duration time windowing function which corresponds to a narrow bandwidth in the frequency window. The convolution with a narrow-bandwidth function will create less strong smearing effects in the frequency domain. The spectrogram of such a type constructed with use of a long-duration window is called a **narrowband spectrogram**. A narrowband spectrogram is capable of showing fine spectral detail of the speech signal including individual spectral harmonics due to glottal vibration. The price to be paid, however, is the loss of sharp localization in time due to the use of the long-duration temporal window.

Trade-off between time and frequency localization

Therefore, if one wants to have fine time localization, a short-duration window in the time domain or the wideband spectrogram should be used. If good frequency localization is desired, a long-duration window in the time domain or the narrowband spectrogram ought be chosen. The dilemma, however, is that the window bandwidth in the frequency domain and the window duration in the time domain cannot be made arbitrarily small simultaneously. This is because they are the Fourier transform pair of each other, and there is an inherent trade-off between time and frequency localization in a spectrogram using a fixed window. Some quantitative analysis on this type of trade-off is provided by the uncertainty principle discussed in [Cohen 95].

Within the framework of STFT discussed in the chapter based on a fixed window, there would be no adequate resolution to the above dilemma. However, researchers in signal processing have actively pursued the use of multiple, adaptive windows in the time-frequency analysis, and joint time-frequency analysis where both time and frequency localization can be carried out concurrently. Examples of such work are the Wigner distribution, the Kernel method, and more generally, wavelet analysis [Cohen 95].

1.7 Summary

In this chapter, we first introdued the basic concept of signal sampling (in time), and then concisely presented the most celebrated **sampling theorem** (as well as the reconstruction theorem) in DSP. The three practical cases of sampling was discussed in detail. Discrete-time systems or digital filters were introduced based on the relationship of the system's input-output signals sampled in the same manner. Fundamental properties of linear discrete-time systems were presented using the mathematical tool of z-transform. The most important property is the transfer function or the impulse response of the digital filter. For almost all speech processing applications, the filter's transfer function can be determined by the "poles" and "zeros" of the filter. Frequency responses of the filter, as related to its pole and zero locations and characterized by the discrete-time Fourier transform of the impulse response, were discussed in some detail with typical examples given.

The chapter then proceeded by introducing discrete Fourier transform (DFT), which is the frequency-domain sampled version of the discrete-time Fourier transform. The related issue of frequency resolution was discussed. This was followed by the topic of short-time Fourier transform (STFT), which was defined equivalent as eigher the time-sample-dependent DFT, or by the array of outputs from a bank of (linear) digital filters. We finally introduced the concept of the spectrogram as the magnitude of the STFT.

Chapter 2

Analysis of Discrete-Time Speech Signals

Equipped with the basic principles and techniques discussed in Chapter 1 for analyzing general discrete-time signals and systems, we are now in a position to apply them to analyzing sampled or discrete-time speech signals. Most speech processing applications utilize certain properties of speech signals in accomplishing their tasks. This chapter describes these properties or features and how to obtain them from a speech signal $s(n)$, i.e., **speech analysis**. This typically requires a transformation of $s(n)$ into a set of parameters, or more generally into a set of signals, with the purpose often being data reduction. The relevant information in speech can often be represented very efficiently. Speech analysis extracts features which are pertinent for different applications, while removing irrelevant aspects of the speech.

In this chapter, we will describe several important speech analysis techniques building directly on the results covered in Chapter 1. These techniques include

- Time-frequency representation of speech signals based on a STFT anaysis;
- Linear predictive speech analysis based on the characterization of all-pole digital filters;
- Cepstral analysis of speech signals based on a specific DSP technique for mixed linear and nonlinear discrete-time system analysis;
- Speech formant tracking based on analysis of digital resonant systems;
- Voicing pitch tracking based on time and frequency analysis techniques for discrete-time signals; and
- Speech analysis using auditory models based on filter-bank signal decomposition.

All these analysis techniques have significant and wide applications in speech technology and in enhancing one's understanding of the fundamental properties of the speech process.

2.1 Time-Frequency Analysis of Speech

In Section 1.6 we provided the mathematical basics for the short-time Fourier transform (STFT), which serves as a commonly used tool for general time-frequency signal analysis.

Applications of the STFT-based time-frequency analysis to speech signals have shed crucial insight into physical properties of the speech signals. Such an analysis enables one to uncover the underlying resonance structure of speech and the vocal tract motion over time, which are responsible for the generation of the directly observable time-domain speech signal.

We also discussed in Chapter 1 the use of the STFT magnitude or **spectrogram** to display the time-frequency representation of speech as an image. This is essentially a way of compressing a three-dimensional function into a two-dimensional display medium exploiting the visual perceptual dimension of darkness. In this section, we will first discuss the separate, *individual* time-domain and frequency-domain properties of speech, as well as the corresponding time-domain and frequency-domain analysis techniques. We will then discuss the *joint* time-frequency properties of speech and related analysis techniques.

2.1.1 Time-domain and frequency-domain properties of speech

Analyzing speech in the time domain often requires simple calculation and interpretation. Among the relevant features found readily in temporal analysis are waveform statistics, power and F0. The frequency domain, on the other hand, provides the mechanisms to obtain the most useful parameters in speech analysis. Most models of speech production assume a noisy or periodic waveform exciting a vocal-tract filter. The excitation and filter can be described in either the time or frequency domain, but they are often more consistently and easily handled spectrally. For example, repeated utterances of the same text by a single speaker often differ significantly temporally while being very similar spectrally. Human hearing seems to pay much more attention to spectral aspects of speech (especially power distribution in frequency) than to phase or timing aspects. In this section, we first examine temporal properties, then spectral ones.

Waveforms

Time-domain speech signals are also called speech **waveforms**. They show the acoustic signals or sounds radiated as pressure variations from the lips while articulating linguistically meaningful information. The amplitude of the speech waveform varies with time in a complicated way, including variations in the global level or intensity of the sound. The probability density function (to be discussed in detail in Chapter 3) of waveform amplitudes, over a long-time average, can be measured on a scale of speech level expressed as sound dB [Sait 85]. This function has a form close to a double-sided (symmetric) exponential at high amplitudes, and is close to Gaussian at low amplitudes. The entire probability density function can be approximated by a sum of exponential and Gaussian functions. Such distibutions can be exploited in speech coding and recognition.

Fundamental frequency

Under detailed examination, a speech waveform can be typically divided into two categories:

- a quasi-periodic part which tends to be repetitive over a brief time interval;
- a noise-like part which is of random shape.

For the quasi-periodic portion of the speech waveform, the average period is called a fundamental period or pitch period. Its inverse is called the **fundamental frequency** or **pitch frequency**, and is abbreviated F0. (Although pitch is actually a perceptual phenomenon, and what is being measured is actually F0, we follow tradition here and consider "pitch" synonymous with F0.) The fundamental frequency corresponds to vocal cord vibrations for vocalic sounds of speech. F0 in a natural speech waveform usually varies slowly with time. It can be 80 Hz or lower for male adults, and above 300 Hz for children and some female adults. F0 is the main acoustic cue for intonation and stress in speech, and is crucial in tone languages for phoneme identification. Many low-bit-rate voice coders require F0 estimation to reconstruct speech.

Overall power

The overall **power** of the speech signal corresponds to the effective sound level of the speech waveform averaged over a long-time interval. In a quiet environment, the average power of male and female speech waveforms measured at 1 cm in front of a speaker's lips is about 58 dB. Male speech is on average about 4.5 dB louder (greater power) than female speech. Under noisy conditions, one's speech power tends to be greater than in a quiet environment (i.e., we speak more loudly in order to be heard). Further, not only the overall power and amplitude is increased, but also the details of the waveform changes in a complicated way. In noisy environments a speaker tends to exaggerate articulation in order to enhance the listener's understanding, thereby changing the spectrum and associated waveform of the speech signal.

Overall frequency spectrum

While the spectral contents of speech change over time, if we take the discrete-time Fourier transform (DFT) of the speech waveform over a long-time interval, we can estimate the overall frequency range that covers the principal portion of the speech power. Such information is important for the design of speech transmission systems since the bandwidth of the systems depends on the overall speech spectrum rather than on the instantaneous speech spectrum. When such an overall frequency spectrum of speech is measured in a quiet environment, it is found that the speech power is concentrated mainly at low frequencies. For example, over 80% of speech power lies below 1 kHz. Beyond 1 kHz, the overall frequency spectrum decays at a rate of about -12 dB per octave. Above 8 kHz, the speech power is negligible. In a telephone channel, due to telephone bandwidth limitations, the overall frequency spectrum becomes negligible above 3.2 kHz, losing some information, mainly for consonants. If the long-time Fourier transform analyzes only a quasi-periodic portion of the speech waveform, we will see the frequency components in harmonic relations, i.e., integer multiples of a common frequency. This common frequency, also called the lowest harmonic component, is the pitch.

Short-time power

Speech is dynamic or time-varying. Sometimes, both the vocal tract shape and pertinent aspects of its excitation may stay fairly constant for dozens of pitch periods (e.g., to 200 ms). On the other hand, successive pitch periods may change so much that their name 'period' is a misnomer. Since the typical phone averages only about 80 ms in duration, dynamic coarticulation changes are more the norm than steady-state sounds. In any event, speech analysis usually presumes that signal properties change relatively slowly over time. This is most valid for short time intervals of a few periods at most. During such a *short-time window* of speech, one extracts parameters or features, which

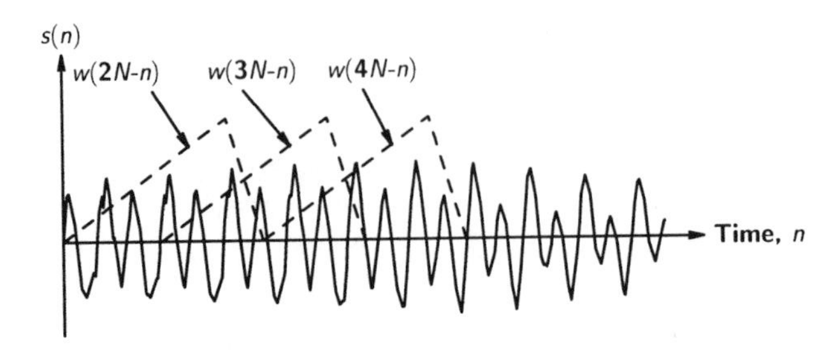

Figure 2.1: Speech signal $s(n)$ with three superimposed windows, offset from the time origin by $2N$, $3N$, and $4N$ samples. (An atypical asymmetric window is used for illustration.)

each represent an average over the duration of the time window. As a result of the dynamic nature of speech, we must divide the signal into many successive windows or *analysis frames*, allowing the parameters to be calculated frequently enough to model dynamic vocal-tract features. Window size is critical to good modelling. Long vowels may allow window lengths up to 100 ms with minimal loss of detail due to the averaging, but stop explosions require much shorter windows (e.g., 5–10 ms) to avoid excess averaging of rapid spectral transitions. In a compromise, typical windows last about 20-30 ms, since one does not know a priori what sound one is analyzing.

Windowing means multiplication of a speech signal $s(n)$ by a window $w(n)$, which yields a set of speech samples $x(n)$ weighted by the shape and duration of the window (Fig. 2.1). By successive shifting of $w(n)$ in time, with spacing corresponding to a suitable update (frame) interval (often 10 ms), we may examine any or all of $s(n)$ via the sliding window. Typically $w(n)$ is a smooth function, so as to properly consider the samples of $s(n)$ under evaluation. The simplest window has a rectangular shape $r(n)$:

$$w(n) = r(n) = \begin{cases} 1 & \text{for } 0 \le n \le N-1 \\ 0 & \text{otherwise.} \end{cases} \tag{2.1}$$

This gives equal weight to all samples of $s(n)$, and limits the analysis range to N consecutive samples. A common alternative to Eq. 2.1 is the Hamming window (Fig. 2.2), a raised cosine pulse:

$$w(n) = h(n) = \begin{cases} 0.54 - 0.46\cos(\frac{2\pi n}{N-1}) & \text{for } 0 \le n \le N-1 \\ 0 & \text{otherwise,} \end{cases} \tag{2.2}$$

or the quite similar Hanning window. Tapering the edges of $w(n)$ allows its periodic shifting (at the update frame rate) along $s(n)$ without having large effects on the resulting speech parameters due to pitch period boundaries.

Short-time average zero-crossing rate

Speech analysis that attempts to estimate spectral features usually requires a Fourier transform (or other major transformation, such as linear prediction). However, a simple measure called the zero-crossing rate (ZCR) provides basic spectral information in some

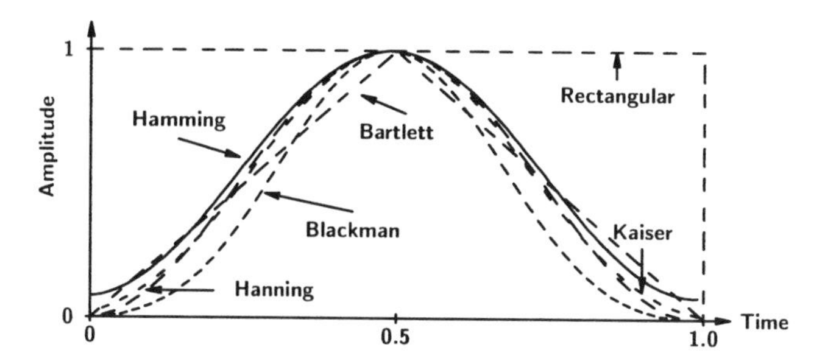

Figure 2.2: Common time windows, with durations normalized to unity.

applications at low cost. For a speech signal $s(n)$, a **zero-crossing** takes place whenever $s(n) = 0$, i.e., the waveform crosses the time axis (changes algebraic sign). Taking the simple case of a sinusoid (instead of speech), ZCR (measured as zero-crossings/s) yields two zero-crossings/period, and thus its F0 = ZCR/2. In DSP terms, we also have ZCR in zero-crossings/sample,

$$F0 = (\text{ZCR} \cdot F_s)/2,$$

for F_s samples/s. For all narrowband signals (e.g., sinusoids), the ZCR can accurately measure the frequency where power is concentrated.

Most short-time processing methods (both temporal and spectral) yield a parameter sequence in the form of a dynamic signal

$$P(n) = \sum_{m=-\infty}^{\infty} T[s(m)]w(n-m), \tag{2.3}$$

where the speech $s(n)$ is subject to a (possibly nonlinear) transformation T and is weighted by the window $w(n)$ to limit the time range examined. The desired parameter $P(n)$ (as specified by the nature of T) appears as a signal with the original sampling rate, representing some speech characteristic averaged over the window duration. $P(n)$ is the convolution of $T[s(n)]$ and $w(n)$. Since $w(n)$ usually behaves as a lowpass filter, $P(n)$ is a smoothed version of $T[s(n)]$.

Thus, Eq. 2.3 serves to calculate the ZCR with

$$T[s(n)] = 0.5|\text{sign}(s(n)) - \text{sign}(s(n-1))|,$$

where the algebraic sign of $s(n)$ is

$$\text{sign}(s(n)) = \begin{cases} 1 & \text{for } s(n) \geq 0 \\ -1 & \text{otherwise.} \end{cases}$$

and $w(n)$ is a rectangular window scaled by $1/N$ (where N is the duration of the window) to yield zero-crossings/sample. The ZCR can be significantly decimated (for data reduction purposes). Like speech power, the ZCR changes relatively slowly with vocal tract movements.

The ZCR is useful for estimating whether speech is voiced. Voiced speech has mostly

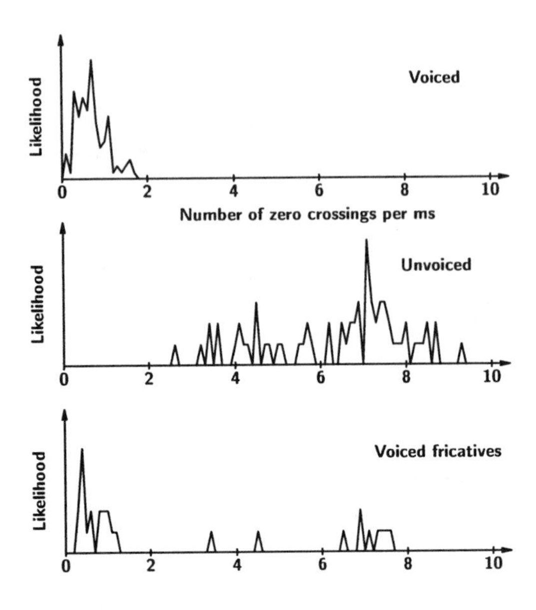

Figure 2.3: Typical distribution of zero-crossings for voiced sonorants, for unvoiced frication, and for voiced frication.

low-frequency power, owing to a glottal excitation spectrum that falls off at about −12 dB per octave. Unvoiced speech comes from broadband noise excitation exciting primarily high frequencies, owing to the use of shorter vocal tracts (anterior to the constriction where noise is produced). Since speech is not narrowband, the ZCR corresponds to the average frequency of primary power concentration. Thus high and low ZCR (about 4900 and 1400 crossings/s) correspond to unvoiced and voiced speech, respectively (Fig. 2.3).

For sonorant sounds, the ZCR follows F1 well, since F1 has more energy than other formants (except in low back vowels, where F1 and F2 are close, and ZCR is thus usually above F1). Voiced fricatives, on the other hand, have bimodal spectra, with voicebar power at very low frequency and frication energy at high frequency; hence, ZCR in such cases is more variable. Unlike short-time power, the ZCR is quite sensitive to noise, especially any low-frequency bias that may displace the zero-amplitude axis.

Short-time autocorrelation function

Another way to estimate certain useful features of a speech signal concerns the short-time autocorrelation function. Like the ZCR, it serves as a tool to access some spectral characteristics of speech without explicit spectral transformations. As such, the autocorrelation function has applications in F0 estimation, voiced/unvoiced determination, and linear prediction. In particular, it preserves spectral amplitude information in the speech signal concerning harmonics and formants, while suppressing (often undesired) phase effects.

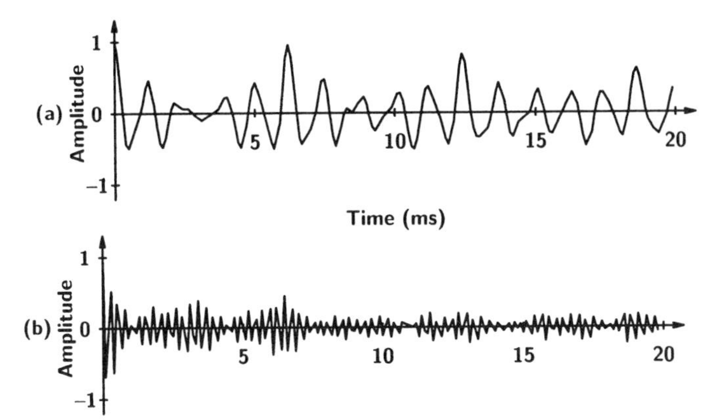

Figure 2.4: Typical autocorrelation function for (a) voiced speech and (b) unvoiced speech, using a 20-ms rectangular window ($N = 201$).

The short-time autocorrelation function is defined as:

$$R_n(k) = \sum_{m=-\infty}^{\infty} s(m)w(n-m)s(m-k)w(n-m+k), \tag{2.4}$$

where the range of summation is limited (i.e., windowed by $w(n)$), as for other speech features. The index n notes that the window slides across the signal $s(n)$. $R_n(k)$ measures the similarity of a signal $s(n)$ with a delayed version of itself, as a function of the time delay. In particular, $R_n(k)$ becomes large at multiples of N if $s(n)$ is quasiperiodic with period N. It is an even function ($R_n(k) = R_n(-k)$), it has maximum value at $k = 0$, and the value $R_n(0)$ gives the average energy in the signal in the vicinity of time n (Fig. 2.4).

Depending on the application, Eq. 2.4 is typically evaluated for different ranges of k. For linear prediction (see later in this chapter), k ranges from 0 to 10 (or so), an order proportional to the signal bandwidth. For F0 estimation, $R_n(k)$ is evaluated for k near the *a priori* estimated number of samples in a pitch period. Compared to power and the ZCR, the autocorrelation can require much more calculation, especially when the window is long (e.g., often a few pitch periods) or the range of k needed is large (as for F0 estimation, when the speaker's pitch can have a wide range). We compromise between minimizing window length (to reduce computation) and retaining enough speech samples in the window (to get a reliable autocorrelation function); e.g., a longer $w(n)$ gives better frequency resolution (see below). Brief windows are sometimes used for spectral envelope estimation (e.g., as in LPC), since the desired spectral detail (to be retained by the autocorrelation) is usually limited to the location of a few formants.

For F0 estimation, an alternative to the autocorrelation is the average magnitude difference function (AMDF). Rather than multiplying speech $s(m)$ by $s(m-k)$, the magnitude of their difference is used:

$$\text{AMDF}(k) = \sum_{m=-\infty}^{\infty} |s(m) - s(m-k)|.$$

Assuming that subtraction is a simpler computer operation than multiplication, the AMDF is faster. The AMDF has minima for values of k near multiples of the pitch period (instead of the peaks for $R_n(k)$).

Filter bank analysis

One traditional speech analysis technique is that of a **filter bank** or set of bandpass filters. A single input speech signal is simultaneously passed through several digital filters. Each analyzes a different range of frequencies, outputting a narrowband signal containing amplitude (and sometimes phase) information about the speech in a narrow frequency range. This can be quite similar to basic DFT analysis, but filter banks are more flexible since each filter bandwidth can be specified separately (unlike the fixed analysis bandwidth of the DFT). The bandwidths normally are chosen to increase with center frequency, thus following decreasing human auditory resolution. They often follow the bark scale, i.e., having equally-spaced, fixed bandwidths below 1 kHz, then logarithmic spacing at higher frequencies (e.g., a fixed Q-factor). Sometimes, one-third-octave filters are used.

As filter-bank outputs, many applications prefer to have a small set of spectral parameters that describe aspects of the power spectrum, especially the smooth (formant) envelope. For example, the amplitude outputs from a dozen bandpass filters furnish a more efficient spectral representation than the much more detailed DFT. For efficiency, one could use two levels of filter banks: a coarse one with few filters for a preliminary classification of sounds (e.g., sonorant versus obstruent), followed by a second round of more precise filters, focussed on the frequency range where most energy was found.

2.1.2 Joint time-frequency properties of speech

Separate time- and frequency-domain characterization of speech as discussed above may be important for some aspects of speech applications. For instance, the overall frequency spectrum of speech determines the bandwidth of communication systems to be allocated for the transmission of speech waveforms. However, such individual, disjoint characterization does *not* reflect the most essential property of speech, that is, the spectral contents (including resonance peaks) which change continuously over time. The physical reason underlying such an essential property is the continual motion of the articulators that forms a time-varying acoustic filter responsible for the generation of the speech waveform. To characterize this type of properties, we need a joint time-frequency representation. As discussed in Chapter 1, the STFT and its magnitude (spectrogram) are good mathematical tools, among other more sophisticated ones, for this purpose.

Window effects

One key factor that determines the time-frequency resolution tradeoffs in a joint time-frequency representation such as a spectrogram is the time span and shape of the window function, used to compute the STFT. In Chapter 1, we discussed the classification of narrowband and wideband spectrograms which use long and short time spans of the window function to focus on different time-frequency properties of the signal. A narrowband spectrogram uses a long time-span window to represent the harmonic structure of speech with fine frequency resolution, at the expense of poor time resolution, which results in smearing resonance peak (formant) change over time. A wideband spectrogram does the reverse, sacrificing frequency resolution.

In addition to controlling the time-frequency resolution tradeoffs, a good choice of window function also would allow error-free recovery of the speech waveform from its spectrogram. It would appear that since a spectrogram discards phase information in the STFT while maintaining only the STFT magnitude, the error-free recovery of the speech waveform would be impossible. However, if the window shape is correctly chosen and sufficient temporal overlap is provided across windows in the STFT, the constraints tacitly imposed by the overlapping windows indeed permit error-free recovery of the speech waveform from which the spectrogram is computed.

We now discuss a number of commonly used window functions (in addition to the simple rectangular and Hamming windows noted above), paying special attention to distortions arising from the use of the windows in constructing the joint time-frequency representation of speech. The main purpose for using a window is to limit the time extent of observation of a speech signal, so that the observed portion displays a relatively steady behavior within the context of dynamic speech. While its time-limiting purpose is explicit, the spectral effects of windowing are also important. Normally a window $w(n)$ changes slowly in time, and as a result has the frequency response of a lowpass filter. For example, the smooth Hamming window $h(n)$ concentrates more energy at low frequencies than a rectangular window $r(n)$ does (owing to the latter's abrupt edges). Owing to this concentration, the spectrum of the windowed signal $x(n) = s(n)w(n)$ corresponds to a convolution of spectra:

$$X(e^{j\omega}) = \frac{1}{2\pi}\int_{\theta=0}^{2\pi} S(e^{j\theta})W(e^{j(\omega-\theta)})\,d\theta.$$

Since the objective here is the average spectrum over the windowed signal, $W(e^{j\omega})$ should ideally have a limited frequency range and a smooth shape, in order to best represent the signal. From one point of view, the best window would correspond to an ideal lowpass filter (i.e., rectangular pulse in frequency), since it strictly limits the frequency range and has constant value. In that case, the output spectrum $X(e^{j\omega})$ is a smoothed version of $S(e^{j\omega})$, where each frequency sample is the average of its neighbors over a range equal to the bandwidth of the lowpass filter. However, a window with a rectangular spectrum may have undesirable edge effects in frequency, with rapidly-fluctuating output as the (ideally) sharp-edged spectrum slides across the signal spectrum in convolution; e.g., for voiced speech, $X(e^{j\omega})$ varies significantly as harmonics are included or excluded under the sliding filter, depending on the interaction between the filter bandwidth and the speech F0.

Furthermore, an ideal lowpass filter is not a feasible window, since it has an infinite duration. In practice, we use finite-duration windows that are not necessarily fully limited in frequency range. Thus each sample in the short-time spectrum $X(e^{j\omega})$ is the sum of the desired average of a range of $S(e^{j\omega})$ plus undesired signal components from frequencies farther away. The latter can be limited by focussing $W(e^{j\omega})$ in a low, narrow frequency range. Since the Hamming $H(e^{j\omega})$ is closer to an ideal lowpass filter than $R(e^{j\omega})$, the former yields a better $X(e^{j\omega})$, more closely approximating the original $S(e^{j\omega})$. Assuming the same window duration for the two cases, $h(n)$ is a lowpass filter spectrally twice as wide as the rectangular $r(n)$, thus smoothing the speech spectrum more (with consequent reduction of spectral detail) (Fig. 2.5).

Many speech applications require a smoothed (i.e., data-reduced) speech spectrum, as seen in wideband spectrograms; e.g., coders and recognizers focussing on the formant detail of vocal-tract response need spectral representations that smooth the rapidly-varying harmonics while retaining the formant envelope. For any specific window type, its duration varies inversely with spectral bandwidth, i.e., the usual compromise between time and frequency resolution. Wideband spectrograms display detailed time resolution

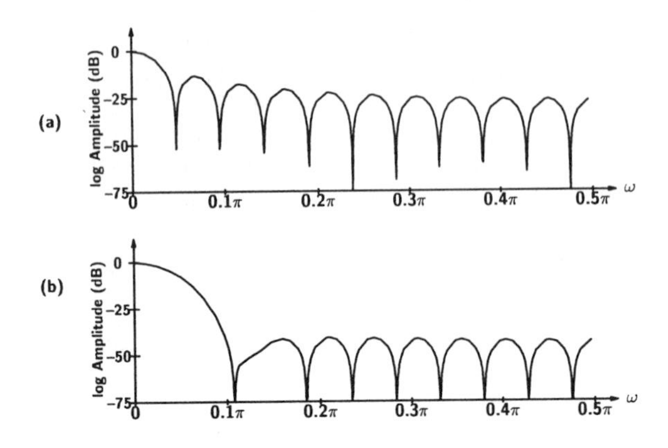

Figure 2.5: Magnitude of Fourier transforms for (a) rectangular window, (b) Hamming window.

(amplitude variations corresponding to vocal cord closures), and typically use a window about 3 ms long. This means a bandwidth of approximately 300 Hz, which smooths away harmonic structure (except for very high-pitched voices). Narrowband spectrograms instead typically use a 20-ms window with a corresponding 45-Hz bandwidth; thus they display individual harmonics, but the time-frequency representation undergoes significant temporal smoothing.

When windowing voiced speech, a rectangular window positioned properly to include exactly one pitch period yields a spectrum similar to that of the vocal-tract impulse response. While close to optimal in output, such **pitch-synchronous** analysis requires reliable location of pitch periods - an extra computation. In practice, most speech analysis uses a fixed window size bigger than a pitch period, e.g., 20-30 ms. Such longer windows require more computation, but reduce edge effects. When such a **pitch-asynchronous** window is time-shifted disregarding pitch periods, the effects of vocal-cord closures are reduced simply because such points represent a smaller proportion in the longer window. Recent attempts to overcome the drawbacks of a fixed window size include more advanced frequency transforms (e.g., wavelets).

The Fourier transform $S(e^{j\omega})$ of speech $s(n)$ provides both spectral magnitude and phase (Fig. 2.6). The autocorrelation $r(k)$ of $s(n)$ is the inverse Fourier transform of the energy spectrum ($|S(e^{j\omega})|^2$). $r(k)$ preserves information about harmonic and formant amplitudes in $s(n)$ as well as its periodicity, while ignoring phase, since phase is less important perceptually.

Other windows

Besides the rectangular and Hamming windows, others include the Bartlett, Blackman, Hann, Parzen, or Kaiser windows [Rab 75, Nut 81]. All are used to represent short-time average aspects of speech signals; they approximate lowpass filters while limiting window duration. Many applications trade off window duration and shape, using larger windows than strictly allowed by stationarity constraints but then compensating by emphasizing the middle of the window.

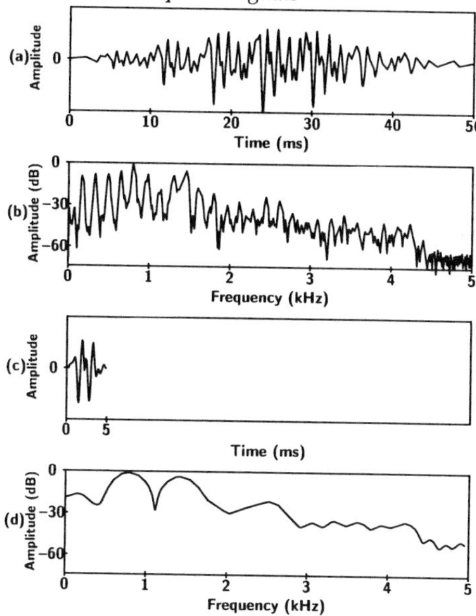

Figure 2.6: Time signals and spectra of a vowel: (a) signal multiplied by a 50-ms Hamming window; (b) the corresponding spectrum (note that harmonic structure is strongest at low frequencies); (c) signal multiplied by a 5-ms Hamming window; (d) its corresponding spectrum.

2.1.3 Spectrogram reading

One insightful use of the joint time-frequency representation of speech is to perform spectrogram reading. This refers to the process, through human efforts, of using visual displays of spectrograms (typically wideband spectrograms) to examine the time-frequency properties of speech signals and to decode the underlying linguistic units. Spectrogram reading has a special educational value since much of the significant acoustic-phonetic properties of speech can be identified directly from the spectrographic display. Efforts have also been made to infer articulation from the spectrographic display of the speech signal.

The importance of spectrogram reading can be appreciated in terms of its possible role for decoding the linguistic message that is encrypted in the speech signal. This coding scheme has been rather indirect, and has been far from being cracked at our current level of understanding. Speech as spoken language contains far more information than written language in communicating thoughts and emotions among human beings. The linguistic message and paralinguistic information including emotions are effectively encoded in the speech signal through several intermediate levels of representation that are all hidden from direct observation in the waveform signal. Spectrograms provide one effective clue to the nature of the speech code since they attempt to examine, in a primitive way, some intermediate levels of speech representation.

A spectrogram translates the speech signal, which we normally process with our ears, into the visual domain as a time- frequency representation. While we as human beings begin to acquire the skill of translating auditory information into meaningful messages in our early lives, it is almost impossible to analyze this process with conscious involvement. In contrast, learning to read a spectrogram allows us to examine the speech "decoding" process. Our conscious mind is able to participate fully in the learning process of reading spectrograms. This conscious participation enables us to explain and analyze the process of translating observations in the spectrogram into linguistic and paralinguistic information. In doing so, we reach a new level of understanding spoken language by the process of analysis, comparison, and learning, and by accumulation of an increasingly large repertoire of time-frequency patterns. These patterns are viewed at a global level, and are reliably linked to the linguistic message as identified in the repetitive learning process in spectrogram reading.

Our understanding of speech as a code, which associates the linguistic message with specific time-frequency sound patterns, is in a continual development phase. Spectrogram reading and analysis serve an extremely useful purpose of continual learning, revision, enhancement, and updating of our understanding of speech as a code. They also serve a useful purpose to provide insight and motivations in modifying architectures for improving the performance of speech technology.

Some illustrations of the association between time-frequency patterns and the linguistic message, categorized by scientific classification of speech sounds, will be presented in Chapter 7. That chapter will be, to a large extent, devoted to the study of acoustic-phonetic characterization of the speech process via the powerful tool of spectrograms discussed here.

A spectrogram provides a three-dimensional spectral representation of speech utterances typically about 2–3 s in duration. The short-time Fourier transform $S_n(e^{j\omega})$ is plotted with time n on the horizontal axis, with frequency ω (from 0 to π) on the vertical axis (i.e., $0 - F_s/2$ on the parallel Hz scale), and with power in the third dimension, often shown graphically as darkness (weak energy below one threshold appears as white regions, while power above a higher threshold is black; levels of grey fill the intervening range). Since the DFT phase is usually of minor interest, only the magnitude of the complex-valued $S_n(e^{j\omega})$ is typically displayed, and usually on a logarithmic (decibel) scale (which follows the perceptual range of audition).

A wideband spectrogram shows individual pitch periods as vertical striations (these correspond to increased power every time the vocal cords close) (Fig. 2.7a). Precise temporal resolution permits accurate estimation of when vocal tract movements occur. The wide filter bandwidth smooths the harmonic amplitudes located in each formant across a range of (usually) 300 Hz, thus displaying a band of darkness (of width proportional to the formant's bandwidth) for each resonance. The center of each band is a good estimate of formant frequency. Narrowband spectrograms, on the other hand, display individual harmonics instead of pitch periods (Fig. 2.7b). They are less useful for segmentation because of poorer time resolution. Instead they are useful for analysis of pitch and vocal tract excitation.

Owing to the typical power fall-off of voiced speech with frequency (about −12 dB per octave), dynamic range in spectrograms is often compressed artificially so that details at the weaker high frequencies are more readily visible. This *pre-emphasis* of the speech, either by differentiating the analog speech $s_a(t)$ prior to A/D conversion or by differencing the discrete-time sampled speech $s(n) = s_a(nT)$, compensates for decreased energy at high frequencies. The most common form of pre-emphasis is

$$y(n) = s(n) - As(n-1), \tag{2.5}$$

where A typically lies between 0.9 and 1.0, reflecting the degree of pre-emphasis. In effect, $s(n)$ passes through a filter with a zero at $z = A$.

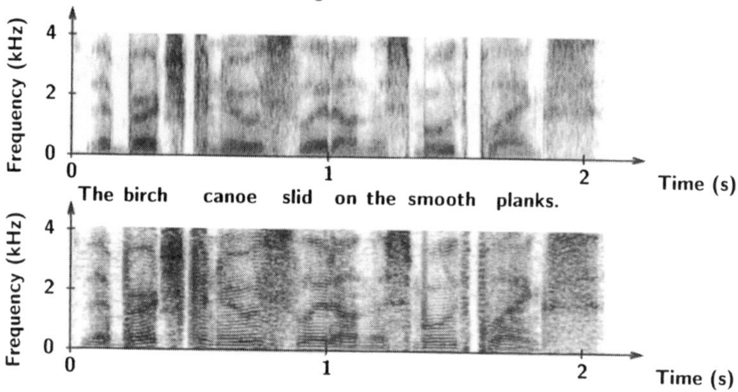

Figure 2.7: (a) Wideband and (b) narrowband spectrograms of a sentence.

2.2 Analysis Based on Linear Predictive Coding

Speech analysis based on **linear predictive coding (LPC)** has had a successful history for more than 30 years. The term "linear prediction" refers to the mechanism of using a linear combination of the past time-domain samples, $s[n-1], s[n-2], \ldots, s[n-M]$, to approximate or to "predict" the current time-domain sample $s[n]$:

$$s[n] \approx \hat{s}[n] = -\sum_{i=1}^{M} a_i s[n-i], \tag{2.6}$$

where $\hat{s}[n]$ is called the predicted sample, and $a_i, i = 1, 2, ..., M$ are called predictor or LPC coefficients. If the prediction is accurate, then a small number of LPC coefficients, $a_1, a_2, ..., a_M$, can be used to efficiently represent or to "code" a long sequence of the signal $s[n]$.

Linear prediction expressed in the time-domain by Eq. 2.6 is the same as the all-pole modeling described in Section 1.3.3, where the "prediction" operation was treated as a discrete-time linear system transformation. In the time-series analysis literature in statistics, the linear prediction of Eq. 2.6 is also called an autoregressive model. The first researchers who directly applied the linear prediction techniques to speech analysis appeared to be Saito and Itakura [Sait 68], and Atal and Schroeder [Atal 79]. A comprehensive coverage of the early research on such applications can be found in [Markel 76, Sait 85].

Define the sampled error for the prediction as

$$e[n] = s[n] - \hat{s}[n] = s[n] + \sum_{i=1}^{M} a_i s[n-i] = \sum_{i=0}^{M} a_i s[n-i], \tag{2.7}$$

where $a_0 = 1$. Taking the z-transform on Eq. 2.7, we obtain

$$E(z) = S(z) + \sum_{i=1}^{M} a_i S(z) z^{-i} = S(z)\left[1 + \sum_{i=1}^{M} a_i z^{-i}\right] = S(z)A(z), \tag{2.8}$$

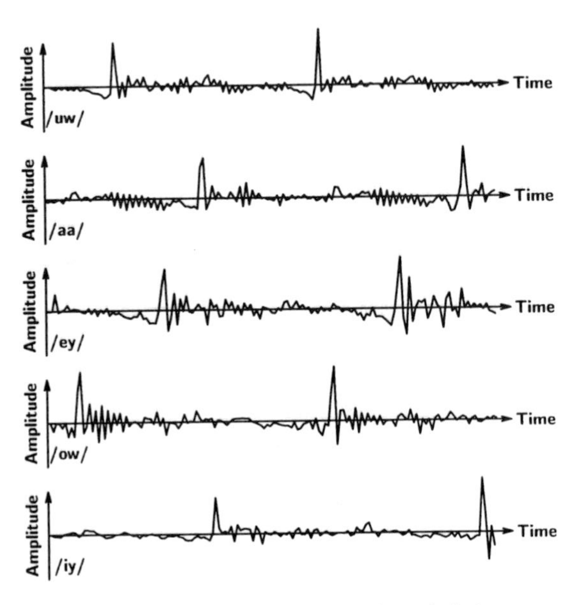

Figure 2.8: Examples of pre-emphasized speech signals and their corresponding prediction error signals for five vowels /uw,aa,ey,ow,iy/.

where

$$A(z) = 1 + \sum_{i=1}^{M} a_i z^{-i} = \sum_{i=0}^{M} a_i z^{-i}.$$

Eq. 2.8 can be equivalently written as

$$S(z) = E(z)\frac{1}{A(z)}.$$

This shows that the speech signal can be viewed as the output of an all-pole digital filter, whose transfer function is

$$\frac{1}{A(z)},$$

and the input to the filter is the LPC error signal $e[n]$. This (forward) filtering view of LPC analysis can be used to interpret Eq. 2.8 as inverse filtering. That is, if we pass the speech signal $s[n]$ as input into an **inverse filter** whose transfer function is $A(z)$, then the output of the inverse filter will be the error signal $e[n]$. $A(z)$ acts as the inverse of the synthesis filter that creates $S(z)$.

2.2.1 Least-squares estimate of LPC coefficients

One key importance of the LPC analysis is that the LPC coefficients, $a_1, a_2, ..., a_M$, which are the parameters of the LPC inverse filter $A(z)$, can be determined directly from the speech signal $s[n]$. The least-squares criterion is mathematically tractable, computationally efficient, and it has been shown to be highly effective in speech analysis.

Under the least-squares optimization criterion, the total squared error is defined by

$$\mathcal{E} \doteq \sum_{n=n_0}^{n_1} e^2[n], \tag{2.9}$$

where n_0 and n_1 are the summation limits, or the sample range over which the optimization is carried out. Substituting Eq. 2.7, we have

$$\mathcal{E} = \sum_{n=n_0}^{n_1} \left[\sum_{i=0}^{M} a_i s[n-i]\right]^2 = \sum_{n=n_0}^{n_1} \sum_{i=0}^{M} \sum_{j=0}^{M} a_i s[n-i] s[n-j] a_j = \sum_{i=0}^{M} \sum_{j=0}^{M} a_i c_{ij} a_j,$$

where

$$c_{ij} \doteq \sum_{n=n_0}^{n_1} s[n-i]s[n-j].$$

To minimize $\mathcal{E}$ with respect to the LPC coefficients, $a_1, a_2, ..., a_M$, we set the partial derivatives to zero and solve the resulting equations. (This is a classical method of optimization; see details of optimization method categories in Chapter 5.) This gives

$$\frac{\partial \mathcal{E}}{\partial a_k} = 2 \sum_{i=0}^{M} a_i c_{ik} = 0.$$

Since $a_0 = 1$, this becomes the following celebrated *normal equation*:

$$\sum_{i=1}^{M} a_i c_{ik} = -c_{0k}, \quad k = 1, 2, ..., M \tag{2.10}$$

Solutions to the $M \times M$ linear system of equations of Eq. 2.10 are straightforward. (Computation complexity of a straightforward solution such as direct matrix inversion is as high as M^3. More efficient solutions are discussed later in this section.) However, the coefficients in Eq. 2.10, c_{ij}, depend on the summation limits n_0 and n_1. The choice of n_0 and n_1 affects the linear prediction properties and the efficiency of the solution methods. Two specific cases pertaining to this issue have been studied in detail, and are presented below.

2.2.2 Autocorrelation and covariance methods

Autocorrelation method

The **autocorrelation method** chooses the summation limits in Eq. 2.8 to be $n_0 = -\infty$ and $n_1 = \infty$ while simultaneously forcing $s[n] = 0$ for $n < 0$ and $n \geq N$, i.e., limit the speech signal to an$N$-sample window. This simplifies c_{ij} to be the symmetric (short-time) autocorrelation function $r(\tau)$:

$$c_{ij} = \sum_{n=-\infty}^{\infty} s[n-i]s[n-j] = \sum_{n=0}^{N-1-|i-j|} s[n]s[n+|i-j|] = r(|i-j|).$$

Due to the truncation of $s[n]$ to zero beyond the N-sample window, the above is equivalent to error minimization over the interval $0 \leq n \leq N + M - 1$.

Summary of autocorrelation method
Seek solution to the linear system of equations:

$$\sum_{i=1}^{M} a_i r(|i-j|) = -r(j), \quad j = 1, 2, ..., M \tag{2.11}$$

where

$$r(\tau) = \sum_{n=0}^{N-1-\tau} s[n]s[n+\tau], \qquad \tau \geq 0.$$

The prediction error signal is

$$e[n] = s[n] + \sum_{i=1}^{M} a_i s[n-i], \qquad n = 0, 1, 2, ..., N+M-1.$$

Covariance method

An alternative, the **covariance method**, chooses the summation limits in Eq. 2.8 to be $n_0 = M$ and $n_1 = N-1$. No signal truncation is forced (the time limitation is imposed on the range of error examined instead) and thus mathematically all speech samples $s[n]$ are used in computing the "covariance matrix" components c_{ij}. This gives

Summary of covariance method
Seek solution to the linear system of equations:

$$\sum_{i=1}^{M} a_i c_{ij} = -c_{0j}, \quad j = 1, 2, ..., M \tag{2.12}$$

where

$$c_{ij} = \sum_{n=M}^{N-1} s[n-i]s[n-j].$$

The prediction error signal is

$$e[n] = s[n] + \sum_{i=1}^{M} a_i s[n-i], \qquad n = M, M+1, ..., N-1.$$

Comparisons

The autocorrelation R and covariance c functions are quite similar, but they differ in windowing effects. The autocorrelation method examines N windowed speech samples, whereas the covariance method uses no explicit window on the speech samples. The former thus imposes a distortion on the spectral estimation (i.e., the explicit windowing of $s(n)$ corresponds to convolving the original short-time $S(e^{j\omega})$ with the frequency response of the window $W(e^{j\omega})$). As noted above, since windows have lowpass spectra, the windowed speech spectrum is a smoothed version of the original, with the extent and type of smoothing dependent on the window shape and duration. The covariance method avoids this distortion.

Both autocorrelation and covariance methods are intended to solve the same **normal**

equation, but the coefficients of the linear equations are not computed the same way in the two methods, giving different results for the estimated parameters $a_1, a_2, ..., a_M$, which characterize the LPC inverse filter $A(z)$. For most applications, the important considerations for LPC analysis of speech are the stability of the forward or synthesis filter $1/A(z)$, the accuracy of the solution, and the efficiency of the solution for real-time computation.

In the autocorrelation method, the M linear equations to be solved are best viewed in matrix form as $\mathbf{Ra} = \mathbf{r}$, where $\mathbf{R}$ is a $M \times M$ matrix of elements $\mathbf{R}(i,j) = r(|i-j|), (1 \leq i,j \leq M)$, $\mathbf{r}$ is a column vector $(r(1), r(2), \ldots, r(M))^{Tr}$, and $\mathbf{a}$ is a column vector of LPC coefficients $(a_1, a_2, \ldots, a_M)^{Tr}$. Solving for the LPC vector requires inversion of the $\mathbf{R}$ matrix and multiplication of the resulting $M \times M$ matrix with the $\mathbf{r}$ vector. The covariance approach is similar, replacing the autocorrelation matrix $\mathbf{R}$ with the $p \times p$ covariance matrix $\mathbf{C}$ of elements $C(i,j) = c_{ij}$ and substituting the $\mathbf{r}$ vector with a $\mathbf{c}$ vector $(c_{01}, c_{02}, \ldots, c_{0M})$.

While matrix inversion is computationally-intensive generally, redundancies in both the $\mathbf{R}$ and $\mathbf{C}$ matrices permit efficient calculation of the LPC coefficients. Both matrices are symmetric (e.g., $c_{ij} = c_{ji}$). In addition, $\mathbf{R}$ is Toeplitz (i.e., all elements along each diagonal are the same); however, $\mathbf{C}$ is not. Thus, the autocorrelation approach (with $2M$ storage locations and $O(M^2)$ math operations) is simpler than the basic covariance method ($M^2/2$ storage locations and $O(M^3)$ operations). ($O(M)$ means "on the order of (or approximately) M".) If $N \gg M$ (e.g., in pitch-asynchronous analysis), then computation of either the $\mathbf{R}$ or $\mathbf{C}$ matrix ($O(pN)$ operations) dominates the total calculation. If the $\mathbf{C}$ matrix is positive definite (which is usually the case for speech signals), its symmetry allows solution through the square root or Cholesky decomposition method [Rab 79], which roughly halves the computation and storage needed for direct matrix inversion techniques.

In comparison with the covariance method, the autocorrelation method's more redundant $\mathbf{R}$ matrix allows the use of more efficient Levinson-Durbin recursive procedure [Makh 75, Rab 79], in which the following set of ordered equations is solved recursively for $m = 1, 2, \ldots, M$:

$$
\begin{aligned}
k_m &= \frac{R(m) - \sum_{k=1}^{m-1} a_{m-1}(k) r(m-k)}{E_{m-1}}, && \text{(a)} \\
a_m(m) &= k_m, && \text{(b)} \\
a_m(k) &= a_{m-1}(k) - k_m a_{m-1}(m-k) \quad \text{for} \quad 1 \leq k \leq m-1, && \text{(c)} \\
E_m &= (1 - k_m^2) E_{m-1}, && \text{(d)}
\end{aligned}
\qquad (2.13)
$$

where initially $E_0 = r(0)$ and $a_0 = 0$. At each cycle m, the coefficients $a_m(k)$ (for $k = 1, 2, \ldots, m$) correspond to the optimal mth-order linear predictor, and the minimum error E_m is reduced by the factor $(1 - k_m^2)$. Since E_m, a squared error, can never be negative, $|k_m| \leq 1$. This condition on the *reflection coefficients* k_m can be related to acoustic tube models, and guarantees a stable LPC synthesis filter $H(z)$ since all the roots of $A(z)$ are then inside (or on) the unit circle in the z-plane. Unlike the covariance method, the autocorrelation method mathematically guarantees a stable synthesis filter (although finite-precision arithmetic may cause an unstable filter).

To minimize the calculation of LPC analysis, both window size N and order M should be minimized. Since M is usually directly specified by how much bandwidth is being modelled in the speech, in practice only the choice of N allows flexibility trading off spectral accuracy and computation. Due to windowing edge effects in the autocorrelation method, N usually includes 2–3 pitch periods to enhance the accuracy of spectral estimates. In the covariance method, the error is windowed (rather than the signal). The formal LPC equations above for the two methods do not explicitly note this windowing difference. Spectral accuracy in both methods increase with larger N, but the lack of direct signal windowing in the covariance method allows an effective window as short as $N = M$; however, such small N risks generating unstable speech synthesis filters.

Most LPC analysis is done with a fixed period (and hence frame rate) and thus without correlation to F0 (i.e., pitch-asynchronously). The basic block analysis methods examine successive sets of N samples, shifted periodically by some fixed fraction of N (e.g., 0.5, leading to a 50% overlap of successive frames). Spectral estimation of the vocal-tract transfer function is less accurate when pitch epochs (the early intense samples in pitch periods, due to vocal-cord closures) are included randomly in an analysis frame. The problem intensifies if N is small, as analysis frames may then be dominated by vocal-cord effects, which are quite poorly handled by LPC models, whose order M is much less than a pitch period length. When the analysis frame contains a few pitch periods (e.g., $N \gg M$), spectral accuracy improves because LPC models the spectral envelope well for speech samples in each period after the inital M points (i.e., after these first samples, $s(n)$ is based on a sample history that includes the effects of the major pitch excitation at vocal-cord closure).

Evaluating the error signal pitch-asynchronously with a rectangular window leads to spectral estimates with undesirable fluctuations, depending on window length. In the autocorrelation method, these are reduced by using a smooth (e.g., Hamming) window to weight the speech signal. These estimation problems can be reduced by pitch-synchronous analysis, where each analysis window is completely within a pitch period. This, however, requires an accurate F0 estimator. In this case, the covariance method is typically used since (unlike the autocorrelation method) it requires no explicit signal window that would distort the speech significantly over short frame analyses of less than one pitch period.

2.2.3 Spectral estimation via LPC

The LPC error that is minimized in choosing the LPC coefficients in Eq. 2.9 was calculated in the time domain, as was all the analysis so far in this section. However, a major use of LPC is for spectral estimation of speech. Thus, we now examine a frequency interpretation for LPC. While we initially defined the error E in the time- domain, Parseval's theorem yields a useful spectral interpretation as well:

$$E = \sum_{n=-\infty}^{\infty} e^2(n) = \frac{1}{2\pi} \int_{\omega=-\pi}^{\pi} |E(e^{j\omega})|^2 d\omega.$$

Furthermore, since $e(n)$ results when filtering speech through its inverse LPC $A(z) = G/H(z)$, the total residual error is

$$E_p = \frac{G^2}{2\pi} \int_{\omega=-\pi}^{\pi} \frac{|S(e^{j\omega})|^2}{|H(e^{j\omega})|^2} d\omega.$$

The usual method to obtain LPC coefficients by minimizing E_p is thus equivalent to a minimization of the average ratio of the speech spectrum to its LPC estimate. In particular, equal weight is accorded to all frequencies (unlike human audition), but $|H(e^{j\omega})|$ models peaks in $|S(e^{j\omega})|$ more accurately than valleys. The latter emphasis on peaks is due to the fact that contributions to the error E_p are greatest for frequencies where the

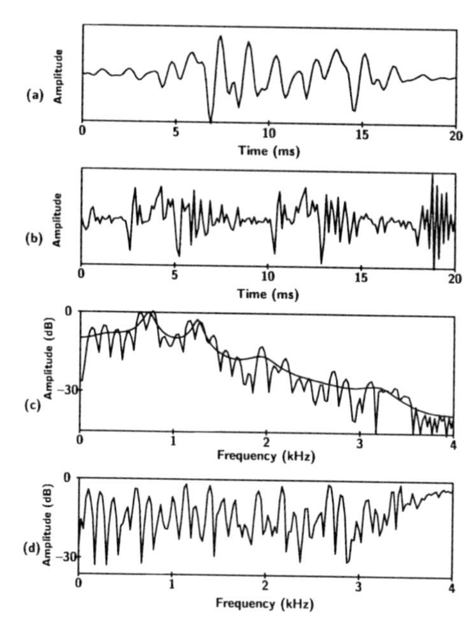

Figure 2.9: Signals and spectra in LPC via the autocorrelation method using 12 poles (a) 20 ms of an $/\alpha/$ vowel from a male speaker at 8000 samples/s (using a Hamming window); (b) residual error signal obtained by inverse LPC filtering the speech (magnified about 3 times); (c) speech spectrum with the smooth LPC spectrum superimposed; and (d) spectrum of the error signal (note the different amplitude scales for parts c and d).

speech spectrum is strong. Larger errors are tolerated in spectral valleys, where both the original speech spectrum and its LPC estimate are small (Fig. 2.9).

We choose the order of the LPC model to be small for coding efficiency, while simultaneously being large enough to model sufficient spectral detail to accomplish coding or recognition applications. The all-pole spectrum $|H(e^{j\omega})|$ models $|S(e^{j\omega})|$ by placement of its p poles; the higher p is, the more spectral detail that LPC can model. A typical choice is $p = 10$, which allows LPC to model four or five resonances well (Fig. 2.10). Obviously, not all relevant spectral detail in a short-time voiced-speech spectrum (rapid frequency variation owing to harmonics, and slower variations of amplitude with frequency owing to formants) can be completely modeled sufficiently by the p-pole $|H(e^{j\omega})|$ of LPC. Use of the minimum-mean-square error criterion forces the slowly-varying LPC spectrum to follow the spectral envelope of $|S(e^{j\omega})|$ at a level just under the harmonic peaks. Any other positioning of $|H(e^{j\omega})|$ (e.g., any attempt to follow the spectral valleys as well as the peaks) would cause larger contributions to the error at spectral peaks. The actual LPC estimation method balances small errors at formant peaks and larger errors in valleys. Happily, this mathematical result owing to the use of the mean-square-error criterion corresponds well with human audition's preference for spectral peaks as more important than valleys. Thus valleys between formants (including valleys due to spectral zeros) are less accurately modeled than formant regions. This leads to less accurate modelling of fomant bandwidths as well.

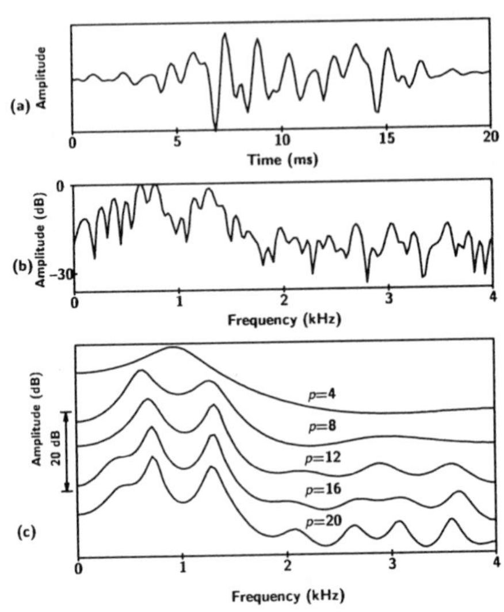

Figure 2.10: Signals and spectra in LPC for 20 ms of an /aa/ vowel at 8000 samples/s: (a) time waveform, (b) speech spectrum, (c) LPC spectra using 4, 8, 12, 16, and 20 poles, respectively.

2.2.4 Pre-emphasis

Following audition, many analysis methods focus their efforts on those parts of the speech spectrum with the highest intensity. If speech perception were to only need the strongest frequencies of speech, then frequencies above 1 kHz (which have much weaker intensity) would be largely ignored. Simple perceptual experiments with such lowpass-filtered speech show this not to be the case. It is thus clear that some aspects of weak energy at higher frequencies in the audio range are also important. In particular, the center frequencies of the second and third formants for sonorant sounds (or correspondingly the lowest broad resonance in obstruent sounds) are very important, and must be modelled well in most speech analysis methods.

To assist LPC to properly model formants of different intensities, a pre- processing technique called **pre-emphasis** is often used as a first step in speech analysis. This pre-processing raises input speech energy by a variable amount that increases as frequency increases. The amount of pre-emphasis is usually specified by α (a constant), determining the cutoff frequency of a single-zero filter used to filter the input speech. Pre-emphasis reduces the large dynamic range of (i.e., "flattens") the speech spectrum by means of adding a zero to the spectrum, which counteracts the -12 dB/octave power falloff owing to the glottal source in voiced speech. Thus, individual zeros owing to pre-emphasis and mouth radiation approximately counterbalance the normal voiced speech falloff, yielding a spectrum with formants of similar heights. This flatter spectrum allows LPC analysis to model all formants equally well. The flattening equalizes formant levels, with little disturbance to formant structure.

Applying pre-emphasis evidently distorts the speech spectrum. So use of the results of pre-emphasized speech must take account of this distortion. In speech coding applications, where the final result is a resynthesized speech signal, the final stage of decoding must contain a de-emphasis filter $1/(1-\beta z^{-1})$ (α and β typically both have values of about 0.94) to counteract the earlier pre-emphasis. Pre-emphasis acts as a differentiator, while de-emphasis does integration. For the speech recognition applications of this book, de-emphasis is unnecessary, as long as pre-emphasis is consistently applied to all speech utterances. It can be thought of simply as a (useful) channel effect.

2.2.5 Choice of order of the LPC model

We noted above that the number p of poles in the all-pole LPC model (i.e., its **order**) reflects the amount of detail in the LPC spectral estimate of the input speech. In practice, the choice of p reflects a compromise of spectral accuracy, computation time and memory. It can be shown that $|H(e^{j\omega})|$ can be made equal to $|S(e^{j\omega})|$ in the limiting case where $p \to \infty$. This is only of theoretical interest, since p rarely goes very high, due to increasingly excessive cost of computation. The order p is chosen to assign enough model poles to represent all formants (at two poles per resonance) in the bandwidth of the input speech signal (for purposes of fixing a single value of p for all speech, vowels are assumed, since some sounds have fewer resonances to model). An additional 2–4 poles are usually assigned (e.g., the standard for 8-kHz sampled speech is 10 poles [Camp 86]) to account for windowing effects and for weaknesses in the all-pole model. The all-pole model ignores spectral zeros and assumes an infinitely-long stationary speech sound; thus assigning only enough poles to model the expected number of formants risks the case that poles may be used by the model to handle non-formant effects in the windowed spectrum (such is seen often in LPC modelling). The non-formant effects derive mostly from the vocal- tract excitation (both glottal and fricative) and from lip radiation. In addition, spectral zeros are regularly found in nasalized sounds. As we will see later, nasal sounds theoretically have more resonances than vowels, but we rarely increase LPC order to handle nasals, because most nasals have more than one resonance with little energy (due to the effects of zeros and losses).

The effect of a zero on a spectrum can be modelled indirectly in an all-pole model. For example, consider the general case of a zero at $z = a$ ($|a| < 1$). At one extreme, any zero can be represented exactly by an infinite number of poles:

$$(1 - az^{-1}) = \frac{1}{1 + \sum_{n=1}^{\infty} (az^{-1})^n}. \tag{2.14}$$

Evaluating on the unit circle ($z = e^{j\omega}$), as is usually done for Fourier analysis of speech, the infinite-order denominator in Eq. 2.14 can be approximated by a finite number M of terms and hence by a finite number of poles. The high-order terms in Eq. 2.14 are reasonably ignored if $a^M \ll 1$. Wide-bandwidth zeros (having small $|a|$) are more accurately modeled with a few poles than are zeros whose bandwidths are comparable to those of the formants.

The prediction error energy E_p can serve as a measure of the accuracy of an LPC model (an equivalent measure is the prediction gain). The **normalized prediction error** (i.e., the energy in the error divided by the speech energy), $V_p = E_p/R(0)$ (Eq. 2.13d), decreases monotonically with predictor order p (i.e., each additional pole in the LPC model improves its accuracy). With voiced speech, poles in excess of the number needed to model all formants (and a few for zero effects) add little to the spectral fit as measured by V_p, but such extraneous poles add increasingly to the computation.

We must note that unvoiced speech yields larger V_p than voiced speech, because the excitation of the former is more spread in time. LPC parameter estimation exploits a history of p prior speech samples, and the LPC model reflects the predictability of each successive speech sample based on the p prior samples. As a result, LPC modelling omits excitation for the basic model in Eq. 2.6. For sonorants, the LPC synthesizer (i.e., the all-pole filter) often admits very simple excitation patterns of one individual impulse per pitch period. Ignoring such a spare excitation in the earlier LPC equations leads to minimal negative effects for voiced speech, because the unpredictable vocal-cord closures (which correspond to the timing of the excitation impulses) are relatively rare occurrences during the typical frame of LPC analysis. (Pitch- synchronous analysis avoids such points when placing the analysis window.) On the other hand, the excitation for unvoiced speech has approximately constant energy over the entire analysis frame however long. As such, the unpredictable, random excitation of unvoiced speech leads to a poorer LPC model for such speech. Luckily, humans have much poorer auditory resolution for spectral differences in such noisy speech, as compared to voiced speech. Some applications of speech processing exploit the fact that V_p is much bigger for unvoiced than voiced speech, by rendering an estimation of the speech frame as voiced or unvoiced based on the relative V_p (high, unvoiced; low, voiced).

2.2.6 Summary

LPC analysis models voiced speech sounds as the output of an all-pole filter in response to a simple sequence of excitation pulses. LPC analysis is based on the characterization of all-pole digital filters. It is the most common technique for low-bit-rate speech coding and is a very important tool in speech analysis and recognition. The popularity of LPC comes from its compact and precise representation of the speech spectral magnitude and its simple computation. LPC is used to estimate F0, vocal tract area functions, and the frequencies and bandwidths of spectral poles (e.g., formants). However, its primary function is to provide a small set of parameters that represent the vocal-tract response.

2.3 Cepstral Analysis of Speech

Speech analysis usually aims toward estimation of parameters of a basic speech-production model, e.g., the output of a linear dynamic system (representing the vocal tract) excited by quasi-periodic (vocal-cord) pulses or random noise (at a vocal-tract constriction). Thus the speech signal is the convolution of an excitation waveform with the vocal-tract response. Many speech applications require separate estimation of these individual components; hence a **deconvolution** of the excitation and envelope components is useful. Producing two signals from one in such a deconvolution is generally nondeterministic, but has some success when applied to speech because the relevant convolved signals forming speech have very different time-frequency behavior.

2.3.1 Principles

As we see below, cepstral deconvolution converts a product of two spectra into a sum of two signals; these may be separated by linear filtering if they are sufficiently different. Let speech spectrum $S = EH$, where E and H represent the excitation and vocal-tract spectra, respectively; then $\log S = \log(EH) = \log(E) + \log(H)$. Since H consists mostly of smooth formant curves (i.e., a spectrum varying slowly with frequency) while E is much more active or irregular (owing to the harmonics or noise excitation), contributions due to E and H can be linearly separated.

To better visualize how cepstral analysis works, take the simple case of a causal exponential signal $x(n) = a^n u(n)$. (If we allow a to be complex, much of speech can be composed of a small number of such exponentials, at least for steady-state sounds.) The z-transform $X(z) = 1/(1 - az^{-1})$ has a pole at $z = a$. Taking a logarithm, we have a power series,

$$\log(X(z)) = \sum_{n=1}^{\infty} \frac{a^n}{n} z^{-n}, \quad \text{if} \quad |z| > |a|.$$

Using the traditional circumflex notation to denote cepstra, the *complex cepstrum* $\hat{x}(n)$ is the inverse transform of $\log(X(z))$. In our example,

$$\hat{x}(n) = -\frac{a^n}{n} u(n-1). \tag{2.15}$$

As a result, $\hat{x}(n)$ is very similar to $x(n)$, retaining the same exponential form except decaying more rapidly, due to the $1/n$ factor.

Since $\log(AB) = \log(A) + \log(B)$, a more complicated speech z-transform $S(z)$ consisting of several poles and zeros results in a complex cepstrum $\hat{s}(n)$ that is a sum of exponential terms, each decaying with the $1/n$ factor. (Since $\log(1/A) = -\log(A)$, the only difference between the effect of a pole and that of a zero in the complex cepstrum is the sign of the power series.) Assuming that speech $S(z)$ has poles p_k and zeros z_k inside the unit circle, the cepstrum has linear contributions of p_k^n/n and $-z_k^n/n$, for $n > 0$. Any zeros b_k outside the unit circle contribute b_k^n/n, for $n < 0$ (no outside poles are allowed, for stability reasons). The overall cepstrum decays very rapidly in time:

$$|\hat{x}(n)| < \alpha \frac{\beta^{|n|}}{|n|}, \quad \text{for} \quad |n| \to \infty,$$

where α is a constant and β is the maximum absolute value among all p_k, z_k, and $1/b_k$ (this maximum corresponds to the closest pole or zero to the unit circle). For speech spectra, the closest pole usually corresponds to the first formant, which has a narrow bandwidth and tends to dominate the shape of the speech waveform. The $1/n$ factor in Eq. 2.15 causes $\hat{s}(n)$ to decay rapidly, i.e., over a few ms. The cepstrum is thus in large contrast to the speech $s(n)$, where individual pitch periods decay slowly over tens of milliseconds.

The cepstrum also contains a component due to the vocal-tract excitation. Voiced speech $s(n)$ can be modelled as the convolution of a unit-sample train $e(n)$ (with spacing N = pitch period) and the vocal-tract response $h(n)$ (in this interpretation, $h(n)$ effectively models all of the speech except its periodicity; i.e., $h(n)$ models both vocal-tract and glottal effects). Modelling $e(n)$ as a periodic train of identical, isolated-sample pulses, $E(e^{j\omega})$ has the form of the same uniform train of impulses, however now in frequency with spacing of $2\pi/N$ radians (the logarithm does not affect the spacing of the impulses). Since $S(z) = E(z)H(z)$, $\log(S(z)) = \log(E(z)) + \log(H(z))$ and $\hat{s}(n) = \hat{e}(n) + \hat{h}(n)$. With $\hat{h}(n)$ lasting only a few milliseconds and $\hat{e}(n)$ for voiced speech being nonzero only at $n = 0, \pm N, \pm 2N, \pm 3N, \ldots$, the two functions are readily separated. An appropriate boundary would be the shortest pitch period, e.g., 3 ms. If no periodic structure is visible in $\hat{s}(n)$, the speech should be considered unvoiced. $\hat{H}(e^{j\omega})$ provides a "cepstrally smoothed" spectrum, mostly removing the undesired $e(n)$ components.

In practice, the real (not complex) cepstrum (the inverse transform of the logarithm of the speech power spectrum) is used:

$$c(n) = \frac{1}{2\pi} \int_{\omega=0}^{2\pi} \log |S(e^{j\omega})| e^{j\omega n} \, d\omega. \tag{2.16}$$

Exploiting the fact that speech $s(n)$ is real-valued, $c(n)$ is the even part of $\hat{s}(n)$ because

$$\hat{S}(e^{j\omega}) = \log(S(e^{j\omega})) = \log |S(e^{j\omega})| + j \arg[S(e^{j\omega})]$$

and the magnitude is real and even, while the phase is imaginary and odd. (In cepstral speech representations, as in other speech modelling methods, phase is often discarded.)

In addition, for digital algorithms, the DFT is used instead of the general Fourier transform in Eq. 2.16:

$$c_d(n) = \frac{1}{N} \sum_{k=0}^{N-1} \log |S(k)| e^{j2\pi kn/N} \qquad \text{for} \quad n = 0, 1, ..., N-1. \tag{2.17}$$

Replacing $S(e^{j\omega})$ with $S(k)$ is like sampling the Fourier transform at N equally spaced frequencies from $\omega = 0$ to 2π. Including the inverse DFT, the total result convolves the original $c(n)$ with a uniform sample train of period N:

$$c_d(n) = \sum_{i=-\infty}^{\infty} c(n + iN).$$

Thus the digital cepstrum $c_d(n)$ contains copies of $c(n)$, at intervals of N samples. Aliasing results, but is unimportant when N is large enough. As the duration of the analysis window, N usually well exceeds a hundred samples, which eliminates most aliasing in $\hat{h}(n)$.

2.3.2 Mel-frequency cepstral coefficients

Cepstral analysis has been widely used as the main analysis method for speech recognition. The most popular use combines the cepstrum with a nonlinear weighting in frequency, following the Bark or mel scale, thus incorporating some aspects of audition. These **mel-frequency cepstral coefficients** c_n (MFCCs) appear to furnish a more efficient representation of speech spectra than other analysis methods such as LPC. A power spectrum S of each successive speech frame is effectively deformed in frequency (according to the bark or critical-band scale) and in amplitude as well (on the usual decibel or logarithmic scale). (The source of the spectrum is sometimes via the DFT and otherwise via LPC.) Then the initial M coefficients c_n of an inverse DFT are obtained ($M \approx 8 - 14$). Usually [Dav 80] critical-band filtering is simulated with a simple set of 20 triangular windows (Fig. 2.11), whose log-energy outputs are designated S_k:

$$c_n = \sum_{k=1}^{20} X_k \cos\left[n\left(k - \frac{1}{2}\right)\frac{\pi}{20}\right] \qquad \text{for} \quad n = 0, 1, 2, \ldots, M. \tag{2.18}$$

The initial value c_0 simply represents the average speech power. Because such power varies significantly with microphone placement and communication channel conditions, it is often not directly utilized in recognition, although its derivative often is. The next MFCC c_1 indicates the balance of power between low and high frequencies, where a

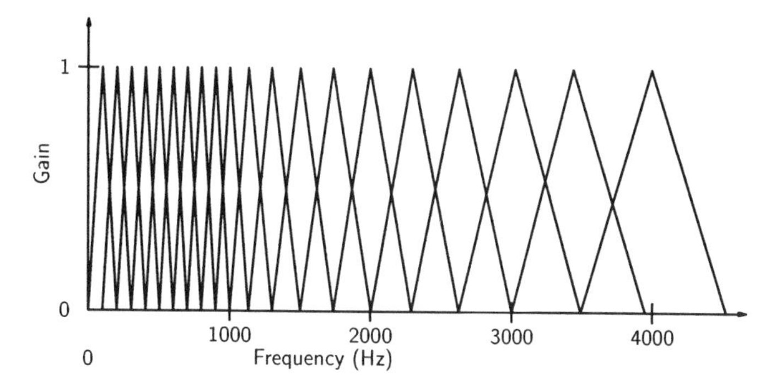

Figure 2.11: Filter bank for generating mel-based cepstral coefficients (after Davis and Mermelstein [Dav 80]).

positive value (owing to the positive half-cycle of the weighting cosine in Eq. 2.18 for the lower half of the full frequency range) indicates a sonorant sound, and a negative c_1 a frication sound. (This reflects the fact that sonorants have most energy at low frequencies, and fricatives the opposite.) For each $i > 1$, c_i represents increasingly finer spectral detail (as a cosine with i periods weights smaller frequency ranges with its alternating oscillations); e.g., c_2 weights the $0 - \alpha$ Hz and $2\alpha - 3\alpha$ Hz ranges positively (and the other two ranges negatively), where α is one-eighth of the sampling rate. Neither MFCCs nor LPC coefficients display a simple relationship with basic spectral envelope detail such as the formants. For example, using a speech bandwidth with four formants (e.g., 0–4 kHz), a high value for c_2 corresponds to high power in the F1 and F3 ranges but low amounts in F2 and F4 regions. Such information is useful to distinguish voiced sounds, but it is difficult to interpret physically.

2.4 Automatic Extraction and Tracking of Speech Formants

The main objective of speech analysis is to automatically extract essential parameters of the acoustic structure from the speech signal. This process serves the purpose of either data reduction or identification and enhancement of information-bearing elements contained in the speech signal. To determine what are such information-bearing elements, one needs to have sufficient knowledge about the physical nature of the speech signal. Such knowledge is often provided by a **synthesis model** that describes how the observed speech signal is generated as the output of a vocal-tract digital filter given an input source. This type of "synthesis" model is often called a source-filter model. This model decomposes the speech waveform into two independent source and filter components. Formants comprise a very important set of the information-bearing elements in view of the source-filter model.

2.4.1 Formants and vocal tract resonances

Formants characterize the "filter" portion of a speech signal. They are the poles in the digital resonance filter or digital resonator. Section 1.4.2 gave the simplest example of a single-formant resonator with one pair of complex poles. The resonator filter for generating the speech signal is far more complex. However, for a subset of speech sounds, voiced sounds with negligible coupling to the nasal and subglottal systems, extension of the all-pole, single-formant resonator discussed in Section 1.4.2 to an all-pole, multiple-formant resonator will give a good approximation to the reality.

Given the source-filter model for voiced speech that is free of coupling, the all-pole filter is characterized by the pole positions, or equivalently by the formant frequencies, $F_1, F_2, \ldots, F_n$, formant bandwidths, $B_1, B_2, \ldots, B_n$, and formant amplitudes, $A_1, A_2, \ldots, A_n$. Among them, the formant frequencies or resonance frequencies, at which the spectral peaks are located, are the most important. As exemplified in Section 1.4.2, a formant frequency is determined by the angle of the corresponding pole in the discrete-time filter transfer function (i.e., in the z-plane). The normal range of the formant frequencies for adult males is $F_1 = 180 - 800$ Hz, $F_2 = 600 - 2500$ Hz, $F_3 = 1200 - 3500$ Hz, and $F_4 = 2300 - 4000$ Hz. These ranges are useful to provide constraints for automatic formant extraction and tracking. The average difference between adjacent formants is 1000 Hz. For adult females, the formant frequencies are about 20% higher than adult males. The relationship between male and female formant frequencies, however, is not uniform and the relationship deviates from a simple scale factor. When the velum is lowered to create nasal phonemes, the combined nasal+vocal tract is effectively lengthened from its typical 17-cm vocal-tract length by about 25%. As a result, the average spacing between formants reduces to about 800 Hz.

Formant bandwidths are physically related to energy loss in the vocal tract, and are determined by the distance between the pole location and the origin of the z-plane in the filter transfer function. Empirical measurement data from speech suggest that the formant bandwidths and frequencies are systematically related and the relationship can be approximated by

$$B_n = 50(1 + \frac{F_n}{1000}) \;\; Hz.$$

Note that this relationship holds for each of the individual formants separately. Formant amplitudes, on the other hand, vary with the overall pattern of formant frequencies as a whole. They are also related to the spectral properties of the voice source.

Vocal-tract resonances

While the formants are defined in terms of the power spectrum of the speech output of the vocal tract and the nasal tract, the actual vocal-tract resonances (**VT-resonances**) are not exactly the same as the formants. Formants and VT-resonances are related but distinct concepts: the former are defined in the acoustic domain (peaks of smoothed spectra in a narrow frequency region), and the latter are associated with the VT properties per se (e.g., the possibility to describe switching VT-cavity affiliations). The acoustic-domain definition of the formant in terms of a local concentration of smoothed spectral energy has been mainly applied to voiced portions of the speech signal, which contain glottal excitation, that excites the entire vocal tract.

Wideband spectrograms display formant patterns which exhibit continuity in time during voiced sequences, i.e., they largely appear as horizontal black bands moving smoothly up and down in frequency, as the organs of the vocal tract move. Continuous excitation of a slowly-changing vocal tract leads to slowly-varying formant patterns. The center-frequencies of the resonance bands consistently correspond to VT poles in such conditions. However, formants become discontinuous when certain major changes occur in the vocal tract. These are either abrupt excitation or VT-shape changes: 1) the vocal cords start or stop vibrating, 2) the velum lowers or raises, or 3) a noisy excitation starts or ceases. The excitation for voiced sounds is more intense than for unvoiced sounds, and excites the full vocal tract; thus a major change in vocal cord status (e.g., cessation of voicing) frequently leads to the appearance or disappearance of most of the formants (especially the higher ones, with correspondingly lower amplitudes in general). A position change in the velum causes the nasal cavity to either participate or not in the VT response; a raised velum excludes its effects, while a lowered velum abruptly introduces the lossy effects of the nasal cavity. Thus at the boundaries of nasal sounds, formants usually change abruptly in terms of their center frequency, intensity, and bandwidth.

The VT-resonances are defined uniformly for all classes of speech sounds regardless of the VT excitation (an analogy can be made here – *poles* can be used (uniformly) to characterize an electrical circuit, regardless of whether zeros exist or not) and are temporally smooth from one speech unit to another during the production of any speech utterance. The relations between the acoustically-defined formants and the VT-resonances can be summarized as follows. For the production of vocalic sounds of speech where the VT-resonances are generally well separated in frequency and the bandwidths of the resonances are reasonably narrow, the VT-resonances manifest themselves as the local peaks of the acoustic spectra and are consistent with the formants. During the production of most consonants, some VT-resonances are either weakly excited or masked/distorted by VT anti-resonants (zeros) and hence will not manifest themselves in the acoustic spectra. The property of temporal smoothness applies to the VT-resonances, but not to the acoustically defined formants as local maxima in the spectra of the speech signal.

2.4.2 Formant extraction and tracking methods

LPC analysis, as discussed in Section 2.2, models voiced speech as the output of an all-pole filter in response to a simple sequence of excitation pulses. In addition to major speech coding and recognition applications, LPC is often used as a standard formant extraction and tracking method. It has limitations in that the vocal-tract filter transfer function, in addition to having formant poles (which are of primary interest for speech analysis), generally also contains zeros due to sources located above the glottis and to nasal and subglottal coupling. Furthermore, the model for the voice source as a simple sequence of excitation impulses is inaccurate, with the source actually often containing local spectral peaks and valleys. These factors often hinder accuracy in automatic formant extraction and tracking methods based on LPC analysis. The situation is especially serious for speech with high pitch frequencies, where the automatic formant-estimation method tends to pick harmonic frequencies rather than formant frequencies. Jumps from a correctly- estimated formant in one time frame to a higher or a lower value in the next frame constitute one common type of tracking error.

The automatic tracking of formants is not trivial, despite several observations that may suggest otherwise: formants exhibit an average spacing in frequency inversely proportional to vocal-tract length (e.g., 1 kHz for a typical man), formant bandwidths are fairly consistent and gradually increasing with frequency, and temporal displacement of formants is gradual most of the time. The factors rendering formant identification complex include the following. The ranges for formant center-frequencies are large, with significant overlaps both within and across speakers. In phoneme sequences consisting only of oral vowels and sonorants, formants smoothly rise and fall, and are easily estimated via spectral peak-picking. However, nasal sounds cause acoustic coupling of the oral and nasal tracts, which lead to abrupt formant movements. Zeros (due to the glottal source excitation or to the vocal tract response for lateral or nasalized sounds) also may obscure formants in spectral displays. When two formants approach each other, they sometimes appear as one spectral peak (e.g., F1-F2 in back vowels). During obstruent sounds, a varying range of low frequencies is only weakly excited, leading to a reduced number of formants appearing in the output speech.

Assuming access to an LPC representation of speech, and hence a spectral representation $S(z)$ (via the LPC coefficients), one could directly locate formants by solving directly for the roots of the denominator polynomial in $S(z)$. Each complex-conjugate pair of roots would correspond to a formant if the roots correspond to a suitable bandwidth (e.g., 100-200 Hz) at a frequency location where a formant would normally be expected. This process is usually very precise, but quite expensive since the polynomial usually requires an order in excess of 10 to represent 4–5 formants. Alternatively, one can use phase to label a spectral peak as a formant [Red 84, Mik 84]. When evaluating $S(z)$ on the unit circle $z = \exp(j\omega)$, a negative phase shift of approximately π radians (i.e., -180 degrees) should occur as ω passes a pole close to the unit circle (i.e., a formant pole).

Two close formants often appear as a single broad spectral peak, a situation that causes many formant estimators difficulty, in determining whether the peak corresponds to one or two resonances. A modified DFT method called the **chirp z-transform** (CZT) (denoting a 'chirp,' like a bird's signal of increasing frequency) may be used to resolve this issue. It calculates the z-transform of a signal (e.g., windowed speech) on any spiral contour inside the unit circle. The traditional DFT (or FFT) samples $S(z)$ at uniform intervals on the entire unit circle, but the generally more computationally-expensive CZT allows much greater freedom to place the evaluation contour anywhere in the z-plane. For formant estimation, the CZT is typically done near a pole corresponding to a specific ambiguous spectral peak, to limit computation. Appropriate placement of a CZT contour can easily resolve two closely-spaced formants. As a more general replacement of the DFT for formant estimation, a complete CZT would have a spiral contour starting near $z = \alpha$ inside the unit circle (e.g., $\alpha = 0.9$), and gradually spiraling inward with increasing frequency $\omega_k = 2\pi k/N$ ($z_k = \alpha\beta^k \exp(j\omega_k)$, with β just less than 1). Such a contour follows a typical path for formant poles, which generally have formant bandwidths that are larger at higher frequencies.

Formant estimation is increasingly difficult for voices with high F0, as in children's voices. In such cases, F0 often exceeds formant bandwidths [Mon 83], and harmonics are so widely separated that only one or two make up each formant. A spectral analyzer, traditionally working independently on one speech frame at a time (for simpler computation), would often equate the strongest harmonics as formants. Human perception, integrating speech over many frames, is capable of properly separating F0 and the spectral envelope (formants), but simpler computer analysis techniques often fail. It is wrong to label a multiple of F0 as a formant center frequency, except for the few cases where the formant aligns exactly with a multiple of F0 (such alignment is common in song, but much less so in speech).

2.5 Automatic Extraction of Voicing Pitch

While the formant parameters just discussed characterize the "filter" portion of the speech signal (in light of the source-filter model), pitch parameters characterize the "source" portion of the signal. Although F0 estimation appears fairly simple at first glance, full accuracy has so far been elusive, owing to speech's nonstationary nature, irregularities in vocal cord vibration, the wide range of possible F0 values, interaction of F0 with vocal tract shape, and degraded speech in noisy environments [Hess 83, Rab 76].

F0 can be estimated either from periodicity in the time-domain or from harmonic spacing in frequency. Spectral approaches generally have higher accuracy than time-domain methods, but need more computation.

Since the major excitations of the vocal tract occur when the vocal cords close for a pitch period, each period starts with high amplitude and then has an amplitude envelope that decays exponentially with time. Since the very lowest frequencies dominate power in voiced speech, the overall rate of decay is usually inversely proportional to the bandwidth of the first formant. The basic method for pitch period estimation is a simple search for amplitude peaks, constraining the peak-to-peak interval to be consistent in time (since F0 varies slowly). Because speakers can range from infants to large men, a large pitch-period range from about 2 ms to 20 ms is possible.

Input speech is often lowpass-filtered to approximately 900 Hz so as to retain only the first formant, thus removing the influence of other formants, and simplifying the signal, while retaining enough harmonics to facilitate peak-picking. F0 estimation in the time domain has two advantages: efficient calculation, and direct specification of pitch periods in the waveform. This is useful for applications when pitch periods need to be manipulated. On the other hand, F0 values alone (without explicit determination of the placement of pitch periods) suffice for many applications, such as vocoders.

When F0 is estimated spectrally, the fundamental itself and, more often, its equally-spaced harmonics can furnish the main clues. In time-domain peak-picking, errors may be made due to peaks corresponding to formants (especially F1), misinterpreting the waveform oscillations due to F1 as F0 phenomena. Spacing between harmonics is usually more reliable as an F0 cue. Estimating F0 directly in terms of the lowest spectral peak in the speech signal can be unreliable because the speech signal is often bandpass filtered (e.g., over telephone lines). Even unfiltered speech has a weak first harmonic when F1 is at high frequency (as in low vowels). While often yielding more accurate estimates than time-domain methods, spectral F0 detectors require much more calculation due to the required spectral transformation. Typical errors include: 1) misjudging the second harmonic as the fundamental and 2) the ambiguity of the estimated F_0 during aperiodicities such as in voice creak. A given estimation method often performs well on some types of voice but less so for other types. Without expressing an opinion on the actual quality of different people's voices, we can say that speakers whose speech in general works well with a given speech processor are called 'sheep' while others are 'goats' (it is, of course, the challenge for speech processor designers to accommodate goats better, while not losing too many sheep).

2.5.1 Basics of pitch estimation methods

We estimate F0 either from periodicity in the time domain or from regularly-spaced harmonics in the frequency domain. Like many pattern recognition algorithms, most pitch estimators have three components: a preprocessor to simplify the input signal (eliminate information in the signal that is irrelevant to F0), a basic F0 extractor to form the F0 estimate, and a postprocessor to correct errors. The preprocessor serves to focus the remaining data towards the specific task of F0 determination, reducing data rates by eliminating much formant detail. Since the basic pitch estimator, like all pattern recognizers, makes errors, a postprocessor may help to 'clean up' the time series of output pitch estimates (one per frame), e.g., imposing continuity constraints from speech production theory, which may not have been applied in the basic F0 extractor, which often operates independently on each speech frame.

The pitch detector algorithm tries to locate one or more of the following features in the speech signal or in its spectrum: the fundamental harmonic F0, a quasi-periodic time structure, an alternation of high and low amplitudes, and signal discontinuities. The intuitive approach of looking for harmonics and periodicities usually works well, but fails too often to be relied upon without additional support. In general, pitch detectors trade off complexity in various components; e.g., harmonic estimation requires a complex preprocessor (e.g., often including a Fourier transform) but allows a simple basic extractor that just does peak picking. The preprocessor is often just a lowpass filter, but the choice of the filter's cutoff frequency can be complicated by the large range of F0 values possible when accepting speech from many different speakers.

Frequency-domain methods for pitch detection exploit correlation, maximum likelihood, and other spectral techniques where speech is analyzed during a short-term window for each input frame. Autocorrelation, average magnitude difference, cepstrum, spectral compression, and harmonic-matching methods are among the varied spectral approaches [Hess 83]. Spectral methods generally have greater accuracy than time-domain methods, but require more computation.

To provide results in real-time for many applications, F0 estimators must work with little delay. Delays normally arise in part from the use of a buffer to accumulate a large frame of speech to analyze, since pitch can only be detected over intervals corresponding to pitch periods (unlike spectral envelope information, like formants, which can succeed with much shorter analysis frames). Time-domain F0 estimators often incur less delay than frequency-domain methods. The latter mostly require a buffer of speech samples prior to their spectral transformation. Many of the F0 detectors with less delay sacrifice knowledge about the timing of the pitch periods; i.e., they estimate the lengths of pitch periods without explicitly finding their actual locations. While most F0 estimators do not need to locate period times, those that do locate them are more useful in certain applications (e.g., to permit pitch-synchrounous LPC analysis).

2.5.2 Time-domain F0 estimation

Most people can easily divide the image of a voiced speech waveform into successive pitch periods. Following the major excitation of the vocal tract for each pitch period, at the time of vocal-cord closure, each period starts with a high amplitude (referred to as an **epoch**) and then has a time envelope that roughly decays exponentially. The initial time peak occurs about 1/2 ms after vocal-cord closure, the time power takes to travel the 17-cm length of the vocal tract (assuming the microphone recording the speech is quite close to one's mouth). The rate of envelope decay is usually inversely proportional to the bandwidth of the first formant, since F1 dominates the speech (when F2 is close to F1 a broader common bandwidth applies). Often, there is sufficient envelope decay in each pitch period to allow epoch determination by the means of simple peak-picking, as long as some simple guidelines are followed as to maximum and minimum durations for pitch periods. Since speakers theoretically may range in vocal-cord size from an infant to a large man, pitch periods are found over a range of almost 20 ms (starting from less than 2 ms in infants). In many practical applications, we can exclude from consideration extremely high F0 voices, which thus simplifies the F0 estimation task. The rate at which F0 can change is finite (excluding the issue of defining F0 for unvoiced speech). In a given section of voiced speech, F0 usually changes slowly with time, and rarely by more than an octave during a 100 ms phone. Continuity constraints must be applied with care, since once a mistake is made continuity would cause errors to propagate. Continuity cannot be invoked across consonants: the resumption of voicing in speech after a silence or unvoiced consonant can have F0 very different from that at the end of the previous voiced section.

Many F0 estimation errors occur when speech switches from voiced to unvoiced excitation. The initial and final pitch periods in a voiced section of speech are often irregular, as the vocal cords change state. To assist the basic peak-picking of time-domain pitch detectors (and also useful for other methods), the input speech is normally lowpass-filtered as part of the preprocessor so that only F1 is retained (e.g., preserving only the 0–900-Hz range). This filtering removes higher formants (which tend just to confound F0 estimation), thus 'flattening' the remaining spectrum, while still retaining the strongest harmonics to assist F0 estimation. In one traditional approach, epoch candidates are retained if they exceed a variable-amplitude threshold: assuming all periods exceed 2 ms (or so), this threshold would remain very high for 2 ms after each estimated epoch, declining exponentially thereafter (following a basic F1-bandwidth time constant) [Gold 69].

One can easily conceive of a more direct approach: filter out all frequencies except that of the fundamental harmonic, then do a zero-crossing analysis (if only the speech fundamental sinusoid remains, then F0 is just half the ZCR (section 2.1.1)). Unfortunately, setting the cutoff frequency for the filter is quite difficult. It would have to be high enough to pass a harmonic from a high-F0 voice yet low enough to reject the second harmonic of a low-F0 voice. In addition, telephone speech does not pass the fundamental in most cases (eliminating the 0–300-Hz range).

Signal anaylsis for speech coding applications requires not losing any relevant information in the speech signal. For F0 estimation (and other selective feature extraction procedures), however, faithful reproduction of the speech is not needed, and thus the speech signal is often distorted during preprocessing so as to reduce calculation and to enhance periodicity parameters. In one procedure called **center clipping**, the speech $s(n)$ has its low-amplitude samples set to zero and the magnitude of each remaining sample is reduced, in an effort to improve F0 estimation [Sond 68]. A clipping threshold that varies with averaged estimated speech power (e.g., 30% of the maximum $|s(n)|$) is used. Extending this idea to **infinite-peak clipping**, which reduces $s(n)$ to a simple binary zero-crossing signal, also can yield good F0 estimation, while also significantly reducing calculation (e.g., all multiplications involve only zeros and ones).

The autocorrelation function is commonly used for F0 estimation, because it eliminates phase variation from the speech signal. It is equivalent to convolving the speech signal with a time-reversed version of itself. This zeros the phase and squares the spectrum, thus emphasizing waveform peaks (both in the time and frequency domains), which allows more reliable peak-picking for F0.

Maximum-likelihood methods provide another time-domain F0 estimator, especially useful for noisy speech [Frie 78]. We add a delayed version $s(n - D)$ to the original speech $s(n)$. When the delay $D = 1/F0$ (i.e., equals one pitch period), then the sum $s(n) + s(n - D)$ has reinforced speech, but any additive noise tends to cancel, because its random phase is uncorrelated with the original signal after any such delay.

2.5.3 Short-time spectral techniques for F0 estimation

The major advantages for doing F0 estimation in the time domain are lower computation and specification of the pitch epochs. Despite these advantages, many pitch detectors are nonetheless based on spectral analysis. The latter require a frame of speech samples to process, and transform it into a spectral representation to make the periodicity in the signal more evident. Periodicity is most clear in a regular spacing of the major peaks in the spectrum at the fundamental frequency (if not using telephone speech) and at certain harmonics (primarily those in the low-frequency formant regions, where intensity is strongest). Peak picking on the speech signal in the time domain risks misinterpreting waveform oscillations due to F1 as being F0-related, whereas the spectral peaks of frequency-domain pitch detectors are usually much easier to relate to F0.

For spectral pitch estimators, we view the spectral transformation (e.g., FFT) as the preprocessor, while the spectral peak detector acts as the basic F0 estimator. The postprocessor then examines estimates from successive frames to correct errors. Such errors can be major, e.g., estimates that are double or half of the actual F0, which often result from confusing the fundamental with its first harmonic. (Half estimates for F0 are more common in time-domain methods, when two periods are mistaken as one.) Fine (small) F0 errors of a few Hz may be more difficult to handle than major errors (e.g., doubling), and tend to occur when analysis frames are chosen to be too short (not retaining enough data to specify F0 accurately) or too long (allowing F0 to change over the course of the frame). Often, simple postprocessing can significantly improve performance [Herm 93].

These example errors illustrate the compromises needed when choosing frame length. As found in all windowing applications of speech (i.e., short-term analysis), we get the most useful parameters if the signal is stationary during the frame chosen. Hence the frame must be short (at most, a few pitch periods), since F0 may change rapidly. On the other hand, the frame must contain more than one period to provide reliable periodicity information. The precision of any F0 measurement is often proportional to the number of samples in the analysis frame; as a result, short frames are more liable to small pitch errors. Since windows always furnish average results, the single F0 estimate that is output from each analyzed frame is an average F0 value for that frame.

In spectral F0 detection, it is tempting to simply look for the fundamental frequency component of the spectrum. However, we noted that this is unreliable because the fundamental often has low energy. The harmonic structure (i.e., spectral peaks at multiples of F0), which appears throughout the spectrum, is a much more reliable cue to F0. Once several strong (harmonic) peaks are identified, the F0 estimate is just the frequency of the greatest common divisor of the located harmonics. Another related approach measures F0 as the common separation of adjacent harmonics. Yet another alternative compresses the spectrum by successive integer factors (i.e., shrink the frequency scale by factors of two, three, four, ...): a sum of such compressed spectra would have its strongest peak at F0, owing to reinforcement of shifted harmonics [Herm 93]. In the **harmonic-sieve** approach, a spectral "sieve" with variable equally-spaced "holes" is aligned with the spectrum, and the frequency spacing where the most harmonics align with holes of the sieve becomes the F0 estimate [Duif 82].

The cepstrum provides a related way to estimate F0. As noted earlier, it is the inverse Fourier transform of a spectral representation of the speech. The MFCC approach (as used for speech recognition) destroys pitch information when calculating the cepstrum, but if one instead uses the basic cepstrum (i.e., with no mel-scale and no triangular filters) and if the original windowed speech contains multiple pitch periods, then the cepstrum will have peaks at intervals of the average pitch period of the windowed speech. Since the effects of the vocal-tract transfer function (i.e., the spectral envelope) compress to the low-time portion of the cepstrum (the first few ms), the upper range, which conveniently matches one's F0 range, has little energy except that corresponding to F0. Thus, peak-picking in that range usually reveals a good estimate for pitch period duration. The cost of the two transforms (the original Fourier transform, and its inverse after taking the log magnitude), however, usually discourages use of the cepstrum for pitch detection.

In general, spectral F0 detectors give more accurate pitch estimates than time-domain methods, but require about 10 times more calculation, owing to the required spectral transformation. This transformation tends to focus information about speech periodicity in a way that time-domain analysis does not. If we assume that **voicing determination** is an inherent part of F0 detection (e.g., the absence of an F0 indicator is a good sign of unvoiced speech), the performance of different pitch detectors can be evaluated objectively by four types of errors: gross F0 errors (e.g., doubling and halving), fine F0 errors (being off by a few Hz), misinterpreting a voiced speech frame as unvoiced, and vice versa. No algorithm has been found to be superior in all four categories [Rab 76].

Rather than judge F0 detectors objectively by how close their estimates are to true values, the detectors can instead be evaluated perceptually by using their results in speech vocoders, if such latter systems represent excitation in terms of F0 and voicing decisions. No single type of objective F0 error correlates well with subjective quality of coded speech, but voiced frames interepreted as unvoiced appear to be the errors most objectionable perceptually, as they cause harsh, noisy sounds where periodic sounds should be produced [McG 77]. For many applications, the computationally simple AMDF ranks high in subjective tests [Pal 84].

2.6 Auditory Models for Speech Analysis

The analysis methods discussed above try to simulate very simple aspects of the human auditory system, often based on the assumption that human signal processing is an efficient way to do speech recognition. Since the inner ear apparently transfers information to the auditory nerve on a spectral basis, the simple DFT (as well as the more complex MFCC) approximates this transfer very roughly. Other speech analysis methods go further in this simulation in the hope that improved analysis may occur. Many of these alternative approaches use auditory models based on filter-bank signal decomposition, where the filtering follows that found in the cochlea more precisely than do other popular techniques.

2.6.1 Perceptual linear prediction

For example, the **perceptual linear prediction** (PLP) method modifies basic LPC, using a critical-band power spectrum with a logarithmic amplitude compression. The spectrum is multiplied by a mathematical curve modelling the ear's behavior in judging loudness as a function of frequency. The output is then raised to the power 0.33 to simulate the power law of hearing [Herm 90], [Herm 98]. Seventeen bandpass filters equally-spaced in Bark z (i.e., critical bands),

$$z = 6\log\left(\frac{f}{600} + \sqrt{\left(\frac{f}{600}\right)^2 + 1}\right),$$

map the range 0–5 kHz, for example, into the range 0–17 Bark. Each band is simulated by a spectral weighting,

$$c_k(z) = \begin{cases} 10^{z-y_k} & \text{for } z \le y_k \\ 1 & \text{for } y_k < z < y_k + 1 \\ 10^{-2.5(z-y_k-1)} & \text{for } z \ge y_k + 1, \end{cases}$$

where z_k are the center frequencies and $y_k = z_k - 0.5$. One advantage of the PLP is that its order is significantly less than orders generally used in LPC (e.g., fifth-order versus LPC orders typically exceeding ten).

PLP has often been combined with the RASTA (RelAtive SpecTrAl) method of speech processing. RASTA bandpasses spectral parameter signals to eliminate steady or slowly-varying components in the speech signal (including environmental effects and speaker characteristics) and rapid noise events. The bandpass range is typically 1–10 Hz, with a sharp zero at 0 Hz and a time-constant of about 160 ms [Herm 94], [Herm 98]. Events changing more slowly than once a second (e.g., most channel effects, except in severe fading conditions) are thus eliminated by the highpass filtering. The lowpass cutoff is more gradual, which smooths parameters over about 40 ms, thus preserving most phonetic events, while suppressing noise.

2.6.2 Other methods

Another yet simpler auditory modification to the basic DFT approach is the **modulation spectrogram** [King 97], which emphasizes slowly-varying speech changes corresponding to a rate of approximately 4 per second. Such 250-ms units conform roughly to syllables, which studies have shown to be important units for perceptual organization. Such spectral displays show less rapid detail than standard wideband spectrograms. In a sense, they follow the idea of wavelets, which allow time and frequency resolution in automatic analysis to follow that of the ear, i.e., variable (rather than fixed, as in the DFT).

Part of the auditory processing of sounds is adjusting for context, or 'normalization.' Since human audition is always active, even when people are not specifically listening, it is normal to ignore ambient sounds. People pay attention to unexpected auditory information, and 'filter' what they hear to 'normalize out' (i.e., ignore) many predictable (and thus useless) aspects of sounds. These adjustments come naturally to humans, as part of the maturation of their hearing systems. In computer sound processing, however, we must explicitly model this behavior. Later sections of this book will discuss adaptations for variations across speakers (e.g., speaker-dependent recognition models). Here it suffices to mention that humans adjust to their acoustic environment (e.g., noise, use of telephone, poor microphone or loudspeaker). Since automatic speech recognition involves comparing patterns or models, such environmental variations can cause major acoustic differences which are superfluous for the ASR decision, and which human audition normalizes for automatically.

Basic speech analysis methods such as the DFT, LPC and MFCC suffer significantly here. The filtering effects of RASTA provide one way to try to do normalization, to improve the results of analysis for noisy speech. In another common approach, a mean spectrum or cepstrum is subtracted from that of each speech frame (e.g., as a form of blind deconvolution), to eliminate channel effects. It is unclear over what time range this mean should be calculated; it is often done over several seconds, on the assumption that environmental conditions do not change more rapidly. This, however, may impose a delay on the speech analysis; so the channel conditions are sometimes estimated from prior frames and imposed on future ones in their analysis (in the hope that prior conditions remain valid over the short term). Calculating such a mean may require a long-term average for efficiency, which is difficult for real-time applications. Often the mean is estimated from a prior section of the input signal that is estimated to be (noisy) silence (i.e., non-speech). This latter approach requires a speech detector and assumes that pauses occur fairly regularly in the speech signal. As the channel changes with time, the mean must be updated.

2.7 Summary

Following the first chapter, which introduced basic mathematical notions of signal analysis, we now passed in this chapter to specific techniques for speech processing. This chapter presented a discussion of several speech analysis methods, assuming that our task is to transform a speech signal into a set of parameters that more economically represents the pertinent information in the original speech. Time-domain analysis typically yields simple speech parameters efficiently, which are especially suitable for segmentation purposes, whereas spectral analysis provides the more common approach to an efficient representation of speech information.

First, a time-frequency representation for speech signals based on the short-time Fourier transform was discussed. Much of signal analysis in many fields derives from this STFT. More specific to speech was the following topic of linear predictive speech analysis, based on a characterization of all-pole digital filters. This all-pole or autoregressive approach is popular in non-speech estimation tasks, but has dominated much of speech analysis for many years. We discussed the mathematical foundations of LP analysis, as well as the practical ways to implement it.

This was followed by a description of the cepstral analysis approach for speech signals. This was based on a specific DSP technique for mixed linear and nonlinear discrete-time system analysis; its value lies in its efficient representation of the spectral envelope of speech, and especially in its ability to accept frequency warping, e.g., in the form of mel-scale weighting. Cepstral analysis is the standard analysis technique for automatic speech recognition. The cepstral coefficients usually include a perceptual mel-scale mapping, as well as a means to produce a small set of parameters for efficient use. For this, we use a fast Fourier transform (FFT) on each speech frame, to get the standard $X(k)$ Fourier coefficients, then we take the log-magnitude $\log X(k)$, multiply by the critical-band triangular filters (Eq. 2.18), and take an inverse FFT. The low-order 10-16 parameters are the mel-scale cepstral coefficients, from which the delta parameters are simply differenced values between two neighboring frames. These cepstral parameters are often weighted by a raised-sine pulse (to de-emphasize low-order values that may relate to channel conditions, as well as high-order values that correspond to less relevant fine spectral detail).

Next, we examined how the spectral peaks of speech known as formants might be estimated and tracked over time, based on an analysis of digital resonant systems. Formants have been shown to be very important from several points of view: speech production, perception, coding, recognition and synthesis. An efficient way to track their dynamic behavior would be very helpful in many areas of speech processing. Fundamental frequency or pitch is another important feature of speech, and we examined how its dynamic behavior could be estimated practically. Voicing pitch tracking is based on time- and frequency-analysis techniques.

The final section of this chapter dealt with speech analysis methods that employ auditory models based on filter-bank signal decomposition. Such auditory models have been the subject of much recent research effort, aimed at improving the performance of speech coders and recognizers, through better simulation of human auditory phenomena.

Chapter 3

Probability and Random Processes

The preceding two chapters were concerned with analysis and processing of deterministic signals, where we managed to dispense with the notion of uncertainty or randomness in the signal description. In the real world we live in, however, experiences of randomness permeate our daily lives. This could be either due to the true nature of our encounters, or due to our admitted lack of understanding of their character, or even due to our deliberate attempt to use randomness (in spite of the understanding) to our advantage. In speech processing, all these three possibilities are common. In particular, the philosophy of building random, statistical, "ignorance" models for the speech data has been prevalent and rooted in much of modern speech processing techniques. We will discuss many such techniques in the remainder of this book.

The fundamental mathematical background for understanding and developing the techniques capitalizing on the notion of randomness is probability theory, statistics, and random process theory. We will review these theories in this chapter, and use some of the results reviewed here to build up further mathematical background in statistical signal processing in the next two chapters.

3.1 Random Variables, Distributions, and Summary Statistics

3.1.1 Random variables and their distributions

A very fundamental concept in probability theory and in statistics is a **random variable**. A scalar random variable is a real-valued function or variable which takes its value based on the outcome of a *random* experiment. A vector-valued random variable is a set of scalar random variables, which may either be related to or be independent of each other. Since the experiment is random, the value assumed by the random variable is random as well.

A random variable can be understood as a mapping (with some arbitrariness) from a random experiment to a variable. Depending on the nature of the experiment and of the design of the mapping, a random variable can take either discrete values, continuous values, or a mix of discrete and continuous values. We hence see the names of **discrete random variable, continuous random variable**, or **mixed (hybrid)**

random variable. All possible values which may be assumed by a random variable are sometimes called its **domain**.

The key characterization of a (scalar) continuous random variable, x, is its **distribution** or the **probability density function (PDF)**, denoted generally by $p(x)$. The PDF for a continuous random variable at $x = a$ is defined by

$$p(a) \doteq \lim_{\Delta a \to 0} \frac{P(a - \Delta a < x \leq a)}{\Delta a} \geq 0,$$

where $P(\cdot)$ denotes the probability of the event $(\cdot)$.

The **cumulative distribution function (CDF)** of a continuous random variable x evaluated at $x = a$ is defined by

$$P(a) \doteq P(x \leq a) = \int_{\infty}^{a} p(x)dx.$$

A PDF has to satisfy the property of **normalization**:

$$P(x \leq \infty) = \int_{-\infty}^{\infty} p(x)dx = 1.$$

If the normalization property is *not* held, we sometime call the PDF an **improper** density.

The characterization of a discrete random variable, analogous to the PDF for a continuous random variable, is the **probability mass function (PMF)**. Let the domain of a discrete random variable, x, be the set $\{d_i, i = 1, \ldots, n\}$. Then, its PMF is defined by

$$P(d_i) \doteq P(x = d_i) = P_i, \quad i = 1, \ldots, n.$$

where P_i are sometimes called **point masses**. The normalization condition for a proper PMF is

$$\sum_{i=1}^{n} P_i = 1.$$

For a continuous random vector $\mathbf{x} = (x_1, x_2, \ldots, x_D)^{\mathrm{Tr}} \in \mathcal{R}^D$ where superscript Tr denotes transpose, we can similarly define their **joint PDF** of $p(x_1, x_2, \ldots, x_D)$. Further, a **marginal PDF** for each of the random variable x_i in the random vector $\mathbf{x}$ is defined by

$$p(x_i) \doteq \int\int_{all\ x_j:\ x_j \neq x_i} \cdots \int p(x_1, \ldots, x_D)dx_1 \ldots dx_{i-1}dx_{i+1} \ldots dx_D \tag{3.1}$$

It has the same properties as the PDF for a scalar random variable.

3.1.2 Summary statistics — expectations, moments, and covariances

The distribution of a random variable — PDF for a continuous random variable and PMF for a discrete random variable — contains all information that is available about the random variable. Some summaries of such information are useful in practice.

One major type of such summary statistical information is the **expectation** (also called expected value, mean, average, or first moment) of the random variable. The expectation of scalar continuous random variable x with PDF of $p(x)$ is defined by

$$E[x] \doteq \int_{-\infty}^{\infty} \xi p(\xi) d\xi = \overline{x}. \tag{3.2}$$

Going beyond the first moment, we can further define higher, n-*th* moments of summary statistics as

$$E[x^n] \doteq \int_{-\infty}^{\infty} \xi^n p(\xi) d\xi = \overline{x^n}. \tag{3.3}$$

The expectation of a function $g(x)$ of the random variable x is

$$E[g(x)] \doteq \int_{-\infty}^{\infty} g(\xi) p(\xi) d\xi = \overline{g(x)}. \tag{3.4}$$

The analogous definition of the expectation of a function for a discrete random variable with PMF of $P(d_i)$ is

$$E[g(x)] \doteq \sum_{i=1}^{n} g(d_i) P(d_i). \tag{3.5}$$

Another commonly used summary statistics for a scalar continuous random variable x is its **variance** (denoted by var(x) or σ_x^2, also called second central moment), defined by

$$\text{var}(x) = \sigma_x^2 \doteq E[(x - \overline{x})^2] = \int_{-\infty}^{\infty} (\xi - \overline{x})^2 p(\xi) d\xi = E[x^2] - \overline{x}^2 \tag{3.6}$$

The square root of the variance of x is called its **standard variation**, conventionally denoted by σ_x. Inverse of variance is sometimes called **precision**.

For a continuous random vector $\mathbf{x} = (x_1, x_2, \ldots, x_D)^{\text{Tr}}$, its expectation is a vector consisting of the expectations of its all component random variables:

$$E[\mathbf{x}] \doteq (E(x_1), E(x_2), \ldots, E(x_D))^{\text{Tr}} = (\overline{x_1}, \overline{x_2}, \ldots, \overline{x_D})^{\text{Tr}} = \overline{\mathbf{x}}.$$

The summary statistics for the random vector $\mathbf{x}$ of dimension D, analogous to variance defined for a scalar random variable, is **covariance matrix** of dimension $D \times D$ defined by

$$\text{cov}(\mathbf{x}) \doteq E[(\mathbf{x} - \overline{\mathbf{x}})(\mathbf{x} - \overline{\mathbf{x}})^{\text{Tr}}].$$

The $(i,j)th$ element of the covariance matrix (which is always symmetric) is defined by

$$\text{cov}(x_i, x_j) = \text{cov}(x_j, x_i) \doteq E[(x_i - \overline{x_i})(x_j - \overline{x_j})^{\text{Tr}}]. \tag{3.7}$$

We often call the diagonal elements of the covariance matrix **variances**, consistent with the definition for a scalar random variable. The off-diagonal elements of the covariance matrix, $\text{cov}(x_i, x_j), (i \neq j)$, is called **covariance**.

The **correlation coefficient** between two random variables, x_1 and x_2, is their covariance normalized to unity by the product of their standard variations:

$$\rho_{x_1,x_2} \doteq \frac{\text{cov}(x_1, x_2)}{\sigma_{x_1}\sigma_{x_2}}.$$

If the two random variables, x_1 and x_2, have zero correlation coefficient (i.e., zero covariance), we say they are **uncorrelated**. It is easy to show that x_1 and x_2 are uncorrelated if and only if

$$E(x_1 x_2) = E(x_1)E(x_2).$$

At the other extreme, when the correlation coefficient between x_1 and x_2 is unity (maximal value), we can prove that they are **linearly dependent**, or x_1 is a linear function of x_2 and vice versa.

In the remainder of this subsection, a number of common PDF's for continuous random variables and PMF's for discrete random variables are listed and elaborated. Most of these distributions have been exploited in statistical speech processing techniques, and some are embedded deeply into speech processing systems. The forms of these distributions should be committed to memory.

3.1.3 Common PDF's

Normal or Gaussian distribution

A scalar continuous random variable x is normally or Gaussian distributed if its PDF is

$$p(x) = \frac{1}{(2\pi)^{1/2}\sigma} \exp\left[-\frac{1}{2}\left(\frac{x-\mu}{\sigma}\right)^2\right] \doteq \mathcal{N}(x;\mu,\sigma^2) \quad (-\infty < x < \infty; \sigma > 0)$$

An equivalent notation for the above is

$$x \sim \mathcal{N}(\mu, \sigma^2),$$

denoting that random variable x obeys a normal distribution with mean μ and variance σ^2. With use of the precision parameter, a Gaussian PDF can also be written as

$$p(x) = \sqrt{\frac{r}{2\pi}} \exp\left[-\frac{r}{2}(x-\mu)^2\right].$$

It is a simple exercise to show that for a Gaussian random variable x, $E(x) = \mu$, $\mathrm{var}(x) = \sigma^2 = r^{-1}$.

The normal random vector $\mathbf{x} = (x_1, x_2, \ldots, x_D)^{\mathrm{Tr}}$, also called multivariate or vector-valued Gaussian random variable, is defined by the following joint PDF:

$$p(\mathbf{x}) = \frac{1}{(2\pi)^{D/2}|\mathbf{\Sigma}|^{1/2}} \exp\left[-\frac{1}{2}(\mathbf{x}-\boldsymbol{\mu})^{\mathrm{Tr}}\mathbf{\Sigma}^{-1}(\mathbf{x}-\boldsymbol{\mu})\right] \doteq \mathcal{N}(\mathbf{x};\boldsymbol{\mu},\mathbf{\Sigma}). \tag{3.8}$$

The likewise equivalent notation is

$$\mathbf{x} \sim \mathcal{N}(\boldsymbol{\mu} \in \mathcal{R}^D, \mathbf{\Sigma} \in \mathcal{R}^{D\times D}).$$

It is also straightforward to show that for a multivariate Gaussian random variable, the expectation and covariance matrix are given by

$$E(\mathbf{x}) = \boldsymbol{\mu}; \quad E[(\mathbf{x}-\overline{\mathbf{x}})(\mathbf{x}-\overline{\mathbf{x}})^{\mathrm{Tr}}] = \mathbf{\Sigma}.$$

The Gaussian distribution is commonly used in many engineering and science disciplines including many areas in speech processing. The popularity arises not only from its highly desirable computational properties, but also from its ability to approximate many naturally occurring real-world data thanks to the law of large numbers.

Mixture Gaussian distribution

A scalar continuous random variable x has a **mixture** Gaussian distribution [Everitt 81] if its PDF is specified by

$$\begin{aligned} p(x) &= \sum_{m=1}^{M} \frac{c_m}{(2\pi)^{1/2}\sigma_m} \exp\left[-\frac{1}{2}\left(\frac{x-\mu_m}{\sigma_m}\right)^2\right] \\ &= \sum_{m=1}^{M} c_m \mathcal{N}(x; \mu_m, \sigma_m^2) \quad (-\infty < x < \infty; \sigma_m > 0; c_m > 0) \end{aligned} \tag{3.9}$$

where the positive mixture weights sum to unity: $\sum_{m=1}^{M} c_m = 1$.

The most obvious property of the mixture Gaussian distribution is its multi-modal one ($M > 1$ in Eq. 3.9, in contrast to the uni-modal property of the Gaussian distribution where $M = 1$). This makes it possible for a mixture Gaussian distribution to adequately describe many types of physical data (including speech data) exhibiting multi-modality poorly suited for a simple Gaussian distribution. The multi-modality in data may come from multiple underlying causes each being responsible for one particular mixture component in the distribution. If such causes are identified, then the mixture distribution can be decomposed into a set of cause-dependent or context-dependent component distributions.

It is easy to show that the expectation of a random variable x with the mixture Gaussian PDF of Eq. 3.9 is $E(x) = \sum_{m=1}^{M} c_m \mu_m$. But unlike a (uni-modal) Gaussian distribution, this simple summary statistic is not very informative unless all the component means, $\mu_m, m = 1, ..., M$, in the mixture distribution are close to each other.

The multivariate generalization of the mixture Gaussian distribution has the joint PDF of

$$\begin{aligned} p(\mathbf{x}) &= \sum_{m=1}^{M} \frac{c_m}{(2\pi)^{D/2}|\mathbf{\Sigma}_m|^{1/2}} \exp\left[-\frac{1}{2}(\mathbf{x}-\boldsymbol{\mu}_m)^{\text{Tr}}\mathbf{\Sigma}_m^{-1}(\mathbf{x}-\boldsymbol{\mu}_m)\right] \\ &= \sum_{m=1}^{M} c_m \mathcal{N}(\mathbf{x}; \boldsymbol{\mu}_m, \mathbf{\Sigma}_m), \qquad (c_m > 0). \end{aligned} \tag{3.10}$$

Use of this multivariate mixture Gaussian distribution has been one key factor contributing to improved performance of many modern speech recognition systems [Deng 91b, Juang 86]. In most applications, the number of mixture components, M, is chosen a priori according to the nature of the problem, although attempts have been made to sidestep such an often difficult problem of finding the "right" number (e.g., [Rasm 99]).

In using the multivariate mixture Gaussian distribution of Eq. 3.10, if the variable x's dimensionality, D, is large (say, 40, for speech recognition problems), then use of full (non-diagonal) covariance matrices ($\mathbf{\Sigma}_m$) would involve a large number of parameters (on the order of $M \times D^2$). To reduce such a number, one can opt to use diagonal covariance matrices for Σ_m's. (When M is large, one can also constrain all covariance matrices to be the same ("tying" Σ) for all mixture components, m.) An additional advantage of using diagonal covariance matrices is significant simplification of computations needed for the applications of the mixture Gaussian distributions (much to be said in later chapters). Reducing full covariance matrices to diagonal ones may have seemed to impose uncorrelatedness among data vector components. This has been misleading, however, since a mixture of Gaussians each with a diagonal covariance matrix can at least effectively describe the correlations modeled by one Gaussian with a full covariance matrix.

Exponential distribution

A scalar continuous random variable x is exponentially distributed if its PDF is

$$p(x) = \frac{1}{\sigma} \exp\left[-\frac{(x-\theta)}{\sigma}\right] \qquad (x \geq \theta; \sigma > 0)$$

The PDF is zero outside the domain $x \geq \theta$. That is, $p(x) = 0$ for $x < \theta$.

Different from the Gaussian and mixture Gaussian distributions which have a two-sided domain $(-\infty < x < \infty)$, the domain of the exponential distribution is one-sided. This makes it suitable to characterize some types of "one-sided" data not well characterized by a Gaussian distribution. For example, the degree of vocal tract constriction cannot be negative; hence its randomness should not be described by a Gaussian distribution but may be adequately described by an exponential distribution.

There are many situations where one expects an exponential distribution to provide a useful characterization of observed pattern of randomness. It can be shown that if a sequence of events occur at random in time, then the interval (which is a random variable!) of the successive events obeys an exponential distribution.

Some summary statistics of an exponential distribution are

$$E(x) = \theta + \sigma; \qquad \text{var}(x) = \sigma^2.$$

Inverse Gaussian distribution

Another one-sided continuous distribution is the inverse Gaussian which has the PDF:

$$p(x) = \sqrt{\frac{\lambda}{2\pi x^3}} \exp\left[-\frac{\lambda(x-\mu)^2}{2\mu^2 x}\right], \qquad (x > 0; \mu > 0; \lambda > 0) \tag{3.11}$$

where μ can be shown to be the mean, $E(x)$, of the distribution. λ is the scale (skewness) factor determining the overall shape of the distribution. The variance can be shown to be $\text{var}(x) = \mu^3/\lambda$.

The name inverse Gaussian arises from the inverse relationship between the cumulant generating functions of these distributions and those of Gaussian distributions [Chhik 89].

Log-normal distribution

Yet another one-sided distribution is the log-normal PDF

$$p(x) = \sqrt{\frac{1}{2\pi\beta x^2}} \exp\left[-\frac{(\log x - \alpha)^2}{2\beta^2}\right], \qquad (x > 0; \beta > 0)$$

The genesis of the log-normal is as follows: If the random variable $\log x$ obeys a Gaussian distribution, then it can be shown that its exponential function (hence always positive), or x, will have the PDF above.

Some summary statistics for the log-normal distribution are

$$E(x) = \exp[\alpha + 0.5\beta]; \qquad \text{var}(x) = [\exp(\beta^2) - 1]\exp(2\alpha + \beta^2).$$

Gamma distribution

The gamma distribution (one-sided again) has the following PDF:

$$p(x) = C\, x^{\alpha-1}\exp\left[-\beta x\right], \qquad (x > 0; \alpha > 0, \beta > 0)$$

with the normalization constant $C = \frac{\beta^\alpha}{\Gamma(\alpha)}$, in which $\Gamma(\alpha)$ is a value of the gamma function defined by

$$\Gamma(\alpha) = \int_0^\infty y^{\alpha-1}\exp(-y)dy.$$

The gamma function can be computed recursively by

$$\Gamma(\alpha) = (\alpha - 1)\Gamma(\alpha - 1).$$

Some summary statistics for the gamma distribution are

$$E(x) = \frac{\alpha}{\beta}; \qquad \text{var}(x) = \frac{\alpha}{\beta^2}.$$

Normal-gamma and normal-Wishart distributions

The normal-gamma distribution is a joint distribution for two scalar random variables, x_1 and x_2, with its PDF closely related the PDF's of gamma and normal distributions:

$$\begin{aligned} p(x_1, x_2) &= \frac{(\omega x_1)^{1/2}}{(2\pi)^{1/2}}\exp\left[-\frac{\omega x_1}{2}(x_2 - \mu)^2\right] \times \frac{\beta^\alpha}{\Gamma(\alpha)}x_1^{\alpha-1}\exp\left[-\beta x_1\right] \\ &= C x_1^{\alpha-0.5}\exp\left[-\frac{\omega x_1}{2}(x_2 - \mu)^2 - \beta x_1\right]. \end{aligned} \tag{3.12}$$

The parameters of the normal-gamma distribution are α, β, μ, and ω. The form of the PDF in Eq. 3.12 is derived from the two requirements:

- the marginal distribution of x_1 (i.e., obtained by integrating out x_2 in the joint PDF $p(x_1, x_2)$) is a gamma distribution with parameters α and β; and

- given x_1, the conditional distribution (a concept to be discussed shortly) of x_2 is a normal distribution with the mean parameter being μ and the variance parameter being $(\omega x_1)^{-1}$.

It was proved in [DeG 70] that these two requirements lead to the natural **conjugate prior** (a concept to be discussed later) for the joint mean and precision (inverse of variance) in a Gaussian distribution. This conjugate prior and the associated normal-gamma distribution have found important applications in adaptive systems of speech recognition (Chapter 13).

A generalization of the normal-gamma distribution to two matrix-valued ($\mathbf{X_1}$) and vector-valued $\mathbf{x_2}$ random variables, respectively, is the normal-Wishart distribution. This is the distribution proved to be the natural conjugate prior for the joint mean vector and full (nondiagonal) precision matrix (inverse of covariance matrix) of a multivariate Gaussian random variable (dimension D). The joint PDF for the normal-Wishart distribution is

$$p(\mathbf{X_1}, \mathbf{x_2}) = K\ |\mathbf{X_1}|^{(\alpha_0 - D)/2} \times \exp\left[-\frac{\omega}{2}(\mathbf{x_2} - \boldsymbol{\phi})^{\mathrm{Tr}}\mathbf{X_1}(\mathbf{x_2} - \boldsymbol{\phi})\right] \exp\left[-\frac{1}{2}tr(\mathbf{B}\mathbf{X_1})\right], \tag{3.13}$$

where $tr(\cdot)$ denotes trace of a matrix. The parameters in the distribution are $\alpha_0 > D - 1, \mathbf{B} \in \mathcal{R}^{D\times D}$ (positive definite), $\boldsymbol{\phi} \in \mathcal{R}^{D}$, and ω. (Note when setting $D = 1$ and $\alpha_0 = 2\alpha$, the normal-Wishart distribution reduces to a normal-gamma distribution of Eq. 3.12.) Like the normal-gamma distribution, the normal-Wishart distribution has also been widely used in adaptive speech recognition.

Laplace distribution

The Laplace distribution, also known as a double exponential, two-tailed exponential, or bilateral exponential distribution, has a PDF

$$p(x) = \frac{1}{2\phi}\exp\left[-\frac{|x-\theta|}{\phi}\right] \qquad (-\infty < x < \infty; \phi > 0).$$

When a histogram of the magnitudes of digitized speech samples is constructed over long periods of time, it has been shown [Paez 72] to approach the Laplace distribution. (The gamma distribution is also a good approximation, but a Gaussian distribution is a poor one.)

Compared with the Gaussian PDF which has a quadratic exponential rate of decay, the Laplace distribution has a much slower rate of decay — linear exponential. Hence, we sometimes say the Laplace distribution has much heavier tails than the Gaussian distribution. This property makes the Laplace distribution more suitable than the Gaussian distribution for physical data which may occasionally manifest *outliers*.

Beta and Dirichlet distributions

A beta distribution is associated with the PDF of

$$p(x) = \frac{\Gamma(\alpha+\beta)}{\Gamma(\alpha)\Gamma(\beta)}x^{\alpha-1}(1x)^{\beta-1}, \quad (0 < x < 1; \alpha > 0; \beta > 0)$$

The following summary statistics can be easily verified:

$$E(x) = \frac{\alpha}{\alpha+\beta}; \qquad \mathrm{var}(x) = \frac{\alpha\beta}{(\alpha+\beta)^2(\alpha+\beta+1)}.$$

The beta distribution can be generalized to its multivariate counterpart, called the Dirichlet distribution, which has found useful applications in speech processing (Chapter 13). A vector-valued discrete random variable, $\mathbf{x} = (x_1, x_2, \ldots, x_k)^{\mathrm{Tr}}$, has a Dirichlet distribution with parameter vector $\alpha = (\alpha_1, \alpha_2, \ldots, \alpha_k)^{\mathrm{Tr}}$ $(\alpha_i > 0)$ if, for $\sum_{i=1}^{k} x_i = 1$ and $x_i > 0$, the joint PMF of α is

$$P(\mathbf{x}) = \frac{\Gamma(\alpha_1 + \alpha_2 + \cdots + \alpha_k)}{\Gamma(\alpha_1)\Gamma(\alpha_2)\ldots\Gamma(\alpha_k)}x_1^{\alpha_1-1}x_2^{\alpha_2-1}\cdots x_k^{\alpha_k-1}. \tag{3.14}$$

Note that Eq. 3.14 is a $(k-1)$–dimensional joint PMF, rather than a k-dimensional one, due to the constraint $\sum_{i=1}^{k} x_i = 1$.

If a scalar random variable x has a beta distribution with parameters α and β, then the random vector $(x, 1-x)^{\text{Tr}}$ will have a Dirichlet distribution with parameter vector $(\alpha, \beta)^{\text{Tr}}$. On the other hand, if a random vector $\mathbf{x} = (x_1, x_2, \ldots, x_k)^{\text{Tr}}$ has a Dirichlet distribution, it can be shown that the marginal distribution for each component, x_i, will have a beta distribution with parameters α_i and $\sum_{j=1}^{k} \alpha_j - \alpha_i$. Also note that the beta distribution for which $\alpha = \beta = 1$ is called the uniform distribution on the interval $(0, 1)$. This is a member of the distribution family to be discussed next.

Uniform distribution

A uniform distribution on the interval (α, β), where $-\infty < \alpha < \beta < \infty$, is associated with the PDF of

$$p(x) = \frac{1}{\beta - \alpha}, \qquad (\alpha < x < \beta)$$

The PDF is zero outside the domain $\alpha < x < \beta$.

The important summary statistics are

$$E(x) = \frac{\alpha + \beta}{2}; \qquad \text{var}(x) = \frac{(\beta - \alpha)^2}{12}.$$

3.1.4 Common PMF's

Bernoulli distribution

A scalar discrete random variable, x, has a Bernoulli distribution (parameter p) if x takes only the values 0 and 1 with the PMF of

$$P(x = 1) = p; \qquad P(x = 0) = 1 - p, \qquad (0 < p < 1)$$

This can also be written in a more compact form:

$$P(x) = p^x q^{1-x}, \qquad for \;\; x = 0, 1$$

where $q = 1 - p$, and $P(x) = 0$ for values of x other than 0 and 1.

The Bernoulli distribution is of fundamental importance — this is a model for any random experiment whose outcome must belong to one of two mutually exclusive classes (e.g., success or failure, correct word recognition or not, etc.).

The main summary statistics of the Bernoulli distribution are

$$E(x) = p; \qquad \text{var}(x) = pq.$$

Poisson distribution

A scalar discrete random variable, x, has a Poisson distribution (with "mean" rate parameter λ) if x has the PMF of

$$P(x) = \frac{\lambda^x \exp(-\lambda)}{x!}, \qquad (x = 0, 1, 2, \ldots,; \;\; \lambda > 0)$$

The main summary statistics of the Poisson distribution are

$$P(x) = \lambda; \qquad \text{var}(x) = \lambda.$$

It can be mathematically derived that under certain conditions, the number of event occurrences in a fixed period of time obeys a Poisson distribution. For example, over a fixed period, the distribution of the number of action potentials (which is a discrete random variable) in an auditory nerve fiber in response to a speech stimulus can be approximated by a Poisson distribution (especially when the refractoriness is ignored). If the fixed period is reasonably short over which the speech spectral information does not change substantially, the Poisson rate λ may be constant. Otherwise, we will have a variable-rate Poisson distribution over the observation time interval. Another example of a Poisson distribution is the number of data packets (such as digitally coded speech) observed over a fixed period of time in a telecommunication network (e.g., [Deng 93d]).

Binomial and multinomial distributions

A scalar discrete random variable, x, has a binomial distribution (parameters n and p) if x has the PMF of

$$P(x) = \binom{n}{x} p^x q^{1-x}, \qquad \text{for } x = 0, 1, 2, \ldots, n. \ (n = 1, 2, \ldots; 0 < p < 1)$$

where $q = 1-p$. The binomial distribution is closely related to the Bernoulli distribution. If $x_1, x_2, \ldots, x_n$ are independent Bernoulli distributed random variables, then their sum, $x_1 + x_2 + \cdots + x_n$, as a new random variable has a binomial distribution. That is, the binomial distribution is for the total number of "success" events, each having two possible outcomes (success or failure) where the success rate is p and failure rate is $q = 1 - p$, in a total of n trials. The main summary statistics of the binomial distribution are

$$P(x) = np; \qquad \text{var}(x) = npq.$$

An immediate generalization of the binomial distribution arises when each trial can have more than two possible outcomes. This gives the vector-valued, discrete multinomial distribution defined below. Let the outcome of a random experiment belong to one of $k(k \geq 2)$ mutually exclusive categories, and let $p_i(0 < p_i < 1)$ be the probability that the outcome belongs to the ith category ($i = 1, 2, \ldots, k$ and $\sum_{i=1}^{k} p_i = 1$). Perform the random experiment independently for n times. If x_i denotes the number of the outcomes that belong to category i, then the random vector $\mathbf{x} = (x_1, x_2, \ldots, x_k)^{\text{Tr}}$ has a multinomial distribution with parameters $p_i(i = 1, 2, \ldots, k)$ and n. A multinomial distribution for $\mathbf{x}$ with $\sum_{i=1}^{k} x_i = n$ can be shown to have the joint PMF of

$$P(\mathbf{x}) = \frac{n!}{x_1!x_2!\ldots x_k!} \, p_1^{x_1} p_2^{x_2} \cdots p_k^{x_k}.$$

The main summary statistics of the multinomial distribution are

$$\begin{aligned} &P(x_i) = np_i; \quad \text{var}(x_i) = np_i(1 - p_i), && \textit{for } \ i = 1, 2, \ldots, k \\ &\text{cov}(x_i, x_j) = -np_ip_j, && \textit{for } \ i, j = 1, 2, \ldots, k; i \neq j. \end{aligned}$$

3.2 Conditioning, Total Probability Theorem, and Bayes' Rule

The notion of **conditioning** has a fundamental importance in probability theory, statistics, and their engineering applications including speech processing. We first discuss several key concepts related to conditioning.

3.2.1 Conditional probability, conditional PDF, and conditional independence

In a random experiment, an **event** or occurrence of an event refers to a set of the random outcomes or to a particular outcome within this set. An event, say A, is associated with a probability, denoted by $P(A)$. The probability of two joint events, A and B, is called the joint probability, denoted by $P(A, B)$.

Conditional probability

The **conditional probability** of event A given event B is defined as

$$P(A|B) \doteq \frac{P(A, B)}{P(B)}. \tag{3.15}$$

Events A and B are **independent** if

$$P(A|B) = P(A). \tag{3.16}$$

That is to say, knowing the event B occurrence tells nothing about the occurrence of event A.

Eq. 3.16 is equivalent to

$$P(A, B) = P(A)P(B)$$

after using Eq. 3.15.

Conditional PDF

Recall that each outcome of a random experiment, or a particular occurrence of an event, assigns a value taken by a random variable. Based on such correspondence or one-to-one mapping between an event and a random variable, we can use the concept of PDF defined already for a random variable (Section 3.1) to prescribe the **conditional PDF** (and similarly the conditional PMF). The conditional PDF of a random variable X given another random variable Y is given by

$$p(x|y) \doteq \frac{p(x, y)}{p(y)}, \tag{3.17}$$

where $p(x, y)$ is the joint PDF of X and Y and $p(y)$ is the marginal PDF of Y.

We can also define the conditional PDF of a random variable X given an event E in a similar way:

$$p(x|E) \doteq \frac{p(x, E)}{p(E)}.$$

Example 1: Let the conditioning event be: $E = \{X > c\}$. Then the conditional PDF becomes:

$$p(x|E) = p(x|X > c) = \frac{P(x, X > c)}{P(X > c)} = \begin{cases} \frac{p(x)}{P(X>c)} & \text{if } x > c \\ 0 & \text{otherwise} \end{cases}$$

Similar to the definition of independence between two events, two random variables X and Y are independent if the relevant PDF's satisfy

$$p(x|y) = p(x),$$

or equivalently

$$p(x, y) = p(x)p(y).$$

Conditional independence

A further concept closely related to those of conditional probability and of independence is that of **conditional independence** [Daw 79]. Two random variables X and Y are independent conditioned on a third random variable Z if the relevant PDF's satisfy

$$p(x, y|z) = p(x|z)p(y|z). \tag{3.18}$$

This can be shown (by multiplying $p(z)$ on both sides of Eq. 3.18) to be equivalent to

$$p(x|y, z) = p(x|z). \tag{3.19}$$

The assumption of conditional independence underlies the most popular model (hidden Markov model) underpinning current speech recognition technology. Many researchers have advocated a need to remove such an assumption in order to achieve technology improvement.

The concepts of conditional PDF's and of independence can be easily generalized to more than two random variables. A simple technique is to appropriately "group" all random variables into two sets and then apply the definitions shown above for two random variables.

Example 2: The "grouping" technique allows us to verify the correctness of the following "chain rule" for the decomposition of the joint PDF for N random variables $\mathbf{x} = (x_1, x_2, \ldots, x_N)$:

$$\begin{aligned} p(\mathbf{x}) &= p(x_1)p(x_2|x_1)p(x_3|x_2, x_1)p(x_4|x_3, x_2, x_1) \ldots p(x_N|x_{N-1} \ldots x_2, x_1) \\ &= \prod_{n=1}^{N} p(x_n|x_{n-1} \ldots x_2, x_1). \end{aligned} \tag{3.20}$$

That is, the first two factors give rise to the joint PDF of $p(x_1, x_2)$. By grouping x_1, x_2 as a "single" random variable set, we obtain a further joint PDF of $p(x_1, x_2, x_3) = p(x_1, x_2)p(x_3|x_2, x_1)$, and so on.

A more direct proof of the "chain rule" follows the steps below by repetitively "grouping" random variable sets (starting with grouping $[x_1, \ldots, x_{N-1}]$):

$$
\begin{aligned}
& p(x_1, \ldots, x_N) = p([x_1, \ldots, x_{N-1}])p(x_N|[x_1, \ldots, x_{N-1}]) \\
= \; & p(x_1, \ldots, x_{N-2})p(x_{N-1}|x_1, \ldots, x_{N-2})p(x_N|x_1, \ldots, x_{N-1}) \\
= \; & p(x_1, \ldots, x_{N-3})p(x_{N-2}|x_1, \ldots, x_{N-3}p(x_{N-1}|x_1, \ldots, x_{N-2})p(x_N|x_1, \ldots, x_{N-1}) \\
= \; & \vdots \\
= \; & p(x_1)p(x_2|x_1)p(x_3|x_1, x_2)p(x_4|x_1, x_2, x_3) \ldots\ldots\ldots\ldots\ldots p(x_N|x_1, \ldots, x_{N-1})
\end{aligned}
$$

The decomposition or factorization technique exemplified in Eq. 3.20 (together with further approximations to be discussed later) form the basis of a most popular "language model" used in many modern speech recognition systems.

Example 3: We use this "grouping" technique to prove that Eq. 3.18 is equivalent to Eq. 3.19. By grouping $[x, y]$, we have

$$p([x, y], z) = p(z)p([x, y]|z).$$

Decomposing $p(x, y, z)$ using Eq. 3.20 (for $N = 3$), we have

$$p(x, y, z) = p(z)p(y|z)p(x|y, z).$$

Equating the above two equations and cancelling out the common $p(z)$, we obtain

$$p(x, y|z) = p(y|z)p(x|y, z). \tag{3.21}$$

Combining Eqs. 3.18 and 3.21, we complete the proof of Eq. 3.19, or combining Eqs. 3.19 and 3.21, we complete the proof of Eq. 3.18.

Eq. 3.21 is universally true. Eqs. that the conditioning random variable (here, z) can be inserted in both right-hand-side factors in the plain definition of the conditional PDF: $p(x, y) = p(y)p(x|y)$. This, in addition to the above steps of proof, can also be straightforwardly verified by left-multiplying $p(z)$ on both sides of Eq. 3.21. The right-hand-side would give $p(z)p(y|z)p(x|y, z) = p(y, z)p(x|y, z) = p(x, y, z)$, the same as the left-hand side: $p(z)p(x, y|z) = p(x, y, z)$.

Graphical representation of conditional independence

The conditional independence (between X and Y given Z) expressed by Eq. 3.19 simplifies the generic decomposition $p(x, y, z) = p(z)p(y|z)p(x|y, z)$ into

$$p(x, y, z) = p(z)p(y|z)p(x|z). \tag{3.22}$$

Since in this case Y depends on Z, and X depends on Z, we can use a directed graph to represent such dependency:

$$X \leftarrow Z \rightarrow Y.$$

The decomposition of the joint probability as in Eq. 3.22 is not unique. If we use a new directed graph to represent the dependency relationship: $X \rightarrow Z \rightarrow Y$ (maintaining the same conditional independence relationship as before), we can then decompose the joint probability in a new way:

$$p(x, z, y) = p(x)p(z|x)p(y|z). \tag{3.23}$$

Eq. 3.23 is an illustration for approximating a "trigram" probability by a "bigram" one by assuming the conditional independence. This form of conditional independence is called the Markov property to be discussed later in this chapter.

Generalizing this "trigram to bigram" case to the "N-gram to trigram" case, we can approximate the "N-gram" probability in Eq. 3.20 by

$$\begin{aligned} p(\mathbf{x}) &= \prod_{n=1}^{N} p(x_n|x_{n-1},\ldots,x_2,x_1) \\ &\approx \prod_{n=1}^{N} p(x_n|x_{n-1},x_{n-2}). \end{aligned} \tag{3.24}$$

Yet another way of decomposing the joint PDF $p(y,z,x)$, maintaining the X and Y independence given Z, is

$$p(y,z,x) = p(y)p(z|y)p(x|z). \tag{3.25}$$

The graphical representation for this decomposition is

$$Y \to Z \to X.$$

In summary, random variables X and Y are independent given Z (i.e, X is dependency separated from Y by Z) if and only if

$$p(x,y|z) = p(x|z)p(y|z),$$

or equivalently,

$$p(x|y,z) = p(x|z) \quad or \quad p(y|x,z) = p(y|z).$$

Under such conditional independence, any one of the three probability decompositions is possible (for the three different ways of dependency among X, Y, and Z):

- $p(x,y,z) = p(z)p(y|z)p(x|z)$
- $p(x,y,z) = p(x)p(z|x)p(y|z)$
- $p(x,y,z) = p(y)p(z|y)p(x|z)$

3.2.2 The total probability theorem

We first define **mutually exclusive** events. The events $E_j, j = 1,2,\ldots,J$ are mutually exclusive if

$$P(E_j, E_i) = 0 \quad \forall i \neq j$$

The events $E_j, j = 1,2,\ldots,J$ are **exhaustive** if

$$\sum_{j=1}^{J} P(E_j) = 1.$$

The **total probability theorem** states that using a set of mutually exclusive and exhaustive events $E_j, j = 1,2,\ldots,J$, we can decompose the probability of any arbitrary event A according to

$$P(A) = \sum_{j=1}^{J} P(A, E_j) = \sum_{j=1}^{J} P(A|E_j)P(E_j).$$

The total probability theorem for continuous random variables is the following decomposition of the PDF with infinitely fine granularity:

$$p(x) = \int_{-\infty}^{\infty} p(x, y)dy = \int_{-\infty}^{\infty} p(x|y)p(y)dy. \tag{3.26}$$

Here, the mutually exclusive and exhaustive "events" are represented by the random variable Y taking a full range of values. (The mutual exclusivity is ensured due to the nature of the mapping from an event occurrence and the assignment of a value taken by the random variable.)

The following equalities are all derived from the above basic total probability theorem:

$$p(x) = \sum_{j=1}^{J} p(x|E_j)P(E_j) \tag{3.27}$$

$$P(A) = \int_{-\infty}^{\infty} P(A|x)p(x)dx \tag{3.28}$$

$$p(x|z) = \int_{-\infty}^{\infty} p(x, y|z)dy = \int_{-\infty}^{\infty} p(x|y, z)p(y|z)dy \tag{3.29}$$

(using also the result of Eq. 3.21)

$$P(A|B) = \sum_{j=1}^{J} P(A, E_j|B) = \sum_{j=1}^{J} P(A|E_j, B)P(E_j|B) \tag{3.30}$$

and

$$P(A|z) = \int_{-\infty}^{\infty} P(A|x, z)p(x|z)dx. \tag{3.31}$$

3.2.3 Bayes' rule and its sequential form

General description

Bayes' rule (or Bayes' theorem, Bayes' formula, etc.) is a direct result of reverse conditioning by plain manipulation of the conditional probabilities:

$$P(H|D) = \frac{P(D|H)P(H)}{P(D)}. \tag{3.32}$$

Often, while using Bayes' rule, the conditional probability $P(H|D)$ is referred to as **posterior probability** (aftering observing Data D), the unconditional one is referred to as **prior probability** (of Hypothesis H), and the $P(D|H)$ is referred to as **data or observation probability** (given the Hypothesis class H) or likelihood function. The prior represents our state of knowledge (or ignorance) about the truth of the hypothesis before we have observed and analyzed the current data. This is modified by the experimental observation of data through the likelihood function. This modification yields the posterior which represents our enhanced state of knowledge about the truth of the hypothesis in light of the observation. We often treat the denominator $P(D)$ in Eq. 3.32

as a normalization factor. If we want to be explicit about $P(D)$ by decomposing it using the total probability theorem, we can rewrite Bayes' rule as

$$P(H_i|D) = \frac{P(D|H_i)P(H_i)}{\sum_{j=1}^{J} P(D|H_j)P(H_j)}.$$

The conventional way of writing the Bayes' rule in Eq. 3.32 tacitly omits an additional conditioning variable I for the relevant background information underlying the hypothesis and data. Incorporating such background information, Bayes' rule has the following precise form:

$$P(H|D,I) = \frac{P(D|H,I)P(H|I)}{P(D|I)}. \tag{3.33}$$

Philosophical arguments aside, Eq. 3.32 is again a direct result of conditional probability manipulation. To verify its correctness, multiply both the numerator and denominator by $P(I)$ to obtain $\frac{P(D|H,I)P(H,I)}{P(D,I)} = \frac{P(D,H,I)}{P(D,I)} = P(H|D,I)$.

Discounting the normalization factor, we commonly express Bayes' rule in a more informative form (based on the notation in [Siv 96]):

$$\text{Prob(Hypothesis|Data, I)} \propto \text{Prob(Data|Hypothesis, I)} \times \text{Prob(Hypothesis, I)}.$$

Bayes' rule for random variables

When both "Data" and "Hypothesis" are expressed in terms of the values taken by two continuous random variables, z and λ, respectively, then Bayes' rule can be rewritten by the prior PDF and posterior PDF:

$$p(\lambda|z) = \frac{p(z|\lambda)p(\lambda)}{p(z)} = \frac{p(z|\lambda)p(\lambda)}{\int_{-\infty}^{\infty} p(z|\lambda)p(\lambda)d\lambda}. \tag{3.34}$$

For the case involving a mixed continuous random variable z and a discrete random variable H_i, Bayes' rule can be worked out to be

$$P(H_i|z) = \frac{p(z|H_i)P(H_i)}{p(z)} = \frac{p(z|H_i)P(H_i)}{\sum_{j=1}^{J} p(z|H_j)P(H_j)}.$$

With more than one conditioning random variable, we can work out the following form of Bayes' rule:

$$P(H_i|z,x) = \frac{p(z|H_i,x)P(H_i|x)}{p(z|x)} = \frac{p(z|H_i,x)P(H_i|x)}{\sum_{j=1}^{J} p(z|H_j,x)P(H_j|x)}.$$

This is as if the second random variable, x, can be inserted into both the conditional and prior probabilities in the normal Bayes' rule involving only one conditioning random variable z.

Sequential Bayes' rule

The following sequential or iterative form of Bayes' rule can be derived from iterative applications of Bayes' rule discussed above. The discussion below follows that in [Spra 72].

Let λ be a continuous random variable with prior PDF $p(\lambda)$ and let $z_1, z_2, \cdots, z_n$ be a sequence of identically distributed random variables which are statistically related to λ. The normal form of Bayes' rule is

$$p(\lambda|z_1, z_2, \cdots, z_n) = \frac{p(\lambda)p(z_1, z_2, \cdots, z_n|\lambda)}{\int_{-\infty}^{\infty} p(\lambda)p(z_1, z_2, \cdots, z_n|\lambda)d\lambda}.$$

If the sequence $z_1, z_2, \cdots, z_n$ (data) are conditionally independent of each other given λ, that is, $z_1, z_2, \cdots, z_n$ are dependency separated by λ, or

$$p(z_1, z_2, \cdots, z_n|\lambda) = p(z_1|\lambda)p(z_2|\lambda)\cdots p(z_n|\lambda),$$

or equivalently

$$p(z_n|\lambda) = p(z_n|z_1, z_2, \cdots, z_{n-1}, \lambda), \tag{3.35}$$

then the a posteriori PDF of λ given $z_1, z_2, \cdots, z_n$ can be computed iteratively by the sequential form of Bayes' rule:

$$p(\lambda|z_1, z_2, \cdots, z_n) = \frac{p(z_n|\lambda)p(\lambda|z_1, z_2, \cdots, z_{n-1})}{P(z_1, z_2, \cdots, z_n)} = \frac{p(z_n|\lambda)p(\lambda|z_1, z_2, \cdots, z_{n-1})}{\int_{-\infty}^{\infty} p(z_n|\lambda)p(\lambda|z_1, z_2, \cdots, z_{n-1})d\lambda}. \tag{3.36}$$

To prove Eq. 3.36, we multiply the denominator and numerator in the right hand side of Eq. 3.36 by the same quantity of $p(z_1, z_2, \cdots, z_{n-1})$. This gives

$$\frac{p(z_n|\lambda)p(\lambda|z_1, z_2, \cdots, z_{n-1})p(z_1, z_2, \cdots, z_{n-1})}{\int_{-\infty}^{\infty} p(z_n|\lambda)p(\lambda|z_1, z_2, \cdots, z_{n-1})p(z_1, z_2, \cdots, z_{n-1})d\lambda} = \frac{p(z_n|\lambda)p(\lambda, z_1, z_2, \cdots, z_{n-1})}{\int_{-\infty}^{\infty} p(z_n|\lambda)p(\lambda, z_1, z_2, \cdots, z_{n-1})d\lambda}.$$

This, after using Eq. 3.35, becomes

$$\frac{p(\lambda, z_1, z_2, \cdots, z_{n-1}, z_n)}{\int_{-\infty}^{\infty} p(\lambda, z_1, z_2, \cdots, z_n)d\lambda} = \frac{p(\lambda, z_1, z_2, \cdots, z_n)}{p(z_1, z_2, \cdots, z_n)} = p(\lambda|z_1, z_2, \cdots, z_n).$$

Eq. 3.36 depicts the iterative nature of the a posteriori probability computation. At $n = 1$, the computation is initialized with the prior distribution:

$$p(\lambda|z_1, z_2, \cdots, z_{n-1}) = p(\lambda|z_0) = p(\lambda).$$

For all $n > 1$, each time a new data point, z_n, is observed, the same computation is executed according to Eq. 3.36, except using the updated PDF's on λ and on z. For the computational complexity analysis of the sequential Bayes' rule and under what conditions the computation remains reasonably simple as n increases, see [Spra 72].

The sequential Bayes' rule has been used successfully for on-line adaptation of speech models in speech recognition, where the vector-valued random variable λ is a collection of the speech model parameters.

3.3 Conditional Expectations

We introduced the (unconditional) expectation in Eq. 3.2 earlier, which is now extended to the **conditional expectation** with respect to a conditional PDF:

$$E(x|y) \doteq \int_{-\infty}^{\infty} xp(x|y)dx. \tag{3.37}$$

The conditional expectation of a function of random variable X, $g(X)$, is given by

$$E[g(x)|y] \doteq \int_{-\infty}^{\infty} g(x)p(x|y)dx. \tag{3.38}$$

More generally, the function $g(\cdot)$ can include arguments of both X and Y. The conditional expectation is given similarly by

$$\boxed{E[g(x,y)|y] \doteq \int_{-\infty}^{\infty} g(x,y)p(x|y)dx.} \tag{3.39}$$

It is essential to keep in mind that the integration is over the regular random variable (x in the above cases), not over the conditioning random variable (y). Hence we say, for the above cases, that the conditional expectation is carried out over the random variable x.

It is also important to note that a conditional expectation is a function of the conditioning random variable (y in the above cases) since y will remain after the integrals of Eqs. 3.37, 3.38, and 3.39. Hence, *a conditional expectation is a random variable itself* (with its own PDF $p(y)$), and as such, we can take (unconditional) expectation of this new random variable.

The result of this expectation (over y) of the conditional expectation (over x) is rather informative:

$$\begin{aligned} E_y[E_x(x|y)] &= \int_{-\infty}^{\infty} [E_x(x|y)]p(y)dy \\ &= \int_{y=-\infty}^{\infty} \left[\int_{x=-\infty}^{\infty} xp(x|y)dx\right] p(y)dy \\ &= \int_{x=-\infty}^{\infty} x \left[\int_{y=-\infty}^{\infty} p(x|y)p(y)dy\right] dx \\ &= \int_{x=-\infty}^{\infty} x \left[\int_{y=-\infty}^{\infty} p(x,y)dy\right] dx \\ &= \int_{x=-\infty}^{\infty} xp(x)dx \\ &= E(x). \end{aligned} \tag{3.40}$$

This result holds when either or both of the regular and conditioning random variables are discrete random variable(s), or when they are mixed discrete and continuous random variables.

Conditional expectation has a fundamental importance in statistical estimation theory (reviewed in Chapter 5). Minimal mean square estimate of a random parameter is precisely the conditional expectation of this parameter. The conditional expectation is also a central quantity in the celebrated Expectation and Maximization algorithm to be covered later in this book.

Example 4: We wish to write an expression of the conditional expectation, analogous to Eq. 3.39, for the random variable X being discrete (the conditioning random variable Y remains continuous). This involves mainly replacing the unconditional PMF $P(d_i)$ in Eq. 3.5 by the conditional PMF $P(d_i|y)$:

$$E_X[g(X,y)|y] = \sum_{i=1}^{n} g(d_i,y)P(d_i|y), \tag{3.41}$$

where $d_i, i = 1, 2, \ldots, n$ are the exhaustive set of values that the random variable X can take.

3.4 Discrete-Time Random Processes

Up until now in this chapter, the random variables we have considered are "static", in the sense that no time or temporal elements are involved. If we let a "static" random variable unfold in time, we obtain a "dynamic" **random process** and stochastic process. If we let a vector-valued (i.e., multivariate) random variable unfold in time, we will have a vector-valued (i.e., multivariate) random process. A random process can be either continuous valued, discrete valued, or mixed valued, depending on the nature of the underlying random variables.

In this book, we will restrict ourselves mainly to the situation where the time axis has been uniformly discretized. A random process under such a condition is called a **random sequence** or **discrete-time random process**. Formally, a random sequence is defined as a time-indexed sequence of random variables:

$$x_1^T = \{x_t\}_{t=1}^T = \{x_t|t = 1, 2, \ldots, T\} = x_1, x_2, \ldots, x_T, \qquad (T = 1, 2, \ldots) \qquad (T = 1, 2, \ldots)$$

The dynamic behavior of a random sequence (continuous valued) X_1^T is fully characterized by the joint PDF with the dimensionality up to infinity:

$$p(\xi_1, \xi_2, \ldots, \xi_T; t_1, t_2, \ldots, t_T); \qquad T = 1, 2, \ldots.$$

The complexity of this joint PDF usually renders it of little practical use. Hence the sequence is often described by its summary statistics or ensemble properties including its expectation, covariance, and correlation function, which we discuss now.

3.4.1 Summary statistics of a random sequence

The **mean**, or the first-order statistics, of a random sequence, $\{x_t\}_{t=1}^T$, is taken across the ensemble at each time t, and is a deterministic function of time:

$$E(x_t) = \overline{x}(t) = \int_{-\infty}^{\infty} \xi p(\xi; t) d\xi,$$

where $p(\xi; t)$ is the PDF for the random variable of x_t, which is written as an explicit function of time t.

The **variance** of the random sequence $\{x_t\}_{t=1}^T$ is also a deterministic function of time t:

$$\sigma_x^2(t) = E[(x_t - \overline{x}(t))^2] = \int_{-\infty}^{\infty} |\xi - \overline{x}(t)|^2 p(\xi; t) d\xi.$$

The **autocorrelation**, or the second-order statistics, of the random sequence is a function of two time indices t and τ:

$$r_x(t, \tau) = E[x_t x_\tau^*] = \int_{-\infty}^{\infty} \xi_t \xi_\tau^* p(\xi_t, \xi_\tau; t, \tau) d\xi_t d\xi_\tau,$$

where the superscript in ξ^* denotes complex conjugate of ξ.

The **autocovariance** is essentially the same function, except being centralized about the mean values:

$$c_x(t, \tau) = E[(x_t - \overline{x}(t))(x_\tau - \overline{x}(\tau))^*] = \int_{-\infty}^{\infty} (\xi_t - \overline{x}(t))(\xi_\tau - \overline{x}(\tau))^* p(\xi_t, \xi_\tau; t, \tau) d\xi_t d\xi_\tau. \tag{3.42}$$

If the product in Eq. 3.42 is expanded, then it follows that the autocovariance and autocorrelation functions are related by

$$c_x(t, \tau) = r_x(t, \tau) - \overline{x}(t)\overline{x}^*(\tau).$$

Therefore, if the random sequence has zero mean, then autocovariance and autocorrelation can be used interchangeably. This results in no loss of generality even for the random sequence $\{x_t\}_{t=1}^T$ with non-zero mean $\overline{x}(t)$ which is known in advance. In this case, a new zero-mean random sequence $\{z_t\}_{t=1}^T$ can always be created by subtracting the mean from $\{x_t\}_{t=1}^T$ according to

$$z_t = x_t - \overline{x}(t).$$

The autocovariance and autocorrelation functions provide information about the statistical relationship between two random variables, x_t and x_τ, that are derived from the same random sequence. When more than one random sequences is involved, it is often of interest to determine the covariance or the correlation between a random variable in one sequence and a random variable in another sequence. Given two random sequences, $\{x_t\}_{t=1}^T$ and $\{y_t\}_{t=1}^T$, the **cross-correlation** function is defined by

$$r_{xy}(t, \tau) = E[x_t y_\tau^*],$$

and the **cross-covariance** function is defined by

$$c_{xy}(t, \tau) = E[(x_t - \overline{x}(t)(y_\tau - \overline{y}^*(\tau))].$$

The two functions satisfy the relation

$$c_{xy}(t, \tau) = r_{xy}(t, \tau) - \overline{x}(t)\overline{y}^*(\tau).$$

3.4.2 Stationary random sequences

The concept of **stationarity** is used to describe "statistical time-invariance" for a random sequence. Several types of stationarity can be defined depending on the precise nature of such statistical time-invariance.

If the first-order PDF, $p(x_t; t)$, of a random sequence, $\{x_t\}_{t=1}^{T}$, is invariant to the time t, then the sequence is said to be **first-order stationary**. The mean function or the first-order statistics of a first-order stationary random sequence is thus constant over its entire time span:

$$E(x_t) = \overline{x}, \qquad 1 \le t \le T.$$

The same is true also for the variance of a first-order stationary random sequence.

Similarly, a random sequence is said to be **second-order stationary** if the second-order joint PDF, $p(x_t, x_\tau; t, \tau)$, of the sequence depends only on the time difference $t - \tau$ and not on the individual times t and τ. If the sequence is second-order stationary, it must be first-order stationary also.

The second-order stationary sequences have second-order statistics, i.e., autocorrelation and autocovariance functions, which are invariant to a time shift of the sequence. That is, for an arbitrary integer k,

$$r_x(t, \tau) = r_x(t + k, \tau + k); \qquad c_x(t, \tau) = c_x(t + k, \tau + k).$$

If we choose $k = -\tau$, then we have

$$r_x(t, \tau) = r_x(t - \tau, 0).$$

That is, the second-order statistics depends only on time lag and not on individual times.

Higher-order stationarity can be similarly defined. A random sequence is said to be **Lth-order stationary** if the sequences x_t and x_τ have the same Lth-order joint PDF. A random sequence that is stationary for all orders $L > 0$ is called a **stationary in the strict sense**.

In most practical applications including nearly all speech processing applications, one is often primarily concerned with the mean and autocorrelation of a random sequence, and not the full PDF's. Restricting ourselves to limited scope, we have the following new form of stationarity:

A random sequence is **wide-sense stationary** if

- The mean of the sequence is a constant: $E(x_t) = \overline{x}$;
- The autocorrelation function $r_x(t, \tau)$ depends only on the difference $t - \tau$; and
- The variance of the sequence is finite: $\sigma_x^2 < \infty$.

A random sequence which is not stationary is called a nonstationary sequence. Natural speech signals are strongly nonstationary. However for mathematical tractability, one often uses random sequences with piecewise stationarity to approximate the speech signals. A **piecewise stationary sequence** is a nonstationary sequence whose local temporal regimes can be each approximated by stationary sub-sequences. These local temporal regimes jointly occupy the entire time span of the overall nonstationary sequence.

3.4.3 White sequence, Markov sequence, Gauss-Markov sequence, and Wiener sequence

A real-valued zero-mean sequence $\{v_t : t = 1, 2, \ldots, \infty\}$ is a **white sequence** or **(discrete-time) white noise** if its autocorrelation function is such that

$$r_x(t, \tau) = E[v_t v_\tau] = Q(t)\delta_{t\tau}, \tag{3.43}$$

where the **Kronecker delta function** $\delta_{t\tau}$ is one if $t = \tau$ and zero otherwise. $Q(t)$ is the variance of this white sequence and is generally time-varying. If the variance is time-invariant, i.e., $Q(t) = Q$, then this white sequence is called a **stationary white sequence**.

In addition, if the PDF at each time in a white sequence has a Gaussian distribution, then we have a **Gaussian white sequence**. A Gaussian white sequence is stationary if its variance, as well as its mean, is time-invariant.

A sequence with the property of Eq. 3.43 is temporally uncorrelated, or weakly independent. However, in most practical applications of speech processing, one has only the statistics up to second order and not full PDF's. Hence, taking the **uncorrelatedness** property as a usual assumption for the property of (weak) independence, one often calls a stationary white sequence an **independent and identically distributed (IID)** sequence.

When a random sequence $\{x_t : t = 1, 2, \ldots, \infty\}$ is generated by a system excited by white sequence $\{v_t : t = 1, 2, \ldots, \infty\}$ according to

$$x_{t+1} = T(t, x_t, v_t), \tag{3.44}$$

then the sequence x_t is called a **Markov sequence** or **(discrete-time) Markov process**. A Markov sequence is generally a time-varying sequence; for example, the mean of x_t is generally a function of time t. However, a Markov sequence is stationary if the system $T(\cdot)$ is time-invariant.

From the above definition, we see that the statistical properties of the random variable, x_{t+1}, are determined only by its immediate past, x_t, and not by its further past. It hence has the following

Markov property expressed as conditional PDF's:

$$p[x_{t+1}|x_t, x_{t-1}, \cdots, x_1] = p[x_{t+1}|x_t].$$

A verbal statement of the above is: The future $(t + 1)$ is independent of the past $(t - 1, t - 2, \cdots, 2, 1)$ if the present (t) is known; or: The future and the past are dependency separated by the present.

The domain of each random variable x_t in the Markov sequence, i.e., all possible values that can be assumed by x_t, is called the **state space** of the Markov sequence. In general, the state space of a Markov sequence according to the generation process Eq. 3.44 is $x_t \in (-\infty, +\infty)$. For this reason, we say the Markov sequence according to Eq. 3.44 is a **continuous-state Markov sequence**.

A special but very popular case of a Markov sequence is a **Gauss-Markov sequence**, which is generated by a linear system excited by a Gaussian white sequence according to

$$x_{t+1} = \phi x_t + v_t. \tag{3.45}$$

A Gauss-Markov sequence is stationary when ϕ is constant, but a Gauss-Markov sequence is generally time-varying. Furthermore, because of the linearity in Eq. 3.45, x_t is a Gaussian sequence, and hence its name (assuming the initial condition is such that x_0 is Gaussian). Finally, it can be easily shown that a stationary Gauss-Markov sequence has an exponential or geometric autocorrelation function.

A special case of a Markov sequence is a **Wiener sequence** or **(discrete-time) Wiener process**, which is generated by a linear system of Eq. 3.45 with $\phi = 1$:

$$x_{t+1} = x_t + v_t.$$

That is, the Wiener sequence is the cumulative temporal sum of the white sequence terms.

3.5 Markov Chain and Hidden Markov Sequence

In this section, we provide basics for two very important random sequences related to the general Markov sequence discussed above.

3.5.1 Markov chain as discrete-state Markov sequence

A **Markov chain** or **discrete-state Markov sequence** is a special case of a general Markov sequence. The state space of a Markov chain is of a discrete nature and is finite: $s_t \in \{s^{(j)}, j = 1, 2, \cdots, N\}$. Each of these discrete values is associated with a **state** in the Markov chain. Because of the one-to-one correspondence between state $s^{(j)}$ and its index j, we often use the two interchangeably.

A Markov chain, $\mathcal{S} = s_1, s_2, \cdots$, is completely characterized by the **transition probabilities**, defined by

$$P(s_t = s^{(j)} | s_{t-1} = s^{(i)}) \doteq a_{ij}(t), \qquad i, j = 1, 2, \cdots, N$$

and by the initial state-distribution probabilities. If these transition probabilities are independent of time t, then we have a **homogeneous Markov chain**.

The transition probabilities of a (homogeneous) Markov chain are often conveniently put into a matrix form:

$$A = [a_{ij}], \quad \text{where } a_{ij} \geq 0 \quad \forall i, j; \quad \text{and } \sum_{j=1}^{N} a_{ij} = 1 \quad \forall i$$

which is called the **transition matrix** of the Markov chain.

Given the transition probabilities of a Markov chain, the state-occupation probability

$$p_j(t) \doteq P[x_t = s^{(j)}]$$

can be easily computed. The computation is recursive according to

$$p_i(t+1) = \sum_{j=1}^{N} a_{ji} p_j(t), \quad \forall i.$$

If the state-occupation distribution of a Markov chain, $\mathcal{S} = s_1, s_2, \cdots$, asymptotically converges: $p_i(t) \to \pi(s^{(i)})$ as $t \to \infty$, we then call $p(s^{(i)})$ a **stationary distribution** of the Markov chain. For a Markov chain to have a stationary distribution, its transition probabilities, a_{ij}, have to satisfy

$$\pi(s^{(i)}) = \sum_{j=1}^{N} a_{ji}\pi(s^{(j)}), \quad \forall i.$$

The stationary distribution of a Markov chain plays an important role in a class of powerful statistical methods collectively named "Markov chain Monte Carlo (MCMC) methods. These methods are used to simulate (i.e., to sample or to draw) arbitrarily complex distributions, enabling one to carry out many difficult statistical inference and learning tasks which would otherwise be mathematically intractable. The theoretical foundation of the MCMC methods is the asymptotic convergence of a Markov chain to its stationary distribution, $\pi(s^{(i)})$. That is, regardless of the initial distribution, the Markov chain is an asymptotically unbiased draw from $\pi(s^{(i)})$. Therefore, in order to sample from an arbitrarily complex distribution, $p(s)$, one can construct a Markov chain, by designing appropriate transition probabilities, a_{ij}, so that its stationary distribution is

$$\pi(s) = p(s).$$

Three other interesting and useful properties of a Markov chain can be easily derived. First, the state duration in a Markov chain is an exponential or geometric distribution:

$$p_i(d) = C\ (a_{ii})^{d-1},$$

where the normalizing constant is $C = 1 - a_{ii}$.

Second, the mean state duration is

$$\overline{d}_i = \sum_{d=1}^{\infty} dp_i(d) = \sum_{d=1}^{\infty}(1 - a_{ii})(a_{ii})^{d-1} = \frac{1}{1 - a_{ii}}.$$

Finally, the probability for an arbitrary observation sequence of a Markov chain, which is a finite-length state sequence $\{\mathcal{S} = s_1, s_2, \cdots, s_T\}$, can be easily evaluated. This is simply the product of the transition probabilities traversing the Markov chain:

$$P(\mathcal{S}) = \pi_{s_1} \prod_{t=1}^{T-1} a_{s_t s_{t+1}},$$

where π_{s_1} is the initial state-occupation probability at $t = 1$.

3.5.2 From Markov chain to hidden Markov sequence

Let us view the Markov chain discussed above as an information source capable of generating observational output sequences. Then we can call the Markov chain an "observable" Markov sequence because its output has one-to-one correspondence to a state. That is, each state corresponds to a deterministically observable variable or event. There is no randomness in the output in any given state. This lack of randomness makes the Markov chain too restrictive to describe many real-world informational sources, such as the speech process, in an adequate manner.

Extension of the Markov chain to embed randomness which overlaps among the states in the Markov chain gives rise to a **hidden Markov sequence**. This extension is accomplished by associating an **observation probability distribution** with each state in the Markov chain. The Markov sequence thus defined is a doubly embedded random sequence whose underlying Markov chain is not directly observable, hence a hidden sequence. The underlying Markov chain in the hidden Markov sequence can be observed only through a separate random function characterized by the observation probability distributions.

Note that if the observation probability distributions do not overlap across the states, then the underlying Markov chain would not be hidden. This is because, despite the randomness embedded in the states, any observational value over a fixed range specific to a state would be able to map uniquely to this state. In this case, the hidden Markov sequence essentially reduces to a Markov chain.

When a hidden Markov sequence is used to describe a physical, real-world informational source, i.e., to approximate the statistical characteristics of such a source, we often call it a **hidden Markov model (HMM)**. One very successful practical use of the HMM has been in speech processing applications, including speech recognition, speaker recognition, speech synthesis, and speech enhancement. In these applications, the HMM is used as a powerful model to characterize the temporally nonstationary, spatially variable, but learnable and regular patterns of the speech signal. One key aspect of the HMM as the speech model is its sequentially arranged Markov states which permit the use of piecewise stationarity for approximating the nonstationary properties of the speech signal. Powerful algorithms have been developed to efficiently optimize the boundaries of the local quasi-stationary temporal regimes.

Characterization of a hidden Markov sequence

We now give a formal characterization of a hidden Markov sequence in terms of its basic elements and parameters.

1. Transition probabilities, $\mathbf{A} = [a_{ij}], i, j = 1, 2, ..., N$, of a homogeneous Markov chain with a total of N states

$$a_{ij} = P(s_t = j | s_{t-1} = i), \qquad i, j = 1, 2, \cdots, N.$$

2. Initial Markov chain state-occupation probability $\pi = [\pi_i], i = 1, 2, \cdots, N$

$$\pi_i = P(s_1 = i), \qquad i = 1, 2, \cdots, N.$$

3. If the observation probability distribution, $P(\mathbf{o}_t) = [b_i], i = 1, 2, ..., N$, is discrete, then the PMF associated with each state gives the probabilities of symbolic observations $\{\mathbf{v}_1, \mathbf{v}_2, \cdots, \mathbf{v}_K\}$:

$$b_i(k) = P[\mathbf{o}_t = \mathbf{v}_k | s_t = i], \qquad i = 1, 2, \cdots, N. \tag{3.46}$$

   If the observation probability distribution is continuous, then the parameters, Λ_i, in the PDF characterize the hidden Markov sequence.

The most common and successful PDF used in speech processing for characterizing the continuous observation probability distribution in the hidden Markov sequence or model (HMM) is a mixture Gaussian distribution (for vectored-valued observation $\mathbf{o}_t \in$

$\mathcal{R}^{\mathcal{D}}$):

$$b_i(\mathbf{o}_t) = \sum_{m=1}^{M} \frac{c_{i,m}}{(2\pi)^{D/2}|\mathbf{\Sigma}_{i,m}|^{1/2}} \exp\left[-\frac{1}{2}(\mathbf{o}_t - \boldsymbol{\mu}_{i,m})^{\mathrm{Tr}}\mathbf{\Sigma}_{i,m}^{-1}(\mathbf{o}_t - \boldsymbol{\mu}_{i,m})\right]. \tag{3.47}$$

In this **mixture Gaussian HMM**, the parameter set Λ_i includes mixture weights, $c_{i,m}$, Gaussian means, $\boldsymbol{\mu}_{i,m} \in \mathcal{R}^{\mathcal{D}}$, and Gaussian covariance matrices, $\mathbf{\Sigma}_{i,m} \in \mathcal{R}^{D\times D}$.

When the number of mixture components is reduced to one: $M = 1$, the state-dependent output PDF becomes a (uni-modal) Gaussian:

$$b_i(\mathbf{o}_t) = \frac{1}{(2\pi)^{D/2}|\mathbf{\Sigma}_i|^{1/2}} \exp\left[-\frac{1}{2}(\mathbf{o}_t - \boldsymbol{\mu}_i)^{\mathrm{Tr}}\mathbf{\Sigma}_i^{-1}(\mathbf{o}_t - \boldsymbol{\mu}_i)\right]. \tag{3.48}$$

and the HMM is commonly called a (continuous-density) **Gaussian HMM.**

Given the model parameters, one convenient way of characterizing a Gaussian HMM is to view it as a generative device producing a sequence of observational data, $\mathbf{o}_t, t = 1, 2, ..., T$. In this view, the data at each time t is generated from the model according to

$$\mathbf{o}_t = \boldsymbol{\mu}_i + \mathbf{r}_t(\mathbf{\Sigma}_\mathrm{i}), \tag{3.49}$$

where state i at a given time t is determined by the evolution of the Markov chain characterized by a_{ij}, and $\mathbf{r}_t(\mathbf{\Sigma}_i) = \mathcal{N}(0, \mathbf{\Sigma}_i)$ is a zero-mean, Gaussian, IID residual sequence, which is generally state dependent as indexed by i.

Because the residual sequence $\mathbf{r}_t(\mathbf{\Sigma}_i)$ is IID, and because $\boldsymbol{\mu}_i$ in Eq. 3.49 is constant (i.e., not time-varying) given state i, their sum which gives the observation $\mathbf{o}_t$ is thus also IID given the state. Therefore, the HMM discussed above would produce locally or piecewise stationary (widesense) sequences. Since the temporal locality in question is confined within state occupation of the HMM, we sometimes use the term **stationary-state HMM** to explicitly denote such a property.

One simple way to extend a stationary-state HMM so that the observation sequence is no longer state-conditioned IID is as follows. We can modify the constant term $\boldsymbol{\mu}_i$ in Eq. 3.49 to explicitly make it time-varying:

$$\mathbf{o}_t = \mathbf{g}_t(\Lambda_i) + \mathbf{r}_t(\mathbf{\Sigma}_i), \tag{3.50}$$

where parameters Λ_i in the deterministic time-trend function $\mathbf{g}_t(\Lambda_i)$ is dependent on state i in the Markov chain. This gives rise to the **trended (Gaussian) HMM** [Deng 92b] a special version of a **nonstationary-state HMM** where the first-order statistics (mean) are time-varying and thus violating a basic condition of wide-sense stationarity discussed in Section 3.4.2.

Simulation of a hidden Markov sequence

When we view a hidden Markov sequence or HMM as an information-source generation device which has been explicitly depicted in Eq. 3.49, we naturally would want to implement such a device. This is the problem of simulating the HMM given appropriate values for all model parameters: $\{A, \pi, B\}$ for a discrete HMM or $\{A, \pi, \Lambda\}$ for a continuous-density HMM. The result of the simulation is to produce an observation sequence, $\mathbf{o}_1^T = \mathbf{o}_1, \mathbf{o}_2, \cdots, \mathbf{o}_T$, which obeys the statistical law embedded in the HMM.

A simulation process is as follows:

1. Select an initial state $s_1 = i$ by drawing from the discrete distribution π (using a random number generator);

2. Set $t = 1$;

3. Draw an observation $\mathbf{o}_t$ based on the observation probability distribution in state i;

4. Make a Markov-chain transition from the current state $s_t = i$ to a new state $s_{t+1} = j$ according to the transition probability a_{ij}.

5. Increment t to $t + 1$; return to step **3** if $t < T$; otherwise exit.

In step **3** above, use Eq. 3.46 if the observation probability distribution in state i is discrete, and use Eq. 3.47 or Eq. 3.48 if the distribution is continuous-density multi-modal mixture of Gaussians or uni-modal Gaussian. For the latter case, one can alternatively simulate a zero-mean Gaussian random variable $\mathbf{r}_t(\mathbf{\Sigma}_i)$, and then add it into the constant $\boldsymbol{\mu}_i$ according to Eq. 3.49. Applying this simulation method, one can also directly simulate a trended HMM according to Eq. 3.50.

Likelihood evaluation of a hidden Markov sequence

Likelihood evaluation is a basic task needed for speech processing applications involving an HMM that uses a hidden Markov sequence to approximate vectorized speech signals.

Let $\mathcal{S} = (s_1, \ldots, s_T)$ be a finite-length sequence of states in a Gaussian HMM, and let $P(O, \mathcal{S})$ be the joint likelihood of the observation sequence $\mathbf{o}_1^T = (\mathbf{o}_1, \ldots, \mathbf{o}_T)$ and the event that $\mathbf{o}_1$ is generated from the model as the Markov chain starts from state s_1 at $t = 1$, $\mathbf{o}_2$ generated when the Markov chain enters into state s_2 at $t = 2$, $\mathbf{o}_3$ generated when the Markov chain enters into s_3 at $t = 3$, and so on. Let $P(\mathbf{o}_1^T|\mathcal{S})$ denote the likelihood that the observation sequence $\mathbf{o}_1^T$ is generated by the model conditioned on the state sequence $\mathcal{S}$.

The state-dependent residual, $\mathbf{r}_t(\mathbf{\Sigma}_i) \in \mathcal{R}^\mathcal{D}$, is assumed to have the zero-mean multivariate Gaussian density function

$$\frac{1}{(2\pi)^{D/2}|\mathbf{\Sigma}_i|^{1/2}} \exp\{-\frac{1}{2}\mathbf{r}_t^{\mathrm{Tr}}\mathbf{\Sigma}_i^{-1}\mathbf{r}_t\}.$$

The IID property for the state-dependent residuals leads to the joint density function of residuals, $\mathbf{r}_1^T = (\mathbf{r}_1, \mathbf{r}_2, ..., \mathbf{r}_T)$, conditioned on $\mathcal{S}$ to be of the following product form:

$$P(\mathbf{r}_1^T|\mathcal{S}) = \prod_{t=1}^{T} \frac{1}{(2\pi)^{D/2}|\mathbf{\Sigma}_i|^{1/2}} \exp\{-\frac{1}{2}\mathbf{r}_t^{\mathrm{Tr}}\mathbf{\Sigma}_i^{-1}\mathbf{r}_t\}.$$

According to Eq. 3.49, the transformation from $\mathbf{r}_1^T$ to $\mathbf{o}_1^T$ has unity Jacobian. This produces the conditional likelihood $P(\mathbf{o}_1^T|\mathcal{S})$ in a simple form

$$P(\mathbf{o}_1^T|\mathcal{S}) = \prod_{t=1}^{T} b_i(\mathbf{o}_t) = \prod_{t=1}^{T} \frac{1}{(2\pi)^{D/2}|\mathbf{\Sigma}_i|^{1/2}} \exp\{-\frac{1}{2}(\mathbf{o}_t - \boldsymbol{\mu}_i)^{\mathrm{Tr}}\mathbf{\Sigma}_i^{-1}(\mathbf{o}_t - \boldsymbol{\mu}_i)\}. \quad (3.51)$$

On the other hand, as discussed in Section 3.5.1 the probability of the state sequence

$\mathcal{S}$ is just the product of the transition probabilities

$$P(\mathcal{S}) = \pi_{s_1} \prod_{t=1}^{T-1} a_{s_t s_{t+1}}. \tag{3.52}$$

The joint likelihood $P(\mathbf{o}_1^T, \mathcal{S})$ can then be obtained by the product of likelihoods in Eqs. 3.51 and 3.52

$$P(\mathbf{o}_1^T, \mathcal{S}) = P(\mathbf{o}_1^T|\mathcal{S})P(\mathcal{S}). \tag{3.53}$$

In principle, the total likelihood for the observation sequence can be computed by summing the joint likelihoods in Eq. 3.53 over all possible state sequences $\mathcal{S}$:

$$P(\mathbf{o}_1^T) = \sum_{\mathcal{S}} P(\mathbf{o}_1^T, \mathcal{S}). \tag{3.54}$$

However, the amount of this computation is exponential in the length of the observation sequence, T, and hence the computation not practical.

A practical algorithm, a version of the **forward-backward algorithm** due to [Baum 72], computes $P(\mathbf{o}_1^T)$ for the HMM linearly in T and is described next.

Algorithm for efficient likelihood evaluation

To describe the algorithm, we first define the forward probabilities by

$$\alpha_t(i) = P(s_t = i, \mathbf{o}_1^t), \quad t = 1, \ldots, T, \tag{3.55}$$

and the backward probabilities by

$$\beta_t(i) = P(\mathbf{o}_{t+1}^T | s_t = i), \quad t = 1, \ldots, T-1, \tag{3.56}$$

both for each state i in the Markov chain. The forward and backward probabilities can be calculated recursively from

$$\alpha_t(j) = \sum_{i=1}^{N} \alpha_{t-1}(i) a_{ij} b_j(\mathbf{o}_t), \quad t = 2, 3, ..., T; \;\; j = 1, 2, ..., N \tag{3.57}$$

and

$$\beta_t(i) = \sum_{j=1}^{N} \beta_{t+1}(j) a_{ij} b_j(\mathbf{o}_{t+1}), \quad t = T-1, T-2, ..., 1; \;\; i = 1, 2, ..., N \tag{3.58}$$

where for the Gaussian HMM, we have Eq. 3.48:

$$b_i(\mathbf{o}_t) = \frac{1}{(2\pi)^{D/2}|\mathbf{\Sigma}_i|^{1/2}} \exp\left[-\frac{1}{2}\left((\mathbf{o}_t - \boldsymbol{\mu}_i)^{\mathrm{Tr}} \mathbf{\Sigma}_i^{-1} (\mathbf{o}_t - \boldsymbol{\mu}_i)\right)\right],$$

and for the mixture Gaussian HMM, we have Eq. 3.47:

$$b_i(\mathbf{o}_t) = \sum_{m=1}^{M} \frac{c_{i,m}}{(2\pi)^{D/2}|\mathbf{\Sigma}_{i,m}|^{1/2}} \exp\left[-\frac{1}{2}\left((\mathbf{o}_t - \boldsymbol{\mu}_{i,m})^{\mathrm{Tr}} \mathbf{\Sigma}_{i,m}^{-1} (\mathbf{o}_t - \boldsymbol{\mu}_{i,m})\right)\right].$$

Proofs of the these recursions are given immediately after this section.

The starting value for the α recursion is, according to the definition in Eq. 3.55,

$$\alpha_1(i) = P(s_1 = i, \mathbf{o}_1) = P(s_1 = i)P(\mathbf{o}_1|s_1) = \pi_i b_i(\mathbf{o}_1), \qquad i = 1, 2, ...N$$

and that for the β recursion is chosen as

$$\beta_T(i) = 1, \qquad i = 1, 2, ...N, \tag{3.59}$$

so as to provide the correct values for β_{T-1} according to the definition in Eq. 3.56.

To compute the total likelihood $P(\mathbf{o}_1^T)$ in Eq. 3.54, note that, for each state i,

$$\begin{aligned} P(s_t = i, O_1^T) &= P(s_t = i, O_1^t, O_{t+1}^T) \\ &= P(s_t = i, O_1^t)P(O_{t+1}^T|O_1^t, s_t = i) \\ &= P(s_t = i, O_1^t)P(O_{t+1}^T|s_t = i) \\ &= \alpha_t(i)\beta_t(i), \end{aligned} \tag{3.60}$$

for any $t = 1, 2, ..., T$ according to the definitions in Eqs. 3.55 and 3.56. (The third step above, $P(\mathbf{o}_{t+1}^T|\mathbf{o}_1^t, s_t = i) = P(\mathbf{o}_{t+1}^T|s_t = i)$, is due to the HMM property that the observations are IID given the state.) Hence,

$$P(\mathbf{o}_1^T) = \sum_{i=1}^{N} P(s_t = i, \mathbf{o}_1^T) = \sum_{i=1}^{N} \alpha_t(i)\beta_t(i). \tag{3.61}$$

Taking $t = T$ in Eq. 3.61 and using Eq. 3.59 lead to

$$P(\mathbf{o}_1^T) = \sum_{i=1}^{N} \alpha_T(i).$$

Thus, strictly speaking, the β recursion is not necessary for the forward scoring algorithm, and hence the algorithm is often called the **forward algorithm**. However, the β computation is a necessary step for solving the model parameter estimation problem, which will be covered in Chapter 5.

Proofs of the forward and backward recursions

Proofs of the recursion formulas, Eqs. 3.57 and 3.58, are provided here, using the total probability theorem, Bayes' rule, and using the Markov property and conditional independence property of the HMM.

For the forward probability recursion, we have

$$\begin{aligned} \alpha_t(j) &= P(s_t = j, \mathbf{o}_1^t) \\ &= \sum_{i=1}^{N} P(s_{t-1} = i, s_t = j, \mathbf{o}_1^{t-1}, \mathbf{o}_t) \\ &= \sum_{i=1}^{N} P(s_t = j, \mathbf{o}_t|s_{t-1} = i, \mathbf{o}_1^{t-1})P(s_{t-1} = i, \mathbf{o}_1^{t-1}) \end{aligned}$$

$$
\begin{aligned}
&= \sum_{i=1}^{N} P(s_t = j, \mathbf{o}_t | s_{t-1} = i)\alpha_{t-1}(i) \\
&= \sum_{i=1}^{N} P(\mathbf{o}_t | s_t = j, s_{t-1} = i)P(s_t = j | s_{t-1} = i)\alpha_{t-1}(i) \\
&= \sum_{i=1}^{N} b_j(\mathbf{o}_t)a_{ij}\alpha_{t-1}(i).
\end{aligned}
$$

For the backward probability recursion, we have

$$
\begin{aligned}
\beta_t(i) &= P(\mathbf{o}_{t+1}^T | s_t = i) \\
&= \frac{P(\mathbf{o}_{t+1}^T, s_t = i)}{P(s_t = i)} \\
&= \frac{\sum_{j=1}^{N} P(\mathbf{o}_{t+1}^T, s_t = i, s_{t+1} = j)}{P(s_t = i)} \\
&= \frac{\sum_{j=1}^{N} P(\mathbf{o}_{t+1}^T | s_t = i, s_{t+1} = j)P(s_t = i, s_{t+1} = j)}{P(s_t = i)} \\
&= \sum_{j=1}^{N} P(\mathbf{o}_{t+1}^T | s_{t+1} = j)\frac{P(s_t = i, s_{t+1} = j)}{P(s_t = i)} \\
&= \sum_{j=1}^{N} P(\mathbf{o}_{t+2}^T, \mathbf{o}_{t+1} | s_{t+1} = j)a_{ij} \\
&= \sum_{j=1}^{N} P(\mathbf{o}_{t+2}^T | s_{t+1} = j)P(\mathbf{o}_{t+1} | s_{t+1} = j)a_{ij} \\
&= \sum_{j=1}^{N} \beta_{t+1}(j)b_j(\mathbf{o}_{t+1})a_{ij}.
\end{aligned}
$$

3.6 Summary

In this chapter, we reviewed the aspects of probability theory, statistics, and random process theory relevant to statistical speech processing. We first introduced the concept of random variable and the associated concepts of probability distribution and summary statistics. We then discussed conditional probability, conditional independence, conditional expectation, and Bayes' rule. We then turned to the general characterization of (discrete-time) random sequences, with a special focus on the Markov sequence as the most commonly used class of the general random sequence. Central to the Markov sequence is the concept of state, which is itself a random variable. When the state of the Markov sequence is confined to be discrete, we have a Markov chain, where all possible values taken by the discrete state variable constitutes the (discrete) state space. When each discrete state value is generalized to be a new random variable (discrete or continuous), the Markov chain is then generalized to the (discrete or continuous) HMM. Some basic properties of the HMM have been discussed in this chapter, where the discussion made use of several more basic concepts introduced earlier in the chapter.

The concepts and results discussed in this chapter will be used to build up further mathematical background in the next two chapters, and will be used in several other chapters when discussing modeling issues for the speech process.

Chapter 4

Linear Model and Dynamic System Model

In this chapter, two types of widely used mathematical models, the **linear model** and the **dynamic system model** will be discussed. We will present the canonical forms of these models as their concise mathematical abstraction. Both types of the models, usually in diversified forms, have found their wide uses in speech processing. Presenting these models in their unified, canonical forms will allow us to conduct mathematical treatment in a more concise manner than otherwise.

We formally treat the linear model as a static or memoryless model. That is, the signal source being modeled is assumed to have no temporal memory or time dependence (correlation), as opposed to the dynamic system model where the temporal memory is explicitly represented. However, such a distinction is used mainly for mathematical convenience. If the signal source is represented by a vector whose elements consist of a block of sequentially organized signal samples, then the linear model associated with this vectorized signal representation will exhibit temporal dependence at the signal-sample level.

The linear model and the dynamic system model (the latter is often formulated in a specific form called the **state-space model**) are two distinct types of mathematical structures, both with popular applications in many disciplines of engineering. While it is possible to write the dynamic system model in the canonical form of the linear model by greatly expanding the dimensionality of the observation vector, the two types of the models have very different mathematical properties and computational structures. Using a high-dimensional linear model to represent the temporal correlation structure of a signal naturally captured by a low-dimensional dynamic system model would result in very inefficient computation.

When the parameters of either the linear model or the dynamic system model are made a function of time, we have the time-varying versions of these two types of models. In this chapter, we will devote special attention to the **time-varying models** due to the general time-varying properties of the speech process that require such a treatment.

4.1 Linear Model

The (static) linear model that we will discuss in this chapter is widely used in speech processing as well as in other areas of signal processing and in control and communication theory. The popularity of the linear model arises from the fact that the mathematics associated with its modeling problems, including representation and estimation, has been well developed and is significantly simpler than the non-linear model [Mend 95].

4.1.1 Canonical form of the model

Let us first define:

- $o[n] \equiv$ observation data at sample time n, and
- $\mathbf{o} = (o[0]\, o[1]\, \ldots\, o[N-1])^{\mathrm{Tr}} \equiv$ vector of N observation samples (N-point data set).

The canonical form of a linear model can be written as

$$\mathbf{o} = \mathbf{H}\boldsymbol{\theta} + \mathbf{v}, \tag{4.1}$$

where $\mathbf{o}$, which has the dimensionality of $N \times 1$, is called the observation or measurement vector); $\boldsymbol{\theta}$, which is $n \times 1$, is called the parameter vector; $\mathbf{H}$, which is $N \times n$, is called the observation matrix; and $\mathbf{v}$, which is $N \times 1$, is called the observation noise vector. $\mathbf{v}$ is a random vector, while $\mathbf{H}$ and $\boldsymbol{\theta}$ are deterministic. The random nature of $\mathbf{v}$ makes observation $\mathbf{o}$ a random vector also.

4.1.2 Examples of the linear model

We provide several common examples of the linear model in its canonical form to illustrate its comprehensiveness. Most of these examples have been discussed in earlier chapters. Putting these common models in a unified form of the linear model permits the use of powerful estimation theory (in Chapter 5) in a concise manner.

Discrete-time linear system

As the first example, we consider the FIR discrete-time linear system discussed in Chapter 1. This system model, in a non-canonical form, can be written as

$$o[n] = \sum_{k=0}^{p-1} h[k]x[n-k] + v[n],$$

where $x[n]$ is the input sequence to the system. Assume that there are N input samples used to yield N output samples. Then, define:

$$\begin{aligned}
\mathbf{o} &= [o(t_0),\, o(t_1),\, o(t_2),\, \ldots,\, o(t_{N-1})]^{\mathrm{Tr}} \\
\mathbf{v} &= [v(t_0),\, v(t_1),\, v(t_2),\, \ldots,\, v(t_{N-1})]^{\mathrm{Tr}} \\
\boldsymbol{\theta} &= [h[0],\, h[1],\, h[2],\, \ldots,\, h[p-1]]^{\mathrm{Tr}}.
\end{aligned}$$

This gives the canonical form of

$$\mathbf{o} = \mathbf{H}\boldsymbol{\theta} + \mathbf{v},$$

where $\mathbf{H}$ is the $N \times p$ matrix:

$$\mathbf{H} = \begin{bmatrix} x[0] & 0 & \cdots & 0 \\ x[1] & x[0] & \cdots & 0 \\ \vdots & \vdots & \ddots & \vdots \\ x[N-1] & x[N-2] & \cdots & x[N-p] \end{bmatrix}.$$

Function approximation

In a function approximation problem, we are interested in fitting a given set of data points by an approximating function that can be easily manipulated mathematically. As an example, we consider fitting the data points $[t, o(t)]$ by a pth order polynomial function of t:

$$o(t) = \theta_0 + \theta_1 t + \theta_2 t^2 + \cdots + \theta_p t^p + v(t).$$

Let there be N samples of the data points. By defining

$$\begin{aligned}
\mathbf{o} &= [o(t_0),\, o(t_1),\, o(t_2),\, \ldots,\, o(t_{N-1})]^{\mathrm{Tr}} \\
\mathbf{v} &= [v(t_0),\, v(t_1),\, v(t_2),\, \ldots,\, v(t_{N-1})]^{\mathrm{Tr}} \\
\boldsymbol{\theta} &= [\theta_0,\, \theta_1,\, \theta_2,\, \ldots .\, \theta_p]^{\mathrm{Tr}},
\end{aligned}$$

we are then able to rewrite the above polynomial fitting function in the canonical form:

$$\mathbf{o} = \mathbf{H}\boldsymbol{\theta} + \mathbf{v},$$

where $\mathbf{H}$ is the $N \times p$ matrix:

$$\mathbf{H} = \begin{bmatrix} 1 & t_0 & t_0^2 & \cdots & t_0^p \\ 1 & t_1 & t_1^2 & \cdots & t_1^p \\ \vdots & \vdots & \vdots & \ddots & \cdots \\ 1 & t_{N-1} & t_{N-1}^2 & \cdots & t_{N-1}^p \end{bmatrix}$$

The type of linear model based on functional fitting to sampled data is sometimes called a **linear regression model**.

Discrete Fourier transform

It is interesting to note that the discrete Fourier transform (DFT) discussed in Chapter 1 can be represented in the canonical form of the linear model as in Eq. 4.1. We use the trigonometric form of the DFT (alternative to the exponential form in Chapter 1) to illustrate this.

DFT can be considered as a function fitting problem, where the N samples of data, $o[n]$, are fitted by a linear combination of *sine* and *cosine* functions at different harmonics of the fundamental frequency. This data-fitting model, not in the canonical form, is

$$o[n] = \sum_{k=1}^{M} a_k \cos\left(\frac{2\pi kn}{N}\right) + \sum_{k=1}^{M} b_k \sin\left(\frac{2\pi kn}{N}\right) + v[n].$$

This can be rewritten in the canonical form $\mathbf{o} = \mathbf{H}\boldsymbol{\theta} + \mathbf{v}$, where

$$\begin{aligned}
\mathbf{o} &= [o[0],\, o[1],\, o[2],\, \ldots,\, o[N-1]]^{\mathrm{Tr}} \\
\mathbf{v} &= [v(t_0),\, v(t_1),\, v(t_2),\, \ldots,\, v(t_{N-1})]^{\mathrm{Tr}} \\
\boldsymbol{\theta} &= [a_1,\, a_2,\, \ldots,\, a_M,\, b_1,\, b_2,\, \ldots,\, b_M]^{\mathrm{Tr}}
\end{aligned}$$

and $\mathbf{H}$ is the $N \times 2M$ matrix:

$$\mathbf{H} = \left[\mathbf{h}_1^a\, \mathbf{h}_2^a \cdots \mathbf{h}_M^a\, \mathbf{h}_1^b\, \mathbf{h}_2^b \cdots \mathbf{h}_M^b\right].$$

The elements of the **H** matrix are

$$\mathbf{h}_k^a = \begin{bmatrix} 1 \\ \cos\left(\frac{2\pi k}{N}\right) \\ \cos\left(\frac{2\pi k 2}{N}\right) \\ \vdots \\ \cos\left(\frac{2\pi k(N-1)}{N}\right) \end{bmatrix}, \qquad \mathbf{h}_k^b = \begin{bmatrix} 0 \\ \sin\left(\frac{2\pi k}{N}\right) \\ \sin\left(\frac{2\pi k 2}{N}\right) \\ \vdots \\ \sin\left(\frac{2\pi k(N-1)}{N}\right) \end{bmatrix}$$

4.1.3 Likelihood computation

Computation of the likelihood of a statistical model for arbitrary samples of the observation data is a fundamental issue for the model development. In the canonical form of the linear model described by Eq. 4.1, since the parameter vector $\boldsymbol{\theta}$ is generally considered to be deterministic, the statistical nature of the noise vector $\mathbf{v}$ is the determining factor for how to compute the likelihood of the linear model for the observation data $\mathbf{o}$. If $\mathbf{v}$ is assumed to follow a zero-mean, multivariate Gaussian distribution:

$$\mathbf{v} \sim \mathcal{N}(\mathbf{v}; \mathbf{0}, \boldsymbol{\Sigma}),$$

or

$$p(\mathbf{v}) = \frac{1}{(2\pi)^{N/2}|\boldsymbol{\Sigma}|^{1/2}} \exp\left[-\frac{1}{2}\left(\mathbf{v}^{\mathrm{Tr}}\boldsymbol{\Sigma}^{-1}\mathbf{v}\right)\right], \tag{4.2}$$

then the rule of random variable transformation gives the likelihood of $\mathbf{o}$:

$$p(\mathbf{o}) = \frac{1}{(2\pi)^{N/2}|\boldsymbol{\Sigma}|^{1/2}} \exp\left[-\frac{1}{2}(\mathbf{o} - \mathbf{H}\boldsymbol{\theta})^{\mathrm{Tr}}\boldsymbol{\Sigma}^{-1}(\mathbf{o} - \mathbf{H}\boldsymbol{\theta})\right]. \tag{4.3}$$

That is, observation vector $\mathbf{o}$ is also a multivariate Gaussian distribution but with a mean vector of $\mathbf{H}\boldsymbol{\theta}$.

4.2 Time-Varying Linear Model

The linear model discussed so far assumes that the parameter vector, $\boldsymbol{\theta}$, does not change over the time index, k. Such a model is also called the time-invariant linear model. The discrete-time time-invariant linear system discussed in Chapter 1 is such a model. This is so because when the coefficients (i.e., the parameter vector in the corresponding linear model) of a finite-difference equation that characterize the linear system are constants, then it can be shown that the shift-invariance property required by a time-invariant linear system is satisfied.

The time-invariant linear model often has limited uses, especially in speech processing, since the properties of the speech change over time (with rare exceptions). Making the parameter vector, $\boldsymbol{\theta}$, of the linear model vary as a function of time results in the **time-varying linear model**, which is the topic of this section.

4.2.1 Time-varying linear predictive model

Recall that the linear predictive (LP) model discussed in Chapter 2, which is also called an autoregressive (AR) model or IIR filter model, has a recursive form for the output sequence $o[n]$ according to

$$o[n] = \sum_{k=1}^{K} \phi_k o[n-k] + v[n].$$

This can be put in the canonical form of a linear model with the parameter vector

$$\boldsymbol{\theta} = [\phi_1, \phi_2 \ldots, \phi_K]^{\mathrm{Tr}}.$$

This is a time-invariant linear model because the parameter vector $\boldsymbol{\theta}$ is not a function of time, n. To turn this linear model into a time-varying one, we impose time dependence on $\boldsymbol{\theta}$, resulting in the time-varying LPC model in the form of

$$o[n] = \sum_{k=1}^{K} \phi_{k,n} o[n-k] + v[n].$$

The time dependence of the parameter vector $\phi_{k,n}$ can take a variety of forms. The simplest form is to allow for non-parametric parameter evolution on a block-by-block basis. That is, we divide the entire time axis into a sequential set of blocks (or frames), and for each block (consisting of many samples) the LPC parameters $\phi_{k,n}$ become independent of time n. However, the $\phi_{k,n}$ change from one block to the next. In essence, the time dependency for this type of non-parametric parameter evolution is such that the parameters are piecewise constant over time. The very popular speech coding technique called LPC coding has been based precisely on this type of non-parametric time-varying LPC model, with the time block being preset to be uniform and consisting of time samples on the order of hundreds [Atal 79, Atal 71]. Under this coding scheme, the parameters of the LPC model need to be estimated for each separate block of speech. Such estimated LPC parameters (together with other residual information) are used to represent the original speech samples on a frame-by-frame basis. These estimated parameters have also been used as speech feature vectors for speech recognition.

A more efficient way to characterize the time dependency of the parameter vector $\boldsymbol{\theta}$ in the LPC model is to provide a parametric form for the parameter evolution. We will then have a new set of parameters, Φ, which gives the time varying function for the LPC parameters over the entire time span, eliminating the need to divide the signal into frames. This type of parametric form of time-varying LPC model can be written as

$$o[n] = \sum_{k=1}^{K} \phi_n(\Phi_k) o[n-k] + v[n],$$

where the LPC parameters ϕ_n are explicitly indexed by discrete time index of n and by the new set of parameters Φ_k.

One common parametric form used for time-varying LPC modeling is that of sinusoids. In this case, the parameters Φ that characterize the time-varying nature of the LPC parameters ϕ_n are the amplitudes, angular frequencies, and phases of the sinusoids. This sinusoidally modulated time-varying LPC model has found uses in very low bit-rate speech coding.

4.2.2 Markov modulated linear predictive model

Another parametric method used for describing time-varying LPC coefficients in LP modeling, which found useful applications in speech recognition [She 94a, Juang 86], is to use a Markov chain to control the time evolution of the LPC coefficients. This is analogous to using the sinusoids for the same purpose, but with a probabilistic structure rather than just a deterministic function.

In this **Markov modulated LP model**, which is also called a **hidden filter model** or autoregressive-HMM, the parameters are distinctively associated with separate Markov states. The present value of a one-dimensional observation sequence, $o[n]$ (e.g., digitized speech waveform), is, as usual, expressed as a linear combination of its past values plus a *driving sequence* that is assumed to be a Gaussian IID process. Consider a Markov chain with N states and a state transition matrix $\mathbf{A} = [a_{ij}]$, $i, j = 1, 2, .., N$. The LP model parameters are made conditional on the state of the Markov chain, i. In this model, with a compact notation, the data-generation mechanism conditioned on state i is described by

$$o[n] = \sum_{k=1}^{K} \phi_i(k) o[n-k] + e_i[n] = \mathbf{b}_i \mathbf{x}[n] + e_i[n] \, , \tag{4.4}$$

where $\mathbf{b}_i = \{\phi_i(1), \phi_i(2), \phi_i(K)\}$ is a vector of LP coefficients, $\mathbf{x}[n] = \{o[n-1], o[n-2], ..., o[n-K]\}^{\text{Tr}}$ is the sequence of past K observations. The driving sequence (also called the *residual error*), $e_i(t)$, is Gaussian and IID with a mean of μ_i and a variance of σ_i^2. Note in Eq. 4.4, if $K = 0$ (zero-th order prediction) then the model would be reduced to the standard HMM discussed in Chapter 3 (with scalar observation sequences); that is, the present observation $o[n]$ is equal to $e_i[n]$ and the output distribution parameters (μ_i and σ_i) would change over time according to the Markov state transition. On the other hand, when $N = 1$, then the model would be reduced to the (time-invariant) LP model discussed earlier in this section.

Since the driving sequence is assumed IID, the likelihood of the observation sequence $o_1^{\text{Tr}} = \{o[1], o[2], ..., o[T]\}$ given the state sequence $\mathcal{S}_1^{\text{Tr}} = \{s_1, s_2, ..., s_T\}$ and K initial observations $\mathbf{x}[1]$ (under the model λ) can be calculated as

$$P(o_1^{\text{Tr}} | \mathcal{S}_1^{\text{Tr}}, \mathbf{x}[1], \lambda) = \prod_{n=1}^{\text{Tr}} \frac{1}{\sqrt{2\pi}\sigma_{s_t}} \exp\left[-\frac{1}{2\sigma_{s_t}^2}(o[n] - \mu_{s_t} - \mathbf{b}_{s_t}\mathbf{x}[n])^2\right].$$

From this, the likelihood with which the model produces the observation sequence can be obtained in a brute-force manner as

$$P(o_1^{\text{Tr}} | \mathbf{x}[1], \lambda) = \sum_{\mathcal{S}_1^{\text{Tr}}} P(o_1^{\text{Tr}} | \mathcal{S}_1^{\text{Tr}}, \mathbf{x}[1], \lambda) P(\mathcal{S}_1^{\text{Tr}}) \, . \tag{4.5}$$

As for the standard HMM discussed in Chapter 3, a more efficient method can be used to compute such a likelihood.

4.2.3 Markov modulated linear regression model

As another example of the time-varying linear model, we consider the **Markov modulated linear regression model**. This is also called a trended HMM, (first-order) nonstationary-state HMM, or mean-trajectory segmental model [Deng 91b, Deng 92b,

Ost 96]. (The name of "trended HMM" was initially coined in [Deng 91b, Deng 92b], using the terminology from time series analysis where a trend describes the average behavior of a time series that shifts with time.) We briefly looked at this model earlier in Section 3.5.2 of Chapter 3 as an extension of the standard HMM where the mean in the PDF of each HMM state is made a function of time. We now view this model from another perspective: we first construct a linear model as function approximation (Section 4.1.2) or linear regression, and then allow the observation matrix of this linear model to vary with time according to the evolution of a Markov chain. These two views are equivalent.

The model consists of the following parameter triplex $[\mathbf{A}, \mathbf{\Theta}, \mathbf{\Sigma}]$:

1. Transition probabilities, $a_{ij}, i, j = 1, 2, ..., N$ of the homogeneous Markov chain (a total of N countable states);

2. Parameters $\mathbf{\Theta}_i$ in the deterministic trend function $\mathbf{g}_t(\Lambda_i)$, as dependent on state i in the Markov chain;

3. Covariance matrices, $\mathbf{\Sigma_i}$, of the zero-mean, Gaussian, IID residual $\mathbf{r}_t(\mathbf{\Sigma_i})$, which are also state dependent.

Given the above model parameters, the observation vector sequences, $\mathbf{o}_t, t = 1, 2, ..., T$ are generated from the model according to

$$\mathbf{o}_t = \mathbf{g}_t(\Lambda_i) + \mathbf{r}_t(\mathbf{\Sigma}_i), \tag{4.6}$$

where state i at a given time t is determined by the evolution of the Markov chain characterized by a_{ij}.

Efficient computation for the likelihood of this trended HMM can be accomplished also by the forward algorithm described in Chapter 3 with the following minor modification. In the forward algorithm, the output probability computation is modified to

$$b_i(\mathbf{o}_t) = \frac{1}{(2\pi)^{D/2}|\mathbf{\Sigma}_i|^{1/2}} \exp\left[-\frac{1}{2}(\mathbf{o}_t - \mathbf{g}_t(\Lambda_i))^{\text{Tr}}\mathbf{\Sigma}_i^{-1}(\mathbf{o}_t - \mathbf{g}_t(\Lambda_i))\right], \tag{4.7}$$

for the (unimodal) trended HMM [Deng 92b], and to

$$b_i(\mathbf{o}_t) = \sum_{m=1}^{M} \frac{c_{i,m}}{(2\pi)^{D/2}|\mathbf{\Sigma}_{i,m}|^{1/2}} \exp\left[-\frac{1}{2}(\mathbf{o}_t - \mathbf{g}_t(\Lambda_{i,m}))^{\text{Tr}}\mathbf{\Sigma}_{i,m}^{-1}(\mathbf{o}_t - \mathbf{g}_t(\Lambda_{i,m}))\right] \tag{4.8}$$

for the mixture trended HMM [Deng 97d].

For a wide class of functions of $\mathbf{g}_t(\Lambda_i)$, the parameter estimation problem (to be elaborated in Chapter 5) has convenient linear solutions. This class includes the functions in the following form:

$$\mathbf{g}_t(\Lambda_i) = \mathbf{B}_i\mathbf{f}_t = \sum_{p=0}^{P} \mathbf{b}_i(p) f_t(p) \tag{4.9}$$

where the basis function $\mathbf{f}_t(p)$ has an arbitrary form of time dependence but the parameters $\Lambda_i = \mathbf{B}_i(p)$ in the function is in a linear form above.

In Eq. 4.9, choose the basis function, in each of its vector components, to be

$$f_t(p) = t^p.$$

Then we have the polynomial trended HMM where the trend function is

$$\mathbf{g}_t(\Lambda_i) = \sum_{p=0}^{P} \mathbf{b}_i(p)t^p. \tag{4.10}$$

4.2.4 Speech data and the time-varying linear models

In this section, we first present a general version of the time-varying linear model of which the time-varying linear models discussed in the earlier sections are special cases. We then show some speech data sequences in the form of cepstral coefficients that can be described by this general model. Different choices of the model give rise to different degrees of modeling accuracy.

The general version of the time-varying linear model is the Markov modulated LP model as discussed earlier but with time-varying LP (i.e., autoregressive) parameters and time-varying residuals. Assuming K autoregressive terms (i.e., LP order), the data-generation equation for the model (scalar output assumed) is

$$o_t = \sum_{k=1}^{K} \phi_t(\Psi_{i,k})o_{t-k} + g_t(\boldsymbol{\theta}_i) + r_t(\sigma_i^2), \tag{4.11}$$

where state i at a given discrete time t (frame) is determined by evolution of the underlying N-state Markov chain characterized by transition matrix $[a_{ij}], i, j = 1, 2, \cdots, N$, $\Psi_{i,k}$ are the state(i)-dependent parameters in the time-varying autoregression coefficients $\phi_t(\Psi_{i,k})$ $\boldsymbol{\theta}_i$ are the state(i)-dependent parameters in the time-varying mean functions, $g_t(\boldsymbol{\theta}_i)$, of the autoregression residual, and finally $r_t(\sigma_i^2)$ is the remaining residual (after mean removal) that is assumed to be IID, white, and Gaussian characterized by state(i)-dependent variance σ_i^2. To be concrete, in this section, we choose a parametric form — a polynomial function of time — for each of the time-varying elements of the autoregression coefficients $\phi_t(\Psi_{i,k})$. We choose the same parametric form also for the mean function of the residual term. The model described above has also been called a hierarchical nonstationary model or second-order nonstationary-state HMM described in various publications [Deng 93c, Deng 95a].

We now show numerical examples for how different choices of the model terms and parameters lead to different degrees of accuracy for fitting the model to the actual speech data. The speech data was taken from several tokens of some monosyllabic words spoken by a native English male speaker. The speech was digitally sampled at 16 kHz. A Hamming window of duration 25.6 ms was applied every 10 ms (the frame length) and for each window, eight mel-frequency cepstral coefficients were computed. Although similar data fitting results were obtained for all cepstral coefficients, we show in this section only the results for the first-order cepstral coefficient (C1).

The parameters of the model used in the data fitting experiments discussed here were trained using the EM algorithm (described in detail in Chapter 5). Given the model parameters, the procedure for fitting the models to the data starts by first finding the optimal segmentation of the data into the HMM states and then fitting the segmented data using the fitting functions associated with the corresponding states. The fitting function, distinct for each HMM state i in general, is defined as

$$f_t(i) = g_t(\boldsymbol{\theta}_i) + \sum_{k=1}^{K} \phi_t(\Psi_{i,k})o_{t-k}. \tag{4.12}$$

In the results shown in this section, $g_t(\boldsymbol{\theta}_i)$ and $\phi_t(\Psi_{i,k})$ are taken to be polynomial functions of time, of an order of M and L, respectively. Then Eq. 4.12 (with o_t being the speech data to be fitted by the model) becomes

$$o_t = f_t(i) + r_t(\sigma_i^2).$$

This determines the state-conditioned data fitting residual:

$$r_t(\sigma_i^2) = o_t - f_t,$$

with $t = \tau_{i-1}, \tau_{i-1}+1, ..., \tau_i - 1$, where $\tau_i, i = 0, 1, 2, ..., N$, are the optimal state segmentation boundaries (the optimal boundaries are determined by an algorithm discussed in a later chapter). The measure we used in our experiments for the accuracy of data fitting by the model is a linear summation of the residual square over the states and over the state-bound time frames:

$$\text{Error} \quad = \quad \sum_{i=1}^{N} \sum_{t=\tau_{i-1}}^{\tau_i - 1} r_t^2(\sigma_i^2). \tag{4.13}$$

Results for the model with a single state

Several tokens of a word utterance are used to train the single-state HMM varying in the number of regression terms (K), in the order of the polynomial for the autoregression coefficients (L), and in the order of the polynomial for the regression-residuals' mean functions (M).

The four plots in Fig. 4.1 compare the data fitting results for different values of M, K with fixed $L = 0$. In these plots, the dotted lines are the actual speech data, o_t, expressed as the $C1$ sequence, from one training token. The vertical axis represents the magnitude of $C1$ and the horizontal axis is the frame number. Superimposed on the four plots as solid lines are the four different fitting functions f_t (Eq. 4.12) varying in the M, K values as shown at the top of each plot. Along with each plot is also shown the data fitting error defined in Eq. 4.13. Comparison among the four plots demonstrates that inclusion of an autoregression term ($K = 1$) has a much more serious effect on improving the closeness of data fitting than the regression-residual terms.

Fig. 4.2 differs from Fig. 4.1 by incorporating a linearly time varying term of the regression coefficients ($L = 1$) and by increasing the number of regression terms from one to two ($K = 2$). Increasing K from one to two has little effect on improving fitting accuracy (e.g., fitting error reduction changes from 70.66 to 68.15). In contrast, inclusion of time varying regression coefficients (from $L = 0$ to $L = 1$) reduces the fitting error by a much greater amount — from 99.43 (Fig. 4.1) to 70.66.

Figs. 4.3 and 4.4 are the counterparts of Figs. 4.1 and 4.2 for the test token of the same word uttered by the same speaker. It can be observed that the overall fitting errors increased somewhat for all cases, as expected since the fitted data sequence was not used in training the models. However, the conclusions drawn above from Figs. 4.1 and 4.2 for the training token case largely hold. Here, significant error reduction occurs from the $L = 0$ case (error = 192.40) to the $L = 1$ case (error = 162.61) only when the correlation residual is also modeled as nonstationary ($M = 1$).

Since many applications of speech models (such as speech recognition) involve not only fitting the model to the "correct" data sequence but also the ability to reject the "wrong" data sequence, Figs. 4.5 and 4.6 show the results of fitting the model from

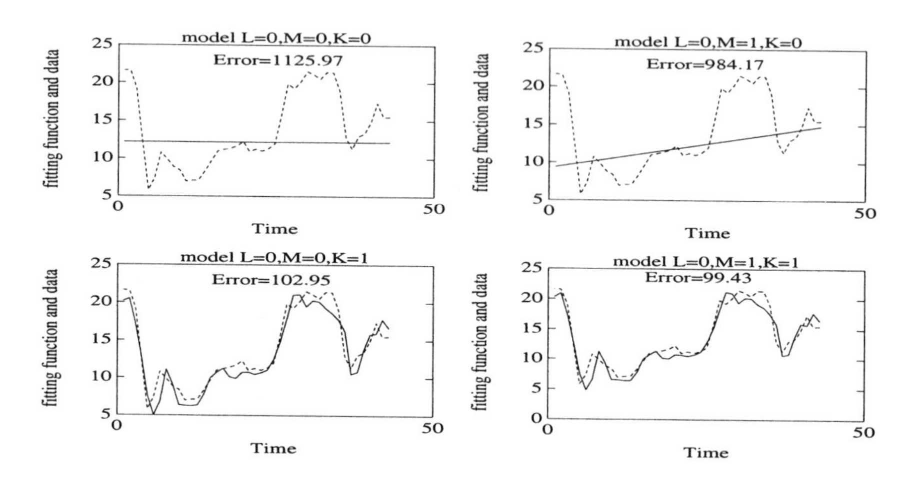

Figure 4.1: Results of fitting models to a training token (A) ($C1$ sequence as data sequence o_t with time-frame rate of 10 ms), with fixed $L = 0$ and two different values of M, K (modified from Deng and Rathinavelu [Deng 95a], @Elsevier).

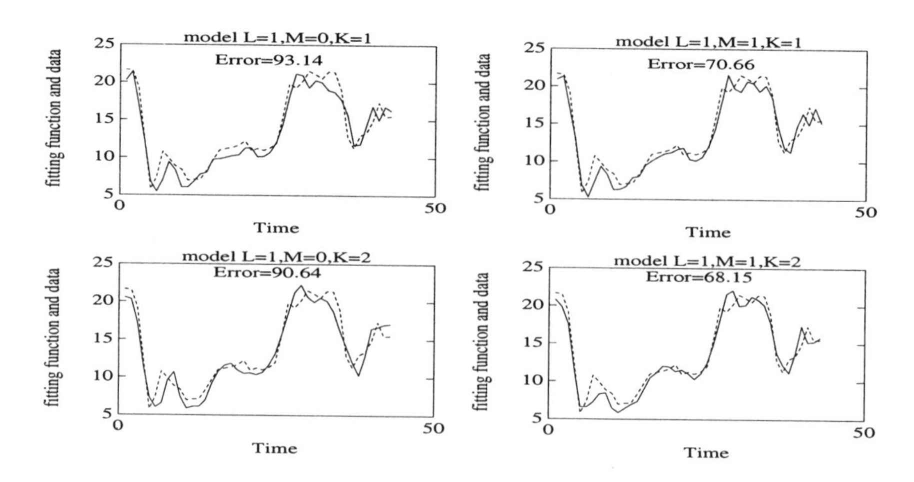

Figure 4.2: Results of fitting models to the training token, incorporating a linearly time varying term of the regression coefficients ($L = 1$) and increasing the number of regression terms to two ($K = 2$) (modified from Deng and Rathinavelu [Deng 95a], @Elsevier).

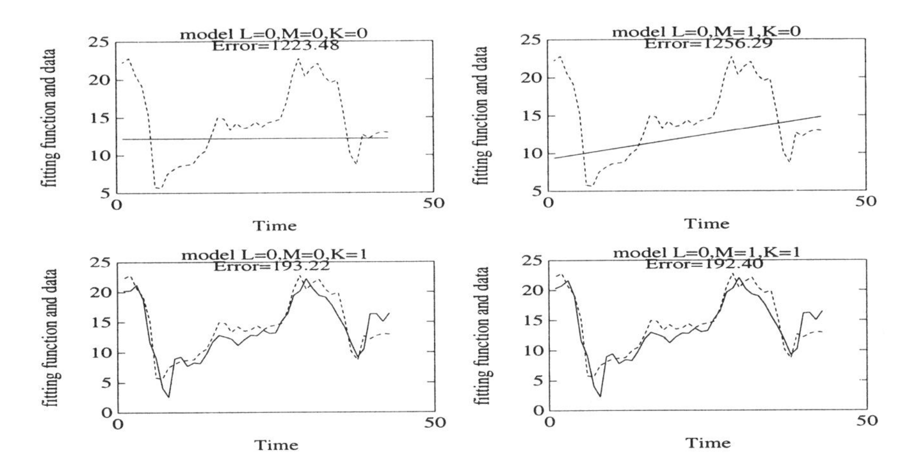

Figure 4.3: Results of fitting models to a *test* token (A) with fixed $L = 0$ and two different values of M, K (modified from Deng and Rathinavelu [Deng 95a], @Elsevier).

data sequences (the same model in Figs. 4.1 to 4.4) to the "wrong" data sequence ($C1$) from the utterance of another word. Figs. 4.5 and 4.6 were obtained with the model parameters $L = 0$ and $L = 1$, respectively, and with several different values of M and K. The significantly larger fitting errors produced from fitting models to "wrong" data than those from fitting to the "correct" data are indicative of the model's discriminative ability.

Results for the model with three states

Here the data fitting results are shown for two models, each having three states but trained with several tokens of two different word utterances as before. These results are intended to illustrate the ability of the constructed models to discriminate different utterances. For Figs. 4.7–4.10 discussed below, only the *test* tokens — disjoint from the training tokens — were used as data sequences for fitting the models. The four plots in each of the figures are arranged with the identical increasing complexity in the model parameter set: the first plot with $L = 0,\ M = 0,\ K = 0$, the second with $L = 0,\ M = 1,\ K = 0$, the third with $L = 0,\ M = 0,\ K = 1$, and the fourth with $L = 1,\ M = 1,\ K = 1$.

Figs. 4.7 and 4.8 show the results of fitting the same word utterance using the "correct" model and the "wrong" model, respectively. Note that in each plot the two breakpoints in the otherwise continuous solid lines correspond to the frames where the "optimal" state transitions occur from state 1 to state 2 and from state 2 to state 3. The "wrong" model apparently produces substantially greater errors in the data fitting than the "correct" model, accounting for the large margin in correctly identifying the utterance and discriminating against the wrong utterance. To appreciate the role of the state-conditioned (local) nonstationarity in the model's discriminability, we examine the data fitting errors in the bottom two plots, one with a stationary-state model ($M = 0, L = 0$) and the other with a nonstationary-state model (nonstationarity in both first and second orders: $M = 1, L = 1$), in Figs. 4.7 and 4.8. For the correct model

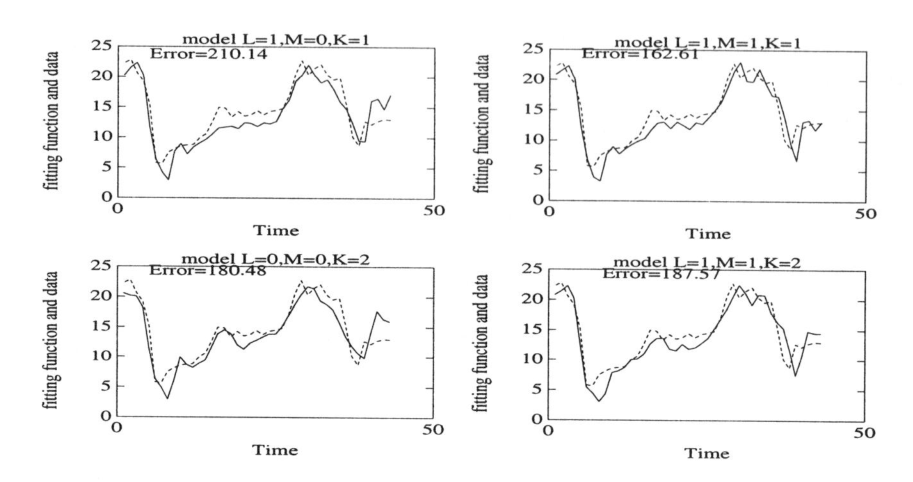

Figure 4.4: Results of fitting models to the *test* token, incorporating a linearly time varying term of the regression coefficients ($L = 1$) and increasing the number of regression terms to two ($K = 2$) (modified from Deng and Rathinavelu [Deng 95a], @Elsevier).

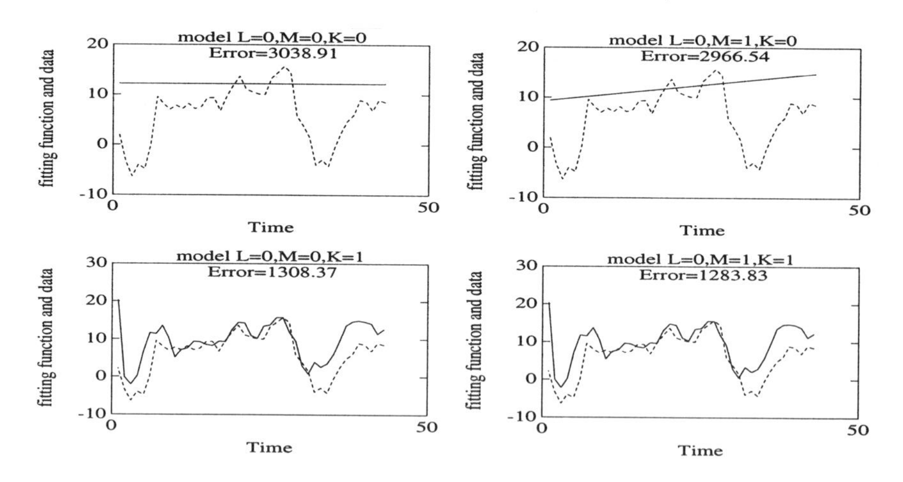

Figure 4.5: Results of fitting models trained from utterance A to a "wrong" data sequence — utterance B, with fixed $L = 0$ (modified from Deng and Rathinavelu [Deng 95a], @Elsevier).

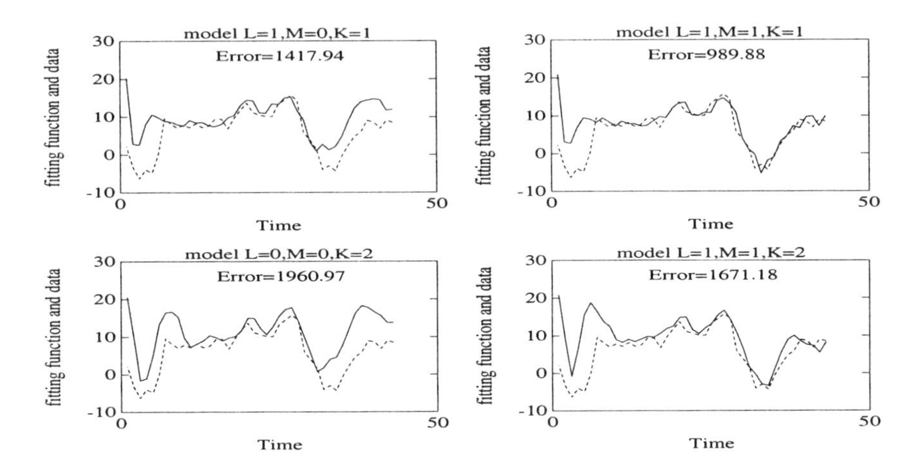

Figure 4.6: Results of fitting models trained from utterance A to a "wrong" data sequence — utterance B, with fixed $L = 1$ (modified from Deng and Rathinavelu [Deng 95a], @Elsevier).

(Fig. 4.7), error reduction in data fitting by incorporating the local nonstationarity goes from 77.84 to 60.86. However, for the purpose of rejecting the wrong model (Fig. 4.8), the local nonstationarity plays a much more significant role — increasing the data fitting error (a measure of the model discrimination power) from 424.74 to as large as 676.23.

The same observations were made by examining the results of fitting the "correct" model (Fig. 4.9) and the "wrong" model (Fig. 4.10) to the other word utterance. Here, the local nonstationarity improves data fitting using the correct model with a reduction of fitting errors only from 139.19 to 110.97. However, use of the local nonstationarity allows the fitting error for the "wrong" model to increase from 602.03 to 696.42.

4.3 Linear Dynamic System Model

The **dynamic system model** has a very different mathematical structure and computational properties from the linear model discussed earlier in this chapter. Although it would be possible to indefinitely increase the dimensionality of the signal vectors in the linear model so as to achieve the same mathematical description as a linear dynamic system model (with an example later in this section), computations associated with the model would become much less efficient. Furthermore, for the signal sources, such as speech, which are endowed with hidden signal dynamics according to their signal generation mechanisms, the dynamic system model representation is much more natural than its high-dimensional linear model counterpart.

Dynamic system models have been an important modeling tool in engineering and sciences including speech processing. The models typically have two essential characteristics. First, the models are statistical in nature, such that the outputs of the systems are characterized not by a deterministic function but by a stochastic process that consists of a rich variety of temporal trajectories. Second, the models are defined with the use of some internal low-dimensional (continuous) state that is hidden from the direct observations. The state summarizes all information there is to know at a given time about the

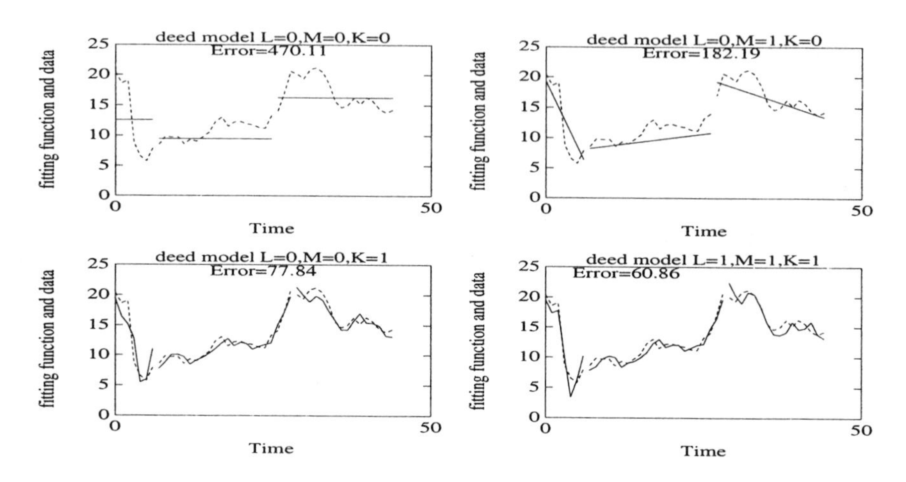

Figure 4.7: Fitting a correct three-state model (A, utterance *deed*) to a data sequence (A).

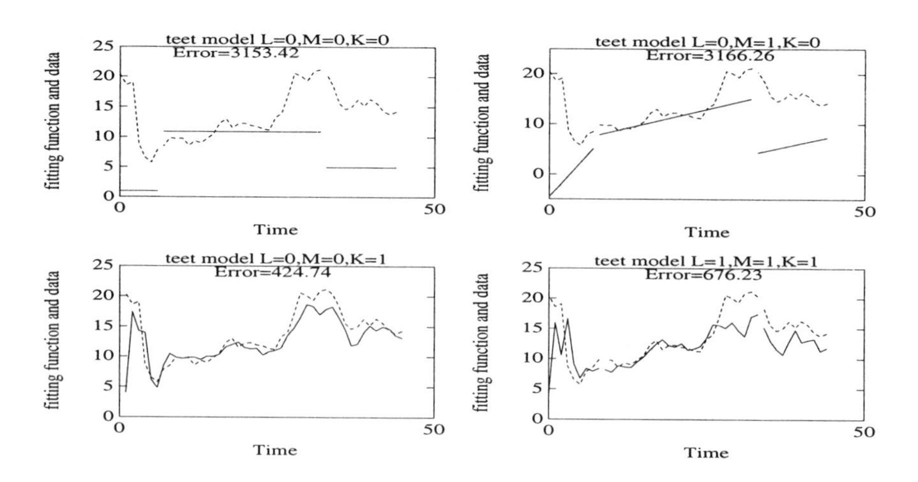

Figure 4.8: Fitting an incorrect three-state model (B, utterance *teet*) to a data sequence (A).

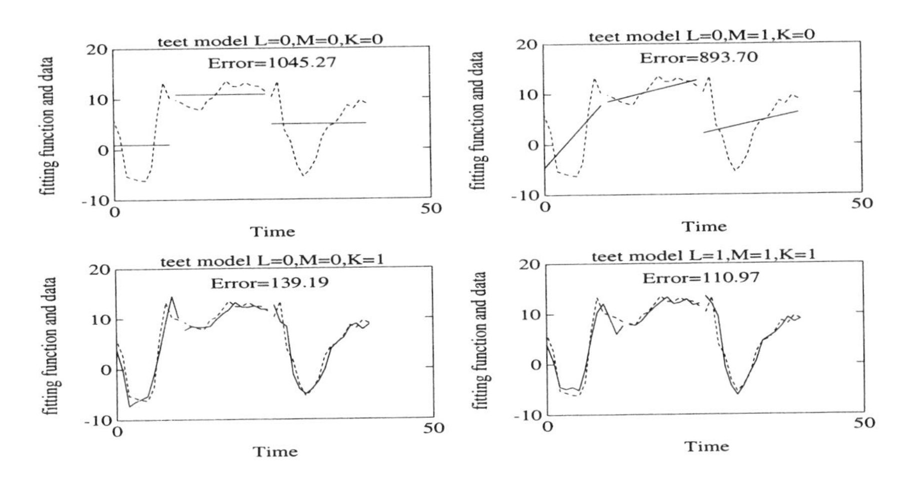

Figure 4.9: Fitting a correct three-state model (B) to a data sequence (B).

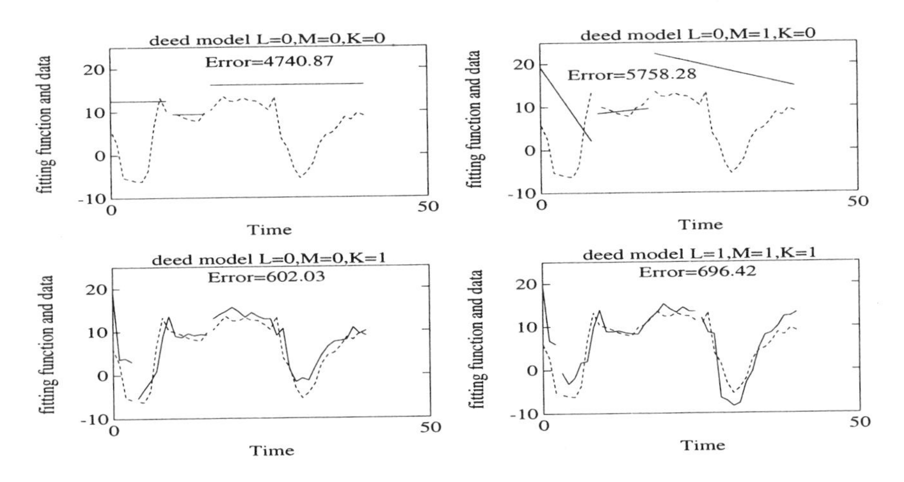

Figure 4.10: Fitting an incorrect three-state model (A) to a data sequence (B).

past behavior for predicting the statistical properties of the model output in the future. Use of the **hidden state** permits decoupling of the internal dynamics from the static relationship between the state variable and the observation variable. The remainder of this chapter is devoted to the dynamic system models with two such essential characteristics. This section will introduce the simplest type of such models — the linear, time-invariant dynamic system model.

4.3.1 State space formulation of the model

The canonical form of the dynamic system model is its state-space formulation. Hence, the names of a dynamic system model and state-space model are often used synonymously. A generative form of the discrete-time, time-invariant, linear state-space model has the following mathematical representation:

$$\mathbf{x}(k+1) = \mathbf{A}\mathbf{x}(k) + \mathbf{u} + \mathbf{w}(k) \tag{4.14}$$

$$\mathbf{o}(k) = \mathbf{C}\mathbf{x}(k) + \mathbf{v}(k), \tag{4.15}$$

initialized by

$$E[\mathbf{x}(0)] = \boldsymbol{\mu}(0), \qquad E[\mathbf{x}(0)\mathbf{x}(0)^{\mathrm{Tr}}] = \Sigma(0).$$

In the above, $\mathbf{x}(k) \in \mathbf{R}^n$ is a (hidden) state vector at time k, $\mathbf{o}(k) \in \mathbf{R}^m$ is an output (or observation or measurement) vector, $\mathbf{w}(k)$ and $\mathbf{v}(k)$ are uncorrelated zero-mean Gaussian noise vectors with covariances $E[\mathbf{w}(k)\mathbf{w}(l)^{\mathrm{Tr}}] = \mathbf{Q}\delta_{kl}$ and $E[\mathbf{v}(k)\mathbf{v}(l)^{\mathrm{Tr}}] = \mathbf{R}\delta_{kl}$, respectively. $\mathbf{u} \in \mathbf{R}^n$ is a deterministic input vector, and is sometimes called the control input. Mathematically, $\mathbf{u} + \mathbf{w}(k)$ could be combined into a single noise vector with the Gaussian mean of $\mathbf{u}$ and with the same covariance $\mathbf{Q}$. The parameter set of the above linear state-space model is

$$\boldsymbol{\theta} = \{\mathbf{A}, \mathbf{u}, \mathbf{Q}, \mathbf{C}, \mathbf{R}\}. \tag{4.16}$$

Note that in the control literature, the control signal is typically different at different time steps, and is hence denoted by $\mathbf{u}(k)$ with time index k. In the model described by Eqs. 4.14 and 4.15 here, however, we assume the control signal $\mathbf{u}$ is constant. We call such a control signal "segmental control", and treat it as a model parameter. As will be seen later, use of this segmental control signal makes the state-space model more appropriate for representing the speech process.

In the above linear state-space model, Eq. 4.14 is often called the state equation, and Eq. 4.15 called the observation (or measurement) equation. The state equation represents linear dynamics of the state variable or vector $\mathbf{x}(k)$ using autoregression (i.e., linear prediction). The noise or error term $\mathbf{w}(k)$ represents the degree of inaccuracy in using the linear state equation to describe the true state dynamics. The greater the determinant of $\mathbf{Q}$ is, the greater degree of inaccuracy there is.

The observation equation Eq. 4.15, on the other hand, is static in nature. It contains no dynamics since the time indices of $\mathbf{o}$ and $\mathbf{x}$ are the same. It represents the noisy relationship between the state vector and the observation vector. Like the noise term in the state equation, the noise term $\mathbf{v}$ in the observation equation represents the degree of inaccuracy in using the linear mapping $\mathbf{o}(k) = \mathbf{C}\mathbf{x}(k)$ to describe the true relationship between the state and observation vectors. Due to the presence of the noise term and due to the possible non-invertibility of matrix $\mathbf{C}$ in the observation equation, the state vector $\mathbf{x}(k)$ cannot be uniquely determined given the observation vector $\mathbf{o}(k)$. In this sense, we say the state vector in the state-space model is hidden, and the state dynamics described by the state equation Eq. 4.14 is hidden dynamics. This is analogous to an HMM, where given an observation it is not possible to uniquely determine which (discrete) HMM state is responsible for generating that observation.

4.3.2 Relationship to high-dimensional linear model

In this subsection, we derive a solution to the state equation Eq. 4.14 in the linear state-space model; that is, we obtain an explicit relationship between input $\mathbf{u}(k)$ and state $\mathbf{x}(k)$ by removing the temporal recursion in k on the state variable $\mathbf{x}(\mathbf{k})$. Based on this solution, we will show representational equivalence of the linear state-space model and a high-dimensional (static) linear model discussed in earlier sections of this chapter, and illustrate the stark differences in their computational properties.

To be concrete, we deal with the scalar observation in the linear, time-invariant state-space model:

$$\mathbf{x}(k+1) = \mathbf{A}\mathbf{x}(k) + \mathbf{u} + \mathbf{w}(k) \tag{4.17}$$

$$o(k) = \mathbf{c}^{\mathrm{Tr}}\mathbf{x}(k) + v(k), \tag{4.18}$$

where the dimensionalities of the observation vector and noise are both $n = 1$. The $n \times m$ observation matrix $\mathbf{C}$ becomes an $m \times 1$ vector $\mathbf{c}$. Its transpose, $\mathbf{c}^{\mathrm{Tr}}$, multiplied by the $m \times 1$ vector $\mathbf{x}(k)$ gives a scalar value.

To solve $\mathbf{x}(k)$ in Eq. 4.17, we write out Eq. 4.17 for each k time steps:

$$\mathbf{x}(1) = \mathbf{A}\mathbf{x}(0) + \mathbf{u} + \mathbf{w}(0) \tag{4.19}$$

$$\begin{aligned}
\mathbf{x}(2) &= \mathbf{A}\mathbf{x}(1) + \mathbf{u} + \mathbf{w}(1) \\
&= \mathbf{A}[\mathbf{A}\mathbf{x}(0) + \mathbf{u} + \mathbf{w}(0)] + \mathbf{u} + \mathbf{w}(1) \\
&= \mathbf{A}^2\mathbf{x}(0) + (\mathbf{A} + \mathbf{I})\mathbf{u} + \mathbf{A}\mathbf{w}(0) + \mathbf{w}(1) \qquad (4.20) \\
\mathbf{x}(3) &= \mathbf{A}\mathbf{x}(2) + \mathbf{u} + \mathbf{w}(2) \\
&= \mathbf{A}[\mathbf{A}^2\mathbf{x}(0) + (\mathbf{A} + \mathbf{I})\mathbf{u} + \mathbf{A}\mathbf{w}(0) + \mathbf{w}(1)] + \mathbf{u} + \mathbf{w}(2) \\
&= \mathbf{A}^3\mathbf{x}(0) + (\mathbf{A}^2 + \mathbf{A} + \mathbf{I})\mathbf{u} + \mathbf{A}^2\mathbf{w}(0) + \mathbf{A}\mathbf{w}(1) + \mathbf{w}(2) \\
&\cdots
\end{aligned}$$

In general, we have the following expression for $\mathbf{x}(k)$ as a function of $\mathbf{x}(0)$, u, and $\mathbf{w}(k)$:

$$\mathbf{x}(k) = \mathbf{A}^k\mathbf{x}(0) + [\sum_{i=1}^{k} \mathbf{A}^{k-i}]\mathbf{u} + \sum_{i=1}^{k} \mathbf{A}^{k-i}\mathbf{w}(i-1).$$

Substituting the above into Eq. 4.18, we have

$$o(k) = \mathbf{c}^{\mathrm{Tr}}\mathbf{A}^k\mathbf{x}(0) + [\sum_{i=1}^{k} \mathbf{c}^{\mathrm{Tr}}\mathbf{A}^{k-i}]\mathbf{u} + \sum_{i=1}^{k} \mathbf{c}^{\mathrm{Tr}}\mathbf{A}^{k-i}\mathbf{w}(i-1) + v(k). \tag{4.21}$$

Creating a high-dimensional vector consisting of the entire sequence of (scalar) observations for $k = 1, 2, \cdots, N-1, N$:

$$\mathbf{o} = \begin{bmatrix} o(N) \\ o(N-1) \\ \vdots \\ o(1) \end{bmatrix},$$

we then fit Eq. 4.21 into the canonical form of the linear model:

$$\mathbf{o} = \mathbf{H}\boldsymbol{\theta} + \mathbf{v}, \tag{4.22}$$

where

$$\mathbf{H} = \begin{bmatrix} \mathbf{c}^{\mathrm{Tr}}\mathbf{A}^{N} \\ \mathbf{c}^{\mathrm{Tr}}\mathbf{A}^{N-1} \\ \vdots \\ \mathbf{c}^{\mathrm{Tr}} \end{bmatrix},$$

$$\boldsymbol{\theta} = \mathbf{x}(0),$$

and $\mathbf{v}$ is a function of $\mathbf{u}$, $\mathbf{w}(0), \mathbf{w}(1), ..., \mathbf{w}(N-1)$, and of $v(1), v(2), ..., v(N)$.

We have thus shown that a (low-dimensional) state-space model has an equivalent mathematical representation to a high-dimensional (static) linear model. Computationally, if one were to estimate the initial state $\mathbf{x}(0)$ or the state at any time frame $\mathbf{x}(k)$, the above equivalence says that one can cast this "state estimation" problem in the dynamic system model into the parameter estimation problem in the linear model setting. However, the latter solution is very inefficient, especially if the state estimates at many time frames are desired. This is so because for each different $k = 0, 1, 2, ...$, $\mathbf{H}$ and $\mathbf{v}$ have to be recomputed. Formal formulation of the various types of estimation problems will be presented in Chapter 5. In particular, we will approach the state estimation problem in the dynamic system model based on recursive algorithms far more efficient than those employed for the high-dimensional linear model.

4.3.3 Likelihood computation

We now discuss the computation of the likelihood for a sequence of observations assuming that these observations are generated as outputs from the linear state-space model of Eqs. 4.14 and 4.15. According to the state-space model definition, however, the observations are no longer statistically independent due to the presence of state dynamics and of the noises; that is,

$$p(\mathbf{o}(1), \mathbf{o}(2), ..., \mathbf{o}(N)|\boldsymbol{\theta}) \neq p(\mathbf{o}(1)|\boldsymbol{\theta})p(\mathbf{o}(2)|\boldsymbol{\theta})...p(\mathbf{o}(N)|\boldsymbol{\theta}). \tag{4.23}$$

So we cannot obtain the likelihood of an output observation sequence by simply computing $p(\mathbf{o}(k)|\boldsymbol{\theta})$ individually. Fortunately, the observations $\mathbf{o}(k)$ and the innovations (a concept to be discussed in Chapter 5) $\tilde{\mathbf{e}}(k|k-1) = (\mathbf{o}(k) - \hat{\mathbf{o}}(k|k-1))$ are causally invertible, and the innovations are all uncorrelated. Hence, it is still relatively easy to determine the likelihood function for the state-space model.

The innovation process $\tilde{e}(k|k-1)$ can be generated by a Kalman filter, which will be discussed extensively in Chapter 5. With the use of the Kalman filter, the output log-likelihood function of the linear state-space model can be computed by [Mend 95]:

$$\log p(\mathbf{o}(1), \mathbf{o}(2), ..., \mathbf{o}(N)|\boldsymbol{\theta}) = -\frac{1}{2}\sum_{k=1}^{N}\{[\tilde{\mathbf{e}}(k|k-1)]^{\mathrm{Tr}}\Sigma_{\tilde{e}}^{-1}(k|k-1)[\tilde{\mathbf{e}}(k|k-1)] + log|\Sigma_{\tilde{e}}(k|k-1)|\} \tag{4.24}$$

where $\Sigma_{\tilde{e}}(k|k-1)$ is the covariance matrix of the innovations process $\tilde{\mathbf{e}}(k|k-1)$; i.e.,

$$\boldsymbol{\Sigma}_{\tilde{e}}(k|k-1) = E\{\tilde{\mathbf{e}}(k|k-1)\tilde{\mathbf{e}}(k|k-1)^{\mathrm{Tr}}\} \tag{4.25}$$

$$= \mathbf{C}\boldsymbol{\Sigma}(k|k-1)\mathbf{C}^{\mathrm{Tr}} + \mathbf{R}. \tag{4.26}$$

4.4 Time-Varying Linear Dynamic System Model

4.4.1 From time-invariant model to time-varying model

The linear state-space model defined earlier by Eqs. 4.14 and 4.15 is time invariant. This is so because the parameters $\boldsymbol{\theta} = \{\mathbf{A}, \mathbf{u}, \mathbf{Q}, \mathbf{C}, \mathbf{R}\}$ that characterize this model do not change as a function of time k. When these model parameters are constant, it can be easily shown that the first-, second-, and higher-order statistics are all invariant with respect to an arbitrary time shift. Therefore, the time-invariant dynamic system model is also stationary according to the definition given in Chapter 3.

In this section, we extend the time-invariant or stationary linear dynamic system model to its time-varying or nonstationary version. That is, we allow the parameters of the dynamic system model to vary as a function of time. This generalizes Eqs. 4.14 and 4.15 into the following state and observation equations where the parameters are indexed by time frames k and k':

$$\mathbf{x}(k+1) = \mathbf{A}_k \mathbf{x}(k) + \mathbf{u}_k + \mathbf{w}(k) \tag{4.27}$$

$$\mathbf{o}(k) = \mathbf{C}_{k'} \mathbf{x}(k) + \mathbf{v}(k). \tag{4.28}$$

In the above, the change of parameters over time is typically slower than the change of state $\mathbf{x}(k)$ and of observation $\mathbf{o}(k)$ over time. In other words, the parameter evolution and the state evolution are at distinct time scales. In this sense, we say that such a model exhibits hierarchical nonstationarity.

We also in Eqs. 4.27 and 4.28 used different time indices to denote the parameter change over time: k in the state equation Eq. 4.27 and k' in the observation equation Eq. 4.28. This difference means that the evolutions of the parameters in the state equation and those in the observation equation may not have to be synchronous in time. (Further, the parameters controlling different components of the state vector $\mathbf{x}(k)$ could also be asynchronously changing over time. For example, components of $\mathbf{u}$ may be indexed by different time subscripts.)

The nature of the parameter evolution over time in the time-varying or nonstationary state-space model can be diverse. One common technique represents the parameter evolution over time as a discrete-state Markov chain. In this case, the Markov chain state transition matrix and the levels associated with each state in the Markov chain become the full set of time-varying state-space model parameters. More sophisticated techniques can use an HMM to represent the parameter evolution over time in the time-varying state-space model. In this case, the full set of model parameters includes additional variance parameters associated with each HMM state.

There is a rather rich literature in statistics, machine learning, control engineering, signal processing (target tracking), and econometrics on various types of time-varying state-space models [Shum 91, Ghah 00, Bar 93, Kim 94, Pavl 99]. Quite a number of different names have been used for the time-varying dynamic system model outlined above, including switching state-space model, jumped dynamic system model, multi-mode model, mixed-state dynamic Bayesian network, etc.

This section gives only a very brief introduction to the time-varying dynamic system model. Many problems associated with computations for this model will be discussed in later chapters. We will also show that depending on the level and the nature of the switching in the model design, various key properties of the speech dynamics can be naturally represented in the time-varying dynamic system model.

4.4.2 Likelihood computation

Likelihood computation for the time-varying linear dynamic system model is in general more complicated than the time-invariant counterpart described earlier. When the parameter evolution follows a Markov chain, for example, then the first step in determining the likelihood for an observation sequence is to segment the sequence into correct regimes, each corresponding to the correct discrete-Markov state associated with the dynamic model parameters. Algorithms used for such segmentation will be presented in a later chapter. Next, the appropriate model parameters according to the segmentation results can be used to perform Kalman filtering. Then Eq. 4.24 is computed for each of the segments and the results are summed to give the log-likelihood for the entire multiple-segment sequence.

4.5 Non-Linear Dynamic System Model

4.5.1 From linear model to nonlinear model

Many physical systems are characterized by nonlinear relationships between various physical variables. Taking an example of the speech process, we can use a dynamic system model to describe the speech production process. The state equation can be used to describe the dynamic articulation process, while the observation equation be used to characterize the relationship between the articulatory variables and the observed speech acoustics. It is well established that the relationship between articulation and acoustics in speech is highly nonlinear, and that the articulatory dynamics can also be (mildly) nonlinear. Therefore, it is inadequate to use a linear dynamic system model to represent the articulatory and acoustic processes in speech production.

When the observation equation is nonlinear, the inversion problem of estimating (i.e., making inference about) the hidden state vector from the observation vector sequence becomes much more difficult than the case where the observation equation is linear. One-to-many (static) mappings from an observation vector to a state vector often arise in the observation equation, and need to be disambiguated using the information contained in the dynamic state equation. This one-to-many mapping problem is often more serious for the nonlinear observation equation than the linear one with the observation matrix being singular or of less than a full rank. Therefore, while the use of a nonlinear dynamic system model often provides an adequate mathematical abstraction for representing real-world physical processes with hidden dynamics, it is a serious challenge to solve the inversion problem associated with the nonlinear dynamic system model.

The most general form of the time-varying nonlinear dynamic system model, with nonlinearity in both state and observation equations, is as follows in its state-space formulation [Tani 96]:

$$\mathbf{x}(k+1) = \mathbf{g}_k[\mathbf{x}(k), \mathbf{u}_k, \mathbf{w}(k)] \quad (4.29)$$

$$\mathbf{o}(k') = \mathbf{h}_{k'}[\mathbf{x}(k), \mathbf{v}(k)]. \quad (4.30)$$

where subscripts k and k' indicate that the nonlinear functions $g[.]$ and $h[.]$ can be time varying and be asynchronous with each other.

A simpler version of the model more frequently used separates out the noise terms from the nonlinear functions, e.g., the model discussed in [Ghah 99]. The time-invariant (stationary) version of such a simplified model has the state-space form of

$$\mathbf{x}(k+1) = \mathbf{g}[\mathbf{x}(k), \mathbf{u}] + \mathbf{w}(k) \quad (4.31)$$

$$\mathbf{o}(k) = \mathbf{h}[\mathbf{x}(k)] + \mathbf{v}(k). \quad (4.32)$$

4.5.2 Nonlinearity and its approximations

Multi-layer perception

One common nonlinear function, which can be used either as the static function $\mathbf{g}[.]$ in the state equation or as the function $\mathbf{h}[.]$ in the observation equation, is called the **multi-layer perceptron** (MLP) neural network [Bishop 97]. The MLP has been shown to possess some strong theoretical properties which sometimes give the MLP an enviable name of "universal nonlinear approximator."

Let us take an example of MLP with three layers (input, hidden and output). Let w_{jl} be the MLP weights from input to hidden units and W_{ij} be the MLP weights from hidden to output units, where l is the input node index, j the hidden node index and i the output node index. Then the output signal at node i can be expressed as a (nonlinear) function $\mathbf{h}(.)$ of all the input nodes (making up the input vector) according to

$$h_i(\mathbf{x}) = \sum_{j=1}^{J} W_{ij} \cdot s(\sum_{l=1}^{L} w_{jl} \cdot x_l), \quad 1 \le i \le I, \quad (4.33)$$

where I, J and L are the numbers of nodes at the output, hidden and input layers, respectively. $s(.)$ is the hidden unit's nonlinear activation function, taken as the standard sigmoid function of

$$s(x) = \frac{1}{1+\exp(-x)}. \quad (4.34)$$

The derivative of this sigmoid function has the following concise form:

$$s'(x) = s(x)(1 - s(x)), \quad (4.35)$$

making it convenient for use in many computations to be discussed later. The parameters that characterize the MLP nonlinearity are the two sets of weights.

Radial basis function

Another popular form of nonlinearity is the **radial basis function** (RBF) neural network [Ghah 99, Bishop 97]. In a vector form, the nonlinearity here can be expressed as:

$$\mathbf{h}(\mathbf{x}) = \mathbf{W}\mathbf{y} \quad (4.36)$$

where W is the weight matrix that consists of the weights connecting the middle layer and the output layer of the RBF network. The (i, j)-th element of $\mathbf{W}$ is the connection weight between node i in the output layer and node j in the middle layer. The middle layer consists of the vector y of the components of

$$\mathbf{y} = [y_1(\mathbf{x}), y_2(\mathbf{x}), \cdots, y_j(\mathbf{x}), \cdots, y_J(\mathbf{x})]^{\mathrm{Tr}}, \quad (4.37)$$

where $y_j(\mathbf{x})$ is the output of the j-th radial-basis-function (or kernel function) in the middle layer. If the Gaussian function is chosen to be the kernel function,

$$y_j(\mathbf{x}) = \exp\{-\frac{1}{2}(\mathbf{x} - \mathbf{c}_j)^{\mathrm{Tr}}\mathbf{\Sigma}_j^{-1}(\mathbf{x} - \mathbf{c}_j)\}, \tag{4.38}$$

then the parameters that characterize the RBF network nonlinearity are the weight matrix $\mathbf{W}$, the kernel centers $\mathbf{c}_j$ and the kernel widths $\mathbf{\Sigma}_j$ $(j = 1, 2, ..., J)$.

The kernel function in the RBF should have the desirable "localized" property. The Gaussian function above satisfies such a property. Another common "localized" kernel function is

$$y_j(\mathbf{x}) = (||\mathbf{x}||^2 + \alpha^2)^{-\alpha}. \tag{4.39}$$

In a scalar form, each element of the RBF nonlinear function in Eq. 4.36 can be written as

$$h_i(\mathbf{x}) = \sum_{j=1}^{J} W_{ij} \cdot y_j(\mathbf{x}), \qquad 1 \le i \le I. \tag{4.40}$$

Truncated Taylor series approximation

Typically, the analytical forms of nonlinear functions, such as the MLP and RBF described above, make the associated nonlinear dynamic systems difficult to analyze and make the estimation problems difficult to solve. Approximations are frequently used to gain computational simplifications while sacrificing accuracy for approximating the nonlinear functions.

One very commonly used technique for the approximation is a truncated **vector Taylor series** expansion. If all the Taylor series terms of order two and higher are truncated, then we have the linear Taylor series approximation that is characterized by the Jacobian matrix $\mathbf{J}$ and the point of Taylor series expansion $\mathbf{x}_0$. For a vector input of dimension n and a vector output of dimension m, this linear approximation can be written as

$$\mathbf{h}(\mathbf{x}) \approx \mathbf{h}(\mathbf{x}_0) + \mathbf{J}(\mathbf{x}_0)(\mathbf{x} - \mathbf{x}_0). \tag{4.41}$$

Each element of the Jacobian matrix $\mathbf{J}$ is a partial derivative of each vector component of the nonlinear output with respect to each of the input vector components. That is,

$$\mathbf{J}(\mathbf{x}_0) = \frac{\partial \mathbf{h}}{\partial \mathbf{x}_0} = \begin{bmatrix} \frac{\partial h_1(\mathbf{x}_0)}{\partial x_1} & \frac{\partial h_1(\mathbf{x}_0)}{\partial x_2} \cdots & \frac{\partial h_1(\mathbf{x}_0)}{\partial x_n} \\ \frac{\partial h_2(\mathbf{x}_0)}{\partial x_1} & \frac{\partial h_2(\mathbf{x}_0)}{\partial x_2} \cdots & \frac{\partial h_2(\mathbf{x}_0)}{\partial x_n} \\ \vdots & & \vdots \\ \frac{\partial h_m(\mathbf{x}_0)}{\partial x_1} & \frac{\partial h_m(\mathbf{x}_0)}{\partial x_2} \cdots & \frac{\partial h_m(\mathbf{x}_0)}{\partial x_n} \end{bmatrix}, \tag{4.42}$$

As an example, for the MLP nonlinearity of Eq. 4.33, the (i, l)-th element of the Jacobian matrix is

$$J_{il} = \sum_{j=1}^{J} W_{ij} \cdot s_j(y) \cdot (1 - s_j(y)) \cdot w_{jl}, \qquad 1 \le i \le I, \;\; 1 \le l \le L, \tag{4.43}$$

where $y = \sum_{l'=1}^{L} W_{jl'} x_{l'}$.

As another example, for the RBF nonlinearity Eq. 4.36 or Eq. 4.40 with Gaussian kernels, the (i, l)-th element of the Jacobian matrix is

$$J_{il} = \sum_{j=1}^{J} W_{ij} \cdot \frac{\partial y_j}{\partial x_l} = \sum_{j=1}^{J} W_{ij} \cdot \exp\{-\frac{1}{2}(\mathbf{x} - \mathbf{c}_j)^{\mathrm{Tr}} \mathbf{\Sigma}_j^{-1} (\mathbf{x} - \mathbf{c}_j)\} [\mathbf{\Sigma}_j^{-1}(\mathbf{x} - \mathbf{c}_j)]_l \quad (4.44)$$

for $1 \le i \le I, \quad 1 \le l \le L$.

If the kernel function of Eq. 4.39 is used in the RBF, then the (i, l)-th element of the Jacobian matrix becomes

$$J_{il} = -2 \sum_{j=1}^{J} W_{ij} \cdot (\alpha_l + 1)((||\mathbf{x}||^2 + \alpha^2)^{-\alpha-1} x_l \qquad 1 \le i \le I, \quad 1 \le l \le L, \qquad (4.45)$$

Quasi-linear approximation

The linear Taylor series approximation discussed above requires evaluation of the Jacobian matrix in an analytical form. If such a form is not available, or some or all elements of the Jacobian matrix do not exist as in the case where discontinuity exists, then a linear approximation to a nonlinear function can be accomplished by the quasi-linear approximation method, which we discuss now.

Consider a nonlinear vector function $\mathbf{h}(\mathbf{x})$ of a vector random vector $\mathbf{x}$ with its PDF denoted by $p(\mathbf{x})$. The quasi-linear approximation gives

$$\mathbf{h}(\mathbf{x}) \approx E[\mathbf{h}(\mathbf{x})] + \mathbf{A}(\mathbf{x} - E[\mathbf{x}]), \qquad (4.46)$$

where the matrix $\mathbf{A}$ is

$$\mathbf{A} = (E[\mathbf{h}(\mathbf{x})\mathbf{x}^{\mathrm{Tr}}] - E[\mathbf{h}(\mathbf{x})]E[\mathbf{x}^{\mathrm{Tr}}])\mathbf{\Sigma}^{-1}.$$

In the above, $\mathbf{\Sigma}$ is the covariance matrix for $\mathbf{x}$:

$$\mathbf{\Sigma} = E[(\mathbf{x} - E[\mathbf{x}])(\mathbf{x} - E[\mathbf{x}])^{\mathrm{Tr}}].$$

Note that in Eq. 4.46, no explicit derivative of the nonlinear function $\mathbf{h}(\mathbf{x})$ is needed in the computation. Rather, some statistics up to the second order need to be computed, using the PDF of $p(\mathbf{x})$ if it is known in advance, or using the estimates of the PDF from empirical data pairs of the input and output of the nonlinear function.

Piecewise linear approximation

A very popular technique for simplifying nonlinear functions is **piecewise linear approximation**. This is essentially an extension of the linear Taylor series approximation from one point to multiple points, where these multiple points are determined using some optimality criteria. The essence of this technique is to linearize the (vector-valued) nonlinear function $\mathbf{h}(\mathbf{x})$ about a finite number of points, and to parameterize these linearized mappings by the set of locations of the linearization points $\mathcal{C}_x = \{\mathbf{x}_c(i)\}, i = 1, 2, \ldots, I\}$, together with the vector values of the function at those points $\mathcal{C}_h = \{\mathbf{h}(\mathbf{x}_c(i)), i = 1, 2, \ldots, I\}$ plus the local Jacobian matrices $\mathcal{C}_J = \{\mathbf{J}(\mathbf{x}_c(i))\}, i = 1, 2, \ldots, I\}$. All these parameters are contained in the form of a **codebook** $\mathcal{C} = \{\mathcal{C}_x, \mathcal{C}_h, \mathcal{C}_J\}$, which may be learned from the available data or may be determined from prior knowledge or by computer simulation.

Given the piecewise linearization parameters (i.e., given the codebook), any value of $\mathbf{h}(\mathbf{x})$ can then be approximated by

$$\begin{aligned}\mathbf{h}(\mathbf{x}) &\approx \mathbf{h}(\mathbf{x}_c) + \mathbf{J}(\mathbf{x}_c)(\mathbf{x} - \mathbf{x}_c) \\ \mathbf{x}_c &= arg \min_{\mathbf{x}_i \epsilon C_x} ||\mathbf{x} - \mathbf{x}_i|| \\ \mathbf{J}(\mathbf{x}_c) &= \frac{\partial \mathbf{h}}{\partial \mathbf{x}}|_{\mathbf{x}=\mathbf{x}_c},\end{aligned} \tag{4.47}$$

where $||.||$ denotes the Euclidean norm.

4.6 Summary

In this chapter, we covered two types of mathematical models relevant to speech processing. As we stated before, a model is a mathematical abstraction of or an approximation to physical reality. Building a model often serves the purpose of facilitating analytical treatment of problems involving physical processes that are usually too complex to analyze without employing a simplified mathematical model.

The first type of model discussed is called the linear model, where in its canonical form the observation vector is linearly related to the observation matrix via the parameter vector. The residual Gaussian noise in the model is used to account for the error left by the above linear relationship. Examples were given for the canonical form of the linear model. Several forms of the time varying linear model, where the parameter vector in the general linear model changes with time, were presented. They included the time-varying linear predictive model (together with an important special case of the Markov modulated linear predictive model), and the Markov modulated linear regression model. The latter is equivalent to the trended HMM, briefly introduced in Chapter 3.

The second type of model discussed is called the dynamic system model, whose canonical form is called the state space model. We first presented the time-invariant, linear dynamic system model. This basic model was then extended along two directions. First, the time-invariant model was extended to the time varying model, where the parameters of the model are not constant but allowed to switch from one set of values to another at different times. Second, linearity was extended to nonlinearity, giving rise to the nonlinear dynamic system model. Some approximation methods to handle the nonlinearity were discussed.

Chapter 5

Optimization Methods and Estimation Theory

In Chapter 4, we introduced a number of standard statistical models used in many engineering and signal processing applications. These include a class of linear models, and a different class of dynamic system models that are characterized by the new concept of (hidden) continuous state. In Chapter 3, while covering random processes, as principal examples we also introduced several statistical models that are characterized by the concept of (hidden) discrete state. All these models have been specified by a set of model parameters as we introduced each of the models. These model parameters can be deterministic, or can be random variables.

At the heart of statistical modeling and its applications is the problem of estimation, including both **parameter estimation** and **state estimation**. Parameter estimation is sometimes also called learning (or training) and state estimation called probabilistic inference (or decoding with confidence measure). For speech processing applications, probabilistic inference on continuous states is often called speech inversion and that on discrete states called recognition (of linguistic units). Estimation theory to be covered in this chapter addresses key issues of how to optimally determine the unknown model parameters and states given a set of observation data, and what are the criteria that can be used to judge the quality of the estimators. Without appropriate estimation procedures, any statistical model, no matter how close its structure may be to the physical reality, cannot be usefully input to a digital processor and applied to solve the related engineering problem. Estimation theory can also be viewed as an extension of classic digital signal processing in dealing with uncertain data in an optimal manner. At the heart of estimation theory is the problem of **optimization**. Estimation theory sets the objective functions and subsequently requires solutions of optimization. Since estimation theory is so closely linked to optimization techniques, the first half of this chapter is devoted to describing three main classes of optimization techniques. As one main theme of this book is the optimization-oriented and computational approach, we focus special attention on computational aspects of the techniques, serving as the principal tools for implementing the estimation algorithms developed from the estimation theory.

Some materials in this chapter are extracted and modified from references [Kay 93], [Mend 95], [Rust 94], and [Rao 84].

5.1 Classical Optimization Techniques

A great deal of speech processing applications use classical optimization techniques discussed in this section. These techniques are applicable to finding optimum (maxima and minima) of continuous and differential functions, with or without constraints. We first introduce various concepts and preliminary results widely used in optimization.

5.1.1 Basic definitions and results

Let a vector $\mathbf{x}$ be in n-dimensional space, $\mathbf{x} \in \mathcal{R}^n$:

$$\mathbf{x} = \begin{pmatrix} x_1 \\ x_2 \\ \vdots \\ x_n \end{pmatrix},$$

and let $f(\mathbf{x})$ be a real-valued function of $\mathbf{x}$. When one wants to optimize the function $f(\mathbf{x})$, we call it the **objective function**. The function $f(\mathbf{x})$ with its domain $\mathbf{x} \subseteq \mathcal{R}^n = D$ is said to have a **global minimum** at $\mathbf{x}^*$ if

$$f(\mathbf{x}^*) \leq f(\mathbf{x})$$

for all $\mathbf{x} \in D$. The function $f(\mathbf{x})$ has a **global maximum** at $\mathbf{x}^{**}$ if

$$f(\mathbf{x}^{**}) \geq f(\mathbf{x})$$

for all $\mathbf{x} \in D$. The function $f(\mathbf{x})$ is said to have a **local minimum** at $\mathbf{x}^*$ if

$$f(\mathbf{x}^*) \leq f(\mathbf{x})$$

for all $\mathbf{x}$ in the neighborhood of $\mathbf{x}^*$. Note since

$$\min f(\mathbf{x}) = -\left[\max(-f(\mathbf{x}))\right],$$

a minimization problem is equivalent to a maximization one. We thus will treat both of these problems as the same **optimization** problem.

The vector of partial derivatives of $f(\mathbf{x})$ is called the **gradient vector**:

$$\nabla f(\mathbf{x}) \doteq \begin{pmatrix} \frac{\partial f}{\partial x_1} \\ \frac{\partial f}{\partial x_2} \\ \vdots \\ \frac{\partial f}{\partial x_n} \end{pmatrix}.$$

The matrix of second-order partial derivatives of $f(\mathbf{x})$ is called the **Hessian matrix**:

$$\mathbf{H}(\mathbf{x}) \doteq \begin{pmatrix} \frac{\partial^2 f}{\partial x_1^2} & \frac{\partial^2 f}{\partial x_1 x_2} & \cdots & \frac{\partial^2 f}{\partial x_1 x_n} \\ \frac{\partial^2 f}{\partial x_2 x_1} & \frac{\partial^2 f}{\partial x_2^2} & \cdots & \frac{\partial^2 f}{\partial x_2 x_n} \\ \vdots & \vdots & \vdots & \vdots \\ \frac{\partial^2 f}{\partial x_n x_1} & \frac{\partial^2 f}{\partial x_n x_2} & \cdots & \frac{\partial^2 f}{\partial x_n^2} \end{pmatrix}.$$

Taylor series expansion of functions is a foundation for many of the classical optimization problems. We thus explicitly state the result here:

Taylor series expansion at $\mathbf{x}^*$:

$$f(\mathbf{x}) = f(\mathbf{x}^*) + (\mathbf{x} - \mathbf{x}^*)^{\text{Tr}} \nabla f(\mathbf{x}^*) + \frac{1}{2}(\mathbf{x} - \mathbf{x}^*)^{\text{Tr}} \mathbf{H}(\xi)(\mathbf{x} - \mathbf{x}^*), \tag{5.1}$$

where $\xi = \mathbf{x}^* + \theta(\mathbf{x} - \mathbf{x}^*)$, and $0 \leq \theta \leq 1$.

When the points $\mathbf{x}$ of interest are sufficiently close to $\mathbf{x}^*$, that is, the vector $\mathbf{x} - \mathbf{x}^*$ is very small in its norm (vector length), then the last term in Eq. 5.1 can be ignored. This gives the *first-order* Taylor series approximation to the function $f(\mathbf{x})$ at $\mathbf{x}^*$:

$$f(\mathbf{x}) \approx f(\mathbf{x}^*) + (\mathbf{x} - \mathbf{x}^*)^{\text{Tr}} \nabla f(\mathbf{x}^*). \tag{5.2}$$

This idea can be generalized to higher orders. The *second-order* Taylor series approximation of the function $f(\mathbf{x})$ at $\mathbf{x}^*$ is

$$\begin{aligned} f(\mathbf{x}) &\approx f(\mathbf{x}^*) + (\mathbf{x} - \mathbf{x}^*)^{\text{Tr}} \nabla f(\mathbf{x}^*) + \frac{1}{2}(\mathbf{x} - \mathbf{x}^*)^{\text{Tr}} \mathbf{H}(\mathbf{x}^*)(\mathbf{x} - \mathbf{x}^*) & (5.3) \\ &= f(\mathbf{x}^*) + \mathbf{h}^{\text{Tr}} \nabla f(\mathbf{x}^*) + \frac{1}{2}\mathbf{h}^{\text{Tr}} \mathbf{H}(\mathbf{x}^*)\mathbf{h}, & (5.4) \end{aligned}$$

where $\mathbf{h} = (h_1 \ h_2 \ ... \ h_n)^{\text{Tr}} = \mathbf{x} - \mathbf{x}^*$. The general *K-th order* Taylor series approximation of the function $f(\mathbf{x})$ is

$$f(\mathbf{x}) \approx f(\mathbf{x}^*) + Df(\mathbf{x}^*) + \frac{1}{2!}D^2 f(\mathbf{x}^*) + \frac{1}{3!}D^3 f(\mathbf{x}^*) + ... + \frac{1}{k!}D^K f(\mathbf{x}^*),$$

where for each $r, 1 \leq r \leq K$,

$$D^r f(\mathbf{x}^*) = \sum_{i=1}^{n} \sum_{j=1}^{n} ... \sum_{l=1}^{n} h_i h_j ... h_l \frac{\partial^r f(\mathbf{x}^*)}{\partial x_i \partial x_j ... \partial x_l},$$

where the number of summations equals r.

Example: Find the third-order Taylor series approximation to the function

$$f(x_1, x_2, x_3) = x_2^2 x_3 + x_1 \exp(x_3), \qquad (n = 3)$$

at point $x^* = (1 \ \ 0 \ \ -2)^{\text{Tr}}$.

Solution: We need to compute each term in

$$f(\mathbf{x}) \approx f(\mathbf{x}^*) + Df(\mathbf{x}^*) + \frac{1}{2!}D^2 f(\mathbf{x}^*) + \frac{1}{3!}D^3 f(\mathbf{x}^*).$$

The computation is as follows:

$$f(\mathbf{x}^*) = x_2^{*2} x_3^* + x_1^* \exp(x_3^*) = 0 + 1 \times \exp(-2) = \exp(-2).$$

$$\begin{aligned}
Df(\mathbf{x}^*) &= h_1 \frac{\partial f(\mathbf{x}^*)}{\partial x_1} + h_2 \frac{\partial f(\mathbf{x}^*)}{\partial x_2} + h_3 \frac{\partial f(\mathbf{x}^*)}{\partial x_3} \\
&= h_1 \exp(x_3^*) + h_2 \times 2x_2^* x_3^* + h_3 \left[x_2^{*2} + x_1^* \exp(x_3^*) \right] \\
&= h_1 \exp(-2) + h_3 \exp(-2).
\end{aligned}$$

$$\begin{aligned}
D^2 f(\mathbf{x}^*) &= \sum_{i=1}^{3} \sum_{j=1}^{3} h_i h_j \frac{\partial^2 f(\mathbf{x}^*)}{\partial x_i \partial x_j} \\
&= h_1^2 \frac{\partial^2 f(\mathbf{x}^*)}{\partial x_1^2} + h_2^2 \frac{\partial^2 f(\mathbf{x}^*)}{\partial x_2^2} + h_3^2 \frac{\partial^2 f(\mathbf{x}^*)}{\partial x_3^2} \\
&\qquad + 2h_1 h_2 \frac{\partial^2 f(\mathbf{x}^*)}{\partial x_1 x_2} + 2h_1 h_3 \frac{\partial^2 f(\mathbf{x}^*)}{\partial x_1 x_3} + 2h_2 h_3 \frac{\partial^2 f(\mathbf{x}^*)}{\partial x_2 x_3} \\
&= h_1^2(0) + h_2^2(2x_3^*) + h_3^2(x_1^* \exp(x_3^*)) + 2h_1 h_2(0) + 2h_1 h_3(2x_2^*) + 2h_2 h_3 \exp(x_3^*) \\
&= -4h_2^2 + \exp(-2) h_3^2 + 2\exp(-2) h_1 h_3.
\end{aligned}$$

$$\begin{aligned}
D^3 f(\mathbf{x}^*) &= \sum_{i=1}^{3} \sum_{j=1}^{3} \sum_{l=1}^{3} h_i h_j h_l \frac{\partial^3 f(\mathbf{x}^*)}{\partial x_i \partial x_j \partial x_l} \\
&= h_1^3 \frac{\partial^3 f(\mathbf{x}^*)}{\partial x_1^3} + h_2^3 \frac{\partial^3 f(\mathbf{x}^*)}{\partial x_2^3} + h_3^3 \frac{\partial^3 f(\mathbf{x}^*)}{\partial x_3^3} \\
&\qquad + 3h_1^2 h_2 \frac{\partial^3 f(\mathbf{x}^*)}{\partial x_1^2 x_2} + 3h_1^2 h_3 \frac{\partial^3 f(\mathbf{x}^*)}{\partial x_1^2 x_3} + 3h_1 h_2^2 \frac{\partial^3 f(\mathbf{x}^*)}{\partial x_1 x_2^2} \\
&\qquad + 3h_1 h_3^2 \frac{\partial^3 f(\mathbf{x}^*)}{\partial x_1 x_3^2} + 3h_2^2 h_3 \frac{\partial^3 f(\mathbf{x}^*)}{\partial x_2^2 x_3} + 3h_2 h_3^2 \frac{\partial^3 f(\mathbf{x}^*)}{\partial x_2 x_3^2} + 6h_1 h_2 h_3 \frac{\partial^3 f(\mathbf{x}^*)}{\partial x_1 x_2 x_3} \\
&= h_1^3(0) + h_2^3(0) + h_3^3(x_1^* \exp(x_3^*)) + 3h_1^2 h_2(0) + 3h_1^2 h_3(0) + 3h_1 h_2^2(0) \\
&\qquad + 3h_1 h_3^2(\exp(x_3^*)) + 3h_2^2 h_3(2) + 3h_2 h_3^2(0) + 6h_1 h_2 h_3(0) \\
&= \exp(-2) h_3^3 + 3\exp(-2) h_1 h_3^2 + 6h_2^2 h_3.
\end{aligned}$$

The final result is

$$\begin{aligned}
f(x_1, x_2, x_3) &= x_2^2 x_3 + x_1 \exp(x_3) \\
&\approx \exp(-2) + \exp(-2)(h_1 + h_3) + \frac{1}{2} \left[-4h_2^2 + \exp(-2) h_3^2 + 2\exp(-2) h_1 h_3 \right] \\
&\qquad \frac{1}{6} \left[(\exp(-2) h_3^3 + 3(\exp(-2) h_1 h_3^2 + 6h_2^2 h_3 \right],
\end{aligned}$$

where $h_1 = x_1 - x_1^* = x_1 - 1$, $h_2 = x_2 - x_2^* = x_2$, and $h_3 = x_3 - x_3^* = x_3 + 2$.

5.1.2 Necessary and sufficient conditions for an optimum

A necessary condition for a function $f(\mathbf{x})$ to have a local optimum at $\mathbf{x}^*$ is that *the gradient vector has all zero components:*

$$\nabla f(\mathbf{x}^*) = 0, \tag{5.5}$$

as long as $\nabla f(\mathbf{x}^*)$ exists and is continuous at $\mathbf{x}^*$. This necessary condition can be directly proved using the Taylor series expansion of Eq. 5.1.

Note that $\nabla f(\mathbf{x}^*) = 0$ is only a necessary condition; that is, a point $\mathbf{x}^*$ satisfying $\nabla f(\mathbf{x}^*) = 0$ may be just a stationary or saddle point, not an optimum point. However, in many optimization problems including those in speech processing, prior knowledge about the nature of the objective function in the problem domain can eliminate the possibility of having a stationary point.

To theoretically guarantee an optimum point (i.e., elimination of the possibility of a stationary point), we need the following sufficient condition:

Sufficient condition for local optimum:
Let there exist continuous partial derivatives up to the second order for objective function $f(\mathbf{x})$. If the gradient vector $\nabla f(\mathbf{x}^*) = 0$ and the Hessian matrix $\mathbf{H}(\mathbf{x}^*)$ is positive definite, then $\mathbf{x}^*$ is a local minimum.
Similarly, if $\nabla f(\mathbf{x}^*) = 0$ and $\mathbf{H}(\mathbf{x}^*)$ is negative definite, then $\mathbf{x}^*$ is a local maximum.

Again, the proof of this condition comes directly from applying the Taylor series expansion of Eq. 5.1.

Many examples of optimization, based on the differential calculus techniques as required by the above necessary and sufficient condition, will be given later in this chapter when we discuss statistical estimation theory. The necessary and sufficient conditions discussed above are applied to optimization problems with no constraints. For the situation where constraints must be imposed, the related optimization problems are discussed next.

5.1.3 Lagrange multiplier method for constrained optimization

The **Lagrange multiplier** method is a popular method in speech processing, as well as in many other optimization problems, which converts constrained optimization problems into unconstrained ones. It uses a linear combination of the objective function and the constraints to form a new objective function with no constraints.

The constrained optimization problem, where the constraints are in the form of equalities, can be formally described as follows: Find $\mathbf{x} = \mathbf{x}^*$ which optimizes the objective function $f(\mathbf{x})$ subject to the M constraints:

$$\begin{aligned} g_1(\mathbf{x}) &= b_1, \\ g_2(\mathbf{x}) &= b_2, \\ &\cdots \\ g_M(\mathbf{x}) &= b_M. \end{aligned}$$

The Lagrange multiplier method solves the above problem by forming

New objective function for equivalent unconstrained optimization :

$$F(\mathbf{x}, \mathbf{\Lambda}) = f(\mathbf{x}) + \sum_{m=1}^{M} \lambda_m (g_m(\mathbf{x}) - b_m),$$

where $\mathbf{\Lambda} = (\lambda_1, \lambda_2, ..., \lambda_M)$ are called the Lagrange multipliers.

Optimization of the new objective function $F(\mathbf{x}, \mathbf{\Lambda})$ proceeds by setting its partial derivatives to zero with respect to $x_1, x_2, ..., x_n$ and with respect to $\lambda_1, \lambda_2, ..., \lambda_M$. (Note that setting the partial derivatives to zero with respect to $\lambda_1, \lambda_2, ..., \lambda_M$ would give the equations which are identical to the original constraints, and hence can be omitted). This produces a set of $n + M$ equations which determine the $n + M$ unknowns including the desired solution for optimization $\mathbf{x} = \mathbf{x}^* = x_1^*, x_2^*, ..., x_n^*$.

When the constraints are in the form of inequalities, rather than of equalities as discussed above, a common method for optimization is to transform the related variables so as to eliminate a common method for optimization is to transform the related variables to eliminate the constraints. For example, if the constraint is $x > 0$ (as required for estimating the variance, which is always positive, in a PDF), then we can transform x into $X = \exp(x)$. Since x and X are monotonically related, optimization of one automatically gives the solution to the other.

5.2 Numerical Methods for Optimization

Most of optimization problems in speech processing and in other disciplines do not have solutions, based typically on setting the gradient vector to zero and solving, in a closed form. This is due either to the difficulty of finding the explicit form of the gradient vector, or to the difficulty of the equation solving. The difficulty often arises in nonlinear optimization. In such cases, numerical solutions are often needed that typically require use of iterations. We first describe a class of methods for which the closed-form gradient vector is available and numerical procedures are used for finding the roots of the gradient vector.

5.2.1 Methods based on finding roots of equations

As discussed earlier, the necessary condition for an optimum of a function $F(\mathbf{x}), \mathbf{x} \in R^n$ is given by the solution of the equation:

$$\nabla f(\mathbf{x}) = \begin{pmatrix} \frac{\partial f}{\partial x_1} \\ \frac{\partial f}{\partial x_2} \\ \vdots \\ \frac{\partial f}{\partial x_n} \end{pmatrix} = \mathbf{g}(\mathbf{x}) = \begin{pmatrix} g_1(x_1, x_2, ..., x_n) \\ g_2(x_1, x_2, ..., x_n) \\ \vdots \\ g_n(x_1, x_2, ..., x_n) \end{pmatrix} = \begin{pmatrix} 0 \\ 0 \\ \vdots \\ 0 \end{pmatrix}.$$

Fixed-point iteration method

A fixed point of a function $\mathbf{h}(\mathbf{x})$ is a real vector $\mathbf{x}^*$ for which $\mathbf{h}(\mathbf{x}^*) = \mathbf{x}^*$. The iteration

$$\mathbf{x}_{k+1}^* = \mathbf{h}(\mathbf{x}_k^*)$$

is called a **fixed-point iteration**. If we wish to solve $\mathbf{g}(\mathbf{x}) = 0$ using the fixed-point iteration method, we can define

$$\mathbf{h}(\mathbf{x}) = \mathbf{g}(\mathbf{x}) + \mathbf{x},$$

and then proceed with the fixed-point iteration given the initial value of $\mathbf{x} = \mathbf{x}_0$. Other ways of defining the function $\mathbf{h}(\mathbf{x})$ are possible, with different and often superior convergent properties. The general procedure is to manipulate the equation $\mathbf{g}(\mathbf{x}) = 0$ into an equivalent form of $\mathbf{h}(\mathbf{x}) = \mathbf{x}$. When the fixed-point iteration is constructed to solve a linear system of equations, the method is called the **Jacobi iteration.**

An improvement to the above fixed-point iteration for solving either linear or nonlinear systems of equations can be made by using partially updated solutions to find the remaining ones within the same fixed-point iteration. This often speeds up the convergence, and is a commonly used technique with the special name of **Gauss-Seidel or Seidel iteration** [Math 99].

Newton-Raphson method

The Newton-Raphson method is one of the most widely used numerical techniques for finding roots of systems of nonlinear equations. To develop the method, we use a first-order Taylor series approximation of Eq. 5.2 for each of the nonlinear equations in

$$\mathbf{g}(\mathbf{x}) = 0$$

around the point $\mathbf{x}^{(k)}$:

$$\begin{aligned}
g_1(\mathbf{x}) &\approx g_1(\mathbf{x}^{(k)}) + (\mathbf{x} - \mathbf{x}^{(k)})^{\mathrm{Tr}} \nabla g_1(\mathbf{x}^{(k)}) \\
&= g_1(\mathbf{x}^{(k)}) + (x_1 - x_1^{(k)}) \frac{\partial g_1(\mathbf{x}^{(k)})}{\partial x_1} + (x_2 - x_2^{(k)}) \frac{\partial g_1(\mathbf{x}^{(k)})}{\partial x_2} + \ldots + (x_n - x_n^{(k)}) \frac{\partial g_1(\mathbf{x}^{(k)})}{\partial x_n}, \\
g_2(\mathbf{x}) &\approx g_2(\mathbf{x}^{(k)}) + (\mathbf{x} - \mathbf{x}^{(k)})^{\mathrm{Tr}} \nabla g_2(\mathbf{x}^{(k)}) \\
&= g_2(\mathbf{x}^{(k)}) + (x_1 - x_1^{(k)}) \frac{\partial g_2(\mathbf{x}^{(k)})}{\partial x_1} + (x_2 - x_2^{(k)}) \frac{\partial g_2(\mathbf{x}^{(k)})}{\partial x_2} + \ldots + (x_n - x_n^{(k)}) \frac{\partial g_2(\mathbf{x}^{(k)})}{\partial x_n}, \\
&\cdots \\
g_n(\mathbf{x}) &\approx g_n(\mathbf{x}^{(k)}) + (\mathbf{x} - \mathbf{x}^{(k)})^{\mathrm{Tr}} \nabla g_n(\mathbf{x}^{(k)}) \\
&= g_n(\mathbf{x}^{(k)}) + (x_1 - x_1^{(k)}) \frac{\partial g_n(\mathbf{x}^{(k)})}{\partial x_1} + (x_2 - x_2^{(k)}) \frac{\partial g_n(\mathbf{x}^{(k)})}{\partial x_2} + \ldots + (x_n - x_n^{(k)}) \frac{\partial g_n(\mathbf{x}^{(k)})}{\partial x_n}.
\end{aligned}$$

Its matrix form is

$$\mathbf{g}(\mathbf{x}) = \mathbf{g}(\mathbf{x}^{(k)}) + \begin{pmatrix} \frac{\partial g_1(\mathbf{x}^{(k)})}{\partial x_1} & \frac{\partial g_1(\mathbf{x}^{(k)})}{\partial x_2} & \cdots & \frac{\partial g_1(\mathbf{x}^{(k)})}{\partial x_n} \\ \frac{\partial g_2(\mathbf{x}^{(k)})}{\partial x_1} & \frac{\partial g_2(\mathbf{x}^{(k)})}{\partial x_2} & \cdots & \frac{\partial g_2(\mathbf{x}^{(k)})}{\partial x_n} \\ \vdots & \vdots & \vdots & \vdots \\ \frac{\partial g_n(\mathbf{x}^{(k)})}{\partial x_1} & \frac{\partial g_n(\mathbf{x}^{(k)})}{\partial x_2} & \cdots & \frac{\partial g_n(\mathbf{x}^{(k)})}{\partial x_n} \end{pmatrix} \begin{pmatrix} x_1 - x_1^{(k)} \\ x_2 - x_2^{(k)} \\ \vdots \\ x_n - x_n^{(k)} \end{pmatrix}, \tag{5.6}$$

or

$$\mathbf{g}(\mathbf{x}) = \mathbf{g}(\mathbf{x}^{(k)}) + \mathbf{J}(\mathbf{x}^{(k)})(\mathbf{x} - \mathbf{x}^{(k)}),$$

where $\mathbf{J}(\mathbf{x}^{(k)})$ is the **Jacobian matrix** evaluated at $\mathbf{x}^{(k)}$:

$$\mathbf{J}(\mathbf{x}^{(k)}) = \begin{pmatrix} \frac{\partial g_1(\mathbf{x}^{(k)})}{\partial x_1} & \frac{\partial g_1(\mathbf{x}^{(k)})}{\partial x_2} & \cdots & \frac{\partial g_1(\mathbf{x}^{(k)})}{\partial x_n} \\ \frac{\partial g_2(\mathbf{x}^{(k)})}{\partial x_1} & \frac{\partial g_2(\mathbf{x}^{(k)})}{\partial x_2} & \cdots & \frac{\partial g_2(\mathbf{x}^{(k)})}{\partial x_n} \\ \vdots & \vdots & \vdots & \vdots \\ \frac{\partial g_n(\mathbf{x}^{(k)})}{\partial x_1} & \frac{\partial g_n(\mathbf{x}^{(k)})}{\partial x_2} & \cdots & \frac{\partial g_n(\mathbf{x}^{(k)})}{\partial x_n} \end{pmatrix}.$$

To obtain the next estimate $\mathbf{x}^{(k+1)}$, we solve the matrix equation:

$$\mathbf{g}(\mathbf{x}^{(k+1)}) = \mathbf{g}(\mathbf{x}^{(k)}) + \mathbf{J}(\mathbf{x}^{(k)})(\mathbf{x} - \mathbf{x}^{(k)}) = \mathbf{0}.$$

This gives the updating equation for the Newton-Raphson method (first-order):

$$\mathbf{x}^{(k+1)} = \mathbf{x}^{(k)} - [\mathbf{J}(\mathbf{x}^{(k)})]^{-1}\mathbf{g}(\mathbf{x}^{(k)}).$$

For the special, scalar case of $n = 1$ where roots of one single nonlinear equation $g(x) = 0$ are sought and the Jacobian matrix reduces to $g'(x)$, we have the simplest updating equation for the Newton-Raphson method:

$$x^{(k+1)} = x^{(k)} - \frac{g(x^{(k)})}{g'(x^{(k)})}. \tag{5.7}$$

If the derivative $g'(x^{(k)})$ above is replaced by a numerical approximation:

$$g'(x^{(k)}) \approx \frac{g(x^{(k)}) - g(x^{(k-1)})}{x^{(k)} - x^{(k-1)}},$$

then Eq. 5.7 becomes

$$x^{(k+1)} = x^{(k)} - \frac{g(x^{(k)})(x^{(k)} - x^{(k-1)})}{g(x^{(k)}) - g(x^{(k-1)})}.$$

This updating method is called the **Secant method**. This method requires two initial points to start, and it does not require a supply of any explicit form of the derivative.

Similar development using second-order Taylor series approximation Eq. 5.4 leads to more efficient but more complicated *second-order* Newton-Raphson method. For the special, scalar case where $n = 1$, the updating equation is

$$x^{(k+1)} = x^{(k)} - \frac{g(x^{(k)})}{g'(x^{(k)}) - \frac{g(x^{(k)})g''(x^{(k)})}{2g'(x^{(k)})}}.$$

It can be proved that near a simple root, the first-order Newton-Raphson method converges at a quadratic rate, and the second-order Newton-Raphson method converges at significantly faster cubic rate.

5.2.2 Methods based on gradient descent

Another popular family of numerical methods for optimization is based on gradient descent. As discussed earlier, the gradient is a vector in an n-dimensional space where the objective function is defined. The effectiveness of these gradient-based methods derives from its important property (see [Rao 84] (pages 302-304) for an interesting theoretical proof of this property):

> The gradient vector represents the direction of steepest ascent of the objective function, and the negative gradient vector represents the direction of steepest descent.

That is, if we move along the gradient direction from any point in the n-dimensional space over which the objective function is defined, then the function value increases at the fastest rate. Note that the direction of steepest ascent is a local and not global property. Hence all the optimization methods based on gradients give only local optimum, and not global optimum.

Due to the steepest ascent or descent property associated with the gradient vector, any method that makes use of it can be expected to find an optimum point faster than the methods without using it. We will discuss in this section a few very popular methods that directly or indirectly make use of the gradient vector in finding the search directions for the objective function optimization.

Steepest descent method

The use of the negative gradient vector, $\nabla f(\mathbf{x})$, as a direction for minimizing an objective function, $f(\mathbf{x})$, was made as early as 1847. An initial point $\mathbf{x}_{(0)}$ is supplied to this method, called the **steepest descent method**, and it iteratively moves towards the minimum point using the updating equation:

$$\mathbf{x}^{(\mathbf{k+1})} = \mathbf{x}^{(\mathbf{k})} - \alpha_{min}^{(k)} \nabla f(\mathbf{x}).$$

$\alpha_{min}^{(k)}$ in the above is called the step size, and in the strict steepest descent method, it is optimized along the search direction $-\nabla f(\mathbf{x})$. That is, in each iteration of the steepest descent, $\alpha_{min}^{(k)}$ is found which minimizes $f(\mathbf{x}^{(\mathbf{k})} - \alpha^{(k)} \nabla f(\mathbf{x}))$.

Example: Use the steepest descent method to minimize $f(x_1, x_2) = x_1 - x_2 + 2x_1^2 + 2x_1x_2 + x_2^2$ with initial point $\mathbf{x}^{(\mathbf{0})} = (0\ 0)^{\mathrm{Tr}}$.

Solution: Denote $f(\mathbf{x}) = f(x_1, x_2)$. The gradient vector is computed by

$$\nabla f(\mathbf{x}) = \begin{pmatrix} \frac{\partial f(\mathbf{x})}{\partial x_1} \\ \frac{\partial f(\mathbf{x})}{\partial x_2} \end{pmatrix} = \begin{pmatrix} 1 + 4x_1 + 2x_2 \\ -1 + 2x_1 + 2x_2 \end{pmatrix}.$$

At the first iteration, when $\mathbf{x}^{(\mathbf{0})} = (0\ 0)^{\mathrm{Tr}}$, this gradient vector becomes

$$\nabla f^{(0)}(\mathbf{x}) = \begin{pmatrix} -1 \\ 1 \end{pmatrix}.$$

To find $\mathbf{x}^{(\mathbf{1})}$, we now need to find the optimal step size $\alpha_{min}^{(0)}$; that is, to minimize

$$f(\mathbf{x}^{(\mathbf{0})} - \alpha^{(0)} \nabla f^{(0)}(\mathbf{x})) = f(-\alpha^{(0)}, \alpha^{(0)}) = [\alpha^{(0)}]^2 - 2\alpha^{(0)},$$

with respect to $\alpha^{(0)}$. Solving $\frac{\partial f(\mathbf{x}^{(\mathbf{0})})}{\partial \alpha^{(0)}}$, we have $\alpha_{min}^{(0)} = 1$. This then gives

$$\mathbf{x}^{(\mathbf{1})} = \mathbf{x}^{(\mathbf{0})} - \alpha^{(0)} \nabla f^{(0)}(\mathbf{x}) = \begin{pmatrix} 0 \\ 0 \end{pmatrix} - 1 \begin{pmatrix} -1 \\ 1 \end{pmatrix} = \begin{pmatrix} 1 \\ -1 \end{pmatrix}.$$

At the next iteration, when $\mathbf{x}^{(\mathbf{1})} = (1\ \ -1)^{\mathrm{Tr}}$, we have

$$\nabla f^{(1)}(\mathbf{x}) = \begin{pmatrix} -1 \\ -1 \end{pmatrix}.$$

(This checks that no minimum occurred at $\mathbf{x}^{(\mathbf{1})}$.) Now,

$$f(\mathbf{x}^{(\mathbf{1})} - \alpha^{(1)} \nabla f^{(1)}(\mathbf{x})) = f(-1 + \alpha^{(1)}, 1 + \alpha^{(1)}) = 5[\alpha^{(1)}]^2 - 2\alpha^{(1)} - 1.$$

Its minimum occurs at $\alpha_{min}^{(1)} = 0.2$ by solving $\frac{\partial f(\mathbf{x}^{(1)})}{\partial \alpha^{(1)}}$. So,

$$\mathbf{x}^{(2)} = \mathbf{x}^{(1)} - \alpha^{(1)} \nabla f^{(1)}(\mathbf{x}) = \begin{pmatrix} -1 \\ 1 \end{pmatrix} - 0.2 \begin{pmatrix} -1 \\ -1 \end{pmatrix} = \begin{pmatrix} -0.8 \\ 1.2 \end{pmatrix}.$$

Continuing on for the next iteration, we have

$$\nabla f^{(2)}(\mathbf{x}) = \begin{pmatrix} 0.2 \\ -0.2 \end{pmatrix}.$$

Now,

$$f(\mathbf{x}^{(2)} - \alpha^{(2)} \nabla f^{(2)}(\mathbf{x})) = f(-0.8 - 0.2\alpha^{(2)}, 1.2 + 0.2\alpha^{(2)}) = 0.04[\alpha^{(2)}]^2 - 0.08\alpha^{(1)} - 1.20.$$

Its minimum occurs at $\alpha_{min}^{(2)} = 1.0$. So,

$$\mathbf{x}^{(3)} = \mathbf{x}^{(2)} - \alpha^{(2)} \nabla f^{(2)}(\mathbf{x}) = \begin{pmatrix} -0.8 \\ 1.2 \end{pmatrix} - 1.0 \begin{pmatrix} 0.2 \\ -0.2 \end{pmatrix} = \begin{pmatrix} -1.0 \\ 1.4 \end{pmatrix}.$$

Further continuation of the iteration will lead to the optimum solution $\mathbf{x}^{(*)} = (-1.0\ 1.5)^{\mathrm{Tr}}$.

Note that the trivial optimization problem above can be most easily solved by the classical technique of Eq. 5.5, via setting the gradient vector to zero and solving. However, this gives a simple example (adopted from [Rao 84]) to demonstrate the steps involved in the steepest descent method. The steps of the steepest descent in the above example are straightforward because the optimal step size α can be easily obtained by setting the derivative to zero, which gives a simple linear equation. If such an equation is nonlinear instead, as in most practical cases, many heuristic methods for determining the optimal step size α will be needed. A common technique is to use the adaptive step size, $\alpha^{(k)}$, which decreases as the iteration proceeds. Another common technique is to approximate the function $f(\mathbf{x}^{(k)} - \alpha^{(k)} \nabla f^{(k)}(\mathbf{x}))$ by a quadratic function in $\alpha^{(k)}$ around the point $\mathbf{x}^{(k)}$. This leads to straightforward optimization of $\alpha^{(k)}$ by solving a linear equation.

While the steepest descent method is a very popular and effective one, since the steepest descent direction is only a local, rather than a global, property, improvements and enhancement of the method are needed. The limitation of the basic steepest descent method discussed above is especially serious for high-dimensional optimization functions. When such functions have significant eccentricity, the steepest descent method will typically settle into steady high-dimensional zigzag patterns that cause the convergence to be extremely slow. Several superior methods that are aimed to overcome this type of limitation are discussed now.

5.3 Dynamic Programming Techniques for Optimization

Dynamic programming (DP) is a special class of optimization techniques, distinct from the ones discussed so far in this chapter, that solve optimization problems concerned with a sequence of interrelated decisions. DP is a procedure aimed to optimize a generally complex objective function that can be simplified to be a sum of objective functions each dependent on individual decisions and the conditions of the decision sequence in the DP procedure. DP is typically based on a recursive relation, sometimes called a functional equation, which is derived by transforming a current condition to a new condition through a decision.

5.3.1 Principle of optimality

The DP technique was originally developed by R. Bellman in 1950s. The foundation of DP was laid by the Bellman optimality principle. The principle stipulates that

> **Principle of Optimality in Dynamic Programming**:
> In an optimization problem concerning multiple stages of interrelated decisions, whatever the initial state (or condition) and the initial decisions are, the remaining decisions must constitute an optimal rule of choosing decisions with regards to the state that results from the first decision.

Example: Optimality principle in **Markov decision processes**. A Markov decision process is characterized by two sets of parameters. The first set is the transition probabilities of

$$P_{ij}^k(n) = P(\text{state j, stage (n + 1)}|\text{state i, stage n, decision k}),$$

where the current state of the system is dependent only on the state of the system at the previous stage and the decision taken at that stage (Markov property). The second set of parameters provide rewards defined by:

$R_i^k(n)$ = reward at stage n and at state i when decision k is chosen.

Define $F(n, i)$ to be the average of the total reward at state n and state i when the optimal decision is taken. This can be computed by DP using the following recursion (functional equation) given by the optimality principle:

$$F(n,i) = \max_k \{R_i^k(n) + \sum_j P_{ij}^k(n) F(n+1, j)\}.$$

In particular, when $n = N$ (at the final stage), the total reward at state i is

$$F(N, i) = \max_k R_i^k(N).$$

The optimal decision sequence can be traced back after the end of this recursive computation.

We see in the above that in applying DP, various stages (e.g., stage $1, 2, ..., n, ...N$ in the above example) in the optimization process must be identified. We are required at each stage to make optimal decision(s). There are several states (indexed by i in the above example) of the system associated with each stage. The decision (indexed by k) taken at a given stage changes the problem from the current stage n to the next stage $n + 1$ according to the transition probability $P_{ij}^k(n)$.

If we apply the DP technique to finding the optimal path, then the Bellman optimality principle can be alternatively stated as follows: the optimal path from nodes A to C through node B must consist of the optimal path from A to B concatenated with the optimal path from B to C. The implication of this principle is tremendous. In order

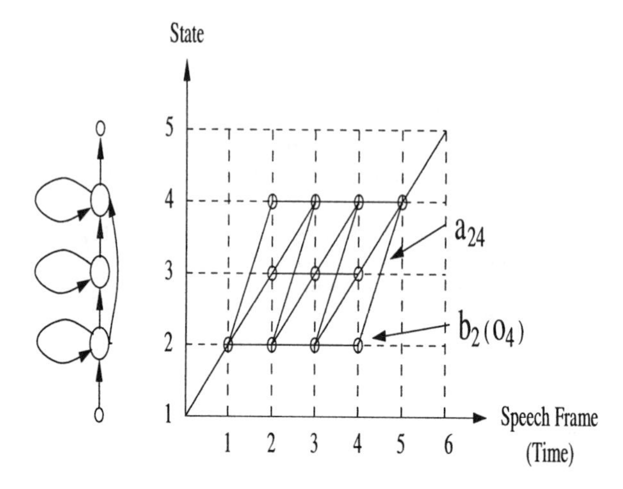

Figure 5.1: Trellis diagram for a left-to-right HMM and the Viterbi algorithm for finding the optimal state sequence using the DP technique.

to find the best path from node A via a "predecessor" node B, there will be no need to reconsider all the partial paths leading from A to B. This significantly reduces the path search effort compared with the brute-force search (i.e., exhaustive search). While it may be unknown whether the "predecessor" node B is on the best path or not, many candidates can be evaluated and the correct one be ultimately determined via a back-tracking procedure in the DP. This optimality principle is the essence of a very popular optimization technique in speech processing applications involving the HMM, which we present below.

5.3.2 Dynamic programming for the hidden Markov model

One fundamental computational problem associated with the HMM introduced in Chapter 3 is finding the best sequence of the HMM states given an arbitrary sequence of observations $\mathbf{o}_1^T = \mathbf{o}_1, \mathbf{o}_2, \cdots, \mathbf{o}_T$. This is a complex T-stage path-finding optimization problem, and is directly suited for the DP solution. The DP technique used for such purposes is also called the **Viterbi algorithm**, developed originally for optimal convolutional-code channel decoding in digital communication.

To illustrate the Viterbi algorithm as an optimal path-finding technique, we draw the two-dimensional grid, also called the trellis diagram, in Fig. 5.1 for a left-to-right HMM. A node in the trellis diagram is associated with both a time frame t on the horizontal axis and an HMM state i on the vertical axis. For a given HMM characterized by the state transitional probabilities a_{ij} and by the state-conditioned output probability distributions $b_i(\mathbf{o}_t)$, let $\delta_i(t)$ represent the maximal value of the joint likelihoods of the partial observation sequence $\mathbf{o}_1^t$ up to time t, and the associated HMM state sequence while in state i at time t. That is,

$$\delta_i(t) = \max_{s_1, s_2, \ldots, s_{t-1}} P(\mathbf{o}_1^t, s_1^{t-1}, s_t = i). \tag{5.8}$$

Note that each $\delta_i(t)$ defined here is associated with a node in the trellis diagram. Each increment of time corresponds to reaching a new stage in DP. At the final stage $t = T$, we have the objective function of $\delta_i(T)$ which is accomplished via all the previous stages of computation for $t \leq T-1$. Based on the DP optimality principle, the optimal partial likelihood of Eq. 5.8 at the processing stage of $t+1$ can be computed using the following functional equation as a recursion:

$$\delta_j(t+1) = \max_i \delta_i(t) a_{ij} b_j(\mathbf{o}_{t+1}), \tag{5.9}$$

for all states j. Each state at this processing stage is a hypothesized "precursor" node in the global optimal path. All such nodes except one will be eventually eliminated after the backtracking operation. The essence of DP used here is that we only need to compute the quantities of $\delta_j(t+1)$ as individual nodes in the trellis, removing the need to keep track of a very large number of partial paths from the initial stage to the current $(t+1)^{th}$ stage, which would be required for the exhaustive search. The optimality is guaranteed, due to the DP optimality principle, with the computation only linearly, rather than geometrically, increasing with the length T of the observation data sequence.

Besides the key recursion of Eq. 5.9, the complete Viterbi algorithm requires additional steps of recursion initialization, recursion termination, and path backtracking. This complete algorithm is described as follows for an N-state HMM with initial state probabilities of π_i:

1. Initialization:

$$\begin{aligned} \delta_i(1) &= \pi_i b_i(\mathbf{o}_1), \quad 1 \leq i \leq N \\ \psi_i(1) &= 0 \end{aligned}$$

(where ψ is used to store the nodes for later backtracking).

2. Forward Recursion:

$$\begin{aligned} \delta_j(t) &= \max_i \delta_i(t-1) a_{ij} b_j(\mathbf{o}_t), \quad && 2 \leq t \leq T \\ & && 1 \leq j \leq N \\ \psi_j(t) &= \arg \max_{1 \leq i \leq N} \delta_i(t-1) a_{ij}, \quad && 2 \leq t \leq T \\ & && 1 \leq j \leq N. \end{aligned}$$

3. Termination of Recursion:

$$\begin{aligned} P^* &= \max_{1 \leq i \leq N} [\delta_i(T)] \\ s^*(T) &= \max_{1 \leq i \leq N} [\delta_i(T)]. \end{aligned}$$

4. Optimal-Path Backtracking:

$$s^*(t) = \psi_{s^*(t+1)}(t+1), \quad t = T-1, T-2, ..., 1. \tag{5.10}$$

The result of the above Viterbi algorithm is the maximal joint likelihood of observation and state sequence P^*, together with the state transition path, $s^*(t)$, which gives such a maximum. This optimal state transition path is equivalent to the information required to determine the optimal state segmentation. The concept of state segmentation is most relevant to a left-to-right HMM, as for the case of speech modeling, because each state in such an HMM is typically associated with a reasonably large number of consecutive time frames in the observation sequence. This is so because the observations cannot be easily assigned back to earlier states due to the left-to-right constraint and due to the fact that the last frame must be accounted for by the right-most state in the left-to-right HMM.

5.3.3 Dynamic programming for the trended hidden Markov model

The same DP optimality principle can be applied to the trended HMM discussed in Chapter 4, to drastically reduce computation time, compared with the exhaustive search, for finding the optimal joint likelihood of observation and state sequences together with the optimal state-transition path. However, a more complex stage for the optimization need be considered. The complexity here is that the output distribution of the trended HMM is not only conditioned on the state, but also on the duration of the state (or state-sojourn time). This is because the mean (for a Gaussian trendend HMM) in the distribution is duration-varying according to the trend function in the model. Therefore, the DP optimization needs to be carried out not only over the time frame t and HMM state j as in the previous case, but also over the additional variable d of state-sojourn time. The Viterbi algorithm based on the DP principle developed for this type of problems has been called the **modified Viterbi algorithm** or 3-D Viterbi algorithm. [Deng 94c]

The partial likelihood for the trended HMM while using the 3-D Viterbi algorithm is now defined to be

$$\delta_i(t,d) = \max_{s_1,s_2,\ldots,s_{t-1}} P(\mathbf{o}_1^t, s_1^{t-1}, s_t = i, d_t = d). \tag{5.11}$$

with the additional variable d of the state-sojourn time compared with Eq. 5.8.

The essence of the 3-D Viterbi algorithm then is to efficiently compute $\delta_t(j,d)$ in an iterative way. To keep track of the optimal state sequence, we use $\psi_t(j,d)$ to trace the most likely state information (state identity and state sojourn time) at time $t-1$ given $s_t = i$ and $d_t = d$.

The complete modified 3-D Viterbi algorithm can now be described below:

1. Initialization:

$$\begin{aligned}
\delta_1(j,d) &= \begin{cases} \pi_j\, b_j(\mathbf{o}_1|d) & d=0,\ 1\le j\le N, \\ 0 & \text{otherwise} \end{cases} \\
\psi_1(j,d) &= (0,0) \qquad d=0,\ 1\le j\le N.
\end{aligned}$$

2. Forward Recursion:

$$\delta_{t+1}(j,d) = \begin{cases} \max\limits_{i<j}\ \max\limits_{0\le\tau\le t-1} \delta_t(i,\tau)\, a_{ij}\, b_j(\mathbf{o}_{t+1}|d) & d=0 \\ \delta_t(j,d-1)\, a_{jj}\, b_j(\mathbf{o}_{t+1}|d) & d>0, \end{cases}$$

$$\psi_{t+1}(j,d) = \begin{cases} \arg\max\limits_{i<j}\ \max\limits_{0\le\tau\le t-1} \delta_t(i,\tau) a_{ij} & d=0 \\ (j,d-1) & d>0 \end{cases}$$

where $1\le t<T$, $1\le j\le N$, $0\le d\le t$, and the explicit form of the output PDF

for the Gaussian trended HMM with M-order linear trend functions is

$$b_j(\mathbf{o}_t|d) = \frac{1}{(2\pi)^{D/2}|\mathbf{\Sigma}_j|^{1/2}} \exp\{-\frac{1}{2}[\mathbf{o}_t - \sum_{m=0}^{M} \mathbf{B}_j(m) f_m(d)]^{\mathrm{Tr}} \mathbf{\Sigma}_j^{-1} [\mathbf{o}_t - \sum_{m=0}^{M} \mathbf{B}_j(m) f_m(d)]\}.$$

3. Termination of Recursion:

$$\begin{aligned} P^* &= \max_i \max_{d_T} [\delta_T(i, d_T)] \\ (s_T^*, d_T^*) &= \arg \max_i \max_{d_T} [\delta_T(i, d_T)]. \end{aligned}$$

4. Optimal-Path Backtracking:

$$(s_t^*, d_t^*) = \psi_{t+1}(s_{t+1}^*, d_{t+1}^*), \; t = T-1, T-2, \cdots, 1.$$

Similar to the case of the conventional HMM, the result of the above modified or 3-D Viterbi algorithm is the maximal joint likelihood of observation and state sequence P^*, together with the optimal state transition path (state segmentation), $s^*(t)$, which also gives the optimal state sojourn durations, d_t^*.

5.4 Preliminaries of Estimation Theory

Equipped with the valuable tools of optimization techniques reviewed up to now in this chapter, we are now in a position to provide an overview of the rich field of estimation theory. We mentioned that the problem of estimation is a central issue in statistical modeling and its applications. Several major classes of estimation techniques with the associated optimization solutions and examples will be given in the remainder of this chapter.

We first introduce the general concepts of estimate and estimator function. Let $\mathbf{o} = (o[0]\, o[1]\, \ldots\, o[N-1])^{\mathrm{Tr}}$ be observation samples or data, and let $p(\mathbf{o}; \theta)$ be a statistical model or its full PDF for this N-sample data set parameterized by θ.

The (parameter) estimation problem, as one central issue addressed by estimation theory, is to find a function of the N-sample data set that provides an estimate of θ, that is:

$$\hat{\theta} = g(\mathbf{o} = \{o[0], o[1], \ldots, o[N-1]\}),$$

where $\hat{\theta}$ is called an **estimate** of θ, and $g(\mathbf{o})$ is known as the estimator function, or **estimator** for short. In other words, the estimate is the resulting functional value after evaluating the estimator function, and estimation is the procedure or process of evaluating the estimator function in order to obtain the value of the estimate. Note that an estimator is itself a random variable, since it is a function of the data as random variables. Hence an estimator is characterized by its PDF, like any random variable, and therefore the quality or performance of an estimator can be completely described and assessed only statistically, not by an absolute criterion.

Once a candidate estimator $g(\mathbf{o})$ is found, two performance issues arise:

1. How *close* will $\hat{\theta}$ be to θ, or how good or optimal is the found estimator?

2. Are there better estimators? If so, by what criteria are they better?

One obvious criterion for measuring goodness of an estimator is whether it has a zero bias, or whether $E(\hat{\theta}) - \theta = 0$. In addition, given the desirable unbiasedness property, an estimator with a smaller variance is more reliable than that with a larger variance. Hence it is also desirable to minimize $\text{var}(\hat{\theta})$ in looking for a high-performance estimator function. This produces the following estimator formally defined as:

Minimum Variance Unbiased (MVU) estimator: An estimator is MVU if and only if 1) it is unbiased, or

$$E(\hat{\theta}) = \theta \text{ for } \quad a < \theta < b$$

where [a,b] is the range of interest; and 2) it has minimum variance, or

$$\hat{\theta}_{MVU} = \arg\min_{\hat{\theta}}\{\text{var}(\hat{\theta})\} = \arg\min_{\hat{\theta}}\{E(\hat{\theta} - E(\hat{\theta}))^2\}.$$

Example: Consider a fixed signal, A, embedded in a zero-mean WGN (White Gaussian Noise) signal, $w[n]$:

$$x[n] = A + w[n] \qquad n = 0, 1, \ldots, N-1,$$

where $\theta = A$ is the parameter to be estimated from the observed data, $x[n]$. Consider the *sample-mean* estimator function:

$$\hat{\theta} = \frac{1}{N}\sum_{n=0}^{N-1} x[n].$$

Is the sample mean an MVU estimator for A? Since

$$E(\hat{\theta}) = E(\frac{1}{N}\sum x[n]) = \frac{1}{N}\sum E(x[n]) = \frac{1}{N}\sum A = \frac{1}{N}NA = A,$$

the estimator is unbiased. Also, we have

$$\text{var}(\hat{\theta}) = \text{var}(\frac{1}{N}\sum x[n]) = \frac{1}{N^2}\sum \text{var}(x[n]) = \frac{1}{N^2}\sum \sigma^2 = \frac{N\sigma^2}{N^2} = \frac{\sigma^2}{N}.$$

However, in order for it to be an MVU estimator, we need to show that $\text{var}(\tilde{\theta}) \geq \frac{\sigma^2}{N}$ for all other unbiased estimator functions $\tilde{\theta}$. This requires the following results.

5.4.1 Cramer-Rao lower bound and minimum variance unbiased estimator

The variance of any unbiased estimator $\hat{\theta}$ must be lower bounded by the **Cramer-Rao Lower Bound(CRLB)**, with the variance of the MVU estimator attaining the CRLB. That is, for a scalar parameter:

$$\text{var}(\hat{\theta}) \geq \frac{1}{-E\left[\frac{\partial^2 \log p(\mathbf{o};\theta)}{\partial\theta^2}\right]},$$

and

$$\text{var}(\hat{\theta}_{MVU}) = \frac{1}{-E\left[\frac{\partial^2 \log p(\mathbf{o};\theta)}{\partial\theta^2}\right]}.$$

Furthermore, if, for some functions g and I:

$$\frac{\partial \log p(\mathbf{o};\theta)}{\partial \theta} = I(\theta)(g(\mathbf{o})-\theta),$$

then we find the MVU estimator as $\hat{\theta}_{MVU} = g(\mathbf{o})$, which will have the minimum variance of $1/I(\theta)$. For a p-dimensional vector parameter, $\boldsymbol{\theta}$, the equivalent condition is:

$$\mathbf{C}_{\hat{\boldsymbol{\theta}}} - \mathbf{I}^{-1}(\boldsymbol{\theta}) \geq \mathbf{0},$$

i.e., $\mathbf{C}_{\hat{\boldsymbol{\theta}}} - \mathbf{F}^{-1}(\boldsymbol{\theta})$ is positive semidefinite where $\mathbf{C}_{\hat{\boldsymbol{\theta}}} = E[(\hat{\boldsymbol{\theta}} - E(\hat{\boldsymbol{\theta}}))^{\mathrm{Tr}}(\hat{\boldsymbol{\theta}} - E(\hat{\boldsymbol{\theta}}))]$ is the covariance matrix. The "Fisher" matrix, $\mathbf{F}(\boldsymbol{\theta})$, is given as:

$$[\mathbf{F}(\boldsymbol{\theta})]_{ij} = -E\left[\frac{\partial^2 \log p(\mathbf{o};\boldsymbol{\theta})}{\partial \theta_i \partial \theta_j}\right].$$

Furthermore if, for some p-dimensional function g and $p \times p$ matrix $\mathbf{I}$:

$$\frac{\partial \log p(\mathbf{o};\boldsymbol{\theta})}{\partial \boldsymbol{\theta}} = \mathbf{I}(\boldsymbol{\theta})(\mathbf{g}(\mathbf{o}) - \boldsymbol{\theta}),$$

then we can find the MVU estimator as: $\boldsymbol{\theta}_{MVU} = \mathbf{g}(\mathbf{o})$ and the minimum covariance will be $\mathbf{I}^{-1}(\boldsymbol{\theta})$.

Example: Continuing the earlier case of a signal embedded in noise:

$$x[n] = A + w[n] \qquad n = 0, 1, \ldots, N-1,$$

where $w[n]$ is a zero-mean WGN with variance σ^2, we have

$$\begin{aligned} p(\mathbf{o};\theta) &= \prod_{n=0}^{N-1} \frac{1}{\sqrt{2\pi\sigma^2}} \exp\left[-\frac{1}{2\sigma^2}(x[n]-\theta)^2\right] \\ &= \frac{1}{(2\pi\sigma^2)^{\frac{N}{2}}} \exp\left[-\frac{1}{2\sigma^2}\sum_{n=0}^{N-1}(x[n]-\theta)^2\right], \end{aligned}$$

where $p(\mathbf{o};\theta)$ is considered a function of the parameter $\theta = A$. Taking the first and then second-order derivatives:

$$\begin{aligned} \frac{\partial \log p(\mathbf{o};\theta)}{\partial \theta} &= \frac{N}{\sigma^2}(\frac{1}{N}\sum x[n] - \theta) = \frac{N}{\sigma^2}(\hat{\theta} - \theta) \\ \frac{\partial^2 \log p(\mathbf{o};\theta)}{\partial \theta^2} &= -\frac{N}{\sigma^2}. \end{aligned}$$

For a MVU estimator the lower bound has to apply, that is:

$$\mathrm{var}(\hat{\theta}_{MVU}) = \frac{1}{-E\left[\frac{\partial^2 \log p(\mathbf{o};\theta)}{\partial \theta^2}\right]} = \frac{\sigma^2}{N}.$$

Since we know from the previous example that $\mathrm{var}(\hat{\theta}) = \frac{\sigma^2}{N}$ we thus prove that the sample-mean is an MVU estimator. We can also show this by considering the first derivative:

$$\frac{\partial \log p(\mathbf{o};\theta)}{\partial \theta} = \frac{N}{\sigma^2}(\frac{1}{N}\sum x[n] - \theta) = I(\theta)(g(\mathbf{o}) - \theta),$$

where $I(\theta) = \frac{N}{\sigma^2}$ and $g(\mathbf{o}) = \frac{1}{N}\sum x[n]$. Thus the MVU estimator is indeed $\hat{\theta}_{MVU} = \frac{1}{N}\sum x[n]$ with minimum variance $\frac{1}{I(\theta)} = \frac{\sigma^2}{N}$.

Example: Consider the linear model (discussed in Chapter 4) with N-sample data vector $\mathbf{o}$:

$$\mathbf{o} = \mathbf{H}\boldsymbol{\theta} + \mathbf{v}. \tag{5.12}$$

Then using the CRLB theorem above, $\boldsymbol{\theta} = \mathbf{g}(\mathbf{o})$ will be an MVU estimator if:

$$\frac{\partial \log p(\mathbf{o};\boldsymbol{\theta})}{\partial \boldsymbol{\theta}} = \mathbf{I}(\boldsymbol{\theta})(\mathbf{g}(\mathbf{o}) - \boldsymbol{\theta})$$

with $\mathbf{C}_{\hat{\boldsymbol{\theta}}} = \mathbf{I}^{-1}(\boldsymbol{\theta})$. So we need to examine whether we can factor

$$\frac{\partial \log p(\mathbf{o};\boldsymbol{\theta})}{\partial \boldsymbol{\theta}} = \frac{\partial}{\partial \boldsymbol{\theta}}\left[-\log(2\pi\sigma^2)^{\frac{N}{2}} - \frac{1}{2\sigma^2}(\mathbf{o} - \mathbf{H}\boldsymbol{\theta})^{\mathrm{Tr}}(\mathbf{o} - \mathbf{H}\boldsymbol{\theta})\right] \tag{5.13}$$

into the form of $\mathbf{I}(\boldsymbol{\theta})(\mathbf{g}(\mathbf{o}) - \boldsymbol{\theta}$. Such a factorization is indeed possible, and we find, from the $\mathbf{g}(\mathbf{o})$ in the factorization, the MVU estimator for $\boldsymbol{\theta}$ to be

$$\hat{\boldsymbol{\theta}} = (\mathbf{H}^{\mathrm{Tr}}\mathbf{H})^{-1}\mathbf{H}^{\mathrm{Tr}}\mathbf{o}, \tag{5.14}$$

with the covariance matrix of $\boldsymbol{\theta}$, from the $\mathbf{I}(\boldsymbol{\theta})$ in the factorization, to be

$$\mathbf{C}_{\hat{\boldsymbol{\theta}}} = \sigma^2(\mathbf{H}^{\mathrm{Tr}}\mathbf{H})^{-1}.$$

The MVU estimator in Eq. 5.14 for the canonical form of the linear system is a very useful one because many models will have the same general form of the estimator after fitting into the same canonical form. Examples are FIR and IIR digital filters, linear regression models (e.g., polynomial curve fitting), discrete Fourier transforms, finite-difference equations, nonlinear function approximations, etc., as discussed in Chapter 4. The matrix $(\mathbf{H}^{\mathrm{Tr}}\mathbf{H})^{-1}\mathbf{H}^{\mathrm{Tr}}$ in the MVU estimator Eq. 5.12 has a special name — pseudo-inverse matrix of $\mathbf{H}$.

5.4.2 Example: MVU estimator for generalized linear model

Extending the canonical form of the linear model Eq. 5.12, the generalized linear model for the observation data has the following canonical form:

$$\mathbf{o} = \mathbf{H}\boldsymbol{\theta} + \mathbf{s} + \mathbf{v}, \tag{5.15}$$

where $\mathbf{s}$ is an $N \times 1$ vector of known signal samples, $\mathbf{v}$ is an $N \times 1$ noise vector with PDF $\mathbf{N}(0, \mathbf{C})$, where $\mathbf{C}$ is non-diagonal matrix. Therefore, the noise $\mathbf{v}$ is no longer white (correlation across samples) as in the original linear model.

The MVU estimator for the earlier linear model where the noise is assumed white can be used after applying a suitable "whitening" transformation. If we factor the noise covariance matrix as:

$$\mathbf{C}^{-1} = \mathbf{D}^{\mathrm{Tr}}\mathbf{D},$$

then the matrix $\mathbf{D}$ is the required transformation since:

$$E[\mathbf{v}\mathbf{w}^{\mathrm{Tr}}] = \mathbf{C} \Rightarrow E[(\mathbf{D}\mathbf{w})(\mathbf{D}\mathbf{w})^{\mathrm{Tr}}] = \mathbf{D}\mathbf{C}\mathbf{D}^{\mathrm{Tr}} = (\mathbf{D}\mathbf{D}^{-1})(\mathbf{D}^{T^{-1}}\mathbf{D}^{\mathrm{Tr}}) = \mathbf{I}.$$

Therefore, $\mathbf{v}' = \mathbf{D}\mathbf{w}$ has PDF $\mathbf{N}(0, \mathbf{I})$. By transforming the general linear model:

$$\mathbf{o} = \mathbf{H}\boldsymbol{\theta} + \mathbf{s} + \mathbf{v}$$

to

$$\begin{aligned} \mathbf{o}' = \mathbf{D}\mathbf{o} &= \mathbf{D}\mathbf{H}\boldsymbol{\theta} + \mathbf{D}\mathbf{s} + \mathbf{D}\mathbf{w} \\ \mathbf{o}' &= \mathbf{H}'\boldsymbol{\theta} + \mathbf{s}' + \mathbf{v}' \end{aligned}$$

or

$$\mathbf{o}'' = \mathbf{o}' - \mathbf{s}' = \mathbf{H}'\boldsymbol{\theta} + \mathbf{v}',$$

we can then write the MVU estimator of $\boldsymbol{\theta}$ given the observed data $\mathbf{o}''$ as:

$$\begin{aligned} \hat{\boldsymbol{\theta}} &= (\mathbf{H}'^{\mathrm{Tr}}\mathbf{H}')^{-1}\mathbf{H}'^{\mathrm{Tr}}\mathbf{o}'' \\ &= (\mathbf{H}^{\mathrm{Tr}}\mathbf{D}^{\mathrm{Tr}}\mathbf{D}\mathbf{H})^{-1}\mathbf{H}^{\mathrm{Tr}}\mathbf{D}^{\mathrm{Tr}}\mathbf{D}(\mathbf{o} - \mathbf{s}). \end{aligned}$$

That is,

$$\hat{\boldsymbol{\theta}} = (\mathbf{H}^{\mathrm{Tr}}\mathbf{C}^{-1}\mathbf{H})^{-1}\mathbf{H}^{\mathrm{Tr}}\mathbf{C}^{-1}(\mathbf{o} - \mathbf{s})$$

and the covariance matrix is:

$$\mathbf{C}_{\hat{\theta}} = (\mathbf{H}^{\mathrm{Tr}}\mathbf{C}^{-1}\mathbf{H})^{-1}.$$

5.4.3 Sufficient statistic

For the cases where the CRLB cannot be established, a more general approach to finding the MVU estimator is required. To do so, we need to first find a **sufficient statistic** for the unknown parameter θ:

Sufficient statistic for an optimal estimator: If it is possible to factor the PDF $p(\mathbf{o};\theta)$ as:

$$p(\mathbf{o};\theta) = g(T(\mathbf{o}), \theta)h(\mathbf{o}),$$

where $g(T(\mathbf{o}), \theta)$ is a function of $T(\mathbf{o})$ and θ only and $h(\mathbf{o})$ is a function of $\mathbf{o}$ only, then $T(\mathbf{o})$ is a sufficient statistic for θ.

Conceptually one expects that the PDF after the sufficient statistic has been observed, $p(\mathbf{o}|T(\mathbf{o}) = T_0;\theta)$, should not depend on θ since $T(\mathbf{o})$ is sufficient for the estimation of θ and no more knowledge can be gained about θ once we know $T(\mathbf{o})$.

Example: Consider again a signal embedded in a WGN signal:

$$x[n] = A + w[n].$$

The PDF is

$$p(\mathbf{o};\theta) = \frac{1}{(2\pi\sigma^2)^{\frac{N}{2}}} \exp\left[-\frac{1}{2\sigma^2}\sum_{n=0}^{N-1}(x[n] - \theta)^2\right],$$

where $\theta = A$ is the unknown parameter we want to estimate. We factor the PDF as follows:

$$\begin{aligned} p(\mathbf{o};\theta) &= \frac{1}{(2\pi\sigma^2)^{\frac{N}{2}}}\exp\left[-\frac{1}{2\sigma^2}(N\theta^2 - 2\theta\sum_{n=0}^{N-1}x[n])\right]\exp\left[-\frac{1}{2\sigma^2}\sum_{n=0}^{N-1}x^2[n]\right] \\ &= g(T(\mathbf{o}),\theta)\cdot h(\mathbf{o}), \end{aligned}$$

where we define $T(\mathbf{o}) = \sum_{n=0}^{N-1} x[n]$ which constitutes a sufficient statistic for θ.

Sufficient statistics $T(\mathbf{o})$ can be used to derive MVU estimators in two different ways:

1. Let $\tilde{\theta}$ be *any* unbiased estimator of θ. Then $\hat{\theta} = E(\tilde{\theta}|T(\mathbf{o})) = \int \tilde{\theta}p(\tilde{\theta}|T(\mathbf{o}))d\tilde{\theta}$ is the MVU estimator.

2. Find some function g such that $\hat{\theta} = g(T(\mathbf{o}))$ is an unbiased estimator of θ (i.e., $E[g(T(\mathbf{o}))] = \theta$), then $\hat{\theta}$ is the MVU estimator.

It can be proved that $\hat{\theta} = E(\tilde{\theta}|T(\mathbf{o}))$ is a valid estimator for θ, is unbiased, and is of lesser or equal variance than that of $\tilde{\theta}$, for all θ. Further, $\hat{\theta} = E(\tilde{\theta}|T(\mathbf{o}))$ is an MVU estimator if the sufficient statistic, $T(\mathbf{o})$, is complete. The sufficient statistic, $T(\mathbf{o})$, is complete if there is only one function $g(T(\mathbf{o}))$ that is unbiased. That is, if $h(T(\mathbf{o}))$ is another unbiased estimator (i.e., $E[h(T(\mathbf{o}))] = \theta$) then we must have that $g = h$.

Example: Consider again the previous example of a signal embedded in a WGN signal:

$$x[n] = A + w[n],$$

where we derived the sufficient statistic to be $T(\mathbf{o}) = \sum_{n=0}^{N-1} x[n]$. Using Method 2 above, we need to find a function g such that $E[g(T(\mathbf{o}))] = \theta = A$. Now:

$$E[T(\mathbf{o})] = E\left[\sum_{n=0}^{N-1} x[n]\right] = \sum_{n=0}^{N-1} E[x[n]] = N\theta.$$

It is obvious that

$$E\left[\frac{1}{N}\sum_{n=0}^{N-1} x[n]\right] = \theta,$$

and thus $\hat{\theta} = g(T(\mathbf{o})) = \frac{1}{N}\sum_{n=0}^{N-1} x[n]$, which is the **sample mean** we have already seen before, is the MVU estimator for θ.

5.4.4 Best linear unbiased estimator

It may occur that the MVU estimator or a sufficient statistic cannot be found or, indeed, the PDF of the data is itself unknown (e.g., only the second-order statistics are known). In such cases one solution is to assume a functional model of the estimator, as being linear in the data, and find the linear estimator which is both *unbiased* and has *minimum variance*. This is called the **best linear unbiased estimator**, or BLUE for short.

For the general vector case we want our estimator to be a linear function of the data, that is:

$$\hat{\theta} = \mathbf{A}\mathbf{o}.$$

Our first requirement is that the estimator be unbiased, that is:

$$E(\hat{\boldsymbol{\theta}}) = \mathbf{A}E(\mathbf{o}) = \theta$$

which can only be satisfied if:

$$E(\mathbf{o}) = \mathbf{H}\boldsymbol{\theta},$$

i.e., $\mathbf{AH} = \mathbf{I}$. The BLUE is derived by finding the $\mathbf{A}$ which minimizes the variance, $\mathbf{C}_{\hat{\boldsymbol{\theta}}} = \mathbf{ACA}^{\mathrm{Tr}}$, where $\mathbf{C} = E[(\mathbf{o} - E(\mathbf{o}))(\mathbf{o} - E(\mathbf{o}))^{\mathrm{Tr}}]$ is the covariance of the data $\mathbf{o}$, subject to the constraint $\mathbf{AH} = \mathbf{I}$. Carrying out this minimization yields the following for the BLUE:

$$\hat{\boldsymbol{\theta}} = \mathbf{Ax} = (\mathbf{H}^{\mathrm{Tr}}\mathbf{C}^{-1}\mathbf{H})^{-1}\mathbf{H}^{\mathrm{Tr}}\mathbf{C}^{-1}\mathbf{o},$$

where $\mathbf{C}_{\hat{\boldsymbol{\theta}}} = (\mathbf{H}^{\mathrm{Tr}}\mathbf{C}^{-1}\mathbf{H})^{-1}$. The form of the BLUE is identical to that of the MVU estimator for the general linear model. The crucial difference is that the BLUE does not make any assumptions on the PDF of the data (or noise), whereas the MVU estimator was derived assuming Gaussian noise. Of course, if the data is truly Gaussian then the BLUE is also the MVU estimator.

Now consider a linear model of the form:

$$\mathbf{o} = \mathbf{H}\boldsymbol{\theta} + \mathbf{v},$$

where $\mathbf{H}$ is known, and $\mathbf{v}$ is noise with covariance $\mathbf{C}$ (the PDF of $\mathbf{v}$ is otherwise arbitrary), then the BLUE of $\boldsymbol{\theta}$ is:

$$\hat{\boldsymbol{\theta}} = (\mathbf{H}^{\mathrm{Tr}}\mathbf{C}^{-1}\mathbf{H})^{-1}\mathbf{H}^{\mathrm{Tr}}\mathbf{C}^{-1}\mathbf{o},$$

where $\mathbf{C}_{\hat{\boldsymbol{\theta}}} = (\mathbf{H}^{\mathrm{Tr}}\mathbf{C}^{-1}\mathbf{H})^{-1}$ is the minimum covariance.

Example: Consider a signal embedded in noise:

$$o[n] = A + w[n],$$

where $w[n]$ is of unspecified PDF with $\mathrm{var}(w[n]) = \sigma_n^2$ and the unknown parameter $\theta = A$ is to be estimated. We assume a BLUE estimate and we derive $\mathbf{H}$ by noting:

$$E[\mathbf{o}] = \mathbf{1}\theta,$$

where $\mathbf{o} = [o[0], o[1], o[2], \ldots, o[N-1]]^{\mathrm{Tr}}$, $\mathbf{1} = [1, 1, 1, \ldots, 1]^{\mathrm{Tr}}$and we have $\mathbf{H} \equiv \mathbf{1}$. Also:

$$\mathbf{C} = \begin{bmatrix} \sigma_0^2 & 0 & \cdots & 0 \\ 0 & \sigma_1^2 & \cdots & 0 \\ \vdots & \vdots & \ddots & \vdots \\ 0 & 0 & \cdots & \sigma_{N-1}^2 \end{bmatrix} \quad \Rightarrow \quad \mathbf{C}^{-1} = \begin{bmatrix} \frac{1}{\sigma_0^2} & 0 & \cdots & 0 \\ 0 & \frac{1}{\sigma_1^2} & \cdots & 0 \\ \vdots & \vdots & \ddots & \vdots \\ 0 & 0 & \cdots & \frac{1}{\sigma_{N-1}^2} \end{bmatrix}$$

and hence the BLUE is:

$$\hat{\theta} = (\mathbf{H}^{\mathrm{Tr}}\mathbf{C}^{-1}\mathbf{H})^{-1}\mathbf{H}^{\mathrm{Tr}}\mathbf{C}^{-1}\mathbf{o} = \frac{\mathbf{1}^{\mathrm{Tr}}\mathbf{C}^{-1}\mathbf{o}}{\mathbf{1}^{\mathrm{Tr}}\mathbf{C}^{-1}\mathbf{1}} = \frac{\sum_{n=0}^{N-1} \frac{o[n]}{\sigma_n^2}}{\sum_{n=0}^{N-1} \frac{1}{\sigma_n^2}} \tag{5.16}$$

and the minimum covariance is:

$$\mathbf{C}_{\hat{\theta}} = \text{var}(\hat{\theta}) = \frac{1}{\mathbf{1}^{\text{Tr}}\mathbf{C}^{-1}\mathbf{1}} = \frac{1}{\sum_{n=0}^{N-1} \frac{1}{\sigma_n^2}}. \tag{5.17}$$

In the above example, for the case of white noise with equal variances $\sigma_n^2 = \sigma^2$ we will obtain the sample mean as the BLUE estimator from Eq. 5.16:

$$\hat{\theta} = \frac{1}{N} \sum_{n=0}^{N-1} o[n].$$

The minimum variance is then $\text{var}(\hat{\theta}) = \frac{\sigma^2}{N}$ from Eq. 5.17.

5.4.5 Method of moments

Method of moments is another useful but rudimentary technique for (deterministic) parameter estimation free of PDFs. While we may not have an expression for the PDF for the observation data, we can use the data to first obtain the empirical k^{th} data-moments, $\mu_k = E(o^k[n])$, by

$$\hat{\mu}_k = \frac{1}{N} \sum_{n=0}^{N-1} o^k[n],$$

and then estimate the model parameters by constructing a match between the model's moments and the data moments.

If we can write an expression for the kth moment as a function of the unknown parameter θ:

$$\mu_k = h(\theta)$$

and assuming h^{-1} exists then we can derive an estimate by:

$$\hat{\theta} = h^{-1}(\hat{\mu}_k) = h^{-1}\left(\frac{1}{N} \sum_{n=0}^{N-1} o^k[n]\right).$$

If $\boldsymbol{\theta}$ is a p-dimensional vector then we require p equations to solve for the p unknowns. That is, we need some set of p moment equations. Using the lowest order p moments what we would like is:

$$\hat{\boldsymbol{\theta}} = \mathbf{h}^{-1}(\hat{\boldsymbol{\mu}}),$$

where:

$$\hat{\boldsymbol{\mu}} = \begin{bmatrix} \frac{1}{N}\sum_{n=0}^{N-1} o[n] \\ \frac{1}{N}\sum_{n=0}^{N-1} o^2[n] \\ \vdots \\ \frac{1}{N}\sum_{n=0}^{N-1} o^p[n] \end{bmatrix}.$$

Example: Consider a two-Gaussian mixture PDF:

$$p(o;\theta) = (1-\theta)g_1(o) + \theta g_2(o),$$

where $g_1 = \mathcal{N}(\mu_1, \sigma_1^2)$ and $g_2 = \mathcal{N}(\mu_2, \sigma_2^2)$ are two different component Gaussian PDFs and θ is the unknown parameter to be estimated. We can write the second moment as a function of θ as follows:

$$\mu_2 = E(o^2[n]) = \int o^2 p(o;\theta)do = (1-\theta)\sigma_1^2 + \theta\sigma_2^2 = h(\theta).$$

This gives the estimate of θ by the method of moments:

$$\hat{\theta} = \frac{\hat{\mu}_2 - \sigma_1^2}{\sigma_2^2 - \sigma_1^2} = \frac{\frac{1}{N}\sum_{n=0}^{N-1} o^2[n] - \sigma_1^2}{\sigma_2^2 - \sigma_1^2}.$$

5.5 Least Squares Estimation

The several estimators discussed in the above preliminary section are, although fundamental to estimation theory, often not widely used in engineering and signal/speech processing applications. This is because they are either difficult to find and to compute (while having highly desirable qualities such as an MVU estimator), or because they are too empirical in nature (such as the method of moments). The requirement for knowing the precise form of PDF is also an important issue.

In the next few sections, we will discuss several very widely used parameter estimation techniques in estimation theory. Their successes are due to desirable tradeoffs between practicality of deriving and computing these estimators, and the strength of the theoretical properties associated with these estimators. We first start with the **least squares estimation (LSE)** procedure.

5.5.1 Basic LSE procedure

The MVU and BLUE estimators discussed previously required an explicit expression for the PDF $p(\mathbf{o};\theta)$ in order to estimate the unknown parameter θ in some optimal fashion. An alternative approach is to assume a signal model (rather than probabilistic assumptions about the data) and achieve a design goal assuming this model. With the least squares approach we assume that the signal model is a function of the unknown parameter θ and produces a signal:

$$s[n] \equiv s(n;\theta),$$

where $s(n;\theta)$ is a function of n and parameterized by θ. Due to noise and model inaccuracies, $w[n]$, the signal $s[n]$ can only be observed as:

$$o[n] = s[n] + w[n].$$

Unlike previous approaches no statement is made on the probabilistic distribution nature of $w[n]$. We only state that what we have is an "error": $e[n] = o[n] - s[n]$ which with the appropriate choice of θ should be minimized in a least-squares sense. That is, we choose $\theta = \hat{\theta}$ so that the criterion:

$$J(\theta) = \sum_{n=0}^{N-1} (o[n] - s[n])^2$$

is minimized over the N observation samples of interest and we call this the LSE of θ. More precisely we have:

$$\hat{\theta} = \arg\min_{\theta} J(\theta)$$

and the minimum LS error is given by:

$$J_{\min} = J(\hat{\theta}).$$

An important assumption made in order to produce a meaningful unbiased estimate is that the noise and model inaccuracies, $w[n]$, have zero-mean. However no other probabilistic assumptions about the data are made (i.e., LSE is valid for both Gaussian and non-Gaussian noise), although by the same token we also cannot make any optimality claims with LSE, since this would depend on the distribution of the noise and modeling errors.

A problem that arises from assuming a signal model function $s(n;\theta)$ rather than knowledge of $p(\mathbf{o};\theta)$ is the need to choose an appropriate signal model. Then again in order to obtain a closed form or "parametric" expression for $p(\mathbf{o};\theta)$ one usually needs to know what the underlying model and noise characteristics are anyway.

Example: Consider observations, $o[n]$, arising from a DC-level signal model, $s[n] = s(n;\theta) = \theta$:

$$o[n] = \theta + w[n],$$

where θ is the unknown parameter to be estimated. Then we have:

$$J(\theta) = \sum_{n=0}^{N-1}(o[n] - s[n])^2 = \sum_{n=0}^{N-1}(o[n] - \theta)^2.$$

Differentiating with respect to θ and setting it to zero gives

$$\left.\frac{\partial J(\theta)}{\partial \theta}\right|_{\theta=\hat{\theta}} = 0 \quad \Rightarrow \quad -2\sum_{n=0}^{N-1}(o[n] - \hat{\theta}) = 0 \quad \Rightarrow \quad \sum_{n=0}^{N-1} o[n] - N\hat{\theta} = 0.$$

Hence $\hat{\theta} = \frac{1}{N}\sum_{n=0}^{N-1} o[n]$ which is the *sample-mean*. We also have the minimum LS error of

$$J_{\min} = J(\hat{\theta}) = \sum_{n=0}^{N-1}\left(o[n] - \frac{1}{N}\sum_{n=0}^{N-1} o[n]\right)^2.$$

5.5.2 Least squares estimator for the linear model

If we assume the signal model is a linear function of the model parameters to be estimated:

$$\mathbf{s} = \mathbf{H}\boldsymbol{\theta},$$

where $\mathbf{s} = [s[0], s[1], \ldots, s[N-1]]^{\mathrm{Tr}}$ and $\mathbf{H}$ is a known $N \times p$ matrix with $\boldsymbol{\theta} = [\theta_1, \theta_2, \ldots, \theta_p]$, then we have the following observation model in the same form of the linear model:

$$\mathbf{o} = \mathbf{H}\boldsymbol{\theta} + \mathbf{w},$$

where $\mathbf{o} = [o[0], o[1], \ldots, o[N-1]]^{\mathrm{Tr}}$. The LS error in this case can be written as

$$J(\boldsymbol{\theta}) = \sum_{n=0}^{N-1}(o[n] - s[n])^2 = (\mathbf{o} - \mathbf{H}\boldsymbol{\theta})^{\mathrm{Tr}}(\mathbf{o} - \mathbf{H}\boldsymbol{\theta}). \tag{5.18}$$

If $\mathbf{o}$ and $\mathbf{y}$ are two $N \times 1$ nonzero vectors, then

$$\frac{d}{d\mathbf{o}}(\mathbf{y}^{\mathrm{Tr}}\mathbf{o}) = \mathbf{y}, \tag{5.19}$$

and

$$\frac{d}{d\mathbf{o}}(\mathbf{o}^{\mathrm{Tr}}\mathbf{o}) = 2\mathbf{o}. \tag{5.20}$$

Using the above matrix-calculus results, we take differentiation of Eq. 5.18 with respect to $\boldsymbol{\theta}$ and set it to zero to obtain

$$\left.\frac{\partial J(\boldsymbol{\theta})}{\partial \boldsymbol{\theta}}\right|_{\boldsymbol{\theta}=\hat{\boldsymbol{\theta}}} = 0 \quad \Rightarrow \quad -2\mathbf{H}^{\mathrm{Tr}}\mathbf{o} + 2\mathbf{H}^{\mathrm{Tr}}\mathbf{H}\hat{\boldsymbol{\theta}} = 0.$$

This yields the required LSE:

$$\hat{\boldsymbol{\theta}} = (\mathbf{H}^{\mathrm{Tr}}\mathbf{H})^{-1}\mathbf{H}^{\mathrm{Tr}} \cdot \mathbf{o},$$

which, surprisingly, is the identical functional form of the MVU estimator for the linear model.

An interesting extension to the LSE for the linear system is the weighted LS, where the contribution to the error from each component of the parameter vector can be weighted in importance by using a different form of the error criterion:

$$J(\boldsymbol{\theta}) = (\mathbf{o} - \mathbf{H}\boldsymbol{\theta})^{\mathrm{Tr}}\mathbf{W}(\mathbf{o} - \mathbf{H}\boldsymbol{\theta}),$$

where $\mathbf{W}$ is an $N \times N$ positive definite (symmetric) weighting matrix. The weighted LS estimator can be similarly proved to be:

$$\hat{\boldsymbol{\theta}} = (\mathbf{H}^{\mathrm{Tr}}\mathbf{W}\mathbf{H})^{-1}\mathbf{H}^{\mathrm{Tr}}\mathbf{W} \cdot \mathbf{o}.$$

Another extension of the LSE is to incorporate constraints. Assume that in the vector parameter LSE problem we impose constraints on the individual parameters. That is, the p parameters $\boldsymbol{\theta}$ are subject to $r < p$ independent linear constraints. The constraints can be expressed by the following system of linear equations:

$$\mathbf{A}\boldsymbol{\theta} = \mathbf{b}.$$

To solve this constrained LS optimization problem, we use the technique of Lagrangian multipliers (cf. Section 5.1.3) to minimize the following modified error criterion:

$$J_c = (\mathbf{o} - \mathbf{H}\boldsymbol{\theta})^{\mathrm{Tr}}(\mathbf{o} - \mathbf{H}\boldsymbol{\theta}) + \boldsymbol{\lambda}^{\mathrm{Tr}}(\mathbf{A}\boldsymbol{\theta} - \mathbf{b}).$$

Let $\hat{\boldsymbol{\theta}}$ be the unconstrained LSE of a linear model, then the expression for the constrained estimate becomes

$$\hat{\boldsymbol{\theta}}_c = \hat{\boldsymbol{\theta}} + (\mathbf{H}^{\mathrm{Tr}}\mathbf{H})^{-1}\mathbf{A}^{\mathrm{Tr}}[\mathbf{A}(\mathbf{H}^{\mathrm{Tr}}\mathbf{H})^{-1}\mathbf{A}^{\mathrm{Tr}}]^{-1}(\mathbf{A}\boldsymbol{\theta} - \mathbf{b}),$$

where $\hat{\boldsymbol{\theta}} = (\mathbf{H}^{\mathrm{Tr}}\mathbf{H})^{-1}\mathbf{H}^{\mathrm{Tr}}\mathbf{o}$ for unconstrained LSE.

5.5.3 Order-recursive least squares

In many cases the signal model is unknown and must be assumed. Obviously we would like to choose the model, $s(\boldsymbol{\theta})$, that minimizes $J_{\min}$, that is:

$$s_{\text{best}}(\boldsymbol{\theta}) = \arg\min_{s(\hat{\boldsymbol{\theta}})} J_{\min}.$$

We can do this arbitrarily by simply choosing models, obtaining the LSE $\hat{\theta}$, and then selecting the model which provides the smallest $J_{\min}$. However, models are not arbitrary and some models are more "complex" (or more precisely have a larger number of parameters or degrees of freedom) than others. The more complex a model the lower the $J_{\min}$ one can expect, but also the more likely the model is to *overfit* the data or be *overtrained* (i.e., fit the noise and not generalize to other data sets).

Example: Consider the problem of "line fitting," where we have observations $o(t)$ plotted against the sample time index t and we would like to fit the "best" line to the data. Many possibilities exist: (1) a constant: $s(t;\boldsymbol{\theta}) = \theta_1$; (2) a straight line: $s(t;\boldsymbol{\theta}) = \theta_1 + \theta_2 t$; (3) a quadratic line: $s(t;\boldsymbol{\theta}) = \theta_1 + \theta_2 t + \theta_3 t^2$; etc. Each possibility represents an increase in the *order* of the model (i.e., order of the polynomial fit), or the number of parameters to be estimated and a consequent increase in the "modeling power". A polynomial fit represents a linear model, $\mathbf{s} = \mathbf{H}\boldsymbol{\theta}$, where $\mathbf{s} = [s(0), s(1), \ldots, s(N-1)]^{\text{Tr}}$. For the three possibilities, we have

$$\textbf{(1) Constant}: \boldsymbol{\theta} = [\theta_1]^{\text{Tr}} \text{ and } \mathbf{H} = \begin{bmatrix} 1 \\ 1 \\ \vdots \\ 1 \end{bmatrix} \text{ is an } N \times 1 \text{ matrix,}$$

$$\textbf{(2) Linear}: \boldsymbol{\theta} = [\theta_1, \theta_2]^{\text{Tr}} \text{ and } \mathbf{H} = \begin{bmatrix} 1 & 0 \\ 1 & 1 \\ 1 & 2 \\ \vdots & \vdots \\ 1 & N-1 \end{bmatrix} \text{ is an } N \times 2 \text{ matrix,}$$

$$\textbf{(3) Quadratic:} \boldsymbol{\theta} = [\theta_1, \theta_2, \theta_3]^{\text{Tr}} \text{ and } \mathbf{H} = \begin{bmatrix} 1 & 0 & 0 \\ 1 & 1 & 1 \\ 1 & 2 & 4 \\ \vdots & \vdots & \vdots \\ 1 & N-1 & (N-1)^2 \end{bmatrix} \text{ is an } N \times 3 \text{ matrix,}$$

and so on. If the underlying model is indeed a straight line then we would expect not only that the minimum $J_{\min}$ results with a straight line model but also that higher order polynomial models (e.g., quadratic, cubic, etc.) will yield the same $J_{\min}$ (indeed higher-order models would "degenerate" to a straight line model, except in cases of overfitting). Thus the straight line model is the "best" model to use.

As an alternative to providing an independent LSE for each possible signal model, a more efficient **order-recursive LSE** is possible if the models are different orders of the same base model (e.g., polynomials of different degrees). In this method the LSE is updated in order (of increasing parameters). Specifically define $\hat{\boldsymbol{\theta}}_k$ as the LSE of order k (i.e., k parameters to be estimated). Then for a linear model we can derive the order-recursive LSE as:

$$\hat{\boldsymbol{\theta}}_{k+1} = \hat{\boldsymbol{\theta}}_k + U_k,$$

where U_k is the updating term for order k. The success of this approach depends on proper formulation of the linear models in order to facilitate the derivation of the recursive update. For example, if $\mathbf{H}$ has orthonormal column vectors, then the LSE is equivalent to projecting the observation $\mathbf{o}$ onto the space spanned by the orthonormal column vectors of $\mathbf{H}$. Since increasing the order implies increasing the dimensionality of the space by just adding another column to $\mathbf{H}$, this allows a recursive update relationship to be derived.

5.5.4 Sequential least squares

In most signal processing applications, the observation samples arrive as a stream of data. All our estimation methods discussed so far have assumed a *batch* or block mode of processing whereby we wait for N samples to arrive and then form our estimate based on these samples. One problem is the delay in waiting for the N samples before we produce our estimate, another problem is that as more data arrives we have to repeat the calculations on the larger blocks of data (N increases as more data arrives). The latter not only implies a growing computational burden but also the fact that we have to buffer all the data we have seen; both will grow linearly with the number of samples we have. Since in signal processing applications samples arise from sampling a continuous process, our computational and memory burden will grow linearly with time.

One solution to the above problem is to use a sequential mode of processing, where the parameter estimate for n samples, $\hat{\theta}[n]$, is derived from the previous parameter estimate for $n-1$ samples, $\hat{\theta}[n-1]$. For linear models we can represent sequential LSE as:

$$\hat{\theta}[n] = \hat{\theta}[n-1] + K[n](o[n] - s[n|n-1]),$$

where $s[n|n-1] \equiv s(n; \theta[n-1])$. The $K[n]$ is the correction gain and $(o[n] - s[n|n-1])$ is the prediction error. The magnitude of the correction gain $K[n]$ is usually directly related to the value of the estimator error variance, $\text{var}(\hat{\theta}[n-1])$, with a larger variance yielding a larger correction gain. This behavior is reasonable since a larger variance implies a poorly estimated parameter (which should have minimum variance) and a larger correction gain is expected. Thus one expects the variance to decrease with more samples and the estimated parameter to converge to the "true" value (or the LSE with an infinite number of samples).

Example: We consider the following linear model with a vector parameter:

$$\mathbf{s}[n] = \mathbf{H}[n]\boldsymbol{\theta}[n].$$

The very interesting example of sequential LS arises with the weighted LS error criterion with $\mathbf{W}[n] = \mathbf{C}^{-1}[n]$, where $\mathbf{C}[n]$ is the covariance matrix of the zero-mean noise, $\mathbf{w}[n]$, which is assumed to be uncorrelated. The argument $[n]$ implies that the vectors are based on n sample observations. We also consider:

$$\begin{aligned} \mathbf{C}[n] &= \text{diag}(\sigma_0^2, \sigma_1^2, \ldots, \sigma_n^2) \\ \mathbf{H}[n] &= \begin{bmatrix} \mathbf{H}[n-1] \\ \mathbf{h}^{\text{Tr}}[n] \end{bmatrix}, \end{aligned}$$

where $\mathbf{h}^{\mathrm{Tr}}[n]$ is the n^{th}row vector of the $n \times p$ matrix $\mathbf{H}[n]$. Note first that:

$$\mathbf{s}[n-1] = \mathbf{H}[n-1]\boldsymbol{\theta}[n-1].$$

We also have that $s[n|n-1] = \mathbf{h}^{\mathrm{Tr}}[n]\boldsymbol{\theta}[n-1]$. So the estimator update becomes:

$$\hat{\boldsymbol{\theta}}[n] = \hat{\boldsymbol{\theta}}[n-1] + \mathbf{K}[n](o[n] - \mathbf{h}^{\mathrm{Tr}}[n]\boldsymbol{\theta}[n-1]).$$

Let $\boldsymbol{\Sigma}[n] = \mathbf{C}_{\hat{\boldsymbol{\theta}}}[n]$ be the covariance matrix of $\hat{\boldsymbol{\theta}}$ based on n samples of data. It can be shown that the optimal correction gain is [Kay 93, Mend 95]

$$\mathbf{K}[n] = \frac{\boldsymbol{\Sigma}[n-1]\mathbf{h}[n]}{\sigma_n^2 + \mathbf{h}^{\mathrm{Tr}}[n]\boldsymbol{\Sigma}[n-1]\mathbf{h}[n]}.$$

Furthermore, it is also possible to derive an optimal **covariance update**:

$$\boldsymbol{\Sigma}[n] = (\mathbf{I} - \mathbf{K}[n]\mathbf{h}^{\mathrm{Tr}}[n])\boldsymbol{\Sigma}[n-1],$$

yielding a wholly recursive procedure requiring only knowledge of the observation data $o[n]$ and the initialization values: $\hat{\boldsymbol{\theta}}[-1]$ and $\boldsymbol{\Sigma}[-1]$, the initial estimate of the parameter and a initial estimate of the parameter covariance matrix.

5.5.5 Nonlinear least squares

So far we have assumed a linear signal model: $\mathbf{s}(\boldsymbol{\theta}) = \mathbf{H}\boldsymbol{\theta}$, where the notation for the signal model, $\mathbf{s}(\boldsymbol{\theta})$, explicitly shows its dependence on the parameter $\boldsymbol{\theta}$. In general the signal model will be an N-dimensional nonlinear function of the p-dimensional parameter $\boldsymbol{\theta}$. In such a case the minimization of:

$$J = (\mathbf{o} - \mathbf{s}(\boldsymbol{\theta}))^{\mathrm{Tr}}(\mathbf{o} - \mathbf{s}(\boldsymbol{\theta}))$$

becomes much more difficult. Differentiating with respect to $\boldsymbol{\theta}$ and setting to zero yields:

$$\frac{\partial J}{\partial \boldsymbol{\theta}} = \mathbf{0} \quad \Rightarrow \quad \frac{\partial \mathbf{s}(\boldsymbol{\theta})^{\mathrm{Tr}}}{\partial \boldsymbol{\theta}}(\mathbf{o} - \mathbf{s}(\boldsymbol{\theta})) = \mathbf{0},$$

which requires solution of N nonlinear simultaneous equations. This is a nonlinear optimization problem that generally requires numerical solutions, some of which were discussed in early Section 5.2.

For example, if the Newton-Raphson method is applied to the nonlinear LS problem, and we let the gradient of the LS error be denoted by

$$\mathbf{g}(\boldsymbol{\theta}) = \frac{\partial \mathbf{s}(\boldsymbol{\theta})^{\mathrm{Tr}}}{\partial \boldsymbol{\theta}}(\mathbf{o} - \mathbf{s}(\boldsymbol{\theta})),$$

then the LS problem becomes that of finding the zero of the nonlinear function $\mathbf{g}(\boldsymbol{\theta}) = \mathbf{0}$. Recall from Section 5.2.1 that the Newton-Raphson method linearizes the function about the initialized value $\boldsymbol{\theta}_k$ and directly solves for the zero of the function to produce the next estimate:

$$\boldsymbol{\theta}_{k+1} = \boldsymbol{\theta}_k - \left(\frac{\partial \mathbf{g}(\boldsymbol{\theta})}{\partial \boldsymbol{\theta}}\right)^{-1} \mathbf{g}(\boldsymbol{\theta})\Bigg|_{\boldsymbol{\theta}=\boldsymbol{\theta}_k}.$$

Note that the gradient $\left(\frac{\partial \mathbf{g}(\boldsymbol{\theta})}{\partial \boldsymbol{\theta}}\right)$ is the second order derivative of the nonlinear function in the signal model $\mathbf{s}(\boldsymbol{\theta})$. So if the function $\mathbf{s}(\boldsymbol{\theta})$ were linear, then the adjustment term would be zero. However, since the function is nonlinear, this will not be the case and the procedure is iterated until the estimates converge.

Another common technique used for solving the nonlinear LS problem is called the iterative LS algorithm [Mend 95] or Gauss-Newton method [Kay 93]. Here, we first linearize the signal model $\mathbf{s}(\boldsymbol{\theta})$ about the known (i.e., initial guess or current estimate) $\boldsymbol{\theta}_k$:

$$\mathbf{s}(\boldsymbol{\theta}) \approx \mathbf{s}(\boldsymbol{\theta}_k) + \left.\frac{\partial \mathbf{s}(\boldsymbol{\theta})}{\partial \boldsymbol{\theta}}\right|_{\boldsymbol{\theta}=\boldsymbol{\theta}_k} (\boldsymbol{\theta} - \boldsymbol{\theta}_k).$$

This approximate LS minimization problem will then be the LS problem for a linear model we discussed earlier, and will have the following familiar LS error function:

$$J = (\hat{\mathbf{o}}(\boldsymbol{\theta}_k) - \mathbf{H}(\boldsymbol{\theta}_k)\boldsymbol{\theta})^{\mathrm{Tr}}(\hat{\mathbf{o}}(\boldsymbol{\theta}_k) - \mathbf{H}(\boldsymbol{\theta}_k)\boldsymbol{\theta}),$$

where $\hat{\mathbf{o}}(\boldsymbol{\theta}_k) = \mathbf{o} - \mathbf{s}(\boldsymbol{\theta}_k) + \mathbf{H}(\boldsymbol{\theta}_k)\boldsymbol{\theta}_k$ is known and

$$\mathbf{H}(\boldsymbol{\theta}_k) = \left.\frac{\partial \mathbf{s}(\boldsymbol{\theta})}{\partial \boldsymbol{\theta}}\right|_{\boldsymbol{\theta}=\boldsymbol{\theta}_k}.$$

The standard LS solution for the linear model can then be applied directly to yield the following update of the LS estimate for the next iteration before convergence:

$$\boldsymbol{\theta}_{k+1} = \boldsymbol{\theta}_k + (\mathbf{H}^{\mathrm{Tr}}(\boldsymbol{\theta}_k)\mathbf{H}(\boldsymbol{\theta}_k))^{-1}\mathbf{H}^{\mathrm{Tr}}(\boldsymbol{\theta}_k)(\mathbf{o} - \mathbf{s}(\boldsymbol{\theta}_k)).$$

5.6 Maximum Likelihood Estimation

Maximum likelihood estimation (MLE) is one very commonly used technique for estimation of deterministic parameters in statistical signal processing, and in speech processing in particular. In addition to its asymptotic optimal properties, the MLE technique is especially powerful in handling the complex situations when the data are partially, rather than fully, observable. In this section, we will discuss the simplest case first for fully observed data, followed by a discussion of the case for partially observed data (i.e., the complete data contain hidden variables). The latter case is most relevant to speech processing because the acoustic signal that is typically observed contains many hidden components not directly observable. This naturally accounts for the popularity of the MLE technique in speech processing.

5.6.1 Basic MLE procedure for fully observed data

When the form of PDF is known for the observed data, then estimation methods making use of the PDF in general can outperform the LS method that does not make use of the PDF information. Furthermore, in many cases when the PDF is known, the theoretically optimal MVU estimator unfortunately may not exist or it cannot be found by any of the methods discussed so far. The MLE approach is an alternative method in cases where the PDF is known. With the MLE, the unknown parameters are estimated by maximizing the PDF. That is, define $\hat{\theta}$ such that:

$$\hat{\theta} = \arg\max_{\theta} p(\mathbf{o}; \theta),$$

where $\mathbf{o}$ is the vector of (fully) observed data (of N samples). It can be shown that $\hat{\theta}$ is asymptotically unbiased:

$$\lim_{N\to\infty} E(\hat{\theta}) = \theta$$

and asymptotically efficient (i.e., the Cramer-Rao Lower Bound):

$$\lim_{N\to\infty} \mathrm{var}(\hat{\theta}) = \mathrm{CRLB}.$$

One important result in estimation theory concerning the optimality of the MLE is that if an MVU estimator exists, then the MLE procedure will produce it. With an infinite amount of data, or $N \to \infty$, the CRLB typically approaches zero. Therefore, if the observations are produced from a true parametric model that is characterized by a set of deterministic parameters, then using a large enough data set so that the CRLB becomes close to zero, then the ML estimate (as a random vector) for the model parameters will have a mean that is equal to the true model parameters while having near-zero variance. That is, the ML estimate will be almost deterministic, and the value will be almost the same as the true model parameters. The implication of this key property of the MLE is that if the model is reasonably accurate, then the MLE results will be highly meaningful in terms of the physical model parameters. Otherwise, the MLE may give irrelevant results.

From the procedure point of view, one practical advantage of the MLE is that unlike the previous estimates, the MLE does not require an explicit, analytical expression for $p(x;\theta)$. Given a histogram plot of the PDF as a function of θ, one can numerically search for the θ that maximizes the PDF.

Example: Consider the signal embedded in noise:

$$o[n] = A + w[n],$$

where $w[n]$ is WGN with zero mean but unknown variance which is also A; that is, the unknown parameter, $\theta = A$, manifests itself both as the unknown signal and the variance of the noise. Although this is a highly unlikely scenario, this simple example demonstrates the power of the MLE approach since finding the MVU estimator by the previous procedures would not be easy. Consider the PDF:

$$p(\mathbf{o};\theta) = \frac{1}{(2\pi\theta)^{\frac{N}{2}}} \exp\left[-\frac{1}{2\theta}\sum_{n=0}^{N-1}(o[n]-\theta)^2\right]$$

as a function of θ; thus it is a likelihood function and we need to maximize it with respect to θ. For a Gaussian PDF, it is easier to find the equivalent maximum of the log-likelihood function:

$$\log p(\mathbf{o};\theta) = \log\left(\frac{1}{(2\pi\theta)^{\frac{N}{2}}}\right) - \frac{1}{2\theta}\sum_{n=0}^{N-1}(o[n]-\theta)^2.$$

According to the classical optimization technique, by partial differentiation, we obtain:

$$\frac{\partial \log p(\mathbf{o};\theta)}{\partial\theta} = -\frac{N}{2\theta} + \frac{1}{\theta}\sum_{n=0}^{N-1}(o[n]-\theta) + \frac{1}{2\theta^2}\sum_{n=0}^{N-1}(o[n]-\theta)^2.$$

Then, setting the derivative to zero and solving for θ produces the MLE estimate:

$$\hat{\theta} = -\frac{1}{2} + \sqrt{\frac{1}{N}\sum_{n=0}^{N-1} x^2[n] + \frac{1}{4}},$$

where we have assumed $\theta > 0$. It can be shown that:

$$\lim_{N\to\infty} E(\hat{\theta}) = \theta$$

and

$$\lim_{N\to\infty} \text{var}(\hat{\theta}) = \text{CRLB} = \frac{\theta^2}{N(\theta + \frac{1}{2})}.$$

Example: Consider a signal embedded in noise:

$$o[n] = A + w[n],$$

where $w[n]$ is WGN with zero mean and known variance σ^2. We know the MVU estimator for θ is the sample mean. To see that this is also the MLE, we consider the PDF:

$$p(\mathbf{o};\theta) = \frac{1}{(2\pi\sigma^2)^{\frac{N}{2}}} \exp\left[-\frac{1}{2\sigma^2}\sum_{n=0}^{N-1}(o[n]-\theta)^2\right]$$

and maximize the log-likelihood function be setting it to zero:

$$\frac{\partial \log p(\mathbf{o};\theta)}{\partial\theta} = -\frac{1}{\sigma^2}\sum_{n=0}^{N-1}(o[n]-\theta) = 0 \quad \Rightarrow \quad \sum_{n=0}^{N-1} o[n] - N\theta = 0.$$

Thus $\hat{\theta} = \frac{1}{N}\sum_{n=0}^{N-1} o[n]$, which is the sample-mean, the same as the MVU estimator.

Example: Suppose the signals $o[n], n = 1, \cdots, N$ are independently drawn from the inverse Gaussian distribution (see Chapter 3):

$$p(x|\lambda,\mu) = \sqrt{\frac{\lambda}{2\pi}} x^{-\frac{3}{2}} \exp\left\{-\frac{\lambda(x-\mu)^2}{2\mu^2 x}\right\}.$$

The logarithm of this PDF is:

$$\log p(x|\lambda,\mu) = \frac{1}{2}\log\frac{\lambda}{2\pi} - \frac{3}{2}\log x - \frac{\lambda(x-\mu)^2}{2\mu^2 x}.$$

Since $o[1], o[2], ..., o[N]$ are independent of each other, the joint likelihood function for them is:

$$l(\lambda,\mu) = p(o[1]|\lambda,\mu)p(o[2]|\lambda,\mu)...p(o[N]|\lambda,\mu),$$

and its logarithm is:

$$\begin{aligned} L(\lambda,\mu) &= \sum_{i=1}^{N} \log p(x_i|\lambda,\mu) = \sum_{i=1}^{N}\left[\frac{1}{2}\log\frac{\lambda}{2\pi} - \frac{3}{2}\log x_i - \frac{\lambda(x_i-\mu)^2}{2\mu^2 x_i}\right] \\ &= \frac{N}{2}\log\frac{\lambda}{2\pi} - \frac{3}{2}\sum_{i=1}^{N}\log x_i - \frac{\lambda}{2\mu^2}\sum_{i=1}^{N}\frac{(x_i-\mu)^2}{x_i}. \end{aligned} \tag{5.21}$$

To find the ML estimate for λ and μ, we differentiate 5.21 with respect to λ and μ, respectively, which gives:

$$\begin{aligned} \frac{\partial L}{\partial \lambda} &= \frac{N}{2}\cdot\frac{2\pi}{\lambda}\cdot\frac{1}{2\pi} - 0 - \frac{1}{2\mu^2}\sum_{i=1}^{N}\frac{(x_i-\mu)^2}{x_i} \\ \frac{\partial L}{\partial \mu} &= 0 - 0 - \frac{\partial}{\partial\mu}\left[\frac{\lambda}{2\mu^2}\sum_{i-1}^{N}\frac{(x_i-\mu)^2}{x_i}\right] \\ &= -\frac{\partial}{\partial\mu}\left(\frac{\lambda}{2\mu^2}\right)\sum_{i=1}^{N}\frac{(x_i-\mu)^2}{x_i} - \frac{\lambda}{2\mu^2}\cdot\frac{\partial}{\partial\mu}\left[\sum_{i=1}^{N}\frac{(x_i-\mu)^2}{x_i}\right] \\ &= \frac{\lambda}{\mu^3}\sum_{i=1}^{N}\frac{(x_i-\mu)^2}{x_i} - \frac{\lambda}{2\mu^2}\sum_{i=1}^{N}\frac{\partial}{\partial\mu}\frac{(x_i-\mu)^2}{x_i} \\ &= \frac{\lambda}{\mu^3}\sum_{i=1}^{N}\frac{(x_i-\mu)^2}{x_i} - \frac{\lambda}{2\mu^2}\sum_{i=1}^{N}\frac{-2(x_i-\mu)}{x_i} \\ &= \frac{\lambda}{\mu^3}\sum_{i=1}^{N}\frac{(x_i-\mu)^2}{x_i} + \frac{\lambda}{\mu^2}\sum_{i=1}^{N}\frac{(x_i-\mu)}{x_i} \\ &= \frac{\lambda}{\mu^3}\left\{\sum_{i=1}^{N}\frac{(x_i-\mu)^2}{x_i} + \mu\sum_{i=1}^{N}\frac{(x_i-\mu)}{x_i}\right\} \\ &= \frac{\lambda}{\mu^3}\sum_{i=1}^{N}(x_i-\mu). \end{aligned} \tag{5.22}$$

Equating these partial derivatives to zero:

$$\frac{N}{2\hat{\lambda}} - \frac{1}{2\hat{\mu}_{ML}^2}\sum_{i=1}^{N}\frac{(x_i-\hat{\mu}_{ML})^2}{x_i} = 0 \tag{5.23}$$

$$\frac{\hat{\lambda}_{ML}}{\hat{\mu}_{ML}^3}\left\{\sum_{i=1}^{N}(x_i-\hat{\mu}_{ML})\right\} = 0. \tag{5.24}$$

Eq. 5.24 gives

$$\hat{\mu}_{ML} = \frac{1}{N}\sum_{i-1}^{N} x_i = \bar{x},$$

and Eq. 5.23 gives

$$\hat{\lambda}_{ML} = \frac{N\bar{x}^2}{\sum_{i=1}^{N}\frac{(x_i-\bar{x})}{x_i}}.$$

We now discuss another interesting and commonly used property of the MLE when it is applied to the transformed parameters $\alpha = g(\theta)$. It can be shown that the MLE in this case is given by

$$\hat{\alpha} = g(\hat{\theta}),$$

where $\hat{\theta}$ is the MLE of θ. If $g(.)$ is not a one-to-one function (i.e., not invertible), then $\hat{\alpha}$ is obtained as the MLE of the transformed likelihood function, $p_1(\mathbf{o}; \alpha)$, which is defined as

$$p_1(\mathbf{o}; \alpha) = \max_{\{\theta:\alpha=g(\theta)\}} p(\mathbf{o}; \theta).$$

5.6.2 MLE for the linear model

Consider the general linear model of the form:

$$\mathbf{o} = \mathbf{H}\boldsymbol{\theta} + \mathbf{w},$$

where $\mathbf{H}$ is a known $N \times p$ matrix, $\mathbf{o}$ is the $N \times 1$ observation vector with N samples, and $\mathbf{w}$ is a noise vector of dimension $N \times 1$ with PDF $\mathcal{N}(\mathbf{0}, \mathbf{C})$. The PDF for the observation data is:

$$p(\mathbf{o}; \boldsymbol{\theta}) = \frac{1}{(2\pi)^{\frac{N}{2}} |\mathbf{C}|^{1/2}} \exp\left[-\frac{1}{2}(\mathbf{o} - \mathbf{H}\boldsymbol{\theta})^{\mathrm{Tr}} \mathbf{C}^{-1} (\mathbf{o} - \mathbf{H}\boldsymbol{\theta})\right]$$

and the MLE of $\boldsymbol{\theta}$ is found by first differentiating the log-likelihood (classical optimization technique):

$$\frac{\partial \log p(\mathbf{o}; \boldsymbol{\theta})}{\partial \boldsymbol{\theta}} = \frac{\partial (\mathbf{H}\boldsymbol{\theta})^{\mathrm{Tr}}}{\partial \boldsymbol{\theta}} \mathbf{C}^{-1} (\mathbf{o} - \mathbf{H}\boldsymbol{\theta}).$$

Then, removing optimization-independent terms and setting the results to zero, we obtain

$$\mathbf{H}^{\mathrm{Tr}} \mathbf{C}^{-1} (\mathbf{o} - \mathbf{H}\boldsymbol{\theta}) = \mathbf{0}.$$

This gives the MLE of $\boldsymbol{\theta}$:

$$\hat{\boldsymbol{\theta}} = (\mathbf{H}^{\mathrm{Tr}} \mathbf{C}^{-1} \mathbf{H})^{-1} \mathbf{H}^{\mathrm{Tr}} \mathbf{C}^{-1} \mathbf{o},$$

which is the same as the MVU estimator.

5.6.3 EM algorithm — Introduction

We now move on to discuss the more complex situation for MLE when only partially observed or incomplete data are available. We will introduce a powerful technique, called the expectation-maximization algorithm, or **EM algorithm** for short, that is a very natural solution to such an ML problem.

Partially observed (o), hidden (h), and complete (y) data

To set up a context for discussion, we consider the use of the MLE procedure to find an estimate for the unknown parameter $\boldsymbol{\theta}$ which requires maximization of the log-likelihood function, $\log p(\mathbf{o}; \boldsymbol{\theta})$. However, we may find that this is either too difficult or there are difficulties in finding an expression for the PDF itself. In such circumstances, where direct expression and maximization of the PDF or log likelihood in terms of the observed data $\mathbf{o}$ is difficult or intractable, an iterative solution is possible if another data set, $\mathbf{y}$, can be found such that the PDF in terms of the $\mathbf{y}$ set is much easier to express in closed form and maximize. We term the data set $\mathbf{y}$ the *complete* data and the original data $\mathbf{o}$ the *incomplete* or partially observed data. In general we can find a mapping from the complete to the incomplete data:

$$\mathbf{o} = \mathbf{g}(\mathbf{y}).$$

However this is usually a many-to-one transformation in that a subset of the *complete* data will map to the same *incomplete* data (e.g., the incomplete data may represent an accumulation or sum of the complete data). This explains the terminology: $\mathbf{o}$ is *incomplete* (or is "missing" something) relative to the data, $\mathbf{y}$, which is complete (for performing the MLE procedure). This is usually not evident, however, until one is able to define what the complete data set is. Unfortunately defining what constitutes the complete data is usually an arbitrary procedure that is highly problem specific and often requires some ingenuity on the part of the algorithm designer.

To summarize the above discussion that is intended to motivate the EM algorithm, we wish to overcome the computational difficulty of direct optimization of the PDF on the (partially) observed or "incomplete" data $\mathbf{o}$. To accomplish this, we supplement the available data $\mathbf{o}$ with "imaginary" missing, unobserved, or hidden data, which we denote as $\mathbf{h}$, to form the "complete" data $\mathbf{y} = \{\mathbf{o}, \mathbf{h}\}$. The hope is that with ingenious choice of the hidden data $\mathbf{h}$, we can work on the complete data $\mathbf{y}$ rather than on the original data $\mathbf{o}$ to make the optimization easier for the log likelihood of $\mathbf{o}$. We now provide a simple example to make such a motivation concrete.

Example: Consider spectral analysis where a known signal, $o[n]$, is composed of an unknown summation of harmonic components embedded in noise:

$$o[n] = \sum_{i=1}^{p} \cos 2\pi f_i n + w[n] \qquad n = 0, 1, \ldots, N-1,$$

where $w[n]$ is WGN with known variance σ^2 and the unknown parameter vector to be estimated is the group of frequencies: $\boldsymbol{\theta} = \mathbf{f} = [f_1\, f_2\, \ldots\, f_p]^{\mathrm{Tr}}$. The standard MLE would require maximization of the log-likelihood of a multi-variate Gaussian distribution, which is equivalent to minimizing the argument of the exponential:

$$J(\mathbf{f}) = \sum_{n=0}^{N-1} \left(o[n] - \sum_{i=1}^{p} \cos 2\pi f_i n \right)^2,$$

which is a difficult, nonlinear, p-dimensional minimization problem. On the other hand if we had access to the individual harmonic signal embedded in noise:

$$y_i[n] = \cos 2\pi f_i n + w_i[n] \qquad \begin{array}{c} i = 1, 2, \ldots, p \\ n = 0, 1, \ldots, N-1, \end{array}$$

where $w_i[n]$ is WGN with known variance σ_i^2, then the MLE procedure would result in minimization of:

$$J(f_i) = \sum_{n=0}^{N-1} (y_i[n] - \cos 2\pi f_i n)^2 \qquad i = 1, 2, \ldots, p,$$

which are p independent one-dimensional minimization problems, and are much easier to solve. Thus $\mathbf{y}$ is the complete data set that we are looking for to facilitate the MLE procedure, and the "hidden" or missing data $\mathbf{h}$ is the information about what harmonic component i is associated with the individual harmonic signal $\mathbf{y}_i[n]$. However, we do not have access to $\mathbf{y}$. The relationship to the known data $\mathbf{o}$ is:

$$\mathbf{o} = \mathbf{g}(\mathbf{y}) \quad \Rightarrow \quad o[n] = \sum_{i=1}^{p} y_i[n] \qquad n = 0, 1, \ldots, N-1$$

and we further assume that:

$$\begin{aligned} w[n] &= \sum_{i=1}^{p} w_i[n] \\ \sigma^2 &= \sum_{i=1}^{p} \sigma_i^2. \end{aligned}$$

However since the mapping from $\mathbf{y}$ to $\mathbf{o}$ is many-to-one we cannot directly form an expression for the PDF $p(\mathbf{y}; \boldsymbol{\theta})$ in terms of the known $\mathbf{o}$ (since we cannot do the obvious substitution of $\mathbf{y} = \mathbf{g}^{-1}(\mathbf{o})$).

The EM algorithm

Once we have identified the complete data set $\mathbf{y}$, even though an expression for $\log p(\mathbf{y}; \boldsymbol{\theta})$ can now easily be derived, we cannot directly maximize $\log p(\mathbf{y}; \boldsymbol{\theta})$ with respect to $\boldsymbol{\theta}$ since $\mathbf{y}$ is unavailable. However, we observed $\mathbf{o}$ and if we further assume that we have a good "guess" estimate for $\boldsymbol{\theta}$ then we can consider the expected value of $\log p(\mathbf{y}; \boldsymbol{\theta})$ conditioned on what we have observed, or the following conditional expectation, denoted by Q:

$$Q(\boldsymbol{\theta}) = E_{h|o}[\log p(\mathbf{y}; \boldsymbol{\theta}) | \mathbf{o}; \boldsymbol{\theta}] = E_{h|o}[\log p(\mathbf{o}, \mathbf{h}; \boldsymbol{\theta}) | \mathbf{o}; \boldsymbol{\theta}], \tag{5.25}$$

and we attempt to maximize this expectation to yield, not the MLE, but the next "best-guess" estimate for $\boldsymbol{\theta}$.

Using Eq. 5.25 for computing the conditional expectation when hidden vector $\mathbf{h}$ is continuous, we have

$$Q(\boldsymbol{\theta}) = \int \log p(\mathbf{y}; \boldsymbol{\theta}) p(\mathbf{h}|\mathbf{o}; \boldsymbol{\theta}) d\mathbf{h}.$$

When the hidden vector $\mathbf{h}$ is discrete (i.e., taking only discrete values as the vector components), then Eq. 5.25 is used to evaluate the conditional expectation:

$$Q(\boldsymbol{\theta}) = \sum_{\mathbf{h}} \log p(\mathbf{y}; \boldsymbol{\theta}) P(\mathbf{h}|\mathbf{o}; \boldsymbol{\theta}),$$

where $P(\mathbf{h}|\mathbf{o}; \boldsymbol{\theta})$ is a conditional PMF, and the summation is over all possible discrete-value vectors that $\mathbf{h}$ may take.

What is needed now is to iterate through both an **E-step** (find an appropriate expression for the conditional expectation and sufficient statistics for its computation) and **M-step** (maximization of the conditional expectation), hence the name EM algorithm. Specifically, after an initialization of the parameters as $\boldsymbol{\theta}_0$, the EM algorithm iterates the following **E-step** and **M-step** until convergence:

E-step and M-step in the EM algorithm:

Expectation (E): Determine the conditional expectation of the log-likelihood of the complete data given the partially observed data and current (k^{th}) estimate of the parameter vector

$$Q(\boldsymbol{\theta}, \boldsymbol{\theta}_k) = \int \log p(\mathbf{y}; \boldsymbol{\theta}) p(\mathbf{h}|\mathbf{o}; \boldsymbol{\theta}_k) d\mathbf{h},$$

for continuous hidden variables $\mathbf{h}$, or

$$Q(\boldsymbol{\theta}, \boldsymbol{\theta}_k) = \sum_{\mathbf{h}} \log p(\mathbf{y}; \boldsymbol{\theta}) P(\mathbf{h}|\mathbf{o}; \boldsymbol{\theta}_k),$$

for discrete hidden variables $\mathbf{h}$.

Maximization (M): Maximize the conditional expectation to obtain the next estimate of the parameter vector:

$$\boldsymbol{\theta}_{k+1} = \arg\max_{\boldsymbol{\theta}} Q(\boldsymbol{\theta}, \boldsymbol{\theta}_k).$$

Convergence of the EM algorithm is guaranteed (under mild conditions) in the sense that the average log-likelihood of the complete data does not decrease at each iteration, that is:

$$Q(\boldsymbol{\theta}, \boldsymbol{\theta}_{k+1}) \geq Q(\boldsymbol{\theta}, \boldsymbol{\theta}_k)$$

with equality when $\boldsymbol{\theta}_k$ is the MLE. For detailed discussions and proof of the convergence of the EM algorithm, see [Demp 77, Wu 83].

Four main attributes of the EM algorithm are:

1. The EM algorithm gives only a local, rather than the global, optimum in the likelihood of partially observed data $\mathbf{o}$.

2. An initial value for the unknown parameter is needed, and as with most iterative procedures a good initial estimate is required for desirable convergence and a good ML estimate.

3. The selection of the complete data set is arbitrary.

4. Although $\log p(\mathbf{y}; \boldsymbol{\theta})$ can usually be easily expressed in closed form, finding the closed-form expression for the expectation is usually harder.

Example: Applying the EM algorithm to the previous example, we have the following EM-algorithm for estimating the individual harmonic frequencies.

E-step We start with finding an expression for $\log p(\mathbf{y}; \boldsymbol{\theta})$ in terms of the complete data:

$$\begin{aligned} \log p(\mathbf{y}; \boldsymbol{\theta}) &\approx K(\mathbf{y}) + \sum_{i=1}^{p} \frac{1}{\sigma_i^2} \sum_{n=0}^{N-1} y_i[n] \cos 2\pi f_i n \\ &\approx K(\mathbf{y}) + \sum_{i=1}^{p} \frac{1}{\sigma_i^2} \mathbf{c}_i^{\mathrm{Tr}} \mathbf{y}_i, \end{aligned}$$

where the terms in $K(\mathbf{y})$ do not depend on $\boldsymbol{\theta}$, $\mathbf{c}_i = [1, \cos 2\pi f_i, \cos 2\pi f_i(2), \ldots, \cos 2\pi f_i(N-1)]^{\mathrm{Tr}}$ and $\mathbf{y}_i = [y_i[0], y_i[1], y_i[2], \ldots, y_i[N-1]]^{\mathrm{Tr}}$. We write the conditional expectation as:

$$\begin{aligned} Q(\boldsymbol{\theta}, \boldsymbol{\theta}_k) &= E[\log p(\mathbf{y}; \boldsymbol{\theta}) | \mathbf{o}; \boldsymbol{\theta}_k] \\ &= E(K(\mathbf{y}) | \mathbf{o}; \boldsymbol{\theta}_k) + \sum_{i=1}^{p} \frac{1}{\sigma_i^2} \mathbf{c}_i^{\mathrm{Tr}} E(\mathbf{y}_i | \mathbf{o}; \boldsymbol{\theta}_k). \end{aligned}$$

Since we wish to maximize $Q(\boldsymbol{\theta}, \boldsymbol{\theta}_k)$ with respect to $\boldsymbol{\theta}$, this is equivalent to maximizing:

$$Q'(\boldsymbol{\theta}, \boldsymbol{\theta}_k) = \sum_{i=1}^{p} \mathbf{c}_i^{\mathrm{Tr}} E(\mathbf{y}_i | \mathbf{o}; \boldsymbol{\theta}_k).$$

We note that $E(\mathbf{y}_i | \mathbf{o}; \boldsymbol{\theta}_k)$ can be thought of as an estimate of the $y_i[n]$ data set given the observed data set $o[n]$ and current estimate $\boldsymbol{\theta}_k$. Since $\mathbf{y}$ is Gaussian then $\mathbf{o}$ is a sum of Gaussians and thus $\mathbf{o}$ and $\mathbf{y}$ are jointly Gaussian and one of the standard results is:

$$E(\mathbf{y} | \mathbf{o}; \boldsymbol{\theta}_k) = E(\mathbf{y}) + \mathbf{C}_{yx} \mathbf{C}_{xx}^{-1} (\mathbf{o} - E(\mathbf{o}))$$

and application of this yields:

$$\hat{\mathbf{y}}_i = E(\mathbf{y}_i | \mathbf{o}; \boldsymbol{\theta}_k) = \mathbf{c}_i + \frac{\sigma_i^2}{\sigma^2} (\mathbf{o} - \sum_{i=1}^{p} \mathbf{c}_i)$$

and

$$\hat{y}_i[n] = \cos 2\pi f_{i_k} n + \frac{\sigma_i^2}{\sigma^2} \left(o[n] - \sum_{i=1}^{p} \cos 2\pi f_{i_k} n \right).$$

Thus:

$$Q'(\boldsymbol{\theta}, \boldsymbol{\theta}_k) = \sum_{i=1}^{p} \mathbf{c}_i^{\mathrm{Tr}} \hat{\mathbf{y}}_i.$$

M-step Maximization of $Q'(\boldsymbol{\theta}, \boldsymbol{\theta}_k)$ consists of maximizing each term in the sum separately or:

$$f_{i_{k+1}} = \arg\max_{f_i} \mathbf{c}_i^{\mathrm{Tr}} \hat{\mathbf{y}}_i.$$

Furthermore, since we assumed $\sigma^2 = \sum_{i=1}^{p} \sigma_i^2$, we still have the problem that we don't know what the σ_i^2 are. However, as long as:

$$\sigma^2 = \sum_{i=1}^{p} \sigma_i^2 \quad \Rightarrow \quad \sum_{i=1}^{p} \frac{\sigma_i^2}{\sigma^2} = 1,$$

then we can choose these values arbitrarily.

5.6.4 EM algorithm example — Markov modulated Poisson process

Here we provide an example from [Deng 93d] to illustrate the application of the EM algorithm to the particular random process called the **Markov modulated Poisson process (MMPP)**. In Chapter 3 when we discussed the Poisson distribution, we mentioned a variable-rate Poisson distribution over the observation time interval and how such a distribution may arise in practical situations. The MMPP is a model for the variable-rate Poisson process, where the temporal change of the Poisson rate is governed by a Markov chain.

Characterization of MMPP

Mathematically, an MMPP consists of an underlying hidden N-state continuous-time Markov chain characterized by the generator

$$P = \begin{bmatrix} -\sigma_{11} & \sigma_{12} & \cdots & \cdots & \sigma_{1N} \\ \sigma_{21} & -\sigma_{22} & \cdots & \cdots & \sigma_{2N} \\ \vdots & \vdots & \vdots & \vdots & \vdots \\ \vdots & \vdots & \vdots & \vdots & \vdots \\ \sigma_{N1} & \sigma_{N2} & \cdots & \cdots & -\sigma_{NN} \end{bmatrix}$$

($\sigma_{ii} = \sum_{j=1,j\neq i}^{N} \sigma_{ij}$), and associated with each state i ($i = 1, 2, ..., N$) of the Markov chain is a Poisson process with the state-conditioned Poisson rate λ_i.

The parameter estimation problem for the MMPP involves estimating P and the $\lambda_i (i = 1, 2, ..., N)$ from a sequence of interarrival interval data from the MMPP process without observing the underlying Markov chain. A direct application of the EM algorithm for solving this problem is very difficult due to the difficulty of expressing the conditional expectation of the log joint likelihood for the observation data and state sequence (required by the E-step) in a simplified form suitable for maximization required by the M-step. To circumvent this difficulty, we use time discretization, to be discussed next, to convert the MMPP to a corresponding discrete-time Markov modulated Bernoulli process (MMBP) for which the efficient EM algorithm becomes immediately applicable. Note that the MMBP is a simple HMM (discussed in Chapter 3), where the output PMF follows a Bernoulli distribution.

Time discretization

Time discretization of the inter-event intervals at a fixed time step h is performed by assigning the value zero to the time slot (length h), where no event occurs and assigning value one to the slot where an event occurs. (h is chosen sufficiently small so that no more than one event occurs in one time slot.) After the time discretization, the original data sequence consisting of interarrival intervals of events is converted to a new binary data sequence, $O_1^T = \{O_1, O_2, ..., O_T\}$, signaling whether or not there is an event within each time slot.

Correspondingly, the parameters of an MMPP can be straightforwardly converted to those of an MMBP. For the underlying Markov chain, we apply the standard technique to reduce the continuous-time Markov chain in the MMPP to the discrete-time Markov chain in the MMBP. That is, use the approximation

$$\sigma_{ij} \approx \frac{a_{ij}}{h}, \quad \text{for} \quad i, j = 1, 2, ..., N; \ i \neq j, \tag{5.26}$$

where a_{ij} is the transition probability from state i to j in the discrete-time Markov chain of MMBP produced from time discretization of the MMPP. For $i = j$, set $a_{ii} = 1 - \sum_{j=1,j\neq i}^{N} a_{ij}$. For the observation process, the state (i)-dependent Poisson rate in the

MMPP can be approximated by

$$\lambda_i \approx \frac{b_i(1)}{h}, \quad \text{for} \quad i = 1, 2, ..., N, \tag{5.27}$$

where $b_i(1)$ is the probability of having one event occur within slot h conditioned on state i in the MMBP. The probability of no event within h is $b_i(0) = 1 - b_i(1)$. The approximation in Eq. 5.27 is a very natural one since λ_i denotes the rate of Poisson events, or the number of events per second.

EM algorithm for MMBP

After time discretization, we are now in the position to apply the EM algorithm for parameter estimation. We have at our disposal a sequence of binary observation data O_1^T, from which estimates of the MMBP parameters, $a_{ij}, i, j = 1, 2, ..., N$ and $b_i(1), i = 1, 2, ..., N$, are sought.

As discussed earlier, each iteration in the EM algorithm consists of two steps. In the E (expectation) step, the following conditional expectation, or the auxiliary function $Q(\boldsymbol{\theta}|\boldsymbol{\theta}_0)$ is sought :

$$Q(\boldsymbol{\theta}|\boldsymbol{\theta}_0) = E[\log P(O_1^T, \mathcal{S}|\boldsymbol{\theta})|O_1^T, \boldsymbol{\theta}_0], \tag{5.28}$$

where the expectation is taken over the "hidden" Markov state sequence $\mathbf{h} = \mathcal{S}$ conditioned on the observations, $P(O_1^T, \mathcal{S})$ is the joint probability density function, $\boldsymbol{\theta}$ stands for the model whose parameters are to be estimated in the present iteration, and $\boldsymbol{\theta}_0$ stands for the model whose parameters had been estimated in the previous iteration. For the MMPP, parameter set $\boldsymbol{\theta}$ entails $a_{ij}'s$ and $b_j(1)'s, i, j = 1, 2, ..., N$. For the EM algorithm to be of utility, $Q(\boldsymbol{\theta}|\boldsymbol{\theta}_0)$ has to be sufficiently simplified so that the ensuing M (maximization) step can be carried out easily. Estimates of the model parameters are obtained in the M step via maximization of $Q(\boldsymbol{\theta}|\boldsymbol{\theta}_0)$.

For the MMBP, the E-step simplification is particularly easy to carry out. By using the IID property for the state-conditioned Bernoulli process, the auxiliary function in (5.28) can be simplified through the following steps:

$$\begin{aligned}
Q(\boldsymbol{\theta}|\boldsymbol{\theta}_0) &= \sum_{\mathcal{S}} P(\mathcal{S}|O_1^T, \boldsymbol{\theta}_0) \log P(O_1^T, \mathcal{S}|\boldsymbol{\theta}) \\
&= \sum_{\mathcal{S}} P(\mathcal{S}|O_1^T, \boldsymbol{\theta}_0) \sum_{t=1}^{T} \log[b_{s_t}(1)^{O_t} b_{s_t}(0)^{1-O_t}] + \sum_{\mathcal{S}} P(\mathcal{S}|O_1^T, \boldsymbol{\theta}_0) \sum_{t=1}^{T-1} \log a_{s_t s_{t+1}} \\
&\cdots \\
&\cdots \\
&= \sum_{i=1}^{N} \sum_{j=1}^{N} \sum_{t=1}^{T-1} \xi_t(i,j) \log a_{ij} + \sum_{i=1}^{N} \sum_{t=1}^{T} \gamma_t(t) \{O_t \log b_i(1) + (1 - O_t) \log[1 - b_i(1)]\},
\end{aligned} \tag{5.29}$$

where $\xi_t(i,j) = P(s_t = i, s_{t+1} = j|O_1^T, \boldsymbol{\theta}_0)$ and $\gamma_t(i) = P(s_t = i|O_1^T, \boldsymbol{\theta}_0)$ are two posterior probabilities, which can be computed efficiently using the forward-backward algorithm discussed in Chapter 3.

The M-step is then carried out to obtain parameter estimates by maximizing the expression in (5.29). The estimation formula for the transition probabilities of the Markov chain of the MMBP is established by setting

$$\frac{\partial Q(\boldsymbol{\theta}|\boldsymbol{\theta}_0)}{\partial a_{ij}} = 0, \quad i, j = 1, 2, ..., N,$$

subject to the constraint $\sum_{j=1}^{N} a_{ij} = 1$. Use of the standard Lagrange multiplier procedure produces the re-estimation formula

$$\hat{a}_{ij} = \frac{\sum_{t=1}^{T-1} \xi_t(i,j)}{\sum_{t=1}^{T-1} \gamma_t(i)}. \tag{5.30}$$

Setting

$$\frac{\partial Q(\boldsymbol{\theta}|\boldsymbol{\theta}_0)}{\partial b_i(1)} = 0, \quad i = 1, 2, ..., N,$$

on the other hand, leads to the re-estimation formula for the probability of state (i)-conditioned IID Bernoulli event:

$$\hat{b}_i(1) = \frac{\sum_{t=1}^{T} \gamma_t(i) O_t}{\sum_{t=1}^{T} \gamma_t(i)}. \tag{5.31}$$

Estimation procedure for MMPP

We now return to the original problem: given the interarrival interval data (with no MMPP state labels attached), how to estimate the MMPP parameters? Equipped with the results in the last two sections, we provide the answer to this problem in the following procedural form:

- Step 1: Discretize the data temporally into a binary sequence O_1^T.
- Step 2: Obtain initial crude MMPP model parameters and use Eqs. 5.26 and 5.27 to convert them into initial MMBP parameters.
- Step 3: Iteratively improve the MMBP parameters by repetitively using formulas 5.30 and 5.31 until convergence is reached.
- Step 4: Use Eqs. 5.26 and 5.27 to convert these optimal MMBP parameters back to the corresponding MMPP parameters.

This estimation procedure based on the EM-algorithm has demonstrated very high accuracy in a comprehensive set of simulation experiments [Deng 93d].

Some useful references for the EM algorithm, including many examples, can be found in [Demp 77, McL 97, Wu 83] and in the reference lists therein. The EM algorithm has been extensively used in speech processing. In Chapter 10 on computational models for acoustic patterns of speech and for speech production, we will provide several classic examples for the use of the EM algorithm in speech modeling. The examples will include some models we discussed in Chapter 4 and their extensions, and include both discrete and continuous hidden variables. Examples of the models with discrete hidden variables are mixture of Gaussians, HMMs, and trended HMMs. Examples of the models with continuous hidden variables are linear and nonlinear dynamic system models.

5.7 Estimation of Random Parameters

The several estimators we have discussed so far all treated the parameters to be estimated as deterministic rather than random variables. The estimation procedures taken so far have often been called the *classic approach*. Hence, the optimal estimator $\hat{\theta}$ is optimal irrespective of and independently of the actual value of θ. However, in many cases where the actual value or prior knowledge of θ could be a factor — for example, an MVU estimator does not exist for certain values or where prior knowledge would improve the estimation — the classic approach would not work effectively.

In the Bayesian philosophy, as opposed to the classic approach, the parameter θ is treated as a random variable. Also, it is generally assumed that the distribution for the parameter is known as a *prior PDF*, $p(\theta)$, for the continuous variable θ, or *prior PMF*, $P(\theta)$, for the discrete variable θ. In such a case, we have a strict Bayesian framework. Otherwise, when the prior PDF or PMF is unknown and needs estimation for its hyper-parameter(s), we have an empirical Bayesian framework. The prior knowledge concerning the distribution of the model parameters usually provides better estimators than the deterministic case when no prior knowledge is used.

5.7.1 Minimum mean square error (MMSE) estimator

In the classic approach, the MVU estimator was derived by minimization of the error function of the **mean square error,** i.e., $\hat{\theta} = \arg\min_{\hat{\theta}} \text{MSE}(\hat{\theta})$

$$\hat{\theta} = \arg\min_{\hat{\theta}} \int (\hat{\theta} - \theta)^2 p(\mathbf{o}; \theta) d\mathbf{o},$$

where $p(\mathbf{o}; \theta)$ is the PDF of $\mathbf{o}$ parameterized by θ. In the Bayesian approach we similarly derive an estimator by minimizing $\hat{\theta} = \arg\min_{\hat{\theta}} \text{MSE}(\hat{\theta})$, where:

$$\text{MSE}(\hat{\theta}) = E[(\theta - \hat{\theta})^2] = \int\int (\theta - \hat{\theta})^2 p(\mathbf{o}, \theta) d\mathbf{o} d\theta = \int (\theta - \hat{\theta})^2 p(\theta|\mathbf{o}) d\theta.$$

is the Bayesian Mean Square Error (MSE) and $p(\mathbf{o}, \theta)$ is the joint PDF of $\mathbf{o}$ and θ (note that θ is now a random variable). The Bayesian MSE $(\theta - \hat{\theta})^2$ and the classic squared error are the same, but since θ is a random variable, additional integration over θ is needed. The minimum $\text{MSE}(\hat{\theta})$ estimator or MMSE estimator is derived by differentiating the expression for $\text{MSE}(\hat{\theta})$ with respect to $\hat{\theta}$ and setting this to zero to yield:

$$\begin{aligned}
\frac{\partial}{\partial \hat{\theta}} \int (\theta - \hat{\theta})^2 p(\theta|\mathbf{o}) d\theta &= \int \frac{\partial}{\partial \hat{\theta}} (\theta - \hat{\theta})^2 p(\theta|\mathbf{o}) d\theta \\
&= \int (-2)(\theta - \hat{\theta}) p(\theta|\mathbf{o}) d\theta \\
&= -2 \int \theta p(\theta|\mathbf{o}) d\theta + 2\hat{\theta} \\
&= 0.
\end{aligned}$$

This gives

Theorem for MMSE Computation:

$$\hat{\theta}_{MMSE} = E(\theta|\mathbf{o}) = \int \theta p(\theta|\mathbf{o}) d\theta, \tag{5.32}$$

where the *posterior PDF*, $p(\theta|\mathbf{o})$, is given by Bayes' rule:

$$p(\theta|\mathbf{o}) = \frac{p(\mathbf{o}, \theta)}{p(\mathbf{o})} = \frac{p(\mathbf{o}, \theta)}{\int p(\mathbf{o}, \theta) d\theta} = \frac{p(\mathbf{o}|\theta) p(\theta)}{\int p(\mathbf{o}|\theta) p(\theta) d\theta}.$$

While the general form for the MMSE computation above is nice and clean, the actual computation is often very difficult; recall the general difficulty of finding the MVU estimator in the classic approach. First, it is usually difficult to derive an analytical expression for the posterior PDF and then evaluate the expectation $E(\theta|\mathbf{o})$. Second, it is also difficult to find an appropriate, realistic prior PDF. To overcome these difficulties, we often sacrifice possible loss of modeling accuracy and assume that the joint PDF, $p(\mathbf{o},\theta)$, is Gaussian and hence both the prior PDF, $p(\theta)$, and posterior PDF, $p(\theta|\mathbf{o})$, are also Gaussian. This property says that the Gaussian PDF is a conjugate prior distribution. Under this assumption, the form of the PDFs remains the same and all that changes are the means and variances.

Example: Consider a signal embedded in noise:

$$o[n] = A + w[n],$$

where as before $w[n] = \mathcal{N}(0,\sigma^2)$ is a WGN process and the unknown parameter $\theta = A$ is to be estimated. However in the Bayesian approach we also assume the parameter A is a random variable with a prior PDF which in this case is the Gaussian PDF $p(A) = \mathcal{N}(\mu_A, \sigma_A^{\in})$. We also have that $p(\mathbf{o}|A) = \mathcal{N}(\mathcal{A}, \sigma^{\in})$ and we can assume that $\mathbf{o}$ and A are jointly Gaussian. Thus, the posterior PDF

$$p(A|\mathbf{o}) = \frac{p(\mathbf{o}|A)p(A)}{\int p(\mathbf{o}|A)p(A)dA} = \mathcal{N}(\mu_{A|\mathbf{o}}, \sigma^2_{A|\mathbf{o}}) \tag{5.33}$$

is also a Gaussian PDF. After simplifying and combining terms in Eq. 5.33, we have

$$\sigma^2_{A|\mathbf{o}} = \frac{1}{\frac{N}{\sigma^2} + \frac{1}{\sigma_A^2}} \qquad \text{and} \qquad \mu_{A|\mathbf{o}} = \left(\frac{N}{\sigma^2}\bar{x} + \frac{\mu_A}{\sigma_A^2}\right)\sigma^2_{A|\mathbf{o}}.$$

Hence, the MMSE is computed as

$$\begin{aligned} \hat{A} = E[A|\mathbf{o}] &= \int A p(A|\mathbf{o}) dA = \mu_{A|\mathbf{o}} && (5.34) \\ &= \alpha\bar{x} + (1-\alpha)\mu_A, && (5.35) \end{aligned}$$

where $\alpha = \frac{\sigma_A^2}{\sigma_A^2 + \frac{\sigma^2}{N}}$. Upon closer examination of this MMSE result, we observe the following (assume $\sigma_A^2 \ll \sigma^2$):

1. With few data (N is small), we will have $\sigma_A^2 \ll \sigma^2/N$ and $\hat{A} \to \mu_A$. That is, the MMSE moves towards the mean of the prior PDF and thus effectively ignores the contribution of the data. Also $p(A|\mathbf{o}) \approx \mathcal{N}(\mu_A, \sigma_A^{\in})$.

2. With a large amount of data (N is large), we will have $\sigma_A^2 \gg \sigma^2/N$ and $\hat{A} \to \bar{x}$. That is, the MMSE tends towards the sample mean $\bar{x}$ and effectively ignores the contribution of the prior information. Also $p(A|\mathbf{o}) \approx \mathcal{N}(\S, \sigma^{\in}/\mathcal{N})$.

We now briefly discuss the relation between the Bayesian estimator and the classic estimators. In classical estimation we cannot make any assumptions on the parameter prior. Thus, all possible θ have to be considered. The equivalent prior PDF would be a flat distribution, essentially $\sigma_\theta^2 = \infty$. This so-called *noninformative* prior PDF will yield the classic estimator where such is defined.

Example: In the earlier problem of a signal embedded in noise, we showed that the MMSE estimator follows Eq. 5.35 with $\alpha = \frac{\sigma_A^2}{\sigma_A^2 + \frac{\sigma^2}{N}}$. If the prior PDF is noninformative then $\sigma_A^2 = \infty$ and $\alpha = 1$ with $\hat{A} = \bar{x}$, which is the classic estimator.

5.7.2 Bayesian linear model

Now consider the Bayesian linear model (for which the MMSE computation is also relatively easy):

$$\mathbf{o} = \mathbf{H}\boldsymbol{\theta} + \mathbf{w},$$

where $\boldsymbol{\theta}$ is the unknown (random) parameter to be estimated with prior PDF $\mathcal{N}(\boldsymbol{\mu}_\theta, \mathbf{C}_\theta)$ and $\mathbf{w}$ is a WGN with $\mathcal{N}(\mathbf{0}, \mathbf{C}_w)$. Note that the linear model introduced in Chapter 4 has the parameter $\boldsymbol{\theta}$ that is a deterministic rather than a random variable. The MMSE for the parameter $\boldsymbol{\theta}$ is provided by $E(\mathbf{y}|\mathbf{o})$. To compute this, we have

$$\begin{aligned} E(\mathbf{o}) &= \mathbf{H}\boldsymbol{\mu}_\theta \\ E(\boldsymbol{\theta}) &= \boldsymbol{\mu}_\theta, \end{aligned}$$

and further

$$\begin{aligned} \mathbf{C}_{oo} &= E[(\mathbf{o} - E(\mathbf{o}))(\mathbf{o} - E(\mathbf{o}))^{\mathrm{Tr}}] = \mathbf{H}\mathbf{C}_\theta\mathbf{H}^{\mathrm{Tr}} + \mathbf{C}_w \\ \mathbf{C}_{\theta o} &= E[(\boldsymbol{\theta} - E(\boldsymbol{\theta}))(\mathbf{o} - E(\mathbf{o})^{\mathrm{Tr}}] = \mathbf{C}_\theta\mathbf{H}^{\mathrm{Tr}}. \end{aligned}$$

Before deriving the final expression for the MMSE, we introduce the following well known property of the multivariate Gaussian PDF (see proof in [Kay 93, Mend 95]): If $\mathbf{o}$ and $\boldsymbol{\theta}$ are jointly Gaussian, where $\mathbf{o}$ is $k \times 1$ and $\boldsymbol{\theta}$ is $l \times 1$, with mean vector $[E(\mathbf{o})^{\mathrm{Tr}}, E(\boldsymbol{\theta}^{\mathrm{Tr}}]^{\mathrm{Tr}}$ and partitioned covariance matrix

$$\mathbf{C} = \begin{bmatrix} \mathbf{C}_{oo} & \mathbf{C}_{o\theta} \\ \mathbf{C}_{\theta o} & \mathbf{C}_{\theta\theta} \end{bmatrix} = \begin{bmatrix} k \times k & k \times l \\ l \times k & l \times l \end{bmatrix},$$

then the conditional PDF, $p(\boldsymbol{\theta}|\mathbf{o})$, is also Gaussian and:

$$E(\boldsymbol{\theta}|\mathbf{o}) = E(\boldsymbol{\theta}) + \mathbf{C}_{\theta o}\mathbf{C}_{oo}^{-1}(\mathbf{o} - E(\mathbf{o})).$$

Applying this property to $\mathbf{o}$ and $\boldsymbol{\theta}$ that are jointly Gaussian in the Bayesian linear model, we obtain the close-form result for the MMSE estimate

$$\hat{\theta} = E(\boldsymbol{\theta}|\mathbf{o}) = \boldsymbol{\mu}_\theta + \mathbf{C}_\theta\mathbf{H}^{\mathrm{Tr}}(\mathbf{H}\mathbf{C}_\theta\mathbf{H}^{\mathrm{Tr}} + \mathbf{C}_w)^{-1}(\mathbf{o} - \mathbf{H}\boldsymbol{\mu}_\theta).$$

5.7.3 General Bayesian estimators and MAP estimator

The mean square error given by:

$$\mathrm{MSE}(\hat{\theta}) = E[(\theta - \hat{\theta})^2] = \int\int (\theta - \hat{\theta})^2 p(\mathbf{o}, \theta) d\mathbf{o} d\theta$$

is one specific case for a general estimator that attempts to minimize the average of the cost function, $\mathcal{C}(\epsilon)$, which is called the Bayes risk $\mathcal{R} = E[\mathcal{C}(\epsilon)]$, where $\epsilon = (\theta - \hat{\theta})$. There are three different cost functions of interest:

1. **Quadratic:** $\mathcal{C}(\epsilon) = \epsilon^2$ which yields $\mathcal{R} = \mathrm{MSE}(\hat{\theta})$. We already showed that the

estimate minimizing $\mathcal{R} = \text{MSE}(\hat{\theta})$ is:

$$\hat{\theta} = \int \theta p(\theta|\mathbf{o}) d\theta,$$

which is the mean of the *posterior PDF*.

2. **Absolute:** $\mathcal{C}(\epsilon) = |\epsilon|$. The estimate, $\hat{\theta}$, that minimizes $\mathcal{R} = E[|\theta - \hat{\theta}|]$ satisfies:

$$\int_{-\infty}^{\hat{\theta}} p(\theta|\mathbf{o}) d\theta = \int_{\hat{\theta}}^{\infty} p(\theta|\mathbf{o}) d\theta \text{ or } \Pr\{\theta \le \hat{\theta}|\mathbf{o}\} = \frac{1}{2},$$

which is the median of the *posterior PDF*.

3. **Hit-or-miss:** $\mathcal{C}(\epsilon) = \begin{cases} 0 & |\epsilon| < \delta \\ 1 & |\epsilon| > \delta \end{cases}$. The estimate that minimizes the Bayes risk can be shown to be:

$$\hat{\theta} = \arg\max_{\theta} p(\theta|\mathbf{o}),$$

which is the mode of the *posterior PDF* (the value that maximizes the PDF).

For the Gaussian posterior PDF it should be noted that the mean, median and mode are identical. Of most interest are the quadratic and hit-or-miss cost functions which, together with a special case of the latter, yield the following three important classes of estimators:

1. **MMSE** (***Minimum Mean Square Error***) estimator which we have already introduced as the conditional expectation:

$$\hat{\theta} = \int \theta p(\theta|\mathbf{o}) d\theta.$$

2. **MAP** (***Maximum A Posteriori***) estimator that is the mode or maximum of the *posterior PDF* resulting from the hit-or-miss cost function:

$$\begin{aligned} \hat{\theta} &= \arg\max_{\theta} p(\theta|\mathbf{o}) \\ &= \arg\max_{\theta} p(\mathbf{o}|\theta) p(\theta). \end{aligned}$$

3. **Bayesian ML** (***Bayesian Maximum Likelihood***) estimator, which is the special case of the MAP estimator where the prior PDF, $p(\theta)$, is uniform or noninformative:

$$\hat{\theta} = \arg\max_{\theta} p(\mathbf{o}|\theta).$$

Since the conditional PDF of $\mathbf{o}$ given θ, $p(\mathbf{o}|\theta)$, is the same as the PDF of $\mathbf{o}$ parameterized by θ, $p(\mathbf{o};\theta)$, the Bayesian ML estimator is equivalent to the classic MLE we introduced earlier.

Comparing the above three estimators, we can make the following general statements: First, the MMSE is preferred due to its very desirable least-squared cost function. However, it is also very difficult to derive and compute due to the need to find an expression or measurements of the posterior PDF $p(\theta|\mathbf{o})$ in order to integrate $\int \theta p(\theta|\mathbf{o}) d\theta$.

Second, the MAP hit-or-miss cost function is more "crude" but the MAP estimate is easier to derive since there is no need to carry out the integration. It involves only the maximization of the posterior PDF $p(\theta|\mathbf{o})$, or equivalently the maximization of the joint PDF $p(\theta, \mathbf{o})$ since $p(\mathbf{o})$ is independent of θ to be maximized. This maximization can be carried out either analytically or numerically using the optimization techniques discussed earlier in this chapter.

Finally, the Bayesian ML is equivalent in performance to the MAP only in the case where the prior PDF is noninformative, otherwise it is a sub-optimal estimator. However, like the classic MLE, the expression for the conditional PDF $p(\mathbf{o}|\theta)$ is usually easier to obtain than that of the posterior PDF, $p(\theta|\mathbf{o})$. Since in most cases knowledge of the prior PDF is unavailable, ML estimators tend to be more prevalent. However, it may not always be prudent to assume the prior PDF is uniform, especially in cases where prior knowledge of the estimate is available even though the exact PDF is unknown. In these cases a MAP estimate may perform better even if an "artificial" prior PDF is assumed (e.g., a Gaussian prior which has the added benefit of yielding a Gaussian posterior).

5.7.4 Linear minimum mean square error (LMMSE) estimator

Up to now, we assume that the parameter θ is to be estimated based on the data set $\mathbf{o} = [o[0]\, o[1]\, \ldots\, o[N-1]]^{\mathrm{Tr}}$ rather than assume any specific form for the MMSE estimator and for the joint PDF $p(\mathbf{o}, \theta)$. The general difficulty of finding the MMSE estimators can often be overcome by constraining the form of the estimator to be linear. We consider the class of all linear estimators of the form:

$$\hat{\theta} = \sum_{n=0}^{N-1} a_n o[n] + a_N = \mathbf{a}^{\mathrm{Tr}}\mathbf{o} + a_N,$$

where $\mathbf{a} = [a_1\, a_2\, \ldots\, a_{N-1}]^{\mathrm{Tr}}$and choose the weight coefficients $\{\mathbf{a}, a_N\}$ to minimize the Bayesian MSE:

$$\mathrm{MSE}(\hat{\theta}) = E\left[(\theta - \hat{\theta})^2\right].$$

The resultant estimator is called the linear minimum mean square error (**LMMSE**) estimator. The LMMSE will be sub-optimal unless the MMSE is also linear. (This would be the case for the Bayesian linear model: $\mathbf{o} = \mathbf{H}\boldsymbol{\theta} + \mathbf{w}$.) Using the classical optimization techniques, the weight coefficients are obtained from

$$\frac{\partial \mathrm{MSE}(\hat{\theta})}{\partial a_i} = 0, \quad \text{for } i = 1, 2, \ldots, N,$$

which yields:

$$a_N = E(\theta) - \sum_{n=0}^{N-1} a_n E(o[n]) = E(\theta) - \mathbf{a}^{\mathrm{Tr}} E(\mathbf{o}) \quad \text{and} \quad \mathbf{a} = \mathbf{C}_{oo}^{-1}\mathbf{C}_{o\theta},$$

where $\mathbf{C}_{oo} = E\left[(\mathbf{o} - E(\mathbf{o}))(\mathbf{o} - E(\mathbf{o}))^{\mathrm{Tr}}\right]$ is the $N \times N$ covariance matrix and $\mathbf{C}_{o\theta} = E\left[(\mathbf{o} - E(\mathbf{o}))(\theta - E(\theta))\right]$ is the $N \times 1$ cross-covariance vector. Thus the LMMSE estimator is:

$$\hat{\theta} = \mathbf{a}^{\mathrm{Tr}}\mathbf{o} + a_N = E(\theta) + \mathbf{C}_{\theta o}\mathbf{C}_{oo}^{-1}(\mathbf{o} - E(\mathbf{o})),$$

where we note $\mathbf{C}_{\theta o} = \mathbf{C}_{o\theta}^{\mathrm{Tr}} = E\left[(\theta - E(\theta))(\mathbf{o} - E(\mathbf{o}))^{\mathrm{Tr}}\right]$.

For the $1 \times p$ vector parameter $\boldsymbol{\theta}$, an equivalent expression for the LMMSE estimator can be similarly derived as:

$$\hat{\boldsymbol{\theta}} = E(\boldsymbol{\theta}) + \mathbf{C}_{\theta o}\mathbf{C}_{oo}^{-1}(\mathbf{o} - E(\mathbf{o})), \tag{5.36}$$

where now $\mathbf{C}_{\theta o} = E\left[(\boldsymbol{\theta} - E(\boldsymbol{\theta}))(\mathbf{o} - E(\mathbf{o}))^{\mathrm{Tr}}\right]$ is the $p \times N$ cross-covariance matrix. The error covariance matrix is:

$$\begin{aligned} \mathbf{M}_{\hat{\theta}} &= E\left[(\boldsymbol{\theta} - \hat{\boldsymbol{\theta}})(\boldsymbol{\theta} - \hat{\boldsymbol{\theta}})^{\mathrm{Tr}}\right] \\ &= \mathbf{C}_{\theta\theta} - \mathbf{C}_{\theta o}\mathbf{C}_{oo}^{-1}\mathbf{C}_{o\theta}, \end{aligned}$$

where $\mathbf{C}_{\theta\theta} = E\left[(\boldsymbol{\theta} - E(\boldsymbol{\theta}))(\boldsymbol{\theta} - E(\boldsymbol{\theta}))^{\mathrm{Tr}}\right]$ is the $p \times p$ covariance matrix.

Example — LMMSE estimator for the Bayesian linear model: Let the data be generated from the Bayesian linear model:

$$\mathbf{o} = \mathbf{H}\boldsymbol{\theta} + \mathbf{w},$$

where $\mathbf{o}$ is the $N \times 1$ data vector, $\mathbf{H}$ is a known $N \times p$ observation matrix, $\boldsymbol{\theta}$ is a $p \times 1$ random vector of parameters with mean $E(\boldsymbol{\theta})$ and covariance matrix $\mathbf{C}_{\theta\theta}$, and $\mathbf{w}$ is an $N \times 1$ random vector with zero mean and covariance matrix $\mathbf{C}_w$, which is uncorrelated with $\boldsymbol{\theta}$. The joint PDF $p(\mathbf{w}, \boldsymbol{\theta})$ and hence also $p(\mathbf{o}, \boldsymbol{\theta})$ are otherwise arbitrary. We now derive the LMMSE estimator of $\boldsymbol{\theta}$ for this model. We have

$$\begin{aligned} E(\mathbf{o}) &= \mathbf{H}E(\boldsymbol{\theta}) \\ \mathbf{C}_{oo} &= \mathbf{H}\mathbf{C}_{\theta\theta}\mathbf{H}^{\mathrm{Tr}} + \mathbf{C}_w \\ \mathbf{C}_{\theta o} &= \mathbf{C}_{\theta\theta}\mathbf{H}^{\mathrm{Tr}}. \end{aligned}$$

Then using the result of Eq. 5.36, we obtain the LMMSE estimator of $\boldsymbol{\theta}$ as

$$\begin{aligned} \hat{\boldsymbol{\theta}} &= E(\boldsymbol{\theta}) + \mathbf{C}_{\theta\theta}\mathbf{H}^{\mathrm{Tr}}(\mathbf{H}\mathbf{C}_{\theta\theta}\mathbf{H}^{\mathrm{Tr}} + \mathbf{C}_w)^{-1}(\mathbf{o} - \mathbf{H}E(\boldsymbol{\theta})) \\ &= E(\boldsymbol{\theta}) + (\mathbf{C}_{\theta\theta}^{-1} + \mathbf{H}^{\mathrm{Tr}}\mathbf{C}_w^{-1}\mathbf{H})^{-1}\mathbf{H}^{\mathrm{Tr}}\mathbf{C}_w^{-1}(\mathbf{o} - \mathbf{H}E(\boldsymbol{\theta})) \end{aligned}$$

and the covariance matrix of the error is:

$$\mathbf{M}_{\hat{\theta}} = (\mathbf{C}_{\theta\theta}^{-1} + \mathbf{H}^{\mathrm{Tr}}\mathbf{C}_w^{-1}\mathbf{H})^{-1}.$$

Example — Wiener smoothing:

We assume N samples of time-series data $\mathbf{o} = [o[0]\, o[1]\, \ldots\, o[N-1]]^{\mathrm{Tr}}$ which are wide-sense stationary (WSS). As such, the $N \times N$ covariance matrix takes the symmetric Toeplitz form:

$$\mathbf{C}_{oo} = \mathbf{R}_{oo} \quad \text{where} \quad [\mathbf{R}_{oo}]_{ij} = r_{oo}[i-j]$$

where $r_{oo}[k] = E(o[n]o[n-k])$ is the autocorrelation function (ACF) of the $o[n]$ process and $\mathbf{R}_{oo}$ denotes the autocorrelation matrix. Note that since $o[n]$ is WSS the expectation $E(o[n]o[n-k])$ is independent of the absolute time index n. In signal processing the estimated ACF is used:

$$\hat{r}_{oo}[k] = \begin{cases} \frac{1}{N}\sum_{n=0}^{N-1-|k|} o[n]o[n+|k|] & |k| \leq N-1 \\ 0 & |k| \geq N. \end{cases}$$

Both the data $\mathbf{o}$ and the parameter to be estimated $\hat{\boldsymbol{\theta}}$ are assumed zero mean. Thus the general form of the LMMSE estimator in Eq. 5.36 is reduced to

$$\hat{\theta} = \mathbf{C}_{\theta o}\mathbf{C}_{oo}^{-1}\mathbf{o}.$$

Applications of this LMMSE estimator to the three signal processing estimation problems of smoothing, filtering, and prediction give rise to the *Wiener filtering* equation solutions. In this example, we discuss Wiener smoothing only and refer the readers for Wiener filtering and prediction to [Kay 93].

The problem is to estimate the signal $\boldsymbol{\theta} = \mathbf{s} = [s[0]\, s[1]\, \ldots\, s[N-1]]^{\mathrm{Tr}}$based on the noisy data $\mathbf{o} = [o[0]\, o[1]\, \ldots\, o[N-1]]^{\mathrm{Tr}}$ where:

$$\mathbf{o} = \mathbf{s} + \mathbf{w}$$

and $\mathbf{w} = [w[0]\, w[1]\, \ldots w[N-1]]^{\mathrm{Tr}}$ is the noise process. (An important difference between smoothing and filtering is that the signal estimate $s[n]$ can use the entire data set: the *past* values ($o[0]$, $o[1]$, ... $o[n-1]$), the *present* $o[n]$ and *future* values ($o[n+1]$, $o[n+2]$, ..., $o[N-1]$).)

We assume that the signal and noise processes are uncorrelated. Therefore, we have

$$r_{oo}[k] = r_{ss}[k] + r_{ww}[k].$$

This leads to

$$\mathbf{C}_{oo} = \mathbf{R}_{oo} = \mathbf{R}_{ss} + \mathbf{R}_{ww}$$

and

$$\mathbf{C}_{\theta o} = E(\mathbf{s}\mathbf{o}^{\mathrm{Tr}}) = E(\mathbf{s}(\mathbf{s}+\mathbf{w})^{\mathrm{Tr}}) = \mathbf{R}_{ss}.$$

Hence the LMMSE estimator (also called the Wiener estimator) is:

$$\hat{\mathbf{s}} = \mathbf{R}_{ss}(\mathbf{R}_{ss} + \mathbf{R}_{ww})^{-1}\mathbf{o} = \mathbf{W}\mathbf{o},$$

and the $N \times N$ matrix:

$$\mathbf{W} = \mathbf{R}_{ss}(\mathbf{R}_{ss} + \mathbf{R}_{ww})^{-1}$$

is referred to as the *Wiener smoothing matrix.*

5.7.5 Sequential LMMSE estimator

We now describe recursive re-estimation of $\hat{\theta}[n]$ (i.e., $\hat{\theta}$ based on n data samples) from:

- $\hat{\theta}[n-1]$, the estimate based on $n-1$ data samples,
- $o[n]$, the nth data sample, and
- $\hat{o}[n|n-1]$, an estimate of $o[n]$ based on $n-1$ data samples.

The procedure is as follows:

1. From the previous iteration we have the estimate, $\hat{\theta}[n-1]$ (or an initial estimate $\hat{\theta}[0]$). We can consider $\hat{\theta}[n-1]$ as the true value of θ projected on the subspace spanned by $\{\theta[0], ,\theta[1], \ldots, \theta[n-1]\}$.
2. We find the LMMSE estimator of $o[n]$ based on the previous $n-1$ samples, which is the one-step linear predictor, $\hat{o}[n|n-1]$, or we have an initial estimate, $\hat{o}[0|-1] =$

$\hat{o}[0]$. We can consider $\hat{o}[n|n-1]$ as the true value $o[n]$ projected on the subspace spanned by $\{o[0], , o[1], \ldots, o[n-1]\}$.

3. We form the *innovation* sequence $o[n] - \hat{o}[n|n-1]$ which we can consider as being orthogonal to the space spanned by $\{o[0], , o[1], \ldots, o[n-1]\}$ and representing the direction of correction based exclusively on the new data sample $o[n]$.

4. We calculate the *correction gain* $K[n]$ by the normalized projection of θ on the innovation $o[n] - \hat{o}[n|n-1]$, that is:

$$K[n] = \frac{E[\theta(o[n] - \hat{o}[n|n-1])]}{E[(o[n] - \hat{o}[n|n-1])^2]}.$$

5. Finally we update the estimator by adding the correction:

$$\hat{\theta}[n] = \hat{\theta}[n-1] + K[n](o[n] - \hat{o}[n|n-1]).$$

Example — Sequential LMMSE estimator for the Bayesian Linear Model:
Consider the scalar-data form (vector parameter) of the Bayesian linear model:

$$o[n] = \mathbf{h}^{\mathrm{Tr}}\boldsymbol{\theta} + w[n].$$

The sequential LMMSE estimation procedure is as follows:

$$\begin{aligned}
\hat{\boldsymbol{\theta}}[n] &= \hat{\boldsymbol{\theta}}[n-1] + \mathbf{K}[n](o[n] - \mathbf{h}^{\mathrm{Tr}}[n]\hat{\boldsymbol{\theta}}[n-1]) \\
\mathbf{K}[n] &= \frac{\mathbf{M}[n-1]\mathbf{h}[n]}{\sigma_n^2 + \mathbf{h}^{\mathrm{Tr}}[n]\mathbf{M}[n-1]\mathbf{h}[n]} \\
\mathbf{M}[n] &= (\mathbf{I} - \mathbf{K}[n]\mathbf{h}^{\mathrm{Tr}}[n])\mathbf{M}[n-1],
\end{aligned}$$

where $\mathbf{M}[n]$ is the error covariance matrix:

$$\mathbf{M}[n] = E[(\boldsymbol{\theta} - \hat{\boldsymbol{\theta}}[n])(\boldsymbol{\theta} - \hat{\boldsymbol{\theta}}[n])^{\mathrm{Tr}}].$$

5.8 State Estimation

Most of the materials presented so far in this chapter have concerned the problem of estimating parameters in statistical models. Some common statistical models used in signal processing, speech processing in particular, have been covered in Chapters 3 and 4. These models can be classified into two groups. The first group has the models with no "hidden" random variables or states. That is, all random variables defined in the model are fully observable, and the only (possibly) unknown elements in the model are the parameters, which can either be deterministic or be random. The Gaussian model, linear model, and Bayesian linear model all belong to this first group. Since there is no "state", parameter estimation constitutes the entire estimation problem. The second group has the models that contain unobserved "hidden" random variables or states. The mixture-of-Gaussians model, HMM, trended HMM, and the dynamic system model belong to this group. In addition to parameter estimation, the estimation problem also includes state estimation or inference, where estimates for the hidden states using the observed data or the posterior probability of the states given the observed data are sought.

The second group of the models can be further classified into two types: 1) the hidden states are discrete random variables (e.g., HMM, and trended HMM); and 2) the hidden states are continuous random variables (e.g., dynamic system models). Section 5.3 already discussed the state estimation problem and solutions for discrete-state models based on the dynamic programming techniques or Viterbi algorithm. This section will thus focus on the state estimation problem for the models with continuous states. Solutions to the state estimation problem usually assume that the parameters in the model are known or have been estimated.

The role of state estimation in signal processing and pattern recognition can be seen as that of a "decoding" device, since the estimated hidden states are typically the hidden information that needs to be discovered (or decoded) by the pattern recognizer. For speech recognition, for example, the hidden information is the discrete linguistic code represented by discrete states that need to be estimated from the observed speech acoustics.

It is also worth noting that in addition to the very crucial role of a decoder, state estimation is also needed as an indispensable step for the parameter estimation problem involving hidden states. While applying the EM algorithm, state estimation is generally required in the E-step to compute the sufficient statistics required by the M-step. We will detail this in Chapter 10 while discussing dynamic models of speech.

In this section, we will first introduce the generic Kalman filter algorithm for a general form of (nonlinear) state-space model. While this generic algorithm is often not computable, when the form of the model is reduced to a linear one, we are then able to compute all necessary terms. This gives a computable form of the Kalman filter specific to the linear state-space model. We will then extend this basic form of the Kalman filter to handle (mildly) nonlinear state-space models.

5.8.1 Generic Kalman filter algorithm

The Kalman filter algorithm is a fundamental technique for state estimation in the state-space dynamic system model involving unobserved hidden states as continuous random variables. The generic Kalman filter algorithm applies to the general form of the time-varying nonlinear dynamic system model, which we presented in Chapter 4 as below:

$$\mathbf{x}(k+1) = \mathbf{g}_k[\mathbf{x}(k), \mathbf{u}_k, \mathbf{w}(k)] \tag{5.37}$$

$$\mathbf{o}(k) = \mathbf{h}_k[\mathbf{x}(k), \mathbf{v}(k)]. \tag{5.38}$$

The Kalman filter is an MMSE technique for estimating the continuous random vector at a fixed discrete-time step k, $\mathbf{x}(k)$, given the observation vector sequence up to time k, $\mathbf{o}_1^k = \{\mathbf{o}(0), \mathbf{o}(1), ..., \mathbf{o}(k)\}$. Using the Theorem for MMSE computation in Eq. 5.32, we have the generic formula for such an estimate, called a filtered state and denoted by $\hat{\mathbf{x}}_{k|k}$:

$$\hat{\mathbf{x}}_{k|k} = E(\mathbf{x}(k)|\mathbf{o}_1^k). \tag{5.39}$$

To describe the generic Kalman filter that efficiently computes $\hat{\mathbf{x}}_{k|k}$ in Eq. 5.39, we first introduce the following new MMSE estimates and their names:

$$\hat{\mathbf{x}}_{k|k-1} = E(\mathbf{x}(k)|\mathbf{o}_1^{k-1}), \quad \text{(predicted state)} \tag{5.40}$$
$$\boldsymbol{\Sigma}_{k|k-1} = E[(\mathbf{x}(k) - \hat{\mathbf{x}}_{k|k-1})(\mathbf{x}(k) - \hat{\mathbf{x}}_{k|k-1})^{\mathrm{Tr}}|\mathbf{o}_1^{k-1}], \quad \text{(predicted-state error covariance)}$$
$$\hat{\mathbf{o}}_{k|k-1} = E(\mathbf{o}(k)|\mathbf{o}_1^{k-1}), \quad \text{(predicted observation)}$$
$$\mathbf{N}_{k|k-1} = E[(\mathbf{o}(k) - \hat{\mathbf{o}}_{k|k-1})(\mathbf{o}(k) - \hat{\mathbf{o}}_{k|k-1})^{\mathrm{Tr}}|\mathbf{o}_1^{k-1}], \quad \text{(innovation covariance)}$$
$$\mathbf{M}_{k|k-1} = E[(\mathbf{o}(k) - \hat{\mathbf{o}}_{k|k-1})(\mathbf{x}(k) - \hat{\mathbf{x}}_{k|k-1})^{\mathrm{Tr}}|\mathbf{o}_1^{k-1}], \quad \text{(error cross-covariance)}$$
$$\mathbf{K}_k = \mathbf{M}_{k|k-1}^{\mathrm{Tr}}\mathbf{N}_{k|k-1}^{-1}, \quad \text{(Kalman gain matrix)}$$
$$\boldsymbol{\Sigma}_{k|k} = E[(\mathbf{x}(k) - \hat{\mathbf{x}}_{k|k})(\mathbf{x}(k) - \hat{\mathbf{x}}_{k|k})^{\mathrm{Tr}}|\mathbf{o}_1^{k}], \quad \text{(filtered-state error covariance).}$$

The description of a generic Kalman filter states that

Generic Kalman filter (state correction): Given the predicted state $\hat{\mathbf{x}}_{k|k-1}$, its error covariance $\boldsymbol{\Sigma}_{k|k-1}$, and the Kalman gain $\mathbf{K}_k = \mathbf{M}_{k|k-1}^{\mathrm{Tr}}\mathbf{N}_{k|k-1}^{-1}$, the filtered state and its error covariance can be computed according to

$$\hat{\mathbf{x}}_{k|k} = \hat{\mathbf{x}}_{k|k-1} + \mathbf{K}_k(\mathbf{o}(k) - \hat{\mathbf{o}}_{k|k-1}), \tag{5.41}$$
$$\boldsymbol{\Sigma}_{k|k} = \boldsymbol{\Sigma}_{k|k-1} - \mathbf{K}_k\mathbf{N}_{k|k-1}\mathbf{K}_k^{\mathrm{Tr}}. \tag{5.42}$$

The error sequence between the actual and the predicted observation sequences in Eq. 5.42 is called the innovation sequence:

$$\tilde{\mathbf{o}}_k \equiv \mathbf{o}(k) - \hat{\mathbf{o}}_{k|k-1},$$

and the computation of $\hat{\mathbf{x}}_{k|k-1}$, $\boldsymbol{\Sigma}_{k|k-1}$, and $\mathbf{K}_k$ requires that the structure of functions $\mathbf{g}_k[.]$ and $\mathbf{h}_k[.]$ in the model be specified. This computation is typically carried out using the filtered state and its error covariance from the immediately previous time step, forming a recursive computation with Eqs. 5.27 and 5.42.

We now offer some insight to understanding the generic Kalman filter algorithm described above. The Kalman filter is essentially a result of implementing Bayes' rule (Chapter 3) in the general setup of the state-space model. To see this, we apply Bayes's rule to compute the posterior probability $p(\mathbf{x}(k)|\mathbf{o}_1^k)$, from which the first-order statistic $\hat{\mathbf{x}}_{k|k}$ in Eq. 5.27 and the second-order statistic $\boldsymbol{\Sigma}_{k|k}$ in Eq. 5.42 are derived. The posterior probability is computed using Bayes' rule and is decomposed as:

$$\begin{aligned} p(\mathbf{x}(k)|\mathbf{o}_1^k) &= p(\mathbf{x}(k)|\mathbf{o}(k), \mathbf{o}_1^{k-1}) \\ &= \frac{p(\mathbf{x}(k), \mathbf{o}(k)|\mathbf{o}_1^{k-1})}{p(\mathbf{o}(k)|\mathbf{o}_1^{k-1})} \\ &= \frac{p(\mathbf{x}(k), \mathbf{o}(k)|\mathbf{o}_1^{k-1})}{\int p(\mathbf{x}(k), \mathbf{o}(k)|\mathbf{o}_1^{k-1})d\mathbf{x}(k)} \\ &= \frac{p(\mathbf{o}(k)|\mathbf{x}(k), \mathbf{o}_1^{k-1})p(\mathbf{x}(k)|\mathbf{o}_1^{k-1})}{\int p(\mathbf{o}(k)|\mathbf{x}(k), \mathbf{o}_1^{k-1})p(\mathbf{x}(k)|\mathbf{o}_1^{k-1})d\mathbf{x}} \\ &= \frac{p(\mathbf{o}(k)|\mathbf{x}(k))p(\mathbf{x}(k)|\mathbf{o}_1^{k-1})}{\int p(\mathbf{o}(k)|\mathbf{x}(k))p(\mathbf{x}(k)|\mathbf{o}_1^{k-1})d\mathbf{x}}. \end{aligned} \tag{5.43}$$

The last equality in Eq. 5.43 is due to the conditional independence between $\mathbf{o}(k)$ and its history $\mathbf{o}_1^{k-1}$ given the current state $\mathbf{x}(k)$, as assumed in the state-space model. The result of Bayes' rule in Eq. 5.43 says that the conditional PDF $p(\mathbf{x}(k)|\mathbf{o}_1^k)$ from which the filtered state is derived can be expressed in terms of the conditional PDF for the predicted state $p(\mathbf{x}(k)|\mathbf{o}_1^{k-1})$. This is exactly the same kind of relation in Eqs. 5.27 and Eq. 5.42 that governs the filtered and predicted states of interest. The remaining information needed to compute the right-hand side of Eq. 5.43 is readily available: $p(\mathbf{o}(k)|\mathbf{x}(k))$ is determined directly from the observation equation Eq. 5.38 as long as the specific distribution of the error term $\mathbf{v}(k)$ is given.

Likewise, prediction of the state using the previous-step filtered state can be carried out using the law of total probability. This results in the following expression of the conditional PDF for the predicted state $p(\mathbf{x}(k)|\mathbf{o}_1^{k-1})$ in terms of the conditional PDF for the previously filtered state $p(\mathbf{x}(k-1)|\mathbf{o}_1^{k-1})$:

$$\begin{aligned} p(\mathbf{x}(k)|\mathbf{o}_1^{k-1}) &= \int p(\mathbf{x}(k), \mathbf{x}(k-1)|\mathbf{o}_1^{k-1}) d\mathbf{x}(k-1) \\ &= \int p(\mathbf{x}(k)|\mathbf{x}(k-1), \mathbf{o}_1^{k-1}) p(\mathbf{x}(k-1))|\mathbf{o}_1^{k-1}) d\mathbf{x}(k-1) \\ &= \int p(\mathbf{x}(k)|\mathbf{x}(k-1)) p(\mathbf{x}(k-1)|\mathbf{o}_1^{k-1}) d\mathbf{x}(k-1). \end{aligned} \tag{5.44}$$

The remaining quantity $p(\mathbf{x}(k)|\mathbf{x}(k-1))$ needed to compute the right-hand side above again can be readily computed from the state equation Eq. 5.37 when the specific distribution of the error term $\mathbf{w}(k)$ is given.

Another insight to understanding the Kalman filter can be gained by carefully interpreting the state correction Eq. 5.27. The "corrected" state estimate $\hat{\mathbf{x}}_{k|k}$ after observing the current data sample $\mathbf{x}(k)$ at time k consists of two terms. The first term, $\hat{\mathbf{x}}_{k|k-1}$, is the state prediction that does not depend on the newly observed data point $\mathbf{x}(k)$. It is based purely on the prior knowledge concerning the state dynamics encoded in the state equation of the model. The second term, on the other hand, is proportional to $\tilde{\mathbf{o}}_k \equiv \mathbf{o}(k) - \hat{\mathbf{o}}_{k|k-1}$ and is directly employing the observed data as "evidence". The proportionality is the Kalman gain, which can be interpreted as giving the optimal balance between the use of prior knowledge and the use of the observed evidence. This situation is a bit like catching a Frisbee. To get an accurate catch, one has to combine the prior knowledge of how the Frisbee may move in terms of the Newtonian dynamics and the current actual position and movement (observed data) of the Frisbee.

5.8.2 Kalman filter algorithms for the linear state-space system

We now fix the structure of the functions $\mathbf{g}_k[.]$ and $\mathbf{h}_k[.]$ in the general form of the state-space model to be of a linear form, and perform all necessary computation to reduce the generic Kalman filter just described into a concrete form. Therefore, the model to be used here will be the linear state-space model presented in Chapter 4:

$$\mathbf{x}(k) = \mathbf{A}\mathbf{x}(k-1) + \mathbf{u} + \mathbf{w}(k-1) \tag{5.45}$$

$$\mathbf{o}(k) = \mathbf{C}\mathbf{x}(k) + \mathbf{v}(k), \tag{5.46}$$

initialized by

$$E[\mathbf{x}(0)] = \boldsymbol{\mu}(0), \qquad E[\mathbf{x}(0)\mathbf{x}(0)^{\mathrm{Tr}}] = \boldsymbol{\Sigma}(0).$$

Note that $\mathbf{w}(k)$ and $\mathbf{v}(k)$ are uncorrelated zero-mean Gaussian noise vectors with known covariances $E[\mathbf{w}(k)\mathbf{w}(l)^{\mathrm{Tr}}] = \mathbf{Q}\delta_{kl}$ and $E[\mathbf{v}(k)\mathbf{v}(l)^{\mathrm{Tr}}] = \mathbf{R}\delta_{kl}$, respectively. $\mathbf{A}$ is the system matrix, $\mathbf{C}$ is the observation matrix, and $\mathbf{u}$ is the input vector. All these model parameters are deterministic and known.

The predicted state $\hat{\mathbf{x}}_{k|k-1} = E(\mathbf{x}(k)|\mathbf{o}_1^{k-1})$ can now be easily expressed as a function of the previous-step filtered state by applying the conditional expectation "operator" $E(.|\mathbf{o}_1^{k-1})$ to both sides of state equation (Eq.5.45). This gives:

$$\hat{\mathbf{x}}_{k|k-1} = E(\mathbf{x}(k)|\mathbf{o}_1^{k-1}) = \mathbf{A}E(\mathbf{x}(k-1)|\mathbf{o}_1^{k-1}) + \mathbf{u} = \mathbf{A}\hat{\mathbf{x}}_{k-1|k-1} + \mathbf{u}.$$

Similarly, applying the conditional expectation "operator" $E(\cdot|\mathbf{o}_1^{k-1})$ to both sides of observation equation (Eq.5.46), we obtain the predicted observation

$$\hat{\mathbf{o}}_{k|k-1} = \mathbf{C}\hat{\mathbf{x}}_{k|k-1}.$$

Using the state equation 5.45 again, we also easily determine the predicted-state error covariance as a function of the previous-step filtered-state error covariance according to

$$\boldsymbol{\Sigma}_{k|k-1} = E[(\mathbf{x}(k) - \hat{\mathbf{x}}_{k|k})(\mathbf{x}(k) - \hat{\mathbf{x}}_{k|k})^{\mathrm{Tr}}|\mathbf{o}_1^k] = \mathbf{A}\boldsymbol{\Sigma}_{k-1|k-1}\mathbf{A}^{\mathrm{Tr}} + \mathbf{Q}.$$

Using the observation equation 5.46, we can derive the expression for the covariance of the innovation as

$$\mathbf{N}_{k|k-1} = E[\tilde{\mathbf{o}}(k)\tilde{\mathbf{o}}^{\mathrm{Tr}}(k)|\mathbf{o}_1^{k-1}] = \mathbf{C}\boldsymbol{\Sigma}_{k|k-1}\mathbf{C}^{\mathrm{Tr}} + \mathbf{R}.$$

To compute the Kalman gain, we first compute the cross-covariance between the state prediction error and the innovation using the observation equation 5.46:

$$\mathbf{M}_{k|k-1} = E[\tilde{\mathbf{o}}(k)(\mathbf{x}(k) - \hat{\mathbf{x}}_{k|k-1})^{\mathrm{Tr}}|\mathbf{o}_1^{k-1}] = \boldsymbol{\Sigma}_{k|k-1}\mathbf{C}^{\mathrm{Tr}}.$$

Thus, the Kalman gain is computed according to

$$\mathbf{K}_k = \mathbf{M}_{k|k-1}^{\mathrm{Tr}}\mathbf{N}_{k|k-1}^{-1} = \boldsymbol{\Sigma}_{k|k-1}\mathbf{C}^{\mathrm{Tr}}(\mathbf{C}\boldsymbol{\Sigma}_{k|k-1}\mathbf{C}^{\mathrm{Tr}} + \mathbf{R})^{-1}.$$

Summarizing all the above results, we have the following one-time-step Kalman filter procedure, which is recursive between the predicted and filtered states:

Kalman filter for the linear state-space model of Eqs. 5.45 and 5.46:
For $k = 1, 2, ..., N$,

Kalman Prediction

$$\hat{\mathbf{x}}_{k|k-1} = \mathbf{A}\hat{\mathbf{x}}_{k-1|k-1} + \mathbf{u} \tag{5.47}$$

$$\boldsymbol{\Sigma}_{k|k-1} = \mathbf{A}\boldsymbol{\Sigma}_{k-1|k-1}\mathbf{A}^{\mathrm{Tr}} + \mathbf{Q} \tag{5.48}$$

Kalman Gain

$$\mathbf{K}_k = \boldsymbol{\Sigma}_{k|k-1}\mathbf{C}^{\mathrm{Tr}}(\mathbf{C}\boldsymbol{\Sigma}_{k|k-1}\mathbf{C}^{\mathrm{Tr}} + \mathbf{R})^{-1} \tag{5.49}$$

Kalman Correction

$$\hat{\mathbf{x}}_{k|k} = \hat{\mathbf{x}}_{k|k-1} + \mathbf{K}_k(\mathbf{o}(k) - \mathbf{C}\hat{\mathbf{x}}_{k|k-1}), \tag{5.50}$$

$$\boldsymbol{\Sigma}_{k|k} = \boldsymbol{\Sigma}_{k|k-1} - \mathbf{K}_k(\mathbf{C}\boldsymbol{\Sigma}_{k|k-1}\mathbf{C}^{\mathrm{Tr}} + \mathbf{R})\mathbf{K}_k^{\mathrm{Tr}}. \tag{5.51}$$

It can be shown that the error covariance correction formula Eq. 5.51 has the following two equivalent forms (but with different computational properties):

$$\mathbf{\Sigma}_{k|k} = [\mathbf{I} - \mathbf{K}_k\mathbf{C}]\mathbf{\Sigma}_{k|k-1}, \tag{5.52}$$

which is called the standard form, and

$$\mathbf{\Sigma}_{k|k} = [\mathbf{I} - \mathbf{K}_k\mathbf{C}]\mathbf{\Sigma}_{k|k-1}[\mathbf{I} - \mathbf{K}_k\mathbf{C}]^{\mathrm{Tr}} + \mathbf{K}_k\mathbf{R}\mathbf{K}_k^{\mathrm{Tr}}, \tag{5.53}$$

which is called the stabilized form.

The Kalman filter is a sequential algorithm. It is initialized usually by $\hat{\mathbf{x}}_{0|0} = E[\mathbf{x}(0)] = \boldsymbol{\mu}(0)$ and $\mathbf{\Sigma}_{0|0} = E[\mathbf{x}(0)\mathbf{x}(0)^{\mathrm{Tr}}] = \mathbf{\Sigma}(0)$, and it incrementally forward-advances time step k until the end of the observation data sequence. For this reason, the above algorithm is sometimes also called the Kalman forward filter or forward recursion.

Among the family of the Kalman filter algorithms, there is also Kalman backward recursion or Kalman smoother. It uses the results of the Kalman forward filter to compute the MMSE estimate of state at time k, using the entire record of observation data $\mathbf{o}_1^N$), rather than only the data $\mathbf{o}_1^k$) up to k. That is, it computes the estimate $\hat{\mathbf{x}}_{k|N} = E(\mathbf{x}(k)|\mathbf{o}_1^N)$ rather than $\hat{\mathbf{x}}_{k|k} = E(\mathbf{x}(k)|\mathbf{o}_1^k)$. This gives a more accurate state estimate since more data are used. Also, to apply the EM algorithm to the parameter estimation problem for the dynamic system models (see Chapter 10 for detail), the rigorous E-step requires the result of Kalman backward recursion following Kalman forward recursion.

Here we summarize the results of Kalman backward recursion or Kalman smoother as follow:

Kalman smoother for linear state-space models:
Given the Kalman forward filter results $\hat{\mathbf{x}}_{k|k-1}, \hat{\mathbf{x}}_{k|k}$, $\mathbf{\Sigma}_{k|k-1}$, and $\mathbf{\Sigma}_{k|k}$, the smoothed state estimate defined as $\hat{\mathbf{x}}_{k|N} = E(\mathbf{x}(k)|\mathbf{o}_1^N)$ can be computed according to the following backward recursion:
For $k = N-1, K-2, ..., 1$

$$\hat{\mathbf{x}}_{k|N} = \hat{\mathbf{x}}_{k|k} + \mathbf{L}_k(\hat{\mathbf{x}}_{k+1|N} - \hat{\mathbf{x}}_{k+1|k}) \tag{5.54}$$

$$\hat{\mathbf{\Sigma}}_{k|N} = \hat{\mathbf{\Sigma}}_{k|k} + \mathbf{A}_k(\hat{\mathbf{\Sigma}}_{k+1|N} - \hat{\mathbf{\Sigma}}_{k+1|k})\mathbf{A}_k^{\mathrm{Tr}}, \tag{5.55}$$

where $\mathbf{L}_k \equiv \mathbf{\Sigma}_{k|k}A^{\mathrm{Tr}}\mathbf{\Sigma}_{k|k-1}^{-1}$.

The initialization of the above backward recursion $\hat{\mathbf{x}}_{N|N}$ and $\mathbf{\Sigma}_{N|N}$ was already computed from the preceding Kalman forward filter. We refer the readers to [Mend 95] for detailed derivations and elaboration of these results.

5.8.3 Extended Kalman filter for nonlinear dynamic systems

While there is a rigorous and straightforward solution to the state estimation problem for the linear state-space model as just described, the problem for the general nonlinear state-space model becomes much more complex. Consider the nonlinear state-space model:

$$\mathbf{x}(k+1) = \mathbf{g}[\mathbf{x}(k)] + \mathbf{w}(k) \tag{5.56}$$

$$\mathbf{o}(k+1) = \mathbf{h}[\mathbf{x}(k+1)] + \mathbf{v}(k+1). \tag{5.57}$$

Even with the simplifying assumption that the noises $\mathbf{w}(k)$ and $\mathbf{v}(k)$ are Gaussian, arbitrary nonlinearities $\mathbf{g}[.]$ and $\mathbf{h}[.]$ generally give rise to arbitrary state distributions. This typically makes the state estimation mathematically intractable. There have been a number of techniques developed that overcome this difficulty by approximating the rigorous solution in one way or another. One simple technique of approximation, called **extended Kalman filter** (EKF), together with its iterative version, is described in this section.

The EKF algorithm requires that the nonlinearities of $\mathbf{g}[.]$ and $\mathbf{h}[.]$ are both differentiable and that the noises $\mathbf{w}(k)$ and $\mathbf{v}(k)$ are both Gaussian. The essence of the EKF is to locally linearize $\mathbf{g}[.]$ and $\mathbf{h}[.]$ (i.e., using the first-order Taylor series expansion) at the point of the current state estimate, $\hat{\mathbf{x}}_{k|k}$. This then linearizes the nonlinear system and hence the conventional Kalman filter as described in the preceding section can be applied. This results in the approximate state distribution that remains to be Gaussian.

We now derive the EKF algorithm. First, we take the first-order Taylor series approximations. The points at which the Taylor series approximation is made differ between the state equation and the measurement equation. For the former, the point is at the filtered value $\hat{\mathbf{x}}_{k|k}$ since this will be used for the predicted estimate $\hat{\mathbf{x}}_{k+1|k}$. For the latter, the point is at the predicted value $\hat{\mathbf{x}}_{k+1|k}$ since we go from it to the next filtered estimate $\hat{\mathbf{x}}_{k+1|k+1}$. Such approximations give

$$\mathbf{g}(\mathbf{x}) \approx \mathbf{g}(\hat{\mathbf{x}}_{k|k}) + \mathbf{G}(\hat{\mathbf{x}}_{k|k})(\mathbf{x} - \hat{\mathbf{x}}_{k|k}) \tag{5.58}$$

$$\mathbf{h}(\mathbf{x}) \approx \mathbf{h}(\hat{\mathbf{x}}_{k+1|k}) + \mathbf{H}(\hat{\mathbf{x}}_{k+1|k})(\mathbf{x} - \hat{\mathbf{x}}_{k+1|k}), \tag{5.59}$$

where each element of the Jacobian matrix $\mathbf{G}$ is a partial derivative of each vector component of the nonlinear output with respect to each of the input vector components:

$$\mathbf{G}(\hat{\mathbf{x}}_{k|k}) = \frac{\partial \mathbf{g}(\hat{\mathbf{x}}_{k|k})}{\partial \mathbf{x}} = \begin{bmatrix} \frac{\partial g_1(\hat{\mathbf{x}}_{k|k})}{\partial x_1} & \frac{\partial g_1(\hat{\mathbf{x}}_{k|k})}{\partial x_2} \cdots & \frac{\partial g_1(\hat{\mathbf{x}}_{k|k})}{\partial x_n} \\ \frac{\partial g_2(\hat{\mathbf{x}}_{k|k})}{\partial x_1} & \frac{\partial g_2(\hat{\mathbf{x}}_{k|k})}{\partial x_2} \cdots & \frac{\partial g_2(\hat{\mathbf{x}}_{k|k})}{\partial x_n} \\ \vdots & & \vdots \\ \frac{\partial g_n(\hat{\mathbf{x}}_{k|k})}{\partial x_1} & \frac{\partial g_n(\hat{\mathbf{x}}_{k|k})}{\partial x_2} \cdots & \frac{\partial g_n(\hat{\mathbf{x}}_{k|k})}{\partial x_n} \end{bmatrix}, \tag{5.60}$$

and likewise,

$$\mathbf{H}(\hat{\mathbf{x}}_{k+1|k}) = \frac{\partial \mathbf{h}(\hat{\mathbf{x}}_{k+1|k})}{\partial \mathbf{x}} = \begin{bmatrix} \frac{\partial h_1(\hat{\mathbf{x}}_{k+1|k})}{\partial x_1} & \frac{\partial h_1(\hat{\mathbf{x}}_{k+1|k})}{\partial x_2} \cdots & \frac{\partial h_1(\hat{\mathbf{x}}_{k+1|k})}{\partial x_n} \\ \frac{\partial h_2(\hat{\mathbf{x}}_{k+1|k})}{\partial x_1} & \frac{\partial h_2(\hat{\mathbf{x}}_{k+1|k})}{\partial x_2} \cdots & \frac{\partial h_2(\hat{\mathbf{x}}_{k+1|k})}{\partial x_n} \\ \vdots & & \vdots \\ \frac{\partial h_n(\hat{\mathbf{x}}_{k+1|k})}{\partial x_1} & \frac{\partial h_n(\hat{\mathbf{x}}_{k+1|k})}{\partial x_2} \cdots & \frac{\partial h_n(\hat{\mathbf{x}}_{k+1|k})}{\partial x_n} \end{bmatrix}. \tag{5.61}$$

Substitution of Eqs. 5.58 and 5.59 into Eqs. 5.56 and 5.57 gives a linearized form of the system:

$$\mathbf{x}(k+1) = \mathbf{g}[\hat{\mathbf{x}}_{k|k}] + \mathbf{G}(\hat{\mathbf{x}}_{k|k})(\mathbf{x} - \hat{\mathbf{x}}_{k|k}) + \mathbf{w}(k) \tag{5.62}$$

$$\mathbf{o}(k+1) = \mathbf{h}(\hat{\mathbf{x}}_{k+1|k}) + \mathbf{H}(\hat{\mathbf{x}}_{k+1|k})(\mathbf{x}(k+1) - \hat{\mathbf{x}}_{k+1|k}) + \mathbf{v}(k+1). \tag{5.63}$$

Now, applying the operator $E[.|\mathbf{o}]$ to both sides of the linearized state equation Eq. 5.62, we obtain the **EKF prediction** equation:

$$\hat{\mathbf{x}}_{k+1|k} = \mathbf{g}[\hat{\mathbf{x}}_{k|k}].$$

To derive the **EKF correction** equation, we first move "constant" terms in the right-hand side of the linearized measurement equation (Eq.5.63), and rewrite it as

$$\mathbf{o}_1(k+1) = \mathbf{H}(\hat{\mathbf{x}}_{k+1|k})\mathbf{x}(k+1) + \mathbf{v}(k+1), \tag{5.64}$$

where

$$\mathbf{o}_1(k+1) = \mathbf{o}(k+1) - \mathbf{h}(\hat{\mathbf{x}}_{k+1|k}) + \mathbf{H}(\hat{\mathbf{x}}_{k+1|k})\hat{\mathbf{x}}_{k+1|k}.$$

For the linearized system Eqs. 5.62 and 5.64, we apply to conventional Kalman correction equation Eq. 5.50 to obtain the EKF correction equation:

$$\begin{aligned}\hat{\mathbf{x}}_{k+1|k+1} &= \hat{\mathbf{x}}_{k+1|k} + \mathbf{K}_{k+1}(\mathbf{o}_1(k+1) - \mathbf{H}(\hat{\mathbf{x}}_{k+1|k})\hat{\mathbf{x}}_{k+1|k}) \\ &= \hat{\mathbf{x}}_{k+1|k} + \mathbf{K}_{k+1}(\mathbf{o}(k+1) - \mathbf{h}(\hat{\mathbf{x}}_{k+1|k})).\end{aligned}$$

Note that in the above EKF correction equation, the term

$$\mathbf{o}(k+1) - \mathbf{h}(\hat{\mathbf{x}}_{k+1|k})$$

is the error between the actual and the "nonlinearly" predicted observations. As in the linear system case, this error is sometimes called **pseudo innovation**.

Also note that the Jacobian $\mathbf{G}(\hat{\mathbf{x}}_{k|k})$ plays the same role as the system matrix $\mathbf{A}$ in the matrix equations for Kalman gain and error covariance matrices of the Kalman filter for the linear dynamic system. Likewise, the Jacobian $\mathbf{H}(\hat{\mathbf{x}}_{k+1|k})$ plays the same role as the observation matrix $\mathbf{C}$ as in the linear dynamic system case.

Putting it all together, we summarize the EKF algorithm as follows:

Extended Kalman filter for the nonlinear state-space model (Eqs. 5.56 and 5.57):
For $k = 1, 2, ..., N$,

EKF Prediction

$$\begin{aligned}\hat{\mathbf{x}}_{k+1|k} &= \mathbf{g}[\hat{\mathbf{x}}_{k|k}] \\ \boldsymbol{\Sigma}_{k+1|k} &= \mathbf{G}(\hat{\mathbf{x}}_{k|k})\boldsymbol{\Sigma}_{k|k}\mathbf{G}^{\mathrm{Tr}}(\hat{\mathbf{x}}_{k|k}) + \mathbf{Q}\end{aligned}$$

EKF Gain

$$\mathbf{K}_k = \boldsymbol{\Sigma}_{k+1|k}\mathbf{H}^{\mathrm{Tr}}(\hat{\mathbf{x}}_{k+1|k})[\mathbf{H}(\hat{\mathbf{x}}_{k+1|k})\boldsymbol{\Sigma}_{k+1|k}\mathbf{H}^{\mathrm{Tr}}(\hat{\mathbf{x}}_{k+1|k}) + \mathbf{R}]^{-1}$$

EKF Correction

$$\begin{aligned}\hat{\mathbf{x}}_{k+1|k+1} &= \hat{\mathbf{x}}_{k+1|k} + \mathbf{K}_{k+1}(\mathbf{o}(k+1) - \mathbf{h}(\hat{\mathbf{x}}_{k+1|k})) \\ \boldsymbol{\Sigma}_{k+1|k+1} &= \boldsymbol{\Sigma}_{k+1|k} - \mathbf{K}_{k+1}[\mathbf{H}(\hat{\mathbf{x}}_{k+1|k})\boldsymbol{\Sigma}_{k+1|k}\mathbf{H}^{\mathrm{Tr}}(\hat{\mathbf{x}}_{k+1|k}) + \mathbf{R})]\mathbf{K}_{k+1}^{\mathrm{Tr}}.\end{aligned}$$

Iterated extended Kalman filter

The above EKF algorithm uses the Taylor series approximation to the first iteration only. The approximation would be poor if the Taylor series expansion point were far away from the actual state variable. That is, the EKF algorithm is expected to work well only if the difference $\mathbf{x} - \hat{\mathbf{x}}_{k|k}$ is small. The **iterated EKF** algorithm is designed to keep this difference as small as possible. The iteration is carried out in the EKF-correction step until the error is small:

$$||\hat{\mathbf{x}}_{k|k}^{(I)} - \hat{\mathbf{x}}_{k|k}^{(I-1)}|| < \epsilon,$$

where I is the iteration number and ϵ is the pre-set convergence criterion.

At the first iteration, the EKF gain $\mathbf{K}_{k+1}$, and the error covariance matrices $\mathbf{\Sigma}_{k+1|k+1}$, $\mathbf{\Sigma}_{k+1|k}$ are computed using $\hat{\mathbf{x}}_{k+1|k}$, just like the conventional EKF. At the ith iteration, these quantities are computed using more accurate state-update information, $\hat{\mathbf{x}}_{k+1|k+k}^{(i-1)}$, which is computed from the previous $(i-1)^{th}$ iteration.

Nonlinear filtering

Both EKF and iterated EKF algorithms are ad-hoc state-estimation algorithms because they in principle do not provide an optimal estimate of the state $\mathbf{x}(k)$. We showed in earlier sections that the optimal state estimate in the MMSE sense is $E[\mathbf{x}(k)|\mathbf{o}]$. This, in general, is very difficult to compute for a nonlinear dynamic system. The difficulty arises due to the generally non-Gaussian nature in the distributions for the state and for the observation. This is so even if both state noise and observation noise in the system are Gaussian. In addition, even if at particular time step, k, the state happens to be Gaussian, the immediate next time step, $k+1$, will turn the system into non-Gaussian due to the nonlinearity in the state or in the observation equation.

The EKF algorithms give a first-order approximation to the optimal state estimate. It does not guarantee to work well in practice, and no convergence results have been known for them. However, there are rich studies in the field of nonlinear filtering devoted to improving state estimation accuracy over the EKF-like algorithms. Some common algorithms in nonlinear filtering include the quasi-linear Kalman filter, high-order Kalman filter, Gaussian sum filter, particle filters, Monte Carlo filter, bootstrap filter, unscented filter, and approximate inference algorithms based on mean-field theory and variational methods.

5.9 Summary

In this chapter we provided elements of estimation theory which forms another key aspect of statistical modeling after the process of model building as discussed in Chapter 4. Intimately related to estimation theory are optimization methods. Estimation theory establishes the criteria for determining model parameters or model states, while optimization methods provide solutions to achieving the criteria. It is thus natural to put optimization methods and estimation theory together in a single chapter.

This chapter began with classical optimization techniques, including the use of Lagrange multiplier method for constrained optimization problems. It then discussed common numerical methods for optimization, including root finding and gradient descent. A relatively modern optimization technique called dynamic programming (DP) was introduced thereafter, where examples were given for applying the DP technique to the HMM and the trended HMM. After the completion of the discussion on the various types of optimization techniques, the chapter started with an introduction to estimation theory. Then, two of the most common types of estimation criteria and solutions, least square estimation and maximum likelihood estimation, were covered for deterministic model parameters. For random parameters in statistical models, new estimation criteria were discussed, including minimum mean square error, maximum a posterior, and their generalization. Finally, moving beyond the parameter estimation problem, the chapter ended with a detailed discussion on the state estimation problem, where Kalman filter and extended Kalman filter techniques were presented.

Chapter 6

Statistical Pattern Recognition

The preceding chapter covered fundamentals of estimation theory, where the focus was the optimal determination of the parameters of given models. Examples of such models are the linear model, the hidden Markov model, and the state-space dynamic system model. The assumption has generally been that the functional forms of the models are a reasonably accurate reflection of the properties of the observation data. Often such properties are expressed as statistical generative properties.

In this chapter, we turn to a rather different class of problems where either the underlying models for the data generation are unknown, or such models are so complex that drastic simplification is used that significantly reduces the accuracy of the models to the observation data. The goal for this class of problems is not to obtain the optimal set of parameters or statistical distributions so as to make the models closely fit the data. Rather, the goal is to discriminate different discrete categories associated with different sets of the observation data using the data with unknown categories. This type of problem is called **pattern classification** or **pattern recognition**, and is the topic of this chapter. (By the tradition commonly followed in speech processing, pattern classification is usually referred to as a simplified pattern recognition problem where the time segments of the individual categories in the overall dynamic pattern are known in advance.)

For speech processing applications such as speech recognition or classification, pattern recognition theory plays an extremely important role. This is because speech recognition deals essentially with discrimination among discrete sets of linguistic units, such as sentences, words, or phonetic units. Although establishing accurate models may be instrumental for high-accuracy speech recognition and thus estimation theory has a significant role to play, this is only a means to an end, rather than the end itself. Pattern recognition theory and estimation theory have very different motivations, philosophies, and traditions. However, they are also intimately linked in several aspects. One aspect is that if indeed accurate models can be established via physical insight to the problem and via estimation theory, then Bayes decision theory (to be discussed as Section 6.1 of this chapter) — fundamentals of pattern recognition theory — provides principled ways of designing classifiers that guarantee minimal classification errors. Another common link between pattern recognition theory and estimation theory is that many techniques employed for solving their respective problems are based on similar optimization principles and on methods such as those discussed in Chapter 5.

For speech science, the concept of discrimination or classification is also highly significant. Theories of phonology have their very objective of accounting for the speech sound patterns that form the distinctive linguistic categories and that are maximally discriminative by the human perceptual system. Theories of phonetics, on the other hand, are required to explain why there is so much variability in either the acoustics or the articulation aspects of the speech process. One key to such accounts, which forms the core of some popular phonetic theories, is the proposed critical role of inherent (linguistic) discriminability among speech sound patterns despite the surface variability at the various observational or measurement levels.

Statistical pattern recognition/classification is a vast area. In this chapter, we only select a few topics that are most relevant to speech processing and that prepare a background for some later chapters. For more comprehensive treatment of the subject of statistical pattern recognition, readers are referred to [Bishop 97, Kay 98, Srin 78].

6.1 Bayes' Decision Theory

Bayes' decision theory is the foundation for optimal pattern classifier design, and provides the "fundamental equation" for modern speech recognition. The theory quantifies the concept of "accuracy" in pattern classification and recognition in statistical terms. That is, it defines the measure of accuracy in terms of the **minimum expected risk**, which can be achieved via the use of the Bayes decision rule.

6.1.1 Bayes' risk and MAP decision rule

To introduce the theory, consider a C-class classification problem, where each (continuous) random variable as an observation sample x (scalar, for simplicity) is to be classified as one of the C classes. The objective of the classifier is to design a mapping or decision function $C(x)$, also called the classification operation, from the observation space $x \in x$ into the discrete set $C_i \in Y, i = 1, 2, ...C$.

Bayes' decision theory assumes that *the joint probability distribution, $P(x, C_i)$, of the observation sample and the class is known to the designer.* It further assumes that each class pair (C_i, C_j) is associated with a risk (or loss, or cost) function r_{ij}, which is also given to the designer. The risk r_{ij} represents the cost of classifying a class-i observation into class j, where $r_{ii} = 0$ represents correct classification invoking a zero risk.

Given x that is mapped into class C_i by the decision function $C(x)$, we can compute its associated **average risk**, R, according to

$$R(C_i|x) = \sum_{j=1}^{C} r_{ji} P(C_j|x). \tag{6.1}$$

Note that x above is a random variable. Since $R(C_i|x)$ is a function of x, it is also a random variable. Hence we can take its expectation, which gives the **expected risk**:

$$\mathcal{L} = \int R(C(x)|x) p(x) dx. \tag{6.2}$$

This is a reasonable measure of the classifier performance, since it averages out all possible random events. Based on this measure — the lower the $\mathcal{L}$, the better the classifier — the best generic classifier is the decision function that gives

$$R(C(x)|x) = \min_i R(C_i|x). \tag{6.3}$$

For specific applications including speech processing, the risk function can be reasonably chosen to be zero or one: $r_{ij} = 1$ for $i \neq j$ and $r_{ii} = 0$. Now, Eq. 6.1 can be simplified to

$$R(C_i|x) = \sum_{j \neq i} 1 \times P(C_j|x) + 0 \times P(C_i|x) = 1 - P(C_i|x). \tag{6.4}$$

Then, the generic decision function design given by Eq. 6.3 becomes

$$C(x) = C_i \qquad \text{if and only if} \quad P(C_i|x) = \max_j P(C_j|x). \tag{6.5}$$

This is often more concisely written as

$$C_i = arg \max_{C_j} P(C_j|x), \tag{6.6}$$

where $arg \max$ is an operation that finds the argument maximizing the function that follows.

The decision rule given by Eq. 6.5 is the celebrated **MAP** or **maximum a posteriori** decision, since the posterior probability $P(C_j|x)$ is used as the basis for choosing the class using the maximum operation.

Note that given the average risk in Eq. 6.4 for which a unit risk is assigned to any error and no risk is assigned to correct classification, the expected risk given by Eq. 6.2 is the same as the error probability of classification.

In summary, we proved in the above that when all posterior probabilities of classes conditioned on the observation samples are known precisely, the classifier designed using the MAP decision rule of Eq. 6.5 is an optimal one in terms of minimizing the error probability of classification.

6.1.2 Practical issues

In practice, however, the posterior probabilities required for the MAP decision rule are either not known precisely, or need to be learned from training data (with known class labels provided). In this sense, the Bayes' decision theory outlined above effectively converts the classifier design problem into the estimation problem.

To elaborate on this further, we rewrite Eq. 6.6 using Bayes' rule:

Practical MAP decision rule for pattern classification:

$$C_i = arg \max_{C_j} \frac{P(C_j)p(x|C_j)}{p(x)} = arg \max_{C_j} P(C_j)p(x|C_j). \tag{6.7}$$

The last step above is due to the fact that $p(x)$ is not a function of the class label and thus can be dropped without affecting the maximization. Now, the posterior probabilities needed in the MAP decision rule can be equivalently represented by the class prior probability $P(C_j)$ and by the class-conditional probability density function $p(x|C_j)$. The estimation problem is typically much easier for these factorized probabilities than for the original posterior probabilities as in Eq. 6.6. Most techniques covered in Chapter 5, in fact, are applied to the estimation problem associated with the class-conditional probabilities.

As stated above, the most important practical issue related to the use of the MAP decision rule is the basic assumption in the Bayes' decision theory that the posterior probabilities, or equivalently the prior and class-conditional probabilities are precisely known. This assumption is often not valid in practice due to at least three reasons. First, for continuous observation x, the parametric form for $p(x|C_j)$ is often needed for computational reasons. The choice of the parametric form is usually limited by tractable computation and is thus at best only an approximation to the true data distribution. Second, solutions to the parameter estimation problem may not always be able to give the most desirable parameter set given the parametric form of the distribution. Third, the training data may not be sufficient or be free of mismatch for the parameter estimation problem. These practical issues suggest that despite the conceptual optimality provided by the MAP decision rule, lack of validity in its underlying assumption guarantees no such optimality in practice.

6.2 Minimum Classification Error Criterion for Recognizer Design

The practical issues discussed above against the theoretical optimality of the MAP-based classifier design motivate the search for alternative designs of classifiers. In this section, we discuss one such alternative, namely the use of the **minimum classification error (MCE)** criterion. This particular approach was originally proposed in [Juang 92], and has been fully described in [Chou 96, Chou 00, Kata 98] for speech recognition applications.

The objective of the classifier design and the associated parameter estimation in the MCE approach is to maximally discriminate the classes for direct and most desirable classification results. It pays no attention to fitting the models (represented by the associated distributions) to data, as required by the MAP decision rule derived from the classic Bayes' decision theory discussed earlier.

6.2.1 MCE classifier design steps

The key insight into the MCE classifier design is the realization that it is possible to estimate the classifier's parameters so that minimizing the expected risk could lead to a minimization of the recognition error rate. This is so even if there is no sufficient knowledge about the posterior probabilities as required by the classic Bayes' decision theory.

The role of the posterior probabilities in the classic theory has been replaced by the new concept of **discriminant function**. Discriminative functions, denoted by $g_\kappa(x, \Phi)$, are associated each class g_κ and determine the decision rule of the classifier. They may or may not be related to class-conditioned probabilities or class posterior probabilities. Using the discriminative functions, we can write down the decision rule of the classifier in the most general form:

$$C(x) \;=\; C^\kappa, \;\text{ if }\; g_\kappa(x, \Phi) \;=\; \max_j \; g_j(x, \Phi).$$

Based on this discriminative function approach, the MCE classifier design consists of the following three steps:

Misclassification measure

Given a discriminant function, the misclassification measure for the l-th input training observation sample x^l from class κ is defined as

$$\begin{aligned} d_\kappa(x^l, \Phi) &= -g_\kappa(x^l, \Phi) + \max_{j \neq \kappa} g_j(x^l, \Phi) \\ &= -g_\kappa(x^l, \Phi) + g_\chi(x^l, \Phi), \end{aligned}$$

where C^χ is the most confusable class. Clearly, $d_\kappa(x^l, \Phi) > 0$ implies misclassification and $d_\kappa(x^l, \Phi) \leq 0$ means correct classification.

Loss function

The loss function is defined as a sigmoid, non-decreasing function of d_κ:

$$\Upsilon_\kappa(x^l, \Phi) = \frac{1}{1 + e^{-d_\kappa(x^l, \Phi)}},$$

which approximates the classification error count.

Overall loss function

The overall loss function for the entire classifier is defined for each class as

$$\Upsilon(x^l, \Phi) = \sum_{\kappa=1}^{\mathcal{K}} \Upsilon_\kappa(x^l, \Phi)\delta[x^l \in C^\kappa], \tag{6.8}$$

where $\delta[\xi]$ is the Kronecker indicator function of a logic expression ξ that gives value 1 if the value of ξ is true and value 0 otherwise. The average loss (or error probability) for the entire training data set is defined as

$$\mathcal{L}(\Phi) = \frac{1}{L} \sum_{l=1}^{L} \Upsilon(x^l, \Phi), \tag{6.9}$$

where L is the total number of training tokens.

The above definitions emulate the classifier performance evaluation, based on the error rate count, in a smooth functional form, and they constitute the MCE classifier design strategy. The smoothness introduced above is necessary for classifier parameter optimization, which we discuss now.

6.2.2 Optimization of classifier parameters

The goal of the training process in the MCE approach, as discussed here, is to search for the parameters Φ in the discriminative functions of the classifier such that the overall loss function (and hence the classification error rate) is minimized.

For sequential optimization, the loss function $\Upsilon(x^l, \Phi)$ as in Eq. 6.9 is to be minimized, each time a training token x^l is presented, by adaptively adjusting the parameter set Φ according to

$$\Phi_{l+1} = \Phi_l - \epsilon \nabla \Upsilon(x^l, \Phi_l), \tag{6.10}$$

where Φ_l is the parameter set at the lth iteration, $\nabla\Upsilon(x^l, \Phi_l)$ is the gradient of the loss function for training sample x^l and ϵ is a small positive learning constant. This gradient-based optimization procedure has been called **generalized probabilistic descent (GPD)** [Kata 98, Chou 00].

In order to carry out the GPD computation, it is highly desirable to be able to compute the gradient $\nabla\Upsilon(x^l, \Phi_l)$ in an analytical form. We now use the chain rule to compute the gradient as follows. Let ϕ^j denote a parameter in the classifier associated with the discriminant function, $g_j(x^l, \Phi)$, for class j. The gradient for this parameter can be computed as

$$\begin{aligned}
\frac{\partial\Upsilon(x^l,\Phi)}{\partial\phi^j} &= \frac{\partial}{\partial\phi^j}\left(\sum_{\kappa'=1}^{\mathcal{K}} \Upsilon_{\kappa'}(x^l,\Phi)\delta[x^l \in C^{\kappa'}]\right) \\
&= \frac{\partial}{\partial\phi^j}\Upsilon_\kappa(x^l,\Phi) \\
&= \frac{\partial\Upsilon_\kappa(x^l,\Phi)}{\partial d_\kappa(x^l,\Phi)}\frac{\partial d_\kappa(x^l,\Phi)}{\partial g_j(x^l,\Phi)}\frac{\partial g_j(x^l,\Phi)}{\partial\phi^j}.
\end{aligned} \tag{6.11}$$

The first factor in the right-hand-side of Eq. 6.11 can be simplified to

$$\begin{aligned}
\frac{\partial\Upsilon_\kappa(x^l,\Phi)}{\partial d_\kappa(x^l,\Phi)} &= \frac{\partial}{\partial d_\kappa(x^l,\Phi)}\left(\frac{1}{1+e^{-d_\kappa(x^l,\Phi)}}\right) \\
&= \frac{e^{-d_\kappa(x^l,\Phi)}}{(1+e^{-d_\kappa(x^l,\Phi)})2} \\
&= e^{-d_\kappa(x^l,\Phi)}\Upsilon_\kappa^2(x^l,\Phi) \\
&= \{\frac{1}{\Upsilon_\kappa(x^l,\Phi)} - 1\}\Upsilon_\kappa^2(x^l,\Phi) \\
&= \Upsilon_\kappa(x^l,\Phi)[1-\Upsilon_\kappa(x^l,\Phi)].
\end{aligned} \tag{6.12}$$

The second factor of the right-hand-side of Eq. (6.11) can be simplified as follows:

$$\begin{aligned}
\frac{\partial d_\kappa(x^l,\Phi)}{\partial g_j(x^l,\Phi)} &= \frac{\partial}{\partial g_\chi(x^l,\Phi)}\left(-g_\kappa(x^l,\Phi) + g_\psi(x^l,\Phi)\right) \\
&= \begin{cases} -1 & \text{if } j=\kappa \\ 1 & \text{if } j=\chi\,. \end{cases}
\end{aligned} \tag{6.13}$$

The third factor of the right-hand-side of Eq. 6.11, $\frac{\partial g_j(x^l,\Phi)}{\partial\phi^j}$, requires that the functional form of the discriminant function be given. We will take up this issue later in this chapter.

6.3 Hypothesis Testing and the Verification Problem

6.3.1 MAP decision rule and hypothesis testing

We learned from Section 6.1 that the MAP decision rule of Eq. 6.7:

$$C_i = arg\max_{C_j} P(C_j)p(x|C_j) \tag{6.14}$$

gives minimum Bayes' risk for the zero-one risk (loss or cost) function r_{ij}. When the total number of classes is two, i.e., $C = 2$, this MAP rule is equivalent to the following **likelihood ratio test**:

$$\frac{P(x|C_1)}{P(x|C_2)} > \frac{P(C_2)}{P(C_1)} = \xi,$$

where ξ is called the **threshold** of the test.

For the general (non-zero or one) risk function r_{ij}, it can be easily shown that the two-class MAP decision rule is also a likelihood ratio test with the test threshold being:

$$\xi = \frac{(r_{12} - r_{22})P(C_2)}{(r_{21} - r_{11})P(C_1)}.$$

The above decision or test has been associated with the following *null* hypothesis:

H₀: Observation x belongs to class C_1,

against the *alternative* hypothesis:

H₁: Observation x belongs to class C_2.

In contrast to the likelihood ratio test discussed above that is based on the criterion of minimal Bayes' risk, a more frequently used formulation of hypothesis testing is based on the **Neyman-Pearson** criterion. The Neyman-Pearson testing aims to minimize the "false-alarm" decision error e_1 (also called the type-I error) while keeping the "miss" error e_2 (also called the type-II error) constant: $e_2 = e_0$. This is a constrained optimization problem, whose solution will give the Neyman-Pearson decision rule. To solve this optimization problem, we use the Lagrange multiplier technique discussed in Chapter 4, and form the objective function of

$$E = e_1 + \lambda(e_2 - e_0),$$

where λ is the Lagrange multiplier.

Using the definitions for the errors e_1 and e_2, it can be shown that the solution to the above constrained optimization problem happens also to be that of the likelihood ratio test:

$$\frac{P(x|C_1)}{P(x|C_2)} > \lambda,$$

with the test threshold being the Lagrange multiplier. In theory, the threshold can be determined analytically according to

$$e_2 \equiv \int_{-\infty}^{\lambda} p(x|C_2)dx = e_0.$$

In practice, an analytical solution to λ above is often impossible, and numerical or experimental methods are needed.

6.3.2 Verification problem in pattern recognition

A typical L-class pattern recognition problem (L is equal to or greater than two) as we have discussed so far can be summarized as follows: Given L classes $\{C_i \mid 1 \leq i \leq L\}$ and an unknown observation y, y is to be classified into one of the above classes C_i. It does not consider the situation that y belongs to none of these classes. This situation arises frequently in practice, where we call the observation an *outlier*. The outlier should be detected and rejected, rather than being misclassified into one of the classes. Most statistical pattern recognition techniques do not have an explicit mechanism to reject y

when it is an outlier.

Now we consider the problem of rejecting outliers in pattern recognition based on a two-pass strategy. In the first phase of *pattern classification*, y is classified into one very likely class, C_l $(1 \leq l \leq L)$. In the second phase of *outlier verification*, we verify whether y really belongs to the class C_l (accept) or y is an outlier (reject). The first stage is a typical pattern recognition problem with a well-accepted solution of the MAP decision rule as discussed earlier. In this section, we discuss some solutions to the second phase, i.e., the outlier verification problem.

We assume that there is no knowledge about each class C_l, except that one can collect a set of training samples for C_l. Suppose we have samples $\mathbf{x}_l = \{x_{l1}, x_{l2}, \cdots, x_{lT_l}\}$ for the class C_l, where T_l means the total number of samples for class C_l. Meanwhile, we assume another outlier class C_0, which represents all outliers. We also assume having another set of samples of outliers: $\mathbf{x}_0 = \{x_{01}, x_{02}, \cdots, x_{0T_0}\}$ for class C_0.

Based on the above assumptions, we formulate the verification problem as hypothesis testing. Here the *null* hypothesis is:

$\mathbf{H_0}$: $\mathbf{x}_l$ and y all come from C_l while $\mathbf{x}_0$ comes from C_0,

and the alternative hypothesis is:

$\mathbf{H_1}$: $\mathbf{x}_l$ comes from C_l while $\mathbf{x}_0$ and y comes from C_0. The solution to the above testing problem is: if

$$\eta = \frac{f(\mathbf{x}_l, y, \mathbf{x}_0 \mid H_0)}{f(\mathbf{x}_l, y, \mathbf{x}_0 \mid H_1)} = \frac{f(\mathbf{x}_l, y \mid C_l) \cdot f(\mathbf{x}_0 \mid C_0)}{f(\mathbf{x}_l \mid C_l) \cdot f(\mathbf{x}_0, y \mid C_0)} < \xi \tag{6.15}$$

then y is rejected as an outlier; otherwise, accept y as belonging to class C_l. ξ is a fixed test threshold, and $f(.|.)$ denotes the conditional PDF.

In the following, we study the above outlier verification problem from both non-Bayesian and Bayesian viewpoints. We take the Gaussian distribution as an example and derive the closed-form solution for each case.

6.3.3 Neymann-Pearson approach to verification

The Neymann-Pearson approach to verification described in this subsection is based on the traditional, non-Bayesian framework. In conventional non-Bayesian statistics, we usually assume the form of a parametric model for each class C_l is known in advance, and leave its parameters Λ_l to be estimated from the training samples $\mathbf{x}_l$. Here we consider the case of the Gaussian distribution $\mathcal{N}(\mu_l, \sigma_l)$, where the model parameters are $\Lambda_l = \{m_l, \sigma_l\}$, that is,

$$f(x \mid \Lambda_l) = \mathcal{N}(m_l, \sigma_l) = \sqrt{\frac{\sigma_l}{2\pi}} e^{-\frac{\sigma_l}{2}(x - m_l)^2} \quad (l = 0, 1, 2 \cdots, L). \tag{6.16}$$

In the non-Bayesian framework (Neymann-Pearson method), the parameter Λ_l is viewed as an estimator from the samples $\mathbf{x}_l$. The above problem of hypothesis testing Eq.6.15 becomes the conventional likelihood ratio test:

$$\eta = \frac{f(\mathbf{x}_l \mid \Lambda_l) \cdot f(y \mid \Lambda_l) \cdot f(\mathbf{x}_0 \mid \Lambda_0)}{f(\mathbf{x}_l \mid \Lambda_l) \cdot f(y \mid \Lambda_0) \cdot f(\mathbf{x}_0 \mid \Lambda_0)} = \frac{f(y \mid \Lambda_l)}{f(y \mid \Lambda_0)} < \xi, \tag{6.17}$$

where Λ_l and Λ_0 are the estimators of the model parameters of the class C_l and C_0 based on the samples $\mathbf{x}_l$ and $\mathbf{x}_0$. Now, we adopt the maximum likelihood (ML) estimator for Λ_l and Λ_0:

$$\mu_l = \frac{1}{T_l} \sum_{j=1}^{T_l} x_{lj} \tag{6.18}$$

$$\sigma_l^2 = \frac{1}{T_l - 1} \sum_{j=1}^{T_l} (x_{lj} - \mu_l)^2 \tag{6.19}$$

$$\mu_0 = \frac{1}{T_0} \sum_{j=1}^{T_0} x_{0j} \tag{6.20}$$

$$\sigma_0^2 = \frac{1}{T_0 - 1} \sum_{j=1}^{T_0} (x_{0j} - \mu_0)^2. \tag{6.21}$$

Based on these, the likelihood test above becomes

$$\eta' = \ln \eta = \log(\frac{\sigma_l}{\sigma_0}) - \frac{\sigma_l}{2}(y - m_l)^2 + \frac{\sigma_0}{2}(y - m_0)^2 < \xi'. \tag{6.22}$$

6.3.4 Bayesian approach to verification

In the Bayesian framework, all knowledge about each class C_l $(l = 0, 1, \cdots, L)$ is contained in a prior PDF $\rho_l(\cdot)$. For simplicity, we assume that the variance σ_l is known and only mean m_l is unknown. If we choose the natural conjugate prior, the prior PDF's will be

$$\rho(m_l) = \sqrt{\frac{\tau_l}{2\pi}} e^{-\frac{\tau_l}{2}(m_l - \mu_l)^2} \quad (l = 0, 1, 2, \cdots, L), \tag{6.23}$$

where $\{\mu_l, \tau_l \mid l = 0, 1, 2, \cdots, L\}$ are the hyperparameters.

The Bayesian solution to the hypothesis testing Eq. 6.15 can be rewritten as:

$$\begin{aligned} \eta &= \frac{\hat{p}_l(\mathbf{x}_l, y) \cdot \hat{p}_0(\mathbf{x}_0)}{\hat{p}_l(\mathbf{x}_l) \cdot \hat{p}_0(\mathbf{x}_0, y)} \\ &= \frac{\int f(\mathbf{x}_l, y \mid m_l) \cdot \rho(m_l) dm_l \cdot \int f(\mathbf{x}_0 \mid m_0) \cdot \rho(m_0) dm_0}{\int f(\mathbf{x}_l \mid m_l) \cdot \rho(m_l) dm_l \cdot \int f(\mathbf{x}_0, y \mid m_0) \cdot \rho(m_0) dm_0} < \xi, \end{aligned} \tag{6.24}$$

where $\hat{p}_l(\cdot)$ denotes the integration over the prior (also called the Bayesian predictive density) of class C_l.

For $l = 0, 1, \cdots, L$, we have

$$\begin{aligned} \hat{p}_l(\mathbf{x}_l) &= \int f(\mathbf{x}_l \mid m_l) \cdot \rho(m_l) dm_l \\ &= \frac{\tau_l^{1/2} \cdot \sigma_l^{T_l/2}}{(2\pi)^{T_l/2} \cdot (\tau_l + \sigma_l T_l)^{1/2}} \cdot \exp\{-\frac{\sigma_l T_l}{2(\tau_l + \sigma_l T_l)} [\tau_l \overline{(x - \mu)_l^2} + \sigma_l T_l (\overline{x_l^2} - \overline{x}_l^2)]\} \end{aligned} \tag{6.25}$$

and

$$\hat{p}_l(\mathbf{x}_l, y) = \int f(\mathbf{x}_l, y \mid m_l) \cdot \rho(m_l) dm_l = \frac{\tau_l^{1/2} \cdot \sigma_l^{(T_l+1)/2}}{(2\pi)^{(T_l+1)/2} \cdot (\tau_l + \sigma_l + \sigma_l T_l)^{1/2}}$$
$$\exp\{-\frac{\sigma_l \tau_l T_l \overline{(x-\mu)_l^2} + \sigma_l \tau_l (y-\mu_l)^2 + \sigma_l^2 (T_l+1)(T_l \overline{x_l^2} + y^2) - \sigma_l^2 (T_l \overline{x}_l + y)^2}{2(\tau_l + \sigma_l + \sigma_l T_l)}\}, \tag{6.26}$$

where

$$\overline{(x-\mu)_l^2} = \frac{1}{T_l} \sum_{i=1}^{T_l} (x_i - \mu_l)^2 \quad (l = 0, 1, 2, \cdots, L) \tag{6.27}$$

$$\overline{x_l^2} = \frac{1}{T_l} \sum_{i=1}^{T_l} x_i^2 \quad (l = 0, 1, 2, \cdots, L) \tag{6.28}$$

$$\overline{x}_l = \frac{1}{T_l} \sum_{i=1}^{T_l} x_i \quad (l = 0, 1, 2, \cdots, L). \tag{6.29}$$

Therefore, the basic hypothesis testing in Eq. 6.24 becomes

$$\begin{aligned}
\eta' &= \ln \eta \\
&= \ln \frac{\tau_l (\tau_l + \sigma_l T_l)^{1/2} (\tau_0 + \sigma_0 + \sigma_0 T_0)^{1/2}}{\tau_0 (\tau_0 + \sigma_0 T_0)^{1/2} (\tau_l + \sigma_l + \sigma_l T_l)^{1/2}} + \frac{\sigma_l T_l}{2(\tau_l + \sigma_l T_l)} [\tau_l \overline{(x-\mu)_l^2} + \sigma_l T_l (\overline{x_l^2} - \overline{x}_l^2)] \\
&- \frac{\sigma_l \tau_l T_l \overline{(x-\mu)_l^2} + \sigma_l \tau_l (y-\mu_l)^2 + \sigma_l^2 (T_l+1)(T_l \overline{x_l^2} + y^2) - \sigma_l^2 (T_l \overline{x}_l + y)^2}{2(\tau_l + \sigma_l + \sigma_l T_l)} \\
&+ \frac{\sigma_0 \tau_0 T_0 \overline{(x-\mu)_0^2} + \sigma_0 \tau_0 (y-\mu_0)^2 + \sigma_0^2 (T_0+1)(T_0 \overline{x_0^2} + y^2) - \sigma_0^2 (T_0 \overline{x}_0 + y)^2}{2(\tau_0 + \sigma_0 + \sigma_0 T_0)} \\
&- \frac{\sigma_0 T_0}{2(\tau_0 + \sigma_0 T_0)} [\tau_0 \overline{(x-\mu)_0^2} + \sigma_0 T_0 (\overline{x_0^2} - \overline{x}_0^2)] < \xi'.
\end{aligned} \tag{6.30}$$

6.4 Examples of Applications

In this section, we provide some examples of applications of the basic statistical pattern recognition theory discussed earlier in this chapter.

6.4.1 Discriminative training for HMM

In this example of application, we take the HMM (discussed in Chapter 3) as a mathematical tool to define the discriminative function for pattern recognition. (This is in contrast to the use of the HMM as a generative model for the observation data viewed in Chapter 3.) Consistent with the maximum-likelihood-based decision rule commonly used for the HMM-based pattern recognition, the discriminant function is accordingly defined as the largest log-likelihood of the observation sequence scored by a complete state path in the HMM. That is, the discriminant function is the log-likelihood score of the input sequence $\mathbf{o}^l$ along the optimal state sequence $\Theta^\kappa = \{\theta_1^\kappa, \theta_2^\kappa \cdots, \theta_{T^l}^\kappa\}$ for the

HMM associated with the κth class Φ^κ:

$$g_\kappa(\mathbf{o}^l, \Phi) = \sum_{t=1}^{T^l} \log b_{\theta_t^\kappa}(\mathbf{o}_t^l) \tag{6.31}$$

The example below in this subsection shows how the MCE-based discriminative learning for the recognizer design can be accomplished using this discriminant function.

As discussed in Section 6.2.2, the optimization procedure for the MCE training centers on the computation of the gradient $\nabla\Upsilon(\mathbf{o}^l, \Phi_l)$ in an analytical form, where $\Upsilon(\mathbf{o}^l, \Phi_l)$ was defined in Eq. 6.8. Use of the chain rule further reduces the computation to that of the gradient of the discriminant function (Eq. 6.31) with respect to the HMM parameters.

Let the HMM parameters to be optimized be mixture (m at state i) weights $c_{i,m}$, Gaussian mean vectors $\boldsymbol{\mu}_{i,m}$, and Gaussian covariance matrices $\boldsymbol{\Sigma}_{i,m}$, for all the model classes $j = 1, 2, ...\mathcal{K}$. Define the set T_i^l for all the time indices for which the state index of the state sequence at time t belongs to state i in the Markov chain, i.e.,

$$T_i^l = \{t|\theta_t = i\}, \quad 1 \leq i \leq N, \quad 1 \leq t \leq T^l.$$

Also, define the *a posteriori* probability of

$$\gamma_{i,m}^j(t) = \frac{c_{i,m} b_{i,m}(\mathbf{o}_t^l)}{b_i(\mathbf{o}_t^l)}.$$

Then, the gradients required for the MCE training can be computed in the following analytical forms:

$$\frac{\partial\Upsilon(\mathbf{o}^l, \Phi)}{\partial\boldsymbol{\mu}_{i,m}} = \psi \sum_{t\in T_i^l} \gamma_{i,m}(t)\boldsymbol{\Sigma}_{i,m}^{-1}(\mathbf{o}_t^l - \boldsymbol{\mu}_{i,m}) \tag{6.32}$$

$$\frac{\partial\Upsilon(\mathbf{o}^l, \Phi)}{\partial\tilde{\boldsymbol{\Sigma}}_{i,m}} = 0.5\psi \sum_{t\in T_i^l} \gamma_{i,m}(t)[(\mathbf{o}_t^l - \boldsymbol{\mu}_{i,m})^{\mathrm{Tr}}(t)\boldsymbol{\Sigma}_{i,m}^{-1}(\mathbf{o}_t^l - \boldsymbol{\mu}_{i,m}) - \mathbf{I}] \tag{6.33}$$

$$\frac{\partial\Upsilon(\mathbf{o}^l, \Phi)}{\partial\tilde{c}_{i,m}} = \psi \sum_{t\in T_i^l(j)} \gamma_{i,m,t}(j) - c_{i,m}(j), \tag{6.34}$$

where in Eq. 6.33 $\tilde{\boldsymbol{\Sigma}}_{i,m}$ are the log-transformed covariance matrices:

$$\tilde{\boldsymbol{\Sigma}}_{i,m} = \log \tilde{\boldsymbol{\Sigma}}_{i,m}.$$

Its use is for the purpose of implementation simplicity with the incorporation of the non-negativity constraint [Chou 00]. For the same reason, in Eq. 6.34 $\tilde{c}_{i,m}$ are also the transformed mixture weights:

$$c_{i,m} = \frac{\exp(\tilde{c}_{i,m})}{\sum_l \exp(\tilde{c}_{i,l})}.$$

The quantity ψ in Eqs. 6.32, 6.33 and 6.34 is

$$\psi = \begin{cases} \Upsilon_\kappa(\mathbf{o}^l, \Phi)[\Upsilon_\kappa(\mathbf{o}^l, \Phi) - 1] & \text{for the correct class } \kappa \\ \Upsilon_\kappa(\mathbf{o}^l, \Phi)[1 - \Upsilon_\kappa(\mathbf{o}^l, \Phi)] & \text{for the most confusable class } \chi\,. \end{cases}$$

6.4.2 Discriminative training for the trended HMM

In this example, we generalize the results above from the conventional HMM to the trended HMM introduced in Chapters 3 and 4. In this section, we will use the mixture version of the trended HMM, which is slightly more general than the unimodal Gaussian-trended HMM discussed in Chapters 3 and 4. The technique described in this section has been successfully applied to speech recognition [Rathi 98].

Recall that the trended HMM is a data-generative type. The mixture version of this model can be described by the following equation for the generation of the observation sequence:

$$\begin{aligned} \mathbf{o}_t &= g_t(i,m) + \mathbf{r}_t(\mathbf{\Sigma}_{i,m}), \\ &= \sum_{p=0}^{P} \mathbf{b}_{i,m}(p)(t-\tau_i)^p + \mathbf{r}_t(\mathbf{\Sigma}_{i,m}), \quad m = 1,2,\cdots,M; \quad i = 1,2,\cdots,N, \quad (6.35) \end{aligned}$$

where $\mathbf{o}_t$, $t = 1,2,\cdots,T$ is a modeled observation data sequence of length T, within the HMM state indexed by i; $B_{i,m}(p)$ are mixture-dependent and state-dependent polynomial regression coefficients of order P indexed by mixture component m and by state i; and the term R_t is the stationary residual (after the data-fitting by the first term F_t) assumed to be independent and identically distributed (IID) and zero-mean Gaussian source characterized by state (i)-dependent, mixture (m)-dependent, but time-invariant diagonal covariance matrix $\mathbf{\Sigma}_{i,m}$.

In the conventional, stationary-state HMM, the first term in Eq. 6.35 is only a function of state i, not a function of time t. Note also that the polynomials for each state depend not only on the coefficients $B_{i,m}(p)$, but also on the time-shift parameter τ_i. The term $t - \tau_i$ represents the sojourn time in state i at time t, where τ_i registers the time when state i in the HMM is just entered before regression on time takes place. Polynomial coefficients $B_{i,m}(p)$ (for state i and mixture component m) are considered as true model parameters and τ_i is merely an auxiliary parameter for the purpose of obtaining maximal accuracy in estimating $B_{i,m}(p)$. In the recognition step, τ_i is again estimated as the auxiliary parameter so as to achieve a maximal score in matching the model to the unknown utterance over all possible τ_i values.

In short, a mixture-trended HMM consists of the following parameter quadruple $[\mathbf{A}, \mathbf{B}, \mathbf{\Sigma}, \mathbf{w}]$:

1. $\mathbf{A} = a_{i,j}$, $i,j = 1,2,\cdots,N$ is the transition probability matrix of the underlying Markov chain with a total of N states.

2. $\mathbf{B} = \mathbf{b}_{i,m}(p)$, $i = 1,2,\cdots,N$, $m = 1,2,\cdots,M$, and $p = 1,2,\cdots,P$ are the polynomial coefficients, of order P and associated with state i and mixture m, in the state-dependent deterministic regression function of time.

3. $\mathbf{\Sigma} = \mathbf{\Sigma}_{i,m}$, $i = 1,2,\cdots,N$ and $m = 1,2,\cdots,M$ are the time-invariant covariance matrices (dimensionality of $n \times n$) of the zero-mean Gaussian IID residual signals $R_t(\mathbf{\Sigma}_{i,m})$. (These matrices are also state and mixture dependent.)

4. $\mathbf{w} = w_{i,m}$, $i = 1,2,\cdots,N$ and $m = 1,2,\cdots,M$ are the mixture weights.

We now describe an MCE-based discriminative training paradigm in the context of the above (mixture) trended HMM for achieving optimal estimation of the state-dependent polynomial coefficients. Let Φ_j, $j = 1,2,\cdots,\mathcal{K}$, denote the parameter set characterizing the trended HMM for the j-th class, where $\mathcal{K}$ is the total number of classes. The classifier based on these $\mathcal{K}$ class models can be characterized by $\Phi = \{\Phi_1, \Phi_2, \cdots, \Phi_{\mathcal{K}}\}$. The purpose of the MCE-based discriminative training is to find the parameter set Φ such that the probability of misclassifying all the training tokens is minimized.

Let $g_j(\mathbf{o}, \Phi)$ denote the log-likelihood associated with the optimal state sequence Θ for the input token $\mathbf{o}$, obtained by applying the Viterbi algorithm using model Φ_j for the j-th class. Then, for the utterance $\mathbf{o}$ (from class c), the misclassification measure $d_c(\mathbf{o}, \Phi)$ is determined by

$$d_c(\mathbf{o}, \Phi) = -g_c(\mathbf{o}, \Phi) + g_\chi(\mathbf{o}, \Phi), \tag{6.36}$$

where χ denotes the incorrect model with the highest log-likelihood (i.e., the most confusable class). In this definition, a negative value of $d_c(\mathbf{o}, \Phi)$ corresponds to a correct classification. The definition in Eq. 6.36 focuses on the comparison between the true model and only the closest-competing wrong model, an approximation that we adopt in this study for computation efficiency. (A more general form of the misclassification measure using the log-likelihoods from all models can be found in [Chou 00, Rathi 97].) A loss function with respect to the input token is defined in terms of the misclassification measure given by

$$\Upsilon(\mathbf{o}, \Phi) = \frac{1}{1 + e^{-d_c(\mathbf{o}, \Phi)}}, \tag{6.37}$$

which projects $d_c(\mathbf{o}, \Phi)$ into the interval [0,1]. Note that the loss function $\Upsilon(\mathbf{o}, \Phi)$ is directly related to the classification error rate and is first-order differentiable with respect to all the model parameters of Φ_j, $j = 1, 2, \cdots, \mathcal{K}$. Once the objective function in Eq. 6.37 is determined, the MCE-based discriminative training is reduced to finding the gradient of the objective function with respect to all the model parameters and to using the computed gradient to update the model parameters in an iterative manner.

Let ϕ be a parameter in the model Φ. Provided that $\Upsilon(\mathbf{o}, \Phi)$ is differentiable with respect to ϕ, that parameter is adjusted in the gradient descent method according to

$$\begin{aligned} \hat{\phi} &= \phi - \epsilon \frac{\partial \Upsilon(\mathbf{o}, \Phi)}{\partial \phi}, \quad \text{or} \\ \hat{\phi} &= \phi - \epsilon \underbrace{\Upsilon(\mathbf{o}, \Phi)(1 - \Upsilon(\mathbf{o}, \Phi))}_{\psi} \frac{\partial d_c(\mathbf{o}, \Phi)}{\partial \phi}. \end{aligned} \tag{6.38}$$

In Eq. 6.38, $\hat{\phi}$ is the new estimate of the parameter and ϵ is a small positive constant that monotonically decreases as the iteration number increases. This gradient descent method is iteratively applied to all training tokens in a sequential manner (for all model parameters) to minimize the loss function during the training process.

Some intuitive explanations for Eq. 6.38 are given here. In the case of near error-free classification (i.e., $\Upsilon(\mathbf{o}, \Phi) \approx 0$), or in the case of a complete loss (very poor classification; i.e., $\Upsilon(\mathbf{o}, \Phi) \approx 1$), the magnitude of ψ in Eq. 6.38 would be close to zero and therefore the change of ϕ would become very small. On the other hand, if $\Upsilon(\mathbf{o}, \Phi) \approx 0.5$ (i.e., the likelihoods for the correct and the best wrong model about the same), then the magnitude of ψ would reach a maximum. Therefore, the training procedure as described in Eq. 6.38 will focus on input tokens which are likely to be misclassified but can be classified correctly after proper adjustment of the model parameters.

In order to determine $\frac{\partial d_c(\mathbf{o},\Phi)}{\partial \phi}$ in Eq. 6.38, we note that in the trended HMM, each mixture of each state is characterized by a multivariate time-varying Gaussian density function in the form of

$$b_{i,m}(\mathbf{o}_t|\tau_i) = \frac{(2\pi)^{\frac{-D}{2}}}{|\mathbf{\Sigma}_i|^{\frac{1}{2}}} \exp\left(\frac{-1}{2}\left[\mathbf{o}_t - \sum_{p=0}^{P} \mathbf{b}_{i,m}(p)(t-\tau_i)^p\right]^{\mathrm{Tr}} \mathbf{\Sigma}_{i,m}^{-1}\left[\mathbf{o}_t - \sum_{p=0}^{P} \mathbf{b}_{i,m}(p)(t-\tau_i)^p\right]\right), \tag{6.39}$$

where $\mathbf{b}_{i,m}(p)$ and $\mathbf{\Sigma}_{i,m}$ denote the polynomial coefficients for the time-varying Gaussian mean and the covariance matrix associated with the m-th mixture of the i-th state, respectively; $(t-\tau_i)$ is the sojourn time in state i at time t, and D is the dimensionality of the observation vector $\mathbf{o}_t$. Based on the trended HMM for speech class j, the optimal state sequence $\Theta^j = \theta_1^j, \theta_2^j, \cdots, \theta_T^j$ and the corresponding mixture sequence $\mathcal{M}^j = m_1^j, m_2^j, \cdots, m_T^j$ for an input token $\mathbf{o} = \mathbf{o}_1, \mathbf{o}_2, \cdots, \mathbf{o}_T$ (T frames in total) is obtained by means of the Viterbi algorithm, with modification by incorporating an additional optimization loop for the state sojourn time. Then, the log-likelihood used as the discriminant function is given by

$$g_j(\mathbf{o},\Phi) = \sum_{t=1}^{T} \log b_{\theta_t^j, m_t^j}(\mathbf{o}_t|\tau_{\theta_t^j}). \tag{6.40}$$

We now derive the expression for the gradient computation of the polynomial coefficients in the trended HMM. By substituting Eqs. 6.36, 6.39 and 6.40 into Eq. 6.38, the gradient calculation of the m-th mixture of the i-th state parameter, $\mathbf{b}_{i,m}^{(j)}(l)$, $l = 0, 1, \cdots, P$, for the j-th model becomes

$$\begin{aligned}
\frac{\partial \Upsilon(\mathbf{o},\Phi)}{\partial \mathbf{b}_{i,m}^{(j)}(l)} &= \psi \frac{\partial d_c(\mathbf{o},\Phi)}{\partial \mathbf{b}_{i,m}^{(j)}(l)} = \psi \frac{\partial}{\partial \mathbf{b}_{i,m}^{(j)}(l)}\left(-g_c(\mathbf{o},\Phi) + g_\chi(\mathbf{o},\Phi)\right) \\
&= \psi \frac{\partial}{\partial \mathbf{b}_{i,m}^{(j)}(l)}\left(-\sum_{t=1}^{T} \log b_{\theta_t^c, m_t^c}(\mathbf{o}_t|\tau_{\theta_t^c}) + \sum_{t=1}^{T} \log b_{\theta_t^\chi, m_t^\chi}(\mathbf{o}_t|\tau_{\theta_t^\chi})\right) \\
&= \psi_j \sum_{t \in T_{i,m}(j)} \frac{\partial}{\partial \mathbf{b}_{i,m}^{(j)}(l)}\left(-\frac{n}{2}\log 2\pi - \frac{1}{2}\log|\mathbf{\Sigma}_{i,m}^{(j)}|\right. \\
&\quad \left. -\frac{1}{2}\left[\mathbf{o}_t - \sum_{p=0}^{P} \mathbf{b}_{i,m}^{(j)}(p)(t-\tau_i)^p\right]^{\mathrm{Tr}} \mathbf{\Sigma}_{i,m}^{(j)^{-1}} \left[\mathbf{o}_t - \sum_{p=0}^{P} \mathbf{b}_{i,m}^{(j)}(p)(t-\tau_i)^p\right]\right) \\
&= \psi_j \sum_{t \in T_{i,m}(j)} \mathbf{\Sigma}_{i,m}^{(j)^{-1}} \left[\mathbf{o}_t - \sum_{p=0}^{P} \mathbf{b}_{i,m}^{(j)}(p)(t-\tau_i)^p\right](t-\tau_i)^l,
\end{aligned} \tag{6.41}$$

where the variable ψ_j is defined as

$$\psi_j = \begin{cases} \psi & \text{if } j = c \text{ (correct class)} \\ -\psi & \text{if } j = \chi \text{ (closest competing class)} \\ 0 & \text{otherwise,} \end{cases}$$

and the set $T_{i,m}(j)$ includes all the time indices such that mixture m and state i are in the optimal Viterbi path determined using the j-class model; that is,

$$T_{i,m}(j) = \{t|\theta_t^j = i, m_t^j = m\}, \quad 1 \le i \le N, \quad 1 \le m \le M.$$

The gradient formula for covariance matrices can be similarly derived, which has the following final form:

$$\frac{\partial \Upsilon(\mathbf{o}, \Phi)}{\partial \tilde{\boldsymbol{\Sigma}}_{i,m}^{(j)}} = 0.5\psi_j \sum_{t \in T_{i,m}(j)} \left[\mathbf{o}_t - \sum_{p=0}^{P} \mathbf{b}_{i,m}^{(j)}(p)(t-\tau_i)^p \right] \left[\mathbf{o}_t - \sum_{p=0}^{P} \mathbf{b}_{i,m}^{(j)}(p)(t-\tau_i)^p \right]^{\mathrm{Tr}} \boldsymbol{\Sigma}_{i,m}^{(j)^{-1}} - \mathbf{I}, \tag{6.42}$$

where $\mathbf{I}$ indicates the $n \times n$ unity matrix and $\tilde{\boldsymbol{\Sigma}}_{i,m}$ are the log-transformed diagonal covariance matrices to automatically impose the constraint that the variances always remain positive definite during training.

6.4.3 Discriminative feature extraction

In this last example, we apply the MCE-based discriminative training to joint optimization of HMM parameters and of feature extraction/transformation. This strategy has been successfully used in speech recognition [Rathi 97].

Let $\mathcal{F} = \{\mathcal{F}^1, \mathcal{F}^2, \cdots, \mathcal{F}^L\}$ denote a set of L D-dimensional vector-valued sequences. These are the raw features subject to further transformation before entering into a pattern classifier. (This transformation is subject to optimization jointly with the classifier parameters.) An example of such raw features is the DFT values computed directly from speech waveforms. Another example is mel-filter-bank log-energy (or mel-warped DFT), as used in the work of [Rathi 97]. Also, let $\mathcal{F}^l = \{\mathcal{F}_1^l, \mathcal{F}_2^l, \cdots, \mathcal{F}_{T^l}^l\}$ denote the l-th sequence having a length of T^l frames.

Based on the basic feature vectors $\mathcal{F}$, we now create a set of new, linearly transformed (static) feature vectors $\mathbf{x}_t^l$. The transformed features are state (i)-dependent, mixture-component (m)-dependent, linear combinations of each row of the transformation matrix with each element of basic features:

$$\mathbf{x}_{p,t}^l = \sum_{q=1}^{n} h_{p,q,i,m} \mathcal{F}_{q,t}^l \qquad p = 1, 2, \cdots, D, \qquad t = 1, 2, \cdots, T^l. \tag{6.43}$$

In the matrix form, Eq. 6.43 can be written as

$$\begin{pmatrix} x_{1,t}^l \\ x_{2,t}^l \\ \vdots \\ x_{D,t}^l \end{pmatrix} = \begin{pmatrix} h_{1,1,i,m} & h_{1,2,i,m} & \cdots & h_{1,n,i,m} \\ h_{2,1,i,m} & h_{2,2,i,m} & \cdots & h_{2,n,i,m} \\ \vdots & \vdots & \vdots & \vdots \\ h_{D,1,i,m} & h_{D,2,i,m} & \cdots & h_{D,n,i,m} \end{pmatrix} \begin{pmatrix} \mathcal{F}_{1,t}^l \\ \mathcal{F}_{2,t}^l \\ \vdots \\ \mathcal{F}_{n,t}^l \end{pmatrix},$$

$$\text{or} \quad \mathbf{x}_t^l = \mathbf{H}_{i,m} \mathcal{F}_t^l,$$

where $h_{p,q,i,m}$ is the pq-th element of the transformation matrix $\mathbf{H}_{i,m}$ associated with the m-th mixture residing in the Markov state i, n is the number of MFB log channel energies for each frame, and D is the vector size of the transformed static feature. Note that the transformed static features constructed above can be interpreted as the output from a slowly time-varying (due to state dependence of the transformation) linear filter with the MFB log energy vector sequence as the input.

Given the transformed static features as described above, the dynamic feature vectors $\mathbf{y}_t^l$ (for frame t of the l-th token) can be constructed as additional state-dependent, trainable linear combinations of the static features stretching over the interval f frames forward and b frames backward according to

$$\mathbf{y}_t^l = \sum_{k=-b}^{f} w_{k,i,m}\mathbf{x}_{t+k}^l, \qquad 1 \leq l \leq L,\ 1 \leq t \leq T^l, \tag{6.44}$$

where $w_{k,i,m}$ is the kth scalar weighting coefficient associated with the mth mixture residing in the Markov state i. In the matrix form, Eq. (6.44) can be written as

$$\begin{pmatrix} \mathbf{y}_{1,t}^l \\ \mathbf{y}_{2,t}^l \\ \vdots \\ \mathbf{y}_{D,t}^l \end{pmatrix} = \begin{pmatrix} \mathbf{x}_{1,t-b}^l & \cdots & \mathbf{x}_{1,t}^l & \cdots & \mathbf{x}_{1,t+f}^l \\ \mathbf{x}_{2,t-b}^l & \cdots & \mathbf{x}_{2,t}^l & \cdots & \mathbf{x}_{2,t+f}^l \\ \vdots & \vdots & \vdots & \vdots & \vdots \\ \mathbf{x}_{D,t-b}^l & \cdots & \mathbf{x}_{D,t}^l & \cdots & \mathbf{x}_{D,t+f}^l \end{pmatrix} \begin{pmatrix} w_{-b,i,m} \\ w_{-b+1,i,m} \\ \vdots \\ w_{f,i,m} \end{pmatrix},$$

where subscript $1, 2, \cdots, D$ denotes the individual element in the feature vector. The static feature matrix above has the dimensionality $D \times (f + b + 1)$. Using Eq. (1), we rewrite Eq. 6.44 as

$$\begin{aligned} \mathbf{y}_t^l &= \sum_{k=-b}^{f} w_{k,i,m}\mathbf{H}_{i,m}\mathcal{F}_{t+k}^l \\ &= \mathbf{H}_{i,m}\sum_{k=-b}^{f} w_{k,i,m}\mathcal{F}_{t+k}^l. \end{aligned} \tag{6.45}$$

According to the definition of Eq. 6.44, the dynamic features can be interpreted as the output from a slowly and step-wise time-varying linear filter with the (optimally transformed) static feature vector sequence serving as the input to the filter. The time-varying filter coefficients are evolving slowly according to the Markov chain in the underlying HMM. In the model discussed here, the jointly transformed static and dynamic features are provided as data input into the modeling stage of the classifier based on a mixture continuous-density HMM.

In order to carry out the MCE training for all the transformations and HMM parameters, we define the following discriminant function consistent with the maximum-likelihood decision rule used for classification:

$$g_\kappa(\mathbf{o}^l, \Phi) = \sum_{t=1}^{T^l} \log b_{\theta_t^\kappa}(\mathbf{o}_t^l),$$

where $\Theta^\kappa = \{\theta_1^\kappa, \theta_2^\kappa \cdots, \theta_{T^l}^\kappa\}$ is the optimal state sequence for the model associated with the κth class Φ^κ, and $b_{\theta_t^\kappa}(\mathbf{o}_t^l)$ is the probability of generating the transformed feature vector $\mathbf{o}_t^l$ (including both static and dynamic features $\mathbf{x}_t^l$ and $\mathbf{y}_t^l$) at time t in state θ_t^κ by the model for the κth class. Assuming the static and dynamic parameters are uncorrelated, the probability in the mixture Gaussian HMM has the familiar form of

$$b_i(\mathbf{o}_t^l) = b_i(\mathbf{x}_t^l, \mathbf{y}_t^l) = \sum_{m=1}^{M} c_{i,m} b_{i,m}(\mathbf{x}_t^l) b_{i,m}(\mathbf{y}_t^l), \quad 1 \le i \le N, \tag{6.46}$$

where $\mathbf{o}_t^l$ is the augmented feature vector (including both static and dynamic features) of the l-th token at frame t, M is the total number of Gaussian mixtures in the HMM's output distribution, and $c_{i,m}$ is the mixture weight for the mth mixture in state i. In Eq. 6.46, $b_{i,m}(\mathbf{x}_t)$ and $b_{i,m}(\mathbf{y}_t)$ are D-dimensional unimodal Gaussian densities for static and dynamic features, respectively:

$$\begin{aligned} b_{i,m}(\mathbf{x}_t^l) &= \frac{1}{(2\pi)^{\frac{d}{2}} |\mathbf{\Sigma}_{x,i,m}|^{\frac{1}{2}}} \exp\left(\frac{-1}{2} [\mathbf{x}_t^l - \boldsymbol{\mu}_{x,i,m}]^{\mathrm{Tr}} \mathbf{\Sigma}_{x,i,m}^{-1} [\mathbf{x}_t^l - \boldsymbol{\mu}_{x,i,m}] \right) \\ b_{i,m}(\mathbf{y}_t^l) &= \frac{1}{(2\pi)^{\frac{d}{2}} |\mathbf{\Sigma}_{y,i,m}|^{\frac{1}{2}}} \exp\left(\frac{-1}{2} [\mathbf{y}_t^l - \boldsymbol{\mu}_{y,i,m}]^{\mathrm{Tr}} \mathbf{\Sigma}_{y,i,m}^{-1} [\mathbf{y}_t^l - \boldsymbol{\mu}_{y,i,m}] \right), \end{aligned}$$

where variables $\mathbf{x}$ and $\mathbf{y}$ indicate the static and the dynamic features, respectively.

Now, the standard MCE learning described in earlier parts of this chapter can be carried out using a set of gradient computations detailed below. Define the time indices

$$T_i^l = \{t | \theta_t = i\}, \quad 1 \le i \le N, \quad 1 \le t \le T^l,$$

and the *a posteriori* probability

$$\gamma_{i,m}^j(t) = \frac{c_{i,m} b_{i,m}(\mathbf{x}_t^l) b_{i,m}(\mathbf{y}_t^l)}{b_i(\mathbf{o}_t^l)}.$$

Also, define the adaptive step size of

$$\psi_j = \begin{cases} \Upsilon_\kappa(\mathbf{o}^l, \Phi)[\Upsilon_\kappa(\mathbf{o}^l, \Phi) - 1] & \text{if } j = \kappa \\ \Upsilon_\kappa(\mathbf{o}^l, \Phi)[1 - \Upsilon_\kappa(\mathbf{o}^l, \Phi)] & \text{if } j = \chi \,. \end{cases}$$

Then the gradient related to learning the transformation matrices for the static features is derived in the following analytical form:

$$\begin{aligned} \frac{\partial \Upsilon(\mathbf{o}^l, \Phi)}{\partial \mathbf{H}_{i,m}} &= \psi \sum_{t \in T_i^l} \gamma_{i,m}(t) \frac{\partial}{\partial \mathbf{H}_{i,m}} \frac{-1}{2} ([\mathbf{H}_{i,m} \mathcal{F}_t^l - \boldsymbol{\mu}_{x,i,m}]^{\mathrm{Tr}} \mathbf{\Sigma}_{x,i,m}^{-1} \\ & \quad [\mathbf{H}_{i,m} \mathcal{F}_t^l - \boldsymbol{\mu}_{x,i,m}] + [\mathbf{H}_{i,m} \sum_{k'=-b}^{f} w_{k',i,m} \mathcal{F}_{t+k'}^l - \boldsymbol{\mu}_{y,i,m}]^{\mathrm{Tr}} \\ & \quad \mathbf{\Sigma}_{y,i,m}^{-1} [\mathbf{H}_{i,m} \sum_{k'=-b}^{f} w_{k',i,m} \mathcal{F}_{t+k'}^l - \boldsymbol{\mu}_{y,i,m}]) \\ &= -\psi \sum_{t \in T_i^l} \gamma_{i,m}(t) (\mathbf{\Sigma}_{x,i,m}^{-1} [\mathbf{x}_t^l - \boldsymbol{\mu}_{x,i,m}] [\mathcal{F}_t^l]^{\mathrm{Tr}} \\ & \quad + \mathbf{\Sigma}_{y,i,m}^{-1} [\mathbf{y}_t^l - \boldsymbol{\mu}_{y,i,m}] [\sum_{k'=-b}^{f} w_{k',i,m} \mathcal{F}_{t+k'}^l]^{\mathrm{Tr}}). \end{aligned} \tag{6.47}$$

The gradient related to learning the weighting vectors for the dynamic features can be computed according to:

$$\frac{\partial \Upsilon(\mathbf{o}^l, \Phi)}{\partial w_{k,i,m}} = \psi \sum_{t \in T_i^l} \gamma_{i,m}(t) \frac{\partial}{\partial w_{k,i,m}} \frac{-1}{2} \left([\sum_{k'=-b}^{f} w_{k',i,m} \mathbf{x}_{t+k'}^l - \boldsymbol{\mu}_{y,i,m}]^{\mathrm{Tr}} \right.$$

$$\left. \boldsymbol{\Sigma}_{y,i,m}^{-1} [\sum_{k'=-b}^{f} w_{k',i,m} \mathbf{x}_{t+k'}^l - \boldsymbol{\mu}_{y,i,m}] \right)$$

$$= -\psi \sum_{t \in T_i^l} \gamma_{i,m}(t) [\mathbf{y}_t^l - \boldsymbol{\mu}_{y,i,m}]^{\mathrm{Tr}} \boldsymbol{\Sigma}_{y,i,m}^{-1} [\mathbf{H}_{i,m} \mathcal{F}_{t+k}^l]. \tag{6.48}$$

The gradient computation for the model parameters are only slightly changed from those for the conventional HMM discussed earlier. We list these formulae here:

$$\frac{\partial \Upsilon(\mathbf{o}^l, \Phi)}{\partial \boldsymbol{\mu}_{x,i,m}} = \psi \sum_{t \in T_i^l} \gamma_{i,m}(t) \boldsymbol{\Sigma}_{x,i,m}^{-1} (\mathbf{H}_{i,m} \mathcal{F}_t^l - \boldsymbol{\mu}_{x,i,m})$$

$$\frac{\partial \Upsilon(\mathbf{o}^l, \Phi)}{\partial \boldsymbol{\mu}_{y,i,m}} = \psi \sum_{t \in T_i^l} \gamma_{i,m}(t) \boldsymbol{\Sigma}_{y,i,m}^{-1} (\mathbf{H}_{i,m} \sum_{k=-b}^{f} w_{k,i,m} \mathcal{F}_{t+k}^l - \boldsymbol{\mu}_{y,i,m})$$

$$\frac{\partial \Upsilon(\mathbf{o}^l, \Phi)}{\partial \tilde{\boldsymbol{\Sigma}}_{x,i,m}} = 0.5\psi \sum_{t \in T_i^l} \gamma_{i,m}(t) [(\mathbf{H}_{i,m} \mathcal{F}_t^l - \boldsymbol{\mu}_{x,i,m}) \boldsymbol{\Sigma}_{x,i,m}^{-1} (\mathbf{H}_{i,m} \mathcal{F}_t^l - \boldsymbol{\mu}_{x,i,m})^{\mathrm{Tr}} - \mathbf{I}]$$

$$\frac{\partial \Upsilon(\mathbf{o}^l, \Phi)}{\partial \tilde{\boldsymbol{\Sigma}}_{y,i,m}} = 0.5\psi \sum_{t \in T_i^l} \gamma_{i,m}(t) [(\mathbf{H}_{i,m} \sum_{k=-b}^{f} w_{k,i,m} \mathcal{F}_{t+k}^l - \boldsymbol{\mu}_{y,i,m}) \boldsymbol{\Sigma}_{y,i,m}^{-1}$$

$$(\mathbf{H}_{i,m} \sum_{k=-b}^{f} w_{k,i,m} \mathcal{F}_{t+k}^l - \boldsymbol{\mu}_{y,i,m})^{\mathrm{Tr}} - \mathbf{I}]$$

$$\frac{\partial \Upsilon(\mathbf{o}^l, \Phi)}{\partial \tilde{c}_{i,m}} = \psi \sum_{t \in T_i^l(j)} \gamma_{i,m,t}(j) - c_{i,m}(j).$$

6.4.4 Bayesian approach to verification using the Gaussian mixture model

In Section 6.3, we provided basic concepts and techniques for the verification problem in statistical pattern recognition. In particular, we introduced the Bayesian approach to verification, where the Gaussian distribution was used as the example data model to illustrate the Bayesian approach to verification. In this section, the same Bayesian approach to the same verification problem will be applied, but now with a more complex and more realistic data model. The Gaussian mixture distribution is used as the example data model in the following discussion. A full description of this application, including evaluation results in speech processing (speaker verification), can be found in [Jia 01].

Due to the missing-data nature in the Gaussian mixture model (GMM), it is necessary to use approximations in computing the basic integrations with the form of

$$\hat{p}(\mathbf{x}) = \int f(\mathbf{x} \mid \Lambda) \cdot \rho(\Lambda)\, d\Lambda.$$

Such kinds of integrals are needed to carry out hypothesis testing as shown in Eq. 6.24.

Given a set of training observations $\mathbf{x} = \{x_1, x_2, \cdots, x_T\}$, one testing observation y, and the prior PDF $\rho(\Lambda)$ of the model parameters Λ (such as the means in the Gaussian mixture distribution), the Bayesian approach to the verification problem consists of the following several steps.

The first step is to collect sufficient statistics. This involves:

- Initialize the 'global' statistics related to all Gaussian mixture components (let there be N in total) in model Λ:

$$\overline{x}_i = \overline{x^2}_i = \overline{(x-\mu)^2}_i = \Upsilon_i = 0 \quad (1 \le i \le N) \tag{6.49}$$

- For each training observation $x_l = \{x_{l1}, x_{l2}, \cdots, x_{lT_l}\}$ $(1 \le l \le L)$ in $\mathbf{x}$, perform a Viterbi or dynamic programming search for x_l based on $\rho(\Lambda)$ to obtain one optimal path $\mathcal{P}_l$ for the Gaussian mixture component over the model Λ.

- Based on the path $\mathcal{P}_l$, collect relevant statistics for each Gaussian mixture component $\mathcal{N}(m_i, \sigma_i)$, $(1 \le i \le N)$. Let the related prior PDF be $\rho(m_i) = \mathcal{N}(\mu_i, \tau_i)$. Then collect the following 'local' statistics for all mixture components $(1 \le i \le N)$:

$$v_i = \sum_{t=1}^{T_l} \delta(\mathcal{P}_{lt} - i) \tag{6.50}$$

$$\overline{x}_i = \frac{1}{v_i} \sum_{t=1}^{T_l} x_{lt} \cdot \delta(\mathcal{P}_{lt} - i) \tag{6.51}$$

$$\overline{x^2}_i = \frac{1}{v_i} \sum_{t=1}^{T_l} x_{lt}^2 \cdot \delta(\mathcal{P}_{lt} - i) \tag{6.52}$$

$$\overline{(x-\mu)^2}_i = \frac{1}{v_i} \sum_{t=1}^{T_l} (x_{lt} - \mu_i)^2 \cdot \delta(\mathcal{P}_{lt} - i), \tag{6.53}$$

where

$$\delta(\mathcal{P}_{lt} - i) = \begin{cases} 1 & \text{if path } \mathcal{P}_l \text{ lies in mixture } i \text{ at time } t \\ 0 & \text{otherwise.} \end{cases} \tag{6.54}$$

- Use these 'local' statistics to update all 'global' statistics for all $1 \le i \le N$ according to

$$\overline{x}_i = \frac{\overline{x}_i \cdot \Upsilon_i + \overline{x}_i \cdot v_i}{\Upsilon_i + v_i}$$

$$\overline{x^2}_i = \frac{\overline{x^2}_i \cdot \Upsilon_i + \overline{x^2}_i \cdot v_i}{\Upsilon_i + v_i}$$

$$\overline{(x-\mu)^2}_i = \frac{\overline{(x-\mu)^2}_i \cdot \Upsilon_i + \overline{(x-\mu)^2}_i \cdot v_i}{\Upsilon_i + v_i}$$

$$\Upsilon_i = \Upsilon_i + v_i.$$

The second step is to calculate the integral (also called Bayesian prediction) $\hat{p}(\mathbf{x})$:

$$\begin{aligned}\hat{p}(\mathbf{x}) &= \int f(\mathbf{x}|\Lambda)\cdot\rho(\Lambda)d\Lambda \approx \prod_{i=1}^{N}\int f(\mathbf{x}|m_i)\mathcal{N}(m_i,\sigma_i)dm_i \\ &= \prod_{i=1}^{N}\frac{\tau_i^{1/2}\cdot\sigma_i^{\Upsilon_i/2}}{(2\pi)^{\Upsilon_i/2}\cdot(\tau_i+\sigma_i\Upsilon_i)^{1/2}} \\ &\quad \exp\{-\frac{\sigma_i\Upsilon_i}{2(\tau_i+\sigma_i\Upsilon_i)}[\tau_i\overline{(x-\mu)^2}_i+\sigma_i\Upsilon_i(\overline{x^2}_i-\overline{x}_i^2)]\}\end{aligned} \tag{6.55}$$

At the same time, also calculate another integral $\hat{p}(\mathbf{x},y)$. To do so efficiently, one needs to keep all 'global' statistics related to $\mathbf{x}$ collected in the above step, and then collect 'local' statistics for y alone, which are denoted as $\{v_i,\overline{y}_i,\overline{y^2}_i,\overline{(y-\mu)^2}_i \mid 1\le i\le N\}$.

Using these collected statistics, $\hat{p}(\mathbf{x},y)$ is computed according to

$$\begin{aligned}\hat{p}(\mathbf{x},y) &\approx \prod_{i=1}^{N}\frac{\tau_i^{1/2}\cdot\sigma_i^{(\Upsilon_i+v_i)/2}}{(2\pi)^{(\Upsilon_i+v_i)/2}\cdot[\tau_i+(v_i+\Upsilon_i)\sigma_i]^{1/2}} \\ &\exp\{-\frac{\sigma_i\tau_i\Upsilon_i\overline{(x-\mu)^2}_i+\sigma_i\tau_i v_i\overline{(y-\mu)^2}_i+\sigma_i^2(\Upsilon_i+v_i)(\Upsilon_i\overline{x^2}_i+v_i\overline{y^2}_i)-\sigma_i^2(\Upsilon_i\overline{x}_i+v_i\overline{y}_i)^2}{2[\tau_i+(v_i+\Upsilon_i)\sigma_i]}\}.\end{aligned} \tag{6.56}$$

The third step is to perform hypothesis testing using the above results. Given the *null* hypothesis that y comes from model Λ_n and the alternative hypothesis that y is an outlier (Λ_0), the Bayesian approach to this testing (verification) problem becomes: If

$$\eta = \frac{\hat{p}_l(\mathbf{x}_l,y)\cdot\hat{p}_0(\mathbf{x}_0)}{\hat{p}_l(\mathbf{x}_l)\cdot\hat{p}_0(\mathbf{x}_0,y)} < \xi \tag{6.57}$$

then reject y as an outlier.

6.5 Summary

This chapter provided background material and selected topics on statistical pattern recognition and classification relevant to speech processing applications. Statistical pattern recognition uses many optimization techniques (Chapter 5) in common with those used in estimation theory, but it focuses on discrimination among classes instead of on the accuracy of fitting the models to the data. In general, the optimization problems associated with pattern recognition are much more complex than those with parameter or state estimation as discussed in Chapter 5.

The chapter began with Bayes' decision theory, which forms the foundation of statistical pattern recognition and classification theory. The principal result of Bayes' decision theory is that if the joint probability of the data and classes is known precisely, then the MAP decision rule will guarantee to minimize the error probability of classification. When this MAP decision rule is applied to speech processing (together with Bayes' rule), we obtain the celebrated fundamental equation of speech recognition as will be discussed in Chapter 12.

Two selected topics in pattern recognition were then discussed. First, discriminative training, which is often called minimum classification error (MCE) learning, was presented. It is motivated by the fact that, in practice, the joint probability of the data and classes is rarely precisely known, nor can it easily be obtained accurately. Detailed discussions on the optimization issues associated with the MCE learning are provided. Second, we discussed the verification problem in pattern recognition. It deals with mechanisms to detect and reject outliers, rather than forcing them to be one of the designated classes.

Finally, this chapter provided four examples of applications of statistical pattern recognition. Three of them applied the MCE learning principle and the associated optimization techniques to solve three different types of discriminative training problems. The last example was a general Bayesian approach to solving the verification problem under the GMM assumption.

Part II

FUNDAMENTALS OF SPEECH SCIENCE

Chapter 7

Phonetic Process

7.1 Introduction

In its physical form, speech is a pressure waveform that travels from a speaking person to one or more listeners. This signal is typically measured (or received by a microphone) directly in front of the speaker's mouth, which is the primary output location for the speech (speech energy also emanates from the cheeks and throat, and nasal sounds leave the nostrils as well). Since the ambient atmosphere in which one speaks imposes a basic pressure (which varies with weather and altitude), it is actually the variation in pressure caused by the speaker that constitutes the **speech signal**. The signal is continuous in nature and is very dynamic in time and amplitude, corresponding to the constantly changing status of the vocal tract and vocal cords. We nonetheless characterize speech as a discrete sequence of sound segments called **phones**, each having certain acoustic and articulatory properties during its brief period of time. Phones are acoustic realizations of **phonemes**, which are the abstract linguistic units that comprise words. Each phoneme imposes certain constraints on the positions for these vocal tract **articulators** or organs: **vocal folds** (or **vocal cords**), **tongue**, **lips**, **teeth**, **velum**, and **jaw**. Speech sounds fall into two broad classes: (a) **vowels**, which allow unrestricted airflow throughout the vocal tract, and (b) **consonants**, which restrict airflow at some point and have weaker intensity than the vowels.

7.2 Articulatory Phonetics and Speech Generation

Speech is generated as one exhales air from the lungs while the articulators move. This sound production will be discussed later as a filtering process in which a speech sound **source** excites the vocal tract **filter**. The source either is periodic, causing **voiced** speech, or is noisy (aperiodic), causing **unvoiced** speech. The source of the periodicity for the former is found in the larynx, where vibrating vocal cords interrupt the airflow from the lungs, producing pulses of air. Both the area of the glottal opening (between the vocal cords) and the volume of the pulse of air can be approximated as half-rectified sine waves in time, except that the glottal closure is more abrupt than its opening gesture. This asymmetry assists the speech signal to be more intense than might otherwise be the case, because abrupt changes in the excitation increase the bandwidth of the resulting speech (e.g., if the glottal pulses were sinusoidal, speech energy would only appear at the single fundamental frequency).

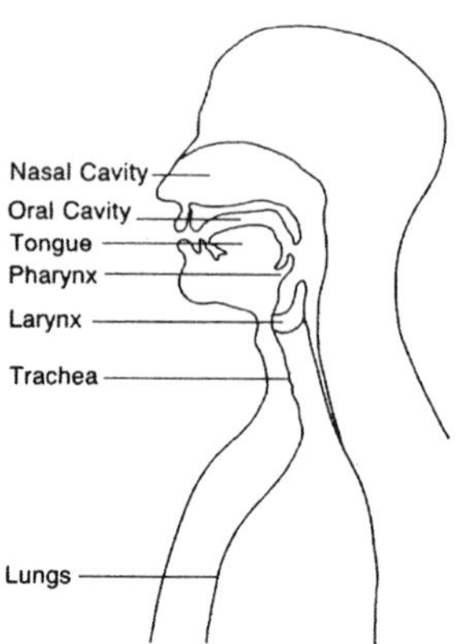

Figure 7.1: A cross-sectional view of the upper portion of a human, showing an outline of the major speech organs.

7.2.1 Anatomy and physiology of the vocal tract

While the mammal auditory system has the sole purpose of understanding sounds (see Sections 7.6-7.8), the human speech production organs are multi-purpose: speech generation, breathing, eating, and sensing odors. Thus, in a communication sense, speech production cannot be as optimal an information source as the ear is a receiver. In radio communication, an optimal source creates a signal with the least energy and shortest duration within the constraints of noisy channel bandwidth. Nonetheless, certain parallels can be made between electronic and human speech communication. Humans minimize effort, in terms of their 'energy' and time, while maximizing the perceptual contrast in listeners (as so anticipated by the speaker). As we will see later, the frequency range occupied by speech is important for many reasons; so we mention here that the bandwidth of human speech communication is from below 100 Hz up to 7 kHz or so (music has a broader range). Both production and perception organs for speech are efficient at these frequencies (much higher frequencies are used for vision and radio signals).

Vocal tract

The lungs provide the airflow and pressure source for speech, and the vocal cords usually modulate the airflow to create many sound variations. However, it is the **vocal tract** that is the most important system component in human speech production (Fig. 7.1). It is a tube-like passageway made up of muscles and other tissues, and enables the production of the different sounds that constitute spoken language. For most sounds, the vocal tract modifies the temporal and spectral distribution of power in the sound waves, which are initiated in the glottis. In addition, the vocal tract generates some sounds directly (e.g., is the source for **obstruent** (stop and fricative) sounds). We distinguish different phones primarily by their periodicity (voiced or unvoiced), spectral shape (mostly, which frequencies have major power) and duration (longer phones are perceived as having greater stress). The state of the vocal folds usually specifies each phone's (binary) voicing feature choice. By far the most important aspect of speech production is the specification of different phones via the filtering actions of the vocal tract.

Because speech perception is dominated by the presence of sound power (i.e., listeners pay primary attention to which sound frequencies dominate), the vocal tract is often described in terms of its resonances, called **formants**, owing to poles in the vocal tract transfer function. The formants are often abbreviated Fi; hence F1 means the formant with the lowest frequency. In voiced phones, the formants often decrease in power as a function of frequency (due to the general lowpass nature of the glottal excitation); thus F1 is usually the strongest formant. For some phones, inverse singularities of the vocal tract transfer function (zeros) exist and cause **antiresonances**, where the speech power dips much more than usual between formants. Displacing the articulators changes the shape of the acoustic tube (i.e., the vocal tract) through which sound passes, and alters its frequency response.

For simplicity, we consider the vocal tract of a typical adult male. Since it is about 17 cm long from glottis to lips, this numerical choice will lead to simpler calculations later (results for different-sized tracts scale easily, although size differences are complex in general across different humans). The cross-sectional area of the vocal tract ranges from 0 (complete occlusion) to about 20 cm^2. After leaving the larynx, air from the lungs passes through the **pharyngeal** and **oral cavities**, then exits at the **lips**. For nasal sounds, air is allowed to enter the **nasal cavity** (by lowering the **velum**), at the boundary between the pharyngeal and oral cavities. The **velum** (or soft palate) is kept in a raised position for most speech sounds, blocking the nasal cavity from receiving air. During nasal sounds, as well as during normal breathing, the velum lowers to allow air through the nostrils.

Later, we will model the vocal tract as a sequence of cylinders, each having a variable cross-sectional area. In reality, its shape is much more complex, but our simple model allows us to obtain data that reasonably simulate empirical results.

In the vocal tract, the tongue, the lower teeth, and the lips undergo significant movements during speech production. In contrast, the upper and rear boundaries of the vocal tract are relatively fixed, but with diverse composition; e.g., the oral cavity roof consists of a soft palate toward the rear, then a stiff hard palate, and in the front the upper lip and teeth. The nasal cavity consists of many passages lined with mucous tissue, and has no movable structures. Its large interior surface area significantly attenuates speech energy. The opening between the nasal and pharyngeal cavities controls the amount of acoustic coupling between the cavities, and hence the amount of energy leaving the nostrils. Increased heat conduction and viscous losses cause formant bandwidths to be wider in nasals than for other sonorants.

During speech production, if either the vocal tract or glottis is completely closed for a time, airflow ceases and typically no sound emerges. (Of course, during strong voiced phones, the glottis closes for a few ms periodically as part of normal voicing, and energy nonetheless continues to leave the system throughout as speech, due to reflections inside the tract. We discuss here instead closures longer than just a few ms.) The class of phones called **stops** or **plosives** uses such closures lasting several tens of ms. Immediately prior to closure and at closure release, stops have acoustics that vary depending on the closure point: in the glottis (closed vocal folds), in the vocal tract (tongue against palate), or the lips. During normal speech from a single breath, the chest muscles continuously attempt to expel air; so pressure builds during these stop phones behind the closure point until it is released by opening the occlusion. **Fricative** phones resemble stops, but have a maintained narrow vocal tract constriction (rather than complete occlusion and then release).

Larynx and vocal folds

The vocal folds are important for speech production because normal breathing (whatever the vocal tract shape) creates little audible sound, as the air expelled by the lungs passes mostly unobstructed throughout the vocal tract. Sound (air pressure variation) is generated in the vocal tract only when the airflow path is narrowly (or totally) constricted, effectively interrupting the airflow, thus creating either turbulent noise (with a steady narrow constriction) or pulses of air (from vibrating vocal folds). Most speech originates in the larynx, which consists of four cartilages (thyroid, cricoid, arytenoid, and epiglottis) joined by ligaments and membranes. The passageway between the lungs and the vocal tract is called the **trachea**, which divides into two bronchial tubes towards the lungs. A nonspeech organ called the epiglottis protects the larynx from food and drink. The vocal folds inside the larynx are typically about 15 mm long, have a mass of about 1 gram each, and have amplitude vibrations of about 1 mm.

When one breathes normally, the vocal folds are far enough apart to avoid sound creation, although increased airflow during exercise leads to loud **whispers** (or sighs). If airflow is strong enough and the folds are close enough, such turbulent noise occurs at the glottis. This speech is called whisper or **aspiration**, and corresponds to the phoneme /h/ (we will place phoneme symbols inside / / to distinguish them from other linguistic units, and will use ARPAbet notation [Shou80]). Similar noise can be generated in the same way higher up in the vocal tract, at a narrow constriction either between the tongue and the palate (the roof of the mouth) or with the lips and teeth. The latter noises are called **frication** (i.e., **fricative** sounds). The main difference between aspiration and frication lies in the range of frequencies: broadband for whisper and high-frequency only for frication, because each noise source excites mostly the portion of the vocal tract in front of the source constriction. In frication, shorter cavities correspond to higher-frequency resonances than with aspiration. All such noise sounds are **aperiodic** due to the random nature of their turbulent source.

Air leaving the lungs during sonorant phones is interrupted by the quasiperiodic closing and opening of the vocal folds. The rate of this vibration is called the **fundamental frequency**. It is often abbreviated F0, in contrast to the formants F1, F2,..., although F0 is not a resonance (power in sonorants appears primarily at **harmonic** multiples of F0, but these harmonics have widely varying intensity depending on the formants). The **fundamental period** or **pitch period** of voiced speech, T_0=1/F0, corresponds to the time between successive vocal fold closures. T_0 has a large dynamic range, but its average value is proportional to the size of the vocal folds; thus men generally have longer periods (about twice as long).

To physically cause the vocal folds to vibrate, they must be close together (i.e., smaller glottis than for normal breathing) and the lungs must generate sufficient pressure (using the diaphragm) so that the difference between the pressure P_{sub} below the glottis (subglottal) and that above (supraglottal) is large enough. Voicing is unlike other rapid movements in speech, which are due to voluntary muscle actions and are thus limited to low frequencies (e.g., under 10 Hz). Vocal fold vibration is caused by aerodynamics and exploits the elasticity of the fold tissue. A simple upward movement of the diaphragm causes a constant P_{sub} and air flows through the narrow glottis. Quantitatively, voicing typically uses P_{sub} = 5−15 cm H_2O and has a peak glottal airflow of 250−750 cc/s [Stev 71].

As the lungs push the folds apart, a negative pressure develops in the glottis due to a **Bernoulli force** and closes the vocal folds via a sucking action. This cycle repeats each pitch period. Air flowing through a tube such as the vocal tract has a constant sum of kinetic and potential energy. Potential energy varies as air pressure, while kinetic energy follows the square of air velocity. The increased velocity in the narrow glottis causes local pressure to drop. When it is low enough, the negative pressure forces the vocal folds to close, which interrupts the airflow. Then a positive pressure develops in the larynx, forcing the vocal folds open again, and the cycle repeats until the diaphragm or vocal cord muscles relax.

Major and secondary articulators

As noted above, vocal tract organs that move to produce speech sounds are called articulators. The tongue and the lips are the major (most important) articulators. In secondary roles, the velum and larynx also contribute to speech production. Through its varying glottal opening, the larynx controls airflow into the vocal tract. The larynx can also be raised or lowered, which alters the tract length, which in turn raises or lowers, respectively, formant frequencies. The jaw (mandible) may be thought of as a secondary articulator because it helps to position the tongue and lips for most sounds.

The lips are a pair of muscular folds that can cause a vocal tract closure or produce a narrow slit at the mouth (when they are pressed together or when the lower lip presses against the upper teeth). The lips can also either round and protrude (pucker) or spread and retract. For all constrictions, the sounds produced are called labials. Closure is typically accomplished by raising the jaw (and consequently the lower lip), while rounding employs the muscles surrounding the lips.

Only the four front upper teeth participate actively in speech production: for dental obstruent phones either the lower lip (i.e., /f,v/) or the tongue tip (e.g., 'th') contacts these teeth. Immediately behind the teeth is the hard palate (the upper wall of the oral tract). Many phones (e.g., /t,s,n,l/) require a constriction between the hard palate and the tongue.

The most important primary articulator is the tongue (consisting mostly of muscle), which has four components: tip, blade, dorsum, and root. Owing to the relative rigidity of the upper and rear walls of the vocal tract, speakers rely on the very flexible tongue to provide the mechanism to create the different vocal tract shapes needed to produce the various speech sounds. The tip is fast and agile, able to move up and then down within about 100 ms. The dorsum is the surface of the tongue, whose frontal portion is the blade; the tongue body (root) positions the dorsum.

Most articulators move towards different target positions for each successive phone, starting as ballistic (rapid and less controlled) motion and then becoming more focused as the target nears or as other muscle commands are issued for a new target. The peak velocity of an articulator is often linearly related to its actual displacement. Movements are often very small (less than 1 cm), at speeds up to 30 cm/s.

7.2.2 Major features of speech articulation

Of main interest in the production of phonemes are the following: the state of the vocal cords (vibrating or not), the degree of any major constriction in the vocal tract, and the location of such constrictions. These correspond to the features of voicing, manner of articulation, and place of articulation, respectively.

Manner of articulation

Manner of articulation deals with airflow in the vocal tract: whether it flows through the oral and/or nasal cavities, and the degree of any major vocal tract constrictions. The manner classes are: vowels (including diphthongs), glides, liquids, nasals, fricatives, and stops. The **vowels** are the most important and largest class of phonemes; air passes relatively freely (at rates of 100–200 cc/s) through the pharynx and oral cavities. (In some languages, nasalized vowels also allow air through the nasal cavity as well.) Vowels have no constrictions narrow enough to cause frication noise (turbulent and random airflow) or to block the airflow completely. To avoid noise generation, the area of minimum constriction (except for the glottis) exceeds 0.3 cm^2.

Glides (also called semivowels) resemble vowels, but have a very high tongue position, which causes a narrow vocal tract constriction barely wide enough to avoid frication. In many languages, there is a glide that closely resembles each high vowel in the language (e.g., glide /y/ resembles vowel /iy/ and /w/ resembles /uw/). In practice, glides are simply very high vowels that are difficult to maintain for more than a few tens of milliseconds; thus they may be thought of as transient vowels (vowels, on the other hand, can easily last for hundreds of ms, if desired).

Liquids also resemble vowels, except for use of part of the tongue as a major obstruction in the oral tract, which causes air to deflect from a simple path. For the liquid /l/ (also called a lateral), the tongue tip is in contact with the alveolar ridge and causes a division of the airflow into two streams on both sides of the tongue. The liquid /r/ (also called rhotic) has several articulations in different languages. In English, /r/ has the tongue tip pointing toward the alveolar ridge and usually curled back (retroflex). There is often a raised dorsum [Boyce 77], and the sides of the tongue also contact the upper molar teeth, all of which tends to restrict the airflow, but not enough to cause frication. In languages other than English, phoneme [r] behaves more like a fricative or trill than a liquid.

Nasal consonants have a lowered velum, which allows airflow into the nasal cavity and through the nostrils, while the oral tract is completely closed. Some languages (e.g., French) have nasalized vowels, where air flows through both the oral and nasal cavities. Such **nasalization** (lowering the velum) often occurs in parts of English vowels, but such sounds are not associated with a separate phonemic category, because English listeners interpret these sounds as normal free variation when vowels occur adjacent to nasal consonants. Thus the distinction is 'allophonic' and not phonemic. In French, on the other hand, words can be distinguished directly on whether a vowel is nasalized.

All phonemes in the preceding four classes (vowel, glide, liquid, nasal) are part of a more general manner class called **sonorants**; they are all voiced and relatively strong in power. The other general class of phonemes is called **obstruents** and is comprised of stops and fricatives, which are noisy and relatively weak, with the primary acoustic excitation at a major vocal tract constriction (rather than at a vibrating glottis).

Stops (or plosives) employ a complete closure in the oral tract, which is then released. Air continues to flow through the glottis throughout a stop (owing to a relatively constant lung pressure, irrespective of vocal tract dynamics for different phonemes). The velum is raised throughout a stop to prevent nasal airflow during oral closure (otherwise a snorting sound would occur, which is not a phoneme in most languages). Pressure builds up behind the oral closure and is then abruptly released. Air flows through the increasing orifice (at a decreasing speed, as the opening grows over a few tens of ms). The initial intense **burst** of noise upon oral tract opening is called an explosion, and is effectively a brief fricative. The phone following a stop is normally a sonorant; so the oral tract widens for that sound and the vocal folds adduct in preparation for voicing. Prior to actual periodicity, an interval of noisy **aspiration** typically occurs, with a duration called the **voice onset time** or VOT. The initial portion of the VOT is frication, produced at the opening constriction, whereas the longer remainder is aspiration noise, created at the constricting glottis.

In voiced stops, on the other hand, vocal folds may continue to vibrate throughout the stop or start to vibrate right after the burst. One difference between voiced and unvoiced stops is that the vocal folds are more widely separated during the vocal-tract closure for unvoiced stops and start to adduct only at the release, hence the longer VOT for unvoiced stops.

Like other phones, stops usually last on the order of 80 ms. Unlike other sounds, stops are sometimes very brief when they occur between two vowels; e.g., an alveolar stop followed by an unstressed vowel (in the same word) often becomes a **flap**, where the tongue tip maintains contact with the palate for as little as 10 ms.

While stops have a complete occlusion in the vocal tract, **fricatives** use a narrow constriction instead. To generate noise, fricatives need sufficient pressure behind the constriction (e.g., 3–8 cm H_2O) with a narrow passage (0.05–0.2 cm^2). This causes sufficiently rapid airflow (41 m/s) to generate turbulence at the end of the constriction [Flan 72]. The Reynolds Number, vh/ν (where v is air particle velocity, h is the width of the orifice, and $\nu = 0.15$ cm^2/s is the viscosity of air) must exceed 1800. Airflow for frication is in the range of 200–500 cc/s, while glottal aspiration is greater (500–1500 cc/s) [Stev 71].

Most speech sounds (i.e., vowels, liquids, nasals and fricatives) each have a specific articulatory position and can be maintained over several seconds, if desired. Stops, on the other hand, are **transient** or dynamic consonants which have a sequence of articulatory events (i.e., closure, then release). Glides may also be considered as transient phonemes, owing to the difficulty of sustaining them.

We did not note above distinct manner classes for diphthongs and affricates, although some linguists view these as separate classes. These sounds are actually phoneme sequences: each **diphthong** consists of a vowel followed by a glide, and each **affricate** consists of a stop followed by a fricative. There are phonological conventions for considering these as individual phonemes, owing to distributional restrictions and durational phenomena (e.g., affricates have less duration than other stop+fricative combinations, and are the only such sequences allowed to start English syllables).

Place of articulation

Place of articulation refers to the location where the vocal tract is the most constricted, if any (other than the glottis). It is usually associated with consonants because they require a significant constriction. We traditionally associate about nine regions along the vocal tract with places of consonant constrictions:

- Bilabial: the lips touch each other.
- Labiodental: the lower lip contacts the upper teeth.
- Dental: the tongue tip touches the upper incisor teeth.
- Alveolar: the tongue tip or blade approaches or contacts the alveolar ridge.
- Palatal: the tongue blade or dorsum constricts against the hard palate.

- Velar: the dorsum constricts with the soft palate.
- Uvular: the dorsum approaches the uvula.
- Pharyngeal: the pharynx constricts.
- Glottal: the vocal folds adduct (since the glottis necessarily constricts periodically for all sonorants, the term **glottal** is reserved for obstruents).

For consonants within each manner class, languages usually exploit only a few of these places. For example, English fricatives occur at five: labiodental, dental, alveolar, palatal, and glottal, and stops occur at four points: bilabial, alveolar, velar, and glottal (although glottal stops are much rarer than the others). Each consonant is primarily specified by its point of narrowest constriction, but some also have a secondary (wider) constriction. In the latter case, consonants are said to be labialized (if lips are involved), and palatalized, velarized, or pharyngealized in other cases; e.g., /w/ is a velarized bilabial glide.

We now examine in more detail the place of articulation for vowels. Vowels in general have an open vocal tract, but individual vowels are distinguished by the degree and location of relative constriction, in terms of tongue position and lip rounding. While **place of articulation** usually refers to a major obstruent constriction, for vowels it means a lip constriction or an upward positioning of part of the tongue. Just as the degree of constriction in consonants distinguishes among stops, fricatives and glides, the tongue height and amount of lip rounding lead to different vowel phonemes. For example, the mouth ranges in opening from 1 to 5 cm^2 between the rounded vowel /uw/ and the open /aa/. Phonology tends to classify many sounds via binary features, and most vowels have only two horizontal tongue positions (front or back) and two lip rounding states (rounded or unrounded). When a given language has only one vowel at a certain tongue height, the front-back position may be neutralized to a central location. For example, the **schwa** vowel is a short and weak centralized vowel, which is not used to contrast with other vowels.

While lip status and lateral tongue position during vowels are mostly binary in nature, tongue height is more varied. Many languages distinguish four levels: high (or close) (e.g., /iy,uw/, mid-high (/ey,ow/), mid-low (/eh,aw/), and low (open) (/aa, ae/). There is often a correlation between lip rounding and tongue position. For example, in English, all front vowels (/iy,ih,ey,eh,ae/) and low vowels (/aa,uh/) are unrounded, while the high back vowels (/ao,ow,uh,uw/) are rounded.

7.2.3 Phonemes, coarticulation, and acoustics

One can analyze speech from different points of view: articulatory, acoustic, phonetic, and perceptual. This section describes articulatory acoustics, which relates speech sound features to articulator positions. With their vocal tract, people can produce a wide range of sounds, only a small number of which are recognized as speech phones as part of a formal language such as English. Each language is characterized by a set of a few dozen abstract linguistic units called **phonemes**. A phoneme is the smallest meaningful contrastive unit in a language's phonology. The actual sounds (or phones) which are produced when a speaker utters each phoneme (a many-to-one relationship) share some articulatory configuration, and correspondingly share certain acoustic features. Each word in a language differs from all other words by at least one phoneme (e.g., cat vs. bat), except for homonyms (where two words sound alike, but can have different meanings depending on lexical context).

While languages vary in their semantic and grammatical structure, they generally use the vocal tract in very similar ways. The 'cardinal' vowels /iy,aa,uw/ appear in almost all languages, while other phonemes are specific to a small minority of languages (e.g., [**th**] in English). Phonemes are often described by interrelated features: phonetic, articulatory, and acoustic. For example, the phoneme /s/ has phonetic features (unvoiced, fricative, alveolar), articulatory features (open glottis, raised velum, narrow alveolar constriction), and acoustic features (noise, aperiodic, strong, high-frequency).

Since the vocal tract can is unlimited in terms of possible speech positions, an infinite number of phones (physical sounds) can correspond to each phoneme (abstract linguistic concept). Attempts by anyone to repeatedly utter a given phoneme yield acoustic versions (phones) that differ, but usually by a lesser amount than versions from different speakers (this fact is a basis of automatic identification of speakers by their voice). We use **allophone** to describe a class of phones that form a certain variant of a phoneme, often with regard to vocal tract shape, in cases where the same phoneme may have different shapes. For example, for the phoneme /k/ the tongue dorsum is in full contact with the palate, but the occlusion may occur along a wide range of the velar region. Prior to a front vowel, the /k/ closure is far forward, whereas it is far enough back in a rear vowel context to cause significant acoustic differences (i.e., energy shifts from F3 to F2).

The phenomenon of allophones leads us to the concept of **coarticulation**, which is the general name for the articulatory and acoustic effects of adjacent phonemes affecting each other's articulation. While designers of speech synthesis and recognition applications would strongly prefer otherwise, natural speech cannot be readily divided into separate phones with a simple invariant correspondence to phonemes. In practical applications, we may divide speech into successive segments (tentatively labeled as phones), whose boundaries are usually times of major spectral change, which in turn correspond to significant changes in the vocal tract. Such changes include onset or offset of occlusion, narrow constriction, and vocal cord vibration (i.e., voicing). Due to coarticulation, the labeled "phoneme boundaries" can be unreliable and sometimes arbitrary. A phoneme's articulatory features often affect the speech signal well beyond such boundaries (see a later discussion of context-dependent speech recognition models).

To better relate articulation to acoustics, it is useful to consider speech units (e.g., words and syllables) more familiar to most people than phonemes. Like phonemes and phones, words exist in both abstract and physical forms (although we will not attempt to distinguish these forms by name, using **word** to denote both). Each word consists of one or more phonological units called **syllables**. A syllable has one vowel (or a diphthong); this central component is the syllable's most intense sound and occurs when the vocal tract is most open. One or more consonants may occur both before and after the vowel in each syllable. (The perceptual importance of speech units larger than phonemes can be demonstrated by **shadowing** experiments, where subjects repeat what they hear as quickly as possible. Typical repetitions are delayed by 270–800 ms. This suggests that listeners process speech in syllable or word units [Mars 85]. Furthermore, using headphones, speech which alternates rapidly between one's ears is most disruptive to perception when switch times align with syllable durations [Wils 84].)

Phoneme	Manner of Articulation	Place of Articulation	Voiced?	Example Word
iy	vowel	high front tense	yes	beat
ih	vowel	high front lax	yes	bit
ey	vowel	mid front tense	yes	bait
eh	vowel	mid front lax	yes	bet
ae	vowel	low front tense	yes	bat
aa	vowel	low back tense	yes	cot
ao	vowel	mid back lax rounded	yes	caught
ow	vowel	mid back tense rounded	yes	coat
uh	vowel	high back lax rounded	yes	book
uw	vowel	high back tense rounded	yes	boot
ah	vowel	mid back lax	yes	but
er	vowel	mid tense (retroflex)	yes	curt
ax	vowel	mid lax (schwa)	yes	about
ay	diphthong	low back → high front	yes	bite
oy	diphthong	mid back → high front	yes	boy
aw	diphthong	low back → high back	yes	bout
j	glide	front unrounded	yes	you
w	glide	back rounded	yes	wow
l	liquid	alveolar	yes	lull
r	liquid	retroflex	yes	roar
m	nasal	labial	yes	maim
n	nasal	alveolar	yes	none
ng	nasal	velar	yes	bang
f	fricative	labiodental	no	fluff
v	fricative	labiodental	yes	valve
th	fricative	dental	no	thin
dh	fricative	dental	yes	then
s	fricative	alveolar strident	no	sass
z	fricative	alveolar strident	yes	zoos
sh	fricative	palatal strident	no	shoe
zh	fricative	palatal strident	yes	measure
h	fricative	glottal	no	how
p	stop	labial	no	pop
b	stop	labial	yes	bib
t	stop	alveolar	no	tot
d	stop	alveolar	yes	did
k	stop	velar	no	kick
g	stop	velar	yes	gig
ch	affricate	alveopalatal	no	church
jh	affricate	alveopalatal	yes	judge

Table 7.1: English phonemes and corresponding features.

Quantal theory of speech production

Speech processing would be easier if there were simple linear relationships between articulation and acoustics, and between acoustics and perception. This would greatly facilitate automatic speech synthesis and recognition, respectively. However, both these relationships are not linear. Some large articulatory changes cause minimal acoustic effects, yet other small movements cause large spectral changes. One theory of speech production [Stev 89] suggests that phones have evolved so as to exploit certain preferred articulatory vocal tract positions which, when varied within a certain range, do not cause significant acoustic variations. By thus having assigned such positions to phonemes, speakers then allow themselves some flexibility in articulation without misleading listeners. Whether due to coarticulation or laziness by the speaker, or even due to noise introduced in speech transmission, many small speech signal variations can be tolerated if listeners can implicitly interpret them as irrelevant to phoneme decisions. As an example, spectra for the basic and opposite vowels /iy, aa/ are modeled well by simple two-tube acoustic models; in these simulations, small perturbations in tube length have little effect on formant locations. As befits the dual (production and perception) nonlinear nature of speech, evidence for this **quantal** theory comes from both speech production and speech perception [Wood 79].

7.2.4 Source-filter description of speech production

We now examine in more detail the traditional electrical and computer model of speech generation. Speech production resembles many other types of sound generation, i.e., a power source (e.g., motor, explosion, lungs) initiates a sound and the immediate environment shapes or 'filters' it. In such cases, it is convenient to view the system as a source 'exciting' a filter, allowing a view of the acoustic results in both time and frequency domains. Such a model also permits an examination of the effects of the excitation source and the vocal tract independently. The source and tract interact acoustically, but their interdependence usually causes only secondary effects. As in most speech analysis discussions, we will assume source-filter independence.

We first examine the production of sonorant phones, which occur more frequently than other phones; sonorant generation is also simpler to analyze. Pulses of air passing through the glottis excite the entire vocal tract, which acts as a filter to shape the speech spectrum. These air puffs occur quasi-periodically, due to the vibration of the vocal folds. (They are not fully periodic, owing to small variations between cycles of the folds and to gradual movement of the vocal tract.) In the second case, we will discuss unvoiced sounds, which result from a noise source that excites only the portion of the vocal tract in front of the source. In both cases, we model the resulting speech signal $s(t)$ as the convolution of an excitation $e(t)$ and an impulse response for the vocal tract $v(t)$. (Naturally occurring excitations are of course not impulses, but the assumed linear system of the vocal tract obeys basic superposition rules, and the impulse response, as the inverse transform of the filter's transfer function, is convenient to use.) Because convolution of two signals corresponds to multiplication of their spectra, the final speech spectrum $S(\Omega)$ is the product of the excitation spectrum $E(\Omega)$ and the frequency response $V(\Omega)$ of the vocal tract.

Glottal excitation

The starting place for most speech production is the glottis, since all sonorant phones have a glottal excitation. We concentrate first on voiced excitation, because it is relevant for most speech and because the naturalness of coded speech and synthesized speech is greatly affected by how well voiced speech is modeled. Actual glottal pressure waveforms are difficult to measure, as are vocal tract shapes, especially in the vast range of variation seen across different speakers and phonetic contexts. Almost all speech processing starts with the pressure signal at the mouth, due to the convenience of capturing such energy and to the great inconvenience of measuring other possibly relevant signals (e.g., X-rays of the vocal tract, facial images, throat accelerometers, ultrasound imaging). To estimate a glottal signal in terms of the mouth output (i.e., the speech signal), we sometimes turn to inverse filtering, i.e., forming an estimate $|\hat{V}(\Omega)|$ of the actual vocal tract transfer function $|V(\Omega)|$, using the spectral amplitude of the observed speech $|S(\Omega)|$. The excitation estimate then is

$$|\hat{E}(\Omega)| = \frac{|S(\Omega)|}{|\hat{V}(\Omega)|}.$$

The major difficulty here (as in cepstral deconvolution, and indeed in automatic speech recognition) is that we estimate two signals from one, and thus need to constrain the task with other knowledge about the speech production system.

To a certain degree, the glottal volume velocity $e(t)$ of voiced speech resembles a half-rectified sine wave: starting with a closed glottis, $e(t)$ gradually increases from a zero value as the vocal folds separate. The folds close more quickly than they separate, skewing each glottal pulse to the right, owing to the Bernoulli force adducting the vocal folds. It is indeed at the rapid glottal closure that the major excitation of the vocal tract occurs in each cycle, owing to the discontinuity in the slope of $e(t)$. We see a sudden increase in $|s(t)|$ (sometimes called a **pitch epoch**) every glottal period about 0.5 ms after glottal closure, the delay owing to the transit time for the sound to travel from the glottis to the lips. The open phase of the glottal cycle is shorter in loud speech (e.g., 40% of the period) and longer in breathy voices (e.g., 70%); the shorter 'duty cycle' causes a sharper closure and thus more energy.

The spectrum of many periods of voiced excitation appears as a line spectrum $|E(\Omega)|$, owing to F0 and its harmonics. Its envelope is lowpass in shape (cutoff near 500 Hz and then falloff of about -12 dB/octave), due to the relatively smooth nature of each glottal air pulse; nonetheless, power does persist into high frequencies, mostly due to the abrupt closure excitation. Consider the theoretical case of an infinitely long voiced excitation $e(t)$, and assume that each period is a glottal pulse $g(t)$. We model $e(t)$ as the convolution of a uniform impulse train $i(t)$ with one pulse $g(t)$, and thus $|E(\Omega)|$ is the product of a uniform impulse train $|I(\Omega)|$ in radial frequency $\Omega = 2\pi f$ and $|G(\Omega)|$. Since $e(t)$ and $i(t)$ are periodic with period $T = 1/\mathrm{F0}$, both $|I(\Omega)|$ and $|E(\Omega)|$ are line spectra spaced by F0 Hz. Typical speakers vary F0 over a general range of about an octave (e.g., about 80–160 Hz for men), although song has often a more extensive F0 range. Average values of F0 for men and women are 132 Hz and 223 Hz, respectively [Pet 52].

Glottal excitation can be aperiodic as well. Indeed for whispered speech, the vocal tract movements are largely the same as for voiced speech, except that the speech is unvoiced. Unvoiced excitation is usually modeled as random noise with an approximately Gaussian amplitude distribution and a flat spectrum over the audio frequencies of concern to speech processors (e.g., 0–10 kHz). Such aperiodic excitation can occur elsewhere in the vocal tract, as well as at the glottis. At the glottis, it is often called aspiration, and is called frication when occurring at a constriction higher up in the vocal tract, or called an **explosion** when occurring at a stop release. White noise, although limited practically to the bandwidth of speech, is a reasonable model. In this case, $|E(\Omega)|$ has no effect on $|S(\Omega)|$. We tend to ignore the phase of $E(\Omega)$ for unvoiced speech as largely irrelevant perceptually. Even in voiced speech, where phase effects are more readily heard, it is often unanalyzed because spectral amplitude is much more important than phase for speech perception and because simple models for $e(t)$ are adequate for acceptable speech quality in coding and synthesis. The ear is much more sensitive to a sound's intensity than to its phase. Unusual phase variations, as can occur in synthetic speech, render sounds unnatural, but many natural phase variations are imperceptible [Schr 75, Opp 81].

When a stop occlusion is released into an ensuing voiced phone, aspiration occurs during the VOT period as the vocal folds move towards a suitable position for voicing. A final comment about glottal excitation is that female speech tends to be more breathy, owing to looser vocal fold vibration, which leads to wider bandwidths and a steeper fall-off in the speech spectrum [Hans 97].

Acoustics of the vocal tract

Except for the glottal effect (speech amplitude decreasing with frequency), the voiced speech spectrum is largely specified by the shape of the vocal tract, i.e., the amplitudes of the spectral lines in $|S(\Omega)|$ are mostly determined by the vocal tract transfer function $|V(\Omega)|$. We calculate $|V(\Omega)|$ below for some simplified models of the vocal tract. To better relate speech articulation and acoustics, we analyze vocal tract acoustics in this section in terms of resonances and antiresonances, as can be obtained from basic articulation models.

We start with the simplest model for the vocal tract: a hard-walled acoustic tube of length l and uniform cross-sectional area A, closed at the glottal end and open at the lips. When quantitative values are needed, we will assume $l = 17$ cm (e.g., for a typical adult male) because that specific value yields very simple formant numbers. Otherwise, we use the general l, since even within a single speaker, length varies (e.g., protruding the lips or lowering the larynx extends the vocal tract). The main acoustic effect of vocal tract size is that speech frequencies increase inversely with l (i.e., smaller tracts have higher values for both F0 and resonant frequencies, although the F0 effect is due to vocal cord size, which only indirectly correlates with vocal tract size).

More complex models (to be discussed below) will relax the assumption of a single area A, since in reality cross-sectional area varies substantially along the length of the vocal tract, even for a neutral phoneme such as the schwa vowel. While the walls of the vocal tract actually move under the pressure of speech, we will treat all effects of energy loss in the vocal tract only in a simple fashion. The assumption of a closed glottal end reflects the fact that, even when allowing air to pass, glottal area is very small relative to typical A values in the vocal tract. Much more variable is the lip end: labial phonemes often cause complete lip closure, which is opposite to our basic open-end assumption (hence other models below must reflect this).

Sound waves, of course, are initiated by some vibration. In the case of speech, it is usually the vocal folds that vibrate (but, for example, the tongue tip does so during **trills** in languages such as Spanish). The other major speech source is that of the random motion of air particles when air passes quickly through a narrow constriction. Once initiated, the sound waves propagate through any physical medium (e.g., air), following the laws of physics, e.g., conservation of mass, momentum, and energy. In free space, sound travels away from a source (e.g., the mouth) in a spherical wave whose radius increases with time, at the speed of sound c. (This is about 340 m/s in the most common conditions, but increases slightly with temperature (0.6 m/s per degree Celsius), and is much higher in media that are denser than air.)

Inside the vocal tract, unlike in free space, sound waves propagate only along the axis of the vocal tract, due to the physical constraints of the vocal tract tube, at least for most audio frequencies (note however, that above 4 kHz this assumption is less valid). This simpler **planar propagation** takes place when sound wavelengths λ are large compared to the diameter of the vocal tract, with $\lambda = \frac{c}{f} \leq \frac{340\,\text{m/s}}{4000/\text{s}} = 8.5\,\text{cm}$, which readily exceeds an average 2-cm vocal tract diameter. Also to simplify our acoustic analysis, we assume that the vocal tract is hard-walled and lossless, thus ignoring actual losses due to viscosity, heat conduction, and vibrating walls.

We now describe vocal tract acoustics in terms of **sound pressure** p and the velocity of air, at a given point in the vocal tract ($\mathbf{v}$ is the vector of velocity in three-dimensional space). Sound wave motion is linear and follows two physical laws: one due to continuity and the other famous law from Newton (force equals mass times acceleration), respectively:

$$\frac{1}{\rho c^2}\frac{\partial p}{\partial t} + \text{div}\,\mathbf{v} = 0, \tag{7.1}$$

$$\rho\frac{\partial \mathbf{v}}{\partial t} + \text{grad}\,p = 0, \tag{7.2}$$

where ρ is the density of air (about 1.2 mg/cc) [Sond 83]. Assuming one-dimensional airflow, we may examine the velocity of a volume of air u, rather than its particle velocity v:

$$u = Av,$$

where A is the cross-sectional area of the vocal tract. $A, u, \mathbf{v}$, and p are all functions of time t and distance x from the glottis ($x = 0$) to the lips ($x = l$) (where, for simplicity, l will usually be assumed to be 17 cm). Thus for **volume velocity** $u(x,t)$ and area $A(x,t)$, we have

$$-\frac{\partial u}{\partial x} = \frac{1}{\rho c^2}\frac{\partial(pA)}{\partial t} + \frac{\partial A}{\partial t}, \tag{7.3}$$

$$-\frac{\partial p}{\partial x} = \rho\frac{\partial(u/A)}{\partial t}. \tag{7.4}$$

We often omit to note the explicit dependence on x and t, where they can be understood from context.

Acoustics of a uniform lossless tube

The initial example of a simple tube with constant area can reasonably be used to describe a single vowel, close to a schwa vowel. We start with this oversimplified case, fixing A in both time and space, and examine more complex $A(x)$ functions later as concatenations of tube sections with uniform cross-sectional area. By developing results using general (but fixed) values for A and l, the analysis for one long uniform tube can be applied to numerous, shorter tube sections as well, as long as we account for the interactions between tube sections. (While an actual vocal tract bends 90°, from a vertical orientation at the lung exit to the horizontal oral tract, we will ignore the small resulting acoustic effects [Sond 86].)

For A constant, Eqs. 7.3 and 7.4 simplify to

$$-\frac{\partial u}{\partial x} = \frac{A}{\rho c^2}\frac{\partial p}{\partial t} \quad \text{and} \quad -\frac{\partial p}{\partial x} = \frac{\rho}{A}\frac{\partial u}{\partial t}. \tag{7.5}$$

Suppose that the vocal tract is excited at its lower (glottal) end by a sinusoidal volume velocity source $u_G(t)$ and that pressure is zero at the lip end. (These assumptions will be relaxed later, but correspond to a typically high impedance at the mostly-closed glottis and to the opposite at the open mouth.) If we use complex exponentials to represent sinusoids (again to simplify the mathematics), we have

$$\begin{aligned} u(0,t) &= u_G(t) = U_G(\Omega)e^{j\Omega t}, \\ p(l,t) &= 0, \end{aligned} \tag{7.6}$$

where Ω is the frequency of the source (in radians) and U_G is its amplitude. The linear Eqs. 7.5 have solutions

$$\begin{aligned} p(x,t) &= P(x,\Omega)e^{j\Omega t} \\ u(x,t) &= U(x,\Omega)e^{j\Omega t}, \end{aligned} \tag{7.7}$$

using P and U to represent complex spectral amplitudes that vary as a function of time and position of the waves in the tube. If we substitute Eqs. 7.7 into Eqs. 7.5, ordinary differential equations result:

$$-\frac{dU}{dx} = YP \quad \text{and} \quad -\frac{dP}{dx} = ZU, \tag{7.8}$$

where $Z = j\Omega\rho/A$ and $Y = j\Omega A/(\rho c^2)$ are the distributed acoustic impedance and admittance, respectively, of the tube. Eqs. 7.8 have solutions of the form

$$P(x,\Omega) = a_1 e^{\gamma x} + a_2 e^{-\gamma x},$$

where the **propagation constant** γ is

$$\gamma = \sqrt{ZY} = j\Omega/c \tag{7.9}$$

in the lossless case. (Losses, which add real components to Z and Y, cause mathematical complexities beyond the scope of this book, and are unnecessary to examine to obtain most insight into speech production.) We apply boundary conditions (Eqs. 7.6) to determine the coefficients a_i, and get steady-state solutions for p and u in a simple tube excited by a glottal source:

$$\begin{aligned} p(x,t) &= jZ_0 \tfrac{\sin(\Omega(l-x)/c)}{\cos(\Omega l/c)} U_G(\Omega)e^{j\Omega t} \\ u(x,t) &= \tfrac{\cos(\Omega(l-x)/c)}{\cos(\Omega l/c)} U_G(\Omega)e^{j\Omega t}, \end{aligned} \tag{7.10}$$

where $Z_0 = \rho c/A$ is the **characteristic impedance** of the tube. As Eqs. 7.10 note, both pressure and volume velocity in an acoustic tube excited by a sinusoid are themselves sinusoids (as expected in a linear system), one being 90° out of phase with respect to the other. The volume velocity at the lips is

$$u(l,t) = U(l,\Omega)e^{j\Omega t} = \frac{U_G(\Omega)e^{j\Omega t}}{\cos(\Omega l/c)}.$$

The vocal tract transfer function, relating lip and glottal volume velocity, is thus

$$V(\Omega) = \frac{U(l,\Omega)}{U_G(\Omega)} = \frac{1}{\cos(\Omega l/c)}. \tag{7.11}$$

Resonances (poles) occur when the denominator is zero at formant frequencies $F_i = \Omega_i/(2\pi)$, i.e., where

$$\begin{aligned} \Omega_i l/c &= (2i-1)(\pi/2) \\ F_i &= (2i-1)c/(4l) \quad \text{for} \quad i = 1,2,3,... \end{aligned} \tag{7.12}$$

Our earlier choice of l = 17 cm means that $V(\Omega)$ is infinite at F_i = 500, 1500, 2500, 3500,... Hz, i.e., vocal tract resonances every 1 kHz, with the first at 500 Hz. Other vocal tract lengths simply scale these values proportionately, following the simple inverse relationship of wavelength to frequency. One is tempted to suppose that people of different sizes would simply have a linear scaling of formants with vocal tract length. However, there are many physical differences among people, and the pharynx tends in general to be disproportionately small in smaller people [Kent 82].

7.3 Acoustic Models of Speech Production

Speech and music production share several parallels. A simple vocal tract tube closed at one end and open at the other much resembles an organ pipe. Both systems have quarter-wavelength resonances, as can be seen from the equations immediately above. At these frequencies, sound waves moving up and down the tube reflect and reinforce at each end of the tube. Within a fixed tube, conditions are static and waves travel unimpeded, but at each end, the impedance changes abruptly, causing major wave reflections. The formant frequencies are those that match boundary conditions on relative pressure P and volume velocity U: a closed tube end obviously renders $U = 0$, while P approaches basic atmospheric pressure at the open lips. This leads to wavelengths λ_i where vocal tract length l is an odd multiple of a quarter-wavelength (Fig. 7.2):

$$l = (\lambda_i/4)(2i-1), \quad \text{for} \quad i = 1,2,3,...$$

7.3.1 Resonances in a nonuniform vocal tract model

Since the various phones of any language result from different shapes of the vocal tract, it is obvious that we must go beyond a simple uniform acoustic tube as a speech production model, to determine the formant frequencies for other phones. A useful heuristic approach views the acoustic effects of modifying a uniform tube to have a single constriction (i.e., a slightly reduced diameter $A_c < A$ over a short length l_c of the tube). As a result, the resonances are 'perturbed' from their original positions, with the amount of formant change being correlated with the degree of the constriction. Any given resonance F_i in a simple uniform vocal tract model has maxima in U and P that alternate periodically along the length of the tube. If we constrict the tube where U is maximum, F_i falls; conversely, if P is maximum, F_i rises due to a constriction. As simple results, we note that a constriction in the upper half of the vocal tract (i.e., in the oral cavity) lowers F1, and a constriction at the lips lowers all formant frequencies. These results come from simple circuit theory, if we compare volume velocity U to electrical current

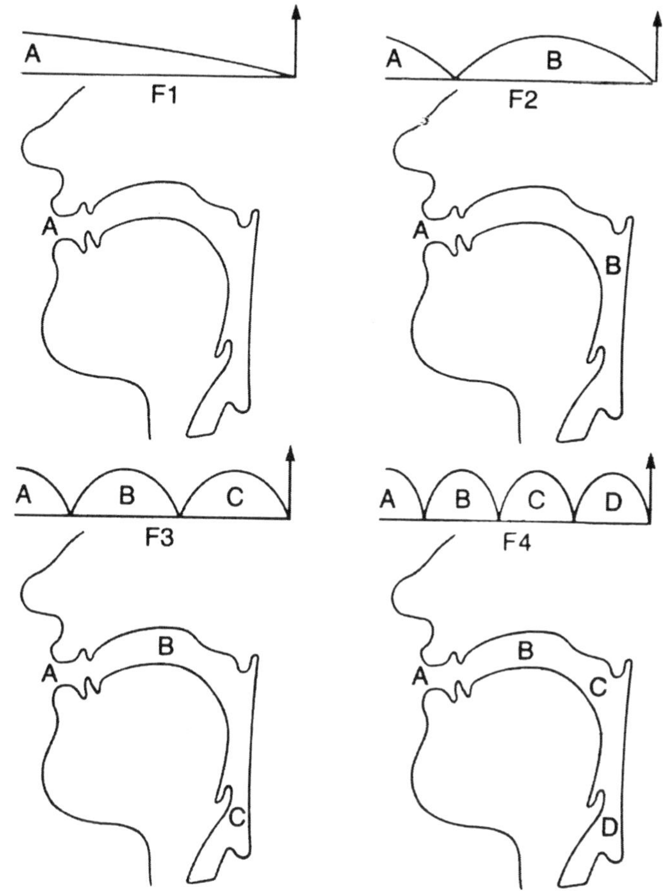

Figure 7.2: Spatial distribution of volume velocity at frequencies of the first four resonances of a uniform cross-sectional vocal tract. Places of maximum volume velocity are noted.

I and pressure P to voltage V, with impedance Z being viewed as acoustic resistance, inductance, and capacitance.

The mass of air in a tube has an **inertance** which opposes acceleration (hence causing an acoustic inductance), and the compressibility of its volume exhibits a **compliance** (capacitance), respectively. Unlike in discrete components on circuit boards, these acoustic (or electrical) parameters are uniformly distributed throughout each tube of interest. Thus the inductance L (per unit length) for a section of tube of area A is ρ/A, while the corresponding capacitance C is $A/\rho c^2$. Following basic electrical theory [Adl 60], we define a **characteristic impedance** for the tube as

$$Z_0 = \sqrt{L/C} = \rho c/A. \tag{7.13}$$

Allowing A to vary as a function of distance x along the tube (e.g., from $x = 0$ at the glottis to $x = l$ at the lips), wide and narrow sections have large values of C and L, respectively.

Continuing this electrical analogy, since many readers are familiar with basic circuits and their resonances, the resonance of a simple **LC** (inductor+capacitor) circuit is $(2\pi\sqrt{LC})^{-1}$. For this very basic filter, an increase in either inductance or capacitance lowers its resonant frequency. When we perturb a uniform tube via a slight constriction (as occurs for several phonemes), we reduce A at one location x along the tube. For each resonance F_i of interest, if U is large prior to the constriction, then kinetic energy dominates there and the change in L (due to the constriction) has the greater effect (i.e., unlike a simple LC-circuit). Similarly, at points where P is large, potential energy dominates and changes in C dominate F_i movement. Since reducing A raises L and lowers C, F_i increases if the constriction occurs where P is large and decreases where U is large [Fant 68].

7.3.2 Two-tube vowel models

The common low vowel /aa/ (found in virtually all languages) can be modeled by a narrow tube (corresponding to the pharynx) which opens abruptly into a wide tube (the oral cavity) (Fig. 7.3a). If we take the simplest model, where each of the two tubes has a length of half the full 17 cm, each produces the same set of resonances, at odd multiples of 1 kHz. This is a simple consequence of halving the length of a 17-cm tube: each tube acts a quarter-wavelength resonator, with its back end relatively closed and its front end open. When the ratio of areas for the two tubes exceeds an order of magnitude, any acoustic coupling between tubes is quite small and the individual tube's resonances have little interaction. However, when the theory predicts identical values, e.g., F1 and F2 for /aa/ both at 1000 Hz, actual acoustic coupling causes the formants to separate by about 200 Hz; thus, F1 = 900, F2 = 1100, F3 = 2900, and F4 = 3100 Hz. Actual observed values for /aa/ deviate somewhat from these numbers since the two-tube model is only a rough approximation, but the two-tube model is reasonably accurate and easy to interpret physically.

The symmetrically-opposite vowel from /aa/ is the high front /iy/ (which is also universal) (Fig. 7.3b). The respective model for /iy/ thus has a wide back tube constricting abruptly to a narrow front tube. The boundary conditions are the opposite of those for /aa/: the back tube is closed at both ends (interpreting the narrow openings into the glottis and the front tube as negligible compared to the large area of the back tube), whereas the front tube has open ends. These tubes are thus **half-wavelength resonators**, owing to their symmetrical boundary conditions. A tube closed at both

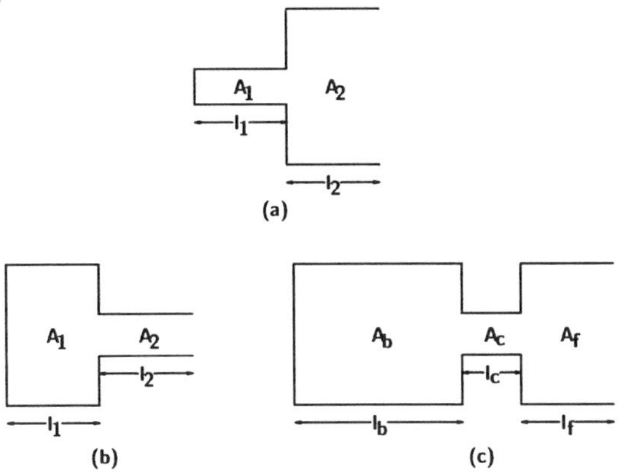

Figure 7.3: Two- and three-tube models for the vocal tract. In the two-tube case, the first tube section of area A_1 and length l_1 may be viewed as modeling the pharyngeal cavity, while the second section of area A_2 and length l_2 models the oral cavity. The first two models represent the vowels: (a) /aa/ and (b) /iy/. The three-tube model (c) has a narrow section corresponding to the constriction for a consonant.

ends forces U to be minimal at its ends (at the reinforcing resonance frequencies), while a tube open at both ends needs minimal P at its ends. For tube lengths l that are multiples of half a resonance wavelength λ_i:

$$l = \frac{\lambda_i}{2} i \quad \text{and} \quad F_i = \frac{c}{\lambda_i} = \frac{ci}{2l}, \quad \text{for} \quad i = 1, 2, 3, \ldots \; . \tag{7.14}$$

Thus, for /iy/, each tube has theoretical resonances at multiples of 2 kHz. An additional resonance occurs at very low frequency for vocal tract models with a large tube (e.g., the back cavity) closed at both ends. F1 in such cases effectively decreases from a nominal value of 500 Hz in our earlier one-tube model as a major constriction occurs in the upper portion of the vocal tract. The theoretical value of 0 Hz is not achieved, but F1 approaches as low as 150 Hz for many consonants with significant constrictions. We cannot go much further in simply modeling other phones using only two tubes, but we end here by noting that such a model for a back rounded vowel (e.g., /ow,uw/) would have two closed tubes with a constricted tube between them, causing both F1 and F2 to be low.

7.3.3 Three-tube consonant modeling

For most consonants there is a single major constriction in the oral tract, which leads to a reduced value for F1, when compared to vowels. An understanding of the behavior of the higher formants is more complex, and can be aided by examining a three-tube vocal tract model, where the middle narrow tube represents the constriction (in an otherwise open system) (Fig. 7.3c). The back and central tubes, being symmetrical (i.e., at both ends having the same configuration: either both closed or both open), are half-wavelength resonators. The front tube is of primary importance and acts as a quarter-wavelength resonator (i.e., asymmetrical: closed at the rear, open at the front). Often excited at its rear end (i.e., at the outlet of the constriction), the front tube acts as a reduced version of the original one-tube system, leading to high-frequency resonances (very common for obstruents). Theoretically the three tubes yield three sets of resonances F_i:

$$\frac{ci}{2l_b}, \frac{ci}{2l_c}, \frac{c(2i-1)}{4l_f}, \quad \text{for} \quad i = 1, 2, 3, ...,$$

where l_b, l_c, l_f are the lengths of the back, central, and front tubes. The front and back cavity resonances display the simple inverse relationship of frequency and wavelength, with one set of resonances increasing (and the other falling) as the constriction position varies. While many of these resonances lie at frequencies above the usual audio range (e.g., 0-10 kHz, at most), the longer cavities tend to generate the more relevant low-frequency formants. The constriction, being of length of a few cm at most, has resonances at multiples of a value above 5 kHz and thus is ignored acoustically in many applications.

Consider the three most common places of articulation for stop consonants: labial, alveolar and velar (ordered by increasing length of the front cavity). The labials /p,b,f,v/ have virtually no front cavity and thus have very weak energy (labial /m/ is an exception due to its glottal excitation). While the front cavity is most important for speech production in obstruents, rear-cavity resonances are of interest for their effects on the temporally-adjacent sonorant formants, as the constriction opens up for the sonorant. In the adjacent sounds, most formants of interest (the strongest ones at low frequencies) are associated with the long back tube, and they generally have center frequencies lower than those for a single uniform open tube (owing to the constriction).

The relative tube lengths are of more interest for non-labial consonants. For example, a typical alveolar might have a 10-cm back tube, a 3-cm constriction, and a 3-cm front tube; the simple formulae of frequency via inverse tube length suggest that such alveolar consonants have a back cavity resonance near 1.7 kHz for F2 and a front cavity resonance for F3 around 2.8 kHz. Empirically, alveolars are often characterized by F3 being higher than in adjacent vowels.

Velar consonants typically have both F2 and F3 around 2 kHz, and these values correspond well to a model with an 8.5-cm back tube and a 4.5-cm front tube. The constriction location for velars is more variable than for labials or alveolars, owing to coarticulation. When the velar consonant has a more forward position (as occurs when adjacent to a forward vowel), F2 originates in a longer back cavity and F3 is due to the correspondingly shorter front cavity. When the velar is near a back vowel, it constricts farther back on the palate, causing a reversal of the F2-F3 roles. This distinction is relevant acoustically because obstruent speech power is primarily caused by front-cavity resonances. Frication is generated just forward of the constriction in an obstruent, and excites the anterior cavity. The resonances of the posterior cavity are largely cancelled by spectral zeros during obstruent production, until the constriction becomes wider. At that point the noise ceases, the tubes interact acoustically, and the sound source is glottal. As a result, alveolars and front-articulated velars are dominated by F3 power, while F2 is prominent in back velars. Labials, with their very short front cavity, have very little power.

7.3.4 Speech production involving both poles and zeros

Acoustic modeling of speech production usually focuses on the effects on the formant frequencies of an acoustic tube, owing to a single tongue constriction and/or a simultaneous labial constriction. However, research on dynamic speech production has revealed that tongue gestures overlap in time, so two or more tongue constrictions simultaneously affect the acoustic output of the model. Understanding the acoustics of natural speech production therefore requires an investigation of the acoustic effects of multiple lingual constrictions of varying degrees and locations. Of specific interest is how the acoustic effects of individual constrictions combine and interact in determining the acoustic output of the whole vocal tract. One could investigate how individual tongue constrictions are dynamically related in speech production. Secondly, one could determine how the modes of gesture combination are acoustically realized.

When a sound source in the vocal tract has only one acoustic path to an oral output, its frequency response has only poles (vocal tract resonances or formants) and no zeros. The spectrum rises and falls with frequency (owing to the poles), but there are no specific frequencies heavily attenuated via zeros; this is the case for all vowels in English. During nasals and obstruents, however, multiple acoustic paths occur and introduce zeros into the vocal tract transfer function. For fricatives, noise generated at a constriction propagates back into the constriction as well as going towards the mouth. There are certain frequencies where the acoustic impedance into the constriction becomes infinite, and as a result no power moves toward the mouth and the final speech thus has zeros. Frication noise is well modeled by a pressure source [Flan 72], and like a voltage source seeing an infinite electrical impedance, no air volume (analogous to current) flows. The same does not occur at the glottis, because the glottal sound source is better modeled as a volume velocity (not pressure) source. (Recall that current sources are usually modeled with high impedance, a situation that the narrow glottis reflects well.)

Electrical circuit theory provides good analogies for both pole and zero estimation. For the simple two-tube model of the vocal tract, at the boundary between the tubes, the acoustic impedance Z_b backward into the posterior tube (pharyngeal cavity) acts in parallel with an impedance Z_f forward into the front tube (oral cavity). Such a parallel network has resonances at all frequencies when

$$Z_b + Z_f = 0. \tag{7.15}$$

For our initial simple model of a tube closed at one end (e.g., glottis) and open at the other (mouth), poles occur when tube length is an odd multiple of a quarter-wavelength. For a tube closed or open at both ends (e.g., a constriction, or a back cavity), poles are found at frequencies where tube length is an even multiple of a half-wavelength. Simple affiliations of resonances to tubes are modeled best when the boundaries between tube sections are abrupt enough to isolate each one's acoustical effects, due to abrupt changes in impedance. In other cases, as occur in most vowels, the end of each tube section cannot be interpreted as fully open or closed, impedances assume intermediate values (neither zero nor infinite), and resonances cannot be associated with individual tubes.

A simple three-tube model with abrupt section boundaries is a reasonable approximation model for many fricatives: a narrow middle tube models the constriction and a wide front tube models the oral cavity. The large rear cavity has little acoustic effect in these models because it is mostly decoupled from the frication excitation (owing to the abrupt boundary between the back cavity and the constriction). (The pharyngeal resonances are canceled by corresponding antiresonances here.) Hence fricatives have poles for frequencies where l_f is an odd multiple of a quarter-wavelength or l_c is a multiple of a half-wavelength. Conversely, zeros appear where l_c is an odd multiple of a quarter-wavelength. Modeling alveolar fricatives /s,z/ with a 2.5-cm constriction and a 1-cm front cavity, a zero occurs at 3.4 kHz and poles at 6.8 and 8.6 kHz; the two lowest frequencies here are due to the constriction, which thus has the greatest acoustic effect on many fricatives. Palatal fricatives have a longer l_c and consequently lower values for this pole-zero pair, while dental and labial fricatives have very low values for l_c and l_f, which leads to poles and zeros well above the usual audio range. Fricatives have quite reduced power at frequencies below that of their first zero, and their power attains a maximum at their first pole.

7.3.5 Transmission line analog of the vocal tract

While obstruents can be reasonably modeled with tubes having abrupt ends, most sonorants need more complicated models in order to understand the behavior of their resonances. One way to obtain theoretical values that correspond well with empirical results is to compare acoustic tubes to transmission lines (T-lines). Eqs. 7.5 have a direct parallel with the behavior of voltage $v(x,t)$ and current $i(x,t)$ in a uniform lossless T-line [Adl 60]:

$$-\frac{\partial i}{\partial x} = C\frac{\partial v}{\partial t} \qquad \text{and} \qquad -\frac{\partial v}{\partial x} = L\frac{\partial i}{\partial t},$$

if we use these analogies: current–volume velocity, voltage–pressure, line capacitance C–acoustic capacitance $A/(\rho c^2)$, and line inductance L–acoustic inductance ρ/A. All effects (impedances, inductances and capacitances) are distributed per unit length of the tube or transmission line. The acoustic impedance $\rho c/A$ follows Z_0, the **characteristic impedance** of a transmission line. In the following, we calculate the frequency response of various vocal tract models in terms of T-line impedances.

A T-line of length l can be viewed as an electrical network with two ports (input and output), having respective voltages V_1 and V_2 and currents I_1 (into the T-line) and I_2 (out):

$$V_1 = Z_{11}I_1 + Z_{12}I_2 \quad \text{and} \quad V_2 = Z_{21}I_1 + Z_{22}I_2, \tag{7.16}$$

where Z_{11} and Z_{22} are the input and output impedances of the T-line, respectively, and Z_{12} and Z_{21} are cross-impedances. Owing to the symmetry of a single uniform tube or T-line,

$$Z_{12} = -Z_{21} \quad \text{and} \quad Z_{11} = -Z_{22}. \tag{7.17}$$

Suppose we set I_2 to zero (i.e., assume the output port to be an open circuit); then Z_{11} is the input impedance of an open-circuited T-line [Adl 60]:

$$Z_{11} = Z_0 \coth(\gamma l), \tag{7.18}$$

where Z_0 and γ are from Eqs. 7.13 and 7.9. If instead V_2 is set to zero (i.e., the output port is a short circuit), Eqs. 7.16 give

$$\begin{aligned} Z_{21} &= -Z_{22}\tfrac{I_2}{I_1} = Z_{11}\tfrac{I_2}{I_1}, \\ V_1 &= [Z_0 \coth(\gamma l)]I_1 + Z_{12}I_2, \end{aligned}$$

and V_1 is obtained in terms of the input impedance of a short-circuited T-line [Adl 60]:

$$V_1 = [Z_0 \tanh(\gamma l)]I_1. \tag{7.19}$$

Thus

$$Z_{21} = Z_0 \operatorname{csch}(\gamma l). \tag{7.20}$$

We often assume a zero acoustic pressure at the lips (Eq. 7.6). Then using Eq. 7.9, we obtain the transfer function for the tube (being modeled as a section of T-line) in terms of its volume velocity (analogous to electrical current):

$$\frac{I_2}{I_1} = \frac{U_2}{U_1} = \frac{Z_{21}}{Z_{11}} = \frac{Z_0\text{csch}(\gamma l)}{Z_0\text{coth}(\gamma l)} = \frac{1}{\cosh(\gamma l)} = \frac{1}{\cos(\Omega l/c)}. \quad (7.21)$$

This is the same result for a lossless model of the vocal tract, whether via T-line (Eq. 7.21) or acoustic analysis (Eq. 7.11).

Practical tube models

We continue with some circuit theory to model the symmetric network of a uniform acoustic tube as a T-section (so called due to its longitudinal and transverse impedance effects). The longitudinal and transverse impedances of a T-section can be easily determined from Eqs. 7.17, 7.20, and 7.18 to be, respectively,

$$Z_a = Z_0\tanh(\gamma l/2) \quad \text{and} \quad Z_b = Z_0\text{csch}(\gamma l). \quad (7.22)$$

Our earlier analysis for a two-tube vocal tract model was limited to cases with tube-sections having very abrupt boundaries and acoustically-uncoupled cavities. Eqs. 7.16 allow a resonance analysis for a general two-tube model whereby concatenated sections are modeled by T-line networks. We view a two-tube model as an electrical network where two such circuits are in parallel where the two sections meet. At this juncture, the impedance Z_1 back into the lower tube and Z_2 up into the second tube model the effects of the two circuits. The poles of a network with N parallel components are the frequencies where their admittances add to zero:

$$\sum_{i=1}^{N} Y_i = \sum_{i=1}^{N} \frac{1}{Z_i} = 0. \quad (7.23)$$

Then the poles of a two-tube model satisfy

$$Z_1 = -Z_2. \quad (7.24)$$

As the glottal opening is much smaller than the pharyngeal A_1, that first tube (pharynx) is modeled as being closed at its lower end. In T-line analogy, an open circuit (i.e., passing no current (volume velocity)) models the closed end of an acoustic tube. We define Z_{0i} as the characteristic impedance of tube section i ($Z_{0i}{=}\rho c/A_i$); then Eq. 7.18 gives

$$Z_1 = Z_{01}\coth(\gamma l_1) = -jZ_{01}\cot(\beta l_1), \quad (7.25)$$

where $\beta = \Omega/c$ in the lossless case. For most phonemes (those with unrounded lips), the mouth ending of the second tube is viewed as open, and is modeled by a T-line short circuit. Thus Eq. 7.19 yields

$$Z_2 = Z_{02}\tanh(\gamma l_2) = jZ_{02}\tan(\beta l_2). \quad (7.26)$$

Combining Eqs. 7.24–7.26 gives

$$\cot(\beta l_1) = \frac{A_1}{A_2}\tan(\beta l_2). \quad (7.27)$$

A number of speech sounds (e.g., /iy,aa/, schwa) allow models with $l_1 = l_2$; symmetry then allows an explicit solution of Eq. 7.27 in terms of formant center frequencies F_i:

$$F_i = \frac{c}{2\pi l} \cot^{-1}\left(\sqrt{\frac{A_1}{A_2}}\right) \quad \text{for} \quad i = 1, 2, 3, \ldots.$$

Cases where $l_1 \neq l_2$ can be solved by simple graphical solutions, to locate the formant frequencies: where the tangent and cotangent curves of Eq. 7.27 intersect. Such intersections occur, on average, every 1 kHz for a 17-cm vocal tract. For the two basic and opposite vowels /iy,aa/, the model allows simple interpretations: for /aa/, $A_2 \gg A_1$ and resonances occur in pairs close to odd multiples of 1 kHz; /iy/ has pairs of resonances near even multiples of 1 kHz (see section 7.3.2).

Three-tube model for nasals

The simple mathematical solution above for a two-tube model is not easily extended to most models having more than two tubes. For example, the resonances for our earlier three-tube fricative model (in the general case where boundaries are not necessarily abrupt) are not as easily determined as for the two-tube case because more than one tube junction needs to be considered. We will study later the filtering behavior of a general model of N concatenated tube sections, but that later method does not easily determine resonance positions. That model will nonetheless be very useful from a speech coding point of view. In this section we examine one final model that allows direct determination of pole positions, one that is useful for an important class of sounds: nasals.

For most speech phones, air flows in the vocal tract up (and back, due to reflections) along a simple path from the source to the output. Thus a direct chain of tube sections models the system well for such phones. Nasal sounds, on the other hand, present a different modeling challenge, owing to the existence of a side tube that affects the sound. The simplest nasal model is a three-tube system (one tube each for the pharyngeal, oral, and nasal cavities), where all three tubes meet at a juncture corresponding to the velum, which is lowered in the natural vocal tract to allow the nasal tract to accept airflow. Thus an analogous electrical circuit for the three tubes links the three in parallel. The poles for such a model are specified by Eq. 7.23:

$$\frac{1}{Z_p} + \frac{1}{Z_m} + \frac{1}{Z_n} = 0, \tag{7.28}$$

where Z_p, Z_n, Z_m are the respective impedances of the pharyngeal, nasal, and oral (mouth) cavities, $Z_p{=}-jZ_{0p}\cot(\beta l_p)$, $Z_m{=}-jZ_{0m}\cot(\beta l_m)$, and $Z_n{=}jZ_{0n}\tan(\beta l_n)$. The mouth and pharyngeal tubes terminate in closed acoustic ends, whereas the nostrils of the nasal tube are open. A graphical solution for this equation is not as simple as in the two-tube case, but we obtain approximate results by assuming that the pharynx and nasal tubes are similar in dimensions. If $l_p = l_n = 10.5$ cm, each of $1/Z_p$ and $1/Z_n$ has periods of about 1.6 kHz and the function $1/Z_p + 1/Z_n$ attains infinity approximately every 800 Hz. The mouth tube, on the other hand, is shorter and varies with the place of articulation of the nasal. Thus nasal consonants have formants spaced approximately every 800 Hz. This smaller spacing is due to the combined pharynx-nasal tube of 21 cm length, compared to the normal 17-cm pharynx-mouth tube.

To obtain the zeros of nasal spectra, we need to determine when $Z_m = 0$; at those frequencies no air flows into the nasal tube and thus no speech leaves the nostrils. Air from the lungs separates into two streams at the velum junction, depending on the impedances looking into the oral and nasal tubes. Solving $Z_m = -jZ_{0m}\cot(2\pi F_i l_m/c) = 0$ for F_i shows zeros at odd multiples of $c/4l_m$. The longest oral tube (about 7 cm) occurs for /m/, yielding zeros at 1.2, 3.6, 6.0, ... kHz. The other nasal stops have oral tube lengths of 3–5 cm; their initial zeros occur at 1.7–2.8 kHz. In addition to the poles spaced by 800 Hz due to the pharynx-nasal cavities, nasal spectra have poles and zeros occurring in close pairs due to the mouth cavity.

While all English nasals are stop consonants, some languages have nasalized vowels, which have much more power than nasal stops. Nasalized vowels have almost all air flowing through the low-impedance oral cavity, and little air flows in the lossy nasal cavity due to its higher impedance. Returning to our three-tube model, the main change for nasalized vowels is that the mouth cavity terminates with open lips. As a result, Z_m is replaced by $jZ_{0m}\tan(\beta l_m)$ in Eq. 7.28. This change causes a shift in pole positions, when comparing oral and nasalized vowels with similar oral tract shapes. There is an additional **nasal formant** near F1, as a result of the nasal cavity [Chen 97].

7.4 Coarticulation: Its Origins and Models

Speech production involves a sequence of gestures by a few articulators that are synchronously timed so that key aspects of vocal tract shape occur in the correct order for the intended phoneme sequence. Gestures for adjacent phonemes overlap temporally, with the result that vocal tract shapes for a phone are highly dependent on its context. We call this behavior coarticulation. The articulation and, consequently, acoustics of a phoneme thus vary according to its phonetic context. Coarticulation can be observed directly in vocal tract movements or in related muscle activity [Honda 94]. However, the difficulty of direct measurements of either has led to more use of indirect acoustic observation.

Each phoneme has its own requirements for certain parts of the vocal tract, but these mostly pertain to only a subset of all the articulators used for speech. Thus there is often considerable freedom for the timing and amount of articulator movements. In a phoneme sequence ABC, when a specific articulator gesture needed for one phoneme (e.g., B) does not oppose the requirements for that articulator for an adjacent phoneme (A or C), the articulator is free to move toward the needed position for B early (i.e., during A) or may remain in the B state later (i.e., during C). The first type of coarticulation (toward a B target and during A) is called **forward, anticipatory**, or **right-to-left** coarticulation, because a target for a phoneme (on the "right") causes motion in an articulator during an earlier phone (on the "left"). It may be that speech production planning implicitly "looks ahead" so that each articulator moves toward its next required state when the last phone that needs it has been completed. As an example, vowel lip rounding usually starts during earlier sounds that do not employ such rounding; this may occur several phones in advance (e.g., during the /s/ of "strew"). The limited acoustic changes that result in these earlier sounds are likely expected by listeners [Sharf 81].

The other type of coarticulation (that of phone B on C, in the explanation above) is called **backward, carryover**, or **left-to-right** coarticulation. In this case some of a phoneme's features persist into later phones. For example, formant movements during that portion of a vowel immediately after a consonant are greatly influenced by the prior consonant [Sharf 81]. Backward coarticulation seems to be a low-level phenomenon, unlike forward coarticulation's origins in high-level speech planning. In a sense, coarticulation is an example of the general trend in human communication to exert the least effort while accomplishing the desired transfer of information. Less effort is used to move an articulator slowly toward a target than to do the motion in a shorter time span.

7.4.1 Effects of coarticulation

Defining coarticulation is at the heart of what distinguishes a phoneme from a phone, i.e., the multitude of ways that the abstract concept of a phoneme can be realized physically as phones. Obviously, more than phonetic context (e.g., speaker differences) is involved in the phoneme-to-phone mapping. In this section, we concentrate solely on the variations of speech production that depend on immediate phonetic context – aspects that can be directly related to the physical movements of articulators.

Each phoneme has a set of requirements about the positioning of various articulators, but the timing of movements are often not simply synchronized together. Researchers usually interpret a phone as the events during the time period when all required articulators for a corresponding phoneme are in appropriate positions. Thus, one phone "ends" when one or more articulators move (often suddenly) toward positions for the following phoneme. Such motion causes acoustic changes in the speech signal, which allow interpretation (by human spectrogram readers, or by automatic speech recognizers) as phone boundaries. Since the articulation needed for each phoneme exceeds the temporal boundaries of its phone realization, it can be said that a phoneme's "articulation period" exceeds its "acoustic period" [Bell 81]; i.e., the gestures for a phoneme start during a preceding phone and finish during an ensuing one. (Usually this articulatory spread extends to the immediately adjacent phonemes, but the effects often extend further.)

Phone boundaries are typically associated with times of large acoustic change, most often changes in manner of articulation, which often involve changes to vocal tract constriction. Often, not all articulators are directly involved in a phoneme's constriction, leaving other articulators free to coarticulate. For example, phonemes with a labial constriction (/p,b,m,f,v/) allow the tongue to coarticulate. Conversely, lingual phonemes permit labial coarticulation. In the latter, the lip-rounding feature of a labial spreads to adjacent lingual consonants, with corresponding acoustic effects. Listeners hearing lower resonant frequencies in lingual phones due to the labial coarticulation subconsciously make the necessary perceptual adjustments to understand the lingual phoneme in such a phonetic context. Coarticulation is particularly helpful to perception in helping listeners understand place of articulation for obstruents and nasals. The formant transitions during vowels before and after oral tract closure for stops and for nasals provide the primary cues to their place of articulation. Place for weak fricatives is mostly recognized from the acoustic effects of coarticulation in adjacent sonorants.

Coarticulation varies with speaking rate. Phoneticians often speak of classical steady-state positions and formant frequency targets for each phoneme, which can be achieved when the phonemes are uttered by themselves (i.e., in isolation, with adjacent silences). In practical speech, due to coarticulation, such target positions and spectra are rarely achieved, with resulting **undershoot** of articulators (and correspondingly, of formant transitions). These effects of articulators' movements toward different phoneme targets in succession during normal speech are greatest when speech is rapid [Moon 94].

Actual measurements of coarticulation vary significantly. One way to quantify the acoustic effects of coarticulation is to measure variability in durations or formant frequencies. When an individual speaker repeatedly utters the same sequence of phonemes, any changes in duration or formants for a given phoneme can be attributed to inherent variability, since any effects of context are controlled. In such identical phonetic contexts, speakers typically produce variations (i.e., standard deviations) for a phoneme on the order of 5–10 ms in phone durations and 50–100 Hz in formant center frequencies. Beyond this, further variations in these parameters when a given phoneme is found in different contexts can be attributed to coarticulation.

7.4.2 Coarticulation effects for different articulators

To better explain coarticulation, consider six articulators that have varying degrees of independence from each other: 1) glottis, 2) velum, 3) tongue body, 4) tongue tip, 5) lips, and 6) jaw (these are the articulators most frequently modeled in articulatory synthesizers). Our first articulator, the glottis, controls the voicing for all sonorant phonemes and, for the /h/ fricative, acts as the major vocal tract constrictor. Major evidence for coarticulation involving the glottis concerns delays, relative to articulator motion for adjacent phonemes, in the onset and offset of voicing. For example, onsets and offsets of voicing and frication often do not coincide.

Of particular interest are variations in voice onset time (VOT) after a stop release, which is another example of glottal coarticulation. VOT tends to be longer for stops articulated further back in the oral tract (e.g., velars have the longest VOTs). VOT also increases with the height of the ensuing vowel [Oha 83] and is longer before sonorant consonants than before vowels [Zue 76, O'Sh 74]. The first of these three effects is due to vocal tract motion away from the stop closure, where the narrow constriction causing frication is maintained longer for velars than for labials, since the tongue body is slower to move toward an ensuing vowel. For the latter two effects, phonemes following the stop that have more vocal tract constriction cause delays in voicing onset, because these phonemes retain more intraoral pressure behind the constriction, which in turn lowers the pressure drop available across the glottis. Lastly, we note that F0 patterns in tone languages (which, like all F0 phenomena, are due to glottal articulation) show significant modifications in phonetic context [Xu 94].

Like the glottis (except for its special role in creating the /h/ phoneme), the velum is not involved in oral tract constrictions. The latter constrictions, which are critical to most phonemes, can often operate relatively independently of the velum, which leaves the velum relatively free to coarticulate in many common cases of speech. Although usually open most of the time when one is not talking, the velum remains closed during most of speech. It lowers specifically and only for nasal sounds, and its lowering often starts well before a nasal consonant (as defined by the closure of the oral tract) and continues afterward as well. This overlap of velum lowering with adjacent phonemes causes **nasalization** to spread into their phones. Nasalization of adjacent phones is avoided only when a raised velum is essential in the adjacent sounds: for obstruents (which require pressure build-up that would be thwarted by a lowered velum) or for French vowels (where inadvertent nasalization would cause a listener to perceive a different phoneme, i.e., a nasalized vowel).

Most vocal-tract constrictions involve tongue and/or lip articulators. When one articulator (or part of an articulator) constricts for a given phoneme, the other articulators are relatively free to coarticulate (as long as they do not cause an additional constriction). Much coarticulation is evident during vowels, where either the tongue body or tip may be displaced toward targets for preceding or ensuing phonemes, while a given phoneme is being uttered. Any phoneme that needs a complicated lingual constriction (e.g., strident fricatives, especially the more complex palatal ones) allows little tongue coarticulation [Reca 97]. On the other hand, the occlusion location for velar stops is sufficiently flexible to allow significant coarticulation (as is evident in the classic front and back allophones for /k,g/).

The jaw is less well studied in speech production, because its role is indirect; it is used to grossly position the tongue and lips (as a unit) so that the latter may attain their appropriate tongue height and lip rounding for each phoneme. As the tongue body and tip are correlated articulators, which consequently limit each other's ability to coarticulate, the jaw too has a limited role in coarticulation owing to its links to the lips and tongue. Nevertheless, the jaw does exhibit coarticulation into one preceding phone and up to two ensuing phones [Sharf 81].

7.4.3 Invariant features

Coarticulation significantly aids human speech perception, although its effects are difficult to model for automatic synthesis and recognition applications. Listeners can identify most phonemes from small portions of the speech signal found in the middle of their phones (i.e., from their most steady-state portion). Certain articulatory gestures have very clear acoustic effects, such that coarticulation is not much needed for accurate perception of the phoneme articulated. For example, an abrupt and major constriction in the vocal tract causes the rapid spectral change of lowering F1 to almost 200 Hz; since it is easy to hear and interpret such F1 behavior (even in noisy conditions), related phonemic information (e.g., vowel height, and manner of articulation) is rarely confused by listeners.

For place of articulation, however, the acoustic cues are often complex in terms of spectral distribution of energy. Coarticulation usually complicates the relationship between speech spectrum and perceived place. Strong sounds (strident fricatives and most sonorants) have simple and reliable place cues; e.g., for strident fricatives, place corresponds to the cutoff frequency of their highpass frication spectra. For each of the stops, nasals, and weak fricatives, however, the speech signal during most of the gesture for the phoneme is inadequate (too weak for obstruents and too similar for nasals) to cue place perception reliably. Place is thus primarily signaled for these phonemes by spectral transitions that occur before and after oral tract closure or constriction (the signal during constriction being less helpful in cuing place). Coarticulation causes these transitions to be complicated functions of phonetic context, however, which has led to the search for **invariant acoustic features** for place of articulation [Mack 83, Perk 86].

For stops, the spectral transitions of F2 and F3, as the stop is either approached or departed, appear to be key cues to place, but the stop burst spectrum is also important. Burst spectra for labial stops tend to be flat or falling, while alveolar stops have "diffusely" rising patterns, and velars have "compact" spectra [Blum 79]. The terms "diffuse" and "compact" refer to the absence or presence, respectively, of a single prominent spectral peak. Being farthest back in the vocal tract among the stops, velars excite a relatively low-frequency front-cavity resonance, which leads to a compact spectral peak near 2 kHz.

Formant transitions appear to be the most important cues to place in stops. Since nasals lack a burst to help cue place and since nasal murmurs vary little with place (the variable oral-tract length, which leads to different placement of spectral zeros, has too slight an effect on the murmur), formant transitions are even more important for nasal place.

The classical search for invariant cues has usually involved stops and nasals, due to their complex production and perception, involving coarticulation. Such ideas could be applied to vowels as well, but simple models of undershoot of targets have generally been sufficient for most applications. In English, there seems to be a trend toward centralization of unstressed vowels, where undershoot leads to many cases of vowels resembling the schwa vowel. As a neutral phoneme at the center of the vowel triangle, schwa is perhaps a prime example of coarticulation [vB 94].

7.4.4 Effects of coarticulation on duration

Coarticulation is usually described in terms of effects on spectra (e.g., undershoot of targets). However, it has important effects on duration as well. Consonants are generally shorter when they occur in clusters (i.e., as part of a sequence of consonants between two vowels). This may be partly explained by the shorter articulatory distances that articulators travel in sequential consonants, all having a relatively closed vocal tract (as opposed to vowel-consonant alternations, which require opening and closing the tract) [O'Sh 74]. As another example, low vowels are typically longer than high vowels because the relatively slow jaw is required to move farther for low vowels. Not all of these durational effects are due to coarticulation, since learned phonological variation (to aid communication) may also be involved.

7.4.5 Models for coarticulation

Since a clear understanding of coarticulation is so important for most speech applications, many models for coarticulation have been proposed. Typically, each phoneme is represented by a set of features or articulatory targets, which are allowed to spread to adjacent phones, if the features/targets for such phones do not conflict with the spreading ones. Taking a syllabic approach, one may view speech production as a series of vowel-to-vowel articulations, upon which consonant articulations are superimposed [Ohm 66]. As justification, we can note that vowels tend to use large muscles, which are less complex and less precise in timing in general than the muscles used to articulate consonants. Other examples of coarticulation that act over fairly long time intervals are lip rounding and nasalization, which also involve relatively slow muscles.

Research on **compensatory speech** supports such models of coarticulation [Kel 84]: when a person speaks with an artificial constraint (e.g., with clenched teeth or with an object in one's mouth), the articulators tend to deviate from normal positions so that the output speech achieves near-normal acoustics. These models, with their view of uniform overlapping articulatory gestures, contrast with other "look-ahead" or "feature-migration" models; the latter view anticipatory coarticulation as allowing features (e.g., lip rounding) to spread far ahead via planning and have a different mechanism for carryover coarticulation (e.g., inertia).

Linear regressions of F2 transitions in consonant+vowel (CV) syllables (called **locus equations** [Sus 97]) have shown recent promise to model coarticulation and invariant features. Labial stops have steeper falling spectral slopes than alveolars, suggesting stronger coarticulatory influence of lip closure on the tongue position for an ensuing vowel. Comparisons with VC syllables suggest that CV F2 transitions are more precisely controlled than for VCs, which corresponds well with the view that CVs are more dominant in speech communication.

There is general support for the idea that auditory feedback is important for production, i.e., that control of speech production is directed toward auditory goals [Lind 90]. People generally minimize their speaking effort whenever the conversational context permits (e.g., we usually employ less clear speech when talking among friends). Under different conditions, we articulate more clearly to assure understanding when talking to less familiar listeners, to avoid having to repeat (which is much less efficient than accomplishing communication in a single utterance). As supporting evidence, we cite the **Lombard effect** (the different acoustical effects of shouted speech) and the deterioration of speech if a speaker develops deafness [Lane 91] (or if self-speech perception is disrupted in other ways). Some researchers think that articulatory goals replace auditory ones after the initial development years of language acquisition as a child [Brow 90]. However, compensation experiments (e.g., putting a bite block or tube in the mouth) show that speakers can radically adjust their articulators to accomplish the acoustic goals needed for speech communication [Kel 86, Sav 95].

7.5 Acoustic-Phonetics and Characterization of Speech Signals

We will now study speech production from the acoustic point of view, relating phonemes to their spectral effects, since the listener often only has access to the sound when interpreting speech (i.e., no visual cues, such as the speaker's face). We will describe different sounds in terms of the classical features of voicing, manner of articulation, and place of articulation. Perhaps the most fundamental feature of speech sounds is whether the phoneme nominally requires vocal fold vibration (we will note later that actual fold vibration is not always needed to achieve the perception of "voicing"). Voiced sounds (i.e., those excited by vibrating folds) appear to be approximately periodic (corresponding, of course, to the F0 rate of the folds). Unvoiced speech, on the other hand, appears as a noisy (or aperiodic) signal, due to the random nature of the sound pressure generated at a narrow constriction in the vocal tract for such sounds. (Voiced fricatives have both periodic and noisy features, in which noise occurs in periodic bursts.) Whether the excitation is voiced or unvoiced, the vocal tract acts as a filter, to amplify some sound frequencies (at and near resonances) while attenuating others.

As a periodic signal, voiced speech signals theoretically have a line spectra consisting of **harmonics** of the fundamental frequency (F0). As we shall see later, F0 is the physical correlate in the audio signal that corresponds to the pitch perceived by listeners. The harmonics consist of energy concentrations at integer multiples of F0. A theoretical periodic signal (i.e., $s(t) = s(t + T_0)$, for all t) has a discrete-line spectrum. Humans are incapable of true periodicity on any time scale since the vocal tract is always changing shape and the fold vibration is irregular. Thus voiced sounds are only locally quasi-periodic in practice. Independent of the vocal tract excitation, the envelope of the speech spectrum is often relatively fixed (quasi-stationary) over brief time intervals (e.g., tens of milliseconds) as each phoneme is produced. Since the average duration of a phone is about 80 ms, speech over any interval longer than a few tens of ms can be expected to vary substantially.

Speech signals are readily characterized by their spectra or by their time waveforms. Observing the time signals directly, the following important aspects of speech can be observed: the quasi-periodicity of sonorants, changes in power, major changes in spectral patterns, and sound durations. For example, vowels and diphthongs are voiced (whispered vowels are the sole exceptions), have the greatest intensity among all the phoneme classes, and average about 100 ms or so in normal speech. Glottal excitations cause an abrupt signal increase once every pitch period; then signal intensity decays exponentially with a time constant that is inversely proportional to the bandwidth of F1 (because that formant dominates a vowel's energy). Like all sonorant sounds, vowel energy is mostly below 1 kHz and falls off at about −6 dB/oct with frequency (owing to the lowpass nature of the glottal excitation). F1 can be identified in the time waveforms of many vowels as the inverse of the period of dominant oscillation within a pitch cycle (when F1 and F2 are close, the period is closer to their average).

7.5.1 Acoustics of vowels

Vowels are usually distinguished by the center frequencies of their first three formants. Fig. 7.4 shows a wideband spectrogram of short sections of the English vowels from a typical speaker. Table 7.2 notes average formant frequencies from a classic study of 60 speakers, who each repeated twice a set of monosyllabic nonsense words /hVd/ (V = one of 10 different vowels listed in the table). There is much variability in any set of acoustic parameters, including formants, due to different vocal tract shapes and sizes for different speakers; Fig. 7.5 plots values for F1 and F2 for the ten vowels spoken by the 60 speakers. Considerable overlap occurs across speakers, with the result that vowels having the same F1 and F2 values can be heard as different phonemes when they are uttered by different speakers. Listeners distinguish such vowels via other acoustic aspects (F0 and higher formants). A traditional plot of average F1 versus F2 values for vowels displays a **vowel triangle** (Fig. 7.6), where the basic vowels /iy,aa,uw/ assume extreme positions on an F1 versus F2 plot. The other vowels tend to have values along one of the lower sides of the triangle, e.g., the /iy/-/aa/ side for the front vowels and the /uw/-/aa/ side for the back vowels. The vowel triangle allows both articulatory and acoustic interpretations: F1 goes down as tongue height increases, and F2 falls as the tongue moves toward the rear of the oral cavity.

		/iy/	/ih/	/eh/	/ae/	/aa/	/ao/	/uh/	/uw/	/ ah /	/r/
F1	male	270	390	530	660	730	570	440	300	640	490
	female	310	430	610	860	850	590	470	370	760	500
F2	male	2290	1990	1840	1720	1090	840	1020	870	1190	1350
	female	2790	2480	2330	2050	1220	920	1160	950	1400	1640
F3	male	3010	2550	2480	2410	2440	2410	2240	2240	2390	1690
	female	3310	3070	2990	2850	2810	2710	2680	2670	2780	1960

Table 7.2: Average formant frequencies (in Hz) for English vowels by adult male and female speakers from a classic study (see also [Hill 95] for a replication). (after Peterson and Barney [Pet 52], @AIP).

As we saw earlier, formants for a 17-cm-long vocal tract occur on average every 1 kHz, and the table shows an approximate average F1 of 500 Hz and F2 of 1500 Hz. The ranges of values for F1, F2 and F3 are quite different, however. Table 7.2 notes a 460-Hz range for F1 (270–730 Hz), but a 1450-Hz range for F2 (840–2290 Hz). The range for F3 is also large (1690–3010 Hz), but F3 for most vowels lies in a much narrower range (2240–2550-Hz). Very low values for F3 (below 2 kHz) are the main aspect of /r/-sounds in English.

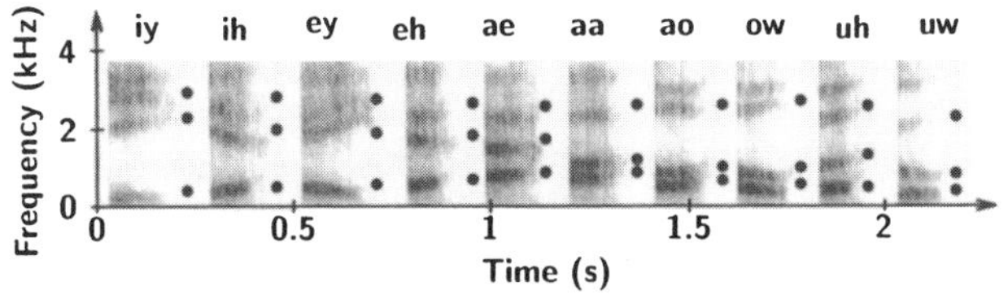

Figure 7.4: Spectrogram of short sections of English vowels from a male speaker. Formants for each vowel are noted by dots.

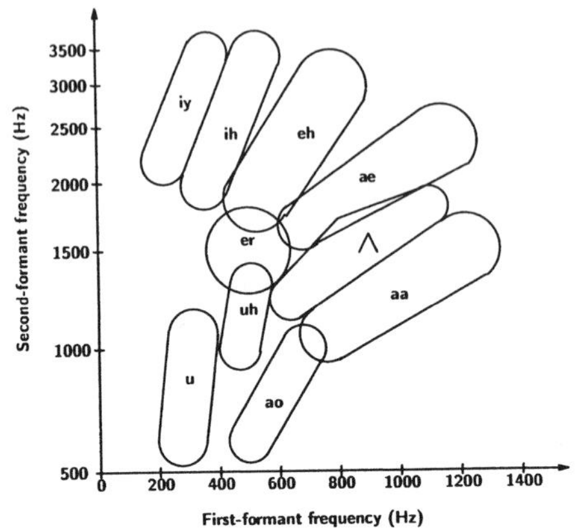

Figure 7.5: Plot of F1 vs. F2 for vowels spoken by 60 speakers (after Peterson and Barney [Pet 52], @AIP).

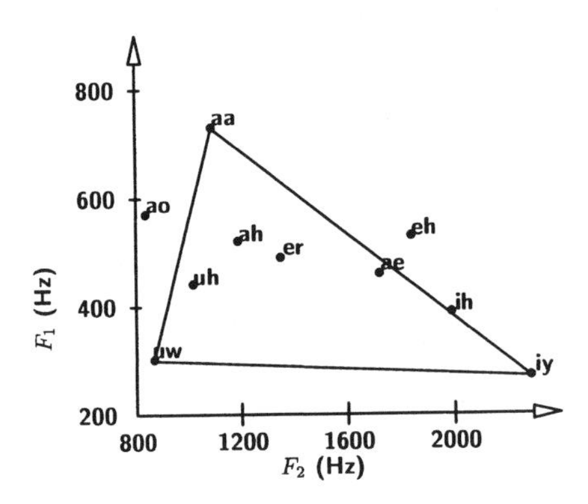

Figure 7.6: The vowel triangle for the vowels of Table 7.2.

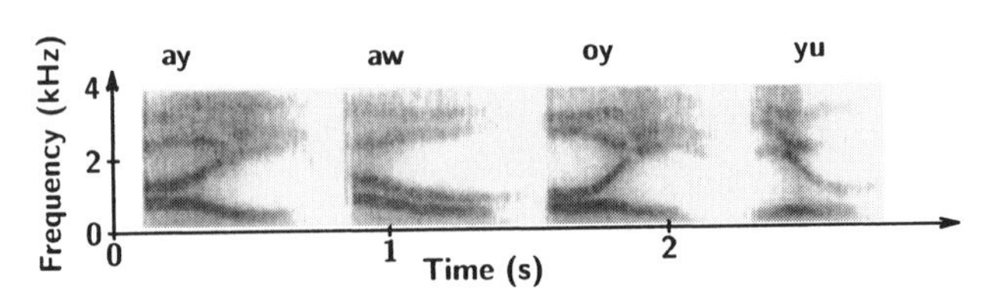

Figure 7.7: Spectrogram of typical English diphthongs from a male speaker.

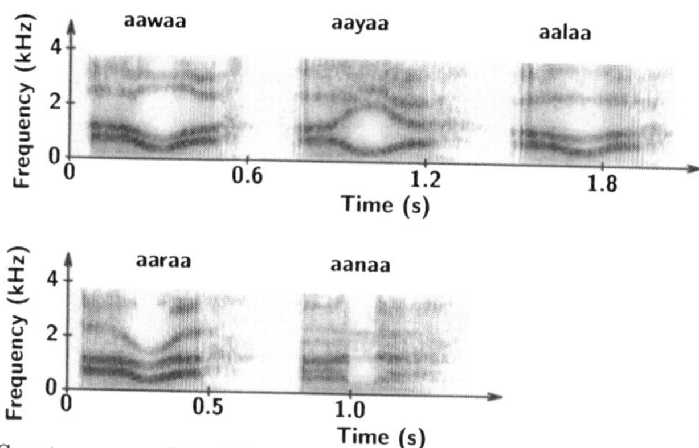

Figure 7.8: Spectrograms of English sonorants in vocalic context: /aa w aa/, /aa y aa/, /aa l aa/, /aa r aa/, /aa n aa/.

7.5.2 Diphthongs and diphthongization

The three diphthongs of Table 7.1 are essentially dynamic vowels where the tongue (and sometimes the lips as well) moves from a low vowel to a high vowel (or corresponding glide) position (Fig. 7.7). In addition, the vowels /ey/ and /ow/ in English are usually articulated so as to end in a related glide, and are thus called **diphthongized**. Diphthongs are quite different from sequences of two vowels; a diphthong has only one syllable. The difference is usually seen in the intensity, which dips between the peaks of two vowels. In a diphthong, amplitude falls during the glide portion.

7.5.3 Glides and liquids

The sonorant consonants called glides and liquids are very similar acoustically to vowels: they all have strong periodic signals with most energy at low frequency (Fig. 7.8a-d). They are slightly weaker than vowels due to their greater vocal tract constriction. Glide /y/ has F1 slightly lower and F2 slightly higher than its cognate /iy/; glide /w/ has both F1 and F2 somewhat lower than /uw/. As noted above, /r/ is distinguished by a very low F3.

Unlike most other sonorants (but like nasals), the liquid /l/ has spectral zeros, caused by the existence of an additional acoustic path for air in the oral cavity: air passes on either side of the tongue tip (which is in contact with the palate). A significant null in the spectrum of /l/ is often seen near 2 kHz. On spectrograms, a slope discontinuity usually appears at the edges of formant trajectories of /l/, at times when the tongue tip starts or ends contact with the palate.

7.5.4 Nasals

Nasal consonants resemble other sonorants acoustically, but are significantly weaker owing to losses in the nasal cavity (Fig. 7.8e). Owing to the long pharyngeal+nasal tract (21 vs. 17 cm), formants occur about every 800 Hz ($\approx 1000 \times 17/21$). During oral tract closure (as occurs for nasal consonants), there is little acoustical change due to coarticulation; so nasal spectra are usually very steady. A low, but strong F1 near 250 Hz dominates the spectrum. Since F2 is typically very weak, an F3 near 2200 Hz is normally

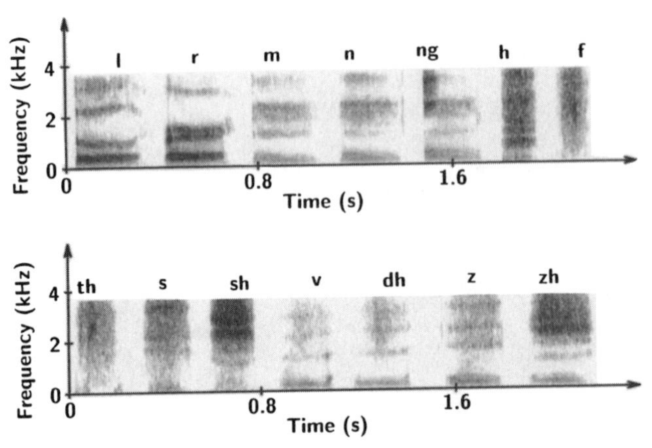

Figure 7.9: Spectrograms of English consonants /l,r,m,n,ng,h,f,th,s,sh,v,dh,z,zh/.

the next strongest formant. A spectral zero, of frequency inversely proportional to the length of the oral cavity, occurs near 1 kHz for /m/ (longest oral cavity), near 2 kHz for /n/, and above 3 kHz for /ng/ (shortened oral cavity). The nasalization of vowels (adjacent to nasal consonants) affects spectra mostly in the F1 region: adding a nasal resonance near F1, as well as weakening and raising the frequency of F1 [Chen 97].

7.5.5 Fricatives

The obstruents (fricatives and stops) are quite different acoustically from sonorants; the obstruents have noisy waveforms, they have much less power, and this power is found primarily at high frequencies (Fig. 7.9). The first feature is due to obstruents' noise excitation at their constriction, the second is caused by the relatively inefficient conversion of airflow to pressure (frication), and the third relates to the shorter vocal tract forward of the noise source at the constriction. Unlike sonorants, which are all voiced (except in whispered speech), the obstruents of many languages (including English) exist in two versions: voiced or unvoiced.

For sonorants, the entire vocal tract is excited, leading to a broadband spectrum. Unvoiced obstruents, on the other hand, have a noise source that excites primarily that part of the vocal tract forward of the constriction. As a result, they have a highpass spectrum, where the cutoff frequency is roughly in inverse proportion to the length of the front cavity. Since the excitation is spectrally flat, obstruent power increases with bandwidth, and thus the palatals (the English fricatives with the most rear constriction) are the most intense, with power from about 2.5 kHz. Moving forward in the vocal tract, the alveolar fricatives lack power below about 3.2 kHz. The **non-strident** fricatives (labiodental and dental) are very weak, having little energy below 8 kHz, due to their extremely small front cavities. One might think that the glottal fricative /h/ would be the most intense fricative since the entire vocal tract is excited, but the glottal noise source (aspiration) is less intense than frication; /h/ is basically a whispered version of an adjacent vowel.

The acoustics are more complicated for voiced obstruents. They have two excitation sources: periodic glottal pulses (as in all voiced phones) and frication noise generated at the oral tract constriction. The weak voiced fricatives /v/ and /dh/ have little actual

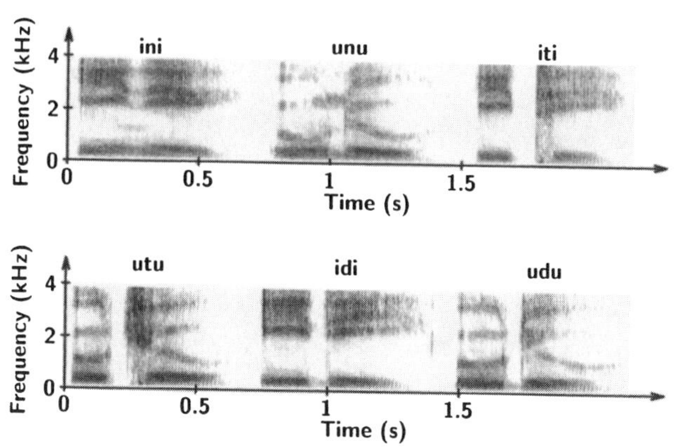

Figure 7.10: Spectrograms of English stops in vocalic context: /iy n iy,iy t iy/.

frication and are almost periodic. The strident voiced fricatives /z/ and /zh/, on the other hand, have significant noise energy at high frequencies as do their unvoiced cognates /s/ and /sh/. However, in addition, the voiced excitation causes significant power at very low frequency (below 200 Hz) called a **voice bar**, owing to its steady appearance resembling a formant on spectrograms. It is actually equivalent to F1, which is not excited in unvoiced obstruents; its low frequency is due to the long back cavity, closed at both ends. The voice bar appears in the output speech signal due to sound radiation through the walls of the vocal tract; losses due to the throat and cheeks heavily attenuate all but the lowest harmonics. Obstruents are not usually characterized in terms of formants because their lowest-frequency formants are cancelled by spectral zeros.

7.5.6 Stop consonants

Stops closely resemble fricatives: they have similar excitations, weak power, option of voicing, and various places of articulation (Fig. 7.10). The major difference is the dynamic nature of stops, versus the steady frication of fricatives. Most phonemes other than stops can be uttered at length (i.e., in steady-state). Stops, however, are transient phonemes: the initial closure portion causes the speech signal to become either silent or to have at most just a voice bar. The second phase of a stop is the release of the vocal tract occlusion, which creates a brief (5–10 ms) explosion or burst of noise frication. This burst often consists of energy over a wide range of frequencies, but primarily matches those of a fricative with the same place of articulation. The third phase is a continuing frication for up to 40 ms, as the vocal tract moves toward an ensuing sonorant. The final phase only appears in unvoiced stops, which have a long VOT (voice onset time), where a glottal aspiration excites the full vocal tract, leading to noisy formants as they continue to move toward positions for the ensuing sonorant [Zue 76].

Since most of each stop is the same acoustically for different places of articulation (i.e., the closure portion), cues to place occur in the transitional intervals of adjacent phones. The formant movements prior to closure and after closure release are key factors. Spectra of the burst and aspiration are also important. The major power concentration of the burst in an alveolar stop lies mostly above 2 kHz. As noted earlier, velar stops vary due to coarticulation; a distinctive peak often occurs at about 2.7 kHz for /k/ before a front vowel whereas a lower-frequency concentration appears before a back vowel [Zue 76].

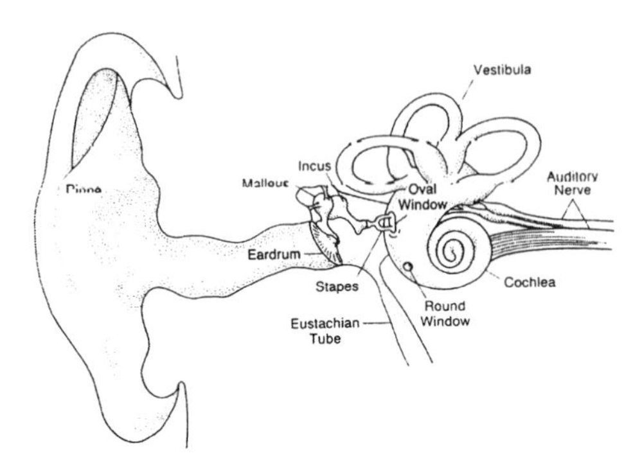

Figure 7.11: The basic structure of the peripheral auditory system.

Burst power is substantially lower than power in adjacent vowels: alveolar and velar bursts are about 16 dB weaker, while labial bursts are very weak. The term **compact** is often used to describe velars because their spectra have concentrated power in a limited frequency region (near the F2-F3 juncture).

7.6 Introduction to Auditory Phonetics

We now examine the auditory process related to the transformation of the speech signal entering a listener's ears into an understood linguistic message in the brain. We distinguish two parts to this process: **audition** (hearing) that converts speech sounds into neural information, and **speech perception** that converts the auditory nerve information into a comprehensible message in the brain. Since much of the latter takes place in the less understood brain, we have more to say technically about audition: the physical details how pressure variations in the outer ear are converted into neural firings in the auditory nerve. In the rest of this chapter, we examine ear behavior and the psychoacoustics of perception.

Humans share with many animals a similar audition system. Indeed most hearing mechanisms are very similar in all mammals and birds, although human brainpower is far superior to that of other animals, which makes humans alone capable of speech communication. There are some differences in the ears of different species; e.g., the dimensions of auditory structures vary with mammal size, leading to ranges of frequency sensitivity inversely proportional to mammal size. Thus small animals can hear much higher frequencies than humans.

The mammal ear has three parts: outer, middle, and inner ear (Fig. 7.11). The outer ear simply funnels incoming speech pressure waves toward the eardrum (at the boundary between the outer ear and middle ear), where these variations are transformed into mechanical vibrations. The inner ear then converts these into electrical patterns in the auditory neurons.

7.6.1 Outer ear

What most people think of as the ear is the readily-visible outer ear (auricle) or **pinna**. Its shape makes listeners more sensitive to sounds from frontal directions. Inside the pinna is an empty cavity that acts as a quarter-wavelength resonator, i.e., a tube open at the pinna end and closed at the eardrum. This canal is about 2.7 cm long, hence causing a resonance near 3 kHz (following earlier results). By amplifying relevant energy in this frequency range, which is often weak in actual speech signals, speech perception is likely aided.

7.6.2 Middle ear

The eardrum (tympanic membrane) transfers power to the middle ear (another air-filled cavity), which contains the ossicular bones: malleus (hammer), incus (anvil), and stapes (stirrup). Among the smallest bones of the body, these three amplify eardrum vibrations and send them to the oval window membrane of the inner ear. They accomplish a match of acoustic impedance (since the lymphatic fluid of the inner ear is about 4000 times denser than air). If not for the ossicles, 99.9% of the sound pressure waves would be reflected and not enter the inner ear. The main amplification effect is due to a large difference in surface area: big eardrum versus small oval window.

7.6.3 Inner ear

The inner ear contains the most important component of the ear, the **cochlea**, a very hard bony tube filled with lymphatic fluid. It converts mechanical vibrations of the oval window into electrical energy for its neural outputs. Tubes inside the cochlea taper off in diameter, from a wide base to a narrow apex. The tubes contain an important structure called the basilar membrane, that is about 32 mm long, and increases in thickness from the base to the apex (despite the narrowing of the tubes). Approximately 30,000 sensory **hair cells** connect the basilar membrane to the **auditory nerve**, which leads to the brain. These cells lie in several rows (inner and outer hair cells, having different functions) along the length of the cochlea. Hair cell compressions (due to basilar membrane vibrations) cause neural **firings** to propagate along the auditory nerve.

Basilar membrane

The basilar membrane does more than simply help convert mechanical acoustics to electrical power. It increases monotonically in thickness along its length, leading to a frequency response that changes from a highpass filter at the base of the basilar membrane to a lowpass filter at the apex. Each point on the basilar membrane has a **characteristic frequency** at which it vibrates the most. One can view each membrane section as a bandpass filter, with a relatively fixed ratio of center frequency to bandwidth (i.e., constant Q). Thus, frequency resolution is best at low frequencies (which corresponds to the membrane apex). Points of maximal response, measured spatially from the apex, are roughly proportional to the logarithm of the characteristic frequency. This logarithmic behavior appears in later sections of this book in terms of the **mel perceptual scale**. (Indeed many measurements in speech production and perception follow better a logarithmic scale than a linear one, which helps greatly to handle the large dynamic ranges of sounds.)

While speech is a complex signal of many frequencies, much of auditory research uses simple signals, as a control in scientific study. So we start by examining ear behavior for one of the simplest signals - a sinusoid or tone. When an acoustic tone excites the oval window (after passing through the outer and middle ears), the varying pressure causes the basilar membrane to vibrate at the frequency of the sinusoid. The actual membrane motion is a traveling wave from the base to the apex, with neither a reflection at the apex nor a standing wave. The traveling wave becomes maximal in size at the point corresponding to a match between the input and characteristic frequency. Each section vibrates at the same sinusoidal rate, except for a phase delay (time for the wave to travel from the oval window). Because the initial basal end is thin and stiff, the amplitude peak occurs there for high-frequency sounds, while the apex responds most to low-frequency tones.

Since speech is very dynamic, we also examine simple transient sounds (the opposite of steady sinusoids, in a sense). For example, in response to an impulsive sound (a **click** of very short duration, thus having a broad spectrum), each point along the basilar membrane responds with a sinusoid of its own characteristic frequency, as would be modeled by the impulse response of a bandpass filter (also after an appropriate **propagation delay**). Such a response decreases with time, with a decay rate that is most rapid at the base (< 1 ms) due to wide filter bandwidths at high frequencies, and the traveling wave gradually loses its high-frequency components as it travels toward the apex.

Activity in the cochlea

The hair cells in the cochlea link to the brain through the VIII-th cranial (auditory) nerve, which is composed of many neurons. The auditory information travels electrically in the form of binary bits, as in digital computer lines, where the line is normally in zero-state (dormant). A series of brief pulses propagates to the brain, which then interprets the frequency and timing of all the positive bits, to understand the speech message. Each individual neuron "fires" (a spike of 0.5–1.0-ms) when it receives enough bending and/or shearing forces from the attached hair cells. The actual mechanical-to-electrical conversion occurs via hair cell tension altering an electrical conductance, which in turn causes the release of a chemical substance [Ratt 98].

Studies of audition in small mammals show distinctive firing patterns in response to simple acoustical stimuli (tones and clicks). With no sound, neurons fire in a random and spontaneous fashion (averaging fewer than 50/s). With sound input, neurons still fire randomly, but the spikes tend to group at certain times as a function of the intensity and frequency of the sound. These firings correspond to the mechanical vibration of the basilar membrane, and match the bandpass filtering action for each point where the neuron contacts the membrane. This frequency-selective nature of auditory fibers in the cochlea provides a way to convert the spectral information of sounds into place information along the BM (i.e., the brain knows approximately what frequencies dominate in a sound by which neurons fire most often).

7.7 Sound Perception

Sound perception deals with what sounds are perceptible, what interpretations a listener makes based on those sounds, and how two or more sounds each interfere with each other's perception. Most of these issues appear to deal with aspects of a sound's power and frequency.

7.7.1 Thresholds

We hear sounds over a wide frequency range (e.g., 20 Hz to 20 kHz) and over a wide power range. The human ear favors the 1–5-kHz range; lower or higher frequencies require increasingly more power to be heard (to the point where very low or high frequencies are inaudible no matter how strong). Sound power is measured as sound pressure level (SPL), with units of decibels (dB). The base reference for SPL (0 dB) is 10^{-16} watt/cm^2 at 1 kHz.

The minimum power for which sound can be perceived is the **hearing threshold**, and a higher threshold corresponds to the strong level of sound intensity that causes pain. Such a pain threshold is approximately 120 dB for most audio frequencies, whereas the hearing threshold varies much more with frequency. Speech typically occupies the 60–70-dB range. For typical sonorants, the strong harmonics centered around the first few formant frequencies are easily audible (but are not equally powerful). However, audibility becomes an important issue for the weaker formants above 3 kHz, as well as for all spectral components in noisy environments.

7.7.2 Just-noticeable differences (JNDs)

Psychophysical experiments generally employ simple sounds (for control purposes), that differ only along a single acoustic dimension (e.g., power, F0, F1, etc.). Listeners are asked whether two sounds differ, as the sounds increase in separation along this dimension; the acoustic value for which 75% of responses are positive is called the **just-noticeable difference** (JND). JNDs are pertinent for perception and coding: they estimate the resolution capacity of audition, as well as the precision needed for speech parameters under quantization. As examples, in the F1-F2 range, two equally-powerful tones must diverge by about 2 Hz to be perceived as different. Power JNDs for tones are about 1 dB. JNDs are often expressed as percentages, because they tend to be proportional to the size of stimuli in general.

7.7.3 Pitch perception

Distinguishing two speech sounds involves aspects of their distribution of power across frequency. Since speech consists of many acoustic aspects, we need to examine how humans interpret simultaneous variations of these factors. While there is theoretically an unlimited number of ways to view information in a complex audio signal, we look first at the perception of pitch, a spectral measure of clear importance. There is much evidence as to how the distribution of power across frequency affects pitch.

Any two periodic sounds (tones, speech sonorants, etc.) can be compared perceptually in terms of their pitch. For tones, perceived pitch is easily related directly to the single frequency of the tone where power occurs. More generally, periodic sounds consist of (many) equally-spaced harmonics (multiples of the fundamental rate F0). In the latter case, pitch is still monotonically related to F0, but the relationship is more complicated by interactions of the effects of the harmonics (which usually have differing amplitudes and phases). A sound is said to have a given pitch if that sound can be 'matched' by a listener to a tone by adjusting the latter's frequency. This perception clearly arises from the amount and timing of neural firings at various locations along the basilar membrane. A sound's loudness is often perceived as a function of the overall rate of neural firings, but other spectral details have a more complex interpretation. According to the **timing** or **volley theory**, low frequencies (including the initial strong harmonics in the vicinity of F1 for speech sounds) are perceived in terms of time-synchronous neural firings occurring toward the apex of the basilar membrane. The alternative **place theory** posits that a sound spectrum is decoded in terms of the basilar membrane locations that cause the most neurons to fire [Pick 82].

As a result, there appear to be two types of pitch. The usual pitch corresponds to the inverse of the fundamental period of the sound (i.e., to F0). A second, **spectral pitch** (or place pitch) corresponds instead to other (more general) aspects of the sound's spectrum. As a simple example, /uw/ sounds generally lower in (spectral) "pitch" than /iy/ does (independently of their F0 values), because /uw/ has a stronger concentration of power at low frequency. The normal pitch, also called a **virtual pitch**, is not due to direct power at the F0 (even though that pitch follows F0, unlike the spectral pitch). This is evident from the fact that it is clearly perceived when F0 is absent [Medd 97]; e.g., F0 and its initial harmonics (those below 300 Hz) are heavily attenuated by telephone lines, yet a pitch corresponding to an F0 of 100 Hz is readily perceived by the higher-frequency harmonics: the relevant information lies apparently in the 100-Hz separation of those strong harmonics.

The human ability to make rapid decisions would seem to imply that the brain could exploit temporal information in the timing of auditory neural firings, thus interpreting a consistent interval between firings on many neurons as the pitch period. This 'timing' theory must necessarily be restricted, however, to the lower end of the range of sound frequencies (e.g., for F0 and F1, but not for higher formants), because neural spikes are only synchronized (even stochastically) at frequencies below 1 kHz. Above that, latency fatigue in each neuron prevents any consistency in firings for individual neurons on successive cycles. Since F0 is always within the lower frequency range, it could be readily handled by neural timing data. Identifying formants, however, often involves higher frequencies, thus apparently necessitating the place theory. The location of maximal vibration on the basilar membrane may be transmitted to the brain by a specific neural pathway (or by the brain learning to associate patterns of certain neural firings with specific sound frequencies).

7.7.4 Masking

Like many types of perception, sound perception is nonlinear. In particular, a listener's interpretation of multiple sounds (or of multiple sound components) cannot be predicted by simply adding the responses to the individual sounds. Thus, while audition of simple sounds such as clicks and tones is readily understood, one's reaction to speech with its complex composition of many harmonics, as well as noise components, is indeed more complicated. In particular, the audibility of one sound can be obscured by the presence of another, if the latter is strong enough. In this phenomenon called **masking**, one sound (called the masker) raises the hearing threshold for another sound (the maskee), when the sounds occur either simultaneously or with a short intervening delay. If simultaneous, we have **frequency masking**, in which a lower-frequency sound has a tendency to mask (i.e., obscure) a higher-frequency one. Sounds separated temporally can cause **temporal masking** of each other, if they are spectrally similar. It is in speech coding that masking has become most important: digital coding necessarily introduces quantization noise, which is often modeled as white noise, but can be distributed non-uniformly in frequency to exploit masking. For example, moving such quantization noise to frequency regions of high speech energy allows some of the noise to be masked, and therefore not to degrade the speech.

The simplest masking experiments show how one tone can obscure another, as a function of the frequency separation between them [Zwi 74]. Even if the two tones lie above the normal threshold of hearing, a listener may hear only one in certain conditions; i.e., the threshold for the weaker tone is elevated by the presence of the stronger tone. As an example, assume a strong 80-dB tone at 1200 Hz: a second weak tone far enough below (e.g., at 800 Hz) is easily heard (as low as 12 dB). On the other hand, if the second tone is close in frequency (e.g., within 100 Hz of 1200 Hz), its audibility threshold rises to 50 dB. The effect is asymmetrical: higher frequencies are much more readily masked than lower ones.

The results of basic masking experiments are not easily applied to speech applications. For example, in speech perception, the listener's task is usually not just to detect a sound in the presence of another, but to identify phonetic aspects of the sound. Thresholds are generally higher in such recognition tasks; e.g., understanding speech in noise (instead of simply hearing its presence) typically requires 10–12 dB more power. **Speech reception thresholds** measure such intelligibility [Nils 94].

7.7.5 Critical bands

The operation of the inner ear is often viewed as a bank of bandpass filters, where the output is a vector of information parameters (indexed on sound frequency) or where the auditory nerve sends such acoustic parameters to the brain on a parallel set of frequency links. The nature of these filters is the subject of this subsection, which explains several aspects of masking. We define the successive ranges of the filters as a set of **critical bands**, which number about 24. Formally, a critical band is the frequency range in acoustic experiments where human perception changes abruptly as a sound (usually constrained temporally and spectrally) is adjusted so as to have frequency components beyond the band. If two sounds lie within a single critical band, the stronger sound typically dominates and thus masks the other sound, rendering it inaudible. We define the **amount of masking** to be the additional power (due to the masker) needed to hear the masked sound.

We can determine the actual shapes of these critical-band filters by using noise of a varying bandwidth to mask a tone. The noise level is fixed, but total power increases as bandwidth does. As the cutoff frequency f_c of, say, a lowpass noise is increased around the tonal frequency f_t, increasing amounts of noise enter the critical-band filter centered at f_t. The masked threshold as a function of f_c provides a definition of the shape of a critical-band filter. The filters are roughly symmetric and have sharp skirts (in excess of 65 dB/oct) [Patt 78]. At low frequencies, critical bandwidth is constant at approximately 100 Hz. Above 1 kHz, critical bandwidths increase logarithmically with frequency (e.g., 700 Hz at 4 kHz).

These bandwidths correspond roughly to 1.5-mm spacings along the BM, which implies that a set of 24 bandpass filters model the BM well along its 32-mm length. In a related measure called the **Bark scale**, one Bark unit corresponds to a critical bandwidth (this is equivalent to the mel scale). Another approach [Glas 90] uses equivalent rectangular bandwidths (ERBs).

7.7.6 Nonsimultaneous (temporal) masking

Sound interference happens in time as well as in frequency [Duif 73]. Sounds that share the same critical band and adjacent time intervals mask each other, with the amount of masking dependent on the sounds' proximity (in both time and frequency). Masking in the 'forward' direction is stronger than **backward masking**. Assume that a potentially masking noise ends abruptly and then a brief sound (in the same frequency band) follows quickly thereafter. Owing to the presence of the first sound, the second is harder to hear, with this forward masking effect decreasing in time over a range of about 200 ms [Zwi 84].

Neuron fatigue is often used to explain forward masking. Neurons gradually adapt to a continuing sound, thereby adjusting the input level which would be needed to cause a certain sound perception after a strong sound. Such a fatigue theory cannot explain backward masking, in which a short sound is not heard if it is followed closely in time by a sufficiently intense noise (again, both must lie within the same critical band). Backward masking only occurs within a 20-ms range, but is nonetheless significant; e.g., a short tone burst followed within 1 ms by a noise undergoes 60 dB of masking (the same burst 1 ms after noise offset only has 30 dB of masking). Instead of fatigue, backward masking likely is due to interference in auditory processing; e.g., perception of an initial tone is interrupted by an ensuing loud noise if the noise arrives before the tone perception has been completed.

7.7.7 Just-noticeable differences (JNDs) in speech

An issue critical to speech coding is how precise we need to measure parameters from speech signals. If listeners cannot distinguish two sounds which differ at a minute level, then such information need not be retained during coding (or sent in transmission). Similarly, if people cannot control their speech production consistently at some level (e.g., if, when asked to carefully repeat an utterance identically, they vary some acoustic parameters at random), then such variation should be ignored (or taken into account as normal free variation in stochastic models). We earlier examined humans' sensitivity to simple tones and clicks. In this subsection, we examine how small acoustic changes in speech must be to render them inaudible to listeners.

Since speech is in general much more complex than tones and clicks, let us first consider a major component of steady vowel sounds – a formant. Suppose a synthesizer varies the center frequency of one formant. The resulting JND for formant frequency is about 1.5–5% [Flan 72, Kew 98], i.e., as little as 14 Hz for low-frequency formants. Listeners are much less sensitive to dynamic vowels; e.g., JNDs are 9–14% for moving formant patterns in simple syllables [Merm 78]. Such JNDs effectively establish lower bounds for formant resolution (and probably for related spectral measures such as LPC parameters or cepstral coefficients). JNDs are likely to increase for cases of more normal speech, as compared to the simple speech stimuli used in these basic experiments.

The most important aspect of spectral envelopes appears to be center frequencies of peaks, but formant amplitudes are also of relevance. Typical JNDs for changes in formant peaks are approximately 1.5 dB for F1 and 3 dB for F2 (higher formants have yet larger JNDs). If we vary just one harmonic (instead of all harmonics in a formant, as would normally occur in natural speech), the corresponding amplitude JNDs vary significantly depending on whether the harmonics are strong or weak: the JNDs for strong harmonics lie in the same range as above (e.g., about 2 dB in F1 or F2 peaks), but are much larger in the weaker valleys between formants (more than 13 dB) [Moo 84]. These large JNDs for low-amplitude harmonics are likely caused by masking. Speech coders and synthesizers must emphasize better spectral modeling of formant peaks than valleys, since changes in valleys are less noticeable. Precise modeling between formants is much less perceptually important than accuracy at spectral peaks.

Formants are generally not perceived in terms of their strongest harmonic alone; the perceived center frequency correlates more with a weighted average of adjacent harmonics under the formant peak. In addition, variation in F0 (which obviously causes much amplitude change of individual harmonics) causes little change in perceived vowel quality [Carl 74]. It further appears that perception of formant center frequencies is largely due to the three strongest harmonics of a formant [Som 97].

A third aspect of formant detail concerns formant bandwidths. Their JNDs are much higher than for center frequencies and amplitudes. Listeners cannot hear differences of less than 20% (at best) in formant bandwidths. Many speech synthesizers use very rudimentary patterns for bandwidth, often fixing their values, especially for higher-frequency formants. Variation of bandwidths seems most important for nasal sounds, where the significantly increased bandwidths due to nasal cavity loss appears to be a major component of perceived nasality.

While the resolution of the spectral envelope is of primary importance for most speech applications, the precision of pitch is also relevant. In steady synthetic vowels, the JND for F0 is 0.3–0.5% [Flan 72], or under 1 Hz [Leh 70]. Since most speech has dynamic F0 or background noise, it is worth noting that JNDs for such cases are an order of magnitude larger [Klatt 73, Sche 84]. As a result, many coders employ 5-6 bits to represent F0.

While formant peak frequencies seem to be the most important perceptual features for sonorants, it is less clear what are relevant parameters for obstruents. Noise in fricatives and stops are usually broadband, but have irregular spectra. In noise sounds, listeners cannot discern spectral peaks with $Q<5$ (and are even less sensitive to spectral notches or zeros). Thus most obstruent spectra can be coded very coarsely, assuming that most spectral irregularities are likely perceptually irrelevant. For example, fricatives can be well modeled acoustically by two high-frequency poles and one lower-frequency zero. The location of the lowest-frequency pole is most important perceptually, as it specifies the cutoff frequency of the highpass frication noise, which largely seems to determine the perception of place of articulation.

7.7.8 Timing

Durations are important for sound perception. We thus explore here the ear's ability to exploit timing information both in simple sounds and in speech. Perhaps the smallest interval of interest is about 2 ms: two clicks are perceived as one sound, unless they are separated by more than that much time. As usual in perception, simply noting a difference may only signal very limited information: we need to have a separation of about 17 ms in order to identify the order of two clicks [Past 82]. Typical critical bands have equivalent window durations of 4–5 ms [Madd 94]. Thus sound differences on the order of a few ms can be relevant in sound perception.

Typical speech contains about 12 phones/s, but rates as high as 50 phones/s [Foul 69] are possible (although it is likely that such fast speech needs much contextual redundancy for accurate understanding). At fast rates, acoustic transitions between sounds become increasingly important. Coarticulation imposes these transitions in natural speech. Synthetic speech at fast rates without appropriate transitions leads to poor perception.

Timing in speech is often important perceptually. The abruptness of a phoneme's start or end is often a cue to phonemic identity. Thus measuring JNDs for transitions is pertinent. If we define the rise time for a phoneme onset as the duration for intensity (in dB) to rise from a 10% level to 90%, we find such JNDs from 10 to 150 ms, or about 25–30%. This suggests that perceptual discrimination may be only accurate enough to reliably cue the distinction between an abrupt and a smooth transition [Broe 83, Howe 83].

Despite the relevance of timing to speech perception, relatively little research has been done for durational JNDs in typical utterances. We know that listeners are more sensitive to changes in vowels than in consonants [Hugg 72a], and that speech timing is based more on syllabic events than phonemic ones [Hugg 72b]. JNDs for natural speech are generally around 10–40 ms [Leh 70]. They depend on segmental duration (the basic Weber's law of perception states that JNDs are directly proportional to overall durations), position in the syllable (smaller in word-initial position), and word position within a sentence [Klatt 75a].

While JNDs are perhaps the most elementary measurements of perception, the concept of a 'distance' between sounds is the next logical step [Carl 79]. How closely related two sounds are is of definite interest in many speech applications. The smallest non-zero perceptual distance would be the JND. We discuss this idea in more detail later in the book.

7.8 Speech Perception

We seem to know less about speech perception than about the mechanics of audition in the ear; i.e., we understand reasonably well how the ear converts sound into auditory nerve firings, but not how the brain converts such firings into a linguistic message [All 94]. Speech perception research often takes a **black box** viewpoint: listeners hear acoustic stimuli and respond to certain questions about the stimuli – detectability, discrimination, and identification. (As with a mysterious box, we may divine its internal operations only through its response to various stimuli.)

Along any relevant dimension (i.e., a useful and measurable aspect of a physical speech sound), speech perception is usually monotonic: increasing that parameter leads to a corresponding increasing (or decreasing) perceptual feature. For example, raising F0 leads to hearing higher pitch. However, the mapping between acoustic cues and perception is nonlinear, as exemplified by **categorical perception**, where equal acoustic changes often lead to widely varying perceptual effects. There is little consensus on how best to model speech perception, owing to the difficulty of designing a test to truly evaluate a model's validity.

Vowel perception appears to be simple to relate to physical aspects of the vowel, because the positions of a vowel's first three formants relate directly to the perception of different vowels. However, listeners place great significance on context. Vowel perception is much affected by coarticulation, speaking rate, intensity, F0, and average formant values: before, during, and after each vowel of interest.

Less is clear about consonant perception, at least with regard to voicing and place of articulation (manner of articulation is perhaps analogous to aspects of vowel perception). In particular, researchers having been looking for **invariant cues** (i.e., clear and consistent acoustic correlates to perception) for the features of voicing and place of articulation for decades. For example, the physical distinction between voiced and unvoiced consonants is often not periodicity, but such parameters as voice onset time and formant transitions.

7.8.1 Physical aspects of speech important for perception

The low-frequency range of about 200–5600-Hz appears to supply almost the needed information for speech perception, as demonstrated by experiments on the intelligibility of bandpass-filtered speech in noisy backgrounds [Pavl 84, Pow 96]. (This range also corresponds to the frequencies of greatest auditory sensitivity and most speech power.) If frequencies below 1 kHz are removed from speech (via a highpass filter), many confusions occur in voicing and manner of articulation for nasals and weak obstruents [Mill 55] (e.g., among /p,b,m,v/). A corresponding lowpass filter, on the other hand, causes errors mostly in place of articulation (e.g., /p/ vs. /k/).

Similar results occur if, instead of filtering, we add noise (to obscure parts of the speech signal). Additive broadband noise causes few voicing errors, but many place errors, as does typical acoustic reverberation (where multiple reflections act as speech-shaped masking noise [Gelf 79]). Noise with a flat spectral level (i.e., white noise) tends to obscure frequencies with low energy, i.e., mostly higher frequencies. It appears that voicing in general (see later discussion about other cues) is perceived through harmonic structure, which is strongest at low frequencies. Manner cues also appear strongest at low frequencies, while place cues reside mostly above 1 kHz (especially in F2 and F3).

7.8.2 Experiments using synthetic speech

In order to demonstrate a direct relationship between any change in an acoustic stimulus and its perceptual consequences, perception experiments have to be carefully controlled. Owing to the many variables in natural speech production, perceptual experiments usually employ synthetic stimuli (i.e., computer-generated), so that all but a few parameters are maintained as constant across stimuli. Unfortunately, perceptual results obtained with synthetic speech are not always directly applicable to understand perception of natural speech. This is obviously true if the synthesis is poorly done, but also can occur with common synthetic speech. If the synthetic stimuli sufficiently resemble natural speech, the perceptual results are nonetheless often assumed to represent a valid model for natural speech. Synthetic speech generally provides the listener with the major acoustic cues needed for perception, but often omits many of the minor cues that enrich natural speech and increase its redundancy.

Natural speech is very redundant and robust. It can tolerate considerable distortion and noise, while remaining comprehensible. The information present in the speech signal can be viewed at the linguistic level as a discrete sequence of phonemes, but with a superimposed layer of other information including emotion, speaker identity, syntax, and emphasis. We initially examine how humans decode the phonemes first, and later investigate the other relevant data. The listener must decode an auditory signal in which the succession of phones exist as acoustic events that overlap in time (owing to coarticulation). Thus the perception of each phoneme is often dependent on its context: the signal prior to and following what one normally thinks of as the phone is usually quite important to understand each phone.

The importance of context goes well beyond the phonetic effects of neighboring phonemes. Listeners anticipate many aspects of a speech communication even before the speech starts. They start with initial expectations based on their knowledge of the speaker, the context of the conversation, and general knowledge. Even when hearing speech from a stranger, we adjust quickly to the new speech patterns, and apply expectations related to a possibly anticipated random encounter with a stranger (as one might have when out in public areas). To accommodate the wide range of possible contexts, our speech adapts to each new situation, instilling extra redundancy in the signal when we (as speakers) expect to have difficulty making ourselves understood. The degree of redundancy (like error protection on noisy digital channels) varies with need. Speakers increase speech redundancy by articulating more clearly [Payt 94] when needed (e.g., in noise or with strangers), but decrease redundancy in informal conversations by speaking rapidly and casually. The nature of such changes, and indeed what quantitatively constitutes redundancy, is complicated, and will be discussed at various points in this book.

7.8.3 Models of speech perception

To better understand human speech perception (as well as to design computer simulations, in the form of recognizers and synthesizers), we often try to model the perception process, based on what we know about perceptual mechanisms. It appears clear that speech perception requires several levels of analysis, including at least the following: auditory, phonetic, phonological, lexical (words), syntactic, and semantic. These levels overlap, in time and in other acoustic dimensions, and are not always readily viewed as a set of sequential steps. Given the quickness of speech perception, some stages must occur simultaneously, for example, correcting mistakes at a low-level (e.g., phonemes) using higher-level knowledge (e.g., at a sentential level). The following subsections examine several models of speech perception.

Categorical perception

A basic task in speech perception is to convert a section of speech signal (i.e., a phone) into the percept of a phoneme. We need to find which acoustic aspects of each sound are relevant to label the stimulus as a member of a class of linguistically relevant sounds (i.e., as phonemes). As in general for physical stimuli, most sounds (e.g., many speech sounds) are perceived on a continuous scale. As the physical characteristics of a sound are gradually changed, the sound is perceived as being different, but the effects are nonlinear. At times, small acoustic changes cause small perceptual changes. With proper consideration (e.g., use of a logarithmic scale), some acoustic-to-perceptual mapping is close to linear. In other cases, however, some changes have little perceptual effect, and others have large effects. The difference frequently is related to one's ability to assign a phonetic label to the stimulus.

Thus certain sounds are perceived in categorical (psycholinguistic) fashion [Stud 70]. For such sounds (often consonants), some changes along a physical continuum yield no perceived difference (i.e., all the stimuli are labeled as identical, both phonetically and acoustically). As such changes move to a different range of values (causing sounds which may be less consistently labeled as phonemes), a small physical change can cause a large perceived difference (a switch between phoneme classes). For example, consider the effect of voice onset time (VOT) on the perception of voicing in syllable-initial stops: with very short or very long VOTs, listeners clearly hear voiced or unvoiced stops, respectively, and poorly distinguish small changes in VOT (presumably because such changes cause no change in phoneme category). However, for stimuli near the voiced-unvoiced boundary (e.g., a 30 ms VOT in English), listeners are much more sensitive to small (e.g., 10-ms) changes, since they cause a labeling change for the stimulus between the features voiced and unvoiced. While consonants may be so modeled as having categorical perception, vowels are generally viewed as not subject to it [Scho 92]. Perhaps the acoustic space for vowels is too tightly packed (along F1-F2 scales) compared to the typical binary (voiced-unvoiced) and ternary (labial-alveolar-velar) scales for consonants.

Distinctive and invariant features

It is often assumed that phonemes are organized, both in perception and in production, according to a set of phonetic features (e.g., voicing, manner, place, etc). If one could reduce every phoneme to a 'bundle' of such features and associate each feature with a consistent set of acoustic characteristics (invariant in various contexts), one would have understood a fundamental process of speech perception. A traditional linguistic approach to phoneme classification uses a set of classical **distinctive features**. This is in contrast to a minimalist theory that might suggest that each phoneme exists as a separate unit of its own, independent of its neighbors. Given much evidence from speech production and perception to the contrary, it is more realistic to consider a small set of orthogonal (and often binary) properties or features that classify phonemes [Jak 61] as well as other aspects of phonology and phonetics [Chom 68]. Most classical features (e.g., voiced, labial-alveolar-velar, nasal, fricative, etc.) have origins in both acoustics and articulation, while perception has usually served in a secondary role to confirm features.

Strong perceptual evidence for the existence of most commonly-used features is found in the nature of typical confusions that occur among phonemes (e.g., in noisy or other difficult listening conditions). Perceptual errors tend to correlate well with features, in the sense that sounds are confused more often in proportion to the number of features they share (e.g., /f/ and /v/ are more often confused than /m/ and /s/, because the first pair share the features of consonant, fricative and labial, whereas the latter pair have only the consonant feature in common).

Given the prominent role that formants have in most models of speech production, it is tempting to characterize many perceptual features in terms of formants. However, the role of formants remains widely debated, perhaps mostly because they have been so difficult to identify consistently for the purpose of automatic speech recognition [McC 74]. While the center frequencies of the first 2-3 formants appear to correlate very well with vowel perception, it may be that other factors of gross spectral shape (rather than explicit formant locations) may eventually provide better explanations of perceptual phenomena.

Active models

To explain the wide range of results from speech perception experiments, researchers have proposed several types of models of speech perception. The primary question is how do listeners use the speech signal to obtain information about the vocal-tract behavior that produced the signal [Fowl 86, Fowl 96]. This subsection examines so-called active models, in which speech analysis employs synthesis as a major component (i.e., listeners 'hear' by thinking, at some subconscious level, about how they would speak the incoming speech). Some researchers suggest that we have a special **speech mode** of listening that switches into action when one starts to hear speech [Lib 67, Scho 80]. In this way, a listener then could pay more attention to speech-specific parameters, such as formants, than occurs in the absence of speech, where humans are more alert to other environmental sounds.

In a related model called the **motor theory of perception**, **motor theory**, listeners decode speech by generating (at some auditory level) an internal articulatory version of the signal, against which to compare the actual speech [Lib 67]. This theory assumes a close relationship between each phoneme and the articulatory commands to achieve it, where invariant patterns for phonemes (at some level, if not at an acoustic level) occur, independent of phonetic context. In this model, the apparent lack of acoustic invariance in the diverse realizations of phonemes, is caused by coarticulation.

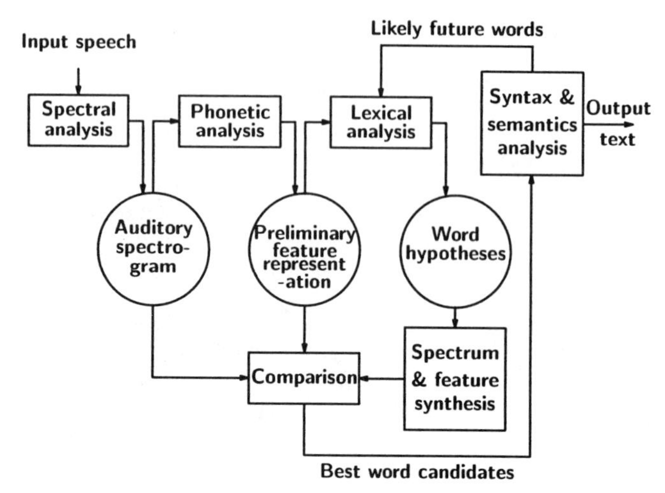

Figure 7.12: Simplified block diagram of an analysis-by-synthesis model of speech perception. Memory buffers are represented by circles.

These approaches often suggest that the listener does an **analysis-by-synthesis**, with the listener transforming the speech by a coarse auditory analysis into a representation that may be 'synthesized' back to a still internal version of the input speech, which would then allow auditory comparison (Fig. 7.12). In this feedback model, if the synthesized version matched the initial auditory one, the perception is complete; otherwise, more detailed perceptual processing is needed. These theories combine **top-down** and **bottom-up** cognitive processes.

Passive models

In the previous subsection, the active models assume that perception requires access to some level of speech production. This subsection discusses passive models, which assume a simpler mapping from acoustic features to phonetic categories (Fig. 7.13). Each speech signal (A) converts into firing patterns on the auditory nerve (B), which are coded directly into auditory features (C), which in turn allow recognition of linguistic units (D), such as phonemes and words. C's auditory patterns exist below the level of phonemes, i.e., they comprise features related to specific acoustic events (such as stop bursts). They might include estimates of periodicity or of power in critical frequency ranges.

Passive models hypothesize that listeners decode speech without any feedback mechanism, e.g., the bottom-up view would mean gradual compression of input speech data to a simple linguistic message, with no mechanism for error feedback. Recognizers that simulate this theory are less successful, which argues for the existence of some feedback in human perception. It is more likely that speech perception combines a bottom-up auditory process and a top-down phonetic process. The former would interpret speech as acoustic features and store them in short-term auditory memory; the latter would suggest such features that would be consistent with linguistic messages that are feasible in the listener's context.

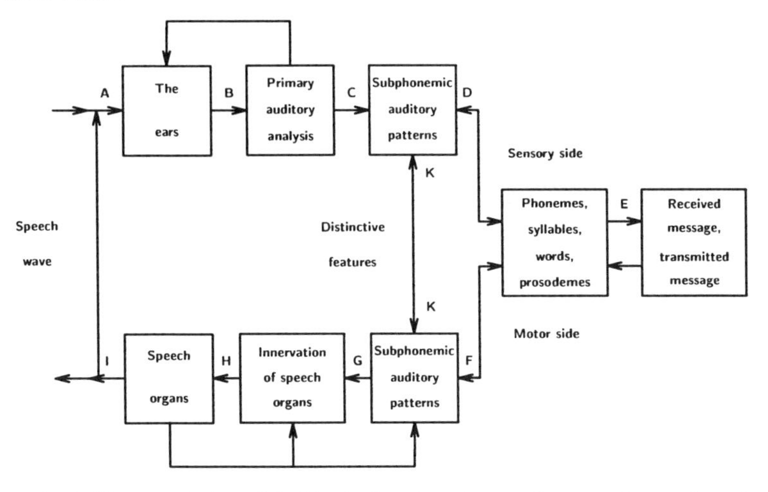

Figure 7.13: Hypothetical model of brain functions in speech perception and production (after Fant [Fant 67]).

7.8.4 Vowel perception

We now examine how specific phones are perceived, starting with the most important class: vowels. As a member of a class of steady-state phones, each vowel can be characterized by its long-term or average spectrum, or perhaps equivalently by a single spectrum in the middle of the vowel. We know, however, that transitional behavior (i.e., coarticulation with adjacent phones) is relevant; such transitions are critical for perception of non-vowel transient sounds, but are important for steady sounds as well. For simplicity, we start by noting how steady vowels are perceived (since they have no transitions), even if such vowels are uncommon in normal speech (isolated vowels bordered by silence, e.g., "oh" and "ahh").

Such static vowels are well described by fixed vocal tract shapes, which presumably form the **targets** for vowels (coarticulated) in words. The perception of a vowel without coarticulation appears to be necessarily based on a single spectrum (selecting either one frame or an average likely makes little difference for steady vowels). Traditionally, the spectrum is described in terms of the center frequencies of the initial three formants (F1-F3) [Hill 93]. As we have noted elsewhere, this formant-based view does not preclude an alternative view of the spectrum (e.g., using other aspects of the spectral envelope), but simply follows the most common and convenient analysis.

Perceived formant location in synthetic vowels

As noted earlier, the use of synthetic speech in listening experiments is essential to be sure which physical property causes which perceptual effect. Earlier in this chapter we showed that vowels can be simply characterized by their F1 and F2 positions (i.e., the vowel triangle), with a specification of F3 being necessary for some vowels (e.g., /er/). Many basic experiments have shown that synthetic vowels using only two or three formants allow easy discrimination (and indeed that, in synthesis using wider bandwidths and more formants, little spectral detail is helpful in formants above F3 for vowel discrimination). In some cases (e.g., rear vowels, which have a low F2), a single broad concentration of power in the F1-F2 region can suffice (rather than two separate formants) to allow good vowel perception, which suggests that individual and separate formants may be unnecessary. Multidimensional scaling analyses show that F1 and F2 are good orthogonal dimensions to separate the vowels perceptually [Klein 70].

To address the issue of the nature of spectral peaks in vowels, listeners can be asked to move the single peak (labeled F1′) in a one-formant synthetic 'vowel' so that it is heard as similar to a two-formant (F1 and F2) vowel. In such cases, listeners locate F1′ about midway between normal F1 and F2 values when they are separated by less than 3.5 critical bands. With greater separation, F1′ is located close to one of the actual formants [Chis 85]. This suggests that audition does a spectral integration, where a **center-of-gravity** of power concentration inside a range of 3.5 bark may be a good perceptual correlate. When listeners adjust the peak frequencies (F1′ and F2′) of two-formant vowels to perceptually match three- and four-formant vowels, they usually locate one peak near the actual F1 and the other near either F2 or F3, depending on where the major power concentration is [Blad 83].

Normalization due to Context

Even steady vowels are affected by context. A given set of formants in a synthetic vowel is not a guarantee that listeners will necessarily agree on what vowel is being spoken. While a given spectral envelope (e.g., as specified by formants) may theoretically have a wide range of possible F0 values, listeners expect a correlation between F0 and formants (e.g., small speakers usually have both small vocal folds and small vocal tracts). If F0 is about 100 Hz, a listener expects the voice to have formants spaced every 1000 Hz [Watk 96]. At F0 = 200 Hz, one expects formants shifted up by 10–15%, corresponding to a typical female voice. Thus, two vowels with the same formants but different F0 values may be perceived as distinct phonemes.

To accommodate many different speakers, speech perception must follow some type of normalization, in order to interpret many different acoustic realizations of a phoneme as the same phonetic unit [Piso 97]. A simple linear scaling of perceived formant frequencies, via a listener's inherent estimation of a speaker's vocal tract length, could provide a first-order approximation to a satisfactory normalization. However, the mechanism by which listeners might accomplish such an estimation is not clear (and is a subject of considerable speculation in automatic speech recognizers, which face the same task). Even an accurate vocal-tract length estimation (based on someone's speech) would not be fully adequate for speaker normalization, because different vocal tracts are not just linearly scaled versions of each another; e.g., there is much more variation across different people in the size of their pharynx than in their oral cavity.

Normalization in listeners is likely to be complex and based on the immediate speech from a speaker. Examples of different vowels from the first few words of each new speaker may well provide a reference formant space for a listener to infer all remaining acoustic patterns from that speaker (e.g., allow the listener to establish an internal estimate of that speaker's vowel triangle). This normalization probably involves cues from both F0 and formant positions [Holm 86]. Since F3 and higher formants tend to be relatively constant for a given speaker, F3 and perhaps F4 provide a simple reference, which has been used in automatic recognizers, although there is little clear evidence from human perception experiments.

Coarticulation effects

All speech sounds are subject to coarticulation (the only exceptions are rare cases of a single phone bordered by silence on both sides, e.g., "oh" or "shh"). In vowel production, formants usually undershoot their 'targets' (values used in isolated utterances, without coarticulation), owing to adjacent phones whose own targets pull the formants in their direction, under pressure of time and economy of effort. Listeners internally 'know' this, and adjust their perceptual interpretations accordingly. As a result, vowel (and similarly, consonant) perception appears to depend on a complex auditory analysis of formant movements before, during, and after the vowel.

If one excises the middle portion of a vowel (from its usual word or sentence context) and presents only that sound to a listener, mistakes are often made in perception. On the other hand, playing short portions of CV (consonant+vowel) and VC transitions often allows proper identification of the vowel despite the removal of a large portion of the vowel [Park 84, Jenk 94]. Hence, spectral transitions in coarticulated vowel perception are very important, although the way listeners exploit the pertinent spectral information is less clear [Nab 97]. When words are excised from normal utterances and presented individually to listeners, typically half the words are misidentified [Poll 63]. These situations lack supporting context.

Formant transitions also assist listeners in auditory stream integration or continuity [Cole 74]. In general, diverse sounds entering one's ears are assigned different sources by the brain. How speech sounds from a single voice (when competing against other background sounds) are associated with each other to form a single perceptual unit is not immediately obvious, although it appears that continuity in formants, pitch and power are important factors. This issue appears with synthetic speech, where sounds poorly coarticulated (i.e., using an inadequate model), tend to be perceived as from different sound streams (i.e., not from a single synthetic voice). For example, inadequate modeling of formant transitions can lead to fricatives being heard as separate hissing noises accompanying a (frication-free) synthetic voice.

7.8.5 Consonant perception

While vowels are evidently perceived in terms of relative formant positioning (or by a related spectral envelope shape), consonant perception is more complicated. The difference is that vowels are completely characterized by F1-F3, but consonants have diverse features (voicing and manner and place of articulation) which do not easily correspond to a few simple parameters extractable from a steady portion of the consonant. Individual features may be as simply perceived as in vowels; e.g., manner appears to be perceived in terms of coarse distribution of power across frequency (with periodicity as a secondary factor). However, the voicing and place features are quite complicated. For example, in fricatives, voicing appears to be simply related to periodicity, but in stops, voicing perception is considerably more complex, with many interrelated acoustic cues.

Perception of the manner of articulation

Manner of articulation is a primary speech feature that distinguishes vowels, glides, nasals, stops, and fricatives. From the production point of view, manner deals with power levels, coarse aspects of speech spectrum, periodicity, and other timing issues. A listener detects a vowel when the associated sound is periodic and has enough power and length, with a spectrum in which the lower formants are excited with sufficiently narrow bandwidths. Glides are distinguished from vowels via less power, less duration, and a more dynamic spectrum. Nasals are distinguished from vowels by the former's lesser overall power, wider bandwidths, and greater concentration of power at low frequencies. Stops are perceived through a silence, accompanied by adjacent appropriate formant transitions, and reinforced often by a following short burst of noise. Sounds with noise of adequate length are heard as fricatives. Whether a sound has harmonic structure without noise is perhaps the basic cue to manner.

Perception of manner of articulation is very robust in noisy conditions. Confusions are relatively rare in normal speech, with the most common errors occurring between the fricatives and the stops. Non-strident fricatives are so weak that they are frequently hard to distinguish from a stop.

The timing of acoustic events is often critical to perception. Each consonant requires a vocal-tract constriction, and the rapidity of the constriction can cue a manner difference. If a sound with steady formant patterns is preceded or followed by linearly rising formants, one often hears /b/ if the transition is short and /w/ if it lasts more than 40 ms [Lib 56]. The directions of formant change cue place of articulation, while the transition durations cue the manner feature. A rapid transition causes a stop perception, while a more gradual articulation signals a glide (and an even slower transition cues a vowel).

Perception of nasality is more subtle than distinguishing other aspects of manner. Wide bandwidths and a relative lack of power except in the F1 range tend to cue nasality. A strong, low, and steady F1 (called a **murmur**) appears to be very important to hear a nasal consonant. A much more subtle aspect of manner concerns nasalization of vowels: a vowel becomes gradually nasalized as a nasal consonant approaches, as cued by widening bandwidths and the gradual introduction of antiresonances.

Perception of place of articulation

The most complex aspect of phonemic perception would appear to be how a listener interprets place cues. For liquids, glides, and fricatives, place perception is similar to that for vowels, i.e., details of the steady spectral envelope are of major importance. As an example, sonorants /r/ and /w/ can be distinguished by the separation of F2 and F3 (/w/ having the wider spacing) [Sharf 81]. Fricatives, however, exploit spectral detail mostly at high frequencies, instead of low-frequency formants. Fricative place seems to be much more coarse in perception than vowel place: small changes in F1 and F2 can signal changes in vowel identity, but the distinction between alveolar and palatal fricatives requires a large change in the cutoff frequency of the highpass frication noise.

Relative power levels, while usually affecting manner perception the most, have an effect on place as well; e.g., distinguishing place among the fricatives is partly based on power [Hed 93], where a strong signal cues a rear place (e.g., /s,sh/), while weak frication cues a forward place (e.g., /f,th/).

For weak consonants (i.e., stops, nasals, and non-strident fricatives), cues to place reside primarily in the spectral transitions during adjacent phones, including burst cues of stops as they are released [Reca 83]. Since burst spectra are somewhat distinct from the ensuing spectral transitions, they form separate but related cues, and experiments can determine which cue is stronger. In any situation where a perceptual feature is normally signaled by multiple cues, synthesizing speech with the cues in conflict (however unnatural), instead of the normal cooperation, can help identify the more important of the cues) [Chris 97]. For example, velar bursts are more effective at signaling place than other bursts, probably because velar bursts are stronger in speech production [Smits 96].

Transition cues are not always symmetric in time. In particular, stops usually end (but never begin) with a release sound. Thus, spectral transitions provide the only place cues for VC (vowel+consonant) contexts where no phoneme follows after the consonant (although word-final stops are occasionally released into the ensuing silence). For plosives in CV syllables, on the other hand, place is cued by the spectrum of the burst release, the aspiration duration, and the spectral transition (especially in F2 and F3) to the ensuing phoneme.

In cases where formant transitions are brief owing to short articulator motion (e.g., a lingual stop followed by a high vowel) or to anticipatory coarticulation (e.g., labial stop, then a rounded vowel), the release burst often is near a major spectral peak for the ensuing vowel and contributes heavily to place perception. On the other hand, large formant transitions (when they occur) tend to dominate over the burst in place perception. As part of speech asymmetry, in VCV stimuli, the CV transition dominates place perception; e.g., conflicting place cues in VC and CV transitions lead to listeners perceiving place from the CV transition [Fuji 78].

Perception of voicing in obstruents

All sonorant phonemes are voiced. So the perceptual contrast of voiced versus unvoiced is only critical for obstruents, where different phonemes from the obstruent class depend directly on correct understanding of voicing. In particular, we distinguish the voiced group /b,d,g,v,dh,z,zh/ from the unvoiced group /p,t,k,f,th,s,sh/, where corresponding pairs of phonemes in the two groups have the same place of articulation. For obstruents, the feature **voiced** does not simply mean 'having vocal cord vibration.' Instead, the actual distinction between the two groups relies on a diverse set of acoustic properties, especially for stops. In the simpler case of fricatives, voicing is indeed generally perceived when the fricative is periodic during most of the sound. However, duration affects voicing perception for all obstruents; e.g., shorter fricatives tend to be heard as voiced [Cole 75] and similarly when the preceding vowel is longer then average.

This last observation refers to the fact that voicing in syllable-final obstruents depends significantly on the duration of the preceding vowel. This may be due to the fact that syllable-final obstruents are often brief, with less time to cue voicing via actual vocal cord vibration. Furthermore, many English voiced stops have little periodicity; i.e., the voice bar, nominally present for all voiced obstruents, occurs more regularly in fricatives than in stops. In sonorant+stop syllables, glottal vibration usually continues into the first part of the stop if voiced, whereas periodicity ends abruptly with oral tract closure for an unvoiced stop [Hill 84].

Perhaps the most research in speech perception has been devoted to the phenomenon of how voicing is cued for syllable-initial stops. The mapping from acoustics to voicing perception for these stops is more complex than for almost all other speech sounds. It involves temporal factors (VOT and the timing of the F1 transition) and spectral factors (intensity and shape of the F1 transition, and the burst). For English stops, the primary cue appears to be VOT: a quick voicing onset after the stop is released causes voiced perception, while a long VOT is heard as an unvoiced stop. If other cues are neutral, perception switches from voiced to unvoiced as VOT increases past 30 ms for an alveolar stop (the boundary for labial or velar stops is about 5–10 ms lower or higher, respectively [Mill 77], which correlates with the production tendency in stops for longer VOT with a longer front cavity).

A secondary cue for initial-stop voicing is the value of F1 at voicing onset [Lib 58]: lower values cue voiced stops. This too correlates with speech production: F1 rises during

CV transitions as the oral cavity opens up (from stop to vowel). The delayed appearance of F1 in unvoiced stops (due to the longer VOT) means the initial F1 rise occurs (silently) during the aspiration. The amount of the F1 transition significantly affects stop voicing perception [Lisk 78], while higher formants seem to have little perceptual effect. This effect too may correlate with production: in stop+vowel sequences, periodicity often appears first in F1 while higher formants still have aperiodic power.

A third cue to voicing is aspiration power. While VOT appears to be the strongest voicing cue, experiments have suggested a trade-off of duration and energy [Repp 83]; i.e., power integrated over time may be more relevant than just duration. Lastly, spectra and F0 can also affect voicing perception [Mass 80]; e.g., rising F0 at stop release helps to cue stop voicing [Hagg 70].

Several of these cues can trade off effects in synthetic experiments; e.g., a change in F1 onset shifts the point along a VOT continuum where perception changes from voiced to unvoiced. In such experiments using open vowels, a 100-Hz change in F1 onset is perceptually equivalent to an 11-ms change in VOT [Summ 77] and a VOT decrease of 4.3 ms is equivalent to a 10-dB increase in aspiration intensity [Repp 79].

7.8.6 Duration as a phonemic cue

We now examine how the length of a sound can directly influence aspects of its phonemic perception. For example, duration is very important in hearing the difference between tense and lax vowels and for detecting voicing in stops.

Manner cues

Many languages (e.g., Swedish and Japanese, but not English) use duration explicitly to distinguish long and short phonemes, where the spectra are essentially the same. In more languages, duration helps to cue subphonemic features, such as VOT for voicing perception. In such cases, duration forms part of a phonemic distinction, along with other acoustic factors. As an example, durational lengthening tends to be associated with unvoiced obstruents, while lengthening of a prior sonorant helps one to hear a succeeding obstruent as voiced.

Typical phonemes last about 80 ms, but sometimes phonemes can be very brief. Lengthening certain sounds or transitions between phonemes can cause perception of additional phonemes (i.e., the lengthened subphonemic segment is transformed by the lengthening into a separate phone). For example, if silence of sufficient duration is inserted between /s/ and /l/ in 'slit,' it can be heard as 'split' (the absence of appropriate formant transitions for a labial stop is overcome perceptually by the primary cue for a stop: silence). When the typical brief silent interval (e.g., 10 ms) occurring between the /s/ frication and the ensuing voicing in 'slit' is abnormally lengthened, the listener reconciles conflicting acoustic cues, at the point where a sound is too long to interpret it as a mere transition. In general, phonemes and their individual features are perceived via several acoustic cues, of which some dominate over others in cases of conflicts. In natural speech, such cues normally act in concert, but experiments with synthetic speech can test each cue's perceptual strength.

Place cues

Place of articulation in consonants is primarily cued spectrally. A secondary cue for stops seems to be VOT duration. As a result of coarticulation, labial stops have a short

VOT, while velar stops have a long one. Labial stops allow substantial tongue movement to anticipate an ensuing vowel, thus allowing more rapid voicing onset than for alveolars or velars. Velars have long VOTs, because the tongue body moves more slowly than the tip and the body is almost always implicated in coarticulation.

Effects due to speaking rate

All acoustic parameters of speech are evaluated by a listener in terms of context. Formants are likely interpreted in terms of an estimated vowel triangle for the speaker, and the loudness of individual phonemes is heard relative to average or typical power levels. As for a phonemic role for duration, perceived lengthening or shortening is judged relative to one's speaking rate [Moo 84, Way 94], both local and global. For example, an acoustic continuum changing perception from /baa/ to /waa/ as the duration of the initial formant transitions lengthen is affected by rate: longer transitions are needed to perceive /waa/ as the overall syllable is lengthened [Mill 81]. In another example, if the /p/ in 'topic' is lengthened, listeners hear 'top pick,' and the amount of lengthening needed to hear the change depends on the local speaking rate.

If a syllable is taken (i.e., excised) from a sentence, listeners hearing that syllable in isolation assume that it should have a long duration and no external coarticulation (as would be appropriate for a short word by itself). As a result, they often misinterpret the word.

7.8.7 Perception of intonational features

Speech perception research usually emphasizes how listeners distinguish different phonemes. Since intonation also affects the way we understand speech, it too warrants discussion. Phonetic context is important for all aspects of speech perception (e.g., coarticulation for phonemes), but the extent of contextual effects is much broader for intonation than for phoneme perception. These extend to syllables, words, phrases, and even sentences. Aspects of intonation such as rhythm and stress assist the listener's understanding by highlighting important words and by indicating syntactic boundaries in each utterance. Such cues to segmentation allow listeners to process speech into small units, for easier perception.

In many languages, a question that solicits a response of yes or no terminates in an F0 rise; perception of such a request often depends upon word order, but F0 alone may signal it. In other aspects of perception of syntactic structure, listeners can determine that a clause is subordinate by its lower F0 (and by durational lengthening of sounds just prior to the clause boundaries), that a word behaves as a vocative (through a low, small F0 rise) or an appositive (larger F0 rise or fall), and that an utterance has been completed (by F0 falling to a very low level).

The global or broader nature of intonation (i.e., affecting context over several seconds) helps both to segment utterances and to group words together. Continuity of the F0 pattern over entire utterances and typically steady speaking rates help listeners to better follow speech in noisy environments, especially in acoustic environments of several competing voices [Broc 82]. Intonation also plays a key role in conveying nonlinguistic information about speakers, such as their emotional state, gender, size, age, etc. [Stre 83, Ladd 85]. Different emotions (e.g., anger, sorrow, fear) alter F0, speaking rates, and precision of articulation [Will 72, Tart 94].

Stress or emphasis

A primary role of intonation is to make certain words stand out as more important than others in each utterance. Speakers generally choose the words to emphasize to be those words that are less predictable from context. Simply put, we stress words which convey new information. A word is usually stressed by intonational changes that occur during what is called the **lexically-stressed** syllable. Thus we must distinguish lexical (word) stress and sentential stress. Lexical stress is an abstract notion: one syllable in every word is designated as **stressed**, as noted in its phonetic description in standard dictionaries (e.g., in boldface, underlined, or capital letters); in long words, another syllable is often designated as having secondary stress. When hearing words, such emphasized syllables appear to stand out owing to local intonational effects. In many languages, lexical stress is straightforward; i.e., the lexically-stressed syllable in every word depends only on its position in the word, e.g., the word's final syllable in French. Other languages allow more freedom for stress placement, but often display a tendency for a default location (e.g., the first syllable in English).

Acoustic correlates of stress

Several acoustic factors affect linguistic stress in a complicated fashion. The major intonational features that can be related to perception of stress are pitch, loudness, and length. An acoustic factor (i.e., a physical correlate) can be readily associated with each of these perceptual features: F0, amplitude, and duration, respectively. As in most cases where one acoustic parameter is readily paired with a perceptual feature, the relationship often has subtleties. Most mappings between the physical and the perceived are monotonic but nonlinear (each increase in the acoustic parameter causes an increase in the corresponding perceptual feature, but the magnitude of the effects can vary widely). Very small physical changes lie below the feature's JND, and very large changes are beyond a saturation range (e.g., extreme intensity overloads the ear). Even within the more typical ranges found in normal speech, nonlinear scales (e.g., logarithmic) are common in the physical-to-perceptual mapping.

F0 is the most direct cause of pitch perception, with amplitude and duration secondarily affecting pitch. F0 is usually measured in the linear scale of Hz, but use of logarithmic tones (where 12 semitones = 1 tone = an octave = doubling of F0) is an attempt to represent F0 changes more closely to a perceptual scale. An even more accurate measure is **ERB-rate** scale, where syllables are heard as equally prominent with F0 movements on this scale [Ghit 92].

A simple way to measure stress is to ask a listener which of two sequential phonemically identical syllables sounds the most emphasized. One is usually heard as having more stress than the other if it has higher amplitude, longer duration, and higher or more varied F0 patterns. Stress may be binary (i.e., a syllable or word is either stressed or unstressed), or it may take several values along a continuum (i.e., one can discern several degrees of stress). The domain of stress appears to be the syllable: one cannot distinguish stress among smaller units (e.g., within a syllable, to hear one phone as stressed and the next as unstressed) and larger units such as words are perceived as stressed depending on whether an appropriate one of their syllables is stressed. A syllable's vowel appears to contribute most to the syllable's stress as it is the most intense part of the syllable (thus the most easily perceived).

Quantifying stress in English is easier than for other languages because it has distinct words with identical phonemic composition but which vary in meaning due to stress (e.g.,

"EXport," which is a noun, versus "exPORT," a verb). Using synthetic speech that controls all acoustic aspects other than specific variations in F0, amplitude, and duration, we can see how stress relates to these acoustic features [Fry 55] [Fry 58] [Bol 58] [Mol 56]. As a result of these and other experiments, it appears that F0 is the most important for stress in English, with duration second and amplitude third. Such rankings are obtained by testing the strength of each cue in cases of conflicting cues (e.g., imposing long duration on a weak syllable); such cases are rarer in natural speech, where cues tend to reinforce each other rather than conflict.

While phones with higher intensity and longer duration obviously cause them to be heard as having greater stress (via perceived loudness and length), the relationship of F0 to stress is considerably more complex. It appears that change in F0, rather than merely high F0 values [Mort 65], correlates better with stress, and that upward changes in F0 cause more stress than downward movements [Bol 58]. There are many ways F0 can vary during a syllable. F0 patterns as complex as a rise+fall+rise (with the movements having different amplitudes) may occur during individual vowels. With level F0 patterns, higher F0 sounds more stressed, but changing F0 usually sounds more stressed than level F0 does. Timing has an effect as well: an F0 rise earlier in a syllable has more stress [Herm 97].

As is often the case in speech perception, context is very important for stress. Phonemes vary inherently in F0, amplitude, and duration, owing to their phonetic features. Furthermore, intonation varies as a function of position within utterances. Thus, the use of intonation to signal stress must be interpreted in the context of what is 'normal' intonation due to phonetics and word positioning. Vowels inherently have greater amplitude and duration than consonants. Low vowels have more intensity than high vowels (hence strong, low vowels do not necessarily sound louder than weaker, high vowels). Vowels with higher tongue positions in stressed syllables have higher F0, and F0 in a vowel starts at a high F0 if the preceding consonant is unvoiced. F0 (and to a lesser extent, amplitude) tends to fall gradually throughout each utterance [Pier 79]; so smaller F0 changes have larger stress effects later in utterances. Syllable-initial consonants are longer than syllable-final consonants [O'Sh 74]; so long final consonants sound emphasized.

Perception of syntactic features

The second major role for intonation is to cue syntactic information, primarily dealing with local grouping of successive words. Since few words in speech are terminated by pauses, organizing the flow of speech into sections of a few seconds at a time (to facilitate perception) is a major role for intonation. It is more efficient for a speaker to pause only at major syntactic boundaries, and allow more subtle intonational changes to signal other syntactic features. Speech with little variation in F0 (i.e., monotonic) and a steady succession of equal-duration syllables would be difficult to understand.

Segmentation

Variation of intonation by a speaker for syntactic purposes obviously interacts with the way stress is signaled. Speech is typically a succession of stressed and unstressed syllables. Certain languages are denoted as **stressed-timed** because their stressed syllables tend to occur in an approximately periodic fashion. For example, English shortens unstressed syllables significantly, which helps to create a sort of rhythm of stresses. **Syllable-timed** languages (e.g., French and Japanese), on the other hand, tend to preserve more constant syllable durations. In both cases, the situation is far from categorical, as the timing regularity (whether of syllables or stressed syllables) is more perceptual than physical, with actual durations varying significantly [Leh 77].

Syllable rhythm is interrupted regularly at many major syntactic breaks, and irregularly when hesitations occur. Distinguishing intentional changes (i.e., intonational variations intended to cue segmentation) from unexpected hesitations can be complicated. We discuss intended changes first. Speaking rate (as measured in syllables or phones per second) often is reduced (i.e., slowed speech) immediately prior to a major syntactic boundary [Klatt 75b]. This speed reduction may precede a pause, but it is still called **prepausal lengthening** even if no pause coincides with the break: the final 1–2 syllables in a major syntactic group usually lengthen, which signals a linguistic break to listeners. Such boundaries are often cued by F0 as well. For example, F0 often rises briefly at the end of the final syllable just prior to a syntactic break. Such a **continuation rise** in F0 (e.g., 10–30 Hz) [Dela 70, Hart 90] informs listeners that the utterance has not yet finished and that the speaker does not wish to be interrupted. Thus, speakers can 'hold the floor' by consistently raising F0 at pauses.

The utility of intonation is exemplified in cases of speech where the structure of the sentence may lead to ambiguity, e.g., where the words are the same phonemically but different meanings are cued by intonation. Examples are 'The good flies quickly passed/past' ('flies' can be a noun or verb in this sentence, which changes the meaning completely) and 'They fed her dog biscuits' (did she or her dog eat?). Coordinate constructions in sentences are often inherently ambiguous, as in 'Joan and Sue or John went.' English noun phrases can consist of several adjectives and nouns, in which medial words may attach syntactically either left or right (e.g., 'light house keeper'). Ambiguity can arise over whether a phone belongs to either of two adjacent words; e.g., 'gray tie' or 'great eye.' While such inherent ambiguities are often resolved by lexical context, examining the intonational behavior in these examples shows how segmentation can be accomplished via intonation.

In all cases here, ambiguity is eliminated if a perceptual break is caused either before or after the word which can associate backward or forward [Kooij 71]; e.g., a break before or after 'dog' signals whether 'dog biscuits' is a syntactic unit. Normal intonation groups words together as phrasal units. Interruptions can override the default groupings, to force perceived boundaries at unexpected places. Among F0, duration, and amplitude, duration is the most reliable segmentation cue [Macd 76, Leh 76], via pauses and prepausal lengthening.

Interrogation

At the end of an utterance, F0 is often used to inform listeners that a response of yes or no is expected next. As in speech immediately preceding sentence-internal major syntactic breaks, the final few syllables of an utterance typically lengthen. In addition, the last few phones often have diminished amplitude. In most cases, F0 falls, often sharply, at the end of an utterance, to a very low value. Such an F0 fall signals finality (the opposite of a continuation rise). For yes/no questions, on the other hand, F0 rises rapidly on the last word in the sentence, often to the highest level of the utterance.

Role of pitch movements in perception

As noted earlier, F0 is a more powerful cue than other intonation factors, owing to its greater degrees of freedom (e.g., rising or falling variable amounts, all within individual syllables). This has led to a complicated relationship between F0 and the linguistic information that it signals, so much so that F0 has not played a significant role in recent speech recognition systems (except when forced to, as when recognizing speech of a tone language). In addition, most speech synthesizers employ very simplistic F0 patterns, e.g., obtrusions for stressed syllables superimposed on a declination line. Despite certain efforts in perception research, it is still unclear what aspects of F0 patterns are of primary perceptual relevance.

Simple experiments have shown that F0 contours can be heavily smoothed (using lowpass filtering) with little perceptual effect [Rosen 71]. Major F0 movements are relevant, and listeners more sensitive to rises than falls [Hart 74]. The rate of an F0 change may be perceptually important, as listeners detect changes in F0 slope as little as 12 Hz/s [Klatt 73]. Pitch is mostly perceived during vowels (i.e., sounds of high intensity), while the more irregular F0 variations during weaker sounds (e.g., most consonants) seem to be largely disregarded [Leon 72]. In particular, interruptions in F0 patterns caused by unvoiced sounds seem to have little effect on pitch perception [Hart 90].

Examining Dutch speech, researchers tried to model F0 contours as successive short F0 patterns selected from a set of 12 prototypes [Hart 90]. Some large F0 movements were found to have little perceptual relevance in certain situations. The declination effect (general trend of F0 to fall) was relevant, although listeners are rarely conscious of it explicitly. Syllable-internal F0 timing was important: stress was heard best when F0 rose early in a syllable or fell late, which correlates well with stresses in typical speech production. Late rises and early falls were not stressed (e.g., a late rise correlates with syntactic grouping, not stress effects).

7.9 Summary

This initial chapter of Part II in this book introduced many of the fundamentals of speech science, specifically those dealing with the phonetics of speech communication. The chapter started with a discussion of the anatomy and physiology of the organs of the vocal tract. The phonemes or sounds of language were introduced, both from the articulatory and acoustic points of view. The phonemes were described in terms of phonetic features, which will be shown to be useful in describing speech applications (recognition and synthesis) later. The acoustic features of phonemes were noted in ways that will be useful for speech analysis methods from different points of view: articulatory, acoustic, and perceptual. The important phenomenon of coarticulation was discussed, noting its usefulness for human speech production and perception, while raising difficult problems for computer speech processing.

Vocal-tract behavior was then analyzed in terms of a source-filter model of speech production, allowing a simple electrical and computer model of speech generation. The model was developed starting from an elementary (but relevant) single-tube model, moving through two- and three-tube models (useful as simple models for many phonemes), and finally to models with more than ten tube sections. The latter models are those used in linear predictive analysis, as found in most speech coders today. The models were developed in both acoustic and electric terms, where the latter lead easily to computer filter simulations. Quantitative descriptions of the phonemes were made for both the time and frequency domains, following the approaches of most speech analysis techniques.

The discussion then turned to human speech perception. First, a physical description of the ear and related internal organs was given, from an acoustical analysis point of view. We focused on the behavior of the inner ear, and especially the vital organ of the cochlea, as the sounds are transformed there to electrical signals for the auditory nerve to the brain. The important perceptual phemomena of masking, critical bands, and just-noticeable differences were analyzed.

The role of intonation, involving the fundamental frequency of the vocal cords (pitch), intensity, and duration, was described in some detail. Intonation provides other mechanisms to help the speaker signal important information to listeners. Simple vocal-tract shape is useful for conveying basic phoentic information to a listener, but intonation provides much other relevant linguistic information.

In short, this chapter examined the mechanisms and acoustics of human speech production, noting the physiology and movements of the vocal tract, as well as the acoustic-phonetic relationships of speech communication and sound perception. Elementary aspects of speech psychoacoustics are fairly well understood, especially at the anatomic level. However, many details remain to be explored to achieve a complete model of speech perception. Much of speech perception research has been motivated by our knowledge of human speech production, i.e., synthetic speech which models natural speech has been used to confirm our production models. Such models have led to a reasonable understanding of the major acoustic cues to perception, but secondary factors remain to be explained. A major task remains the search for invariant cues to phoneme perception, especially for voicing and place of articulation features. Reflecting the less advanced state of production models of coarticulation (as opposed to a more solid understanding of isolated phone production), much research in speech perception continues to better understand the effects of context.

Chapter 8

Phonological Process

8.1 Introduction

The phonetic process covered in the preceding chapter can be regarded as a subject of study by a set of physical sciences, including anatomy and physiology of speech, acoustics of speech, and auditory perception of speech. However, beyond these physical characterizations, speech is also a systematically organized process for the purpose of conveying linguistically and cognitively meaningful messages. This "higher level" view of speech organization forms the subject of **phonology**. That is, phonology is concerned with those aspects of speech that deal with the systems and patterns of speech sounds that occur within specific languages. As examples, the study of acoustic or perceptual properties of vowels (such as vowel duration) is considered as phonetic research, while the study concerning identifying the total number of vowels in English or Chinese, or the study concerning how to formulate rules to predict patterns for vowel duration increase (lengthening before voiced consonants), belong to the realm of phonology. Therefore, phonetics researchers tend to draw on scientific techniques used in physical sciences such as measurement, sampling, and analysis, while phonology researchers, on the other hand, are more likely to examine the cognitive or mental organization of language.

The focus of phonology on linguistic structure and mental organization does not imply that the researchers are free to speculate at will. In fact, studies on any aspect of speech, physical or mental, must build upon empirical observation on a large scale. While the empirical nature of phonetic subjects such as speech articulation and acoustics is obvious, the same standards also apply to phonology, where linguistic systems, structures, and distinctions are equally subject to empirical justification.

Another important characteristic that differentiates phonetics and phonology involves their nature: continuous (i.e., graded) versus discrete (i.e., symbolic). Phonetics concerning physical, concrete reality of speech (such as articulation, acoustic waves, and auditory response) is obviously dealing with continuous variables, either deterministic or random. Phonology, on the other hand, deals with symbolic, discrete categories of speech sound structures and classes, and deals with the distinctions of these sound categories.

As a summary of the introduction to phonology versus phonetics, phonetics is associated with concrete, physical, and continuous aspects of speech such as measurement of articulator movements and capturing/analysis of the speech signal, while phonology concerns abstract, cognitive, and symbolic aspects of speech, such as sound patters of a language and linguistic labels (pronunciation) for words. The division, however, is not absolute; hence we will discuss an important subject called the interface between phonetics and phonology.

In this chapter, basic concepts, models, and theories in traditional and more recent phonology developments are outlined. The topics covered will include phonemes and allophones, distinctive feature theory and associated phonological rules, feature geometry theory, articulatory phonology, and syllables in phonological theory. For more details, readers are referred to a number of books dedicated to both traditional and modern phonology [Chom 68, Katam 89, Clark 95, Golds 90].

8.2 Phonemes: Minimal Contrastive Units of Speech Sounds

In this section, we explore the phonemic organization of speech, which constitutes the segmental units of spoken language. Phonemic organization offers a profound insight into the structure of spoken language. The basic concept of phonology in terms of systems and structures of speech sound patterns and in terms of distinctions of these patterns is fundamentally illustrated by exploring phonemes for particular languages.

When we examine phonetic properties of speech such as articulatory motions, there do not appear to be discrete events separating one part of speech from another in the temporal dimension. These phonetic properties tend to flow continuously in time. However, when we listen to speech, we do sense discrete units arranged as a temporal sequence. Such discrete units are called "segments." The linguistic notion and mental reality of phonemes, which we discuss below, are responsible for this cognitive sense of segments.

8.2.1 Phonemes and allophones

To discuss the phonemic approach to studying speech sound patterns, we need to consider speech sounds not from the viewpoint of individual physical sounds, but from the viewpoint of families of sounds that are regarded as functionally equivalent, in terms of conveying meaning, in the language in question. Such a family of sounds is called a **phoneme**. A phoneme is an abstract or underlying form of sound in a language; it is a unit within the sound system of a language. Members of the same phoneme family, or the various individual physically distinct sounds which are surface realizations of a given underlying phoneme, are called **allophones** of that phoneme. We can say that phonetics is mainly concerned with surface forms, and phonology is concerned with underlying forms as well as the transition from the surface forms to the underlying forms.

Given a particular sound heard in speech, how do we know which phoneme it belongs to? This depends on the language where the sound occurs. For example, the sound of the initial consonant in the Mandarin Chinese word /ba ba/ (father) belongs to the phoneme /b/ in Mandarin, but the same sound belongs to the phoneme /p/ in English, e.g., realized as one of many /p/ allophones that occur in the English word "speech."

Let us pronounce two words "cool" and "key." In uttering "cool," the back of the tongue is in contact with the part of the soft palate near the uvula at the very back of the mouth. In contrast, in "key," the more front part of the soft palate is where the tongue makes the contact to create the constriction. These two varieties of "k" are physically very different, both in terms of articulation and acoustics. However, these differences are not used to distinguish word meaning. Hence, they are two allophones of the same phoneme /k/. Which allophone to use in a given utterance depends on what the adjacent sounds happen to be. From this example, we can give a refined description of a phoneme — it is the smallest unit of sound in a language that is contrastive or distinctive.

The two allophones of /k/ above are said to be in **complementary distribution**. When two sounds are in complementary (i.e., mutually exclusive) distribution, they are forbidden from occurring in identical phonetic contexts. One sound appears in certain contexts and the other in some different ones. In the example above, the back allophone [k] of phoneme /k/ as in "cool" occurs before back vowels. The front allophone [k] in "key" occurs before front vowels, which is a different, clearly defined phonetic environment. Another example of complementary distribution is the several /t/ allophones as in the words "tea," "too," "tree" and "eight." Sounds in complementary distribution are generally allophonic.

Based on the above notion of "complementary distribution," where allophones often occur, one might be tempted to think that the [ng] and [h] sounds in English could be classified as the same phoneme. [h] always occurs in syllable-initial contexts in English, and [ng] always in syllable-final contexts. However, there is an additional principle in phoneme theory that prevents interpreting [ng] and [h] as one phoneme. This principle is called the "phonetic similarity" of allophones. [ng] and [h] lack phonetic similarity — one is a voiced velar nasal, and the other is voiceless glottal fricative. Hence, they cannot be grouped together as allophones of the same phoneme, despite the complementary distribution they have.

A related notion to complementary distribution is **free variation**, where different sounds can appear in the same position in two otherwise identical words but without creating a difference in meaning. Examples of free variation are allophones [t] and the glottal stop in the words "don't," "bottle," and "button." Sounds in free variation can also be phonemic, such as the two pronunciations of the word "either."

The principal approach to phonemic analysis is to examine its linguistic function of contrasting or distinguishing word meaning. Two sounds are contrastive when they can appear in the same position in two otherwise identical words and create a difference in meaning between the words. The phoneme is a minimal sound unit capable of contrasting word meaning, and contrastive sounds are always phonemically distinct. For example, in the word pair "tip" and "dip," /t/ and /d/ are two separate phonemes because a substitution of one by the other would cause a change in the word meaning, and hence they are contrastive.

8.2.2 Phoneme identification

How do we identify phonemes of a language? That is, how do we know two sound segments belong to two separate phonemes or they belong to two allophones of the same phoneme? The first basic procedure for this type of phonemic analysis is called the "minimal pair" test. If two words are identical in all aspects except for one segment, they are referred to as a minimal pair. In the above example, "tip" and "dip" form a minimal pair. The minimal pair test is to determine whether there is a single segment difference that distinguishes the meanings of two words. The segments or sounds are classified as separate phonemes if they are responsible for a difference in meaning in a minimal pair. In other words, the segments are separate phonemes if they contrast in identical environments, where environments refer to all other segments in the words.

Passing the minimal pair test is only a sufficient condition for identifying separate phonemes. Often, it is not possible to find minimal pairs that contrast each single phoneme in a language. There comes the second, less rigorous, and somewhat subjective procedure for phonemic analysis, which is based on contrast in analogous environments, rather than in identical environments. Analogous environments are phonetically very similar environments. To pass this test in identifying two separate phonemes, one needs to show that the two similar environments cannot be attributed to the difference in the two segments in question that provide contrast in meaning. This is a very subjective test, and is used only when minimal pairs cannot be found for the two segments.

Of course, one needs to perform the above tests only if there is no clear lack of phonetic similarity between the two segments in question. If phonetic similarity cannot be identified, then the two are separate phonemes. In the earlier example, [ng] and [h] are in complementary distribution and hence no minimal pair can be found for them. There is no need here to search for analogous environments since the lack of phonetic similarity is already sufficient to classify them as separate phonemes.

Every sound in a language belongs to some phoneme, and phonemes form a discrete set; i.e., the set members take symbolic, categorical values rather than continuous, numerical values. Some phonemes have more allophones than other phonemes. In English, phoneme /t/ has probably more allophones than any other phoneme.

In summary, phoneme and allophone are two separate but related concepts. A phoneme serves the linguistic function of contrasting word meaning, and allophones do not. A phoneme is a linguistic unit that represents the segmental structure and system of spoken language, and an allophone is a phonetic realization of such a unit. Phonemes are abstract units, and allophones tend to have a large number of variations. In speech technology, we often refer to a phoneme as a base-form of the speech unit, and refer to an allophone as a surface-form of the unit.

8.3 Features: Basic Units of Phonological Representation

8.3.1 Why a phonemic approach is not adequate

When we talked about phonemic identification above, we introduced the principle of phonetic similarity. That is, in order to be considered allophones of an underlying phoneme, two sounds must be phonetically "similar." But how do we measure this similarity? How do we know [k] is not "phonetically similar" to [h]? To provide a satisfactory answer to questions such as these, we need to look beyond the phoneme and introduce the new notion of "features." Features are the basic units of internal phonological structure that make up phonemes as sound segments, and they represent linguistic properties of individual sounds, whose values identify that sound uniquely. Features are the members of a small set of elementary categories that combine in various ways to form the speech sounds of the world's languages.

Much of phonological analysis shows that there is only a relatively small inventory of features from which the world's languages select different combinations in order to construct their individual phonemic systems. Since it is clear from phonetic studies that humans are endowed with very similar articulatory and auditory capabilities, it should not be surprising that they will only be able to produce and utilize speech sounds obtained from the economical inventory of features that is pre-determined by their biological endowment. If the phonemic systems of different languages do not share features from a common inventory, a desirable economical system of phonology for the world's languages could hardly be imagined within human capabilities.

The need for features is also clearly revealed when we examine phonological processes in a language, whereby one sound changes to become more like some other sound in its adjacent environment. For any language, the phonological behavior of phonemes is largely determined by the features they have. Consider the behavior of the nasalized vowels in the words "pam," "pan" and "pang," as well as in "pim," "pin" and "ping."

This is in contrast to the unnasalized vowels in the words "pass," "pat" and "pack," as well as "pitch," "pit" and "pick." If we were to treat phonemes as indivisible units, there would be no insightful way to account for the fact that the vowel only assimilates the property of a nasality feature from the following consonant that has that same feature. The spreading of nasality from the nasal consonant to the vowel can be due to the anticipatory lowering of the velum during the articulation of the vowel. Use of features facilitates the straightforward explanation of assimilation processes by highlighting the various separate articulatory gestures involved in speech production. If phonemes were treated as unanalyzable, indivisible entities, no natural way of stating assimilation processes would be available.

The need for features is further made clear when we examine their crucial role of highlighting the internal structure of a sound segment. A phoneme segment consists of a bundle of features, but this bundle is not unordered and unstructured phonetic properties. We will discuss this internal structure of a sound segment via the use of the features later in this chapter.

8.3.2 Feature systems

After showing the need for features beyond phonemes in phonological analysis, we now explore the various systems for features and their properties. Features often act independently in phonological processes. A complete and accurate inventory of features in any feature system should be no larger or smaller than necessary to uniquely define all phonemes and to describe all phonological alternations in a language. It should also be able to provide an accurate representation of the way in which humans store and organize the sound systems of their languages.

In addition to the above "linguistic" requirements, the feature systems are based typically on articulatory, acoustic, or perceptual realities. The best known and most widely used feature system is that of "distinctive features" which combines all these realities in defining its feature inventory. The distinctive feature system was established initially in the work of [Jaco 52]. A number of shortcomings of this system were overcome with a major revision in the book of [Chom 68], named *The Sound Pattern of English* (SPE). The SPE approach to the distinctive feature system is outlined below.

The inventory of features in the SPR system consists of the following groups of distinctive features:

Major class features

1. Consonantal and Nonconsonantal [± cons]: Consonantal sounds are produced with a sustained vocal tract constriction. Nonconsonantal sounds are produced without such constriction. Obstruents, nasals, and liquids are consonantal; vowels and glides are nonconsonantal.

2. Sonorant and Nonsonorant [± son]: Sonorant sounds are produced with a vocal tract configuration sufficiently open so that the air pressure inside and outside the mouth is approximately equal. Nonsonorant sounds, also called obstruents, are produced with enough constriction to increase the air pressure inside the mouth so that it is significantly higher that the pressure of the ambient air. Vowels, nasals, and liquids are sonorant; stops, fricatives, and affricates are obstruents.

3. Syllabic and Nonsyllabic [± syl]: Syllabic sounds function as syllable nuclei; nonsyllabic sounds occur at syllabic margins. Syllabic sounds are auditorily more salient

than adjacent nonsyllabic sounds. Vowels, as well as syllabic glides and syllabic nasals, have the feature [+syl].

Manner features

1. Continuant and Noncontinuant [± cont]: Continuant sounds are produced with a vocal tract configuration that allows the air stream to flow through any midsaggital occlusion of the oral tract. Noncontinuant sounds are made by completely blocking the flow of air through the center of the vocal tract. Nasals, oral stops, laterals, and affricates are noncontinuant. All other sounds (fricatives, vowels, glides) are continuant.

2. Nasal and Nonnasal [± nasal]: Nasal sounds are produced by lowering the velum, allowing the air to pass outward through the nose. Nonnasal or oral sounds are produced with the velum raised to prevent air passing through the nasal cavity. In English, there are three nasal sounds: /m/, /n/, and /ng/.

3. Lateral and Nonlateral [± lat]: Lateral sounds are produced with the tongue placed so as to prevent the air stream from flowing outward through the center of the mouth, while allowing it to pass over one or both sides of the tongue. Nonlateral (or central) sounds do not involve such a constriction. In English, /l/ is the only lateral sound.

4. Strident and Nonstrident [± strid]: Strident sounds are characterized acoustically by stronger random noise than their nonstrident counterparts.

Laryngeal (source) features

1. Voiced and Voiceless [± voice]: Voiced sounds are produced with a laryngeal configuration causing periodic vibration of the vocal cords. Voiceless sounds are produced without such periodic vibration.

2. Spread Glottis and Nonspread Glottis [± spread]: Spread (or aspirated) sounds are produced with the vocal cords drawn apart, producing a non-periodic noise component in the acoustic signal. Spread-glottis sounds include [h], aspirated stops, breathy voiced sounds, voiceless vowels, and voiceless glides.

3. Constricted Glottis and Nonconstricted Glottis [± constr]: Constricted (or glottalized) sounds are produced with the vocal cords drawn together, preventing normal vocal cord vibration. Constricted-glottis sounds include glottalized vowels and glides, implosives, ejectives, and creaky voice.

Place-of-Articulation (cavity) features

These features specify where modifications of the air stream in the vocal tract take place in the speech sound production.

1. Labial and Nonlabial [± labial]: Labial sounds are formed using the lip(s) as the articulator, that is, when there is narrowing made with the lower (and upper) lips. An alternative feature Round [± round] is often used to refer to many of the labial sounds. Rounded labial sounds are produced with a protrusion of the lips. There is a high degree of overlap between the sounds covered by the features

[+labial] and [+round]. [w], [uw], [uh], and [o] with [+round] are a subset of sounds with [+labial]. [m], [p], and [b] are with [+labial] (bilabial) but with [-round]. In addition to bilabial consonants, labial-dental consonants [f] and [v] are also with feature [+labial].

2. Coronal and Noncoronal [± coronal]: Coronal sounds are formed using the tongue tip or blade as the articulator. To produce a coronal consonant sound, the tip or blade is raised towards the front teeth, alveolar ridge, or the hard palate. Sounds with [-coronal] are produced with the tongue tip or blade in a neutral position. All dental, alveolar, alveo-palatal, retroflex, and palatal consonants are with feature [+coronal].

3. Anterior and Nonanterior [± anterior]: For producing anterior consonants, the main constriction of the air stream is at a point no further back in the mouth than the alveolar ridge. For example, anterior coronals are produced with a primary constriction at or in front of the alveolar ridge. All labial, dental, and alveolar sounds are with [+anterior], and all other consonants with [-anterior].

4. Distributed and Nondistributed [± distributed]: Distributed sounds are produced with the main constriction extending over a considerable area along the middle-line of the oral tract and there is a large area of contact between the articulators.

Tongue-body features

1. Dorsal and Nondorsal [± dorsal]: Dorsal sounds are formed using the front or back of the tongue body as the articulator. In English, all vowels, glides, and three consonants [k], [g], [ng] are [+dorsal]. All other sounds are [-dorsal]. All the remaining features in this tongue-body feature category assume [+dorsal] and are used to provide finer classification of vowels/glides based on tongue body position, relative to the neutral position, during vowel production.

2. High and Nonhigh [± high]: High sounds are produced by raising the body of the tongue from the neutral position toward the palate. Nonhigh sounds are produced without such raising.

3. Low and Nonlow [± low]: Low sounds are produced with the body of the tongue drawn down away from the roof of the mouth.

4. Back and Nonback [± back]: Back sounds are produced with the tongue body relatively retracted.

Tongue-root features

1. Advanced-tongue-root and Non-advanced-tongue-root [± ATR]: Vowel sounds with feature [+ATR] are produced by drawing the root of the tongue forward, lowering and backing the tongue body as well. This expands the resonating chamber of the pharynx. The vowels [iy], [ey], [aa], [o], and [uw] are [+ATR], and other vowels are [-ATR].

2. Tense and Lax [± tense]: Vowel sounds with feature [+tense] are produced with the tongue root configuration having a greater degree of constriction than that in their lax counterparts with [-tense]. All English "long" vowels and diphthongs are [+tense], and "short" vowels [-tense].

Prosodic features

1. Long and Short [± long]: This feature distinguishes English "long" vowels versus "short" vowels.

2. Stress [± stress]: This feature refers to the prominence of the syllables, which is due to one or more of the following phonetic properties: raised pitch, longer duration, and greater loudness resulting from increased signal intensity.

3. Tone [± high] [± mid] [± low] [± rising] [± falling] [± fall-rise] : In tonal languages such as Mandarin, the several binary tone features use the differences in pitch and pitch contours to distinguish phonemically between word meanings.

8.3.3 Natural classes

Given the knowledge of features outlined above, we can now come to a deeper understanding as to why the feature approach is superior to the phonemic approach in a variety of issues such as phonetic similarity. We first introduce the notion of **natural class**. A set of sound segments constitutes a natural class if fewer features are needed to specify the set as a whole than to specify any single member of the set. We then can define allophones of a given phoneme as those sounds segments that occur between members of a natural class, without having to reply on the ambiguous concept of phonetic similarity.

For example, all the allophones of phoneme /l/, including unvoiced, dental syllabic allophones, belong to the natural class characterized by the minimum feature set [+lateral]. Likewise, all allophones of phoneme /t/, including aspirated, unaspirated, released, unreleased, dental, and glottal allophones, form the natural class characterized by the minimal feature set [+coronal, -cont, -voice].

We now can also succinctly account for why the sounds segments [h] and [ng] are not allophones of a phoneme. Since [h] is characterized by the laryngeal feature [+spread] and [ng] by the completely feature set [+dorsal, +nasal], no fewer features can specify both [h] and [ng] than to specify either [h] or [ng]. The required feature set for specifying both would be [+cons, -lateral, -constr]. Therefore, no natural class exists for [h] and [ng] and hence they cannot be allophones of a phoneme.

8.4 Phonological Rules Expressed by Features

Equipped with the feature representation of speech segments, we now discuss how to go from the baseform of phonemes to the surface-form of allophones in words and in word sequences. The baseform of each phoneme is assumed to have been stored in the speaker's mental lexicon as the knowledge source. The relationship between the phonemic representations and the allophonic representations that reflect the pronunciation of the words and word sequences is governed by "rules." These **phonological rules** relate the minimally specified phonemic representation to the allophonic representation and are also part of the speaker's knowledge of language. The phonemic representations of words in the mental grammar are minimally specified because the predictable features or their values are not included. This shows the redundancy of such features.

The phonemic representation need only include the unpredictable distinctive features of the phoneme sequence that represent words. The allophonic representations derived by applying the phonological rules include all the linguistically (contrastively) relevant phonetic aspects of the sounds. It may not include all the physical properties of the sounds of an utterance, since the physical signal of speech may vary in many ways that are not relevant to the phonological system. The result of applying the phonological rules to a phoneme sequence is an allophonic sequence (surface form), which is also called the phonetic transcription or "pronunciation." The phonetic transcription is also an abstraction from the physical signal, but it is closer to the physical signal than the phonemic representation. The phonetic transcription includes the nonvariant phonetic aspects of the utterances; i.e., those features that remain relatively the same from speaker to speaker and from one time to another.

8.4.1 Formalization of phonological rules

We now discuss the formalization of feature-based phonological rules that derive the phonetic transcription from the phonemic representation. This formalization constitutes the core of the generative phonology established in SPE.

Each phonological rule requires three types of information: 1) segment to be modified; 2) phonetic change; and 3) phonemic environment specified by features. This gives:

The standard form of a phonological rule:

$$S \rightarrow T \;/\; C \,\_\_\, D; \tag{8.1}$$

Source segment S becomes target segment T in the environment where S is preceded by C and followed by D.

The three types of information are S, change from S to T, and the environment indicated by / and by __, which is placed before or after the relevant segments that determine the phonetic change.

Using this standard form, the well-known nasalization rule that "A vowel (V) becomes nasalized in the environment before a nasal segment" can be written as

$$V \rightarrow [nasal] \;/\; \_\_ \; [+nasal] \tag{8.2}$$

Another well-known phonological rule, called the **aspiration rule**, can be written as

$$\begin{bmatrix} -continuant \\ -voiced \end{bmatrix} \rightarrow [+spread]/\$ \,\_\_\, \begin{bmatrix} -cons \\ +stress \end{bmatrix} \tag{8.3}$$

where the sign $ stands for a syllable boundary. This rule says that all voiceless stops, which are specified by features [-continuant, -voiced], change to aspirated ones (with the added feature [+spread]) if the following vowel is stressed and if the stop starts the syllable. These conditions preclude the application of the rule for words "compass" and "display."

Below we summarize some common phonological rule types that occur in the world's languages using the above rule formalization.

8.4.2 Common phonological rule types

Assimilation rules

Assimilation is the phonological process where one or more features of a segment change their values to match those of a neighboring segment. Such a feature change is also called feature "spreading." The above vowel nasalization rule in (8.2) is an example of assimilation rules as the vowel acquires the feature [+nasal] from the adjacent consonant. Some of the assimilations are pre-planned linguistically, but most of them are caused by articulatory processes. When we speak, there is tendency to make it easier to move articulators, a general phenomenon called **ease of articulation**.

Assimilation rules are phonological, but they reflect the phonetic process of coarticulation that we discussed in the preceding chapter. Coarticulation creates "sloppiness" in speaking while facilitating ease of articulation. When this tendency becomes regularized, it becomes a part of the assimilation rules of the language.

Another example of assimilation rules is vowel devoicing in Japanese when the vowel is adjacent both left and right to voiceless non-sonorant (stop or fricative) consonants:

$$\begin{bmatrix} -cons \\ +syl \end{bmatrix} \rightarrow [-voice]/ \begin{bmatrix} -son \\ -voice \end{bmatrix} \text{ — } \begin{bmatrix} -son \\ -voice \end{bmatrix} \tag{8.4}$$

This rule makes the vowel /u/ in "sukiyaki" realized as a voiceless allophone.

Other examples of assimilation rules in English are those that devoice the nasals and liquids in the environment of prior adjacent voiceless non-sonorant consonants. The rules make the nasal consonants in "smart" and "snow" voiceless, as well as the liquids in "slow" and "price." The phonetic basis of these rules is that the vocal cords become sluggish in changing to the vibration mode because voiceless nasals and liquids do not contrast with their voiced counterparts.

Additional examples of assimilation rules in English are the change of the place-of-articulation feature for coronal sounds. These rules render the /n/ in "in" of the following phrases as [m], [n], and [ng] respectively: "in position," "in decision" and "in complete." This phenomenon is often called **homorganic nasal assimilation**. The well-known **palatalization rule** also belongs to this category of assimilation rules: /s/ is realized as [sh] in "his shoes," "gas shortage," "miss you," etc.

Dissimilation rules

Dissimilation is the phonological process where one or more features of a segment change their values to differentiate it from a neighboring segment. Contrary to the assimilation rules which permit a greater degree of ease of articulation on the part of speaker, dissimilation rules inform the listener about a greater degree of sound contrast.

Dissimilation rules are rare. An example is the derivational suffix added to nouns to form adjectives. Consider the word pairs:

```
Nouns          Adjectives

culture          cultural
orbit            orbital
region           regional
electric         electrical
digit            digital
...              ...
```

It is clear that the suffix *-al* is the baseform added to the nouns to form the adjectives. Now, for a noun that ends with /l/ such as "table," use of this baseform would result in a perceptually low contrast between the noun and the corresponding adjective. After applying dissimilation rules, the contrast is greatly enhanced as shown in the following word pairs:

Nouns	Adjectives
angle	angular
table	tabular
title	titular
single	singular
circle	circular
...	...

Insertion and deletion rules

The aspiration rule of English formalized in (8.3) inserts (i.e., adds) a new non-distinctive feature [+spread]. Phonological rules can also insert a whole segment instead of just one feature. For example, in Spanish, [e] is inserted by a rule at the beginning of a word that would otherwise begin with an [s] followed by another consonant. This rule creates the phonetic forms of "*escuela*," (school), "*España*," (Spain), and "*estampa*" (stamp) from the phonemic representations that begin with /sk/, /sp/, and /st/.

A common **epenthetic stop insertion rule** states that between a coronal nasal and fricative is often inserted an alveolar stop consonant. Thus, "prince" is pronounced identically with "prints." They have the same phonetic realization with different phonemic representations.

A very well-known example for deletion rules is the French one where word-final consonants are deleted when the following word begins with a consonant (including liquid), but are retained when the following word begins with a vowel or a glide. This is formalized as

$$[+cons] \rightarrow \emptyset \;\; / \;\; \_\_ \, \# \;\; \# \, [+cons] \tag{8.5}$$

where the symbol $\emptyset$ denotes the null unit, and # denotes word boundary. The deletion rule in (8.5) can be stated as: A consonant segment (first [+*cons*]) becomes null ($\rightarrow \emptyset$) in the environment at the end of a word (___ #) followed by a word beginning with a consonant (# [+*cons*]).

Another common example of deletion rules comes from casual speech in English. The rule deletes the second vowel in "he is" to create the contracted form of "he's." The deletion rule also creates the contracted form of "I'll" from "I will."

Deletion of schwa in the following English words in casual speech is also an example of deletion rules:

*myst**e**ry*, *mem**o**ry*, *vig**o**rous*, *Barb**a**ra*, *gen**e**ral*, etc.

Other rules

Besides the common classes of phonological rules discussed above, there are some individual and less common rules. Lenition rules weaken a segment, along the scale of *voiceless* > *voiced* or the scale of *stop* > *fricative* > *liquid* > *glide*. Compensatory-lengthening rules increase the duration of a segment in response to deletion of the segment following it. Metathesis rules create the reversal of two neighboring segments.

In English, we have the following **flap rule**:

$$\begin{bmatrix} -continuant \\ -nasal \\ +coronal \end{bmatrix} \rightarrow [+son] / \begin{bmatrix} -cons \\ +stress \end{bmatrix} \_\_ \begin{bmatrix} -cons \\ -stress \end{bmatrix} \tag{8.6}$$

This can be translated to: A coronal ([$+coronal$]) stop ([$-continuant, -nasal$]) is changed to a voiced flap if preceded by a stressed vowel ([$-cons, +stress$]) and followed by an unstressed vowel ([$-cons, -stress$]).

Another individual vowel-lengthening rule in English says that the vowel has a greater duration when followed by a voiced consonant than when followed by a voiceless consonant.

8.5 Feature Geometry — Internal Organization of Speech Sounds

8.5.1 Introduction

The discussion in the preceding section illustrates that the basic units of phonological representation are not phonemes but features. Features play the same linguistically meaningful distinctive role as phonemes, but the use of features offers straightforward accounts for many potentially unrelated observations not explicable by phoneme theory. For example, feature theory explains that the world's languages draw on a small set of speech properties in constructing their phonological systems and share a large portion of common elements that are finer than phonemes. Feature theory also explains that speech sounds are perceived and stored in memory in a categorical manner, because the features that constitute the basic units of speech are binary and their combinations must be discretely valued thereof.

We also saw that phonological processes and rules operate on feature representations, rather than on phoneme representations. Therefore, the rules are typically applied to "natural classes" of sounds that can be uniquely defined in terms of the conjunction of a minimal number of features.

Most importantly, feature theory in generative phonology outlined in this section provides support to the view that sound systems of language reflect a single general pattern rooted in the physical and cognitive capacities of the human beings. It is fair to say that feature theory is one of the major research outcomes of all of linguistic science.

The traditional feature theory discussed so far has addressed questions of what the features are, how they are defined, and what operations can apply to them as phonological rules. More recent advances in feature theory have moved further toward answering the more important question: How are features internally organized in providing phonological representations of speech sounds? This forms the core subject of the feature-geometry theory, which we outline now.

8.5.2 From linear phonology to nonlinear phonology

As discussed earlier, a phoneme is represented as an unstructured set of features, or **feature bundles**. Likewise, a sequence of phonemes is characterized by a sequence of feature bundles, resulting in a **feature matrix** which arranges the feature bundles into columns. The feature matrix does not concern how features might be organized or structured. Because phonemes as feature bundles in a word or in word sequences follow each other in strict succession, this feature-matrix approach is called the sequential linear model of phonology. In this regard, the speech units in the "linear" order are often likened to "beads on a string." Note here the term "linear" refers to the strict sequential order, instead of the superposition property of "linear" systems discussed in Chapter 1.

While this sequential model of phonological representation is conceptually simple and analytically tractable, it has a number of serious inadequacies. All features are assumed to have a positive or negative value for every segment, regardless of whether such features could conceivably apply to that segment or not. It also makes no sense, and serves no practical purpose, to denote labial consonants as [-lateral] or vowels as [-distributed, -strident]. Further, the phonological rules in linear phonology do not explain why some phonological processes are more natural than others. Also, many phonological rules can become so complex that they are almost incomprehensible (such as the stress rule in SPE).

The most important inadequacy of the linear phonological model is that it prevents features from extending over domains greater or lesser than one single phoneme, because each feature value can characterize only one phoneme and vice versa. This is contrary to ample phonological evidence that demonstrates "nonlinear" behavior, where strict sequential order is broken and one feature can occupy a domain significantly greater than a phoneme, or a domain less than a full phoneme. The clearest examples come from tone features in tone languages, where two or more tones may be forced into a single syllable, forming contour tones (e.g., the "third" tone in Mandarin Chinese). Segmental properties also have this type of "nonlinear" behavior. For example, the [+nasal] feature in some languages may occupy only a fraction of a segment, or it can spread across more than one segment or syllable. In languages that exhibit vowel harmony (such as Turkish) (see Section 8.5.4), tongue-body and tongue-root features [+back], [+round], and [+ATR] often extend across many syllables. This type of inadequacy of the linear phonology model has been overcome by the theory of autosegmental phonology where the features that go beyond the segmental limits set by the linear model are extracted from feature matrices and are placed in separate, independent tiers of their own, hence the term **autosegmental**. In autosegmental phonology, tones are represented in a separate tier from vowel and consonant segments. Elements in the same tier are still sequentially ordered, but elements on different tiers are unordered and are related to each other via association lines exhibiting patterns of overlap and alignment across tiers. Autosegmental phonology establishes a "nonlinear" model of phonological representation, where the strict "linear" order is replaced by a multi-tiered representation where feature elements in different tiers often do not follow the linear order but overlap with each other temporally.

The second inadequacy of the linear sequential model of phonological representation is its implicit assumption that feature bundles in the feature matrix have no internal structure; i.e., each feature is equally related to any other feature. This, again, is against a considerable amount of evidence that suggests that features are grouped into higher-level functional units. In many languages including English, all place features function together as a unit. For example, in the traditional linear model of phonology, three separate rules are required to capture place assimilation for "in + position," "in + decision" and "in + complete." However, when the place features are grouped as one larger functional unit, then these separate rules become essentially one rule for place assimilation. This suggests that all features defining place of articulation have a special status in phonological representation and that this grouping relationship constitutes some structure among the features. To overcome this challenge, a tree-like model of feature organization is developed, where segments are represented in terms of hierarchically-organized node configurations with terminal nodes being the feature values and non-terminal nodes being the feature classes resulting from functional feature groupings. This tree-like feature organization, together with a set of general properties associated with the organization and related to phonological rules, is also called feature geometry.

Now, rather than putting features in matrices as in the traditional theory, the features are placed as terminal nodes in a tree-like diagram, where these terminal nodes are unordered and are on separate tiers depending on their parent feature classes. This organization permits nonlinear behavior of feature overlap, as in autosegmental phonology. It also permits strong constraints on the form and functioning of phonological rules.

Feature geometry is a substantial extension of autosegmental phonology, principally in the segmental, rather than the prosodic, aspect. It is one very well established formal model of nonlinear phonology.

8.5.3 Feature hierarchy

We now formally describe how the features are hierarchically organized into a tree-like structure in feature geometry theory. Compared with traditional SPE feature theory, feature geometry is more heavily based on articulators and their functional roles. Central to feature geometry theory is the idea that speech is produced using several independently controlled and functioning articulators, which comprise the lips, the front part of the tongue (tip and blade), the tongue dorsum (body and root), the soft palate (velum), and the larynx. The articulators determine one or more constrictions in the vocal tract, which shape the speech acoustics in a principled way. It is also well known that the articulators follow a hierarchically tiered structure that plays a fundamental role in the organization of segments. Combining these facts, a new phonological representation emerges, building on a tree-like structure consisting of the articulatory nodes associated with their naturally occurring tiers. Fig. 8.1 provides an overall look at the nodes and tiers for this type of tree-like phonological representation for one segment.

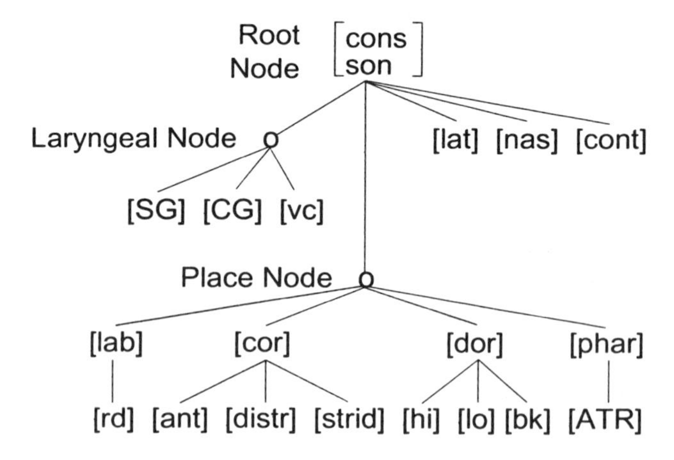

Figure 8.1: Feature geometry expressed as a tree-like structure (with nodes and tiers) for one segment.

The root node

In the **root node** (top of Fig. 8.1), which dominates all features, of this articulator-based feature hierarchy, the coherence of the global segmental properties is expressed. Popular proposals assign features [cons] and [son] in this root node.

The laryngeal node

The tier below the root node in the hierarchical tree contains a non-terminal node called the **laryngeal node**. This is shown as the left branch of the root node in Fig. 8.1. Three laryngeal features [spread] (i.e., spread glottis, SG), [constr] (or constricted glottis, CG), and [voice] (vc) are grouped under the laryngeal node. The motivation for this grouping is not only that the common active articulator responsible for implementing these three features is the larynx, but also that these features in many languages may spread or stop spreading often as a unit.

The place node

We have seen in the previous description of the assimilation rules that the place features [labial], [coronal] (together with its dependent features [anterior], [distributed], [strident]), and [dorsal] (together with its dependent features [high], [low], and [back]) often spread as a unit. This spread is independent of other non-place features such as [cont], [voice], [nasal], etc. This regularity is naturally captured by grouping all these place features under a single place node as shown in the lower part of Fig. 8.1.

The remaining features [lateral], [nasal], and [cont] do not form sub-groups within themselves or with other features. They are listed separately under the root node, as shown at the right side of Fig. 8.1.

It should be noted that while Fig. 8.1 is one very common feature hierarchy, there are alternative proposals for the hierarchy, some of which are far more complex and detailed than Fig. 8.1. See a review in [Clem 95].

When several segments form a sequence, each of which has its feature hierarchy as shown in Fig. 8.1, we obtain a three-dimensional picture where the feature hierarchy unfolds in time. This picture as an expanded structure is schematized in Fig. 8.2 where each sectional cut gives the two-dimensional hierarchy of Fig. 8.1, and where RN, LN, and PN denotes Root Node, Laryngeal Node, and Place Node, respectively. While the root node dominates all features for each segment as shown in Fig. 8.1, for a sequence of segments, all the individual root nodes are linked in a sequence as well, forming the now top tier which is often referred to as the CV (Consonant-Vowel) tier.

One example of the expanded feature geometry for a three-segment sequence, consisting of /p/, /i/, and /n/, is crafted in Fig. 8.3.

8.5.4 Phonological rules in feature geometry

In the feature geometry approach where the feature organization is made explicit in a tree-like structure, the phonological rules are expressed in a very different way from that in the traditional feature-matrix approach discussed earlier. Explicit feature organization makes it possible to impose strong and sometimes more flexible constraints in the form and functioning of phonological rules. This has been either impossible, difficult, or unnatural to accomplish in the traditional linear model of phonology. In addition, as is pointed out in [Chom 68], traditional phonology has not been able to provide an intrinsic way of distinguishing plausible, cross-linguistically attested rules from highly

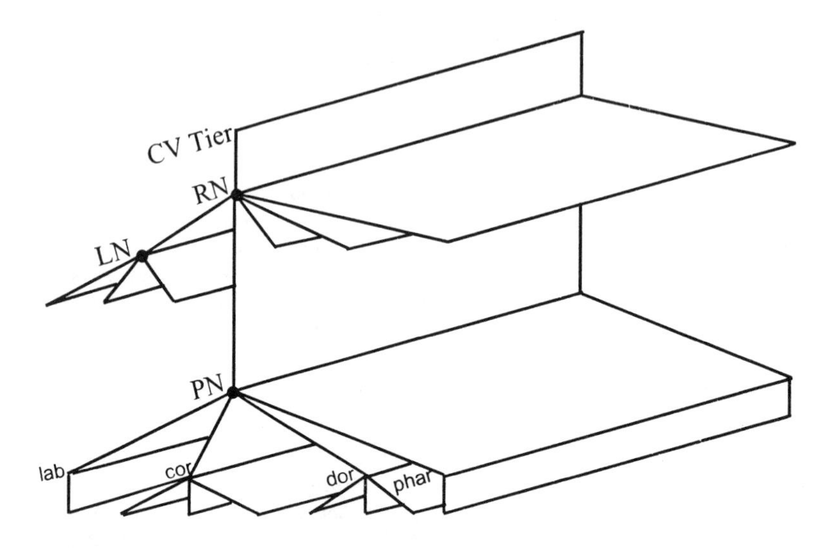

Figure 8.2: Expanding feature geometry along the time direction for a sequence of segments. The CV tier is added at the top to inform the consonant/vowel constituents.

improbable ones. This difficulty can be naturally overcome using the hierarchical feature organization. Armed with this new, powerful representation, a small number of elementary phonological rule types can be projected to a large class of "natural" rule types. They can also simultaneously exclude rare and unnatural rule types.

In formulating phonological rules, feature geometry theory relies heavily on the hierarchical structure of feature grouping, and is based on some new principles. First, the phonological rules perform only single operations on the tree-like feature structures across different segments. Capitalizing on the feature grouping, this principle claims that only feature sets that form a group under a common non-terminal node may function together in phonological rules. Such feature sets form a very small portion of all the logical possibilities, this claim provides a strong constraint on the nature and the possible classes of phonological rules. Second, feature organization is universally determined, independently of languages, thereby offering cross-linguistic predictions. This universality is not affected by phonological rules, and is thus common for both the underlying base-form feature structure and the surface-form structure. Hence there comes the notion of well-formedness, which all steps in the phonological rule applications (i.e., derivations) must obey.

We now take up the issue of assimilation again, which was discussed earlier as a very common type of phonological rule in traditional feature theory, but from the new perspective of feature geometry. Assimilation was characterized in the linear, feature-matrix approach in terms of feature copying; that is, one segment copies or spreads feature specifications from its adjacent segment. In the new, feature geometric approach, however, assimilation rules are now characterized by the association (spreading) of a feature (at the terminal node of the tree) or a non-terminal node to that of an adjacent

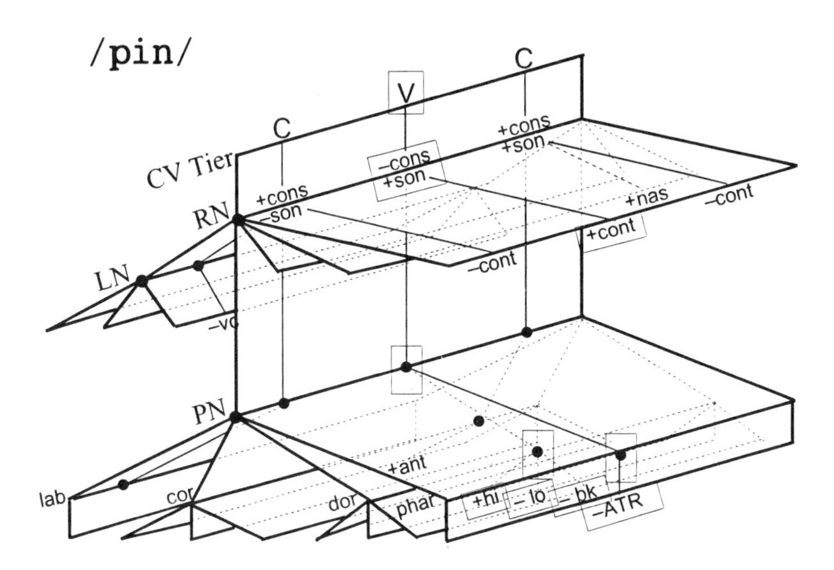

Figure 8.3: Example of the expanded feature geometry for the segment sequence /pin/, where the CV tier takes the values of C, V, and C.

segment.

In this new view of assimilation, there can be classified two main types of assimilation rules depending on the identity of the spreading node. If the node spreads at a level lower than the root node, then the target segment that is affected by the assimilation rule will acquire one or more, but not all, features of the trigger, source segment. This type of assimilation rule is called a partial or incomplete assimilation rule. Fig. 8.4 shows one example of the partial assimilation rule, spreading feature [+labial] leftward from segment /p/ to /n/, in the word "impose." Feature hierarchies for /p/ and /n/ are shown in full. The assimilation rule is characterized as a dashed association (spreading or linking) line, which is accompanied by a simultaneous delinking (cut) line from the root node to the place note in segment /n/. These simple tree operations accomplish the labial assimilation rule.

If the root node spreads, the affected, target segment will acquire all the features of the triggering segment. This rule type is called complete or total assimilation. Fig. 8.5 gives one example of total assimilation where all the features in the target, underlying segment /n/ are completely replaced by those in the neighboring triggering segment /l/ in the word "illicit" (/I n - l I s I t/ – > [I l - l I s I t]). The total assimilation rule in this case is expressed simply by the dashed association or linking line from the root node of the triggering segment /l/ to the root node of the target segment /n/, while delinking all the children nodes and features under the latter root node, both shown in Fig. 8.5. This linking-delinking operation at the very top root-node tier naturally replaces all features from the triggering segment to the target one.

Fig. 8.6 provides another interesting example of partial place assimilation for a well known phonological phenomenon called vowel harmony in languages such as Turkish.

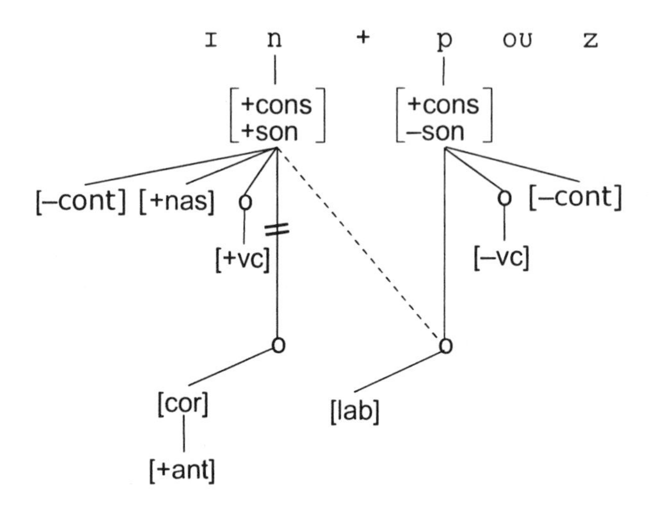

Figure 8.4: Phonological rule of partial assimilation represented as a linking-delinking operation in a lower tier of the feature trees. The affected, target segment /n/ is phonetically realized with the added [+labial] feature that is spread from the following. trigger segment /p/. The assimilation is called partial because the target segment still retains the original features [+nasal], [+cons], and [+son].

Vowel harmony is a regular pattern in these languages that all surface-form vowels in a word have the same [back] feature and the same labial [round] feature. Partial assimilation is responsible for creating this "harmonized" vowel pattern from the baseform that is often inconsistent with this pattern (this situation occurs when a suffix is added to a root word). Using feature geometry, the assimilation rule creating vowel harmony can be most succinctly represented as the two association lines shown in Fig. 8.6 between place features that are across the vowels in the adjacent syllables. The fundamental hypothesis in linear phonology that speech can be segmented into a linear sequence of discrete events would break down here.

8.6 Articulatory Phonology

In the preceding sections of this chapter, we have provided introductions to a number of phonology theories, ranging from traditional phoneme theory and distinctive feature theory (linear phonology) to more recent autosegmental theory and feature geometry theory (nonlinear phonology). All these theories claim a strong division between phonology and phonetics, in terms of the objectives, the domains of applications, and of the representational frameworks. In this section, we provide an introduction to another recent nonlinear phonology, which advocates strong integration of phonology and phonetics on the basis of dynamic articulatory representation. The introduction to the articulatory phonology presented in this section is a summary of the original literature on this new

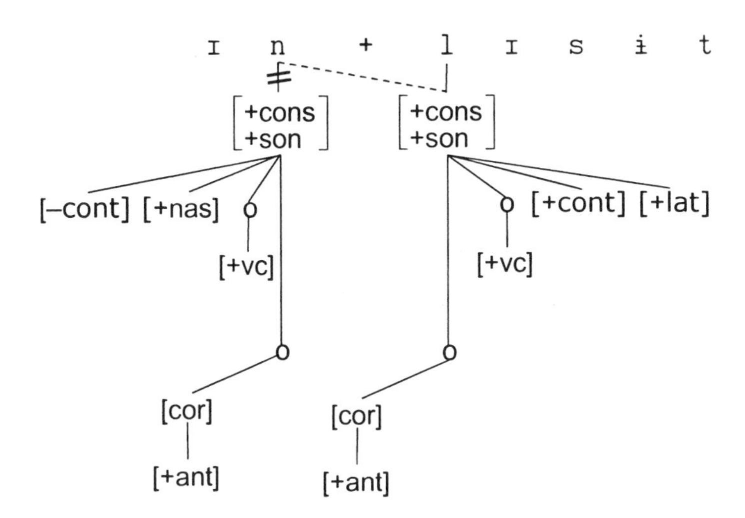

Figure 8.5: Phonological rule of full assimilation represented as a linking-delinking operation in the top tier of the feature trees. /n/ is phonetically realized as /l/ with all its features spread from the following segment /l/.

theory that appeared in 1980's and 1990's [Brow 86, Brow 90, Brow 92].

Speech has been treated in traditional phonology theories as consisting of two structures, one physical or phonetic, and the other cognitive or linguistic. The relationship between the two structures is generally not an intrinsic part of either phonetic or linguistic description. To form a complete account of speech, one needs to translate between the two intrinsically different domains. This has been a key issue at the heart of many phonetic and phonology theories, an issue popularly called the "interface between phonology and phonetics."

Articulatory phonology aims to bridge the boundary between phonology and phonetics, amounting to eliminating the need for their interface. This is a drastic change and it begins with the following very different assumption from traditional phonology theories. That is, the two apparently different physical and cognitive domains of speech are simply the high- and low-dimensional descriptions of the same complex system. Crucial to this approach is identification of basic phonological units, not with the traditional, purely symbolic distinctive features, but with a new set of dynamically specified units of articulatory action, called **articulatory gestures** or vocal tract gestures. We begin the introduction of articulatory phonology below with a discussion on this essential notion of articulatory gesture.

8.6.1 Articulatory gestures and task dynamics

In contrast to defining distinctive features as the basic, "atomic" phonological units in all types of feature theory, articulatory phonology proposes articulatory gestures, or gestures for short, as the atomic phonological units. Articulatory gestures are primitive

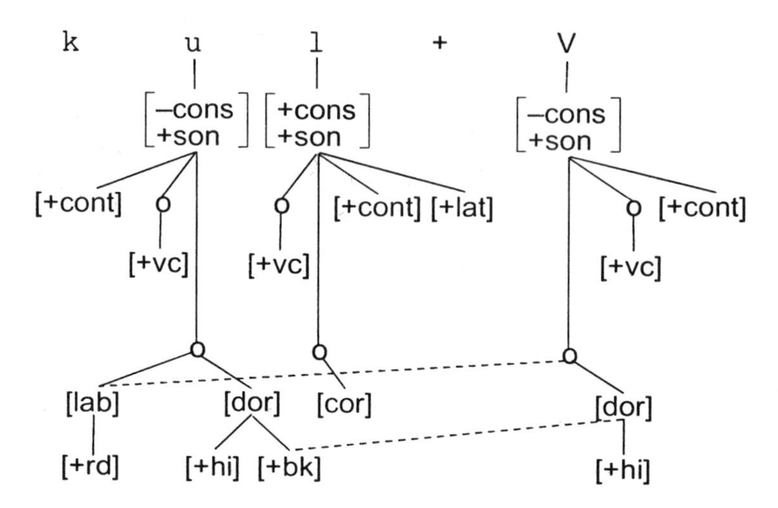

Figure 8.6: Vowel harmony as partial place assimilation.

articulatory actions in the vocal tract. A gesture is associated with two key defining elements. First, a gesture is identified with the formation and release of a characteristic constriction within one of the five relatively independent articulatory subsystems of the vocal tract: three oral subsystems consisting of lips, tongue front, and tongue body, plus the laryngeal and velic subsystems. This notion of independent subsystems (which evolve independently over time) is analogous to the independent tiers in autosegmental phonology, and the classification of the subsystems into oral and non-oral ones is reminiscent of feature geometry. Second, a gesture, as an action, has some intrinsic "time" associated with it. A gesture is a characterization of a specific type of movement through space and over time.

According to articulatory phonology, a speech utterance is described as an abstract action that can be decomposed into a small number of primitive units of articulatory gestures in a particular spatio-temporal configuration. Since these gestures are defined on vocal tract subsystem constrictions, rather than on the articulators themselves, this phonological description of the utterance as an action is a low-dimensional description. Importantly, this same description also provides an intrinsic specification of the high-dimensional properties of the action, which entail the various bio-mechanical consequences of the action as the physically measurable articulators' movements. This equivalence of descriptions in both the low-dimensional, phonological "task" space and in the high dimensional, phonetic "physical articulatory" space is accomplished via a nonlinear transformation, which we will later discuss, from the "task variables" to articulatory variables. A tight integration between phonology and phonetics stems from this equivalent description of the speech production actions in both phonological and phonetic spaces.

Since gestures as the units of speech production are actions, they are dynamic rather than static. Moreover, articulatory phonology considers phonological functions, such as contrast as in any phonological theory, to be low-dimensional descriptions of such actions in the "task" coordinate of vocal tract constriction that induces articulatory movements. Therefore, the gestural units are not neutral between articulation and acoustics, but rather are articulatory in nature.

A gesture in articulatory phonology is specified using a set of related **tract variables** in the low-dimensional **task space**. Each gesture is precisely defined in terms of the parameters of a set of equations for a **task-dynamic model** that was documented in detail in [Salt 89]. Briefly, this is a nonlinear dynamic system consisting of two components. The first component is a target-directed, vector-valued, linear state equation characterized by the "target" parameters and by the "stiffness" parameters related to the "time constants" of the dynamics. The vector in this state equation, which spans the task space, is a set of eight low-dimensional "tract variables," also called the task variables. The "target" parameters are in this same task space. These eight tract variables, as in most articulatory literature, are listed in the left column of Table 8.1. They are essentially the vocal tract constriction degrees and locations created by a set of principal articulators: Lips, Tongue Tip/Blade, Tongue Body, Velum, and Glottis. Since the constriction locations of both velum and glottis, relative to the whole length of the vocal tract, are largely fixed (i.e., much less variable than other articulators) during speech production and they are not responsible for any phonological contrast, only the constriction degree (aperture) is used as the tract variable for the velum or glottis.

Tract variables	Corresponding articulatory variables
LP (lip protrusion)	upper-lip(x), lower-lip(x), jaw(x, y)
LA (lip aperture)	upper-lip(y), lower-lip(y), jaw(x, y)
TTCL (tongue-tip constriction location)	tongue-tip(x, y), tongue-body(x, y), jaw(x, y)
TTCD (tongue-tip constriction degree)	tongue-tip(x, y), tongue-body(x, y), jaw(x, y)
TBCL (tongue-body constriction location)	tongue-body(x, y), jaw(x, y)
TBCD (tongue-body constriction degree)	tongue-body(x, y), jaw(x, y)
VEL (velic aperture)	velum(y)
GLO (glottis aperture)	glottis(x)

Table 8.1: Eight tract variables and related articulatory variables in the task dynamic model. The distinct parameter set of the task dynamic model completely specifies an articulatory gesture.

The second component in the task-dynamic model is a static, nonlinear mapping from the low-dimensional task variables to the high-dimensional articulatory variables. The latter are in the physical, phonetic space of speech production, which can be measured experimentally. There is some freedom in the choice of articulatory variables, as long as they can satisfactorily produce realistic vocal tract area functions in speech production. Phonetic studies have shown that it is not necessary to use a large number of points for all the articulators to form a complete articulatory vector. These points can be analyzed and some principal components extracted to represent all the points rather faithfully. In articulatory phonology and task dynamics literature, twelve articulatory variables are typically used — normalized physical positions in the x and y coordinates for the five articulators of the upper lip, lower lip, jaw, tongue tip, and tongue body, plus the apertures of the velum and glottis. Importantly, in the mapping from task variables to articulatory variables, this is a highly constrained, one-to-many relationship. For example, the lip-protrusion LP tract variable is mapped to four articulatory variables:

the positions of the upper and lower lips in the x coordinate, and the jaw positions in both x and y coordinates. It will not be mapped to the remaining articulatory variables. The right column of Table 8.1 lists the subset of articulatory variables being mapped to by each of the eight tract variables. Each of the Tongue-Tip related tract variables (TTCL and TTCD) has six articulatory variables associated with it, including tongue-body ones. This is because physical movements of the tongue body are required to support tongue tip movements in order to create the constriction made with the tongue tip. It is important to distinguish the Tongue-Tip tract variables as those that form a vocal tract constriction and the physical tongue tip as an articulator.

Existing publications on articulatory phonology have not provided full sets of gestures for particular languages. However, based on the notion of articulatory gestures defined above, it is not difficult to construct the gesture inventory for each individual language. For English, the size of such an inventory is roughly on the order of 20, smaller than the inventory size of distinctive features and that of phonemes. In Table 8.2, we list 13 consonantal gestures in English. They include: 1) three closure gestures, which are complete constrictions in the vocal tract (zero value in constriction degree) as the movement targets in the task dynamic model, 2) four incomplete constriction gestures corresponding to four types of English fricative place of articulation, 3) one lip rounding gesture, 4) two gestures for liquids, 5) one gesture for nasality, and 6) two gestures for the glottis. The related tract variables and the articulatory variables for each of these consonantal gestures are also shown in Table 8.2. The English phonemes that typically contain each of the gestures are listed in the final column.

Conson. Gesture	Tract Variables	Articulatory Variables	Found in phonemes
labial-closure	LA	upper-lip, lower-lip, jaw	/p/,/b/,/m/
alveolar-closure	TTCL, TTCD	tongue-tip, tongue-body, jaw	/t/,/d/,/n/
velar-closure	TBCL, TBCD	tongue-body, jaw	/k/,/g/,/ng/
lab.-dent.-constr.	LA	upper-lip, lower-lip, jaw	/f/,/v/
dental-constr.	TTCL,TTCD	tongue-tip, tongue-body, jaw	/th/,/dh/
alveolar-constr.	TTCL,TTCD	tongue-tip, tongue-body, jaw	/s/,/z/
alv.-pal.-constr.	TTCL,TTCD	ton.-tip/body, upp./low.-lips	/sh/,/zh/
lip-rounding	LP, LA	upper-lip, lower-lip	/w/ (/uw,uh,o,ao/)
laterality	TTCL, TTCD	tongue-tip, tongue-body	/l/
rhoticity	TTCL, TTCD	tongue-tip, tongue-body	/r/,/er/
nasality	VEL	velum	/m/,/n/,/ng/
voicing	GLO	glottis	/m,n,ng,b,d,g,w,j,l,r/(V)
aspiration	GLO	glottis	/p/,/t/,/k/,/h/

Table 8.2: English consonantal gestures as primitive speech units in articulatory phonology, and as tract-variable constriction/closure targets in the task dynamics

There are about the same number of vowel gestures in English as for consonants. They are listed in Table 8.3, denoted by the corresponding vowel name. Note all vowel gestures are associated with tongue-body related tract variables (TBCL, TBCD), which are in turn associated with tongue body and jaw articulatory variables. Also note that these vowel gestures are not further divided in terms of the feature categories such as [high], [low], and [back]. Because of the relative independence among the variable sets of tract variables, their temporally independent organization that follows the same principle in autosegmental phonology as in articulatory phonology permits a high degree of flexibility to account for many phonological phenomena using these vowel gestures denoted simply by the vowel name.

Vowel gesture	Tract Variables	Articulatory Variables
gesture-iy	TBCL, TBCD	tongue-body, jaw
gesture-ih	TBCL, TBCD	tongue-body, jaw
gesture-eh	TBCL, TBCD	tongue-body, jaw
gesture-ae	TBCL, TBCD	tongue-body, jaw
gesture-aa	TBCL, TBCD	tongue-body, jaw
gesture-e	TBCL, TBCD	tongue-body, jaw
gesture-er	TBCL, TBCD	tongue-body, jaw
gesture-uw	TBCL, TBCD, LA, LP	tongue-body, upper-lip, lower-lip, jaw
gesture-uh	TBCL, TBCD, LA, LP	tongue-body, upper-lip, lower-lip, jaw
gesture-o	TBCL, TBCD, LA, LP	tongue-body, upper-lip, lower-lip, jaw
gesture-ao	TBCL, TBCD, LA, LP	tongue-body, upper-lip, lower-lip, jaw
gesture-j	TBCL, TBCD	tongue-body, jaw
gesture-w	TBCL, TBCD, LA, LP	tongue-body, upper-lip, lower-lip, jaw

Table 8.3: English vowel gestures in articulatory phonology. All vowel gestures are associated with tongue-body related tract variables (TBCL, TBCD) and tongue body and jaw articulatory variables.

While articulatory gestures as the underlying tasks of articulatory movements have been known and accepted for a long time, these gestures that are being endowed with a much more important status of primitive phonological units are established only in the relatively recent articulatory phonology. As a result, the analysis of gestures in terms of their contrastive roles in linguistics and in terms of their distributional properties (e.g., complementary distribution vs. free variation) has not been as comprehensive as compared with the analysis of the more traditional phonological units of phonemes and features. Some of this type of analysis, as well as a detailed account of how gestures are used to bridge phonology to phonetics, has been built upon a further concept in articulatory phonology related to the organization of gestures, which we discuss below.

8.6.2 Gestural scores

In articulatory phonology, the basic notion of articulatory gestures introduced above has been expanded to form a new but closely related concept — **gestural score**, also known as gestural constellation or gestural structure. A gestural score is a temporal pattern of organization of gestures for a specific speech utterance. It is often depicted as a two-dimensional graph, where the horizontal dimension is time, and the vertical dimension displays five sets of tract variables consisting of LIPS, TT(CD/CL), TB(CD/CL), VEL, and GLO. In this graph, the temporal pattern of gestural organization is shown as a number of boxes, one for each gesture that is present in the utterance. The vertical placement of the box for a gesture shows which out of five tract variable sets constitutes the gesture according to Tables 8.2 and 8.3. The horizontal length of the box indicates the interval of active control for the individual gesture. The absolute temporal starting point of each gesture depends on the rate of speech. However, the relative starting times of the gestures that are associated with different tract variable sets (sometimes called different articulatory tiers) depend only on the underlying linguistic structure (e.g., phonological contrast). This relative timing across tiers is also called the **phasing relationship**, which quantifies the degree of **intergestural overlapping**.

During the act of speaking, more than one gesture is activated, often in an overlapping fashion across different tract variable sets/tiers. Sometimes different gestures belonging to the same tier overlap in time; in this case, we say the two gestures are blended or mixed. This situation occurs mostly between two consecutive vowels, one in a word-final position and the other in word-initial position, in fast speech. We thus can also define a gestural score as a recurrent, regular pattern of gestural overlap (across tiers) and of blending (within the same tier).

An example of a gestural score for the English word "paw" is shown in Fig. 8.7, which provides the kind of information contained in the gestural structure. Each row, or tier, shows the gestures that control the distinct tract variable sets: LIPS, TT, TB, VEL, and GLO. The gestures are represented as boxes with descriptors taking the names of the gestures from Tables 8.2 and 8.3. Each gesture is assumed to be active for a fixed temporal activation interval. This is shown in Fig. 8.7 as the horizontal extent of each box. Note that there is substantial overlap among the gestures. This kind of overlap can result in certain types of context dependence in the articulatory trajectories of the invariantly specified gestures. In addition, overlap can cause the kinds of acoustic variation that have been traditionally described as allophonic variation. In this example, we see the substantial overlap between the velic lowering gesture called [nasality], and the TB [gesture-ao] for the vowel. This will result in an interval of time during which the velo-pharyngeal port is open and the vocal tract is in position for the vowel; this produces the nasalized vowel. In the traditional linear phonology based on the feature-matrix approach, the fact of nasalization has been represented by a rule that changes an oral vowel into a nasalized one before a (final) nasal consonant. However, in the current new view in terms of the gestural score, this nasalization is just the natural consequence of how the individual gestures are coordinated and organized. The vowel gesture itself has not changed in any way (it has the same specification in this word and in the word "pawed," which is not nasalized). Also, the overlap between the gesture [aspiration] and [gesture-ao] is responsible for the aspirational portion of the acoustics at the junction between [p] and [ao].

In Fig. 8.7, we also show some degree of "blending" of the [voicing] and [aspiration] gestures that occurs at the same GLO tier. In Fig. 8.8 is shown another, more complex example of gestural score in the phrase "ten themes." Note that the gesture blending in the TT tier is responsible for the phonetic realization of /n/ into the dental place of articulation (partial assimilation). There does not need any special phonological rule here. The gestural score automatically predicts this type of phenomena.

From the above examples, we see that one major difference between the gesture and the feature is that the former has intrinsic temporal structure which the latter does not. The gestures' temporal structure is reflected not only in the activation intervals (durations) of the individual gestures, but more importantly in the relative phasing relationship among these gestures expressed in terms of a gestural score. This phasing is typically over a fraction of the activation duration associated with a gesture; in contrast, in the feature approach, when a rule changes a feature, the change usually occurs over the entire segment. Articulatory phonology views this phasing relationship or overlapping pattern as part of phonology; that is, it forms phonological contrasts in meanings.

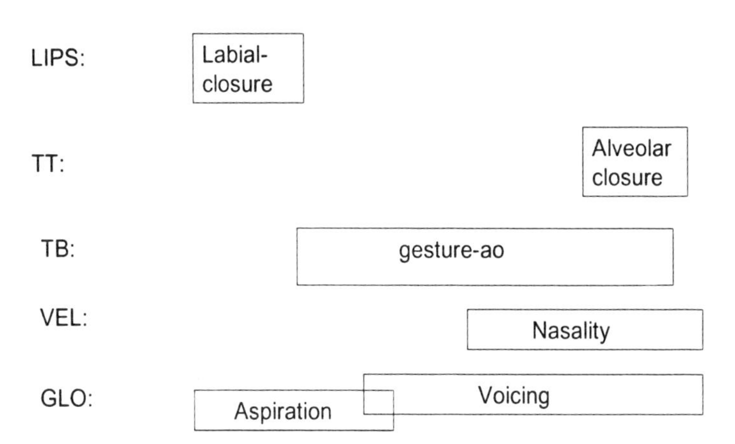

Figure 8.7: Gestural score for English word "pawn" /p ao n/.

8.6.3 Phonological contrast in articulatory phonology

The gestural score discussed above not only characterizes the microscopic properties of speech utterances, but also uses the systematic differences among the gestural scores to form phonological contrast in a language. Articulatory phonology claims that phonological contrast among utterances can be defined in terms of different gestural scores.

Given the nature of gestural scores, the set of possible ways in which they may differ from one another is quite constrained. Examples and analysis have been given in the articulatory phonology literature to show that gestural scores are suitable for characterizing the major phonological function of contrast in meaning [Brow 86, Brow 92] . Relation is also shown between the phonological structure implicit in gestural scores and the hierarchical structure in feature geometry [Clem 95]. Here we give one example for four ways in which the notion of phonological contrast is encoded in a system based on articulatory gestures, using the schematic gestural scores in Fig. 8.9 for eight contrasting words.

The simplest way in which gestural scores for phonological contrastive utterances may differ is in the presence vs. absence of a gesture. This kind of difference is illustrated by two pairs of subfigures in Fig. 8.9(a) vs. (b) and Fig. 8.9(b) vs. (d). The gestural score in Fig. 8.9(a) for "pan" differs from that in (b) for "ban" in having an additional [aspiration] in the GLO tier. Note that all the remaining gestures in the two gestural scores are identical between the two words. Fig. 8.9(b) for "ban" differs from (d) "Ann" in having a [labial-closure] gesture in the TT tier (for the initial consonant).

The second way in which gestural scores for phonological contrastive utterances may differ is in the particular tract variable tier controlled by a gesture within the gestural score. This is illustrated by Fig. 8.9(a) for "pan" vs. Fig. 8.9(c) for "tan," which differ in terms of whether the initial closure gesture occurs in the LIPS or the TT tier.

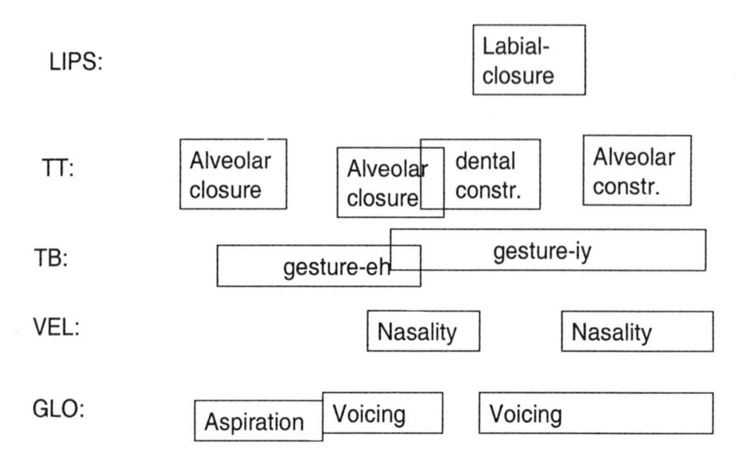

Figure 8.8: Gestural score for English phrase "ten themes" /t en n th iy m z/.

The third way in which gestural scores may differ is in the symbolic value(s) of the tract variable(s). This is illustrated when comparing Fig. 8.9(e) for "sad" to (f) "shad" where the value of the constriction location tract variable for the initial tongue tip constriction (TTCL) is the only difference between the two utterances: one takes the value of [alveolar], while the other takes the value of [alveolar-palatal].

Finally, two gestural scores may contain the same gestures and differ simply in how they are coordinated. This is shown by comparing Fig. 8.9(g) for "dab" vs. (h) "bad." Note that these two gestural scores contain identical gestures.

8.7 Syllables: External Organization of Speech Sounds

The role of the syllable in phonological theory, as well as in speech technology, has become more significant recently. All major schools of phonology have recognized the syllable as a fundamental unit in phonological analysis. The syllable is a larger unit beyond the phoneme, and it plays an important role of organizing phonological systems at this "suprasegmental" level. The main phonological function of the syllable is its role of constraining what speech sound segments are legal or illegal in part of a word. Because of this, it is often said that syllables serve as an external organizer of speech sounds.

8.7.1 The representation of syllable structure

It has been commonly accepted that a syllable consists of two constituents: 1) the **onset**, which comes at the beginning of a syllable, and 2) the **rhyme**, the remaining portion of the syllable which follows the onset. The onset of English syllables may consist of zero, one, two, or three consonants. The rhyme of English syllables must have a vowel, followed by zero, one, or more consonants. Since the rhyme is the only compulsory

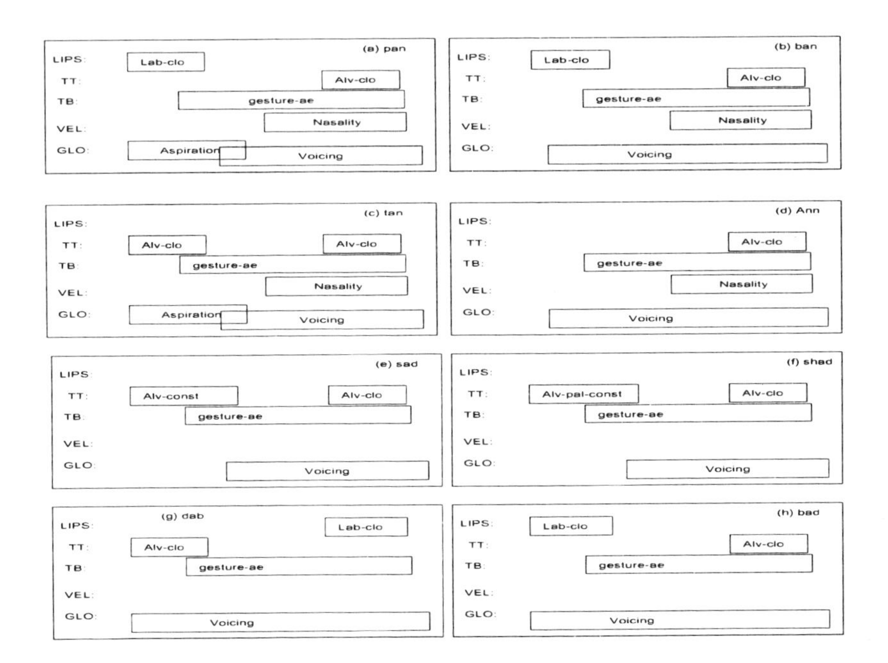

Figure 8.9: Four gestural scores for eight English words: (a) "pan" /p ae n/; (b) "ban" /b ae n/; (c) "tan" /t ae n/; (d) "Ann" /ae n/; (e) "sad" /s ae d/; (f) "shad" /sh ae d/; (g) "dab" /d ae b/; and (h) "bad" /b ae d/. They show phonological contrast among them in various ways (set text).

constituent, it is also called the head constituent. The rhyme can be further divided into the **nucleus** and **coda** sub-constituents, one consisting of the compulsory vowel, and the other consisting of the trailing consonant(s). This basic constituent structure of a syllable is shown in Fig. 8.10, with an example for the word *strengths*, where both the onset and the coda consist of three consonants.

An alternative representation to the above basic onset-nucleus-coda structure of the syllable is the CV-tier (Consonant-Vowel tier) representation, which give the canonical forms of morphemes. This representation is used in mostly theoretical studies of phonology, where a syllable is assumed to have a three-tiered structure consisting of a syllable node, a CV-tier whose C and V elements dominate consonantal and vowel segments, and a segmental tier consisting of bundles of features which may be organized according to feature geometry. Roughly speaking, the V element in the CV-tier representation corresponds to the nucleus in the onset-nucleus-coda structure, and the C element corresponds to a consonant occurring in either the onset or the coda.

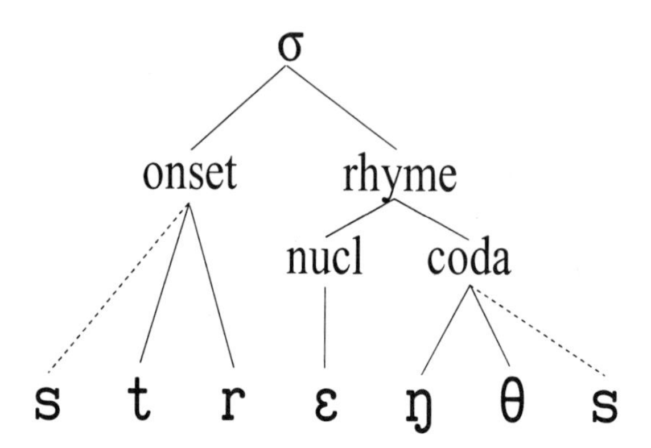

Figure 8.10: Syllable structure template with an example for English word *strengths*.

8.7.2 Phonological function of the syllable: basic phonotactic unit

The syllable serves a number of important functions in phonological theory and practice. The most basic function is to regulate the ways in which lower-level units, including consonants and vowels (as well as the features and gestures), in the phonological hierarchy can legally combine. The rules that govern the legal combinations of segments (consonants and vowels) are called the phonotactic rules, or morpheme structure conditions. Since the syllable is the principal structure which determines these legal combinations (i.e., well-formedness), we say that the syllable is the basic phonotactic unit. In articulatory phonology, the syllable is also used as the unit to organize the ways in which various gestures are allowed to overlap among each other. Therefore, in this view, much of coarticulation is organized around the syllable structure.

In terms of the syllable unit, the phonotactic rules can be most easily stated. For example, in English, the consonant sequence [tl] is not allowed within a syllable, but allowed when the [t] and [l] belong to different syllables as in "partly" [p aa r t - l i]. Constraints on syllable structure serve as a filter to permit only certain phone sequences to occur. These constraints are specific to a particular language.

To begin the discussion of the English phonotactic rules, we list in Fig. 8.11 some **basic** (not exhaustive) legal, well-formed syllable onsets in English. "+" signs denote well-formedness, and "-" signs otherwise. For example, [pr] is legal, and [pm] is not. Is there any regular pattern hiding in these legal onsets? Yes, but to illustrate the pattern, we need to introduce an important concept in syllable theory called the **sonority**, and the related notion of sonority hierarchy.

Sonority, defined rather elusively, is a particular type of "prominence" associated with a segment that represents the loudness aspect of its intrinsic articulation. It measures the output of acoustic energy associated with the production of the segment. Similarly, it can also measure the degree of aperture of the vocal tract in the production of the segment. A complete ranking of segment types in order of their intrinsic sonority is called the sonority hierarchy. A common ranking is, in decreasing order:

- Vowels;

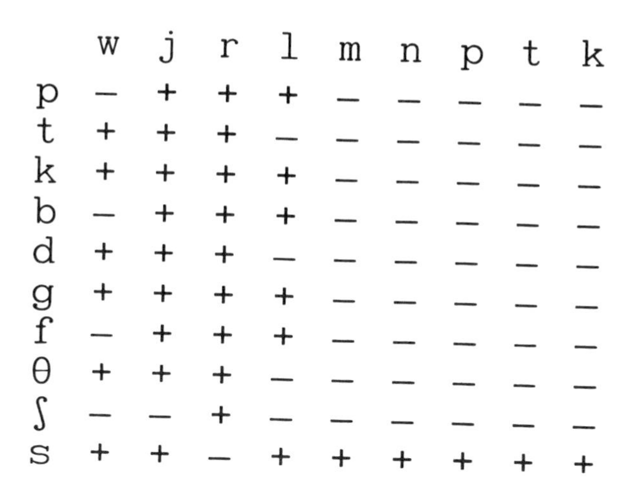

	w	j	r	l	m	n	p	t	k
p	–	+	+	+	–	–	–	–	–
t	+	+	+	–	–	–	–	–	–
k	+	+	+	+	–	–	–	–	–
b	–	+	+	+	–	–	–	–	–
d	+	+	+	–	–	–	–	–	–
g	+	+	+	+	–	–	–	–	–
f	–	+	+	+	–	–	–	–	–
θ	+	+	+	–	–	–	–	–	–
ʃ	–	–	+	–	–	–	–	–	–
s	+	+	–	+	+	+	+	+	+

Figure 8.11: Basic legal syllable onsets in English.

- Glides;
- Liquids;
- Nasals;
- Fricatives; and
- Oral stops.

Also, a voiced consonant has a higher sonority value than its voiceless counterpart within the same segment type in the above list. In Fig. 8.12 is listed the commonly accepted relative sonority of some segments of English.

Given the sonority hierarchy, the regular pattern hiding in the well-formed syllable onsets of Fig. 8.10 is clear: The second segment (top row) has a lower sonority value than the first segment (left column) in the onset. That is, the sonority value is increasing in the syllable onset towards to the syllable nucleus. More generally, the most important phonotactic rule in English is stated as the

> **Sonority Sequencing Principle**: The sonority value of the segments in the syllable onset or coda must rise when proceeding from the beginning of the syllable to the syllable nucleus, and fall when proceeding from the nucleus to the end of the syllable, both according to the sonority hierarchy.

In other words, the sonority profile of the syllable must slope outwards from the syllable nucleus. We also more simply state this phonotactic constraint as: syllables rise in sonority from the onset to the nucleus, and they fall in sonority from the nucleus to the coda.

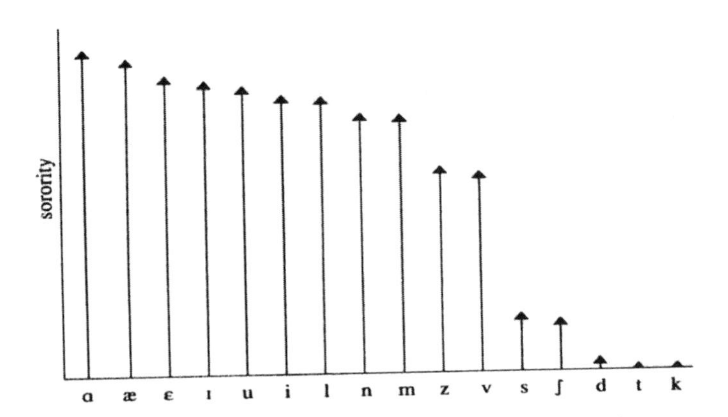

Figure 8.12: The commonly accepted relative sonority of some segments of English.

Due to reasons arising from historical phonology, there are two exceptions to the sonority sequencing principle in English phonotactics. We call these exceptions the adjunction rules. They are

- **Adjunction rule 1:** Segment /s/ may be adjoined to the beginning of an otherwise well-formed syllable onset. According to this, /sw/, /sj/, /sl/, /sm/, /sn/, /sp/, /st/, /sk/, /sh r/, /spj/, /spr/, /spl/, /stw/, /stj/, /str/, /skw/, /skj/, /skr/, and /skl/ can form a legal syllable onset while not strictly rising in sonority.

- **Adjunction rule 2:** An anterior coronal obstruent /t/, /d/, /th/, /s/, and /z/ may be adjoined to the end of a well-formed syllable coda. Examples are in the words of "depth, fifth, apt, strength," etc.

Incorporating both of the adjunction rules, the syllable structure for the word *strengths* was shown in Fig. 8.10, where the application of each of the adjunction rules is denoted by each of the dashed lines.

It should be noted that a syllable onset or coda that follows the sonority sequencing principle may not automatically be a well-formed onset or coda. There are additional language-specific constraints on the well-formedness within the constraints placed already by the sonority sequencing principle. For English, these additional constraints are:

- Voiced fricatives /v, dh, z, zh/ and [-anterior] coronals /sh, zh, tsh, dsh/ may not share the syllable onset with another consonant;

- Segments in the syllable onset tend to favor consonants that are maximally distant in the sonority hierarchy. Thus, /pl/ and /pr/ are favored, but not /pm/, /tf/, and /ml/ since the two segments are close in sonority values. No such a constraint applies to the syllable coda; for example, /lp/, /lm/, /mp/, and /ft/ are all well-formed codas.

- With the exception of [+coronal] followed by /r/, the segments in the syllable onset may not have the same place of articulation.

The above discussion has focused on the major phonological function of the syllable as the basic phonotactic unit, providing powerful constraints on what speech sound segments are well formed in part of a word. This function thus gives the syllable the role of external organizer of speech sounds. Another phonological function of the syllable structure is its role in conditioning the application of phonological rules. Further, the syllable is the most natural framework to account for the phonological structure of complex segments. Finally, the syllable forms a necessary building block for higher levels of phonological structure including: foot, prosodic word, and phonological phrase. We will not discuss these other phonological functions here, but refer interested readers to [Katam 89, Golds 90].

8.8 Summary

This chapter on the phonological process was devoted to the linguistically oriented speech science that pertains to the structure and contrast in language. Phonology is concerned with the symbolic, discrete categories of speech sound classes and their distinction, in contrast to phonetics, which is concerned with continuous variables in the speech process (such as articulation and acoustics).

The chapter started with the most fundamental concept of phonemes or segments, as the minimal contrastive units of speech sounds, in traditional phonology. The related concepts of allophones and of complementary distribution versus free variation were also introduced. We then proceeded to discuss the notion of distinctive features as the basic (symbolic) units for phonological representation. A collection (bundle) of such features form a phoneme. The advantages of features over that of phonemes as phonological representation were first discussed, and then a particular English feature system based on the work of SPE was presented in detail. This SPE feature system consists of a number of categories of features.

Given a full definition of the feature system for a language, phonological rules can be naturally and conveniently expressed in terms of the features. We discussed formalization of phonological rules and some common rule types in English.

Discussions on more recent developments of phonological theory followed traditional phonology presented earlier. First, the notion of feature geometry — internal organization of speech sounds — was motivated. This advanced the traditional, "linear" phonology towards the modern, "nonlinear" phonology. One specific proposal on the feature hierarchy, based on articulators' natural organization, was presented in some detail. Phonological rules were shown to be more natural using such a feature hierarchy than by using unstructured feature bundles. Next, we studied another branch of the modern phonology called articulatory phonology. It intended to form a close bridge between phonological and phonetic processes. Central to articulatory phonology are the concepts of articulatory gesture, gestural score, and task dynamics. We provided a full set of involved gestural units and task dynamic variables for English. Several examples were given to illustrate the gestural scores for some English utterances, and to illustrate phonological contrast as the most important character in any phonological theory.

Finally, we introduced the notion of syllable as external organization of speech sounds. We discussed the representation of syllable structure and its phonological significance in terms of phonotactic constraints governing what phonemic sequences are valid or invalid in a language. The sonority sequence principle and two adjunction rules were outlined.

A recent trend in phonology development is to move from rule-based systems, as we have covered in this chapter, to constraint-based systems. The main idea is to eliminate the generally complex phonological rules, which are typically language specific and sensitive to complex rule ordering, by a new mechanism of language-specific ranking of a set of language-universal constraints. Minimal violation of the constraints gives rise to the surface form of the phonological representation given the underlying form of the phonological representation. **Optimality theory** represents such new phonology development [Arch 97]. More recently, some computational framework for optimality theory has also been developed (e.g., [Eisn 02] and references therein).

Part III

COMPUTATIONAL PHONOLOGY AND PHONETICS

Chapter 9

Computational Phonology

Phonology as a linguistic science of speech sounds and as a discipline of study for systematic patterns of the sounds was introduced in Chapter 8. Typically, the approach in phonology is to provide general frameworks, principles, and rules (generative as well as constraining rules) by which the speech sound patterns of a language or across languages are studied. It is clear that the complexity of phonological patterns, which is due to an astronomically large number of symbolic entities in languages as well as their interactions, necessarily makes phonology a very complex subject. To provide a comprehensive coverage of phonological phenomena in a language would be very difficult without resorting to some sort of computational techniques, to which this chapter is devoted.

Computational phonology can be either a discipline of phonology that applies computing to solve phonology problems, or a discipline of computing that applies the principles and theories of phonology to solve some artificial intelligence problems, such as speech recognition and natural language processing. There have been many motivations for the study of computational phonology. Among them is the current, widely accepted weaknesses of the symbolic pronunciation modeling components in most state-of-the-art speech recognition systems. Much of these pronunciation modeling components are based on traditional linear phonology and typically use the 1960's SPE phonological rule system. Contemporary nonlinear phonological models, such as those from autosegmental phonology, feature geometry, and articulatory phonology discussed in Chapter 8, bear a closer resemblance to the continuous speech streams than linear phonological models. This is so especially for spontaneous speech with a casual speaking style, which has presented formidable challenges to current speech recognition systems. These nonlinear phonological models offer promise to provide a fresh source of techniques to aid the design of symbolic phonological components in the speech recognition systems.

A general coverage of the field of computational phonology can be found in some dedicated books [Bird 95, Cole 98], as well as in a special issue of the journal *Computational Linguistics* (Vol. 20, No. 3, 1994), dedicated to computational phonology. This chapter will be relatively narrowly focused — not aiming at an even coverage of the field, but at a particular approach in computational phonology that has been developed mainly to solve speech recognition problems. This particular approach is called the **overlapping articulatory feature** approach, built on the theoretical foundations of autosegmental phonology, feature geometry, articulatory phonology, and syllable theory.

In this chapter, we will first introduce the concept of multi-tiered overlapping articulatory features and a system for their specification. The development of overlapping articulatory features has been motivated by nonlinear theories of phonology, mainly feature geometry theory and articulatory phonology. They capture essential linguistic properties of the gestures and features in nonlinear phonology, but have taken a substantially simplified form and have offered distinctive advantages from computational considerations. We then discuss the temporal overlapping and mixing (blending) of the articulatory features across tiers and within a tier. The overlapping as applied to articulatory features is analogous to spreading and assimilation as applied to distinctive features in phonology theories, but key differences exist in that the overlapping may be partial (i.e., fractional) within a segment, rather than being "all-or-none" for each segment. This offers a more fine-grained description of the phonological process than standard phonology theory. Motivated by improving pronunciation modeling which is known to be a key component in speech recognition, we further discuss how a pattern emerging from the overlapping of articulatory features can be transformed into a finite state-transition graph or network. We call this equivalent network representation to the feature-overlapping pattern the (discrete) articulatory state graph. Such a graph can be used as the topology for a hidden Markov model (HMM) for speech recognition. Finally, in this chapter we present some recent research work on how high-level linguistic or prosodic information can be used to effectively constrain the overlapping of articulatory features. In particular, a parser for English syllable structure is described which provides some essential linguistic information above the phoneme level. The results of the parser are used to guide the lookup of phonological operators which act as phonological rules to generate overlapping feature bundles. This specific implementation method that transforms a phonotactically valid phonemic sequence (marked by some linguistic tags) for English to a discrete articulatory state graph will be described in detail.

The materials presented in this chapter have been reorganized and rewritten based on a series of published work in [Deng 94e, Deng 96, Deng 97a, Deng 97b, Deng 97c, Deng 98, Sun 00, Sun 01, Deng 99d].

9.1 Articulatory Features and a System for Their Specification

Central to the overlapping articulatory feature approach, on which we will focus in this chapter, is the notion of **articulatory feature**, or overlapping articulatory feature, which has been developed in a series of works published in [Deng 94e, Deng 96, Deng 97a, Deng 97c, Deng 98, Sun 01, Deng 99d]. Articulatory features are designed based on speech processing engineering considerations and on theories of phonology. From the theoretical side, they are a mix of, but different from, the (multi-valued) distinctive features in feature-geometry theory and the gestures in articulatory theory, both discussed in Chapter 8. Compared with the gestures, the articulatory features share the same key property of overlapping or blending across tiers. In fact, the tiers are essentially the same between the two representations. However, one very crucial difference is that unlike the gestures which are defined in terms of the parameters of the abstract dynamics in the "tract variables," the articulatory features are entirely symbolic with no specific reference and association to any continuous variables. A separate module, which is called the interface between phonology and phonetics and which can take many different forms, is responsible for the mapping from a symbolic articulatory feature to some continuous, phonetic variables. The notion of this type of interface simply does not exist in articulatory phonology and in the gesture. The second important difference between the gesture and the articulatory feature is that the former is associated with intrinsic duration due to its connection to an underlying, abstract task-dynamic system, while the latter is not. The temporal characterization of the articulatory feature is only by way of relative timing among the tiers.

Compared with the distinctive features in feature theory, the (spatial) hierarchical tier structure of articulatory features is not nearly as rich, but the temporal structure is much richer. While in feature geometry the linking and delinking operations apply across segments, the temporal precision is rather crude — in the units of segmental duration. That is, there is always a quantal jump, with the size of the duration of segment(s), over which the features spread temporally. In contrast, the spread or overlap of the articulatory features, like the gestures, has a much finer temporal precision in that feature overlapping across tiers is allowed over not only a duration of one or more segments but also over a graded fraction of duration within a segment. This is much more powerful and flexible than feature geometry in accounting for phonological patterns.

Further, the articulatory features are designed to be phonological units. That is, when serving as underlying units for describing word pronunciation, they play contrastive roles in distinguishing meanings of the words. To ensure this property, in the design of the articulatory features which we will discuss shortly, we use phoneme-like units as the basis and make it explicit that the different phonemes have distinctive values of the articulatory features associated with them.

From speech processing engineering considerations, mainly those of speech recognition, the articulatory features are designed with the additional requirement of economy. The articulatory features, with their spatial and temporal structures in place, are used as the nonlinear atomic speech units for lexical representation. This is aimed at providing a superior alternative to the popular linear phonetic representation often called "beads-on-a-string" [Ost 00]. An efficient lexical representation of this sort requires the choice of a small set of feature units so that in terms of these symbols each lexical item can be compactly specified while being made distinguishable from each other at the same time.

Most desirable choices for lexical representation and for pronunciation mapping should simultaneously satisfy the following two requirements: (1) Each speech unit is mapped into a (relationally) invariant, "consistent" acoustic region. Consequently, distinct speech units can be mapped into maximally separate acoustic regions; and (2) The inventory of the speech units is sufficiently small such that different lexical items are allowed to maximally (and beneficially) share the available acoustic data. Consequently, the demand for training data can be kept minimal. Speech units based on entire words satisfy the first requirement well but satisfy the second requirement poorly. Phonemic speech units are just the reverse. The use of overlapping articulatory features, when properly designed, can do well on both.

In the articulatory feature design in the work of [Deng 94e], five multi-valued features, Lips, Tongue Blade (TB), Tongue Dorsum (TD), Velum, and Larynx, are assigned uniquely to each phonemic unit, with intended "minimal" redundancy and "maximal" separability. Then the major contextual variations in speech are modeled as a natural result of overlap or asynchronous spread of the "intrinsic" values of one or more of these features across adjacent phonetic units. Given a fixed feature-overlap pattern, a one-to-one mapping can be made from such a pattern to a state-transition graph which forms the topology of the underlying Markov chain of the HMM (detailed techniques for accomplishing such a mapping will be presented later in this chapter). The graph is constructed in such a way that each node in the graph (or the state in the HMM) represents a unique composition of the five features. Each individual lexical item is represented by a distinct state-transition graph, formed by concatenating a sequence of sub-graphs associated with the phone sequence in the phonetic transcription according to the feature-overlap patterns constructed specifically from these phones-in-context. Since each node in this global graph contains a distinct composition of the features, we can also view the representation of a lexical item described above as an organized set of feature-bundle collections.

The above choice of feature set is motivated by articulatory phonology that we reviewed in Chapter 8, enhanced by practical considerations for a feature-based speech recognizer implementation. The five articulatory features chosen (Lips, TB, TD, Velum, and Larynx) are essentially the same as the five tiers in articulatory phonology, except those tiers were explicitly associated with the tract variables (Lip protrusion/aperture, Tongue Tip constrict location/degree, Tongue Body constrict location/degree, Velic aperture, and Glottal aperture) while the articulatory features are not.

The complete articulatory feature specification system is shown in Table 9.1 for the consonantal segments and in Table 9.2 for the non-consonantal segments in English. In the following, we provide several notes on the conventions used and key properties of the articulatory feature specification system.

type	symbol	example	Lips	TB	TD	Velum	Larynx
Closure:	bcl,dcl,gcl		0	0	0	1	1
	pcl,tcl,kcl		0	0	0	1	2
Stops:	b	bee	1	0	0	1	1
	d	day	0	1	0	1	1
	g	geese	0	0	1	1	1
	p	pea	1	0	0	1	2
	t	tea	0	1	0	1	2
	k	key	0	0	1	1	2
	dx	muddy,dirty	0	2	0	1	1
	q	at, bat	0	0	0	1	3
Fricatives:	s	sea	0	3	0	1	2
	sh	she	0	3	2	1	2
	z	zone	0	3	0	1	1
	zh	azure	0	3	2	1	1
	f	fin	2	0	0	1	2
	th	thin	0	4	0	1	2
	v	van	2	0	0	1	1
	dh	then	0	4	0	1	1
Nasals:	m	mom	1	0	0	2	1
	n	noon	0	1	0	2	1
	ng	sing	0	0	1	2	1
	em	bottom	1	0	17	2	1
	en	button	0	1	17	2	1
	eng	Washington	0	0	20	2	1
	nx	winner	0	2	0	2	1
Affricates:	jh	joke	dcl + d + zh				
	ch	choke	tcl + t + sh				

Table 9.1: Multi-valued articulatory feature specification for consonantal segments of English (using TIMIT labels) (after Deng and Sun [Deng 94e], @AIP).

First, the special feature value denoted by “0” for the phonetic or quasi-phonemic segment in Tables 9.1 and 9.2 indicates that in attempting to produce a sound this particular articulatory dimension is unmarked (or irrelevant). For instance, the Lips feature is unmarked for all vowel segments except for those with lip rounding (feature value "3" for /uw/ and "4" for /ao/), and labial consonants are all unmarked in both TB and TD features. Unmarked features are most susceptible to feature spreading (assimilation) from adjacent phonetic segments; that is, segments having unmarked feature(s) tend to inherit these features from adjacent segments.

type	symbol	example	Lips	TB	TD	Velum	Larynx
Semivowels:	l	lay	0	5	3	1	1
	r	ray	5	6	4	1	1
	w	way	3	0	5	1	1
	y	yacht	0	7	6	1	1
Vowels:	iy	beet	0	0	7	1	1
	ih	bit	0	0	8	1	1
	eh	bet	0	0	9	1	1
	ey	bait	0	0	9/7	1	1
	ae	bat	0	0	10	1	1
	aa	bottom	0	0	11	1	1
	aw	bout	0/3	0	11/15	1	1
	ay	bite	0	0	11/7	1	1
	ah	but	0	0	12	1	1
	ao	bought	4	0	13	1	1
	oy	boy	4/0	0	13/7	1	1
	ow	boat	4/3	0	13/15	1	1
	uh	book	3	0	14	1	1
	uw	boot	3	0	15	1	1
	ux	toot	0/3	0	8/15	1	1
	el	bottle	0	5	17	1	1
	er	bird	5	6	16	1	1
	ax	about	0	0	17	1	1
	ix	debit	0	0	18	1	1
	axr	butter	0	0	19	1	1
	ax-h	suspect	0	0	19	1	2
Aspirations:	hh	hay	0	0	0	1	2
	hv	ahead	0	0	0	1	1

Table 9.2: Multi-valued articulatory feature specification for non-consonantal segments of English (using TIMIT labels) (after Deng and Sun [Deng 94e], @AIP).

Second, for each of the five articulatory features, we provide a range of feature values (i.e., multiple quantization levels) — five for Lips, seven for TB, 20 for TD, two (binary) for Velum, and three for Larynx. The feature assignments should be interpreted as being of a symbolic nature, only without ordering and relational information. These feature values can be equivalently represented by other, non-numerical symbols.

Third, the specified features for each phoneme-like segment are context-independent, and are related to the intended target articulatory gestures or the abstract "hidden" messages to be conveyed through the speech production process. For example, the intended gestures for phoneme /t/ are: to move the tongue blade in contact with the alveolar ridge anteriorly (value "1" in the TB feature in Table 9.1), to close the velopharyngeal port (value "1" in the Velum feature), and to abduct the vocal cords (value "2" in the Larynx feature), all independent of the contexts. Despite allophonic variations, the above intended gestures for /t/ tend to be invariant.

Fourth, the systematic nonstationary segments (or complex segments) are decomposed into relatively stationary sub-segments. Specifically, affricates are decomposed into their associated stop and fricative sub-segments (last two rows in Table 9.1), and diphthong specification is accomplished by sequential connection of the two target vowels (see double labels in either the Lips or TD feature for diphthongs in Table 9.2). Such specifications for the complex segments imply that their feature spread is extended over only the left or right, rather than both, adjacent segments.

Fifth, all vowels are marked in the TD feature since the tongue dorsum is always active in producing these sounds; for lip-rounding vowels, the Lips feature is marked as well. The TB feature is unmarked for vowels except for /el/ and /er/. Semi-vowels, on the other hand, are marked by two or all three principal articulator features.

Finally, we follow the TIMIT convention and adopt the TIMIT labels to denote all the phonetic segments in Tables 9.1 and 9.2. The TIMIT labels have a small degree of deviation from the standard phonemic labels; for example, the closure interval of a stop is distinguished from the stop release, and some rather systematic allophones — flaps /dx/ and /nx/, glottal stop /q/, voiced-h /hv/, fronted-u /ux/, and devoiced-schwa /ax-h/ — are added as separate labels. This expanded set of phonetic segments may be called the quasi-phonemic ones.

9.2 Cross-Tier Overlapping of Articulatory Features

Once the feature-specification system above is established, one establishes an "abstract" lexical representation, which is in a similar form to the feature-matrix representation in feature theory. The next step in the overlapping articulatory feature approach is to "compile" such an abstract representation, based on the concept of probabilistic and fractional overlap of articulatory features which we describe in this section. This description and the examples provided in this section below have been taken from the work published in [Deng 94e, Deng 96].

9.2.1 Major and secondary articulatory features

In performing this "compilation," notions about hierarchical feature structure from feature geometry theory are used to provide constraints on what feature tiers can spread simultaneously and what tiers may spread independently. The overlapping feature approach described in [Deng 96] links the manipulation of the articulatory structure, as symbolized by the articulatory feature bundles, to the dynamic patterns of physical articulators and of the resulting speech signal. This linkage has been based on the critical notion of *major (*or *active) articulator* proposed in [Key 94, Sag 86]. The notion of major articulator, as originally appearing in [Sag 86], is connected to the view that for each consonant there is one (major) articulator forming the main vocal tract constriction. In the later work of [Key 94], the major articulator is defined from a more general principle; i.e., the major articulator is the most anterior articulator below the supra-velar node (for English) that dominates terminal features in the hierarchical feature structure. Under the supra-velar node, there are three possible major articulatory features: Lips (L), Tongue Blade (B), and Tongue Body or Dorsum (D). According to the theory outlined in [Key 94], for a speech segment with the dominant node being supra-velar, the acoustic correlates of the corresponding major articulatory features that define the segment are spread over a region near the time when there is a rapid spectral change or a strongly dynamic pattern in speech. This dynamic pattern in acoustics is correlated with the lawful motion of the major articulator(s) in forming and releasing the vocal tract constriction that defines the corresponding major articulatory feature(s). In contrast, for a segment with the dominant node being supralaryngeal, i.e., the Velum (V) and Larynx (X) features in the feature-specification system of [Deng 94e], the acoustic correlates of these "secondary" articulatory features are determined by relatively slow spectral changes or approximately static patterns showing no abrupt spectral discontinuity. Fig. 9.1 is the hierarchical feature tree structure (revised from Fig. 5 in [Key 94]), which shows the major and the secondary articulatory features as leaves in the tree in relation to the corresponding supra-velar and supralaryngeal nodes. (A similar tree on the vocal tract hierarchy related to articulatory and tube geometry also appeared in Fig. 13 of [Brow 89].) Note that for American English, only two among the four secondary articulatory features in Fig. 9.1, Velum and Glottis, appear to play very important phonological roles. These two secondary articulatory features, together with the three major ones, constitute the complete feature set in the feature-specification system designed in [Deng 94e] and listed in the preceding section.

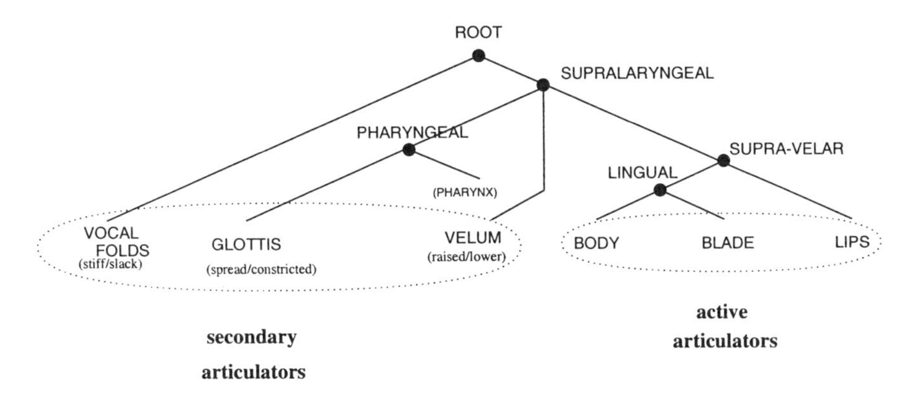

Figure 9.1: Tree structure of articulatory-feature organization emphasizing the division between major and secondary articulatory features (after Deng and Sameti [Deng 96], ©IEEE).

According to the notion of major and secondary articulatory features outlined above and in view of their respective roles in determining dynamic and largely static acoustic correlates, a new set of feature-based, subphonemic speech units can be defined. Given the feature-specification system and given the feature assimilation or spreading process, if the resulting feature bundle contains one or more assimilated feature value(s) marked in the *major* articulatory feature dimension(s), then this feature bundle defines a *Transitional Speech Unit.* On the other hand, if the assimilated feature(s) are all in the *secondary* articulatory feature dimension(s), then the corresponding feature bundle defines a *Static Speech Unit.*

9.2.2 Feature assimilation and overlapping examples

Now we use two example English utterances to illustrate the abstract concepts of major/secondary articulatory features and of feature assimilations, and the consequent definitions of transitional and static speech units, all introduced in the preceding subsection.

Table 9.3 is a (re-coded) subset of Tables 9.1 and 9.2, and it extracts the relevant phonemes and features used in the examples provided here. Note that a different, more mnemonic set of symbols is used here, where the multi-valued features in all five tiers (L, B, D, V, X) are denoted with the prefix taken as the name of the tier. That is, the first letter in the non-"U" entries denotes the feature name (one out of five), and the second letter (in Greek) denotes the symbolic feature value. Symbol "U" denotes feature underspecification, the same as feature value "0" in Tables 9.1 and 9.2.

Segment	Example word	Lips (L)	TongueBlade (B)	TongueDorsum (D)	Velum (V)	Larynx (X)
b	bee	Lα	U	U	Vα	Xα
d	day	U	Bα	U	Vα	Xα
g	geese	U	U	Dα	Vα	Xα
p	pea	Lα	U	U	Vα	Xβ
t	tea	U	Bα	U	Vα	Xβ
k	key	U	U	Dα	Vα	Xβ
s	sea	U	Bβ	U	Vα	Xβ
f	fin	Lβ	U	U	Vα	Xβ
v	van	Lβ	U	U	Vα	Xα
m	mom	Lα	U	U	Vβ	Xα
n	noon	U	Bα	U	Vβ	Xα
ng	sing	U	U	Dα	Vβ	Xα
r	ray	Lγ	Bγ	Dβ	Vα	Xα
iy	beet	U	U	Dγ	Vα	Xα
aa	hot	U	U	Dδ	Vα	Xα
ao	bought	Lδ	U	Dϵ	Vα	Xα

Table 9.3: Five-tuple (L B D V X) articulatory features for some segments in English used for the illustrative examples in this section (after Deng, Ramsay, and Sun [Deng 97c], @Elsevier).

The first example, shown in Fig. 9.2, uses word *strong* (/s t r ao ng/), which contains extensive feature assimilation or overlapping. It goes across far beyond the immediately adjacent phonetic segments. In Fig. 9.2, the feature values corresponding to each phoneme in /s t r a ng/ are directly taken from the feature specification in Table 9.3. This feature-vector sequence forms a multi-valued feature matrix, subject to the following feature assimilation or overlapping process. Feature assimilation is largely independent across some feature tiers, but the features are all associated and asynchronously aligned somewhere (indicated by the vertical dashed lines) between each pair of adjacent phonemes unless feature underspecification ("U" value) is encountered. We now discuss several striking cases of the feature assimilation and the associated acoustic consequences exhibited in this example.

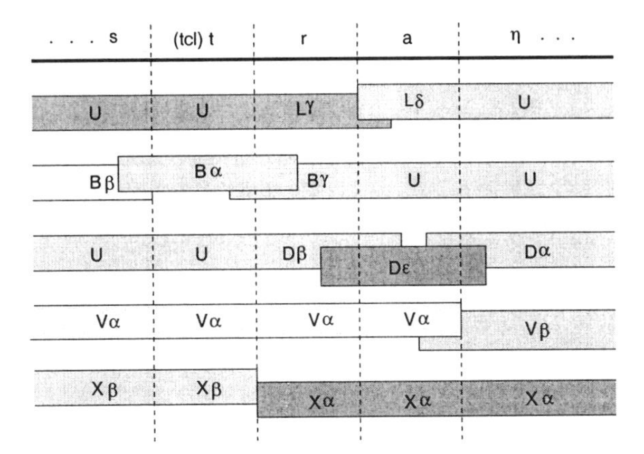

Figure 9.2: Feature assimilation pattern for example word *strong*. Note that feature assimilation takes place beyond immediately adjacent segments and that the assimilation in the major articulatory feature dimensions produces transitional speech units (after Deng [Deng 97a], @Elsevier).

First, for the Lips feature (L, major), the feature value Lα associated with the home segment /r/ is spread in the leftward direction all the way to /s/, because the Lips feature is underspecified for both /t/ and /s/. The same cross-phone feature assimilation takes place in the Tongue Dorsum (D, major) dimension for the same reason. One clear acoustic consequence (also serving as evidence) of these feature assimilations is to lower the low-frequency edge of the /s/ frication noise from the nominal value of about 4 kHz to as low as 1 kHz [Zue 91]. The effect is unusually conspicuous because the Lips-feature assimilation from /r/ results in a lip-rounding gesture in /s/ and the Dorsum-feature assimilation from /r/ results in a palatal-constriction gesture in /s/, both contributing to lengthening the front cavity in the vocal tract and thus producing the /sh/-like /s/ allophone. Second, for the Tongue Blade (B, major) feature, there is a fractional feature spread left from segment /r/ to its adjacent segment /t/. This, together with the left D-feature spread discussed earlier, produces relatively strong frication noise in segment /t/, lowers the low-frequency edge of the noise, and turns /t/ into a /tS/-like allophone. Note that both of the above B-feature and D-feature spreads are due to phonetic (as opposed to phonological) causes. The resulting /tS/-like acoustic sound is generated from the articulatory structure between that of /t/ (with vocal-tract constriction more anterior) and that of /r/ (with the constriction more posterior); this intermediate articulatory structure is created in transit from the two sequentially related gestures. Third, the fractional D-feature spread left from the /a/ segment to the /r/ segment contributes to F3 rising from the characteristically low value of about 1700 Hz associated with /r/ to over 2000 Hz toward the F3 target value for /a/. Finally, the Velum (V, secondary) feature spread left from /ng/ to /a/ produces vowel nasalization. Note the nasalization does not spread further into /r/ due to the blocking non-"U" feature value Vα in the adjacent segment /a/.

In drawing the feature assimilation pattern in Fig. 9.2, in the Larynx (X, secondary) feature dimension, a constraint has been imposed to forbid the feature value Xα (– Voice feature) from spreading in the right direction from /t/ into /r/. This constraining rule applied to the /t/ segment is conditioned on its left context of /s/. Articulatory phonology has provided an elegant account for such a constraint [Brow 86]. Without such a constraint, $X\beta$ would spread into /r/ and produce aspiration, which does not happen acoustically. In general, as a language-dependent phenomenon, there are many feature-assimilation patterns that are not permissible in reality. This can be either due to phonological factors or simply due to physical constraints in vocal tract acoustics. As an example for the latter case, the Velum feature value $V\beta$ (i.e., +nasal) is generally forbidden from spreading into a voiceless stop or fricative segment, because production of that stop or fricative segment requires occlusion or constriction in the vocal tract (by a major articulator) that is accompanied by pressure buildup behind the vocal tract constriction. If the $V\beta$ feature were simultaneously in place, then the corresponding velum gesture (lowered) would open the velopharyngeal port and release the pressure buildup behind the constriction. Therefore, one major component in designing overlapping feature-based models for speech is to devise a comprehensive set of rules to specify possible constraints in feature assimilations for each phonetic segment in sentence-level contexts and for each of the five feature dimensions. Different ways of deriving and representing such rules will be described later in this chapter.

Given the feature-assimilation pattern shown in Fig. 9.2 for the example word, an automatic recursive procedure can be devised to construct a state-transition graph, which has the result of Fig. 9.3. (Details of an algorithm for such construction will be presented in the next section.) This construction is unique and encompasses all allowable constellations, shown as the state contents, of the five-tuple features expressed in Fig. 9.3 that respect all the feature-assimilation constraints imposed therein. Each node (or state) in the graph defines an atomic, subphonemic speech unit, which may be potentially shared across all (English) lexical items. Applying the definitions of the transitional and static speech units made in the preceding subsection to the state-transition graph of Fig. 9.3, we identify a total of ten transitional units, with the remaining seven being static. Each of the transitional units is marked by one or two filled circles in the respective major articulatory feature(s) whose assimilatory nature defines them as being transitional. Each filled circle left (or right) of the corresponding transitional unit signifies the left (or right) origin of the assimilation that creates the transitional unit. One key aspect of the construction of the speech units described here is that each state (or unit) is associated with unambiguous interpretations expressed as the unique articulatory feature bundle that ultimately leads to acoustic interpretations.

The second example provided here is intended to illustrate how the well-known coarticulation of vowels across consonants [Ohm 66] can be elegantly accounted for in the construction of the feature assimilation/overlapping pattern and of the associated state-transition graph. Note that, as in the previous example, the coarticulation across phones in this example cannot be easily captured by the popular tri-phone modeling method employed by the conventional HMM-based speech recognition technology. The study in [Ohm 66] showed many pencil tracings of the formant center lines in spectrograms of VCV utterances containing the medial consonant /b/. If there were no coarticulation of vowels across /b/, all the formants would drop their values towards the medial /b/ according to the perturbation theory. The increase of F2 values of the pre-consonantal /a/ and of the post-consonantal /a/ is a clear evidence for across-consonantal coarticulation. In Fig. 9.4, the feature-assimilation pattern for the utterance /a b i/ is shown. The rising of F2, a dynamic pattern, is a direct result of feature value $D\delta$ spreading (fractionally only) from the /i/ segment, across the "U"-valued /b/ segment, to the home segment of /a/. The key point to note is that the state constructed that contains the assimilated

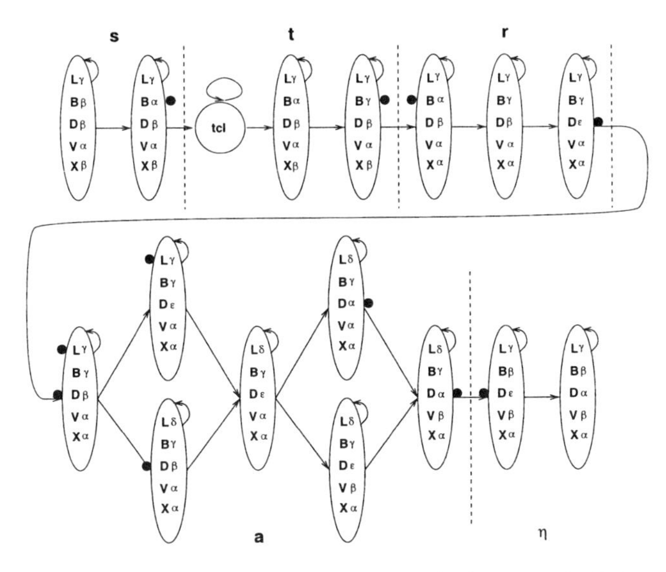

Figure 9.3: State-transition graph constructed from the feature-assimilation pattern of Fig. 9.2 for the word *strong*, containing 10 transitional units, marked by thick dots, and 7 static states. The position of the circle denotes the left or right origin of the directional assimilation of the corresponding major articulatory feature (after Deng [Deng 97a], @Elsevier).

feature value Dδ as a sub-component in the /a/ segment is transitional, because the involved articulatory feature D is a major articulatory feature.

9.2.3 Constraining rules for feature overlapping

In the above examples illustrating the feature overlapping across tiers, we implicitly assumed that the overlaps cannot be arbitrary and that they have to be constrained somehow. Otherwise, it would not be possible to produce the sensible state-transition graphs that may lead to realistic acoustic interpretation. In this subsection, we provide one set of constraining rules for English, elaborated in the work of [Deng 96] where these rules have been successfully applied to speech recognition based on this particular computational phonological approach of overlapping features.

The constraining rules that govern the possibilities of feature overlapping in [Deng 96] had been motivated mainly empirically, and were derived from studies of a large body of speech science literature. These rules are schematized in Figs. 9.5-9.10. They are organized, one for each figure, by the manner classes of English sounds in the segments, towards which the features in the adjacent segments are spreading. These segments are sometimes called the "home" segments. Fig. 9.5 illustrates the rules of feature overlapping for the VOWEL class as the home segment (middle column). The left and right columns, respectively, contain the feature bundles for the specified classes of segments obtained from an analysis of *all* left/right segments, which are generally not just the immediately adjacent left/right segment. Note that the rules are stipulated for each of the five articulatory features (Lips, TongueBlade, TongueDorsum, Velum, and Larynx) separately. The horizontal boxes that intrude into the middle column from the

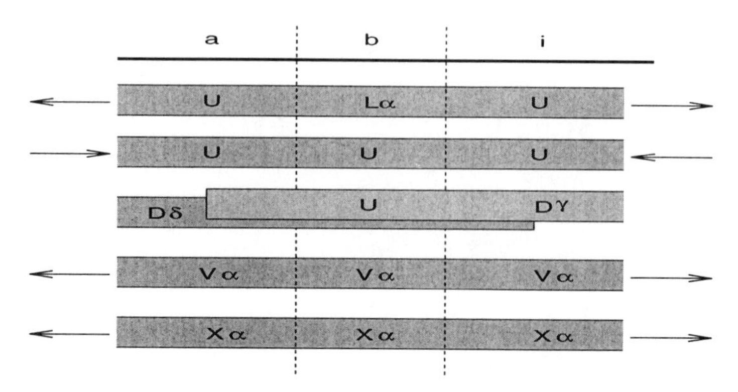

Figure 9.4: Feature assimilation pattern for utterance /a b i/ (after Deng [Deng 97a], @Elsevier).

left/right column express *fractional and probabilistic* feature spreading, with the depth of the spread following a probability distribution, such as an exponential distribution. The boxes that do not intrude specify blocking of the feature overlapping in context. Several example words or phrases, displayed in Fig. 9.5 for each feature tier separately, are given for feature overlapping and/or for blocking of feature overlapping. For example, the Lips feature from both the left and right segments are allowed to spread or be assimilated into a vowel home segment. Since the vowel home segment in this example contains only the underspecified Lips feature, the feature spreading here is called overlapping, rather than mixing. The example of feature spreading that is mixing rather than overlapping can be found in the TongueDorsum tier in Fig. 9.5. Since the vowel home segment also contains a specified TongueDorsum feature, any other specified TongueDorsum feature from its adjacent segments can spread into or be mixed with the home segment. The example for the Velum feature tier in Fig. 9.5 shows that the home vowel segment can be partially or completely nasalized if either or both of its left/right segments contain the specified [+nasality] feature. In the Larynx feature tier, the examples in Fig. 9.5 illustrate blocking of the [+aspiration] feature and of the [-voice] feature in the adjacent segment(s) to the home vowel segment in the same tier.

Fig. 9.6 shows the feature spreading and blocking constraining rules for the semivowel or liquid home segments. The example in the Velum tier says that [w] in the word "with" may be nasalized due to feature spreading (or overlapping) from the left adjacent segment [n]. In the Larynx tier, the [-voice] feature is blocked from spreading (or mixing) into the [l] home segment.

Fig. 9.7 shows similar constraining rules for the fricative home segments. In the Lips tier, with the example word "strong," the specified Lips feature in segment [r] is spread left, across the intervening segment [t] that has the underspecified Lips feature, into the home segment [s]. Since [s] has the underspecified Lips feature, this feature spreading is called overlapping. In the TongueDorsum tier, the home segment [s] in the phrase "gas shortage" is assimilated into [sh] due to feature spreading from the right segment. This naturally gives rise to the palatalization rule discussed in Chapter 8. In the Velum tier, the example of Fig. 9.8 shows that the home voiced fricative segment [dh] may be nasalized due the [+nasality] feature spreading from the left segment. In contrast, if the home fricative segment is voiceless as in [f], the same feature spreading will be blocked.

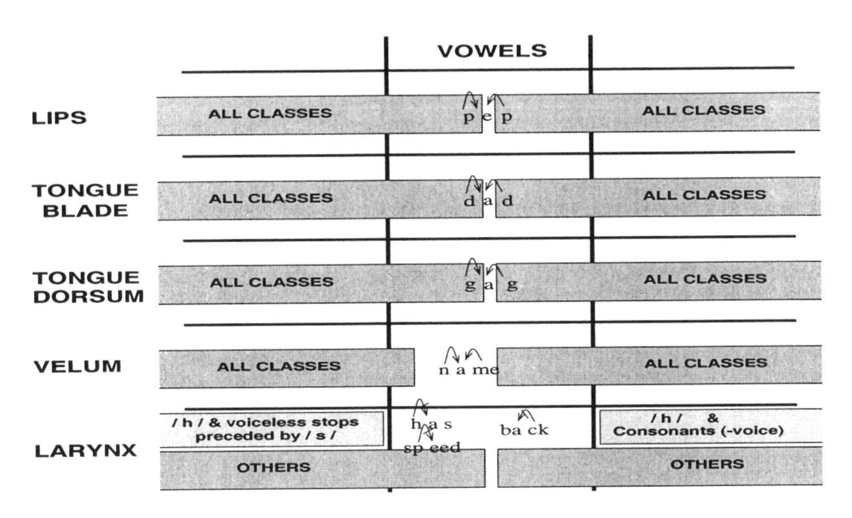

Figure 9.5: Feature spreading and blocking constraining rules for the vowel home segments; examples are provided (after Deng and Sameti [Deng 96], ©IEEE).

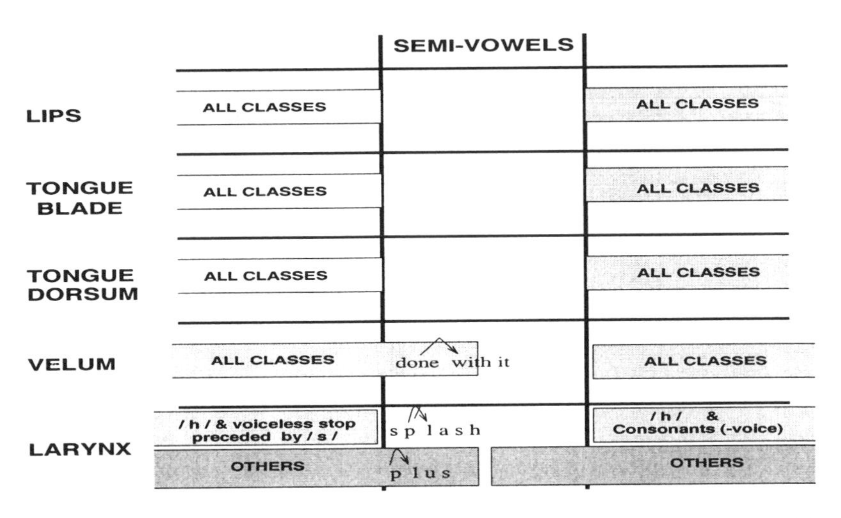

Figure 9.6: Feature spreading and blocking constraining rules for the semivowel and liquid home segments; examples are provided (after Deng and Sameti [Deng 96], ©IEEE).

This type of rule has a strong phonetic basis: since the airstream is typically weaker for the [-voice] sound than for the [+voice] sound, it is more difficult to open up the velum to allow partial airstream to leak through the nose in order to implement the [+nasality] feature. Finally, in the Larynx tier, the home voiced segment [z] in "his" may become devoiced due to the right adjacent voiceless segment [sh] in "shoe."

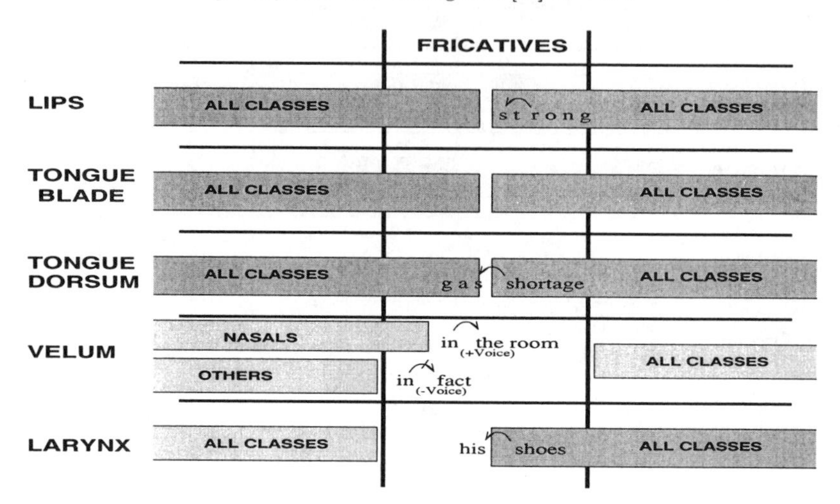

Figure 9.7: Feature spreading and blocking constraining rules for the fricative home segments; examples are provided (after Deng and Sameti [Deng 96], ©IEEE).

Consider a case where the home segment is a nasal consonant as depicted in Fig. 9.8. The examples show that both [n]'s in the word "noon" become rounded by the Lips feature in the adjacent segment [uw]. Likewise, the [n] in "ten" becomes labialized by inheriting the labial feature in the adjacent [m]. In the TongueBlade tier, the example shows that the blade feature in [n] is mixed with that in the adjacent [s]. This is the mixing type of feature spreading, not the overlapping type, because the tongue blade feature in both segments is fully specified. In the TongueDorsum tier, the home segments [n] in both "run" and "ten" are assimilated essentially to [ng].

When the home segment is a stop consonant as depicted in Fig. 9.9, the examples show many types of phonological consequences. For instance, the [k] in "clue" may become rounded due to the Lips feature spreading (overlapping) from its right segment [uw] across the intervening segment [l]. [t] in "strong" has its TongueBlade feature mixed with that of its right segment [r], resulting in more frication than otherwise. The second [d] in "did you" may become palatalized due to the TongueDorsum feature spreading (overlapping) from its right segment [j].

Finally, Fig. 9.10 shows the constraining rules for the separate [h] home segment. The [h] in "who" becomes rounded due to the [+round] Lips feature spreading from its adjacent segment [uw]. In the TongueDorsum tier, the [h] in "he" ([h iy]) inherits the TongueDorsum feature from [iy], making this [h] allophone essentially a voiceless or whispered vowel. In the Larynx tier, the [+voice] feature from both left and right vowel segments is spread into [h], making it a voiced allophone.

After applying all the above feature-spreading (overlapping and mixing) constraining rules to all TIMIT sentences, the acoustic space of the feature-based speech recognizer

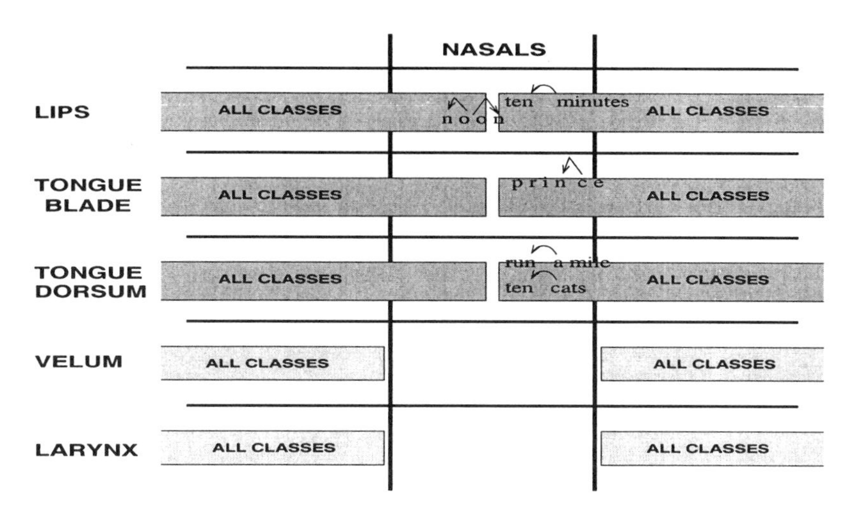

Figure 9.8: Feature spreading and blocking constraining rules for the nasal home segments; examples are provided (after Deng and Sameti [Deng 96], ©IEEE).

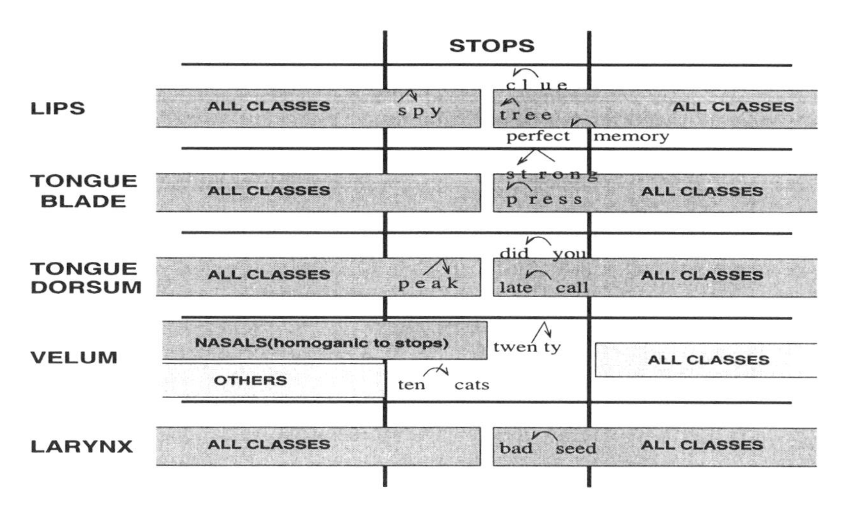

Figure 9.9: Feature spreading and blocking constraining rules for the oral stop home segments; examples are provided (after Deng and Sameti [Deng 96], ©IEEE).

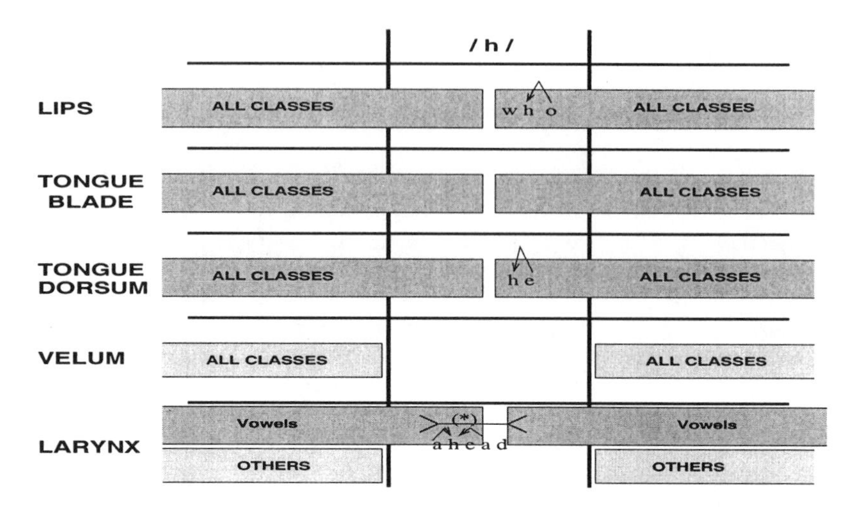

Figure 9.10: Feature spreading and blocking constraining rules for the aspiration home segment; examples are provided (after Deng and Sameti [Deng 96], ©IEEE).

as reported in [Deng 96] is characterized by a total of 5153 states in the state-transition graph, among which 4479 are transitional states and the remaining 674 static. The state-transition graph, which we also call the (discrete) articulatory HMM state topology, is constructed using an automatic algorithm that we describe next.

9.3 Constructing Discrete Articulatory States

9.3.1 Motivations from symbolic pronunciation modeling in speech recognition

A principal motivation of the overlapping articulatory feature model described so far in this chapter comes from our recognition of the weakness of the conventional linear "beads-on-a-string" phonological model for speech recognition. We will discuss many aspects of speech recognition in great detail in Chapter 12, and in this section will only touch on the (symbolic) phonological or pronunciation aspect of speech recognition as the motivation for some techniques of computational phonology.

When we discussed the Markov chain and the hidden Markov sequence (or HMM) in Chapter 3.5, we emphasized the discrete-state nature of this type of mathematical construct. Since the underlying Markov chain of an HMM is an ensemble of discrete or symbolic temporal events, it is naturally suited to represent the symbolic phonological process. The essential issue here is: *what entities of the phonological process can be appropriately made associated with the HMM state?* The currently dominant technique in speech recognition has been to assign the HMM state to the allophone, or a portion of the allophone whose linear, sequential connection constitutes an allophone [Huang 01], hence a linear phonological or pronunciation model. However, there is also a new trend emerging that intends to break away from this linear phonological model by advocating the use of modern versions of nonlinear phonology as computational machinery for pronunciation modeling in speech recognition [Ost 00, Deng 94e, Deng 96, Deng 95b, Deng 97a,

Deng 97c, Deng 98, Sun 01, Deng 99d, Deng 97b, Deng 97c, Cars 01, Kirc 98, Deng 00a]

The following is a quote from an early seminal paper which significantly contributed to the popularity of the HMM in speech recognition: "[...] It is quite natural to think of the speech signal as being generated by such a (HMM) process. We can imagine the vocal tract as being in one of a finite number of articulatory configurations or (HMM) states. [...]" [Lev 83]. It has become apparent nowadays that the mechanism described above that associates HMM states to articulatory configurations has been highly superficial. The phoneme-sized HMM is essentially a flexible piece-wise data-fitting device and describes mere surface phenomena of speech acoustics rather than any underlying mechanisms of the speech process. This is the reason why an increase in size of the HMM state-space and in the amount of training data appear to be the only possibilities for more accurate representation of speech if one is to build more robust speech recognizers for tasks with fewer constraints [Young 95].

The overlapping articulatory feature model that was presented above in this chapter aims at constructing a multi-dimensional HMM whose states can be made to directly correspond to the symbolically-coded, phonologically-contrastive articulatory structure responsible for generating acoustic observations from the states. The very nature of multiple dimensionalities, separate for each phonologically significant articulatory gesture tier, of the HMM allows embodiment of the asynchronous articulatory feature/gesture overlaps (coarticulation) in a natural way. In the work described in [Deng 00a, Deng 97b], a compact set of universal phonological/articulatory features across several languages (English, French, and Mandarin Chinese) is designed. The resulting feature specification systems, one for each language, share intensively among the component features. Through appropriate combinations of component features, new sounds or segments in new target languages from the sounds in the source language(s) can be reliably predicted. The phonological model uses hierarchically organized articulatory features as the primitive phonological units motivated by articulatory phonology and feature-geometry theory. The phonological model further entails a statistical scheme to allow probabilistic, asynchronous but constrained temporal overlapping among components (symbols) in the sequentially-placed feature bundles. A set of feature overlapping and constraining rules are designed based on syllable structure and other prosodic factors. Examples of the rules for English are: 1) overlap among consonant clusters in syllable onset and coda, and in consonant sequences across connected words; 2) overlap between syllable onset and nucleus; 3) overlap between syllable nucleus and coda; 4) overlap of the tongue-dorsum feature between two adjacent syllable nuclei; 5) except for the Lips and Velum features, no overlap between onset and coda within the same syllable. Details of these syllable-based constraining rules for feature overlapping will be described in the next section.

The overall design of the speech recognizer based on overlapping articulatory features can be cast in a probabilistic analysis-by-synthesis framework; no direct inversion operation is necessary and, to perform speech recognition, there is no requirement for the articulatory measurement data. The recognition process involves top-down hypothesizing sentence-level solutions, together with scoring each hypothesis with the acoustic data using the assumption that the data are produced from a sequence of discrete multi-dimensional articulatory states. These discrete articulatory states are constructed in advance, as will be described in the remaining portion of this section, using a phonemic transcription. (More recent work on incorporating prosodic information and syllabic structure in the state construction will be presented in the next section.) At the heart of the recognizer is the algorithm for automatic conversion of any probabilistic and fractional articulatory feature overlap pattern into a Markov state transition graph, which is described below.

9.3.2 Articulatory state construction

Two key components are required for constructing the discrete articulatory states: 1) an articulatory feature specification system and 2) constraints on feature spreads (overlaps and mixes). A portion of the feature specification system (a re-coded subset of the overall system in Table 9.2) is listed in Table 9.3, where symbolic feature value *'U'* denotes feature underspecification.

Notations

To describe the procedure for the articulatory state construction, we first introduce the following notation. Let

$$\mathbf{\Phi} = (\phi_1, \cdots, \phi_m)$$

be the phonetic transcription of a sentence, where m is the number of phonetic segments and ϕ_i takes a discrete value of phonetic symbols. Let

$$\mathbf{f}(\phi_i) = (f_1(\phi_i), \cdots, f_D(\phi_i))^{\mathrm{Tr}}$$

be the vector of articulatory features of target segment ϕ_i in the phonetic transcription. For example, $\mathbf{f}(/b/) = (L\alpha, U, U, V\alpha, X\alpha)$ shown in Table 1. Similarly, we let

$$\mathbf{g}(\mathbf{f}(\phi_{i+\delta}), \phi_i)$$

be the vector of contextual articulatory features of target segment ϕ_i assimilated by the features at segment $i + \delta$, where δ takes integer values ($\delta > 0$ for anticipatory coarticulation and $\delta < 0$ for carry-over coarticulation). Obviously, $\mathbf{g}(\mathbf{f}(\phi_i), \phi_i) \triangleq \mathbf{f}(\phi_i)$ when $\delta = 0$.

Algorithm Description

Input:

phonetic transcription of a given utterance (part or whole sentence): $\mathbf{\Phi} = (\phi_1, \cdots, \phi_m)$.

Output:

articulatory state transition graph.

Algorithm:

1. Attaching D-tuple features to each target segment ($\mathbf{f}(\phi_i)$) according to the feature specification system (D=5).

2. Feature specification of contextual segments ($\mathbf{g}(\mathbf{f}(\phi_{i+\delta}), \phi_i)$):

   With the notation introduced above, the articulatory feature specification of each target segment ϕ_i *in context* can be written as:

$\cdots$	$g_1(\mathbf{f}(\phi_{i-1}), \phi_i)$	$f_1(\phi_i)$	$g_1(\mathbf{f}(\phi_{i+1}), \phi_i)$	$\cdots$
$\vdots$	$\vdots$	$\vdots$	$\vdots$	$\vdots$
$\cdots$	$g_D(\mathbf{f}(\phi_{i-1}), \phi_i)$	$f_D(\phi_i)$	$g_D(\mathbf{f}(\phi_{i+1}), \phi_i)$	$\cdots$

   The values of the contextual features of the segment ϕ_i are determined by the values of $\mathbf{f}(\phi_i), \cdots, \mathbf{f}(\phi_{i+\delta})$ together with feature overlap and spread rules.

Initially, we set

$$\mathbf{g}(\mathbf{f}(\phi_{i+\delta}), \phi_i) = \mathbf{f}(\phi_{i+\delta}).$$

Then, the values of $\mathbf{g}(\mathbf{f}(\phi_{i+\delta}), \phi_i)$ are modified according to a set of rules including:

(a) Maximum feature spread constraint:
$g_d(\mathbf{f}(\phi_{i+\delta}), \phi_i) =$ 'U' for $d = 1, \cdots, D$, and $|\delta| \leq \Delta$, where Δ is a constant indicating the maximum amount of spread.
This rule specifies the constraint on the maximum span of feature spreading.

(b) Discontinuity with underspecified features:
If $g_d(\mathbf{f}(\phi_{i+\delta}), \phi_i) =$ 'U', then $g_d(\mathbf{f}(\phi_{i+\delta'}), \phi_i) =$ 'U', for $|\delta'| \geq |\delta|$.
This rule prevents feature spreading across some segments with underspecified features.

3. Construction of states S:

According to Step-2 results, which give the feature specifications for all the segments *in context*, enumerate all D-tuple feature bundles (states):

$$S \triangleq (g_1(\mathbf{f}(\phi_{i+\delta_1}), \phi_i), \cdots, g_D(\mathbf{f}(\phi_{i+\delta_D}), \phi_i))^{\mathrm{Tr}}$$

which satisfy these two conditions:

(a) $\delta_1, \cdots, \delta_D$ are either all positive or all negative,

(b) $g_d(\mathbf{f}(\phi_{i+\delta_d}), \phi_i) \neq$ 'U' for $\delta_d \neq 0$.

4. Determination of state transitions:

Enumerate all transitions from state (a)

$$\begin{aligned} S(a) = \; & (g_1(\mathbf{f}(\phi_{i+\delta_1(a)}), \phi_i), \cdots, \\ & g_D(\mathbf{f}(\phi_{i+\delta_D(a)}), \phi_i))^{\mathrm{Tr}} \end{aligned}$$

to state (b)

$$\begin{aligned} S(b) = \; & (g_1(\mathbf{f}(\phi_{i+\delta_1(b)}), \phi_i), \cdots, \\ & g_D(\mathbf{f}(\phi_{i+\delta_D(b)}), \phi_i))^{\mathrm{Tr}}, \end{aligned}$$

which satisfy

$$\delta_d(a) \leq \delta_d(b), \quad \text{for } d = 1, \cdots, D,$$

where at least one inequality holds strictly, and

$$\delta_d(b) - \delta_d(a) \leq \Delta_{\text{skip}} \quad \text{for } d = 1, \cdots, D.$$

A speech recognizer that uses the above method of constructing discrete articulatory states has been implemented and evaluated on standard tasks commonly used in speech recognition research. The studies reported in [Deng 94e, Deng 96, Deng 97a] suggest that a great deal of care needs to be taken in incorporating feature-overlap constraint rules, with the degree of rule relaxation made dependent on the amount of training data. Once this care is taken, the results have demonstrated consistently superior performance of the recognizer in comparison with several benchmark systems.

9.4 Use of High-Level Linguistic Constraints

In the earlier sections of this chapter, the feature overlapping rules were constructed based only on the information about the phoneme or its constituents' identity in the utterance. It is well established [Brow 89, Church 87, Cole 98, Golds 90] that a wealth of linguistic factors beyond the level of phoneme, in particular prosodic information (syllable, morpheme, stress, utterance boundaries, etc.), directly control the low-level feature overlapping. Thus, it is desirable to use such high-level linguistic information to control and to constrain feature overlapping effectively. As an example, in pronouncing the word *display*, the generally unaspirated [p] is constrained by the condition that an [s] precedes it in the same syllable onset. On the other hand, in pronouncing the word *displace*, *dis* is a morphological unit of one syllable and the [p] in the initial position of the next syllable subsequently tends to be aspirated.

In order to systematically exploit high-level linguistic information for constructing the overlapping feature-based phonological model in speech recognition, it is necessary to develop a computational framework and methodology in a principled way. Such a methodology should be sufficiently comprehensive to cover a wide variety of utterances (including spontaneous speech) so as to be successful in speech recognition and related applications. Development of such a methodology is the major thrust of some recent research reported in [Sun 01], which we will outline in the remaining portion of this chapter.

9.4.1 Types of high-level linguistic constraints

The computational phonology technique we have introduced here is based on the assumption that high-level (i.e., beyond the phonemic level) linguistic information controls, in a systematic and predictable way, feature overlapping across feature tiers through long-span phoneme sequences. The high-level linguistic or prosodic information used in the work of [Sun 01] for constraining feature overlapping includes

- Utterance, word, morpheme, and syllable boundaries (which are subject to shifts via resyllabification);
- Syllable constituent categories: onset, nucleus and coda; and
- Word stress and sentence accents.

Morpheme boundary and syllabification are key factors in determining feature overlapping across adjacent phonemes. For example, aspiration of voiceless stops in *dis-place* and in *mis-place* versus non-aspiration of the stop in *di-splay* are largely determined by morpheme boundaries and syllabification in these words. In the former case, the feature overlapping occurs at the Larynx tier. Utterance and word boundaries condition several types of boundary phenomena. Examples of the boundary phenomena are glottalized word onset and breathy word ending at utterance boundaries, and the affrication rule at word boundaries (e.g., compare *at right* with *try*) [Jens 93]. Likewise, association of a phoneme with its syllable constituent influences pronunciation in many ways. For example, stops are often unreleased in coda but not so in onset. An example of the effect of word-stress information on feature overlapping is the alveolar-flap rule which only applies to the contextual environment where the current syllable is unstressed and the preceding syllable is stressed within the same word.

This kind of high-level linguistic constraints is applied to the feature-overlapping framework through a predictive model which parses the training sentences into accent groups at the sentence level and into syllabic constituents at the word level. The accent group identification is mainly through part-of-speech tagging information. The syllabic constituent identification is mainly through a context-free-grammar syllable parser based on rules of syllable composition by phonemes (see details later in this section). After this analysis, a sentence is represented by a sequence of symbolic vectors, each containing the phoneme symbol and its syllabic, boundary and accent information, which governs the pronunciation of each phoneme in continuous speech. For example, the utterance "The other one is too big" will be represented as:

[dh ons ub] (ons = syllable onset, ub = utterance beginning)
[iy nuc we ust] (nuc = syllable nucleus, we = word end, ust = unstressed)
[ah nuc wb] (wb = word beginning)
[dh ons]
[ax nuc we ust]
[w ons wb]
[ah nuc ust]
[n cod we] (cod = syllable coda)
[ih nuc wb ust]
[z cod we]
[t ons wb]
[uw nuc we str] (str = stressed)
[b ons wb]
[ih nuc str]
[g cod ue] (ue = utterance end).

In a later part of the chapter we will explain how the high-level information above can be used effectively to constrain feature overlapping.

9.4.2 A parser for English syllable structure

The syllable structures of words are obtained by a recursive transition network-based phonological parser [Church 87], using a pronunciation dictionary. The transition network is derived from a set of context-free grammar (CFG) rules describing the syllable structure of English words. The CFG rules are obtained by reorganizing and supplementing several lists found in [Jens 93]. These rules have been tested for all 6110 words in the TIMIT dictionary. The CFG rules used for constructing the transition network are as follows:

Word → [Init-Onset] V [CvCluster] [Final-Coda]

Init-Onset → C | p,l | p,r | p,w | p,y | b,l | b,r | b,w | b,y | t,r | t,w | t,y | d,r | d,w | d,y | k,l | k,r | k,w | k,y | g,l | g, r | g,w | g,y | f,l | f,r | f,y | v,l | v,r | v,y | th,r | th,w | th,y | s,p | s,p,y | s,t | s,t,y | s,k | s,k,y | s,f | s,m | s,n | s,l | s,w | s, y | s h,m | sh,l | sh,r | sh,w | hh,y | hh,w | m,y | n,y | l,y | s,p,l | s,p,r | s,t,r | s,k,l | s,k,r | s,k,w

CvCluster → [MidC] V [CvCluster]

MidC → MidC41 | MidC31 | MidC32 | MidC20 | MidC21 | C

MidC41 → C, s, C, C

MidC31 → s, C, C | C, s, C | Nas, Fri, Lqd | Nas, Stp, Gld | Nas, Obs, r | Lqd, Fri, Lqd | Lqd, Obs, r | Gld, Fri, Lqd | Gld, Obs, r | Stp, Stp, Lqd | Stp, Stp, Gld | Stp, Fri, Lqd | Fri, Stp, Lqd | Fri, Stp, Gld

MidC32 → Nas, Stp, Lqd | Nas, Stp, Nas | Nas, Stp, Fri | Nas, Stp, Stp | Nas, Stp, Afr | Lqd, Fri, Stp | Lqd, Fri, Nas | Lqd, Fri, Fri | Lqd, Stp, Stp | Lqd, Stp, Lqd | Lqd, Stp, Fri | Lqd, Stp, Gld | Lqd, Stp, Afr | Fri, Fri, hh | r, C, C

MidC20 → p,l | p,r | p,w | p,y | b,l | b,r | b,w | b,y | t,r | t,w | t,y | d,r | d,w | d,y | k,l | k,r | k,w | k,y | g,l | g,r | g,w | g ,y | f,l | f,r | f,y | v,l | v,r | v, y | th,r | th,w | th, y | s,p | s,t | s,k | s,f | s,m | s,n | s,l | s,w | s, y | sh,p | sh,m | sh,l | sh,r | sh,w | hh,y | hh,w | m,y | n,y | l,y

MidC21 → C, C

Final-Coda → C | p, th | t, th | d, th | d,s,t

| k,s | k,t | k,s,th | g, d | g , z | ch, t | jh, d | f, t | f, th | s, p | s, t | s, k | z, d | m, p | m, f | n, t | n, d | n, ch | n, jh | n, th | n, s | n, z | ng, k | ng, th | ng , z | l, p | l, b | l, t | l, d | l, k | l, ch | l , jh | l, f | l, v | l, th | l, s | l, z | l , sh | l, m | l, n | l,p | l,k,s | l,f,th | r, Stp | r,ch | r,jh | r,f | r,v | r,th | r,s | r,z | r,sh | r,m | r,n | r,l

The phoneme type categories are C (consonants), V (vowels), Nas (nasals), Gld (glides), Fri (fricatives), Afr (affricates), Obs (obstruents), Stp (stops), Lqd (liquids). The MidC categories are used for assigning word-internal consonant clusters to either the previous syllable's coda or the next syllable's onset according to one of the following four possibilities:

MidC41 – 1 coda consonants, 3 onset consonants,

MidC31 – 1 coda consonants, 2 onset consonants,

MidC32 – 2 coda consonants, 1 onset consonants,

MidC20 – 0 coda consonants, 2 onset consonants,

MidC21 – 1 coda consonants, 1 onset consonants.

This grammar above is not fully deterministic. The fifth rule of MidC31 and the first rule of MidC32, for example, can result in ambiguous analyses for certain input sequences. For example, the phoneme sequence in *Andrew* can be parsed either by rule MidC31 (Nas Obs r) or by rule MidC32 (Nas Stp Lqd). How to deal with this problem is a practical issue. For parsing a large number of words automatically, the solution described in [Sun 01] is to use this parser to first parse a pronunciation dictionary and then resolve the ambiguities through hand checking. The parsed pronunciation dictionary is then used to provide syllable structures of the words. This procedure was carried out on the TIMIT pronunciation dictionary. The results showed that the rules are a fairly precise model of English syllable and phonotactic structures: Out of 7905 pronunciations, only 135 or 1.7% generated multiple parses.

As an illustration, Fig. 9.11 shows the parse tree for word *display*, which denotes that the word *display* consists of two syllables. The category 'CvCluster' is used for dealing with multiple syllables recursively; 'MidC' and 'MidC31' are categories of intervocalic consonant clusters. The category 'Init-Onset' denotes the word-initial syllable onset. The

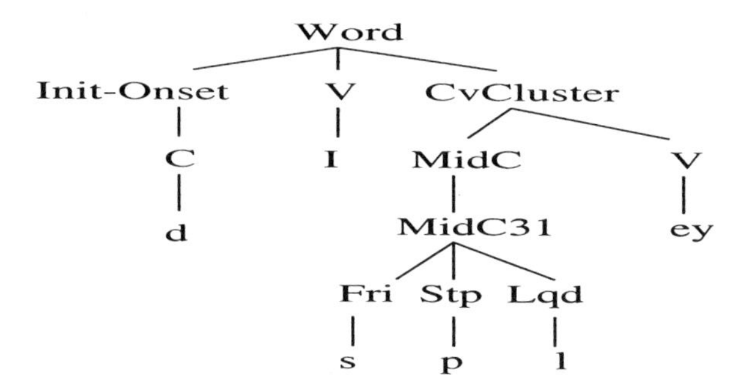

Figure 9.11: The parse tree for the word *display* (after Sun and Deng [Sun 01], @AIP).

separation of syllable-internal consonants into coda and onset is based on the consonant types according to phonotactic principles [Jens 93].

The parser output of an unambiguous tree is transformed into subsegmental feature vectors [Bird 95, Cole 98, Deng 94e] with high-level linguistic information. This is illustrated in Fig. 9.12 for the above example. Here the word *display* is parsed as a single word utterance with **ub** standing for utterance beginning and **ue** for utterance end. Stress is denoted by 0 (unstressed syllable) and 1 (stressed syllable) at the syllable node. The subsegmental feature structure is viewed as an autosegmental structure [Bird 95, Golds 90] with skeletal and articulatory feature tiers and a prosodic structure placed on top of it. There is a resyllabification by which /s/ is moved to the second syllable.

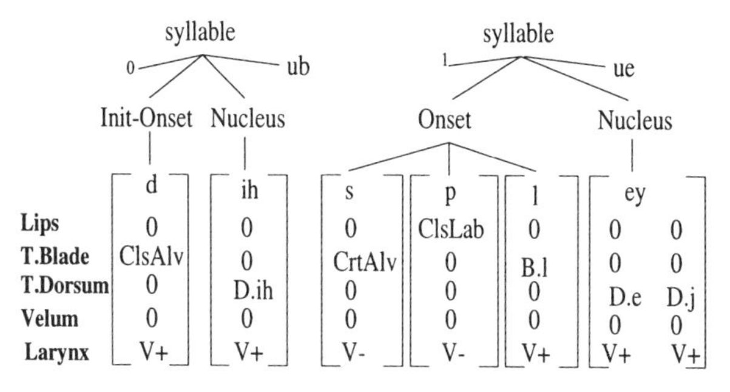

Figure 9.12: Subsegmental feature structure for the word *display* (after Sun and Deng [Sun 01], @AIP).

9.5 Implementation of Feature Overlapping Using Linguistic Constraints

Based on the information obtained from the above syllable parser, as well as other types of high-level linguistic information discussed earlier, we in this section describe a specific implementation of the feature-overlapping engine based on the work published in [Sun 01]. We first outline the feature specification used in this implementation. Then, an overview of a generator for overlapping feature bundles is given, with examples of its outputs provided. We then describe internal operations of the generator. This includes the use of demi-syllables as the organizational units for formulating feature-overlapping rules, and a set of operators used to produce the overlapping feature bundle outputs.

9.5.1 Feature specification

In the specific implementation described in this section, a consistent feature specification system for transforming segment symbols to feature bundles is used. This transformation, also called a generator, is carried out after syllable parsing and before the application of feature overlapping rules. This feature specification system is characterized by the following key aspects:

- Five feature tiers are specified, which are: Lips, Tongue-Blade, Tongue-Dorsum, Velum, and Larynx.
- The feature specification of segments is context independent; it shows canonical articulatory properties coded in symbolic forms. (The total repertoire of the feature values we have designed is intended for all segments of the world languages. For a particular language, only a subset of the repertoire is used.)
- Open (underspecified) feature values are allowed in the feature specification system. These underspecified feature values may be partially or fully filled by temporally adjacent (specified) features during the rule-controlled feature overlapping process.

The feature specification system for American English has the following specific properties. A total of 45 phonemes are classified into eight categories: stops, fricatives, affricates, nasals, liquids, glides, (monophthong) vowels and diphthongs. Each phoneme is specified with a five-tiered feature bundle, corresponding to the five articulators: Lips, Tongue-blade, Tongue-body, Velum, and Larynx. The values for each tier are symbolic, generally concerning the place and manner of articulation (which are distinct from other phonemes) for the relevant articulator. The feature values for any (canonically) irrelevant articulator are underspecified (denoted by the symbol "0" in the example below).

Continuing with the example provided in the previous section, after the phonemes are replaced by their associated articulatory features (before the overlapping operation which will be described later), the utterance "The other one is too big" becomes (the explanations of the prosodic symbols are given in the previous section):

[dh(0 ClsDen 0 0 V+) ons ub] (ClsDen = dental closure, V+ = voiced)
[iy(0 0 D.iy 0 V+) nuc we ust] (D.iy = tongue dorsum position of /iy/)
[ah(0 0 D.ah 0 V+) nuc wb]
[dh(0 ClsDen 0 0 V+) ons]
[ax(0 0 D.ax 0 V+) nuc we ust]
[w(Rnd.u 0 D.w 0 V+) ons wb] (Rnd.u = lip rounding of /u/)
[ah(0 0 D.ah 0 V+) nuc ust]
[n(0 ClsAlv 0 N+ V+) cod we] (ClsAlv = alveolar closure, N+ = nasal)
[ih(0 0 D.ih 0 V+) nuc wb ust]
[z(0 CrtAlv 0 0 V-) cod we] (CrtAlv = alveolar critical, V- = unvoiced)
[t(0 ClsAlv 0 0 V-) ons wb]
[uw(Rnd.u 0 D.uw 0 V+) nuc we str]
[b(ClsLab 0 0 0 V+) ons wb] (ClsLab = labial closure)
[ih(0 0 D.ih 0 V+) nuc str]
[g(0 0 ClsVel 0 V+) cod ue] (ClsVel = velum closure)

9.5.2 A generator of overlapping feature bundles: Overview and examples of its output

The overlapping feature bundle generator is a program which 1) scans the input sequence of feature bundles with high-level linguistic information, 2) matches them to corresponding overlapping rules, 3) executes overlapping (or mixing) operations specified in the overlapping rules during two separate, leftward-scan and rightward-scan processes. The execution starts from the right-most phoneme for the leftward-scan process, and it starts from the left-most phoneme for the rightward-scan process; and 4) integrates the results of leftward-scan and rightward-scan to produce a state-transition network. A block diagram that gives an overview of the overlapping feature bundle generator is shown in Fig. 9.13.

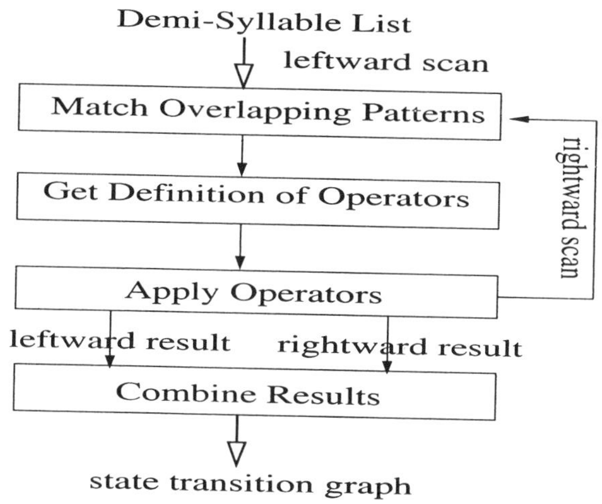

Figure 9.13: Overview of the overlapping feature bundle generator (after Sun and Deng [Sun 01], @AIP).

The feature-overlapping rules, which will be discussed in a much greater detail in this section, contain two types of information: possibility information and constraint information. The possibility component specifies what features can overlap and to what extent, regardless of the context. The constraint component specifies various contexts to block feature overlapping. In the following are some examples of possibility and constraint:

- Possibility of Velum Feature Overlapping: A velum-lowering feature can spread left and right to cause the phenomenon of nasalization in some phones, such as vowels.

- Possibility of Lip Feature Overlapping: A lip rounding feature can spread mainly to the left to cause the phenomenon of lip-rounded allophones.

- Possibility of Tongue Body Feature Overlapping: A tongue body feature can spread to cause such phenomenon in stops as advanced or retracted tongue body closures (as in /g iy/ versus /g uh/).

- Possibility of Larynx Feature Overlapping: A voicing/unvoicing feature can spread to cause such phenomena as voiced/unvoiced allophones.

- Possibility of Tongue Tip Feature Overlapping: The tongue tip feature of /y/ can spread into the release phase of a stop to cause the phenomenon of palatalization (as in "did you").

- Constraint rule: A stop consonant blocks feature spreading of most features, such as lip feature, larynx feature, etc.

- Constraint rule: A vowel usually blocks tongue body features from spreading through it.

The above spreading-and-blocking model can account for many types of pronunciation variation found in continuous speech, but there are some other common phenomena that cannot be described by feature spreading only. The most common among these are the reductive alternation of vowels (into schwa) and consonants (flapping, unreleasing, etc.). Therefore, our model needs to include a control mechanism that can utilize high-level information to "impose" feature transformation in specific contexts. We give some examples below:

- Context-controlled transformation: A stop consonant undergoes a flap transformation in such contexts as: [V stressed] * [V unstressed] (where '*' marks the position of the consonant in question).

- Context-controlled transformation: A stop consonant deletes its release phase in a coda position.

- Context-controlled transformation: A vowel undergoes a schwa transformation in an unstressed syllable of an unaccented word in the utterance.

The output of the generator is a state-transition network consisting of alternative feature bundle sequences as the result of applying feature-overlapping rules to an utterance. This structure directly corresponds to the state topologies of HMMs for speech. Each distinctive HMM state topology can be taken as a phonological representation for a word or for a (long-span) context-dependent phone. (The HMM parameters, given the topology, can be subsequently trained using the feature parameters extracted from the speech signal.)

Examples of feature bundles from the generator's output

Here we give two examples to illustrate typical applications of the feature overlapping rules utilizing high-level linguistic information before details of the rules are formally described. The first example shows how the words *display* and *displace* are endowed with different feature structures in the stop consonant /p/, despite the same phoneme sequence embedding the /p/. The difference is caused by different syllable structures. After syllable parsing and feature overlapping, the results of feature bundles, accompanied by the spectrograms of the two words, are shown in Fig. 9.14. Due to different syllable structures: (/d ih s . p l ey s/ versus /d ih . s p l ey/), different overlapping rules are applied. This simulates the phonological process in which the phoneme /p/ in *displace* tends to be aspirated but in *display* unaspirated.

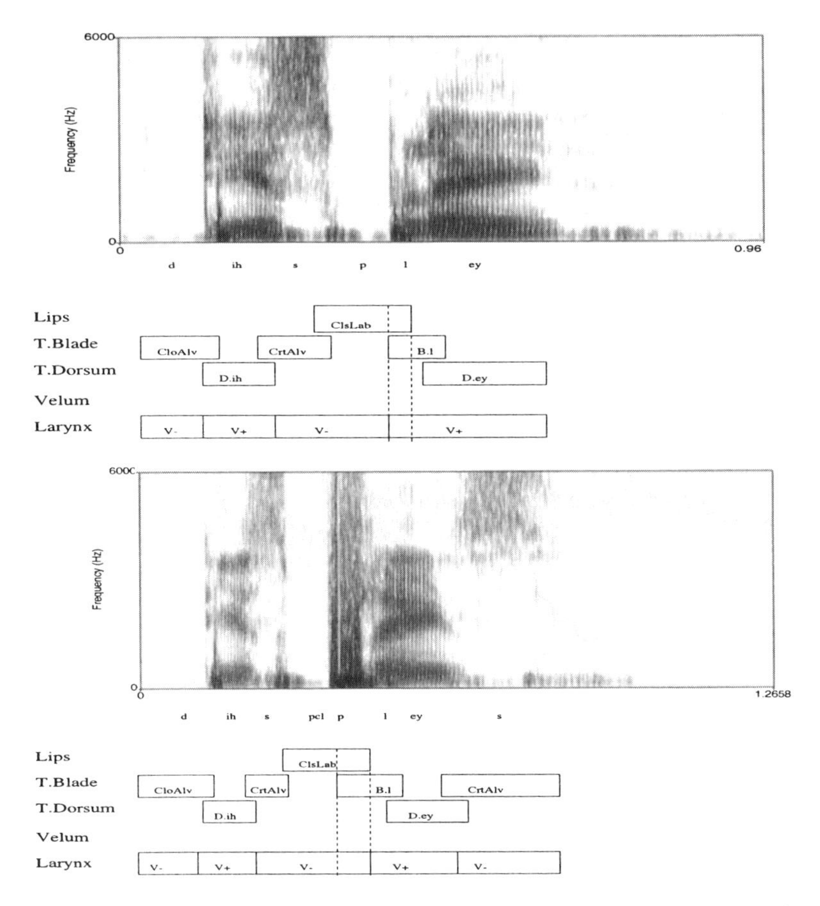

Figure 9.14: Feature overlaps for the words "display" (upper) and "displace" (lower) (after Sun and Deng [Sun 01], @AIP).

The two relevant feature bundles are shown in the figure by dashed vertical lines. The difference lies in the voicing feature at the larynx feature tier. The aspiration is indicated by a V-feature in the feature bundle of the word *displace* between /p/ and /l/. Phonologically, this is called delayed voicing in the onset of /l/. In the model, this is realized through asynchronous leftward spreading of the tongue blade and larynx features of /l/, which overlap with the features of /p/.

The second example in Fig. 9.15 uses the word *strong*, which contains several feature overlaps and mixes; recall that feature mixes are defined as feature overlaps at the same feature tier. Some features have variable (relative) durations (in lip-rounding and nasalization), represented by the dashed boxes. Such variability in the duration of feature overlapping gives rise to alternative feature bundle sequences. By merging identical feature bundles, a network can be constructed, which we call the "state transition network." Each state in the network corresponds to a feature bundle.

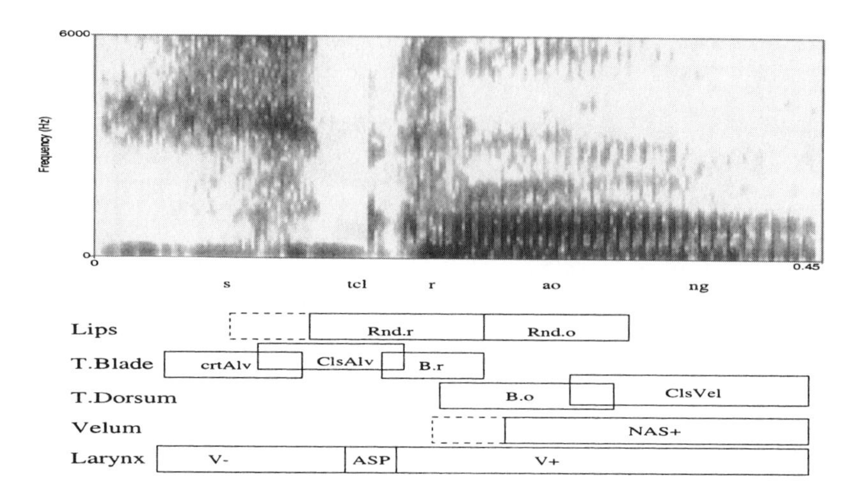

Figure 9.15: Feature overlaps and mixes for the word *strong* (after Sun and Deng [Sun 01], @AIP).

The network constructed by the overlapping feature bundle generator for the word *strong* is shown in Fig. 9.16, where each state is associated with a set of symbolic features. The branches in the network result from alternative overlapping durations specified in the feature overlapping rules.

Generally, a derived feature bundle with overlapping features from adjacent segments represents a transitional phase (coarticulation) between phonemes in continuous speech. Overlapping in real speech can pass several phonemes and our feature-overlapping model effectively simulates this phenomenon. For example, in *strong* /s t r ao ng/, the lip rounding feature of /r/ can spread through /t/ to /s/, and the nasal feature of /ng/ can also pass through /ao/ to /r/, as shown in Fig. 9.15. This ability to model long-span phonetic context is one of the key characteristics of the feature-overlapping approach presented in this section.

9.5.3 Demi-syllable as the rule organizational unit

We now discuss in more detail the generator for overlapping feature bundles. Based on the information obtained by the syllable parser and the feature specification (including underspecification) of phonemes described already, demi-syllables are constructed first, and they are operated upon by the feature-overlapping rules to generate transition networks of feature bundles. A demi-syllable in the system of [Sun 01] is a sequence of broad phoneme categories encompassing the phonemes in either syllable-onset plus nucleus, or nucleus plus syllable-coda formations, together with high-level linguistic information. When a syllable has no onset or coda consonants, that demi-syllable will be only a vowel. The broad phonetic categories used in [Sun 01] are defined as follows:

- V — vowel,
- GLD — glide,
- LQD — liquid,

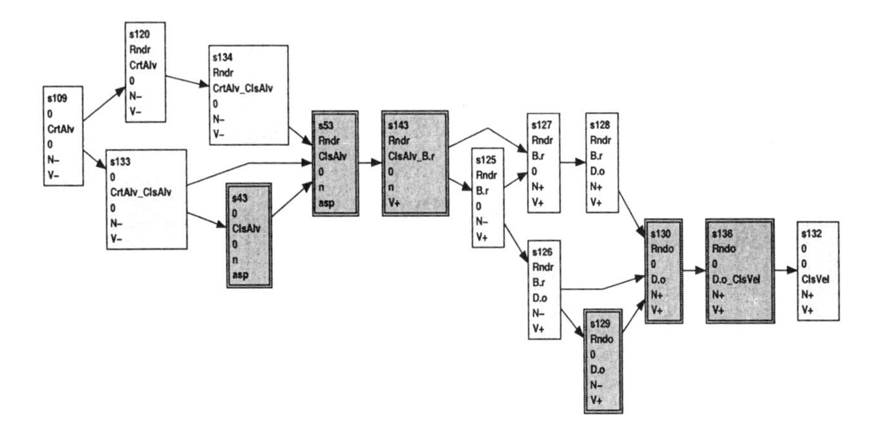

Figure 9.16: State-transitional graph for word *strong* (after Sun and Deng [Sun 01], @AIP).

- NAS – nasal,
- AFR – affricate,
- FRI1 – voiced fricative,
- FRI2 – voiceless fricative,
- STP1 – voiced stop,
- STP2 – voiceless stop.

Other elements included in a demi-syllable are related to higher-level linguistic information. These include:

- ons – syllable onset,
- nuc – syllable nucleus,
- cod – syllable coda,
- ub – utterance beginning,
- ue – utterance end,
- wb – word beginning,
- we – word end,
- str – stressed syllable in the utterance,
- ust – unstressed syllable.

For instance, the demi-syllables of the utterance "The other one is too big," including high-level linguistic information, are as follows:

[FRI1 ons ub] [V nuc we ust] (dh-iy)
[[V, nuc, wb]] (ah)
[FRI1 ons] [V nuc we ust] (dh-ax)
[GLD ons wb] [V nuc ust]] (w-ah)
[V nuc ust] [NAS cod we]] (ah-n)
[V nuc wb ust] [FRI2 cod we] (ih-z)
[STP2 ons wb] [V nuc we str] (t-uw)
[STP1 ons wb] [V nuc str]] (b-ih)
[V nuc str] [STP1 cod ue] (ih-g)

Demi-syllables split a full syllable (one with both onset and coda consonants) into two halves. The purpose of this splitting is to make a small set of units for practical rule development. Contextual constraints specified in the phonological rules are defined on the demi-syllables. In the study of [Sun 01], after parsing all 6110 words in the TIMIT corpus dictionary, a total of 291 distinct word-based demi-syllables (that is, without specifying utterance boundaries and utterance accents) were obtained. This is a very compact set, thus facilitating the development of the feature overlapping rule system.

9.5.4 Phonological rule formulation

We now give a detailed description of the phonological rules that govern articulatory feature overlapping. The phonological rules have been formulated systematically based on the general expected behavior of articulatory features, especially that under the influence of high-level linguistic structures. These rules are used to map any utterance from its demi-syllable representation into its corresponding feature bundle network (i.e., the state transition graph).

The data structure of feature-overlapping rules consists of "overlapping patterns" and "overlapping operators." Each overlapping pattern is defined with respect to a demi-syllable and contains the names of a number of overlapping operators. The demi-syllable contains both segmental information (in the form of broad phonetic categories) and high-level linguistic information (boundaries, accents and syllable constituents). The construction of overlapping patterns starts from the word-based demi-syllables. Based on phonological knowledge concerning coarticulation and phonetic alternations, necessary boundary and accent requirements are added. Further, a number of overlapping operators' names are added to form an overlapping pattern. Each operator corresponds to a broad phonetic category in the demi-syllable.

The overlapping operators are defined on the phonemes based on phonological theories that describe how the relevant articulatory features may overlap (or mix) in speech. When an overlapping pattern is applied, an operator's name will point to the actual definition, which is then applied to the corresponding phoneme that matches a broad phonetic category. One definition of an operator may be pointed to by more than one overlapping pattern. Thus, the overlapping operators realize the possibilities while the overlapping patterns realize the "constraints" on the "possibilities."

Let us denote a broad phone category in a demi-syllable by DSC (standing for demi-syllable constituent). Then a phonological rule can be formulated by a list of DSC's in a demi-syllable, together with all possible operators allowed to operate on each DSC. The overall data structure of a phonological rule is in the following form:

[DSC-1: operator1.1, operator1.2, operator1.3 ... (high-level information)]

[DSC-2: operator2.1, operator2.2, operator2.3 ... (high-level information)]
[DSC-3: operator3.1, operator3.2, operator3.3 ... (high-level information)]
...

An operator describes how feature overlapping could happen on different articulatory tiers, as is described in phonological theory, such as "lip rounding," "jaw lowering," "palatalization," etc. Each operator consists of four components: 1) action, 2) tier-specification, 3) feature-value constraint, and 4) relative-timing. Below we discuss each of these components. First, there are three choices for describing an action:

- L or R: For leftward (look-ahead) or rightward (carry-over) feature spread from an adjacent phoneme onto an underspecified tier of the phoneme.
- M or N: For leftward or rightward mixture of a feature from an adjacent phoneme on the same tier.
- S: For substitution of a feature value by a different feature value.

Second, a tier-indicator specifies at which feature tier an action takes place. A tier indicator is given by an integer as follows:

- 1: the Lips tier,
- 2: the Tongue-Blade tier,
- 3: the Tongue-Dorsum tier,
- 4: the Velum tier,
- 5: the Larynx tier.

Third, a value constraint can optionally be given to stipulate that a feature spread from an adjacent phoneme must have a specified value. If this value constraint is not given, the default requirement is that on this tier of an adjacent phoneme there must be a specified feature in order for the operator to be applicable.

Fourth, a relative-timing indicator is used to specify the temporal extent of a feature spreading. In the current implementation of the model, we use four relative-timing levels: 25%, 50%, 75%, and 100% (full) with respect to the entire duration of the phoneme.

Long-span effects are realized in this model by full (100%) feature spreading. Once an adjacent phoneme's feature is spread to the entire duration of the current phoneme, that feature is visible to the adjacent phoneme on the other side and may spread further. For example, a nasal feature from a right adjacent phoneme may be allowed to spread to the full duration of a vowel. The phoneme to the left of the vowel can "see" this feature and may allow it to spread into itself. This is the mechanism used by the model to pass a feature over several phonemes until it is blocked.

The naming of an operator follows a syntax which reflects its internal definition. The syntax for an operator name is given as:

$$\text{Operator-Name} := Op\, N^{+}\ [\ @\ N^{+}\ ]$$
$$N := 1 \mid 2 \mid 3 \mid 4 \mid 5$$

where the numbers after '*Op*' reflect the tier-indicators in the definition, and the optional numbers after the symbol @ stand for the tiers at which feature-value constraints are imposed.

A phoneme can be given a number of operators. Whether an operator is allowed to apply to a phoneme depends on whether it is listed in a DSC of an overlapping pattern. Furthermore, whether an operator listed in a DSC can be fired or not depends on if the conditions in the operator definition are met. For example, for the operator with the name *Op2* to fire, the second tier of the adjacent phoneme must have a specified feature value. As another example, for the operator of the name *Op12@2* to fire, the adjacent phoneme (whether it is to the left or right depends on the action type of the operator) must have specified features at tier 1 and 2 and the feature value at tier 2 must match the value specified in its definition.

As an illustration, Fig. 9.17 shows the result of applying an operator named *Op*125@15 to the feature bundle of /t/ when it is followed by /r/. The operator is defined as

(125@15, tier_1.L.rnd, tier_2.M, tier_5.L.V+, time:(.5,.25,.25; 1,.25,.25)).

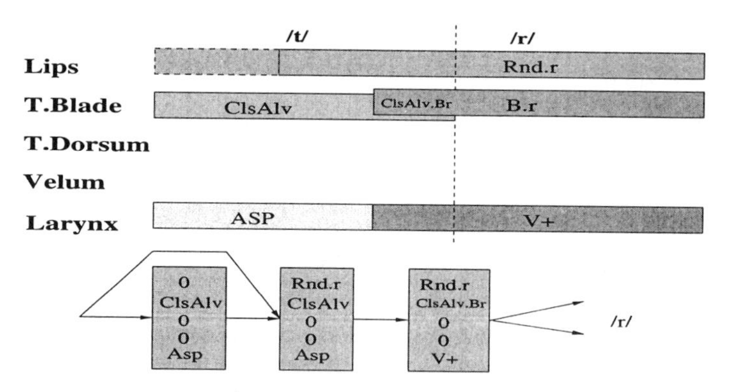

Figure 9.17: Results of applying operator *Op125@15* to /t/ before /r/ and the corresponding state transition graph of /t/ (after Sun and Deng [Sun 01], @AIP).

According to this definition, the three tiers of the phoneme – Lips (1), Tongue-Blade (2) and Larynx (5) – have actions M or L. Tiers 1 and 5 constrain the spreading feature values as *rnd* and *V+* that come from a right neighbor. There are two alternative timing specifications (.5,.25,.25) and (1,.25,.25). Feature spreading at the three tiers will enter the feature bundle of /t/ in two possible ways: 1) Lips feature spreading to 50% of the entire duration, and Tongue-Blade and Larynx features spreading to 25%, or 2) Lips feature spreading to the entire duration and the Tongue-Blade and Larynx feature spreading to 25%. As a consequence, two new feature bundles are derived. The two possible ways for state transitions are shown in Fig. 9.18, which is automatically derived by a node-merging algorithm accepting parallel state sequences. Note how long-distance feature overlapping can be realized by the rule mechanism: Once a feature spreading covers an entire duration, this feature will be visible to the next phoneme. Now we give an example of a phonological rule, which is defined on the demi-syllable with high-level linguistic structure:

[FRI2 ons wb] [STP2 ons] [LQD ons] [V nuc str]

This demi-syllable can match the first four phonemes of the word *strong*. This rule is expressed as:

[FRI2 (Op2, Op3, Op13@1) ons wb]
[STP2 (Op2, Op125@15) ons]
[LQD (Op3, Op34@4) ons]
[V (Op3, Op34@4) nuc str].

Each DSC in this rule is given a number of operators which can operate on the phonemes that are matched by the demi-syllable. Notice the high-level linguistic structures (ons, wb, etc.) which constrain the application of the rule to certain prosodic contexts. In the current implementation of the feature-based model, we have the following operator inventory, which consists of a total of 26 operators defined for the 44 English phonemes for leftward scanning. A corresponding set of operators for rightward scanning are similarly defined. We list the leftward operators as follows:

1. `(Op1,1.M,(.25))` (transitional phase)
2. `(Op1,1.L,(.25))`
3. `(Op2,2.M,(.25))`
4. `(Op2,2.L,(.25))`
5. `(Op3,3.M,(.25))`
6. `(Op3,3.L,(.25))`
7. `(Op5,5.S,())` (glottal substitution)
8. `(Op2,2.S,())` (tongue-blade substitution)
9. `(Op4,4.L.N+,(.5;1))` (nasalization)
10. `(Op12@1,1.L.rnd,2.M,(.5,.25;1,.25))` (transition with lip rounding)
11. `(Op13@1,1.M.rnd,3.L,(.5,.25;.25,.25))`
12. `(Op13@1,1.L.rnd,3.M,(.5,.25;1,.25))`
13. `(Op13@1,1.L.rnd,3.L,(.5,.25;1,.25))`
14. `(Op14@4,1.L,4.L.N+,(.25,.5;.25,1))` (transition with nasalization)
15. `(Op24@4,2.L,4.L.N+,(.25,.5;.25,1))`
16. `(Op34@4,3.M,4.L.N+,(.25,.5;.25,1))`
17. `(Op23@2,2.S.TapAlv,3.L,(.25,.75;1,.25))`
18. `(Op34@4,3.M,4.l.N+,(.25,.5;.25,1))`
19. `(Op34@4,3.L,4.L.N+,(.25,.5;.25,1))`

20. (Op35@5,3.M,5.L.V+,(.25,.25)) (transition with unaspiration)

21. (Op35@5,3.L,5.L.V+,(.25,.25))

22. (Op125@15,1.L.rnd,2.M,5.L.V+,(.5,.25,.25;1,.25,.25)) (more combinations)

23. (Op134@14,1.M.rnd,3.L,4.L.N+,(.5,.25,.5;.5,.25,1;1,.25,.5))

24. (Op134@14,1.L.rnd,3.L,4.L.N+,(.5,.25,.5;.5,.25,1;1,.25,.5))

25. (Op135@15,1.M.rnd,3.L,5.L.V+,(.5,.25,.25;1,.25,.25))

26. (Op135@15,1.L.rnd,3.L,5.L.V+,(.5,.25,.25;1,.25,.25))

To illustrate the use of overlapping phonological rules and how high-level linguistic information is incorporated, we demonstrate with the example utterance "a tree at right" (the corresponding phoneme sequence is /ax t r iy ae t r ay t/). After prosodic processing, where part-of-speech tagging and shallow syntactic parsing is used for deriving the boundary and accent information, and following syllable parsing, the utterance is represented by a sequence of demi-syllables:

1. [V nuc ub ust] (ax)
2. [STP2 ons wb] [FRI1 ons] [V nuc we str] (t-r-iy)
3. [V nuc wb ust] [STP2 cod we] (ae-t)
4. [FRI1 ons wb] [V nuc str] (r-ay)
5. [V nuc str] [STP2 cod ue] (ay-t).

Each demi-syllable is matched by a phonological rule. The overlapping operators in each DSC are examined to see which apply. If the conditions are met, an operator is applied to derive feature bundles. During the derivation process, segment and word boundaries are recorded to divide the derived network into word networks or phone networks, which are used to build word- or phone-based hidden Markov models.

In this example, we illustrate the use of syllable information to realize the "affrication rule" discussed earlier. The utterance's waveform, spectrogram and relevant features concerning the use of the affrication rule are shown in Fig. 9.18. To realize the affrication rule, the phonological rule matching the second demi-syllable: [STP2 ons wb] [FRI1 ons] [V nuc we str] will have its first DSC assigned an operator: (Op2,2.L,(.25)) which allows feature overlapping on the tongue blade tier. The overlapping phonological rule matching the third demi-syllable, on the other hand, will not assign this operator to the second DSC: [STP2 cod we], blocking affrication.

As another example of applying high-level linguistic information, consider the use of a substitution action in an operator at the utterance beginning. For the above utterance, a rule matching the first demi-syllable: [V nuc ub ust] can have an operator with a glottal substitution action. This simulates an utterance with a glottal stop at the outset. Similarly, an unreleased stop consonant at the end of a word or utterance can be simulated by the phonological rule mechanism as well.

We have illustrated how "possibilities" and "constraints" can be implemented by the overlapping patterns and operators. With each DSC within a rule there may be a number of operators available for application. When more than one operator can be applied, it is the more specific ones that are applied first. Depending on how complex we expect the generated network to be, the system is able to control how many operators are applied.

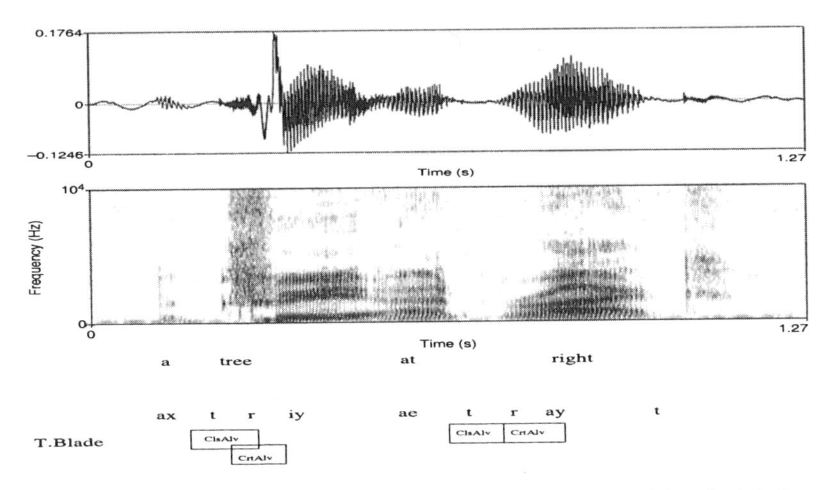

Figure 9.18: Example for using phonological rules and high-level linguistic information (after Sun and Deng [Sun 01], @AIP).

9.6 Summary

This chapter built upon the previous chapter on phonological processes and introduced related computational techniques. It focused on a particular approach based on the overlapping articulatory features, which in turn had their origin in autosegmental phonology, feature geometry, articulatory phonology, and syllable theory.

The chapter started by introducing the concept of multi-tiered overlapping articulatory features and a system for the feature specification for American English. The development of such a concept and system has been motivated by nonlinear theories of phonology. They capture essential linguistic properties of the gestures and features in nonlinear phonology, but have taken a substantially simplified form and have offered special advantages based on computational considerations. We then discussed the temporal overlapping and mixing of the articulatory features across tiers and within a tier, where we introduced the concepts of major and secondary articulatory features, and provided a number of examples for feature assimilation and overlapping. A set of empirical rules that govern feature overlapping were also presented.

The chapter then proceeded to discuss how to construct the discrete articulatory state network or graph from the pattern of overlapping articulatory features. This construction was aimed at improving pronunciation modeling, which has been a key component in speech recognition. This graph can be used as the topology of a sentence-level HMM as the pronunciation model for speech recognition.

Finally, built on top of rule-governed feature overlapping pattern generation and articulatory state network construction, recent research on the exploitation of high-level linguistic or prosodic constraints was presented in detail. A parser for English syllable structure was described first, which provided some essential linguistic information above the phoneme level. The results of the parser were used to guide the lookup of phonological operators, which acted as phonological rules to generate overlapping feature bundles. This specific implementation method that transformed a phonotactically valid phonemic sequence (marked by some linguistic tags) for English to a discrete articulatory state graph has been described in detail in this chapter.

Chapter 10

Computational Models for Speech Production

Phonetics, as the physical and biological sciences of speech sounds, was introduced in Chapter 7. It studies facts and mechanisms of speech production and speech perception, and expands a diverse scope of physiology, acoustics, and psychology as specifically related to the unique human speech process. Computational phonetics is a sub-discipline of phonetics that applies computational techniques to solve complex problems in phonetics; it is also a discipline of computing that applies the principles and theories of phonetics to solve problems in speech technology, speech recognition and speech synthesis.

Phonetic science and computational phonetics have, in general, rather different approaches towards solutions of speech production and perception problems. They also have different objectives. The former typically uses scientific approaches, both theoretical and experimental, with the main objective of understanding the principles and mechanisms underlying the human speech process. The latter, on the other hand, emphasizes an engineering and computational approach, where optimization and statistical techniques tend to be heavily used. The main objectives of computational phonetics include not only providing solutions to complex problems in phonetics that would be otherwise difficult to solve without using powerful computational techniques, but also building intelligent machines that would exhibit some aspects of human speech processing capabilities.

This chapter and the following chapter will cover two main areas of computational phonetics — speech production and auditory speech processing. As in the previous chapter on computational phonology, we will not aim at an even coverage of the field, but instead at some particular approaches in computational phonetics most familiar to and within the research areas of the authors. This chapter is devoted to computational models and techniques for speech production. We will first provide an introduction to and overview of speech production modeling from the perspectives of both speech science and technology. We will introduce multiple levels of dynamics in the speech production process. Then the bulk of this chapter will cover the most peripheral or acoustic level of the dynamics, as well as the hidden level(s) of the dynamics internal to the observable acoustic level. Emphasis will be placed on the parameter estimation aspects of statistical modeling for both the acoustic dynamics and hidden dynamics. One particular way of implementing the hidden dynamic model of speech based on the piecewise approximation technique will be presented in detail. Finally, we will present one comprehensive computational model that covers all levels of speech production dynamics.

The materials presented in this chapter have been reorganized and rewritten based on a series of published work in [Deng 92b, Deng 94c, Deng 97c, Deng 98, Deng 99c, Ma 02, Ma 99, Deng 00c, Deng 97e, Diga 93, Ost 96]

10.1 Introduction and Overview of Speech Production Modeling

10.1.1 Two types of speech production modeling and research

In the past thirty years or so, the same physical entity of human speech has been studied and modeled using drastically different approaches undertaken by largely distinct communities of speech scientists and speech engineers. Models for how speech is generated in the human speech production system developed by speech scientists typically have rich structures [Kent 95, Perr 96b]. The structures have embodied detailed multi-level architectures that transform the high-level symbolic phonological construct into acoustic streams via intermediate stages of phonetic task specification, motor command generation, and articulation. However, these models are often underspecified due to 1) their deterministic nature which does not accommodate random variability of speech and only weakly accommodates systematic variability; 2) a lack of comprehensiveness in covering all classes of speech sounds (with some exceptions), and 3) a lack of strong computational formalisms allowing for automatic model learning from data and for an optimal choice of decision variables necessary for high-accuracy speech classifications.

On the other hand, models for how speech patterns are characterized by statistical generative mechanisms, which have been developed in speech technology, notably by speech recognition researchers, typically contain weak and poor structures. These models often simplistically assume direct, albeit statistical, correlates between phonological constructs and surface acoustic properties. This causes recognizers built from these models to perform poorly on unconstrained tasks, and to break down easily when porting from one task domain to another, from one speaking mode to another, and from one language to another. Empirical system tuning and use of ever-more increasing data appear to be the only options for making the systems behave reasonably if no fundamental changes are made to the speech models underlying the recognizers. However, a distinct strength associated with these models is that they are equipped with powerful statistical formalisms, based solidly on mathematical optimization principles, and are suitable for implementation with a flexible, integrated architecture. The precise mathematical framework, despite poor and simplistic model structure, gives rise to ease of automatic parameter learning (training) and to optimal decision rules for speech-class discrimination.

Logically, there are reasons to expect that a combination of the strengths of the above two styles of speech production models, free from the respective weaknesses, will ultimately lead to superior speech technology. This motivation gives rise to the recent development of several statistical models for dynamic speech production that will be covered in this chapter. In presenting these models, the theoretical motivation, mathematical formulation, and algorithms for parameter estimation will be the main topics.

Here we provide an overview of speech production modeling from both speech science and technology literatures, with an emphasis on drawing parallels and contrasts between these two styles of models, developed by largely separate research communities in the past. The purpose of scientific speech production models is to provide adequate representations to account for the conversion of a linguistic (mainly phonological) message to the actions of the production system and to the consequent articulatory and acoustic outputs. Critical issues addressed by these models are the serial-order problem, the degrees-of-freedom problem, and the related context-sensitivity or coarticulation problem in both articulatory and acoustic domains. The models can be roughly classified into categories of global and component models. Within the category of the global production models, the major classes (modified from the classification of [Kent 95], where a large number of references are listed) are: 1) Feedback-feedforward models, which enable both predictive and adaptive controls to operate, ensuring achievement of articulatory movement goals; 2) Motor program and generalized motor program (schema) models, which use preassembled forms of speech movement (called goals or targets) to regulate the production system; 3) Integrated motor program and feedback models, which combine the characteristics and modeling assumptions of 1) and 2); 4) Dynamic system and gestural patterning models, which employ coordinative structures with a small number of degrees of freedom to create families of functionally equivalent speech movement patterns, and use dynamic vocal-tract constriction variables (in the "task" space) to directly define speech movement tasks or goals; 5) Models based on equilibrium point hypothesis, which use shifts of virtual target trajectories, arising from interactions among central neural commands, muscle properties, and external loads, in the articulatory space ("body" rather than "task" space) to control and produce speech movement; and finally 6) Connectionist models, which establish nonlinear units interconnected in a large network to functionally account for prominent speech behavior, including serial order and coarticulation.

Within the category of the component or subsystem models of speech production are the models for the respiratory subsystem, the laryngeal subsystem, and the supralaryngeal (vocal tract) subsystem. In addition, composite models have also been developed to integrate multiple subsystem models operating in parallel or in series [Kent 95].

All of the scientifically motivated speech production models briefly surveyed above have focused mainly on the explanatory power for speech behavior (including articulatory movements and their relations to speech acoustics), and paid relatively minor attention to computation issues. Further, in developing and evaluating these models, comprehensiveness in covering speech classes is often seriously limited (e.g., CV, CVC, VCV sequences only). In stark contrast, speech models developed by technologists usually cover all classes of speech sounds, and computation issues are given a high priority of consideration with no exception. Another key character of the technology-motivated speech models is the rigorous statistical frameworks for model formulation, which permit automatic learning of model parameters from realistic acoustic data of speech. On the negative side, however, the structures of these models tend to be oversimplified, often deviating significantly from the true stages in the human speech generation mechanisms which the scientific speech production models are aiming to account for. To show this, let us view the HMM (which forms the theoretical basis of modern speech recognition technology) as a primitive (very inaccurate) generative model of speech. To show such inaccuracy, we simply note that the unimodal Gaussian HMM generates its sample paths, which are piecewise constant trajectories embedded in temporally uncorrelated Gaussian noise. Since variances of the noise are estimated from time-independent (except with the HMM state-bound) speech data from all speakers, the model could freely allow generation of speech from different speakers over as short as every 10 ms. (For the mixture Gaussian HMM, the sample paths are erratic and highly irregular due to lack of temporal constraints forcing fixed mixture components within each HMM state; this deviates significantly from speech data coming from heterogeneous sources such as multiple speakers and multiple speech collection channels.)

A simplified hierarchy of statistical generative or production models of speech developed in speech technology is briefly reviewed below. Under the root node of a conventional HMM, there are two main classes of its extended or generalized models: 1) nonstationary-state HMM, also called the trended HMM, segmental HMM, stochastic trajectory model, etc. (with some variation in technical detail according to whether the parameters defining the trend functions are random or not); the sample paths of this type of model are piecewise, explicitly defined stochastic trajectories (e.g., [Deng 92b, Deng 94c, Gish 93, Ghit 93a]); 2) multiple-region dynamic system model, whose sample paths are piecewise, recursively defined stochastic trajectories (e.g., [Diga 93, Ost 96]). The parametric form of Class-1 models typically uses polynomials to constrain the trajectories, with the standard HMM as a special case when the polynomial order is set to zero. This model can be further divided according to whether the polynomial trajectories or trends can be observed directly (this happens when the model parameters — e.g., polynomial coefficients — are deterministic) or the trends are hidden due to assumed randomness in the trend function's parameters. For the latter, further classification gives discrete and continuous mixtures of trends depending on the assumed discrete or continuous nature of the model parameter distribution [Deng 97d, Holm 95, Holm 96, Gales 93]. For Class-2, recursively defined trajectory models, the earlier linear model aiming at dynamic modeling at the acoustic level [Diga 93, Ost 96] has been generalized to nonlinear models taking into account detailed mechanisms of speech production. Subclasses of such nonlinear dynamic models are 1) the articulatory dynamic model, and 2) the task-dynamic model. They differ from each other by distinct objects of dynamic modeling, one at the level of biomechanical articulators (body space) and the other at the level of more abstract task variables (task space). Depending on the assumptions about whether the dynamic model parameters are deterministic or random, and whether these parameters are allowed to change within phonological state boundaries, further subclasses can be categorized.

10.1.2 Multiple levels of dynamics in human speech production

The statistical speech production models discussed above can also be functionally classified in terms of the level of dynamics in the speech production process. Functionally speaking, human speech production can be described at four distinctive levels of dynamics. The top level of dynamics is symbolic or phonological. The linear sequence of speech units in linear phonology or the nonlinear (or multilinear) sequence of the units in autosegmental or articulatory phonology demonstrate the discrete, time-varying nature of the speech dynamics at the mental motor-planning level of speech production, before articulatory execution of the motor commands. The next level of the speech production dynamics is the functional, "task-variable" level. At this level, the goal or "task" of speech production is defined, which could be either the acoustic goal such as spectral prominence or formants, or the articulatory goal such as vocal-tract constrictions, or their combination. This task level can be considered as the interface between phonology and phonetics, since it is at this level that each symbolic phonological unit (feature or gesture) is mapped to a unique set of the phonetic (continuously valued) parameters. These parameters are often called the correlates of the phonological units. The third level of dynamics occurs at the physiological articulators. Finally, the last level of the speech production dynamics is the acoustic one, where speech measurements or observations are most frequently made, both in scientific experiments and in speech technology applications.

The several different types of computational models for speech production covered in the remaining portion of this chapter will be organized in view of the above functional levels of speech production dynamics. The dynamic speech production model at the phonological level (the top level of the production chain) was discussed in some detail in Chapter 9, and will be only lightly touched on in this chapter. The dynamic speech production models at the acoustic level (the most peripheral level of the production chain for human or computer speech perception) will be discussed first. The remaining phonetic levels of the speech dynamics will be grouped into what is called the **hidden dynamic model** of speech production and then be discussed. The model is called "hidden dynamic" because either the task dynamics or the articulatory dynamics of speech production are not directly observable from the usual speech measurements at the acoustic level. The basic mathematical backgrounds for these models were provided in Chapters 3 and 4, and we will build on these backgrounds and focus on speech modeling in this chapter.

10.2 Modeling Acoustic Dynamics of Speech

In this section, we review several types of mathematical models that are used to represent the statistical properties of the acoustic feature sequences extracted from the speech signal. We refer to these feature sequences as the acoustic dynamics. These features are usually in the form of short-time Fourier transforms (discussed in Chapter 1) and their further linear or nonlinear transformations, and are usually employ a uniform time scale.

The models for the acoustic dynamics we will discuss in this section form one major class of the statistical speech production or generative models, because the models are intended to provide statistical characterizations of the acoustic observation data and can be used to physically "generate" some sort of acoustic observation feature sequences. Given a linguistic (phonological) description of the speech, such as a phoneme or allophone sequence, the models are able to 1) select the appropriate sets of parameters corresponding to each of these discrete phonological units, and 2) generate the observation sequence by sampling the appropriate statistical distributions that are characterized by the model parameters.

We will in this section cover two types of the acoustic dynamic models of speech, differentiated by the use of either a discrete state or a continuous state to condition the statistical distributions in the models.

10.2.1 Hidden Markov model viewed as a generative model for acoustic dynamics

The hidden Markov model (HMM) and its generalized versions, which were briefly introduced in Sections 3.5 and 4.2, form one principal type of models for the acoustic dynamics of speech. In order to reflect the temporal flow of the acoustic features, the topology of the HMM is typically arranged in a left-to-right manner. The discrete states in the HMM arranged in terms of the left-to-right sequence are then used to represent the underlying symbolic phonological sequence for the speech utterance, and the sampling of the distributions associated with these states forms the temporal sequence of acoustic features. The HMM with such a particular topology is referred to as the "left-to-right" HMM. This is a special case of HMM where the transition probability matrix is constrained to be upper triangle. The general HMM with no such a constraint is sometimes called the "ergodic" or fully connected HMM. Fig. 10.1 shows a three-state left-to-right

HMM. The arrow lines represent state transitions with non-zero probabilities a_{ij}. It also shows that sequential sampling from the distributions $b_j(o_t)$ associated with these states in a left-to-right order produces a sequence of five observation data points $o_1, o_2, ..., o_5$.

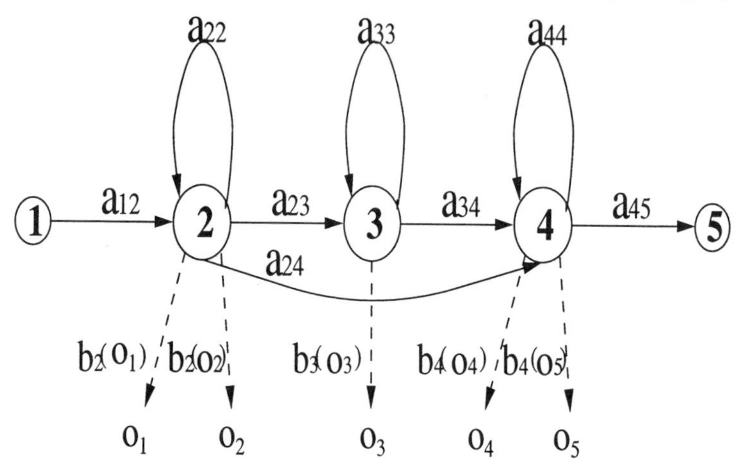

Figure 10.1: A left-to-right HMM and the sampling from its distributions to produce a sequence of observation data points.

The (left-to-right) HMMs are currently responsible for the most successful technology in many speech recognition and other speech processing applications, which we will cover in later chapters. This success is not only due to the rigorous mathematical formulation of the HMM itself, but more importantly due to its conformity to the optimization principle applied to the statistical pattern recognition framework discussed in Chapter 6. In presenting Bayesian decision theory and the related MAP decision rule for pattern recognition in Section 6.1, we discussed the need to compute the class-conditioned "data generation" (or production or synthesis) probability via Bayes' rule. Applying this to the speech modeling and recognition problem, we have the speech observation generation probability $P(\mathbf{o}_1^T|W)$, where $\mathbf{o}_1^T$ is the sequence of the acoustic features, and W is the word (for isolated word recognition) or the word sequence (for continuous speech recognition) as the object or the discrete class of recognition. The MAP decision which evaluates all such production probabilities (in conjunction with prior probabilities) for all the word classes followed by optimization can be interpreted as a probabilistic framework of analysis-by-synthesis. Implicit in Bayesian decision theory is also the need to learn the "production model" parameters so as to achieve high accuracy in evaluating $P(\mathbf{o}_1^T|W)$. HMMs are amenable to highly efficient computation and parameter learning thanks to Baum's original work [Baum 72], and thus would fit naturally into the probabilistic analysis-by-synthesis framework. This is consistent with the qualification of an HMM as a speech generator or production model, because embedded in the HMM there is a mechanism for converting a word sequence W directly into acoustic data $\mathbf{o}_1^T$. One simple way to view an HMM as a speech production or synthesis device is to run Monte Carlo simulation on the HMM and regard the outcome of the simulation as the synthetic speech features.

The theoretical treatment of the HMM as a production model is one thing; how reasonably and effectively it behaves as a production model is another. To examine this latter issue, let us first separate the production probability $P(\mathbf{o}_1^T|W)$ into two factors:

$$\begin{aligned} P(\mathbf{o}_1^T|W) &= \sum_{\mathcal{P}} P(\mathbf{o}_1^T|\mathcal{P})P(\mathcal{P}|W) \\ &\approx \max_{\mathcal{P}} P(\mathbf{o}_1^T|\mathcal{P})P(\mathcal{P}|W), \end{aligned} \tag{10.1}$$

where $\mathcal{P}$ is a discrete-valued *phonological model* and specifies, according to probability $P(\mathcal{P}|W)$, how words and word sequences W can be expressed in terms of a particular organization of a small set of "atomic" phonological units; $P(\mathbf{o}_1^T|\mathcal{P})$ is the probability that a particular organization $\mathcal{P}$ of phonological units produces the acoustic data for the given word sequence W. We shall call this latter mapping device from phonological organization to speech acoustics the *phonetic interface model.*

In view of the factoring of Eq. 10.1, state-of-the-art speech recognizers based on phonetic HMMs can be analyzed as follows. The phonological model $\mathcal{P}$ in most current successful speech recognizers is essentially a linearly-organized multiple-state phonetic sequence governed by a left-to-right Markov chain, and the phonetic interface model is simply a temporally independent random sampling from a set of (trainable) acoustic distributions associated with the states in the Markov chain. Therefore, the following straightforward decomposition in computing the probabilities associated with the phonological model and with the interface model is possible:

$$\begin{aligned} P(\mathcal{P}|W) &= \prod_{t=0}^{T-1} P(s_{t+1}|s_t); \\ P(\mathbf{o}_1^T|\mathcal{P}) &= \prod_{t=1}^{T} b_{s_t}(\mathbf{o}_t), \end{aligned}$$

where $P(s_{t+1}|s_t)$ and $b_{s_t}(\mathbf{o}_t)$ are the transition probabilities of the Markov chain and the state-dependent output distributions of the HMM, respectively.

It is obvious from this discussion that the conventional phonetic HMM outlined above is a poor and naive speech generator or production model: the use of linearly-organized units as the phonological model ignores developments in modern phonological theory as we presented in Chapter 8, whereas the use of an independent and identically distributed (IID) stochastic process (conditioned on the HMM state sequence) as the acoustic interface model discards many of the key temporal correlation properties in the acoustic features resulting from relatively smooth motion of the articulatory structures. This presents many opportunities for improving both phonological and interface components of the HMM.

Given the above rejection of the phonetic HMM as a *good* generator/production model, one may ask why it has achieved so much success in present speech recognition technology. Does such success imply that good production models have no role to play in speech recognition? Our answer is just the opposite. It appears that the current (limited) success of the HMM is to a large degree due to the many constraints imposed on the recognition tasks. These constraints create an artificially sparse phonetic space and limit many possible phonetic confusions. Task constraints are often so strong that even a simple HMM as the production model is able to do a reasonably good job in disambiguating different phonetic classes in many practically useful speech recognition tasks. The important point to note is that such limited success can be largely attributed to the probabilistic "analysis-by-synthesis" framework and to the use of automatic learning for the HMM parameters, which is an implicit component of the framework. There is thus a compelling need for developing production models superior to conventional phonetic HMMs, which we will cover in this and the remaining sections of this chapter, for greater success in speech recognition technology with fewer constraints on the task domain. Recent evaluation experiments on real-world speech recognition tasks using a telephone switchboard database demonstrate poor performance of state-of-the-art technology based on conventional HMMs. The superior performance achieved by discrimination-motivated HMM parameter learning over the conventional maximum-likelihood learning further attests to the poor quality of the latter HMMs as a production model for speech recognition.

In pursuing the development of high-quality global speech production models that can theoretically guarantee superiority in speech recognition tasks as argued above, two key modeling requirements must be emphasized. First, the models must take into account critical mechanisms in the human speech communication process that describe systematic variability in speech acoustics as a necessary and natural means to convey phonologically meaningful information from speaker to listener; much of such systematic variability has been detrimental to the current HMM-based speech technology. Second, the current success of the HMM in technology demonstrates that any good speech production model for use in speech recognition should be compatible with the computational requirements imposed by the probabilistic analysis-by-synthesis framework.

10.2.2 From stationary state to nonstationary state

The above critical view of the HMM as a production model for the acoustic dynamics of speech shows a compelling need to generalize the HMM's capability to better represent speech dynamics. It also shows the importance of parameter estimation or learning for any statistical production model in the probabilistic analysis-by-synthesis framework. In this subsection, we provide one simplistic generalization of the conventional Gaussian HMM in characterizing the acoustic dynamics of speech. This is the trended HMM, or nonstationary-state HMM, developed in [Deng 91c, Deng 92b, Gish 93, Deng 94c, Ost 96], and introduced in Sections 3.5 and 4.2 with its mathematical background. The trended HMM partially overcomes the IID assumption of the conventional HMM, in that the conditional distributions on the observation sequence given the state sequence are no longer identically distributed, although the conditional distributions are still independent across time. In this subsection, we will elaborate on why the trended HMM as a generalized model, which gives piecewise explicit trajectories for the speech feature sequence, is capable of providing a better characterization of speech dynamics than the conventional, stationary-state HMM.

As discussed earlier, an HMM with a left-to-right topology provides a representational form for many non-stationary time series encountered in practical applications, such as the output sequences from the preprocessor in speech recognition systems [Rab 89]. The standard HMM formulation [Baum 72, Lipo 82, Pori 88] assumes that the observation data sequence is state-conditioned IID. It relies solely on the (hidden) Markov chain to fit the manifestly nonstationary behavior of the acoustic feature sequence of speech. This becomes an apparent weakness since, when applying these models to speech processing, the states of the models can only be used to represent piecewise stationary segments of speech, but the acoustic realization of many types of speech sounds is highly dynamic and varies in a continuous manner even during a short segment. Glides, diphthongs, consonant-to-vowel and vowel-to-consonant transitions are typical instances of these sounds; for example, consider the highly regular, non-random trends in the formant transitions in these sounds. It might be argued that use of many states in the standard HMM can approximate such continuously varying, nonstationary acoustic patterns in a piecewise constant fashion. Indeed, it has been found that as many as ten states are needed to model strongly dynamic speech segments in order to achieve a reasonable recognition performance [Deng 91b]. This solution, however, has the obvious shortcoming: many free and largely independent model parameters (e.g., Gaussian mean vectors) — requiring a large amount of data for reliable training — are used to describe just simple non-constant trends. Such trends can be much more efficiently and accurately described by simple deterministic functions of time that require a very small number of parameters, as opposed to using many HMM states to approximate them piecewise constantly.

Our nonstationary time-series model, the trended HMM, accomplishes the above objective. The basic element of the trended model was shown earlier to be the "deterministic trend plus stationary residual" nonstationary time-series model, taking the form of

$$\mathbf{o}_t = \mathbf{g}_t(\mathbf{\Lambda}) + \mathbf{r}_t, \tag{10.2}$$

where $\mathbf{g}_t(\mathbf{\Lambda})$ is the deterministic function of time frame t, parameterized by $\mathbf{\Lambda}$, and $\mathbf{r}_t$ is the stationary residual. Such a model is suitable for a acoustic feature sequence corresponding to a speech segment, either stationary or nonstationary, with a consistent trend. To piece together speech segments with different trends to form a long speech stream, we allow the trend parameters and the residual parameters in the above basic model element Eq. 10.2 to vary over time. Motivated by the general success of the HMM for speech processing applications, we assume that such a parameter variation obeys a Markov chain. We thus arrive at a nonstationary — both locally (i.e., conditioned on the state) and globally — time-series model with deterministic trends and with Markov modulated parameters. This model can be alternatively viewed as a generalized HMM with an underlying Markov chain and with the trend-plus-residual model (Eq. 10.2) as the state-dependent output stochastic process. For simplicity, the residual, $\mathbf{r}_t$, is assumed to be an IID, zero-mean Gaussian source throughout this chapter. (For use of a more general stationary autoregressive moving average source as the residual, the model development follows similar steps but the final parameter estimation formulas are more complex.) In summary, the trended HMM consists of the following parameter sets: 1) transition probabilities, $a_{ij}, i, j = 1, 2, ..., N$; 2) parameters $\mathbf{\Lambda}_i$ in the deterministic trend function $\mathbf{g}_t(\mathbf{\Lambda}_i)$, as dependent on state i in the Markov chain; and 3) covariance matrices, $\mathbf{\Sigma}_i$, of the zero-mean, Gaussian, IID residual $\mathbf{r}_t(\mathbf{\Sigma}_i)$, which are also state dependent. Given these model parameters, the acoustic feature sequence, $\mathbf{o}_t, t = 1, 2, ..., T$ can be "generated" from the model in a probabilistic manner. We now discuss the core problem for the trended HMM — how to optimally determine the model parameters from a set of training data in the form of the acoustic feature sequences.

10.2.3 ML learning for the trended HMM via the EM algorithm

We now describe the work of [Deng 92b], which uses the maximal likelihood (ML) learning (parameter estimation) discussed in Section 5.6, in particular the EM algorithm introduced in Section 5.6.3, to solve the optimal parameter estimation problem for the trended HMM. Since the stationary-state HMM is a special case of the trended HMM, the same ML estimate reduces naturally to that for the standard, stationary-state HMM.

The EM algorithm is a general iterative technique for ML estimation, with local optimality, when hidden variables exist. When such hidden variables take the form of a Markov chain, the EM algorithm becomes the Baum-Welch algorithm, due to the original work in [Baum 72]. For the trended HMM, the hidden random process is indeed a Markov chain; hence the Baum-Welch form of the EM algorithm is developed and discussed below.

Each iteration in the EM algorithm consists of two steps. In the E (expectation) step, the following conditional expectation, or the auxiliary function $Q(\Phi|\Phi_0)$ (we adopt the notation in [Demp 77]), is obtained:

$$Q(\Phi|\Phi_0) = E[\log P(\mathbf{o}_1^T, \mathcal{S}|\Phi)|\mathbf{o}_1^T, \Phi_0], \tag{10.3}$$

where the expectation is taken over the "hidden" state sequence $\mathcal{S}$, Φ stands for the model whose parameters are to be estimated in the present iteration, and Φ_0 stands for the model whose parameters were estimated in the previous iteration. For the EM algorithm to be of utility, $Q(\Phi|\Phi_0)$ has to be sufficiently simplified so that the ensuing M (maximization) step can be carried out easily. Estimates of the model parameters are obtained in the M step via maximization of $Q(\Phi|\Phi_0)$, which is in general much simpler than direct procedures for maximizing $P(\mathbf{o}_1^T|\Phi)$.

As we discussed in Section 5.6 on the EM algorithm, that an iteration of the above two steps will lead to maximum likelihood estimates of model parameters with respect to the objective function $P(\mathbf{o}_1^T|\Phi)$ is a direct consequence of Baum's inequality [Baum 72, Demp 77], which asserts that

$$\log\left(\frac{P(\mathbf{o}_1^T|\Phi)}{P(\mathbf{o}_1^T|\Phi_0)}\right) \geq Q(\Phi|\Phi_0) - Q(\Phi_0|\Phi_0).$$

We now carry out the E and M steps for the trended HMM below.

The E step

The goal of the E step is to simplify the conditional expectation Q into a form suitable for direct maximization in the M step. To proceed, we first explicitly write out the Q function in Eq. 10.3 in terms of the expectation over state sequences $\mathcal{S}$ in the form of a weighted sum

$$Q(\Phi|\Phi_0) = \sum_{\mathcal{S}} P(\mathcal{S}|\mathbf{o}_1^T, \Phi_0) \log P(\mathbf{o}_1^T, \mathcal{S}|\Phi). \tag{10.4}$$

To simplify the writing, denote by $N_t(i)$ the quantity

$$-\frac{D}{2}\log(2\pi) - \frac{1}{2}\log|\mathbf{\Sigma}_i| - \frac{1}{2}(\mathbf{o}_t - \boldsymbol{g}_t(\mathbf{\Lambda}_i))^T \mathbf{\Sigma}_i^{-1}(\mathbf{o}_t - \boldsymbol{g}_t(\mathbf{\Lambda}_i)). \tag{10.5}$$

Using

$$P(\mathcal{S}) = \prod_{t=1}^{T-1} a_{s_t s_{t+1}}$$

and

$$P(\mathbf{o}_1^T, \mathcal{S}) = P(\mathbf{o}_1^T|\mathcal{S})P(\mathcal{S}),$$

we have

$$\log P(\mathbf{o}_1^T, \mathcal{S}|\Phi) = \sum_{t=1}^{T} N_t(s_t) + \sum_{t=1}^{T-1} \log a_{s_t s_{t+1}}.$$

Then the Q function in Eq. 10.4 can be rewritten as

$$Q(\Phi|\Phi_0) = \sum_{\mathcal{S}} P(\mathcal{S}|\mathbf{o}_1^T, \Phi_0) \sum_{t=1}^{T} N_t(s_t) + \sum_{\mathcal{S}} P(\mathcal{S}|\mathbf{o}_1^T, \Phi_0) \sum_{t=1}^{T-1} \log a_{s_t s_{t+1}}. \quad (10.6)$$

To simplify Q, we write the first term in Eq. 10.6 as

$$Q_1 = \sum_{i=1}^{N} \{\sum_{\mathcal{S}} P(\mathcal{S}|\mathbf{o}_1^T, \Phi_0) \sum_{t=1}^{T} N_t(s_t)\}\delta_{s_t,i}, \quad (10.7)$$

and the second term as

$$Q_2 = \sum_{i=1}^{N} \sum_{j=1}^{N} \{\sum_{\mathcal{S}} P(\mathcal{S}|\mathbf{o}_1^T, \Phi_0) \sum_{t=1}^{T-1} \log a_{s_t s_{t+1}}\}\delta_{s_t,i}\delta_{s_{t+1},j}, \quad (10.8)$$

where δ indicates the Kronecker delta function. Let's examine Eq. 10.7 first. By exchanging summations and using the obvious fact that

$$\sum_{\mathcal{S}} P(\mathcal{S}|\mathbf{o}_1^T, \Phi_0)\delta_{s_t,i} = P(s_t = i|\mathbf{o}_1^T, \Phi_0),$$

Q_1 can be simplified to

$$Q_1 = \sum_{i=1}^{N} \sum_{t=1}^{T} P(s_t = i|\mathbf{o}_1^T, \Phi_0) N_t(i). \quad (10.9)$$

Carrying out similar steps for Q_2 in Eq. 10.8 results in a similar simplification

$$Q_2 = \sum_{i=1}^{N} \sum_{j=1}^{N} \sum_{t=1}^{T-1} P(s_t = i, s_{t+1} = j|\mathbf{o}_1^T, \Phi_0) \log a_{ij}. \quad (10.10)$$

Three comments are provided below. First, in maximizing $Q = Q_1 + Q_2$, Q_1 and Q_2 can be maximized independently—Q_1 contains only the parameters in the trend functions and in the residuals, while Q_2 involves just the parameters in the Markov chain. Second, in maximizing Q, the weights in Eq. 10.9 and Eq. 10.10, $P(s_t = i|\mathbf{o}_1^T, \Phi_0)$ and $P(s_t = i, s_{t+1} = j|\mathbf{o}_1^T, \Phi_0)$ which we denote by $\gamma_t(i)$ and $\xi_t(i,j)$ respectively, are treated as known constants due to their conditioning on Φ_0. They can be computed efficiently via the use of the forward and backward probabilities. The posterior state transition probabilities are

$$\xi_t(i,j) = \frac{\alpha_t(i)\beta_{t+1}(j)a_{ij}\exp(N_{t+1}(j))}{P(\mathbf{o}_1^T|\Phi_0)}, \quad (10.11)$$

for $t = 1, 2, ..., T-1$. (Note that $\xi_T(i,j)$ has no definition.) The posterior state occupancy probabilities can be obtained by summing $\xi_t(i,j)$ over all the destination states j

$$\gamma_t(i) = \sum_{j=1}^{N} \xi_t(i,j), \quad (10.12)$$

for $t = 1, 2, ..., T-1$. $\gamma_T(i)$ can be obtained by its very definition:

$$\gamma_T(i) = P(s_T = i|\mathbf{o}_1^T, \Phi_0) = \frac{P(s_T = i, \mathbf{o}_1^T|\Phi_0)}{P(\mathbf{o}_1^T|\Phi_0)} = \frac{\alpha_T(i)}{P(\mathbf{o}_1^T|\Phi_0)}.$$

(For the left-to-right model, $\gamma_T(i)$ has a value of one for $i = N$ and of zero otherwise.) Third, the summations in Eqs. 10.9 and 10.10 are taken over states i or over state pairs i, j, which is significantly simpler than the summations over state sequences $\mathcal{S}$ as in the unsimplified forms of Q_1 and Q_2 in Eq. 10.6. Eqs. 10.9 and 10.10 are the simplistic form of the auxiliary objective function, which can be maximized in the M step discussed next.

The M step

(a) Transition probabilities:

The re-estimation formulas for the transition probabilities of the Markov chain are established by setting

$$\frac{\partial Q_2}{\partial a_{ij}} = 0,$$

for Q_2 in Eq. 10.10 and for $i, j = 1, 2, ..., N$, subject to the constraint $\sum_{j=1}^{N} a_{ij} = 1$. Use of the standard Lagrange multiplier procedure produces the re-estimation formula

$$\hat{a}_{ij} = \frac{\sum_{t=1}^{T-1} \xi_t(i,j)}{\sum_{t=1}^{T-1} \gamma_t(i)},$$

where $\xi_t(i,j)$ and $\gamma_t(i)$ are computed from Eqs. 10.11 and 10.12.

(b) Parameters in the trend functions:

By removing optimization-independent terms and factors in Q_1 expressed in Eq. 10.9, an equivalent objective function is obtained as

$$Q_e(\mathbf{\Lambda}_i) = \sum_{i=1}^{N} \sum_{t=1}^{\mathrm{Tr}} \gamma_t(i)(\mathbf{o}_t - \boldsymbol{g}_t(\mathbf{\Lambda}_i))^{\mathrm{Tr}} \mathbf{\Sigma}_i^{-1} (\mathbf{o}_t - \boldsymbol{g}_t(\mathbf{\Lambda}_i)).$$

The re-estimation formulas are obtained by solving

$$\frac{\partial Q_e}{\partial \mathbf{\Lambda}_i} = 0, \tag{10.13}$$

or

$$\sum_{t=1}^{T} \gamma_t(i)(\mathbf{o}_t - \boldsymbol{g}_t(\hat{\mathbf{\Lambda}}_i)) \frac{\partial \boldsymbol{g}_t(\hat{\mathbf{\Lambda}}_i)}{\partial \mathbf{\Lambda}_i} = 0. \tag{10.14}$$

for $i = 1, 2, ..., N$. (In obtaining Eq. 10.14 from Eq. 10.13, $\mathbf{\Sigma}_i^{-1}$ is dropped out when no constraints, which we assume, are imposed on $\mathbf{\Lambda}_i$ for different states. If such constraints were imposed as with use of the popular technique of parameter tying, Eq. 10.14 would assume a more complex form, which we omit.)

For a very wide class of functions of $\boldsymbol{g}_t(\mathbf{\Lambda}_i)$, Eq. 10.14 is a set of linear systems of equations. Obvious instances of these functions are those with arbitrary forms of time dependence, f_t, but the parameters themselves are in such a linear form,

$$\boldsymbol{g}_t(\boldsymbol{\Lambda}) = \sum_{m=0}^{M} \mathbf{B}(m) f_t(m), \tag{10.15}$$

that derivatives with respect to them contain only time functions. The exact form of Eq. 10.14 can be written once the parametric form of $\boldsymbol{g}_t(\boldsymbol{\Lambda}_i)$ is given. In a later section, we will provide the solution equations for polynomial trend functions.

(c) Covariance matrices of the residuals:

Directly setting

$$\frac{\partial Q_1}{\partial \boldsymbol{\Sigma}_i} = 0$$

would create algebraic difficulties in solving $\boldsymbol{\Sigma}_i$. We employ the trick of setting $\mathbf{K} = \boldsymbol{\Sigma}^{-1}$ (we omit the state index i for simplicity) and treating Q_1 as a function of $\mathbf{K}$. Now, the derivative of $\log|\mathbf{K}|$ (a term in Q_1, see Eqs. 10.9 and 10.5) with respect to $\mathbf{K}$'s lm-th entry, k_{lm}, is the lm-th entry of $\boldsymbol{\Sigma}$, $\boldsymbol{\Sigma}_{lm}$ [Lipo 82]. Using this fact we reduce

$$\frac{\partial Q_1}{\partial k_{lm}} = 0$$

to

$$\sum_{t=1}^{T} \gamma_t(i)\{\frac{1}{2}\hat{\boldsymbol{\Sigma}}_{lm} - \frac{1}{2}(\mathbf{o}_t - \boldsymbol{g}_t(\hat{\boldsymbol{\Lambda}}))_l(\mathbf{o}_t - \boldsymbol{g}_t(\hat{\boldsymbol{\Lambda}}))_m\},$$

for each entry: $l, m = 1, 2, ..., D$. Writing this result in a matrix form, we obtain the compact re-estimation formula

$$\hat{\boldsymbol{\Sigma}}_i = \frac{\sum_{t=1}^{T} \gamma_t(i)(\mathbf{o}_t - \boldsymbol{g}_t(\hat{\boldsymbol{\Lambda}}_i))(\mathbf{o}_t - \boldsymbol{g}_t(\hat{\boldsymbol{\Lambda}}_i))^T}{\sum_{t=1}^{T} \gamma_t(i)}$$

for each state: $i = 1, 2, ..., N$.

10.2.4 Example: Model with state-dependent polynomial trends

Let $f_t(m)$ in Eq. 10.15 be t^m; we then have polynomial trend functions (with a given order M)

$$\boldsymbol{g}_t(\boldsymbol{\Lambda}_i) = \sum_{m=0}^{M} \mathbf{B}_i(m) t^m. \tag{10.16}$$

Here, the trend parameters $\boldsymbol{\Lambda}_i$ are the polynomial coefficients, $\mathbf{B}_i(m), m = 0, 1, ..., M$, each having D dimensions and being a function of state i in the Markov chain.

Substituting Eq. 10.16 into the general formula Eq. 10.14, we obtain the linear system of equations for the re-estimate of the polynomial coefficients

$$\sum_{t=1}^{T} \gamma_t(i)(\mathbf{o}_t - \sum_{l=0}^{M} \hat{\mathbf{B}}_i(l) t^l) t^m = 0, \qquad m = 0, 1, ..., M.$$

This can be re-organized as

$$\sum_{l=0}^{M} C_i(l+m)\hat{\mathbf{B}}_i(l) = \mathbf{V}_i(m), \qquad m = 0, 1, ..., M, \tag{10.17}$$

where the scalar constants are

$$C_i(k) = \sum_{t=1}^{T} \gamma_t(i) t^k, \quad k = 0, 1, ..., 2M, \tag{10.18}$$

the right-hand-side constant vectors are

$$\mathbf{V}_i(m) = \sum_{t=1}^{T} \gamma_t(i) \mathbf{o}_t t^m, \quad m = 0, 1, ..., M,$$

and $\hat{\mathbf{B}}_i(l), l = 0, 1, ..., M$ are the unknowns to be solved.

Three notes are in order. First, no constraints are assumed on the model relating the polynomial parameters associated with different states, and hence Eq. 10.17 is solved for $\mathbf{B}_i(l), l = 0, 1, ..., M$ for each state i independently. Otherwise, a larger set of systems of equations would be encountered to solve $\mathbf{B}_i(l)$ jointly for all states: $i = 1, 2, ..., N$, and jointly with covariance matrices $\mathbf{\Sigma}_i$.

Second, the vector form of the linear system of equations Eq. 10.17 can be decomposed component-by-component into D independent linear systems of equations. Each of the systems, for individual state $i = 1, 2, ..., N$ independently as well, can be put into the matrix form

$$\begin{pmatrix} C_i(0) & C_i(1) & \cdots & \cdots & C_i(M) \\ C_i(1) & C_i(2) & \cdots & \cdots & C_i(M+1) \\ \vdots & \vdots & \vdots & \vdots & \vdots \\ \vdots & \vdots & \vdots & \vdots & \vdots \\ C_i(M) & C_i(M+1) & \cdots & \cdots & C_i(2M) \end{pmatrix} \begin{pmatrix} \hat{b}_i^{(d)}(0) \\ \hat{b}_i^{(d)}(1) \\ \vdots \\ \vdots \\ \hat{b}_i^{(d)}(M) \end{pmatrix} = \begin{pmatrix} v_i^{(d)}(0) \\ v_i^{(d)}(1) \\ \vdots \\ \vdots \\ v_i^{(d)}(M) \end{pmatrix}$$

for $d = 1, 2, ..., D$, where $C_i(k)$ is defined in Eq. 10.18, $\hat{b}_i^{(d)}(k)$ and $v_i^{(d)}(k)$ are the d-th components of $\hat{\mathbf{B}}_i(k)$ and $\mathbf{V}_i(k)$, respectively.

Third, as a special case, if M is set to zero, then Eq. 10.17 reverts to the re-estimation formula for the Gaussian mean vectors in the standard HMM [Lipo 82]. This is expected since, as the time-varying component in the trend function is removed, the resulting degenerated trended HMM is no different from the standard HMM.

10.2.5 Recursively-defined acoustic trajectory model using a linear dynamic system

The trended HMM discussed above explicitly represents the acoustic feature trajectories on a segment-by-segment basis. An alternative way of representing these trajectories is to use recursion on time to define a model for the dynamics. This gives rise to the linear dynamic system, whose state space formulation was introduced in Section 4.3. We now use this recursively defined trajectory model to characterize statistical dynamic properties of the acoustic feature sequence of speech. We note that the recursion in time for this type of model gives rise to a Markov process with its state being continuously valued, in contrast to the HMM, whose state is discretely valued.

In the remaining portion of this section, we will describe the formulation of the linear dynamic system for representing speech feature trajectories, and a unified technique for parameter estimation (learning) of the system. In addition to speech processing, the interest in the dynamic system and in the associated parameter estimation problem has arisen from several different disciplines. In statistics, linear regression techniques can be generalized to include temporal evolution of the input variable, obeying laws governed by an autoregressive process. This generalization of the regression model unfolding in time gives rise to a linear dynamic system model. In control theory, the dynamic system has been widely used as a model for noisy observations on stochastic, linearly-behaved physical plants. In signal processing, a dynamic system can be viewed as a continuous-state counterpart of the conventional discrete-state HMM, which we have discussed extensively in this book.

The success of the dynamic system model, when applied to the various disciplinary areas, depends strongly on the degree to which the model represents the true physical process being modeled. The linear form of the dynamic system model has the advantage of mathematical tractability and (arguably) is a reasonably good representation of the global characteristics of many physical processes. Once the structure of a dynamic system is established (a linear form in many practical applications), then the only factors determining how well the model represents a particular physical process being modeled are the parameters of the system.

Similarly to the trended HMM for nonstationary time series, where discrete states control the global nonstationarity at a higher level than the local trend function within each state, we can impose the same high-level control of the nonstationarity for the linear dynamic system using discrete states. Let s denote the distinctive temporal region (also called mode or regime) corresponding to each discrete state. Then we can use the following nonstationary linear dynamic system model (state-space formulation) to represent the acoustic trajectories of speech in a recursive manner:

$$\begin{aligned} \boldsymbol{x}_{k+1} &= \boldsymbol{A}(s)\boldsymbol{x}_k + \boldsymbol{u}(s) + \boldsymbol{w}_k(s) \\ \boldsymbol{o}_k &= \boldsymbol{C}(s)\boldsymbol{x}_k + \boldsymbol{v}_k(s), \qquad E(\boldsymbol{x}_0) = \boldsymbol{\mu}_0(s), \qquad E(\boldsymbol{x}_0\boldsymbol{x}_0^T) = \boldsymbol{P}_0(s), \end{aligned} \tag{10.19}$$

where $\boldsymbol{x}_k \in \boldsymbol{R}^n$ is the "acoustic state" vector, $\boldsymbol{o}_k \in \boldsymbol{R}^m$ is the observed acoustic output (feature) vector, $\boldsymbol{w}_k$ and $\boldsymbol{v}_k$ are uncorrelated, zero-mean Gaussian vectors with covariance matrices $E(\boldsymbol{w}_k\boldsymbol{w}_k^{\mathrm{Tr}}) = \boldsymbol{Q}(s)$ and $E(\boldsymbol{v}_k\boldsymbol{v}_k^{\mathrm{Tr}}) = \boldsymbol{R}(s)$. The elements of the matrices $\boldsymbol{A}(s)$, $\boldsymbol{C}(s)$, $\boldsymbol{Q}(s)$, and $\boldsymbol{R}(s)$, and vector $\boldsymbol{u}(s)$ are random signals depending on the distinctive region (mode), indexed by s, which serves as the high-level discrete linguistic control of the global nonstationarity for the observed dynamic acoustic features $\boldsymbol{o}_k$.

In the state equation of Eq. 10.20 above, the time k recursion on the state vector $\boldsymbol{x}_k$ is used to temporally smooth the acoustic feature vector $\boldsymbol{o}_k$. The vectors $\boldsymbol{u}(s)$ serve as the mode-dependent acoustic targets for the smoothing. This is one common type of stochastic segment models described in [Diga 93, Ost 96, Deng 97e].

One very important aspect of any statistical model for speech is how to automatically learn or estimate the model parameters given a fixed set of observation (training) data. The next subsection is devoted to this aspect for the linear dynamic system, exploited to recursively represent acoustic trajectories.

10.2.6 ML learning for linear dynamic system

We now describe maximum likelihood (ML) parameter estimation for the linear dynamic system, based on the work published and notations used in [Deng 97e]. For simplicity we only consider the special case with a fixed region s in Eq. 10.20. This corresponds to fixing the two acoustic boundaries (in time frames) for the segment (such as an allophone) when the linear dynamic system is used to represent the dynamics of that segment. Without including the distinct regions, we can drop the region index s in the general nonstationary linear dynamic system Eq. 10.20, and consider the stationary linear dynamic system of

$$\begin{aligned} \boldsymbol{x}_{k+1} &= \boldsymbol{A}\boldsymbol{x}_k + \boldsymbol{u} + \boldsymbol{w}_k \\ \boldsymbol{o}_k &= \boldsymbol{C}\boldsymbol{x}_k + \boldsymbol{v}_k, \qquad E(\boldsymbol{x}_0) = \boldsymbol{\mu}_0, \quad E(\boldsymbol{x}_0\boldsymbol{x}_0^{\mathrm{Tr}}) = \boldsymbol{P}_0, \end{aligned} \tag{10.20}$$

where $\boldsymbol{x}_k \in \boldsymbol{R}^n$ is a state vector, $\boldsymbol{o}_k \in \boldsymbol{R}^m$ is an output vector, $\boldsymbol{w}_k$ and $\boldsymbol{v}_k$ are uncorrelated zero-mean Gaussian vectors with covariances $E(\boldsymbol{w}_k\boldsymbol{w}_k^{\mathrm{Tr}}) = \boldsymbol{Q}$ and $E(\boldsymbol{v}_k\boldsymbol{v}_k^{\mathrm{Tr}}) = \boldsymbol{R}$, respectively.

In parameter estimation problems based on maximum likelihood, it is usually more convenient to work with the negative of the logarithm of the likelihood function as the objective function for optimization. It is possible to do so because the logarithm is a monotonic function. This objective function for the system Eq. 10.20 with respect to $\boldsymbol{\theta}$ is

$$J(\boldsymbol{\theta}) = -\log L(\boldsymbol{x}, \boldsymbol{o}, \boldsymbol{\Lambda}), \tag{10.21}$$

where $\boldsymbol{\theta}$ is a vector of unknown parameters in $\boldsymbol{A}, \boldsymbol{u}, \boldsymbol{C}, \boldsymbol{Q}$, and $\boldsymbol{R}$.

In order to develop a procedure for estimating the parameters in Eq. 10.20, we note first that the joint log likelihood of $\boldsymbol{x} = [\boldsymbol{x}_0, \boldsymbol{x}_1, ..., \boldsymbol{x}_N]$, $\boldsymbol{o} = [\boldsymbol{o}_0, \boldsymbol{o}_1, ..., \boldsymbol{o}_N]$ can be written in the form of

$$\begin{aligned} \log L(\boldsymbol{x}, \boldsymbol{o}, \boldsymbol{\theta}) &= -\frac{1}{2}\sum_{k=1}^{N} \log \boldsymbol{Q} + (\boldsymbol{x}_k - \boldsymbol{A}\boldsymbol{x}_{k-1} - \boldsymbol{u})^{\mathrm{Tr}}\boldsymbol{Q}^{-1}(\boldsymbol{x}_k - \boldsymbol{A}\boldsymbol{x}_{k-1} - \boldsymbol{u}) \\ &\quad - \frac{1}{2}\sum_{k=0}^{N} \log \boldsymbol{R} + (\boldsymbol{o}_k - \boldsymbol{C}\boldsymbol{x}_k)^{\mathrm{Tr}}\boldsymbol{R}^{-1}(\boldsymbol{o}_k - \boldsymbol{C}\boldsymbol{x}_k) + \text{constant}. \end{aligned}$$

We assume that there are no constraints on the structure of the system matrices. When the states are unobservable, the forecasting and smoothing forms of the Kalman filter are used to estimate the unobserved (continuously valued) states $\boldsymbol{x}_k$. The forecast and smoothed values in the Kalman filter estimator depend on the initial values assumed for the system parameters. The new estimates for the system parameters can be obtained again by an iterative technique using the EM algorithm. In the E step, the conditional expectation of the log joint likelihood of $\boldsymbol{x}$ and $\boldsymbol{o}$, given observation $\boldsymbol{o}$, is computed. In the case of hidden state $\boldsymbol{x}$, sufficient statistics containing data series $\boldsymbol{x}_k$, $k = 1, 2, \ldots, N$, are estimated which are conditioned also on $\boldsymbol{o}$. The results of this E step are then used to obtain a new estimate of $\boldsymbol{\theta}(\boldsymbol{A}, \boldsymbol{u}, \boldsymbol{C}, \boldsymbol{Q}, \boldsymbol{R})$ in the M step. The new estimate is then fed back to the E step, and the E-step and the M-step iterate themselves until convergence. We now apply the EM algorithm to the dynamic system model of Eq. 10.20.

The E step

The E step of the EM algorithm computes the conditional expectation of $\log L(\boldsymbol{x}, \boldsymbol{o})$ given $\boldsymbol{o}$. The expectation is defined over the joint space $\boldsymbol{x}$ and $\boldsymbol{o}$. However, the conditioning on $\boldsymbol{o}$ eliminates the $\boldsymbol{o}$ space in the expectation, and hence the conditional expectation is effectively taken over the (hidden) space $\boldsymbol{x}$ only. The conditional expectation can be written as

$$\begin{aligned} U(\boldsymbol{x}, \boldsymbol{o}, \boldsymbol{\theta}) &= E\{\log L(\boldsymbol{x}, \boldsymbol{o}) | \boldsymbol{o}\} \\ &= -\frac{N}{2}\log|\boldsymbol{Q}| - \frac{N+1}{2}\log|\boldsymbol{R}| - \frac{1}{2}\sum_{k=1}^{N} E_N[\boldsymbol{e}_{k1}^{\mathrm{Tr}}\boldsymbol{Q}^{-1}\boldsymbol{e}_{k1}|\boldsymbol{o}] \\ &\quad - \frac{1}{2}\sum_{k=0}^{N} E_N[\boldsymbol{e}_{k2}^{\mathrm{Tr}}\boldsymbol{R}^{-1}\boldsymbol{e}_{k2}|\boldsymbol{o}], \end{aligned} \tag{10.22}$$

where $\boldsymbol{e}_{k1} = \boldsymbol{x}_{k+1} - \boldsymbol{A}\boldsymbol{x}_k - \boldsymbol{u}$, $\boldsymbol{e}_{k2} = \boldsymbol{o}_k - \boldsymbol{C}\boldsymbol{x}_k$, and E_N denotes the conditional expectation based on N samples of data. The estimates $\boldsymbol{Q}$ and $\boldsymbol{R}$ must satisfy

$$\frac{\partial U(\boldsymbol{x}, \boldsymbol{o}, \boldsymbol{\theta})}{\partial \boldsymbol{Q}} = 0$$

and

$$\frac{\partial U(\boldsymbol{x}, \boldsymbol{o}, \boldsymbol{\theta})}{\partial \boldsymbol{R}} = 0.$$

By

$$\begin{aligned} \frac{\partial}{\partial \boldsymbol{Q}}\log|\boldsymbol{Q}| &= \boldsymbol{Q}^{-1}, \\ \frac{\partial}{\partial \boldsymbol{R}}\log|\boldsymbol{R}| &= \boldsymbol{R}^{-1}, \\ \frac{\partial}{\partial \boldsymbol{Q}}\boldsymbol{e}_{k1}^{\mathrm{Tr}}\boldsymbol{Q}\boldsymbol{e}_{k1} &= \boldsymbol{e}_{k1}\boldsymbol{e}_{k1}^{\mathrm{Tr}}, \\ \frac{\partial}{\partial \boldsymbol{R}}\boldsymbol{e}_{k2}^{\mathrm{Tr}}\boldsymbol{R}\boldsymbol{e}_{k2} &= \boldsymbol{e}_{k2}\boldsymbol{e}_{k2}^{\mathrm{Tr}}, \end{aligned}$$

and after setting the partial derivatives of Eq. 10.22, with respect to $\boldsymbol{Q}$ and $\boldsymbol{R}$, to zero we have

$$\begin{aligned} \frac{\partial U}{\partial \boldsymbol{Q}^{-1}} &= -\frac{N}{2}\frac{\partial}{\partial \boldsymbol{Q}^{-1}}\log \boldsymbol{Q}^{-1} + \frac{1}{2}\sum_{k=1}^{N}\frac{\partial}{\partial \boldsymbol{Q}^{-1}}E_N(\boldsymbol{e}_{k1}^{\mathrm{Tr}}\boldsymbol{Q}^{-1}\boldsymbol{e}_{k1}|\boldsymbol{o}) \\ &= \frac{N}{2}\boldsymbol{Q} - \frac{1}{2}\sum_{k=1}^{N}E_N(\boldsymbol{e}_{k1}\boldsymbol{e}_{k1}^{\mathrm{Tr}}|\boldsymbol{o}) \\ &= 0 \\ \frac{\partial U}{\partial \boldsymbol{R}^{-1}} &= -\frac{N+1}{2}\frac{\partial}{\partial \boldsymbol{R}^{-1}}\log \boldsymbol{R}^{-1} + \frac{1}{2}\sum_{k=0}^{N}\frac{\partial}{\partial \boldsymbol{R}^{-1}}E_N(\boldsymbol{e}_{k2}^{\mathrm{Tr}}\boldsymbol{R}^{-1}\boldsymbol{e}_{k2}|\boldsymbol{o}) \\ &= \frac{N+1}{2}\boldsymbol{R} - \frac{1}{2}\sum_{k=0}^{N}E_N(\boldsymbol{e}_{k2}\boldsymbol{e}_{k2}^{\mathrm{Tr}}|\boldsymbol{o}) \\ &= 0. \end{aligned}$$

Then, the estimates Q and R become

$$\bar{Q} = \frac{1}{N}\sum_{k=1}^{N} E_N(e_{k1}e_{k1}^{\mathrm{Tr}}|o), \qquad \bar{R} = \frac{1}{N+1}\sum_{k=0}^{N} E_N(e_{k2}e_{k2}^{\mathrm{Tr}}|o). \tag{10.23}$$

Further, using equality $\mathbf{a}^{\mathrm{Tr}}\mathbf{V}\mathbf{a} = \mathrm{trace}\{\mathbf{V}\mathbf{a}\mathbf{a}^{\mathrm{Tr}}\}$, we have

$$\begin{aligned}
\sum_{k=1}^{N} E_N(e_{k1}^{\mathrm{Tr}}\bar{Q}^{-1}e_{k1}|o) &= \sum_{k=1}^{N} E_N\{e_{k1}^{\mathrm{Tr}}[\frac{1}{N}\sum_{k=1}^{N} E_N(e_{k1}e_{k1}^{\mathrm{Tr}}|o)]^{-1}e_{k1}|o\} \\
&= \mathrm{trace}\{[\frac{1}{N}\sum_{k=1}^{N} E_N(e_{k1}e_{k1}^{\mathrm{Tr}}|o)]^{-1}[\sum_{k=1}^{N} E_N(e_{k1}e_{k1}^{\mathrm{Tr}}|o)]\} \\
&= N\cdot\mathrm{trace}(I) = \mathrm{constant}
\end{aligned} \tag{10.24}$$

$$\begin{aligned}
\sum_{k=0}^{N} E_N(e_{k2}^{\mathrm{Tr}}\bar{R}^{-1}e_{k2}|o) &= \sum_{k=0}^{N} E_N\{e_{k2}^{\mathrm{Tr}}[\frac{1}{N}\sum_{k=1}^{N} E_N(e_{k2}e_{k2}^{\mathrm{Tr}}|o)]^{-1}e_{k2}|o\} \\
&= \mathrm{trace}\{[\frac{1}{N+1}\sum_{k=0}^{N} E_N(e_{k2}e_{k2}^{\mathrm{Tr}}|o)]^{-1}[\sum_{k=0}^{N} E_N(e_{k2}e_{k2}^{\mathrm{Tr}}|o)]\} \\
&= (N+1)\cdot\mathrm{trace}(I) = \mathrm{constant}.
\end{aligned}$$

Finally, substitution of the above back into Eq. 10.22 for $U(x, o, \theta)$ leads to the following objective function for the E step:

$$\begin{aligned}
U(x, o, \theta) &= -\frac{N}{2}\log\{\frac{1}{N}\sum_{k=1}^{N} E_N[|x_k - Ax_{k-1} - u|^2|o]\} \\
&\quad - \frac{N+1}{2}\log\{\frac{1}{N+1}\sum_{k=0}^{N} E_N[|o_k - Cx_k|^2|o]\} + \mathrm{constant}.
\end{aligned} \tag{10.25}$$

The M step

Given the conditional expectation in Eq. 10.25, the M-step aims to minimize the following two separate quantities (both being the expected values of a standard least-squares criterion):

$$\begin{aligned}
U_1(A, u) &= \frac{1}{N}\sum_{k=1}^{N} E_N[|x_k - Ax_{k-1} - u|^2|o] \quad \text{and} \\
U_2(C) &= \frac{1}{N+1}\sum_{k=0}^{N} E_N[|o_k - Cx_k|^2|o].
\end{aligned} \tag{10.26}$$

Since the order of expectation and differentiation can be interchanged, we obtain the parameter estimates by solving

$$\frac{\partial U_1(A, u)}{\partial A} = -\sum_{k=1}^{N} E_N[\frac{\partial}{\partial A}|x_k - Ax_{k-1} - u|^2|o] = 0$$

$$\begin{aligned}\frac{\partial U_1(\boldsymbol{A},\boldsymbol{u})}{\partial \boldsymbol{u}} &= -\sum_{k=1}^{N} E_N[\frac{\partial}{\partial \boldsymbol{u}}|\boldsymbol{x}_k - \boldsymbol{A}\boldsymbol{x}_{k-1} - \boldsymbol{u}|^2|\boldsymbol{o}] = 0\\ \frac{\partial U_2(\boldsymbol{C})}{\partial \boldsymbol{C}} &= -\sum_{k=1}^{N} E_N[\frac{\partial}{\partial \boldsymbol{C}}|\boldsymbol{x}_k - \boldsymbol{A}\boldsymbol{x}_{k-1}|^2|\boldsymbol{o}] = 0,\end{aligned}$$

or, after rearranging terms,

$$\begin{aligned}\bar{\boldsymbol{A}}\sum_{k=1}^{N} E_N(\boldsymbol{x}_{k-1}\boldsymbol{x}_{k-1}^{\mathrm{Tr}}|\boldsymbol{o}) + \bar{\boldsymbol{u}}\sum_{k=1}^{N} E_N(\boldsymbol{x}_{k-1}^{\mathrm{Tr}}|\boldsymbol{o}) &= \sum_{k=1}^{N} E_N(\boldsymbol{x}_k\boldsymbol{x}_{k-1}^{\mathrm{Tr}}|\boldsymbol{o})\\ \bar{\boldsymbol{A}}\sum_{k=1}^{N} E_N(\boldsymbol{x}_{k-1}|\boldsymbol{o}) + N\bar{\boldsymbol{u}} &= \sum_{k=1}^{N} E_N(\boldsymbol{x}_k|\boldsymbol{o}).\\ -\bar{\boldsymbol{C}}\sum_{k=0}^{N} E_N(\boldsymbol{x}_k\boldsymbol{x}_k^{\mathrm{Tr}}|\boldsymbol{o}) + \sum_{k=0}^{N} E_N(\boldsymbol{o}_k\boldsymbol{x}_k^{\mathrm{Tr}}|\boldsymbol{o}) &= 0 \qquad (10.27)\end{aligned}$$

In the matrix form, Eq. 10.27 becomes

$$\begin{aligned}[\bar{\boldsymbol{A}}\ \ \bar{\boldsymbol{u}}] &= [\sum_{k=1}^{N} E_N(\boldsymbol{x}_k\boldsymbol{x}_{k-1}^{\mathrm{Tr}}|\boldsymbol{o})\ \ \sum_{k=1}^{N} E_N(\boldsymbol{x}_k|\boldsymbol{o})]\begin{bmatrix}\sum_{k=1}^{N} E_N(\boldsymbol{x}_{k-1}\boldsymbol{x}_{k-1}^{\mathrm{Tr}}|\boldsymbol{o}) & \sum_{k=1}^{N} E_N(\boldsymbol{x}_{k-1}|\boldsymbol{o})\\ \sum_{k=1}^{N} E_N(\boldsymbol{x}_{k-1}^{\mathrm{Tr}}|\boldsymbol{o}) & N\end{bmatrix}^{-1}\\ \bar{\boldsymbol{C}} &= [\sum_{k=0}^{N} E_N(\boldsymbol{o}_k\boldsymbol{x}_k^{\mathrm{Tr}}|\boldsymbol{o})][\sum_{k=0}^{N} E_N(\boldsymbol{x}_k\boldsymbol{x}_k^{\mathrm{Tr}}|\boldsymbol{o})]^{-1}. \qquad (10.28)\end{aligned}$$

Substituting Eq. 10.28 into Eq. 10.23, we obtain

$$\begin{aligned}\bar{\boldsymbol{Q}} &= \frac{1}{N}\sum_{k=1}^{N} E_N[|\boldsymbol{x}_k - \bar{\boldsymbol{A}}\boldsymbol{x}_{k-1} - \bar{\boldsymbol{u}}|^2|\boldsymbol{o}]\\ &= \frac{1}{N}\sum_{k=1}^{N} E_N\{(\boldsymbol{x}_k\boldsymbol{x}_k^{\mathrm{Tr}}|\boldsymbol{o}) - [\bar{\boldsymbol{A}}\ \ \bar{\boldsymbol{u}}][\boldsymbol{x}_k\boldsymbol{x}_{k-1}^{\mathrm{Tr}}|\boldsymbol{o}\ \ \boldsymbol{x}_k|\boldsymbol{o}]^{\mathrm{Tr}}\}\\ \bar{\boldsymbol{R}} &= \frac{1}{N+1}\sum_{k=0}^{N} E_N[|\boldsymbol{o}_k - \bar{\boldsymbol{C}}\boldsymbol{x}_k|^2|\boldsymbol{o}]\\ &= \frac{1}{N+1}[\sum_{k=0}^{N} E_N(\boldsymbol{o}_k\boldsymbol{o}_k^{\mathrm{Tr}}|\boldsymbol{o}) - \bar{\boldsymbol{C}}(\sum_{k=0}^{N} E_N(\boldsymbol{o}_k\boldsymbol{x}_k^{\mathrm{Tr}}|\boldsymbol{o}))^{\mathrm{Tr}}. \qquad (10.29)\end{aligned}$$

To perform each iteration of the EM algorithm, it remains to evaluate all the conditional expectations, as sufficient statistics for parameter estimation, in Eqs. 10.28 and 10.29. Such evaluations (which form part of the E step) are described next.

Computing sufficient statistics for a single observation sequence

(a) Observable state $\boldsymbol{x}_k$:

For the special case where the continuous state $\boldsymbol{x}_k$, $k = 0, 1, \ldots, N$, in the dynamic system is observable (non-hidden), then we will have

$$E_N(\boldsymbol{x}_k|\boldsymbol{o}) = \boldsymbol{x}_k,$$

$$E_N(\boldsymbol{x}_k\boldsymbol{x}_k^{\mathrm{Tr}}|\boldsymbol{o}) = \boldsymbol{x}_k\boldsymbol{x}_k^{\mathrm{Tr}},$$

and

$$E_N(\boldsymbol{o}_k\boldsymbol{x}_k^{\mathrm{Tr}}|\boldsymbol{o}) = \boldsymbol{o}_k\boldsymbol{x}_k^{\mathrm{Tr}}.$$

Therefore, Eqs. 10.28 and 10.29 can be directly applied to estimate the parameters $\boldsymbol{A}$, $\boldsymbol{u}$, $\boldsymbol{C}$, $\boldsymbol{Q}$, and $\boldsymbol{R}$.

(b) Hidden state $\boldsymbol{x}_k$:

In the general case when the state $\boldsymbol{x}_k$ of the dynamic system is unobservable or hidden, the log likelihood function to be optimized will depend on the unobserved data series $\boldsymbol{x}_k$, $k = 0, 1, \ldots, N$. In this case, we need to evaluate the various conditional expectations on the observation sequence $\boldsymbol{o}_k$, $k = 0, 1, \ldots, N$, which appear in Eqs. 10.28 and 10.29 and are sufficient statistics for the parameter estimation. For convenience, we use the following notation to denote the conditional expectation

$$E_N\{\boldsymbol{x}_k^{\mathrm{Tr}}|\boldsymbol{o}\} \equiv \hat{\boldsymbol{x}}_{k/N}^{\mathrm{Tr}}, \tag{10.30}$$

where the subscript k/N indicates that up to N frames of the observation sequence $\boldsymbol{o}_1^N$ and the estimate is for the k-th frame ($k \leq N$).

Then, to evaluate $E_N[\boldsymbol{x}_k\boldsymbol{x}_k^{\mathrm{Tr}}|\boldsymbol{y}]$, we use the standard formula

$$E_N\{[\boldsymbol{x}_k - E_N(\boldsymbol{x}_k)][\boldsymbol{x}_k - E_N(\boldsymbol{x}_k)]^{\mathrm{Tr}}|\boldsymbol{o}\} = E_N[\boldsymbol{x}_k\boldsymbol{x}_k^{\mathrm{Tr}}|\boldsymbol{o}] - E_N[\boldsymbol{x}_k|\boldsymbol{o}]E_N[\boldsymbol{x}_k^{\mathrm{Tr}}|\boldsymbol{o}],$$

or

$$\boldsymbol{P}_{k/N} = E_N[\boldsymbol{x}_k\boldsymbol{x}_k^{\mathrm{Tr}}|\boldsymbol{o}] - \hat{\boldsymbol{x}}_{k/N}\hat{\boldsymbol{x}}_{k/N}^{\mathrm{Tr}},$$

to obtain

$$E_N[\boldsymbol{x}_k\boldsymbol{x}_k^{\mathrm{Tr}}|\boldsymbol{o}] = \hat{\boldsymbol{x}}_{k/N}\hat{\boldsymbol{x}}_{k/N} + \boldsymbol{P}_{k/N}. \tag{10.31}$$

In using the *EM* algorithm to obtain ML estimates of the parameters in Eq. 10.20, i.e., to evaluate the right-hand sides of Eqs. 10.28 and 10.29, it is necessary to compute the following quantities at each iteration of the algorithm:

$$\begin{aligned}
E\{\boldsymbol{x}_k^{\mathrm{Tr}}|\boldsymbol{o}\} &= \hat{\boldsymbol{x}}_{k/N}^{\mathrm{Tr}} \\
E\{\boldsymbol{x}_k\boldsymbol{x}_k^{\mathrm{Tr}}|\boldsymbol{o}\} &= \boldsymbol{P}_{k/N} + \hat{\boldsymbol{x}}_{k/N}\hat{\boldsymbol{x}}_{k/N}^{\mathrm{Tr}} \\
E\{\boldsymbol{x}_k\boldsymbol{x}_{k-1}^{\mathrm{Tr}}|\boldsymbol{o}\} &= \boldsymbol{P}_{k,k-1/N} + \hat{\boldsymbol{x}}_{k/N}\hat{\boldsymbol{x}}_{k-1/N}^{\mathrm{Tr}} \\
E\{\boldsymbol{o}_k\boldsymbol{x}_k^{\mathrm{Tr}}|\boldsymbol{o}\} &= \boldsymbol{o}_k\hat{\boldsymbol{x}}_{k/N}^{\mathrm{Tr}} \\
E\{\boldsymbol{o}_k\boldsymbol{o}_k^{\mathrm{Tr}}\} &= \boldsymbol{o}_k\boldsymbol{o}_k^{\mathrm{Tr}}.
\end{aligned} \tag{10.32}$$

That is, we can use the fixed interval smoothing form of the Kalman filter which we described in Chapter 5 to compute the required statistics. It consists of a backward pass that follows the standard Kalman filter forward recursions. In addition, in both the forward and the backward passes, we need some additional recursions for the computation of the cross covariance. All the necessary recursions are summarized below:

Forward Recursions:

$$\begin{aligned}
\hat{\boldsymbol{x}}_{k/k} &= \hat{\boldsymbol{x}}_{k/k-1} + \boldsymbol{K}_k \boldsymbol{e}_k \\
\hat{\boldsymbol{x}}_{k+1/k} &= \boldsymbol{A}\hat{\boldsymbol{x}}_{k/k} + \boldsymbol{u} \\
\boldsymbol{e}_k &= \boldsymbol{o}_k - \boldsymbol{C}\hat{\boldsymbol{x}}_{k/k-1} \\
\boldsymbol{K}_k &= \boldsymbol{P}_{k/k-1}\boldsymbol{C}^{\mathrm{Tr}}\boldsymbol{P}_{e_k}^{-1} \\
\boldsymbol{P}_{e_k} &= \boldsymbol{C}\boldsymbol{P}_{k/k-1}\boldsymbol{C}^{\mathrm{Tr}} + \boldsymbol{R} \\
\boldsymbol{P}_{k/k} &= \boldsymbol{P}_{k/k-1} - \boldsymbol{K}_k\boldsymbol{P}_{e_k}\boldsymbol{K}_k^{\mathrm{Tr}} \\
\boldsymbol{P}_{k,k-1/k} &= (\mathbf{I} - \boldsymbol{K}_k\boldsymbol{C})\boldsymbol{A}\boldsymbol{P}_{k-1/k-1} \\
\boldsymbol{P}_{k+1/k} &= \boldsymbol{A}\boldsymbol{P}_{k/k}\boldsymbol{A}^{\mathrm{Tr}} + \boldsymbol{Q}
\end{aligned} \tag{10.33}$$

Backward Recursions:

$$\begin{aligned}
\boldsymbol{\Gamma}_k &= \boldsymbol{P}_{k-1/k-1}\boldsymbol{A}^{\mathrm{Tr}}\boldsymbol{P}_{k/k-1}^{-1} \\
\hat{\boldsymbol{x}}_{k-1/N} &= \hat{\boldsymbol{x}}_{k-1/k-1} + \boldsymbol{\Gamma}_k[\hat{\boldsymbol{x}}_{k/N} - \hat{\boldsymbol{x}}_{k/k-1}] \\
\boldsymbol{P}_{k-1/N} &= \boldsymbol{P}_{k-1/k-1} + \boldsymbol{\Gamma}_k[\boldsymbol{P}_{k/N} - \boldsymbol{P}_{k/k-1}]\boldsymbol{\Gamma}_k^{\mathrm{Tr}} \\
\boldsymbol{P}_{k,k-1/N} &= \boldsymbol{P}_{k,k-1/k} + [\boldsymbol{P}_{k/N} - \boldsymbol{P}_{k/k}]\boldsymbol{P}_{k/k}^{-1}\boldsymbol{P}_{k,k-1/k}
\end{aligned} \tag{10.34}$$

Using Eqs. 10.32–10.34, the parameter estimates of system Eq. 10.20 for unobservable states $\boldsymbol{x}_k$ can be completely obtained by Eqs. 10.28 and 10.29.

Computing sufficient statistics for multiple observation sequences

In general, multiple observation sequences are needed and are available for estimating the parameters of the linear dynamic system model for acoustic trajectories of speech. The necessity is especially strong for the nonstationary dynamic system, which typically demands more training data to make reliable estimates of the system parameters than for the stationary system. The extension from the single-sequence approach we just described to the multiple-sequence one is straightforward: it involves simply summing the appropriate sufficient statistics over the different sequences. In order to show that this extension is correct, we first write down the corresponding objective function for optimization:

$$\begin{aligned}
\log L(\boldsymbol{x}, \boldsymbol{o}, \boldsymbol{\theta}) &= -\frac{1}{2}\sum_{l=1}^{L}\{\sum_{k=1}^{N_l}[\log \boldsymbol{Q} + (\boldsymbol{x}_{kl} - \boldsymbol{A}\boldsymbol{x}_{(k-1)l} - \boldsymbol{u})^{\mathrm{Tr}}\boldsymbol{Q}^{-1}(\boldsymbol{x}_{kl} - \boldsymbol{A}\boldsymbol{x}_{(k-1)l} - \boldsymbol{u})]\} \\
&\quad -\frac{1}{2}\sum_{l=1}^{L}\{\sum_{k=0}^{N_l}[\log \boldsymbol{R} + (\boldsymbol{o}_{kl} - \boldsymbol{C}\boldsymbol{x}_{kl})^{\mathrm{Tr}}\boldsymbol{R}^{-1}(\boldsymbol{o}_{kl} - \boldsymbol{C}\boldsymbol{x}_{kl})]\} + \text{Const.}
\end{aligned} \tag{10.35}$$

where L is the total number of sequences and N_l is the number of observations in the l-th training sequence. It is assumed that the discrete region (mode) s for each training sequence is fully observed. Carrying out similar procedures in the E and M steps for the single-sequence case using the more general objective function in Eq. 10.35, we can show that the new parameter estimates are

$$
\begin{aligned}
[\hat{\boldsymbol{A}}\ \hat{\boldsymbol{u}}]_s &= [\sum_{l=1}^{L}\sum_{k=1}^{N_l} E_N(\boldsymbol{x}_{kl}\boldsymbol{x}_{(k-1)l}^{\text{Tr}} \ \sum_{l=1}^{L}\sum_{k=1}^{N_l} E_N(\boldsymbol{x}_{kl}|\boldsymbol{o})]_s \\
&\cdot \begin{bmatrix} \sum_{l=1}^{L}\sum_{k=1}^{N_l} E_N(\boldsymbol{x}_{(k-1)l}\boldsymbol{x}_{(k-1)l}^{\text{Tr}}|\boldsymbol{o}) & \sum_{l=1}^{L}\sum_{k=1}^{N_l} E_N(\boldsymbol{x}_{(k-1)l}|\boldsymbol{o}) \\ \sum_{l=1}^{L}\sum_{k=1}^{N_l} E_N(\boldsymbol{x}_{(k-1)l}^{\text{Tr}}|\boldsymbol{o}) & \sum_{l=1}^{L} N_l \end{bmatrix}_s^{-1} \\
\hat{\boldsymbol{C}}_s &= [\sum_{l=1}^{L}\sum_{k=1}^{N_l} E_N(\boldsymbol{o}_{kl}\boldsymbol{x}_{kl}^{\text{Tr}}|\boldsymbol{o})]_s[\sum_{l=1}^{L}\sum_{k=0}^{N_l} E_N(\boldsymbol{x}_{kl}\boldsymbol{x}_{kl}^{\text{Tr}}|\boldsymbol{o})]_s^{-1} \\
\hat{\boldsymbol{Q}}_s &= \frac{1}{\sum_{l=1}^{L} N_l}\sum_{l=1}^{L}\sum_{k=1}^{N_l} E_N[|\boldsymbol{x}_{kl} - \hat{\boldsymbol{A}}\boldsymbol{x}_{(k-1)l} - \hat{\boldsymbol{u}}|^2|\boldsymbol{o}]_s \\
&= \frac{1}{\sum_{l=1}^{L} N_l}\sum_{l=1}^{L}\sum_{k=1}^{N_l}\{E_N(\boldsymbol{x}_{kl}\boldsymbol{x}_{kl}^{Tr}|\boldsymbol{o}) - [\hat{\boldsymbol{A}}\ \hat{\boldsymbol{u}}][E_N(\boldsymbol{x}_{kl}\boldsymbol{x}_{(k-1)l}^{\text{Tr}}|\boldsymbol{o})\ \ E_N(\boldsymbol{x}_{kl}|\boldsymbol{o})]^{\text{Tr}}\}_s \\
\hat{\boldsymbol{R}}_s &= \frac{1}{\sum_{l=1}^{L}(N_l+1)}\sum_{l=1}^{L}\sum_{k=0}^{N_l} E_N[|\boldsymbol{o}_{kl} - \hat{\boldsymbol{C}}\boldsymbol{x}_{kl}|^2|\boldsymbol{o}]_s \\
&= \frac{1}{\sum_{l=1}^{L}(N_l+1)}\sum_{l=1}^{L}[(\sum_{k=0}^{N_l}\boldsymbol{o}_{kl}\boldsymbol{o}_{kl}^{\text{Tr}}) - \hat{\boldsymbol{C}}\sum_{k=0}^{N_l} E_N(\boldsymbol{o}_{kl}\boldsymbol{x}_{kl}^{\text{Tr}}|\boldsymbol{o})^{\text{Tr}}]_s,
\end{aligned}
\tag{10.36}
$$

where $s = 1, 2, \ldots, m$ is the s-th mode of the nonstationary model.

The EM algorithm for the nonstationary version of the dynamic system involves the computation, at each iteration, of the sufficient statistics (i.e., all the conditional expectations in the right-hand side of Eq. 10.37 using the Kalman filter recursions with the previous estimates of the model parameters (E-step). The new estimates for the system parameters are then obtained using the sufficient statistics according to Eq. 10.37 (M-step).

10.3 Modeling Hidden Dynamics of Speech

This linear dynamic model just described uses the continuous state in the state-space model formulation to heuristically smooth the acoustic features. It thus does not represent any hidden mechanisms responsible for generating the acoustic features. This body of work, however, has recently been significantly generalized into carefully structured nonlinear dynamic system models where the continuous state explicitly represents either the task dynamics, articulatory dynamics, or vocal-tract-resonance dynamics that form various underlying mechanisms for the observed dynamics of acoustic features. Because these underlying mechanisms are "hidden" from the observed acoustics due to the many-to-one nonlinear relationship and due to acoustic measurement "noise," we call this class of nonlinear dynamic systems the **hidden dynamic model**.

There are many kinds of hidden dynamic models, depending on the different aspects of the underlying speech production mechanisms to be modeled and on the different ways of representing these mechanisms. In particular, different statistical assumptions made on the various portions of the dynamic model structures differentiate the various kinds of hidden dynamic models.

We focus in this section on the statistical hidden dynamic models, where the likelihood for any acoustic observation sequence can be computed using the models and the model parameters can be learned using the optimization techniques and estimation theory described in Chapter 5. We will also discuss several kinds of the statistical hidden dynamic models with different kinds of distributional assumptions made on the model parameters, and in particular the target parameters in the state equation of the dynamic system.

10.3.1 Derivation of discrete-time hidden-dynamic state equation

We first derive a general form of the target-directed state equation, in the discrete-time domain, of the statistical hidden dynamic model. The concept of **target directedness** which we present below is central to all classes of the hidden dynamic computational models for speech production discussed in this chapter.

The derivation we provide below originated from the deterministic, continuous-time task-dynamic model established in the speech science literature [Salt 89]. Starting with the original model in [Salt 89] but incorporating random, zero-mean, Gaussian noise vector $\mathbf{w}(t)$, we have the following state equation, in the continuous-time domain, of a statistical task-dynamic model:

$$\frac{d^2\boldsymbol{x}(t)}{dt^2} + 2\mathbf{S}(t)\frac{d\boldsymbol{x}(t)}{dt} + \mathbf{S}^2(t)(\boldsymbol{x}(t) - \boldsymbol{x}_0(t)) = \mathbf{w}(t), \tag{10.37}$$

where $\mathbf{S}^2$ is the normalized, gesture-dependent stiffness parameter (which controls fast or slow movement of vector-valued tract variable $\boldsymbol{x}(t)$), and $\boldsymbol{x}_0$ is the gesture-dependent point-attractor parameter of the dynamical system (which controls the target and hence direction of the tract-variable movement). Here, for generality, we assume that the model parameters are (slowly) time-varying and hence are indexed by time t.

We can rewrite Eq. 10.37 in a canonical form of

$$\frac{d}{dt}\begin{pmatrix} \boldsymbol{x}(t) \\ \dot{\boldsymbol{x}}(t) \end{pmatrix} = \begin{pmatrix} 0 & 1 \\ -\mathbf{S}^2(t) & -2\mathbf{S}(t) \end{pmatrix}\begin{pmatrix} \boldsymbol{x}(t) \\ \dot{\boldsymbol{x}}(t) \end{pmatrix} + \begin{pmatrix} 0 \\ \mathbf{S}^2(t)\boldsymbol{x}_0(t) \end{pmatrix} + \begin{pmatrix} 0 \\ \mathbf{w}(t) \end{pmatrix},$$

where $\dot{\boldsymbol{x}}(t) = \frac{d\boldsymbol{x}(t)}{dt}$. This, in the matrix form, becomes:

$$\frac{d}{dt}\boldsymbol{x}(t) = \mathbf{F}(t)\boldsymbol{x}(t) - \mathbf{F}(t)\ \mathbf{t}(t) + \bar{\mathbf{w}}(t),$$

where the composite state in the dynamic system is defined by

$$\boldsymbol{x}(t) \equiv \begin{pmatrix} \boldsymbol{x}(t) \\ \dot{\boldsymbol{x}}(t) \end{pmatrix},$$

the system matrix is defined by

$$\mathbf{F}(t) \equiv \begin{pmatrix} 0 & 1 \\ -\mathbf{S}^2(t) & -2\mathbf{S}(t) \end{pmatrix},$$

and the attractor or target vector for the discrete-time system is defined by

$$\mathbf{t} \equiv -\mathbf{F}^{-1}(t)\begin{pmatrix} 0 \\ \mathbf{S}^2(t)\boldsymbol{x}_0(t) \end{pmatrix}.$$

An explicit solution to the above task-dynamic equation can be found to be

$$\boldsymbol{x}(t) = \boldsymbol{\Phi}(t, t_0)\boldsymbol{x}(t_0) + \int_{t_0}^{t} \boldsymbol{\Phi}(t, \tau)[-\mathbf{F}(\tau)\ \mathbf{t}(\tau) + \bar{\mathbf{w}}(\tau)]d\tau,$$

where $\boldsymbol{\Phi}(t, t_0)$ (state transition matrix) is the solution to the following matrix homogeneous differential equation:

$$\dot{\boldsymbol{\Phi}}(t, \tau) = \mathbf{F}(t)\boldsymbol{\Phi}(t, \tau); \quad \text{with initial condition} : \boldsymbol{\Phi}(t, t) = \mathbf{I}.$$

The solution can be written in the following matrix exponential form during a short time interval $t_k \le t \le t_{k+1}$ over which $F(t) = F_k$:

$$\boldsymbol{\Phi}(t, \tau) = \exp[(t - \tau)\mathbf{F}_k].$$

Now, setting $t_0 = t_k, t = t_{k+1}$, we have

$$\boldsymbol{x}(t_{k+1}) \approx \boldsymbol{\Phi}(t_{k+1}, t_k)\boldsymbol{x}(t_k) - [\int_{t_k}^{t_{k+1}} \boldsymbol{\Phi}(t_{k+1}, \tau)\mathbf{F}(\tau)d\tau]\mathbf{t}(t_k) + \int_{t_k}^{t_{k+1}} \boldsymbol{\Phi}(t_{k+1}, \tau)\bar{\mathbf{w}}(\tau)d\tau.$$

This leads to the discrete-time form of the state equation:

$$\boldsymbol{x}(k+1) = \boldsymbol{\Phi}(k)\boldsymbol{x}(k) + \boldsymbol{\Psi}(k)\mathbf{t}(k) + \bar{\mathbf{w}}_d(k), \tag{10.38}$$

where

$$\begin{aligned}
\boldsymbol{\Phi}(k) &\approx \boldsymbol{\Phi}(t_{k+1}, t_k) = \exp(\mathbf{F}_k \Delta t), \qquad \Delta t \equiv t_{k+1} - t_k \\
\boldsymbol{\Psi}(k) &\approx -[\int_{t_k}^{t_{k+1}} \boldsymbol{\Phi}(t_{k+1}, \tau)F(\tau)d\tau] \approx -\int_{t_k}^{t_{k+1}} \exp[\mathbf{F}_k(t_{k+1} - \tau)]\mathbf{F}(\tau)d\tau \\
&\approx -\mathbf{F}_k \exp(\mathbf{F}_k\ t_{k+1}) \int_{t_k}^{t_{k+1}} \exp[-\mathbf{F}_k\tau]d\tau = \boldsymbol{I} - \exp(\mathbf{F}_k\Delta t) = \boldsymbol{I} - \boldsymbol{\Phi}(k),
\end{aligned}$$

and $\mathbf{w}_d(k)$ is a discrete-time white Gaussian sequence which is statistically equivalent through its first and second moments to $\int_{t_k}^{t_{k+1}} \boldsymbol{\Phi}(t_{k+1}, \tau)\bar{\mathbf{w}}(\tau)d\tau$.

Summarizing the above results, we started from the continuous-time dynamic system's state equation and obtained its counterpart in the discrete-time system as a quantitative model for the hidden dynamic variables of speech. The target-directed property in the continuous-time system of Eq. 10.37, i.e., when $t \to \infty, \boldsymbol{x}(t) \to \boldsymbol{x}_0$, is retained for the discrete-time system. That is, in Eq. 10.38, when the effect of noise $\bar{\mathbf{w}}_d(k)$ is negligibly small and when $k \to \infty$ in discrete values, we have $\boldsymbol{x}(k+1) \approx \boldsymbol{x}(k)$ and thus $\boldsymbol{x}(k) \to \mathbf{t}$.

Target-directed state equation for hidden speech dynamics (discrete-time):

$$\boldsymbol{x}(k+1) = \boldsymbol{\Phi}(k)\boldsymbol{x}(k) + [\boldsymbol{I} - \boldsymbol{\Phi}(k)]\mathbf{t}(k) + \bar{\mathbf{w}}_d(k), \tag{10.39}$$

where $\boldsymbol{x}(k) \to \mathbf{t}$ if $k \to \infty$.

10.3.2 Nonlinear state space formulation of hidden dynamic model

The discrete-time state equation derived above serves as a statistical description for the target-directed hidden dynamic variables responsible for generating the observed speech acoustic features. The description for how the acoustic features are actually generated from these hidden dynamic variables belongs to the "observation" or "measurement" equation of the state space formulation of the hidden dynamic model, which we discussed in some detail in Chapter 4.

In all types of statistical hidden dynamic models, whether the hidden dynamics is caused by the task variables, articulatory variables, or the vocal-tract resonance variables, these hidden variables are usually related to the acoustic variables in a highly nonlinear manner. For practical purposes, the nonlinear relationship can be approximated as a static mapping; i.e., not depending on the past history. For the sake of conciseness, we use a general form of static nonlinear function to represent such a mapping. Further, an additive measurement noise $\mathbf{v}(k)$ is assumed to reflect the possible approximation errors introduced by using the nonlinear function for the true relationship between the hidden dynamic variables and the observed acoustic features.

Combining both the state equation and the observation equation described above, we have the following complete state space formulation of the target-directed hidden dynamic model of speech:

State space formulation of a discrete-time hidden dynamic model of speech:

$$\boldsymbol{x}(k+1) = \boldsymbol{\Phi}(k)\boldsymbol{x}(k) + [\boldsymbol{I} - \boldsymbol{\Phi}(k)]\mathbf{t}(k) + \mathbf{w}(k), \tag{10.40}$$

$$\mathbf{o}(k) = \mathbf{h}[\boldsymbol{x}(k)] + \mathbf{v}(k). \tag{10.41}$$

10.3.3 Task dynamics, articulatory dynamics, and vocal-tract resonance dynamics

We now describe three types of statistical hidden dynamic models for speech, differentiated by the use of task variables, articulatory variables, or the vocal-tract-resonance (VTR) variables as the state variable $\boldsymbol{x}(k)$ in Eqs. 10.40 and 10.41.

Task dynamics

During our introduction to Articulatory Phonology in Chapter 8, we mentioned that the parameters in the dynamics of tract variables associated with the vocal-tract (VT) constrictions serve to define the gesture as the atomic phonological unit. The dynamics of the VT constriction variables (including both the aperture and location) is called the **task dynamics**. The task dynamics are closely linked (interfaced) to the discrete articulatory gestures. The gestures define the regions or modes of the dynamics for the task variables, often in an asynchronous way across different components in the task variable vector.

The main property of the task dynamics is its close connection to the gesture-based phonological unit defined in articulatory phonology. In order to give concrete examples of the nature of the task variable $\boldsymbol{x}(t)$'s dynamics mathematically described by Eq. 10.40, a hypothetical utterance for the two-word phrase *ten pin* is shown in Fig. 10.2. The task variable $\boldsymbol{x}(t)$ has five dimensions, including the Tongue-Blade constriction degree associated with [t] and [n], and the Velum constriction degree (aperture) associated with the [n]'s in two different words. Note that the two features in Tongue-Body which control two different temporal regions of the Tongue-Body task variable are distinct, one for each distinct vowel in the utterance. This results in two different target positions, for the same Tongue-Body task variable, at two different temporal regions. In contrast, all three Tongue-Blade features (and the associated target positions of the corresponding task-variable component) in the utterance are identical, and so are the two Velum features and the two Glottal features. An interesting phenomenon shown in this example is a "hiding" of the feature or articulatory gesture of Tongue-Blade of [n] in word *ten* (drawn as the shaded box in Fig. 10.2). The "hiding" is a result of the complete temporal feature overlapping with the Lips feature in [p]. However, according to the task-dynamic model, the Tongue-Blade task variable exhibits its dynamics preceding the feature onset of complete closure (target position) of the Lip-constriction component of the task variable. The Lip-constriction variable also follows a dynamic path towards the complete closure after the feature onset. Therefore, the predicted acoustics from this model will be very different from that predicted simply by assimilation of [n] to [m] (a complete change of phones in pronunciation) according to the conventional linear phonology. The task dynamics is responsible for this "phonological" change.

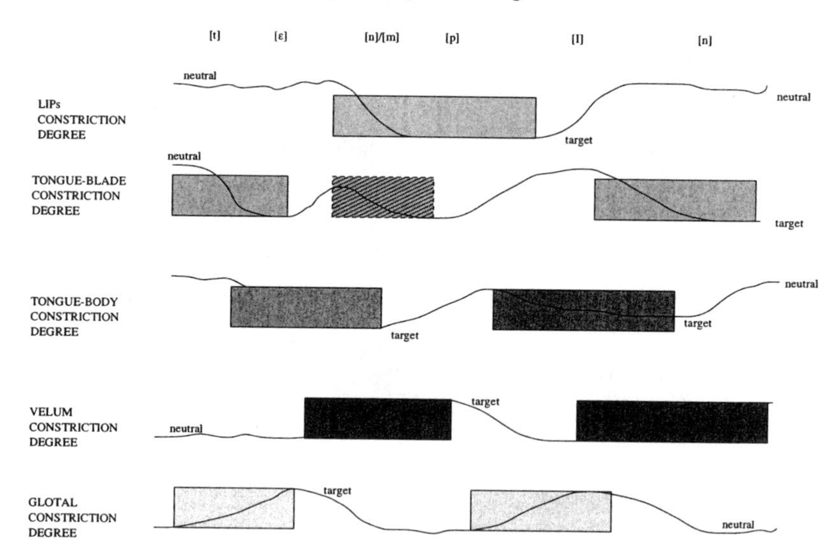

Figure 10.2: Illustration of task dynamics for the phrase *ten pin* (after Deng [Deng 98], @Elsevier).

The second key property of the task dynamics is the high complexity with which the task dynamic vectors (i.e., the tract variables defined in articulatory phonology) are transformed into the acoustic feature vectors. There are two separate steps in this transformation, both captured by the general nonlinear observation function $\mathbf{h}[\boldsymbol{x}(k)]$ in Eq. 10.41 in the state-space model formulation. The first transformation step is from the VT-constriction-based tract variables to the physical articulators' positions (or their principal components, which we call articulatory variables). This transformation is highly nonlinear and one-to-many, and can be determined by the geometric relationship in the VT between the constriction variables and the articulatory variables [Merm 73]. The second transformation step is from the articulatory variables to the acoustic features.

This transformation is often many-to-one. During this step, the articulatory variables are first converted to the VT area function, which, together with the excitation source in the VT, subsequently produces acoustic waveforms and features of speech.

After introducing the intermediate articulator variables $\mathbf{y}(k)$ to link the task variables $\boldsymbol{x}(k)$ to the acoustic features $\boldsymbol{o}(k)$ via static nonlinear functions, $\boldsymbol{o} = \mathcal{O}(\mathbf{y})$ and $\boldsymbol{x} = \mathcal{X}(\mathbf{y})$, we can decompose the general form of nonlinearity in Eq. 10.41 into these two separate forms of nonlinearity. Thus, viewing the hidden task-variable dynamics as observed through a noisy (IID noise $\boldsymbol{v}(k)$) nonlinear relation $\mathbf{h}(\cdot)$, we have

$$\boldsymbol{o}(k) = \mathcal{O}(\mathbf{y}(k)) + \boldsymbol{v}(k) = \mathcal{O}[\mathcal{X}^{-1}(\boldsymbol{x}(k))] + \boldsymbol{v}(k) = \mathbf{h}[\boldsymbol{x}(k)] + \boldsymbol{v}(k). \tag{10.42}$$

The third property of the task dynamics concerns the statistical description of the target vector $\mathbf{t}(k)$ of Eq. 10.40 in the space of VT-constriction tract variables. As we discussed in Section 8.6 on articulatory phonology, the tract variables consist of both the constriction degree (aperture) and the constriction location components. While the target vector $\mathbf{t}$ has been treated traditionally as deterministic parameters in the speech production literature, advantages can be gained if the target vector is treated as the more general random vectors in the computational task dynamic model. In determining the form of the probability distribution suitable for the aperture components of the tract variables, special attention must be paid to the non-negativity of these components. Some of the common PDF's discussed in Section 3.1 can be made to be always positively distributed; e.g., the exponential distribution, inverse Gaussian distribution, and Gamma distribution, etc. They can thus be usefully applied as a statistical model for the task-dynamic target vector. In practice, some critical apertures associated with stop consonants have a strong tendency to have complete closures as their targets. In this case, the target distribution needs to have a very small variance. It can also be shifted to allow for taking slightly negative values in order to ensure that the target of complete closure is reached most of the time.

Articulatory dynamics

When the hidden dynamics in the general discrete-time state-space model Eqs. 10.40 and 10.41 takes the form of articulatory dynamics, the state vector $\boldsymbol{x}(k)$ will no longer be the tract variables (as in the task dynamics), but be the articulatory variables that characterize a set of articulators' geometric positions. The choice of the articulatory variables should be such that they are sufficient to describe all possible VT shapes or "area functions," in addition to being economical. Some commonly used articulatory variables include jaw radius and angle, tongue body radius, tongue tip angle, lip protrusion and height, velar opening, etc.

In the articulatory dynamic model, the relationship between the phonological units and the hidden dynamic variables is not as straightforward as for the task dynamic model just discussed. If we still use the gesture as the phonological unit, then its phonetic interface as the target $\mathbf{t}(k)$ (in the space of articulatory variables) in Eq. 10.40 must be statistical distributions which show strong correlations among various articulatory variables. This is so partly because of the one-to-many nonlinear relationship between the VT constrictions and the articulatory positions.

More generally, in the articulatory dynamic model, one can assume that a phonological unit is associated with a number of physical correlates, which may be acoustic, articulatory or auditory in nature — the requirement for choosing the correlates being that each set of correlates be expressible in terms of quantities that can be measured from suitably-detailed models of the production and auditory systems [Deng 97c]. Further, the region defined by the correlates for each phonological unit in the above can presumably be projected onto a space of the articulatory variables, thereby defining an equivalent target distribution for that unit in the articulatory space.

Under the above assumptions, any hypothesized sequence of phonological units thus induces a succession of statistical target distributions on the articulatory space, which are sampled randomly, as each new unit appears, to construct a control trajectory for the articulators that lasts until the occurrence of a new unit. These control trajectories can be assumed for simplicity to be piecewise-constant functions, representing essentially static spatial targets. The probability distribution of possible control trajectories on the articulatory space is intended to represent the statistical ensemble of idealized articulatory movements that would produce the required sequence of distributions of phonetic correlates for the phonological sequence in question.

Use of the state equation in Eq. 10.40 to represent the true biomechanical articulatory dynamics is necessarily simplistic. At present, it is difficult to speculate how the conversion of higher-level control trajectories into articulator movement takes place. Ideally, modeling of articulatory dynamics and control would require detailed neuromuscular and biomechanical models of the vocal tract, as well as an explicit model of the control objectives and strategies realized by a speaker's motor control system. This is clearly too complicated to implement at present for any reasonable computational model of speech production. It is possible to assume, however, that the combined (non-linear) control system and articulatory mechanism behave macroscopically as a stable linear system that attempts to track the control input as closely as possible in the articulatory parameter space. Articulator motion can then be approximated as the response of a dynamic vocal-tract model driven by a random control trajectory, producing a time-varying tract shape that modulates the acoustic properties of the speech signal.

In the articulatory dynamic model outlined in [Deng 97c], it is assumed that the speech production system can be divided into a motor control module, an articulatory mechanism, and articulatory dynamics. The motor control module generates a single time-varying articulatory target trajectory from a given sequence of phonological states by randomly sampling the sequence of target distributions induced by those states. Each time a new control region is activated or deactivated (when a new phonological state is entered), the motor system chooses a new target point in articulatory space according to the probability law induced by the current target distribution, and maintains that target point until further changes occur in the phonological state. The articulatory mechanism acts as a servomechanism which attempts to track the time-varying articulatory target from the control module as best it can, subject to biomechanical constraints due to articulators' inertia. Finally, the result of the articulatory mechanism is a simple stable, linear articulatory dynamic system driven by the target process produced by the motor control module. The effective parameters of the system are dependent on the phonological state. The dependence of the parameters of the articulatory dynamic system on the phonological state is justified by the fact that the functional behavior of an articulator depends on the particular goal it is trying to implement, and on the other articulators with which it is cooperating in order to produce compensatory articulation.

Again, use of the general observation equation in Eq. 10.41 to represent the true relationship between the articulatory variables and the acoustic features is also highly simplified. The assumption for the simplification is that the articulatory and acoustic states of the vocal tract can be adequately described by low-order finite-dimensional vectors of variables representing, respectively, the relative positions of the major articulators, and the corresponding time-averaged spectral features derived from the acoustic signal. Given further that an appropriate time scale is chosen, it will also be assumed that the relationship between articulatory and acoustic representations can be modeled by a static memoryless transformation, converting a vector of articulatory variables into a vector of acoustic measurements, and that this can be calculated explicitly by computational models.

This final stage in the speech production model based on articulatory dynamics involves accounting for the generation of the observed acoustic features from the time-varying parameters characterizing articulatory geometry as a result of the articulatory dynamics. While the transformation between the articulatory-geometry parameters and the observations can be simulated using a computational model, any observation space may be used, as long as a computational model is available to represent the relationship between articulatory variables and the acoustic measurements.

The common assumptions made on the articulatory geometry and on the acoustic feature vector generation are as follows. First, the geometry of the articulatory system can be described in terms of a finite set of variables, chosen to represent approximate degrees of freedom available to the articulators. These variables measure the positions of a number of reference points fixed on the articulatory organs. Any parameter vector can be converted geometrically into a mid-sagital outline of the vocal tract and hence into the corresponding area function, using a model derived from physical data. Second, generation of speech sound in the vocal tract can be adequately described by the standard model describing the propagation of acoustic waves in a compressible gas with suitable losses. Articulatory movements can be converted into an acoustic pressure signal at the mouth by solution of the partial differential equations defined by this model, under the boundary conditions associated with the sequence of static area functions defined for the sampled articulatory-geometry parameter trajectories. The speech signal generated by the model for any static tract shape can be adequately parameterized by a finite vector of acoustic representations using spectral analysis techniques.

Vocal-tract-resonance dynamics

The hidden dynamic model based on articulatory dynamics just described has the main advantage of being "physical," and thus rich knowledge gained from scientific research on biomechanical and acoustic aspects of speech production and their relations can make potentially useful contributions to the modeling accuracy. However, the use of multivariate distributions on the articulatory targets with component correlation is a rather indirect way to interface with the phonological units. Further, the complexity of realistic biomechanical properties of articulation makes the functional linear state model of Eq. 10.40 a rather poor approximation. On the other hand, the VT-constriction-based hidden dynamic model discussed earlier in this section provides a very direct interface with the phonological units based on gesture and overlapping features. However, the complex mapping from the hidden tract variables to the articulatory variables and further to the acoustic features makes it difficult to accurately characterize the functional observation model with Eq. 10.41. Moreover, the quantal nature of speech [Stev 89] demands great precision in controlling some critical constriction-degree components of $\mathbf{x}(k)$ so that glides, fricatives, and stops can be clearly separated because their VT constriction-degree values are close to each other.

The above weaknesses of the VT-constriction-based and articulation-based hidden dynamic models motivate the use of the new VT resonances (VTRs) as the hidden dynamic state variables $\mathbf{x}(k)$ in the computational model of Eqs. 10.40 and 10.41. The major VTRs with low dimensionality, denoted by F1, F2, and F3, are chosen to be the (vector-valued) state variables. As discussed earlier in Sections 2.4.1 and 7.3, the VTRs are pole locations of the VT configured to produce speech sounds, and have acoustic correlates of formants that may be measurable for vocalic sounds, but are hidden or perturbed for consonantal sounds due to concurrent spectral zeros, turbulence noises, wide resonance bandwidths, weak VT source excitation(s), and/or blocking of sound radiation by full closure of the lips.

The VTR-based hidden dynamic model discussed here is far from the plain use of formants in speech recognition attempted in the past [Deng 94a, Holm 97]. Formants and VTRs are related but distinct concepts: one is defined in the acoustic domain (peaks of smoothed spectra in a narrow frequency region), and the other is associated with the VT properties per se (e.g., the possibility to describe switching VT-cavity affiliations). The acoustic-domain definition of the formant in terms of a local concentration of smoothed spectral energy has been mainly applied to voiced portions of the speech signal that contain glottal excitation. This raises several problematic issues such as how to define formants for consonant spectra and whether the same definition applies to voiced and voiceless consonants. Since any statistical model useful for speech recognition requires consistent definitions of variables across all speech classes, the above problematic issues prevent the use of formants defined in the acoustic domain as the state variables in the hidden dynamic model. Use of the VTRs is free from this consistency problem. The VTRs are defined uniformly for all classes of speech sounds regardless of the VT excitation, and are temporally smooth from one speech unit to another during the production of any speech utterance. The relations between the acoustically defined formants and the VTRs can be summarized as follows. For the production of vocalic sounds of speech where the VTRs are generally well separated in frequency and the bandwidths of the resonances are reasonably narrow, the VT-resonances manifest themselves as the local peaks of the acoustic spectra and are consistent with the formants. During the production of most consonants, some VTRs are either weakly excited or masked/distorted by VT anti-resonances (zeros) and hence will not manifest themselves in the acoustic spectra. The property of temporal smoothness applies to the VTRs, but not to the acoustically defined formants defined as local maxima in the spectra of the speech signal. Therefore, the basic assumption in the hidden dynamic model that the state variables are a continuous flow from one phonological unit to its adjacent one will be untenable if formants, rather than VTRs, were to be used as the state variables $\boldsymbol{x}(k)$ in Eq. 10.40.

According to goal-based speech production theory, articulatory structure and the associated VTRs necessarily manifest asymptotic behavior in their temporally smoothed dynamics. Such behavior, however, is often significantly distorted in the observable acoustic signal due to movements of the VT excitation source(s) and to the intervening aerodynamic events. This distortion, in the current computational framework, is modeled by a static nonlinear functional mapping, $\mathbf{h}[\boldsymbol{x}(k)]$ in Eq. 10.41, from the underlying VTRs to directly measurable acoustic features.

The interface of the VTR-based hidden dynamic model to the overlapping feature-based phonological model is much simplified from that of the VT-constriction-based hidden dynamic model. Feature overlaps are essential to account for look-ahead, target-directed acoustic (e.g., formant) transitions ubiquitously observed in speech data, but due to a less direct correspondence between the feature "contents" and the physical meaning of the VTR variables, the number of feature components can be reduced and feature-overlapping rules need not be as elaborate. The powerful phonological mechanism of feature overlaps can be made to act in concert with the smoothing mechanism inherent in the VTR dynamics to provide an ultimate structural account for the large amount of contextual and speaking style-related variation in speech. The smooth dynamics in the VTR state variables is not only limited within a discrete phonological state (defined by a feature bundle), but more importantly, it also occurs across adjacent phonological states. This model thus incorporates the physical constraints on the relatively slow motion of the speech-production structure and the associated VTRs (both within and across phonological units). These constraints limit how fast the VTRs can change over time across entire speech utterances, enabling the model to exhibit a greater degree of coarticulation for fast, spontaneous, and more casual speech.

One principal advantage of the VTR-based hidden dynamic model over the articulation-based model is the relative ease with which to specify the targets $\mathbf{t}$ in Eq. 10.40 (and possibly their distributions), and to verify the refined target values (or their distributional parameters) after the training (to be discussed later in this chapter). As an illustrating example, we show here typical VTR target values, from a typical male speaker, of some phonemes in English, together with some context-dependent phones. In this example, each VTR target consists of a three-dimensional (F1, F2, and F3), vector-valued target (T^j) in the VTR-frequency domain. Some typical VTR-frequency target values are shown in Table 10.1 for context-independent phones, and in Table 10.2 for some context-dependent phones.

The eight phones in Table 10.2 are context-dependent because their target VTRs are known to be affected by the anticipatory tongue position associated with the following phone. The targets of a phone with subscript f (such as b_f) are conditioned on the immediately next phones being front vowels (/iy, ih, eh, ae, y/). The targets of a phone without the subscript are conditioned on the immediately next phones being the remaining phones (with the phonetic feature of non-front). The VTR target values in Tables 10.1 and 10.2 are related to the Klatt synthesizer setup [Stev 93], and are slightly adjusted by examining spectrograms of some speech data.

For the speech utterances produced by a fixed speaker, the VTR target vector in Eq. 10.40 may be adequately modeled as a deterministic one for each phonological unit, because there is relatively minor variation in the possible range of the target values given the unit. This is in contrast to the articulation-based hidden dynamic model, where the articulatory target vector in Eq. 10.40 must be described in terms of a statistical distribution (with strong component correlations) in order to adequately represent the correlates for the phonological units and to reflect compensatory articulation. However, for the speech utterances produced by a large population of speakers (including men, women, and children, for example), the variation in the range of the VTR target values given a fixed phonological unit will be large. In this case, a deterministic VTR target vector in Eq. 10.40 may not be appropriate. Rather, the VTR target distribution should be used but the VTR vector components may be made uncorrelated to simplify the form of the distribution. For a limited number of speakers, a discrete distribution for the VTR target may be sufficient, while for a large number of speakers, a continuous distribution (such as Gaussian) is more appropriate.

10.4 Hidden Dynamic Model Implemented Using Piecewise Linear Approximation

In this section, we will describe a particular computational implementation of the hidden dynamic model in Eqs. 10.40 and 10.41, where a deterministic target vector is assumed and the general nonlinear function $\mathbf{h}[\boldsymbol{x}(k)]$ in Eq. 10.41 is approximated by a set of

Units	F1 VTR target	F2 VTR target	F3 VTR target
aa	730	1090	2440
ae	660	1720	2410
ah	640	1190	2390
ao	570	840	2410
ax	500	1500	2500
d	180	1800	2700
dh	250	1300	2500
eh	530	1840	2480
el	450	1000	2700
en	500	1500	2500
er	490	1350	1690
ih	390	1990	2550
iy	270	2290	3010
l	450	1060	2640
n	250	1800	2700
r	460	1240	1720
s	250	1900	2700
sh	250	1700	2500
t	180	1800	2700
th	250	1300	2500
uh	440	1020	2240
uw	300	870	2240
w	350	770	2340
y	360	2270	2920

Table 10.1: Context-independent units and their VTR target values (in unit of Hz).

Units	F1 VTR target	F2 VTR target	F3 VTR target
b	180	1100	2300
b_f	180	1800	2300
g	180	1500	2200
g_f	180	2200	2800
k	180	1500	2200
k_f	180	2200	2800
p	180	1100	2300
p_f	180	1800	2300
f	250	1100	2300
f_f	250	1800	2300
m	250	1100	2300
m_f	250	1800	2300
ng	250	1500	2200
ng_f	250	2200	2800
v	250	1100	2300
v_f	250	1800	2300

Table 10.2: Context-dependent units and their VTR target values (in unit of Hz).

optimally derived piecewise linear functions. The relevant model parameters are also represented by a set of parameters related to these piecewise linear functions.

10.4.1 Motivations and a new form of the model formulation

One principal motivation for using the piecewise linear approximation in the implementation of the general hidden dynamic model for speech production is the general difficulty associated with the original nonlinear model that requires gross approximation for developing and implementing the model parameter learning algorithms. (An extensive study based on nonlinear learning of the VTR dynamic model can be found in [Deng 00c] and will not be described here.) In addition, the large amount of computation for implementing the approximate learning algorithms for the original nonlinear model makes it difficult to extensively tune the speech recognition system constructed based on such a model.

To overcome the difficulties inherent in the nonlinear methods for nonlinear models, it is appropriate to seek piecewise linear methods which may possibly provide adequate approximations to the nonlinear relationship between the hidden and observational spaces in the formulation of the speech model while gaining computational effectiveness and efficiency in model learning. A very straightforward method is to use a set of linear regression functions to replace the general nonlinear mapping in Eq. 10.41, while keeping intact the target-directed, linear state dynamics of Eq. 10.40. That is, rather than using a single set of linear-model parameters to characterize each phonological state, multiple sets of the linear-model parameters are used. This gives rise to the mixture linear dynamic model to be described in this section. When we further use separate state equations in the dynamic system model for different phonological states, the hidden space will be rather finely partitioned and hence the degree of nonlinearity of each individual linear model is expected to be relatively minor. Therefore, the use of a small number of mixture components in the piecewise linear model is expected to be adequate for representing the state-dependent nonlinear function between the hidden and observational spaces.

The piecewise or mixture linear dynamic system model motivated above can be written succinctly in the following state-space form (for a fixed phonological state):

$$\boldsymbol{x}(k) = \boldsymbol{\Phi}_m \boldsymbol{x}(k-1) + (\mathbf{I} - \boldsymbol{\Phi}_m)\mathbf{t}_m + \mathbf{w}_m(k-1), \tag{10.43}$$

$$\mathbf{o}(k) = \dot{\mathbf{H}}_m \dot{\boldsymbol{x}}(k) + \mathbf{v}_m(k), \quad m = 1, 2, ..., M \tag{10.44}$$

where $\dot{\mathbf{H}}_m = [\mathbf{a} \mid \mathbf{H}]$ and $\dot{\boldsymbol{x}}(k) = [1 \mid \boldsymbol{x}(k)^{\mathrm{Tr}}]^{\mathrm{Tr}}$ are an extended matrix and vector (where $\mid$ denotes appending an extra unit). In the above, M is the total number of mixture components in the piecewise linear dynamic model for each phonological state (e.g., phone). As before, $\mathbf{w}_m(k)$ and $\mathbf{v}_m(k)$ are the discrete-time state noise and measurement noise, modeled by uncorrelated, IID, zero-mean, Gaussian processes with covariance matrices $\mathbf{Q}_m$ and $\mathbf{R}_m$, respectively. $\mathbf{o}$ represents the sequence of acoustic (MFCC) vectors, $\mathbf{o}(1), \mathbf{o}(2), ..., \mathbf{o}(k)...$, and $\boldsymbol{x}$ represents the sequence of VTR vectors, $\boldsymbol{x}(1), \boldsymbol{x}(2), ..., \boldsymbol{x}(k)...$. In the current simplified version of the model, the VTR target vectors $\mathbf{t}_m$ are considered having a zero covariance matrix and hence are treated as fixed rather than random parameters. The full set of model parameters for each phonological state (not indexed for clarity) are $\Theta = \{\boldsymbol{\Phi}_m, \mathbf{t}_m, \mathbf{Q}_m, \mathbf{R}_m, \mathbf{H}_m, \quad m = 1, 2, ..., M\}$. Each of the parameters is indexed by the mixture component m.

An important constraint, which we call a mixture-path constraint, is imposed on the above piecewise linear dynamic system model. That is, for each sequence of acoustic observations associated with a phonological state, being either a training token or a test token, the sequence is constrained to be produced from a fixed mixture component, m, of the dynamic system model. This means that the target of the hidden continuous variables for each phonological state is not permitted to switch from one mixture component to another at the frame level. The constraint of such a type is motivated by the physical nature of speech production — the target that is correlated with its phonetic identity is defined at the segment level, not at the frame level.

After the general nonlinear mapping in Eq. 10.41 from hidden dynamic variables to acoustic observational variables is approximated by a piecewise linear relationship, a new, discrete random variable m is introduced to provide the "region" (i.e., the mixture-component index) in the piecewise linear mapping. The conditional PDF for the output acoustic variables $\mathbf{o}(k)$ given the input hidden variables $\boldsymbol{x}(k)$ (for the fixed region m at time frame k) is:

$$p[\mathbf{o}(k)|\boldsymbol{x}(k), m_k] = N[\mathbf{o}(k); \dot{\mathbf{H}}_{m_k}\dot{\boldsymbol{x}}(k), \boldsymbol{\Sigma}_v]. \tag{10.45}$$

This is due to the Gaussian assumption for the observational noise $\mathbf{v}_m(k)$. The conditional PDF related to the state equation for the fixed region m at time frame k is

$$p[\boldsymbol{x}(k+1)|\boldsymbol{x}(k), \mathbf{t}(k), \mathbf{s}_k, m_k] = N[\boldsymbol{x}(k+1); \boldsymbol{\Phi}_{s_k,m_k}\boldsymbol{x}(k) - (\mathbf{I} - \boldsymbol{\Phi}_{s_k,m_k})\mathbf{t}(k), \boldsymbol{\Sigma}_w]. \tag{10.46}$$

This is due to the Gaussian assumption for the state noise $\mathbf{w}_m(k)$.

10.4.2 Parameter estimation algorithm

We now describe the parameter estimation or learning algorithm for the hidden dynamic model using the above piecewise linear approximation. Again, the algorithm developed is based on the EM principle for maximum likelihood.

To proceed, we first define a discrete variable $\mathcal{M}$, which indicates the observation-to-mixture assignment for each sequence of observation vectors. For example, for a given sequence of observation vectors corresponding to a phonological state (e.g., phone), $\mathcal{M} = m$ $(1 \leq m \leq M)$ means that the m-th mixture component is responsible for generating that observation vector. The EM algorithm described here treats $\boldsymbol{x}(k)$ and $\mathcal{M}$ as the missing data, and $\mathbf{o}(k)$ as the observed training data.

For generality, we discuss the case with multiple sequences as the training data and let N denote the total number of training sequences (also called tokens) for a given phonological state. Let $\{\mathbf{o}, \mathbf{x}, \mathcal{M}\}^N$ represent the joint random variables

$$\{(\mathbf{o}^1, \boldsymbol{x}^1, \mathcal{M}^1), (\mathbf{o}^2, \boldsymbol{x}^2, \mathcal{M}^2), ..., (\mathbf{o}^n, \boldsymbol{x}^n, \mathcal{M}^n), ..., (\mathbf{o}^N, \boldsymbol{x}^N, \mathcal{M}^N)\},$$

where $\mathbf{o}^n, \boldsymbol{x}^n$ and $\mathcal{M}^n$ are the n-th observed training token, its corresponding hidden dynamic vector, and its corresponding mixture component, respectively. All discrete random variables $\mathcal{M}^n, (1 \leq n \leq N)$ are assumed to have an identical distribution. We assume further that the N tokens are independent of each other.

The development of the EM algorithm described below consists of several steps. First, we develop an explicit expression for the joint PDF of the observation and missing data. We also develop an expression for the mixture weighting factor. These expressions are then used to compute the conditional expectation as required in the E-step of the EM algorithm. The conditional expectation is expressed in turn as a function of a set of sufficient statistics computed from a linear Kalman filter. Finally, re-estimation formulas are derived using the conditional expectation in the M-step of the EM algorithm.

Computation of the joint PDF of observation and missing data

Due to the token-independence assumption, the joint PDF of observation and missing data $\{\mathbf{o}, \boldsymbol{x}, \mathcal{M}\}^N$, given the parameter set Θ, can be written as

$$\begin{aligned} p(\{\mathbf{o}, \boldsymbol{x}, \mathcal{M}\}^N|\Theta) &= \prod_{n=1}^{N} p(\mathbf{o}^n, \boldsymbol{x}^n, \mathcal{M}^n|\Theta) \\ &= \prod_{n=1}^{N} p(\mathbf{o}^n, \boldsymbol{x}^n|m^n, \Theta)\ P(m^n|\Theta). \end{aligned} \tag{10.47}$$

In Eq. 10.47, $p(\mathbf{o}^n, \boldsymbol{x}^n|\mathcal{M}^n, \Theta)$ is the conditional joint PDF of $\mathbf{o}^n$ and $\boldsymbol{x}^n$ given that the mixture component is fixed. It can be further expressed as

$$p(\mathbf{o}^n, \boldsymbol{x}^n|m^n, \Theta) = p(\boldsymbol{x}_0^n|m^n, \Theta) \prod_{k=1}^{K_n} p(\boldsymbol{x}_k^n|\boldsymbol{x}_{k-1}^n, m^n, \Theta)\ p(\mathbf{o}_k^n|\boldsymbol{x}_k^n, \mathcal{M}^n, \Theta), \tag{10.48}$$

where K_n is the total number of frames of the n-th training token, $p(\boldsymbol{x}_0^n|\mathcal{M}^n, \Theta)$ is the distribution of initial value of the hidden dynamics at time $k = 0$ given the mixture component.

Substituting Eq. 10.48 into Eq. 10.47, we obtain the conditional joint PDF $\{\mathbf{o}, \boldsymbol{x}, \mathcal{M}\}^N$ with the explicit form of

$$\begin{aligned} p(\{\mathbf{o}, \boldsymbol{x}, \mathcal{M}\}^N|\Theta) &= \prod_{n=1}^{N} p(\mathbf{o}^n, \boldsymbol{x}^n|m^n, \Theta)P(m^n|\Theta) \qquad (10.49) \\ &= \prod_{n=1}^{N} \{p(\boldsymbol{x}_0^n|m^n)[\prod_{k=1}^{K_n} p(\boldsymbol{x}_{k,m^n}^n|\boldsymbol{x}_{k-1,m^n}^n)\ p(\mathbf{o}_k^n|\boldsymbol{x}_{k,m^n}^n)]P(m^n|\Theta)\}. \end{aligned}$$

Computation of the mixture weighting factor

The conditional joint PDF for $\{\mathbf{o}, \mathcal{M}\}^N$ is

$$p(\{\mathbf{o}, \mathcal{M}\}^N|\Theta) = \prod_{n=1}^{N} p(\mathbf{o}^n|m^n, \Theta)P(m^n|\Theta). \tag{10.50}$$

The PDF for the observation sequence is

$$\begin{aligned} p(\{\mathbf{o}\}^N|\Theta) &= \prod_{n=1}^{N} \sum_{\mathcal{M}^n=1}^{M} p(\mathbf{o}^n|\mathcal{M}^n, \Theta)P(\mathcal{M}^n|\Theta) \\ &= \prod_{n=1}^{N} \sum_{l=1}^{M} p(\mathbf{o}^n|l, \Theta)P(l|\Theta). \end{aligned} \tag{10.51}$$

In the above equation, all $\mathcal{M}^n (1 \le n \le N)$ have an identical distribution so we use one common variable l to replace them. The conditional PDF of $\{\mathcal{M}\}^N$ given $\{\mathbf{o}\}^N$ is

$$\begin{aligned} p(\{\mathcal{M}\}^N|\{\mathbf{o}\}^N, \Theta) &= \frac{p(\{\mathbf{o}, \mathcal{M}\}^N|\Theta)}{p(\{\mathbf{o}\}^N|\Theta)} \\ &= \frac{\prod_{n=1}^N p(\mathbf{o}^n|m^n, \Theta)P(m^n|\Theta)}{\prod_{n=1}^N \sum_{l=1}^M p(\mathbf{o}^n|l, \Theta)P(l|\Theta)} \\ &= \prod_{n=1}^N \omega_{m^n}^n, \end{aligned} \tag{10.52}$$

where we define the (token-dependent) mixture weighting factors to be

$$\omega_{m^n}^n = \frac{p(\mathbf{o}^n|m^n, \Theta)P(m^n|\Theta)}{\sum_{l=1}^M p(\mathbf{o}^n|l, \Theta)P(l|\Theta)}. \tag{10.53}$$

The mixture weighting factors satisfy $\sum_{m^n=1}^M \omega_{m^n}^n = 1$.

The E-step

Given the various PDF's computed above, we are now in a position to derive an iterative EM algorithm for parameter estimation. For the piecewise linear model presented in this section, both $\{\boldsymbol{x}\}^N$ and $\{\mathcal{M}\}^N$ are treated as missing data. The Q-function in the E-step of the EM algorithm is computed below as the conditional expectation over the missing data:

$$\begin{aligned} Q(\Theta|\bar{\Theta}) &= \sum_{\{\mathcal{M}\}^N} \int \log p(\{\mathbf{o}, \boldsymbol{x}, \mathcal{M}\}^N|\Theta) \cdot p(\{\boldsymbol{x}, \mathcal{M}\}^N|\{\mathbf{o}\}^N, \bar{\Theta})\, d\{\boldsymbol{x}\}^N \qquad (10.54) \\ &= \sum_{\{\mathcal{M}\}^N} \int \log p(\{\mathbf{o}, \boldsymbol{x}, \mathcal{M}\}^N|\Theta) \cdot p(\{\boldsymbol{x}\}^N|\{\mathbf{o}, \mathcal{M}\}^N, \bar{\Theta})\, d\{\boldsymbol{x}\}^N \cdot p(\{\mathcal{M}\}^N|\{\mathbf{o}\}^N, \bar{\Theta}), \end{aligned}$$

where $\bar{\Theta}$ denotes the model parameters associated with the immediately previous iteration of the EM algorithm. Since $\{\boldsymbol{x}\}^N$ is continuously distributed and $\{\mathcal{M}\}^N$ is discretely distributed, they are integrated and summed, respectively, in the evaluation of the conditional expectation of Eq. 10.54.

Substituting Eq. 10.50 into Eq. 10.54 above, we have

$$\begin{aligned} Q(\Theta|\bar{\Theta}) &= \sum_{\{\mathcal{M}\}^N} \int \sum_{n=1}^N \{\log p(\boldsymbol{x}_0^n|m^n) + \sum_{k=1}^{K_n} [\log p(\boldsymbol{x}_{k,m^n}^n|\boldsymbol{x}_{k-1,m^n}^n) + \log p(\mathbf{o}_k^n|\boldsymbol{x}_{k,m^n}^n)] \\ &\quad + \log p(m^n|\Theta)\} \cdot p(\{\boldsymbol{x}\}^N|\{\mathbf{o}, \mathcal{M}\}^N, \bar{\Theta})\, d\{\boldsymbol{x}\}^N \cdot p(\{\mathcal{M}\}^N|\{\mathbf{o}\}^N, \bar{\Theta}) \\ &= \sum_{\{\mathcal{M}\}^N} \{\sum_{n=1}^N \int \{\log p(\boldsymbol{x}_0^n|m^n) + \sum_{k=1}^{K_n} [\log p(\boldsymbol{x}_{k,m^n}^n|\boldsymbol{x}_{k-1,m^n}^n) + \log p(\mathbf{o}_k^n|\boldsymbol{x}_{k,m^n}^n)] \\ &\quad + \log p(m^n|\Theta)\} \cdot p(\boldsymbol{x}^n|\mathbf{o}^n, m^n, \bar{\Theta})\, d\boldsymbol{x}^n\} \cdot p(\{\mathcal{M}\}^N|\{\mathbf{o}\}^N, \bar{\Theta}) \\ &= \sum_{\{\mathcal{M}\}^N} \{\sum_{n=1}^N \int \{\log p(\boldsymbol{x}_0^n|m^n) + \sum_{k=1}^{K_n} [\log p(\boldsymbol{x}_{k,m^n}^n|\boldsymbol{x}_{k-1,m^n}^n) \qquad (10.55) \\ &\quad + \log p(\mathbf{o}_k^n|\boldsymbol{x}_{k,m^n}^n)]\} \cdot p(\boldsymbol{x}^n|\mathbf{o}^n, m^n, \bar{\Theta})\, d\boldsymbol{x}^n + \sum_{n=1}^N \log p(m^n|\Theta)\} \cdot p(\{\mathcal{M}\}^N|\{\mathbf{o}\}^N, \bar{\Theta}). \end{aligned}$$

This can be simplified to

$$Q(\Theta|\bar{\Theta}) = \sum_{n=1}^{N}\sum_{m=1}^{M}\int\{\log p(\boldsymbol{x}_0^n|m) + \sum_{k=1}^{K_n}[\log p(\boldsymbol{x}_{k,m}^n|\boldsymbol{x}_{k-1,m}^n) + \log p(\mathbf{o}_k^n|\boldsymbol{x}_{k,m}^n)]\} \cdot p(\boldsymbol{x}^n|\mathbf{o}^n, m, \bar{\Theta})\, d\boldsymbol{x}^n\} \cdot \bar{\omega}_m^n + \sum_{n=1}^{N}\sum_{m=1}^{M}\log p(m|\Theta)\cdot\bar{\omega}_m^n, \tag{10.56}$$

after substituting Eq. 10.52 into Eq. 10.56, changing the order of the summations in Eq. 10.56, and using the common variable m to replace all m^n's. In Eq. 10.56, $\bar{\omega}_m^n$ has the same expression as ω_m^n except that the Θ in the expression is replaced by $\bar{\Theta}$ of the previous EM iteration. We express $Q(\Theta|\bar{\Theta})$ above as

$$Q(\Theta|\bar{\Theta}) = Q_x + Q_p, \tag{10.57}$$

where

$$Q_x = \sum_{n=1}^{N}\sum_{m=1}^{M}\int\{\log p(\boldsymbol{x}_0^n|m) + \sum_{k=1}^{K_n}[\log p(\boldsymbol{x}_{k,m}^n|\boldsymbol{x}_{k-1,m}^n) + \log p(\mathbf{o}_k^n|\boldsymbol{x}_{k,m}^n)]\} \cdot p(\boldsymbol{x}^n|\mathbf{o}^n, m, \bar{\Theta})\, d\boldsymbol{x}^n\} \cdot \bar{\omega}_m^n \tag{10.58}$$

and

$$Q_p = \sum_{n=1}^{N}\sum_{m=1}^{M}\log p(m|\Theta)\cdot\bar{\omega}_m^n. \tag{10.59}$$

From the model definition by Eqs. 10.43 and 10.44 and according to Eqs. 10.45 and 10.46, $p(\boldsymbol{x}_{k,m}^n|\boldsymbol{x}_{k-1,m}^n)$ is a Gaussian with mean $\mathbf{\Phi}_m\boldsymbol{x}^n(k-1)+(\mathbf{I}-\mathbf{\Phi}_m)\mathbf{t}_m$ and covariance $\mathbf{Q}_m$, and $p(\mathbf{o}_k^n|\boldsymbol{x}_{k,m}^n)$ is a Gaussian as well with mean $\dot{\mathbf{H}}_m\dot{\boldsymbol{x}}^n(k)$ and covariance $\mathbf{R}_m$. Fixing $p(\boldsymbol{x}_0^n|m)$ as a Gaussian with zero mean and a given covariance, we simplify Q_x to

$$\begin{aligned} Q_x &= -\frac{1}{2}\sum_{n=1}^{N}\sum_{m=1}^{M}\{K_n\log|Q_m| + K_n\log|\mathbf{R}_m| + \int\sum_{k=1}^{K_n}[\mathbf{e}_{k1,m}^n{}^{\mathrm{Tr}}(\mathbf{Q}_m)^{-1}\mathbf{e}_{k1,m}^n \\ &\quad + \mathbf{e}_{k2,m}^n{}^{\mathrm{Tr}}(\mathbf{R}_m)^{-1}\mathbf{e}_{k2,m}^n]\cdot p(\boldsymbol{x}^n|\mathbf{o}^n, m, \bar{\Theta})\, d\boldsymbol{x}^n\}\cdot\bar{\omega}_m^n + \text{constant} \\ &= -\frac{1}{2}\sum_{n=1}^{N}\sum_{m=1}^{M}\{K_n\log|\mathbf{Q}_m| + K_n\log|\mathbf{R}_m| + \sum_{k=1}^{K_n}\int[\mathbf{e}_{k1,m}^n{}^{\mathrm{Tr}}(\mathbf{Q}_m)^{-1}\mathbf{e}_{k1,m}^n \\ &\quad + \mathbf{e}_{k2,m}^n{}^{\mathrm{Tr}}(\mathbf{R}_m)^{-1}\mathbf{e}_{k2,m}^n]\cdot p(\boldsymbol{x}^n|\mathbf{o}^n, m, \bar{\Theta})\, d\boldsymbol{x}_k^n\}\cdot\bar{\omega}_m^n + \text{constant} \\ &= -\frac{1}{2}\sum_{n=1}^{N}\sum_{m=1}^{M}\{K_n\log|\mathbf{Q}_m| + K_n\log|\mathbf{R}_m| + \sum_{k=1}^{K_n}E[\mathbf{e}_{k1,m}^n{}^{\mathrm{Tr}}(\mathbf{Q}_m)^{-1}\mathbf{e}_{k1,m}^n|\mathbf{o}^n, m, \bar{\Theta}] \\ &\quad + \sum_{k=1}^{K_n}E[\mathbf{e}_{k2,m}^n{}^{\mathrm{Tr}}(\mathbf{R}_m)^{-1}\mathbf{e}_{k2,m}^n|\mathbf{o}^n, m, \bar{\Theta}]\}\cdot\bar{\omega}_m^n + \text{constant}, \end{aligned} \tag{10.60}$$

where $\mathbf{e}_{k1,m}^n = \boldsymbol{x}^n(k) - \mathbf{\Phi}_m\boldsymbol{x}^n(k-1) - (\mathbf{I}-\mathbf{\Phi}_m)\mathbf{t}_m$ and $\mathbf{e}_{k2,m}^n = \mathbf{o}^n(k) - \dot{\mathbf{H}}_m\dot{\boldsymbol{x}}^n(k)$. For notational simplicity, we will use $E_n[\cdot]$ to denote $E[\cdot|\mathbf{o}^n, m, \bar{\Theta}]$ below.

The M-step

(a) Re-estimating $P(m|\Theta)$:

We use π_m to denote $P(m|\Theta)$, $m = 1, 2, ..., M$, which we call mixture weighting probabilities. Since in the Q-function only Q_p is related to π_m, we can obtain the re-estimation formula by setting the partial derivative of Q_p with respect to π_m to zero, and then solving it subject to the constraint:

$$\sum_{m=1}^{M} \pi_m = 1.$$

To proceed, we define the Lagrangian of

$$L_p = Q_p + \lambda(1 - \sum_{m=1}^{M} \pi_m). \tag{10.61}$$

Taking the derivative of L_p with respect to π_m, we obtain

$$\frac{\partial L_p}{\partial \pi_m} = \sum_{n=1}^{N} \frac{1}{\pi_m} \bar{\omega}_m^n - \lambda. \tag{10.62}$$

Setting the derivative equal to zero, we have the re-estimate for π_m:

$$\hat{P}(m|\Theta) = \frac{1}{\lambda} \sum_{n=1}^{N} \bar{\omega}_m^n. \tag{10.63}$$

Taking $\sum_{m=1}^{M}$ on both sides of Eq. 10.63 and using $\sum_{m=1}^{M} \pi_m = 1$, we obtain

$$\lambda = \sum_{n=1}^{N} \sum_{m=1}^{M} \bar{\omega}_m^n. \tag{10.64}$$

This gives the re-estimation formula for π_m:

$$\hat{\pi}_m = \frac{\sum_{n=1}^{N} \bar{\omega}_m^n}{\sum_{n=1}^{N} \sum_{m=1}^{M} \bar{\omega}_m^n} \quad \text{for } 1 \le m \le M. \tag{10.65}$$

(b) Re-estimating $\mathbf{\Phi}_m$ and $\mathbf{t}_m$:

Before deriving the re-estimation formula for these new parameters, we adopt the following notations first:

$$\begin{array}{ll}
\mathbf{A}_m^n = \sum_{k=1}^{K_n} E_n[\boldsymbol{x}^n(k)\boldsymbol{x}^n(k)^{\mathrm{Tr}}|\mathbf{o}^n, m], & \mathbf{b}_m^n = \sum_{k=1}^{K_n} E_n[\boldsymbol{x}^n(k)|\mathbf{o}^n, m] \\
\mathbf{c}_m^n = \sum_{k=1}^{K_n} E_n[\boldsymbol{x}^n(k-1)|\mathbf{o}^n, m], & \mathbf{B}_m^n = \sum_{k=1}^{K_n} E_n[\boldsymbol{x}^n(k-1)\boldsymbol{x}^n(k-1)^{\mathrm{Tr}}|\mathbf{o}^n, m] \\
\mathbf{C}_m^n = \sum_{k=1}^{K_n} E_n[\boldsymbol{x}^n(k)\boldsymbol{x}^n(k-1)^{\mathrm{Tr}}|\mathbf{o}^n, m], & \mathbf{d}_m = (\mathbf{I} - \hat{\mathbf{\Phi}}_m)\hat{\mathbf{t}}_m \\
\mathbf{D}^n = \sum_{k=1}^{K_n} \mathbf{o}^n(k)(\mathbf{o}^n(k))^{\mathrm{Tr}}, & \mathbf{F}_m^n = \sum_{k=1}^{K_n} \mathbf{o}^n(k) E_n[\dot{\boldsymbol{x}}^n(k)|\mathbf{o}^n, m]^{\mathrm{Tr}} \\
\mathbf{G}_m^n = \sum_{k=1}^{K_n} E_n[\dot{\boldsymbol{x}}^n(k)(\dot{\boldsymbol{x}}^n(k))^{\mathrm{Tr}}|\mathbf{o}^n, m]. &
\end{array}$$

where $\hat{\mathbf{\Phi}}_m$ and $\hat{\mathbf{t}}_m$ stand for the newly re-estimated values, and where

$$E_n[\dot{\boldsymbol{x}}^n(k)] = [1 \; | E_n[\boldsymbol{x}^n(k)]^{Tr}]^{Tr},$$

and

$$E_n[\dot{\boldsymbol{x}}^n(k)(\dot{\boldsymbol{x}}^n(k))^{Tr}] = \begin{bmatrix} 1 & E_n[\boldsymbol{x}^n(k)]^{Tr} \\ E_n[\boldsymbol{x}^n(k)] & E_n[\boldsymbol{x}^n(k)(\boldsymbol{x}^n(k))^{Tr}] \end{bmatrix}.$$

To re-estimate $\mathbf{\Phi}_m$ and $\mathbf{t}_m$ we note that they are related only to Q_x. Furthermore, only $\mathbf{e}_{k1,m}$ includes $\mathbf{\Phi}_m$ and $\mathbf{t}_m$. The relevant partial derivatives are

$$\begin{aligned} \frac{\partial Q_x}{\partial \mathbf{\Phi}_m} &= \mathbf{Q}_m^{-1} \sum_{n=1}^{N} (-\mathbf{C}_m^n + \mathbf{b}_m^n \mathbf{t}_m^{\mathrm{Tr}} + \mathbf{t}_m \mathbf{c}_m^{n\ \mathrm{Tr}} - K_n \mathbf{t}_m \mathbf{t}_m^{\mathrm{Tr}}) \cdot \bar{\omega}_m^n \\ &\quad + \mathbf{Q}_m^{-1} \mathbf{\Phi}_m \sum_{n=1}^{N} (\mathbf{B}_m^n - \mathbf{c}_m^n \mathbf{t}_m^{\mathrm{Tr}} - \mathbf{t}_m \mathbf{c}_m^{n\ \mathrm{Tr}} + K_n \mathbf{t}_m \mathbf{t}_m^{\mathrm{Tr}}) \cdot \bar{\omega}_m^n \quad (10.66) \end{aligned}$$

and

$$\frac{\partial Q_x}{\partial \mathbf{t}_m} = -\mathbf{Q}_m^{-1}(\mathbf{I} - \mathbf{\Phi}_m) \sum_{n=1}^{N} \{\mathbf{b}_m^n - \mathbf{\Phi}_m \mathbf{c}_m^n - K_n(\mathbf{I} - \mathbf{\Phi}_m)\mathbf{t}_m\} \cdot \bar{\omega}_m^n. \quad (10.67)$$

Setting the above derivatives to zero, we obtain the re-estimate for $\mathbf{\Phi}_m$ and $\mathbf{t}_m$:

$$\begin{aligned} \hat{\mathbf{\Phi}}_m &= \{\sum_{n=1}^{N} (\mathbf{C}_m^n - \mathbf{b}_m^n \hat{\mathbf{t}}_m^{\mathrm{Tr}} - \hat{\mathbf{t}}_m \mathbf{c}_m^{n\ \mathrm{Tr}} + K_n \hat{\mathbf{t}}_m \hat{\mathbf{t}}_m^{\mathrm{Tr}}) \cdot \bar{\omega}_m^n\} \\ &\quad \cdot \{\sum_{n=1}^{N} (\mathbf{B}_m^n - \mathbf{c}_m^n \hat{\mathbf{t}}_m^{\mathrm{Tr}} - \hat{\mathbf{t}}_m \mathbf{c}_m^{n\ \mathrm{Tr}} + K_n \hat{\mathbf{t}}_m \hat{\mathbf{t}}_m^{\mathrm{Tr}}) \cdot \bar{\omega}_m^n\}^{-1} \quad (10.68) \end{aligned}$$

and

$$\hat{\mathbf{t}}_m = \frac{(\mathbf{I} - \hat{\mathbf{\Phi}}_m)^{-1} \sum_{n=1}^{N} \{\mathbf{b}_m^n - \hat{\mathbf{\Phi}}_m \mathbf{c}_m^n\} \cdot \bar{\omega}_m^n}{\sum_{n=1}^{N} K_n \cdot \bar{\omega}_m^n}. \quad (10.69)$$

In the above re-estimation, the parameters $\mathbf{\Phi}_m$ and $\mathbf{t}_m$ are updated alternatively at separate EM iterations. This gives rise to the generalized EM algorithm, which gives local optimization in the M-step, rather than global optimization.

(c) Re-estimating $\dot{\mathbf{H}}_m$:

To re-estimate $\dot{\mathbf{H}}_m$, we note that it is included only in $\mathbf{e}_{k2,m}$. The relevant partial derivative is

$$\frac{\partial Q_x}{\partial \dot{\mathbf{H}}_m} = -\sum_{n=1}^{N} \bar{\omega}_m^n \sum_{k=1}^{K_n} \mathbf{R}_m^{-1} E_n[(\dot{\mathbf{H}}_m \dot{\boldsymbol{x}}^n(k) - \mathbf{o}_k^n)(\dot{\boldsymbol{x}}^n(k))^{\mathrm{Tr}}]. \quad (10.70)$$

Setting the above to zero, we have the re-estimate:

$$\hat{\dot{\mathbf{H}}}_m = \{\sum_{n=1}^{N} \bar{\omega}_m^n \mathbf{F}_m^n\}\{\sum_{n=1}^{N} \bar{\omega}_m^n \mathbf{G}_m^n\}^{-1}. \quad (10.71)$$

(d) Re-estimating $\mathbf{Q}_m$ and $\mathbf{R}_m$:

Since the noise covariances $\mathbf{Q}_m$ and $\mathbf{R}_m$ are included only in Q_x, we compute the following derivatives:

$$\frac{\partial Q_x}{\partial \mathbf{Q}_m^{-1}} = \frac{1}{2}\sum_{n=1}^{N} K_n \mathbf{Q}_m \bar{\omega}_m^n - \frac{1}{2}\sum_{n=1}^{N}\sum_{k=1}^{K_n} E_n[\mathbf{e}_{k1,m}^n \mathbf{e}_{k1,m}^n{}^{\mathrm{Tr}}] \cdot \bar{\omega}_m^n,$$

and

$$\frac{\partial Q_x}{\partial \mathbf{R}_m^{-1}} = \frac{1}{2}\sum_{n=1}^{N} K_n \mathbf{R}_m \bar{\omega}_m^n - \frac{1}{2}\sum_{n=1}^{N}\sum_{k=1}^{K_n} E_n[\mathbf{e}_{k2,m}^n \mathbf{e}_{k2,m}^n{}^{\mathrm{Tr}}] \cdot \bar{\omega}_m^n.$$

Setting the derivatives equal to zero, we obtain the estimates for $\mathbf{Q}_m$ and $\mathbf{R}_m$:

$$\hat{\mathbf{Q}}_m = \frac{\sum_{n=1}^{N}\sum_{k=1}^{K_n} E_n[\mathbf{e}_{k1,m}^n \mathbf{e}_{k1,m}^n{}^{\mathrm{Tr}}] \cdot \bar{\omega}_m^n}{\sum_{n=1}^{N} K_n \bar{\omega}_m^n}$$

and

$$\hat{\mathbf{R}}_m = \frac{\sum_{n=1}^{N}\sum_{k=1}^{K_n} E_n[\mathbf{e}_{k2,m}^n \mathbf{e}_{k2,m}^n{}^{\mathrm{Tr}}] \cdot \bar{\omega}_m^n}{\sum_{n=1}^{N} K_n \bar{\omega}_m^n}.$$

In $\hat{\mathbf{Q}}_m$ above, $\sum_{k=1}^{K_n} E_n[\mathbf{e}_{k1,m}^n \mathbf{e}_{k1,m}^n{}^{\mathrm{Tr}}]$ is calculated according to

$$\begin{aligned}\sum_{k=1}^{K_n} E_n[\mathbf{e}_{k1,m}^n \mathbf{e}_{k1,m}^n{}^{\mathrm{Tr}}] &= \mathbf{A}_m^n + \hat{\mathbf{\Phi}}_m \mathbf{B}_m^n \hat{\mathbf{\Phi}}_m^{\mathrm{Tr}} - \mathbf{C}_m^n \hat{\mathbf{\Phi}}_m^{\mathrm{Tr}} - \hat{\mathbf{\Phi}}_m (\mathbf{C}_m^n)^{\mathrm{Tr}} - \mathbf{b}_m^n (\mathbf{d}_m)^{\mathrm{Tr}} \\ &\quad - \mathbf{d}_m (\mathbf{b}_m^n)^{\mathrm{Tr}} + \hat{\mathbf{\Phi}}_m \mathbf{c}_m^n (\mathbf{d}_m)^{\mathrm{Tr}} + \mathbf{d}_m (\mathbf{c}_m^n)^{\mathrm{Tr}} \hat{\mathbf{\Phi}}_m^{\mathrm{Tr}} + K_n \mathbf{d}_m \mathbf{d}_m^{\mathrm{Tr}}.\end{aligned}$$

In $\hat{\mathbf{R}}_m$ above, $\sum_{k=1}^{K_n} E_n[\mathbf{e}_{k2,m}^n \mathbf{e}_{k2,m}^n{}^{\mathrm{Tr}}]$ is calculated according to

$$\sum_{k=1}^{K_n} E_n[\mathbf{e}_{k2,m}^n \mathbf{e}_{k2,m}^n{}^{\mathrm{Tr}}] = \mathbf{D}^n - \mathbf{F}_m^n \hat{\hat{\mathbf{H}}}_m^{\mathrm{Tr}} - \hat{\hat{\mathbf{H}}}_m (\mathbf{F}_m^n)^{\mathrm{Tr}} + \hat{\hat{\mathbf{H}}}_m \mathbf{G}_m^n \hat{\hat{\mathbf{H}}}_m^{\mathrm{Tr}}.$$

Computing the sufficient statistics

In order to obtain the re-estimate for the model parameters according to the formulas derived above as the M-step of the EM algorithm, a set of conditional expectations needs to be calculated. Essentially, three conditional expectations (all conditioned on the observation sequences), $E_n[\boldsymbol{x}^n(k)]$, $E_n[\boldsymbol{x}^n(k)\boldsymbol{x}^n(k)^{\mathrm{Tr}}]$ and $E_n[\boldsymbol{x}^n(k)\boldsymbol{x}^n(k-1)^{\mathrm{Tr}}]$, are required by the M-step. The conditional expectation $E_n[\cdot]$, which denotes $E[\cdot|\mathbf{o}^n, m, \bar{\Theta}]$, is precisely the Kalman smoother for the m-th mixture component and for the n-th observation (token). All conditional expectations required in the M-step can be calculated by using the results of the Kalman smoothing algorithm. We now list the computational steps of the Kalman smoothing algorithm below. These steps have been designed, based on the general description of Kalman filter and smoother presented in Section 5.8, for the specific structure of the current piecewise linear hidden dynamic model.

Forward recursion (Kalman filtering):

$$\begin{aligned}
\hat{\boldsymbol{x}}^n_{k|k-1,m} &= \boldsymbol{\Phi}_m \hat{\boldsymbol{x}}^n_{k-1|k-1,m} + (\mathbf{I} - \boldsymbol{\Phi}_m)\mathbf{t}_m \\
\boldsymbol{\Sigma}^n_{k|k-1,m} &= \boldsymbol{\Phi}_m \boldsymbol{\Sigma}^n_{k-1|k-1,m} \boldsymbol{\Phi}_m + \mathbf{Q}_m \\
\tilde{\mathbf{o}}^n_{k,m} &= \mathbf{o}^n(k) - \dot{\mathbf{H}}\dot{\hat{x}}^n_{k|k-1,m} \\
\boldsymbol{\Sigma}^n_{\tilde{\mathbf{o}}_{k,m}} &= \mathbf{H}_m \boldsymbol{\Sigma}^n_{k|k-1,m} \mathbf{H}^{\mathrm{Tr}}_m + \mathbf{R}_m \\
\mathbf{K}_{k,m} &= \boldsymbol{\Sigma}^n_{k|k-1,m} \mathbf{H}^{\mathrm{Tr}}_m (\boldsymbol{\Sigma}^n_{\tilde{\mathbf{o}}_{k,m}})^{-1} \\
\hat{\boldsymbol{x}}^n_{k|k,m} &= \hat{\boldsymbol{x}}^n_{k|k-1,m} + \mathbf{K}_{k,m}\tilde{\mathbf{o}}^n_{k,m} \\
\boldsymbol{\Sigma}^n_{k|k,m} &= \boldsymbol{\Sigma}^n_{k|k-1,m} - \mathbf{K}_{k,m}\boldsymbol{\Sigma}^n_{\tilde{\mathbf{o}}_{k,m}}\mathbf{K}^{\mathrm{Tr}}_{k,m}
\end{aligned} \tag{10.72}$$

Backward recursion (Kalman smoothing):

$$\begin{aligned}
\mathbf{M}^n_{k,m} &= \boldsymbol{\Sigma}^n_{k|k,m}\boldsymbol{\Phi}^{\mathrm{Tr}}_m(\boldsymbol{\Sigma}^n_{k|k-1,m})^{-1} \\
\hat{\boldsymbol{x}}^n_{k|K_n,m} &= \hat{\boldsymbol{x}}^n_{k|k,m} + \mathbf{M}^n_{k,m}[\hat{\boldsymbol{x}}^n_{k+1|K_n} - \hat{\boldsymbol{x}}^n_{k+1|k,m}] \\
\boldsymbol{\Sigma}^n_{k|K_n,m} &= \boldsymbol{\Sigma}^n_{k|k,m} + \mathbf{M}^n_{k,m}[\boldsymbol{\Sigma}^n_{k+1|K_n,m} - \boldsymbol{\Sigma}^n_{k+1|k,m}]\mathbf{M}^{\mathrm{Tr}}_k
\end{aligned} \tag{10.73}$$

Based on the Kalman filtering and smoothing results above, the three required conditional expectations are computed as follows:

$$\begin{aligned}
E_n[\boldsymbol{x}^n(k)] &= \hat{\boldsymbol{x}}^n_{k|K_n,m} \\
E_n[\boldsymbol{x}^n(k)\boldsymbol{x}^n(k)^{\mathrm{Tr}}] &= \boldsymbol{\Sigma}^n_{k|K_n,m} + E_n[\boldsymbol{x}^n(k)]E_n[\boldsymbol{x}^n(k)]^{\mathrm{Tr}} \\
E_n[\boldsymbol{x}^n(k)\boldsymbol{x}^n(k-1)^{\mathrm{Tr}}] &= \boldsymbol{\Sigma}^n_{k|K_n,m}(\boldsymbol{\Sigma}^n_{k|k-1,m})^{-1}\boldsymbol{\Phi}_m\boldsymbol{\Sigma}^n_{k-1|k-1,m} \\
E_n[\boldsymbol{x}^n(k)\boldsymbol{x}^n(k-1)^{\mathrm{Tr}}] &= \boldsymbol{\Sigma}^n_{k,k-1|K_n,m} + E_n[\boldsymbol{x}^n(k)]E_n[\boldsymbol{x}^n(k-1)]^{\mathrm{Tr}},
\end{aligned} \tag{10.74}$$

where $\boldsymbol{\Sigma}^n_{k,k-1|K_n,m}$ is recursively calculated by

$$\boldsymbol{\Sigma}^n_{k,k-1|K_n,m} = \boldsymbol{\Sigma}^n_{k|k,m}(\mathbf{M}^n_{k-1,m})^{\mathrm{Tr}} + \mathbf{M}^n_{k,m}(\boldsymbol{\Sigma}^n_{k+1,k|K_n,m} - \boldsymbol{\Phi}_m\boldsymbol{\Sigma}^n_{k|k,m})(\mathbf{M}^n_{k-1,m})^{\mathrm{Tr}}$$

for $k = K_n, \cdots, 2$, where

$$\boldsymbol{\Sigma}^n_{K_n,K_n-1|K_n,m} = (\mathbf{I} - \mathbf{K}_{K_n,m}\mathbf{H}_m)\boldsymbol{\Phi}_m\boldsymbol{\Sigma}^n_{K_n-1|K_n-1,m}.$$

Updating ω:

To update $\bar{\omega}^n_m$ according to Eq. 10.53, $p(\mathbf{o}^n|m,\bar{\Theta})$ must be calculated. The calculation proceeds as follows:

$$\begin{aligned}
p(\mathbf{o}^n|m,\bar{\Theta}) &= p(\mathbf{o}^n_1|m,\bar{\Theta})p(\mathbf{o}^n_2|\mathbf{o}^n_1,m,\bar{\Theta}) \\
&\quad \cdots p(\mathbf{o}^n_k|\mathbf{o}^n_{k-1},\ldots,\mathbf{o}^n_1,m,\bar{\Theta})\cdots p(\mathbf{o}^n_{K_n}|\mathbf{o}^n_{K_n-1},\ldots,\mathbf{o}^n_1,m,\bar{\Theta}),
\end{aligned}$$

where $p(\mathbf{o}^n_k|\mathbf{o}^n_{k-1},\ldots,\mathbf{o}^n_1,m,\bar{\Theta})$ is the PDF of the innovation sequence. This PDF has the Gaussian form of

$$(2\pi)^{-\frac{D}{2}}\left|\boldsymbol{\Sigma}^n_{\tilde{\mathbf{o}}_{k,m}}\right|^{-\frac{1}{2}}\exp\{-\frac{1}{2}(\tilde{\mathbf{o}}^n_{k,m})^{\mathrm{Tr}}[\boldsymbol{\Sigma}^n_{\tilde{\mathbf{o}}_{k,m}}]^{-1}\tilde{\mathbf{o}}^n_{k,m}\}, \tag{10.75}$$

where the innovation sequence $\tilde{\mathbf{o}}_{k,m}$ is computed directly from the (forward) Kalman Filtering step Eq. 10.72.

10.4.3 Likelihood-scoring algorithm

The piecewise linear hidden dynamic model discussed so far combines M different linear state-space models using the mixture weighting probabilities. After the weighting probabilities and all other models' parameters are estimated as just described, the likelihood of the model for each phonological state, given a sequence of acoustic observations, can be computed directly. We describe this computation below.

Based on the estimation theory for dynamic systems presented in Chapter 5, the log-likelihood function for each individual component of the mixture model is calculated from its innovation sequence $\tilde{\mathbf{o}}_{k,m}$ according to

$$l_m(\mathbf{o}|\Theta) = \exp\left[-\frac{1}{2}\sum_{k=1}^{K}\{\log|\mathbf{\Sigma}_{\tilde{\mathbf{o}}_{k,m}}| + \tilde{\mathbf{o}}_{k,m}\mathbf{\Sigma}^{-1}_{\tilde{\mathbf{o}}_{k,m}}\tilde{\mathbf{o}}^{\mathrm{Tr}}_{k,m}\} + \text{constant}\right], \qquad (10.76)$$

where the innovation sequence $\tilde{\mathbf{o}}_{k,m}$ and its covariance $\mathbf{\Sigma}_{\tilde{\mathbf{o}}_{k,m}}$ are computed from the Kalman filtering recursion. Then we sum up the component models' log-likelihoods according to their mixture weighting probabilities to obtain the log-likelihood for the entire mixture model:

$$L(\mathbf{o}|\Theta) = \log\left[\sum_{m=1}^{M}\pi_m l_m(\mathbf{o}|\Theta)\right]. \qquad (10.77)$$

For a speech utterance that consists of a sequence of phonological states (e.g., phones) with the dynamic regimes (i.e., segmentations) given, the log-likelihoods for each phonological state in the sequence as defined in Eq. 10.77 are summed to give the total log-likelihood score for the entire utterance.

In general, the dynamic regimes associated with the phonological states are not known. Finding the optimal likelihood for the speech utterance using optimized dynamic regimes is an intractable computational problem, with the computational cost increasing exponentially with the length of the utterance. Several approximation techniques that overcome this intractability have been developed for the speech model described in this section and the readers are referred to the publications of [Ma 99, Ma 00, Ma 02] for details. A simpler way of overcoming the computational burden than the approximation techniques described in these publications is to use a high-quality HMM system to produce the segmentation of phonological units for each hypothesized word sequence (e.g., the word sequences in the N-best list). Then, Eq. 10.77 can be used efficiently to compute the needed likelihood for evaluating the recognizer.

10.5 A Comprehensive Statistical Generative Model of the Dynamics of Casual Speech

In this section we describe a comprehensive computational model of speech production which uses functional hidden articulatory dynamics to characterize speech with a casual style. This serves to review and extend many of the concepts in computational models of speech production already discussed in this chapter.

Casual speech is a very pervasive form of human-human communication employed in verbal conversations among two or more speakers either over the telephone, or in meetings, or face to face. Casual speech is also common in non-conversational, spontaneous speech such as natural voice mail. If a computer system can be constructed to automatically decode the linguistic messages contained in conversational speech, one will have vast opportunities for the applications of such speech technology.

The casual style of speaking in conversational or spontaneous speech causes the following two key consequences that make the acoustics of conversational speech significantly differ from that of "read-style" speech. First, in casual speech, which is also called "hypo-articulated" speech [Lind 90], phonological reorganization occurs where the relative timing or phasing relationship across different articulatory feature dimensions in the "orchestrated" feature strands are modified. One obvious manifestation of this modification is that the more casual or relaxed the speech style is, the greater overlapping will result across the feature dimensions. Second, phonetic reduction occurs where articulatory targets as phonetic correlates to the phonological units may shift toward a more neutral position due to the use of reduced articulatory efforts associated with casual "hypo-articulated" speech. Another, perhaps more important, form of phonetic reduction is that the realized articulatory trajectories tend to be further away from reaching their respective targets due to physical inertia constraints in the articulatory movements, which occur during a generally shorter time duration in casual-style speech than in formal-style speech.

The current HMM-based speech recognition systems do not provide effective mechanisms to embrace the huge new acoustic variability in casual conversational speech arising either from phonological organization or from phonetic reduction. Importantly, the new variability due to phonetic reduction is scaled continuously, resulting in many phonetic confusions in a predictable manner. Due to this continuous nature of variability scaling, prohibitively large amounts of labeled conversational speech data would be needed in order to capture this variability if no knowledge about phonetic reduction and about its effects on speech dynamic patterns is incorporated into the speech model underlying casual speech recognition systems.

The comprehensive generative model for casual speech described in this section has been motivated by the desire of incorporating knowledge of both phonological reorganization and phonetic reduction in an integrative fashion. To describe this model, we break up the model into several inter-related components, where the output, expressed as the probability distribution, of one component serves as the input to the next component in a truly "generative" spirit. The top-level component is the phonological model that specifies the discrete (symbolic) pronunciation units of the intended linguistic message in terms of multi-tiered, overlapping articulatory features. The first intermediate component consists of articulatory control and target, which provides the interface between the discrete phonological units to the continuous phonetic variable and which represents the "ideal" articulation and its inherent variability if there were no physical constraints in articulation. The second intermediate component consists of articulatory dynamics, which explicitly represents the physical constraints in articulation and gives the output of "actual" trajectories in the articulatory variables. The bottom component in this comprehensive model is the process of speech acoustics being generated from the vocal tract whose shape and excitation are determined by the articulatory variables as the output of the articulatory dynamic model.

The comprehensive generative model presented here consists of the largely sequentially connected components of 1) Multi-tiered phonological construct (non-observable or hidden; discrete valued); 2) Articulatory target (hidden; continuous); 3) Articulatory dynamics (hidden; continuous); and 4) Acoustic pattern formation from articulatory variables (hidden; continuous). We now describe each of these components in some detail.

10.5.1 Overlapping model for multi-tiered phonological construct

As we discussed in Chapters 8 and 9, phonology is concerned with sound patterns of speech and with the nature of discrete or symbolic units that form such patterns. Traditional theories of phonology differ in the choice and interpretation of the phonological units. Early distinctive feature-based theory and subsequent autosegmental, feature-geometry theory, both discussed in Sections 8.3-8.5, assumed a rather direct link between phonological features and their phonetic correlates in the articulatory or acoustic domain. Phonological rules for modifying features represented changes not only in the linguistic structure of the speech utterance, but also in the phonetic realization of this structure. This weakness has been recognized by more recent theories, e.g., articulatory phonology discussed in Section 8.6, which emphasize the importance of accounting for phonetic levels of variation as distinct from those at the phonological levels.

In the phonological model described here, it is assumed that the linguistic function of phonological units is to maintain linguistic contrasts and is separate from phonetic implementation. It is further assumed that the phonological unit sequence can be described mathematically by a discrete-time, discrete-state, multi-dimensional homogeneous Markov chain. How to construct sequences of symbolic phonological units for any arbitrary speech utterance and how to build them into an appropriate Markov state (i.e., phonological state) structure are the key issues in the model construction.

Motivated by articulatory phonology, the feature-based phonological model outlined here, which is a simplified version of a model in computational phonology discussed in detail in Section 9.5, uses multi-tiered articulatory features that are overlapping with each other in separate tiers temporally, with learnable relative-phasing relationships. This contrasts with nearly all current popular speech recognition systems, where the representation is based on phone-sized units with a single tier for the phonological sequence acting as "beads-on-a-string".

Mathematically, the L-tiered, overlapping model can be described by the "factorial" Markov chain, where the state of the chain is represented by a collection of discrete component state variables:

$$\mathbf{s}_t = s_t^{(1)}, ..., s_t^{(l)}, ..., s_t^{(L)}.$$

Each of the component states can take $K^{(l)}$ values. In implementing this model for American English, we have $L = 5$, and the five tiers are Lips, Tongue Blade, Tongue Body, Velum, and Larynx, respectively. For "Lips" tier, we have $K^{(1)} = 6$ for six possible linguistically distinct Lips-configurations; i.e., those for */b/*, */r/*, */sh/*, */u/*, */w/*, and */o/*. For the remaining tiers, we have $K^{(2)} = 6$, $K^{(3)} = 17$, $K^{(4)} = 2$, and $K^{(5)} = 2$.

The state space of this factorial Markov chain consists of all $K_L = K^{(1)} \times K^{(2)} \times K^{(3)} \times K^{(4)} \times K^{(5)}$ possible combinations of the $s_t^{(l)}$ state variables. If no constraints are imposed on the state transition structure, this would be equivalent to the conventional one-tiered Markov chain with a total of K_L states and a $K_L \times K_L$ state transition matrix. This is an uninteresting case since the model complexity is exponentially (or factorially) growing in L. It is also unlikely to find any useful phonological structure in this huge Markov chain. Further, since all the phonetic parameters in the lower-level components of the comprehensive model (to be seen later) are conditioned on the phonological state, the total number of model parameters would be unreasonably large, presenting a well known sparseness difficulty for parameter learning.

Fortunately, rich sources of phonological and phonetic knowledge are available to constrain the state transitions of the above factorial Markov chain. One particularly useful set of constraints comes directly from the phonological theories that motivated the construction of this model. Both autosegmental phonology and articulatory phonology treat the different tiers in the phonological features as being largely independent of each other in their evolving dynamics over time. This thus allows the a-priori decoupling among the L tiers:

$$P(\mathbf{s}_t|\mathbf{s}_{t-1}) = \prod_{l=1}^{L} P(s_t^{(l)}|s_{t-1}^{(l)}).$$

The transition structure of this constrained (uncoupling) factorial Markov chain can be parameterized by L distinct $K^{(l)} \times K^{(l)}$ matrices. This is significantly simpler than the original $K_L \times K_L$ matrix as in the unconstrained case.

This model is called an overlapping feature model because the independent dynamics of the features at different tiers cause many ways in which different feature values associated with their respective tiers occur simultaneously at a fixed time point. These are determined by how the component features overlap with each other as a consequence of their independent temporal dynamics. In contrast, in the phone-based phonological model, there is only a single tier of phones as the "bundled" component features, and hence there is no concept of overlapping component-features.

The model parameters in this phonological model component includes the Markov chain transition probability for each separate tier, l, defined by

$$P(s_k^{(l)} = j|s_{k-1}^{(l)} = i) = a_{ij}^{(l)}. \tag{10.78}$$

10.5.2 Segmental target model

After a phonological model is constructed, the processes for converting abstract phonological units into their phonetic realization need to be specified. This is a central problem in speech production literature. It concerns the nature of invariance and variability in the processes interfacing phonology and phonetics. Specifically, it needs to address the issue of whether the invariance is more naturally expressed in the articulatory or acoustic/auditory domains. A number of theories assumed a direct link between abstract phonological units and physical measurements. For example, the "quantal theory" [Stev 89] proposed that phonological features possessed invariant acoustic correlates that could be measured directly from the speech signal. The "motor theory" [Lib 85] proposed instead that articulatory properties are directly associated with phonological symbols. No conclusive evidence supporting either hypothesis has been found without controversy (cf. [Lind 96]).

In the current generative model of speech, one commonly held view in phonetics literature is adopted; that is, discrete phonological units are associated with a temporal segmental sequence of phonetic targets or goals. In this view, the function of the articulatory motor control system is to achieve such targets or goals by manipulating the articulatory organs according to some control principles subject to the articulatory inertia and possibly minimal-energy constraints [Lind 90].

Compensatory articulation has been widely documented in the phonetics literature where trade-offs between different articulators and non-uniqueness in the articulatory-acoustic mapping allow for the possibility that many different articulatory target configurations may be able to "equivalently" realize the same underlying goal. Speakers typically choose a range of possible targets depending on external environments and their interactions with listeners. In order to account for compensatory articulation, a complex phonetic control strategy need be adopted. The key modeling assumptions adopted regarding such a strategy are as follows. First, each phonological unit is correlated (i.e., interfaced) to a number of phonetic parameters. These measurable parameters may be acoustic, articulatory or auditory in nature, and they can be computed from some physical models for the articulatory and auditory systems. Second, the region determined by the phonetic correlates for each phonological unit can be mapped onto an articulatory parameter space. Hence the target distribution in the articulatory space can be determined simply by stating what the phonetic correlates (formants, articulatory positions, auditory responses, etc.) are for each of the phonological units (many examples are provided in [Stev 98, Flan 72]), and by running simulations in detailed articulatory and auditory models.

A convenient mathematical representation for the distribution of the articulatory target vector $\mathbf{t}$ is a multivariate Gaussian distribution, denoted by

$$\mathbf{t} \sim \mathcal{N}(\mathbf{t}; \mathbf{m}(\mathbf{s}), \mathbf{\Sigma}(\mathbf{s})), \tag{10.79}$$

where the covariance matrix $\mathbf{\Sigma}(\mathbf{s})$ is non-diagonal. This allows for correlation among the articulatory vector components. Because such a correlation is represented for the articulatory target (as a random vector), compensatory articulation is naturally incorporated in the model.

Since the target distribution as specified in Eq. 10.79 is conditioned on a specific phonological unit (e.g., a bundle of overlapped features represented by the composite state $\mathbf{s}$ consisting of component feature values in the Markov chain) and since the target does not switch until the phonological unit changes, the statistics for the temporal sequence of the target process follows that of a segmental HMM [Rus 93].

For the single-tiered (L=1) phonological model (e.g., phone-based model), the segmental HMM for the target process will be the same as that described in [Rus 93], except the output is no longer the acoustic parameters. The output of this segmental HMM is the random articulatory target vector $\mathbf{t}(k)$ that is constrained to be constant until the phonological state switches its value. This segmental constraint for the dynamics of the random target vector $\mathbf{t}(k)$ represents the adopted articulatory control strategy that the goal of the motor system is to try to maintain the articulatory target's position (for a fixed corresponding phonological state) by exerting appropriate muscle forces. That is, while random, $\mathbf{t}(k)$ remains fixed until the phonological state s_k switches. The switching of target $\mathbf{t}(k)$ is synchronous with that of the phonological state, and only at the time of switching is $\mathbf{t}(k)$ allowed to take a new value according to its probability density function. This segmental constraint can be described mathematically by the following conditional probability density function:

$$p[\mathbf{t}(k)|s_k, s_{k-1}, \mathbf{t}(k-1)] = \begin{cases} \delta[\mathbf{t}(k) - \mathbf{t}(k-1)] & \text{if } s_k = s_{k-1}, \\ \mathcal{N}(\mathbf{t}(k); \mathbf{m}(s_k), \mathbf{\Sigma}(s_k)) & \text{otherwise.} \end{cases}$$

10.5.3 Functional model for hidden articulatory dynamics

At the present state of knowledge, it is difficult to speculate how the conversion of higher-level motor control into articulator movement takes place. Ideally, accurate modeling of articulatory dynamics and control would require detailed neuromuscular and biomechanical models of the vocal tract, as well as an explicit model of the control objectives and strategies. This would be extremely complicated to implement as a useful computational model. A reasonable, simplifying assumption is that the combined (non-linear) control system and articulatory mechanism behave, at the functional level, as a linear dynamic system. It attempts to track the control input equivalently represented by the articulatory target in the physical articulatory parameter space. Articulatory dynamics can then be approximated as the response of a dynamic vocal-tract model driven by a random target sequence as represented by the segmental HMM just described. The statistics of the random target sequence approximates that of the muscle forces that physically drive motions of the articulators.

This simplifying assumption then reduces the generally intractable nonlinear state equation:

$$\boldsymbol{x}(k+1) \quad = \quad \mathbf{g}_s[\boldsymbol{x}(k), \mathbf{t}_s, \mathbf{w}(k)]$$

into the following mathematically tractable, linear, autoregressive (AR) model which we discussed earlier in this chapter:

$$\boldsymbol{x}(k+1) \quad = \quad \boldsymbol{\Phi}_s \boldsymbol{x}(k) + (\mathbf{I} - \boldsymbol{\Phi}_s)\mathbf{t}_s + \mathbf{w}(k), \tag{10.80}$$

where $\boldsymbol{x} \in \mathcal{R}^n$ is the articulatory-parameter vector, $\mathbf{I}$ is the identity matrix, $\mathbf{w}$ is the IID and Gaussian noise, $\mathbf{t}_s$ is the HMM-state-dependent, target vector expressed in the same articulatory domain as $\boldsymbol{x}(k)$, and $\boldsymbol{\Phi}_s$ is the HMM-state-dependent system matrix. The dependence of the $\mathbf{t}_s$ and $\boldsymbol{\Phi}_s$ parameters of the above dynamic system on the phonological state is justified by the fact that the functional behavior of an articulator depends both on the particular goal it is trying to implement, and on the other articulators with which it is cooperating in order to produce compensatory articulation.

Note that the linear AR model of Eq. 10.80 has a special, target-directed property; that is, the articulatory vector $\boldsymbol{x}(k)$ asymptotically approaches the mean of the target random vector $\mathbf{t}$ for artificially lengthened speech utterances. For natural speech, especially for conversational speech with a casual style, the generally short duration associated with each phonological state forces the articulatory dynamics to move away from the target of the current state (and towards the target of the following phonological state) long before it reaches the current target. This gives rise to phonetic reduction, and is one key source of speech variability that cannot be directly captured by a conventional HMM.

The conditional PDF for the continuous articulatory state in this model can be determined directly from Eq. 10.80 to be

$$p_{\boldsymbol{x}}[\boldsymbol{x}(k+1)|\boldsymbol{x}(k), \mathbf{t}(k), \mathrm{s}_k] = p_{\mathbf{w}}[\boldsymbol{x}(k+1) - \boldsymbol{\Phi}_{\mathrm{s}_k}\boldsymbol{x}(k) - (\mathbf{I} - \boldsymbol{\Phi}_{\mathrm{s}_k})\mathbf{t}(k)]. \tag{10.81}$$

This gives rise to a switching, target-directed AR model driven by a segmental HMM.

10.5.4 Functional model for articulatory-to-acoustic mapping

The final component in the generative model moves the speech generative process from articulation to speech acoustics. A simplifying assumption made for the feasibility of model implementation is that an appropriate time scale is chosen which is common for both the articulatory and acoustic vectors. Further, the relationship between articulatory and acoustic representations can be modeled by a static memoryless transformation, converting a vector of articulatory parameters into a vector of acoustic ones (with the same time scale).

This static memoryless transformation can be mathematically represented by the following "observation" equation in the state-space model:

$$\mathbf{o}(k) \quad = \quad \mathbf{h}[\boldsymbol{x}(k)] + \mathbf{v}(k), \tag{10.82}$$

where $\mathbf{o} \in \mathcal{R}^m$ is the observation vector, $\mathbf{v}$ is the IID observation noise vector uncorrelated with the state noise $\mathbf{w}$, and $\mathbf{h}[.]$ is the static memoryless transformation from the articulatory vector to its corresponding acoustic observation vector.

The conditional PDF for the acoustic features as the output of the nonlinear mapping function can be determined directly from Eq. 10.82 to be

$$p_{\mathbf{o}}[\mathbf{o}(k)|\boldsymbol{x}(k)] = p_{\mathbf{v}}[\mathbf{o}(k) - \mathbf{h}(\boldsymbol{x}(k))]. \tag{10.83}$$

There are many ways of choosing the static nonlinear function for $\mathbf{h}[\boldsymbol{x}]$ as in Eq. 10.82. Let us take an illustrative example of a multi-layer perceptron (MLP) with three layers (input, hidden and output), capitalizing on the highly desirable property of the MLP as a universal nonlinear function approximator [Bishop 97]. Let w_{jl} be the MLP weights from input to hidden units and W_{ij} be the MLP weights from hidden to output units, where l is the input node index, j the hidden node index and i the output node index. Then the output signal at node i can be expressed as a (nonlinear) function of all the input nodes (making up the input vector) according to

$$h_i(\boldsymbol{x}) = \sum_{j=1}^{J} W_{ij} \cdot s(\sum_{l=1}^{L} w_{jl} \cdot x_l), \quad 1 \le i \le I, \tag{10.84}$$

where I, J and L are the numbers of nodes at the output, hidden and input layers, respectively, and $s(.)$ is the hidden unit's nonlinear activation function, taken as the standard sigmoid function of

$$s(x) = \frac{1}{1 + \exp(-x)}. \tag{10.85}$$

The derivative of this sigmoid function has the following concise form:

$$s'(x) = s(x)(1 - s(x)), \tag{10.86}$$

making it convenient for use in many computations.

Typically, the analytical forms of nonlinear functions, such as the MLP, make the associated nonlinear dynamic systems difficult to analyze and make the estimation problems difficult to solve. Approximations are frequently used to gain computational simplifications while sacrificing accuracy for approximating the nonlinear functions.

One very commonly used technique for the approximation is a truncated (vector) Taylor series expansion. If all the Taylor series terms of order two and higher are truncated, then we have the linear Taylor series approximation that is characterized by the Jacobian matrix $\mathbf{J}$ and by the point of Taylor series expansion $\boldsymbol{x}_0$:

$$\mathbf{h}(\boldsymbol{x}) \approx \mathbf{h}(\boldsymbol{x}_0) + \mathbf{J}(\boldsymbol{x}_0)(\boldsymbol{x} - \mathbf{x}_0). \tag{10.87}$$

Each element of the Jacobian matrix $\mathbf{J}$ is a partial derivative of each vector component of the nonlinear output with respect to each of the input vector components. That is,

$$\mathbf{J}(\boldsymbol{x}_0) = \frac{\partial \mathbf{h}}{\partial \boldsymbol{x}_0} = \begin{bmatrix} \frac{\partial h_1(\boldsymbol{x}_0)}{\partial x_1} & \frac{\partial h_1(\mathbf{x}_0)}{\partial x_2} \cdots & \frac{\partial h_1(\boldsymbol{x}_0)}{\partial x_n} \\ \frac{\partial h_2(\boldsymbol{x}_0)}{\partial x_1} & \frac{\partial h_2(\mathbf{x}_0)}{\partial x_2} \cdots & \frac{\partial h_2(\boldsymbol{x}_0)}{\partial x_n} \\ \vdots & & \vdots \\ \frac{\partial h_m(\boldsymbol{x}_0)}{\partial x_1} & \frac{\partial h_m(\mathbf{x}_0)}{\partial x_2} \cdots & \frac{\partial h_m(\boldsymbol{x}_0)}{\partial x_n} \end{bmatrix}. \tag{10.88}$$

As an example, for the MLP nonlinearity of Eq. 10.84, the (i, l)-th element of the Jacobian matrix is

$$\sum_{j=1}^{J} W_{ij} \cdot s_j(y) \cdot (1 - s_j(y)) \cdot w_{jl}, \qquad 1 \leq i \leq I, \;\; 1 \leq l \leq L, \tag{10.89}$$

where $y = \sum_{l'=1}^{L} W_{jl'} x_{l'}$.

Besides the MLP, the radial basis function (RBF) is an attractive alternative choice as another form of universal function approximator for implementing the articulatory-to-acoustic mapping.

10.6 Summary

This chapter is a core chapter in the book, and has been devoted to functional modeling of the human speech production process at the phonetic level. It provided a much greater depth of coverage than the largely descriptive treatment of the phonetic process in Chapter 7, and was focused on computational issues. This chapter started out by providing an overview of speech production modeling, where two general types of modeling and research approaches, scientifically and engineering oriented ones, were described. In this overview, we also outlined a key dynamic nature of the speech production process, and classified it into three different levels — the acoustic dynamic level, articulatory dynamic level, and task dynamic level. The acoustic dynamic level is also called the surface level of dynamics since in most relevant computational problems and applications (such as speech recognition), the level of speech acoustics is where the (most peripheral) measurement took place. The levels of articulatory and task dynamics are also called the hidden dynamics, because they are internal to the direct acoustic measurement.

The chapter proceeded then to cover acoustic dynamic modeling of speech. First, the (stationary-state) hidden Markov model (HMM) studied earlier in Chapter 3 was viewed as a primitive statistical generative model of the acoustic dynamics of speech. Then this view was generalized for the nonstationary-state or trended HMM. The EM algorithm for learning the trended HMM parameters was presented in detail, together with a detailed example for the use of polynomials as the trended or state-bound trajectory function. The learning algorithm was naturally reduced to the form appropriate for the stationary-state HMM when setting certain polynomial coefficients to zero. A different form of the acoustic dynamic model of speech was then discussed, based on continuously valued rather than discretely valued Markov processes (cf. Chapter 3). This falls into the category of the linear dynamic system model presented in Chapter 4. The EM algorithm for learning the parameters of such a model was then presented in great detail.

The core part of this chapter is on the models for hidden dynamics of speech. It generalizes the linear dynamic system model for the acoustic dynamics to carefully structured nonlinear dynamic system models where the continuous state explicitly represents either the task dynamics, articulatory dynamics, or vocal tract resonance dynamics. We first derived a general form of the discrete-time hidden-dynamic state equation, which embedded a key concept of target directedness. This provided the basis for the state-space formulation of a discrete-time hidden dynamic model of speech. Three kinds of hidden dynamics of speech were discussed. This was followed by a simplified version of the hidden dynamic model where the nonlinear observation equation that governs the complex relationship between the hidden dynamic variables (state vectors) was approximated by piecewise linear equations. The EM-based parameter learning algorithm was presented in great detail.

Finally, synthesizing from many discussions in this and previous chapters, we presented a comprehensive statistical generative model of the dynamics for natural (in particular, casual) speech. This comprehensive model has a number of layered structures, starting from the phonological construct to the (distortion-free) speech acoustics. In this comprehensive generative model, Eqs. 10.80 and 10.82 form a special version of the switching state-space model appropriate for describing multi-level speech dynamics. The top-level dynamics of the model occurs at the level of discrete-state phonology, represented by the state transitions of $\mathbf{s}$ with a relatively long time scale. The next level is the target ($\mathbf{t}$) dynamics; it has the same time scale and provides systematic randomness at the segmental level. At the level of articulatory dynamics, the time scale is significantly shortened. This is continuous-state dynamics driven by the target process as input. The state equation (Eq. 10.80) explicitly describes this dynamics in $\boldsymbol{x}$, with index of s (which takes discrete values) implicitly representing the switching process. At the lowest level of the acoustic dynamics, there is no switching process. Since the observation equation Eq. 10.82 is static, this simplified acoustic generation model assumes that acoustic dynamics results solely from articulatory dynamics. Improvement of the model that overcomes this simplification is unlikely until better modeling techniques are developed for representing multiple time scales in the dynamic aspects of speech acoustics in relation to the speech articulation process.

Speech recognition applications of some components of the various computational models discussed in this chapter will be presented in Chapter 12.

Chapter 11

Computational Models for Auditory Speech Processing

The preceding chapter covers the first major area of computational phonetics — speech production modeling. This chapter will cover the second major area of computational phonetics — modeling for auditory speech processing. The phonetic science for how speech sounds are represented and processed in the auditory systems of human and other animals was discussed in Chapter 7. This chapter will build on that fundamental knowledge and present some related computational techniques and analysis.

Common techniques for auditory speech processing involve constructing computational models that simulate auditory functions for the speech sounds as the input to the models. There are several levels at which auditory functions are invoked to process speech sounds, including the outer and middle ears, inner ear or cochlea (basilar membrane, inner and outer hair cells, and inner-hair-cell to auditory-nerve synapses), cochlear nuclei, and the auditory neural pathways above the cochlear nuclei towards and at the auditory cortex. The processing stages at and below the auditory nerves are commonly referred to as the peripheral auditory processing. Sometimes the processing at the cochlear nucleus is also included as the peripheral auditory processing. Any processing above the auditory nerve (or the cochlear nuclei) is commonly referred to as the central auditory processing. This chapter concerns mainly the peripheral auditory processing.

The computational modeling approach to auditory processing of speech and other sounds offers a number of distinct advantages over other approaches such as neurophysiological, psychoacoustic, and analytical modeling ones [Ains 02, Gree 88a, Flet 95, Medd 91, Cooke 93, Ghit 93a, Heinz 01a, Heinz 01b, Gree 01]. Several kinds of auditory models have been developed that are capable of reproducing fairly accurate predictions for the realistic auditory responses to speech and other sounds [Deng 87a, Deng 92c, Carn 93, Gig 94b, Dau 96a, Patt 95, Rob 99, Zha 01, Heinz 01a]. In contrast to the physiological approach, which is typically limited to a small number of speech classes and to a small portion of the auditory system, the computational modeling approach can simulate the entire system's response to arbitrary acoustic stimuli, including all types of speech sounds. It also contrasts to the analytical modeling approach, which is also limited to a narrow range of parameters in the acoustic stimuli.

This chapter is devoted to the computational modeling approach to auditory processing of speech, and is outlined as follows. We first present a computational model, based on a partial differential equation, for one-dimensional nonlinear cochlear mechanics. The details of the simulation steps for the model's output for an arbitrary acoustic stimulus including a speech waveform are described. The simulation steps are based on finite-difference solutions to the partial differential equation. We present both frequency-domain and time-domain solutions, and further present stability analysis for the time-domain solution. This analysis contributes to the implementational efficiency of the solution by allowing for optimal choices of the time and spatial discretization steps with minimal loss of simulation accuracy. We then present computational models for inner hair cells (sensors) and for the synapses from the inner hair cells to the auditory nerve fibers. The output of the synapse model, which is the final stage of the cochlear model, is an array (ensemble) of the deterministic signals of instantaneous firing rates for each of the fibers. There are two separate analyses branching from this array of signals as the cochlear model's output. First, temporal interval statistics are computed over this ensemble of signals and are used as a frame-based feature for the speech waveform as the input into the cochlear model. It is shown that these types of features, which capitalize on detailed temporal information provided by the cochlear model, preserve and enhance much of the acoustic phonetic properties of the original speech input, especially under noisy environments. Second, the instantaneous firing rate in the auditory nerve is transformed into the action potential spike sequence that occurs randomly in time. This serves as the input to a layered neural network model for the cochlear-nucleus-like structure. The neural mechanisms represented in the model, such as lateral inhibition and coincidence detection, account for the temporal-to-rate code conversion in the higher levels of the auditory system. There has been a recent trend in auditory research aimed at accounting for human auditory performance, using the results of physiological auditory modeling and statistical decision/detection theory [Heinz 01a, Heinz 01b]. Use of a model for the auditory nerve is shown to have accounted successfully for the rudimentary human auditory performance (e.g., frequency and intensity discrimination) with respect to simple acoustic stimuli, such as single tones. Use of higher-level models and of speech stimuli has the potential to ultimately account for human speech perception performance and to contribute to understanding the auditory mechanisms for achieving such performance.

This chapter has been reorganized from publications in [Deng 86, Deng 87a, Deng 92c, Deng 93b, Deng 93a, Deng 99b, She 98, She 95].

11.1 A Computational Model for the Cochlear Function

11.1.1 Introduction

After the relatively minor signal transformation (largely linear) carried out by the outer ear and middle ear, a very important step in auditory speech processing is the filtering, compression, and other nonlinear operations on the signal performed by the cochlea (including basilar membrane, hair cells, and auditory nerves), the principal structure in the inner ear. Due to the complexity and the importance of the cochlear function, mathematical modeling efforts devoted to understanding the cochlear function have been active for many years [Keid 76, All 85, Deng 87a, Deng 92c, Carn 93, Hawk 96a]. Computational techniques for large-scale simulation of the cochlear responses to complex sounds including speech have also been developed during a long time span [Cooke 93, Deng 87a, Deng 92c, Gig 94a, Gig 94b, Dau 96, Dau 96a, Patt 95, Ains 02]. There are several purposes underlying these computational studies, including: 1) further understanding of the non-measurable cochlear mechanisms at work for processing complex speech sounds; 2) uncovering emerging properties of auditory processing; 3) experimenting on speech feature extraction by higher-level processing based on the parallel-channel output in cochlear processing; and 4) applying the extracted speech features to speech technology applications.

A review of the many types of models for cochlear mechanics can be found in [Hub 96]. Traditionally, the broadest classification of these models is mechanistic models vs. signal processing (or functional) models. The former seeks to explain the mechanism of the cochlea while the latter is aimed at producing cochlear-like output signals to be used in various applications. Among the mechanistic models, the next level of classification is macromechanical models vs. micromechanical models. The macromechanical models are biophysical models in which the fluid of the cochlea plays a major role in shaping the cochlear response, while the detailed structure of the cochlear partition is minimized into a few abstract parameters. The micromechanical models, in contrast, assign the detailed structure of the cochlear partition with a prominent role. Among the macromechanical models, a further subdivision is one-dimensional vs. high-dimensional (two or three) models, where the dimensionality refers to the representation in terms of the cochlear encasement, with the most important dimension being along the length of the cochlea. Finally, the one-dimensional models can be either active or passive, and either linear or nonlinear. Passive models dissipate energy that comes in via the input signal from the middle ear, and active models produce energy in the cochlea. The biological basis for the active models is the motile outer hair cells capable of both reverse electrical-to-mechanical and mechanical-to-electrical transduction. Linear models do not compress the input signals and are linear systems that obey the superposition principle, as discussed in Chapter 1. Nonlinear models reduce the system's gain when the input signal is large, and conditionally suppress the response to a tone when there is another tone present (violation of the superposition principle). Nonlinearity in the models is typically implemented by making some parameters of the models dependent on the model outputs. Later in this section, we will present a particular model which somewhat breaks away from the above traditional classification. It belongs to a nonlinear, active, one-dimensional, macromechanical model, but the computation has been optimized so that it is capable of producing cochlear-like output signals on a massive scale for arbitrary acoustic inputs, including all classes of speech sounds.

Models for the wave motions in the basilar membrane (BM), which is the principal structure in the cochlea, based on the hydrodynamic principle and on the biophysical mechanisms of the cochlear function have been developed by many auditory researchers in the past. These models have shown a relatively high degree of success in replicating existing neurophysiological data on cochlear responses to acoustic signals ranging from pure tones to complex sounds such as speech. The significance of such faithful models may be projected to be their potential application in designing front-end components of advanced speech processing systems, as well as in serving as a powerful tool for studying information processing in central parts of the auditory system.

While analytical solutions to cochlear models are most desirable, they can be obtained only for relatively simple models, for simple input stimuli, and with an unknown degree of accuracy (due to various simplifying assumptions) [Keid 76, All 85]. On the other hand, numerical methods such as the finite-difference scheme have proved to be direct, versatile, and highly effective means for obtaining accurate solutions of complicated cochlear models [Neely 81, Deng 92c, Deng 99b, Deng 93b, Deng 93a]. Frequency-domain solutions by the finite-difference method are generally efficient in computation, but are limited to linear models and to sinusoidal inputs only. With nonlinear components built into the model and with the desire to deal with arbitrary inputs (speech-like sounds), use of the finite-difference scheme in time domain appears necessary. In general, computer simulations of the finite-difference equations in time-domain solutions are slow, which may prevent use of the model in many possible speech-processing applications. In addition, the outputs may suffer from an instability problem, which would render the solution meaningless. Such practical issues and some solutions will be discussed later in this chapter.

11.1.2 Mathematical formulation of the cochlear model

The computational BM model described here is based on an active, nonlinear, one-dimensional transmission-line model developed previously in [Deng 86, Deng 87a, Deng 92c] for speech processing applications. The model is aimed at describing realistic wave motions along the length (i.e., the principal dimension) of the cochlea. The physical structure that the model represents is an uncoiled cochlea that is approximated by two fluid-filled rigid-walled compartments separated by an elastic sheet. The input to the model is the acoustic waveform that can be an arbitrary function of time (e.g., the speech waveform). The output of the model is the vibration pattern of the BM as a function of time and of a one-dimensional space along the cochlea. The model, in the form of a partial differential equation, is obtained by considering the principal hydrodynamic and biophysical properties of the cochlea.

The model for the BM wave motion can be succinctly described by (detailed derivation of the model can be found in [Deng 92c]):

$$\frac{\partial^2}{\partial x^2}\left(m\frac{\partial^2 u}{\partial t^2} + r(u,x)\frac{\partial u}{\partial t} + s(x)u - k(x)\frac{\partial^2 u}{\partial x^2}\right) - \frac{2\rho\beta}{A}\frac{\partial^2 u}{\partial t^2} = 0, \tag{11.1}$$

where the meanings of the variables and the constants with their selected values in the equation are listed below:

$u(x,t)$	BM displacement averaged over the basilar membrane width,
A	cross sectional area of the oval and round windows, A=0.012 cm^2,
m	BM mass per unit area, m=0.05 g/cm^2,
β	BM width, β=0.1 cm,
ρ	density of the fluid inside the cochlea, $\rho = 1$ dyne/cm^3,
x	distance along the BM ($x = 0$ at the stapes and $x = L$ at the helicotrema),
t	time variable (continuous),
$s(x)$	BM stiffness per unit area,
$r(u,x)$	BM damping per unit area,
$k(x)$	stiffness coupling coefficient.

The following boundary conditions can be obtained by considering the constraints imposed on the structure of the cochlea:

$$\left(\frac{\partial p_d}{\partial x}\right)|_{x=0} = -2\rho a_{sp}, \tag{11.2}$$

$$p_d|_{x=L} = 0, \tag{11.3}$$

with p_d being the pressure difference across the BM expressed by

$$p_d = m\frac{\partial^2 u}{\partial t^2} + r(u,x)\frac{\partial u}{\partial t} + s(x)u - k(x)\frac{\partial^2 u}{\partial x^2}; \tag{11.4}$$

L is the length of the BM truncated up to 8 kHz in the characteristic frequency (CF), chosen to be of value 2.1 cm, and a_{sp} is the stapes acceleration which is related to the input stimulus to the BM.

Due to the second-order coupling term in the x dimension, two additional boundary conditions are needed:

$$\frac{\partial^2 u}{\partial x^2}|_{x=0} = 0,$$
$$\frac{\partial^2 u}{\partial x^2}|_{x=L} = 0. \tag{11.5}$$

The natural initial conditions for solving the differential equation can also be easily established as

$$u|_{t=0} = 0,$$
$$\frac{\partial u}{\partial t}|_{t=0} = 0. \tag{11.6}$$

In the BM model described here, the damping parameters in Eq. 11.1 is a function of the model output $u(x,t)$ (hence a nonlinear model), as well as a function of the spatial variable x, according to

$$r(u,x) = r_0(x) - \Delta r + R(x,t,u). \tag{11.7}$$

The individual components in Eq. 11.7 have distinct interpretations in terms of the cochlear function. The first term r_0 is the "passive" hydrodynamic BM damping. The second term $-\Delta r$ is the effective damping reduction resulting from active outer hair cells' motility, assuming the motility does not enter into its saturation region. In the study of [Deng 92c], $-\Delta r$ is shown to be a function of (a) phase lag between the outer hair cell contraction force and its depolarizing voltage; (b) ratio of upward displacement of the outer hair cell cuticular plate over centrifugal displacement of the cilia; and (c) proportionality constant between the magnitudes of the outer hair cell contraction force and of the ciliary displacement. The final term in Eq. 11.7 represents the saturation effect of the outer hair cell motility, which is taken as an integral or "laterally coupled" form:

$$R(x,t,u) = \gamma \int_0^L \mid u(s,t) \mid e^{-|x-s|/\lambda} ds, \tag{11.8}$$

where λ is the space constant for the lateral coupling, whose value is chosen to be the electrical space constant of the cochlea, 0.2 cm; γ is a proportionality constant, set at 10,000 so as to fit quantitative aspects of two-tone suppression data. It has been postulated that this form of lateral coupling can be mediated by cochlear field potentials and by the efferent system of the cochlea [Deng 92c].

In model implementation, the BM is divided into N discrete sections. In each section a serial combination of mass-stiffness-damping constitutes a tuned circuit. The resonant frequency (CF) decreases from the basal ($x = 0$) to the apical end ($x = L$). To achieve this, the stiffness parameter was selected according to the formula 2π CF$=\sqrt{\text{stiffness}/\text{mass}}$, decreasing from the basal to the apical end. The value of the characteristic frequency of a particular point on the partition is based on the cochlear frequency map, based on the work in [Lib 82].

Both the passive damping coefficient (r_0) and the stiffness coupling coefficient (k) are made dependent on the position along the BM, x, according to the empirical relation

$$k(\text{CF}) = a \times e^{-b \times x} + c, \tag{11.9}$$

$$r_0(x) = d \times (x + f), \tag{11.10}$$

where constants a, b, c, d, and f are chosen such that the transfer functions of the modeled BM sections, obtained using low-intensity clicks, approximate those of the physiological data. The empirical values chosen are $a = 55, b = 2, c = 5, d = 2$, and $f = 21$.

11.2 Frequency-Domain Solution of the Cochlear Model

If the partial differential equation Eq. 11.1 is made linear by eliminating the term $R(x,t,u)$ in the general damping expression Eq. 11.7, then an efficient frequency-domain solution of the model can be obtained using the following transformations:

$$\begin{aligned} u(x,t) &\longleftrightarrow u(x,j\omega), \\ \frac{\partial u(x,t)}{\partial t} &\longleftrightarrow jwu(x,j\omega), \\ \frac{\partial^2 u(x,t)}{\partial t^2} &\longleftrightarrow -\omega^2 u(x,j\omega), \end{aligned} \tag{11.11}$$

where $\omega = 2\pi f$ is the angular frequency of the input signal, which is assumed to be sinusoidal.

In the frequency domain, Eq. 11.1 can be transformed to

$$\frac{d^2}{dx^2}\left\{\left(-m\omega^2 + s(x) + j\omega r(x)\right)u - k(x)\frac{d^2 u}{dx^2}\right\} + \frac{2\rho\beta}{A}\omega^2 u = 0. \tag{11.12}$$

The frequency-domain expressions of the boundary conditions are

$$\begin{aligned} \frac{d}{dx}\left\{\left(-m\omega^2 + s(x) + j\omega r(x)\right)u\right\}|_{x=0} &= 2\omega^2\rho, \\ \left\{\left(-m\omega^2 + s(x) + j\omega r(x)\right)u\right\}|_{x=L} &= 0, \\ \frac{d^2u}{dx^2}|_{x=0} &= 0, \\ \frac{d^2u}{dx^2}|_{x=L} &= 0. \end{aligned} \tag{11.13}$$

A numerical solution of the above frequency-domain model by the finite-difference method requires that the spatial dimension be represented by a finite number (N) of discrete points. The solution is obtained for the volume displacement of the BM as a function of the distance from the stapes, x, for selected input frequencies.

To discretize the frequency-domain equations, the derivatives in Eqs. 11.12-11.13 are replaced by the conventional central differences:

$$\begin{aligned} \frac{du}{dx} &= \frac{u_{i+1} - u_{i-1}}{2\Delta x}, \\ \frac{d^2u}{dx^2} &= \frac{u_{i+1} - 2u_i + u_{i-1}}{(\Delta x)^2}, \\ \frac{d^4u}{dx^4} &= \frac{u_{i+2} - 4u_{i+1} + 6u_i - 4u_{i-1} + u_{i-2}}{(\Delta x)^4}. \end{aligned} \tag{11.14}$$

Define $Y(x, j\omega)$ as $j\omega Z(x, j\omega)$, with $Z(x, j\omega)$ being the BM impedance:

$$Z(x, j\omega) = j\omega m + \frac{s(x)}{j\omega} + r(x).$$

The derivatives of $Y(x, j\omega)$ with respect to x can be approximated by

$$\frac{dY}{dx} = \frac{Y_{i+1} - Y_{i-1}}{2\Delta x},$$

$$\frac{d^2Y}{dx^2} = \frac{Y_{i+1} - 2Y_i + Y_{i-1}}{(\Delta x)^2}.$$

For the boundary point at $i = 1$, the forward-difference approximation to the derivative is used, which gives

$$\frac{du}{dx} = \frac{u_2 - u_1}{\Delta x}$$

$$\frac{dY}{dx} = \frac{Y_2 - Y_1}{\Delta x},$$

For the derivatives of the products

$$\frac{d^2}{dx^2}\left(uY(x, j\omega)\right),$$

and

$$\frac{d^2}{dx^2}\left(k(x)\frac{d^2u}{dx^2}\right)$$

no simplification was made and all the terms involved are taken into account. Further, because there is no coupling at the edges, the longitudinal stiffness coupling coefficient is set at zero for the points neighboring the boundaries.

After replacing each derivative by its finite-difference approximation, the following $N - 1$ equations are obtained:

$$[Y_2 - 2Y_1]u_1 + Y_1u_2 = 2\,\omega^2\rho\Delta x \qquad (i = 1), \tag{11.15}$$

$$\begin{aligned} &\left[\frac{-2k_i + k_{i+1} - k_{i-1}}{(\Delta x)^2}\right]u_{i-2} + \left[Y_{i-1} + 2Y_i - Y_{i+1} + \frac{12k_i - 4k_{i+1}}{(\Delta x)^2}\right]u_{i-1} \\ +\; & 2\left[Y_{i-1} - 4Y_i + Y_{i+1} + 2\frac{\rho\beta}{A}\omega^2(\Delta x)^2 + \frac{-10k_i + 2(k_{i-1} + k_{i+1})}{(\Delta x)^2}\right]u_i \\ +\; & \left[-Y_{i-1} + 2Y_i + Y_{i+1} + \frac{12k_i - 4k_{i-1}}{(\Delta x)^2}\right]u_{i+1} + \left[\frac{-2k_i - k_{i+1} + k_{i-1}}{(\Delta x)^2}\right]u_{i+2} = 0 \end{aligned}$$

$$(i = 2, \ldots, N - 2).$$

The boundary condition Eq. 11.13 is equivalent to $u_N = 0$; thus for $i = N - 1$ we have

$$[Y_{N-2} + 2Y_{N-1} - Y_N] + 2\left[Y_{N-2} - 4Y_{N-1} + Y_N + 2\frac{\rho\beta}{A}\omega^2(\Delta x)^2\right]u_{N-1} = 0$$

$$(i = N - 1).$$

This system can be written in a matrix form as **Au**=**B**, where **A** has the following structure:

$$\begin{bmatrix} a_{11} & a_{12} & 0 & 0 & 0 & \cdots & 0 \\ a_{21} & a_{22} & a_{23} & 0 & 0 & \cdots & 0 \\ a_{31} & a_{32} & a_{33} & a_{34} & a_{35} & & \\ \ddots & \ddots & \ddots & \ddots & \ddots & & \vdots \\ 0 & a_{i,i-2} & a_{i,i-1} & a_{i,i} & a_{i,i+1} & a_{i,i+2} & 0 \\ \vdots & \ddots & \ddots & \ddots & \ddots & \ddots & \\ & & a_{N-3,N-5} & a_{N-3,N-4} & a_{N-3,N-3} & a_{N-3,N-2} & a_{N-3,N-1} \\ 0 & \cdots & 0 & 0 & a_{N-2,N-3} & a_{N-2,N-2} & a_{N-2,N-1} \\ 0 & \cdots & 0 & 0 & 0 & a_{N-1,N-2} & a_{N-1,N-1} \end{bmatrix}.$$

This matrix is sparse and band-diagonal (band $= 5$). The matrix does not require storage of order N^2; instead, storage of order N is adequate. To obtain the solution, first, matrix $\mathbf{A}$ is factorized as $\mathbf{A} = \mathbf{PLQ}$, where $\mathbf{P}$ is a permutation matrix, $\mathbf{L}$ is a lower triangular matrix with at most two non-zero sub-diagonal elements per column, and $\mathbf{Q}$ is an upper triangular band matrix with four super-diagonals. The factorization is obtained by Gaussian elimination with partial pivoting. After factorization, the solution vector is calculated for different values of ω. The computation time required is of order N.

11.3 Time-Domain Solution of the Cochlear Model

The finite-difference method for solving the BM model in the time domain requires discretization in both time and spatial dimensions. The numerical solutions are obtained for the volume displacement at selected positions along the BM length as a function of time.

For the derivatives in Eqs. 11.1-11.6, forward differences are used for the first-order derivatives and central differences for the second-order ones. These finite-difference expressions are

$$\frac{\partial u}{\partial t} = \frac{u_i^{n+1} - u_i^n}{\Delta t},$$

$$\frac{\partial^2 u}{\partial t^2} = \frac{u_i^{n+1} - 2u_i^n + u_i^{n-1}}{(\Delta t)^2},$$

$$\frac{\partial u}{\partial x} = \frac{u_{i+1}^n - u_i^n}{\Delta x},$$

$$\frac{\partial^2 u}{\partial x^2} = \frac{u_{i+1}^n - 2u_i^n + u_{i-1}^n}{(\Delta x)^2},$$

$$\frac{\partial^4 u}{\partial x^4} = \frac{u_{i+2}^n - 4u_{i+1}^n + 6u_i^n - 4u_{i-1}^n + u_{i-2}^n}{(\Delta x)^4},$$

$$\frac{\partial^3 u}{\partial t \partial x^2} = \frac{u_{i+1}^{n+1} - 2u_i^{n+1} + u_{i-1}^{n+1} - u_{i+1}^n + 2u_i^n - u_{i-1}^n}{\Delta t\ (\Delta x)^2},$$

$$\frac{\partial^4 u}{\partial t^2 \partial x^2} = \frac{u_{i+1}^{n+1} - 2u_i^{n+1} + u_{i-1}^{n+1} - 2u_{i+1}^{n}}{(\Delta t)^2 \ (\Delta x)^2} \\ \frac{+4u_i^n - 2u_{i-1}^n + u_{i+1}^{n-1} - 2u_i^{n-1} + u_{i-1}^{n-1}}{(\Delta t)^2 \ (\Delta x)^2}, \tag{11.16}$$

where $i = 1, \ldots, N$ and $n = 0, \ldots, M$. N is the total number of points along the BM length and M the total number of time steps. Δt and Δx are the mesh sizes in time and spatial dimensions, respectively. As there is no coupling at the edges, the longitudinal stiffness coupling coefficient is set at zero for the points neighboring the boundaries.

After substituting each derivation by its corresponding approximation, the following $N-2$ equations are obtained:

$$\begin{aligned}
& \left[\frac{m}{(\Delta x)^2 \ (\Delta t)^2} + \frac{r(i)}{(\Delta x)^2 \ \Delta t}\right] u_{i+1}^{n+1} - \left[2\frac{m}{(\Delta x)^2 \ (\Delta t)^2} + 2\frac{r(i)}{(\Delta x)^2 \ \Delta t} + 2\frac{\beta\rho}{A \ (\Delta t)^2}\right] u_i^{n+1} \\
+ & \left[\frac{m}{(\Delta x)^2 \ (\Delta t)^2} + \frac{r(i)}{(\Delta x)^2 \ \Delta t}\right] u_{i-1}^{n+1} \\
= & \left[2\frac{m}{(\Delta x)^2 \ (\Delta t)^2} + \frac{r(i)}{(\Delta x)^2 \ \Delta t} - \frac{2s(i+1) - s(i)}{(\Delta x)^2}\right] u_{i+1}^{n} \\
+ & \left[-4\frac{m}{(\Delta x)^2 \ (\Delta t)^2} - 2\frac{r(i)}{(\Delta x)^2 \ \Delta t} - \frac{-s(i+1) - 2s(i) + s(i-1)}{(\Delta x)^2}\right] u_i^n \\
+ & \left[2\frac{m}{(\Delta x)^2 \ (\Delta t)^2} + \frac{r(i)}{(\Delta x)^2 \ \Delta t} - \frac{-s(i)}{(\Delta x)^2}\right] u_{i-1}^n \\
+ & \left[2\frac{m}{(\Delta x)^2 \ (\Delta t)^2} + 2\frac{\beta\rho}{A \ (\Delta t)^2}\right] u_i^{n-1} - \frac{m}{(\Delta x)^2 \ (\Delta t)^2}\left[u_{i+1}^{n-1} + u_{i-1}^{n-1}\right] \\
+ & \frac{k(i)}{(\Delta x)^4}\left[u_{i+2}^n - 4u_{i+1}^n + 6u_i^n - 4u_{i-1}^n + u_{i-2}^n\right]
\end{aligned} \tag{11.17}$$

$$(i = 2, \ldots, N-1) \qquad (n = 0, \ldots, M).$$

The numerical form of the boundary condition (Eq. 11.2) at the stapes is

$$\begin{aligned}
& \left[\frac{m}{\Delta x \ (\Delta t)^2} + \frac{r(1)}{\Delta x \ \Delta t}\right]\left[u_2^{n+1} - u_1^{n+1}\right] = \frac{-2\rho}{(\Delta t)^2}\left[\epsilon(n+1) - 2\epsilon(n) + \epsilon(n-1)\right] \\
+ & \left[2\frac{m}{\Delta x \ (\Delta t)^2} + \frac{r(1)}{\Delta x \ \Delta t} - \frac{s(1)}{\Delta x}\right]\left[u_2^n - u_1^n\right] - \frac{s(2) - s(1)}{\Delta x} u_1^n \\
- & \frac{m}{\Delta x \ (\Delta t)^2}\left[u_2^{n-1} - u_1^{n-1}\right]
\end{aligned} \tag{11.18}$$

$$(n = 0, \ldots, M),$$

where $\epsilon(n)$ is the displacement of the stapes at time $t = n\Delta t$, which is related to the input stimulus that can take an arbitrary time function. The discrete form of the other boundary condition (Eq. 11.3) is

$$\left[\frac{m}{(\Delta t)^2} + \frac{r(N)}{\Delta t}\right] u_N^{n+1} = \left[2\frac{m}{(\Delta t)^2} + \frac{r(N)}{\Delta t} - s(N)\right] u_N^n - \frac{m}{(\Delta t)^2} u_N^{n-1} \tag{11.19}$$

$$(n = 0, \ldots, M).$$

Thus, for each time step, we have a system of N linear equations with N unknowns. Again, such a system can be put into a matrix form **Au**=**B**, with **A** having a tri-diagonal structure:

$$\begin{bmatrix} a_{11} & a_{12} & 0 & \cdots & 0 \\ a_{21} & a_{22} & a_{23} & \ddots & 0 \\ \ddots & \ddots & \ddots & & \vdots \\ 0 & a_{i,i-1} & a_{i,i} & a_{i,i+1} & 0 \\ \vdots & \ddots & \ddots & \ddots & \\ 0 & \ddots & a_{N-1,N-2} & a_{N-1,N-1} & a_{N-1,N} \\ 0 & \cdots & 0 & 0 & a_{N,N} \end{bmatrix} \begin{bmatrix} u_1^{n+1} \\ u_2^{n+1} \\ \vdots \\ u_i^{n+1} \\ \vdots \\ u_{N-1}^{n+1} \\ u_N^{n+1} \end{bmatrix} = \begin{bmatrix} b_1 \\ b_2 \\ \vdots \\ b_i \\ \vdots \\ b_{N-1} \\ b_N \end{bmatrix},$$

where u_i^{n+1} $(1 \le i \le N)$ is the value of the BM volume displacement at position i at the next time index (i.e., index $n+1$), b_1 is the right hand side of Eq. 11.18, b_i is the right hand side of Eq. 11.17, b_N is the right hand side of Eq. 11.19, $a_{i,i-1}, a_{i,i}, a_{i,i+1}$ are the coefficients of u_{i-1}^{n+1}, u_i^{n+1}, and u_{i+1}^{n+1} in Eqs. 11.18, 11.17, and 11.19, respectively.

The structure of the matrix A is again very sparse. The procedures of LU decomposition and forward/backward substitution provide efficient methods to solve this trigonal system of equations. The storage required is of order N; the computation time is also of order N for each time step.

11.4 Stability Analysis for Time-Domain Solution of the Cochlear Model

One goal of developing the computational techniques described above is to understand numerical properties of the finite-difference approximation to the BM model, and according to such an understanding to improve computation efficiency of the solution to the model. In pursuing this goal, the mesh sizes, both in time and in spatial dimensions, in the finite-difference approximation can be identified as the most direct and important factors in determining the computation efficiency. The second, related goal is to investigate the conditions under which optimal time and spatial mesh sizes can be chosen to minimize the amount of computation while producing stable model outputs. Since stability is only a necessary condition for accuracy of the numerical solution, computer simulations need to be conducted to provide experimental evidence that stable model outputs in practice always lead to rather accurate model outputs.

In this section, we introduce a detailed study published in [Deng 93a] on numerical properties of a finite-difference scheme for the solution of a one-dimensional nonlinear transmission line model for the basilar membrane wave motions just described. Specifically, a sufficient condition is derived and described under which the finite-difference solution to the model is guaranteed to be stable. The significance of this condition is that it determines the optimal time and spatial mesh sizes that would invoke a minimal amount of computation for obtaining stable and accurate model outputs.

11.4.1 Derivation of the stability condition

This section is concerned with conditions that must be satisfied if the solution of the finite-difference equations presented in Section 4 is to be a reasonably accurate approximation to the solution of the corresponding partial differential equations. The conditions to be derived are sufficient ones, which, when satisfied, will prevent unstable growth of errors resulting from the arithmetical operations needed to solve the finite-difference equations.

If it were possible to perform all calculations to an infinite number of decimal places, we would obtain the exact solution of our finite-difference equations. In practice, however, each calculation is carried out to a finite number of decimal places, which necessarily introduces rounding errors. If the cumulative effect of the rounding errors is not negligible, then the errors will soon obliterate the desired solution and lead to entirely spurious results. Any numerical scheme that allows the growth of error, eventually "swamping" the true solution, is unstable. These numerical phenomena must be avoided by some restrictive action, such as limiting the mesh sizes.

A standard way of investigating the growth of errors in the operations needed to solve the finite-difference equations is the von Neumann stability analysis [8]. This method starts by expressing the error between the exact solution and the calculated solution as a Fourier series. Then, assuming an error was introduced at time t, it finds the condition which will prevent this error from growing. Such a condition is referred to as the stability condition.

To apply the von Neumann stability analysis to a variable coefficient partial differential equation, as is the case for the BM model of concern (Eq. 11.1), we "freeze" the variable coefficients to fixed values to begin the analysis. This enables us to derive the stability condition as a function of the parameters of the model as well as of the mesh sizes. Then, investigation is done to determine which set of parameters would give the worst case for selecting the mesh sizes.

We formally describe application of the von Neumann stability analysis to the BM model expressed by Eq. 11.1. The three positive-valued variable coefficients (space-dependent, and for damping, output-dependent as well) are first "frozen" at fixed values:

$$\begin{aligned} s(x) &= s, \\ k(x) &= k, \\ r(u,x) &= r. \end{aligned}$$

Then the corresponding difference equation becomes

$$\begin{aligned}
& \left[\frac{m}{(\Delta x)^2\ (\Delta t)^2} + \frac{r}{(\Delta x)^2\ \Delta t}\right] u_{i+1}^{n+1} \\
- & \left[2\frac{m}{(\Delta x)^2\ (\Delta t)^2} + 2\frac{r}{(\Delta x)^2\ \Delta t} + 2\frac{\beta\rho}{A\ (\Delta t)^2}\right] u_i^{n+1} \\
+ & \left[\frac{m}{(\Delta x)^2\ (\Delta t)^2} + \frac{r}{(\Delta x)^2\ \Delta t}\right] u_{i-1}^{n+1} \\
= & \left[2\frac{m}{(\Delta x)^2\ (\Delta t)^2} + \frac{r}{(\Delta x)^2\ \Delta t} - \frac{s}{(\Delta x)^2}\right] u_{i+1}^{n} \\
- & 2\left[2\frac{m}{(\Delta x)^2\ (\Delta t)^2} + \frac{r}{(\Delta x)^2\ \Delta t} - \frac{s}{(\Delta x)^2} + 2\frac{\beta\rho}{A\ (\Delta t)^2}\right] u_i^{n} \\
+ & \left[2\frac{m}{(\Delta x)^2\ (\Delta t)^2} + \frac{r}{(\Delta x)^2\ \Delta t} - \frac{s}{(\Delta x)^2}\right] u_{i-1}^{n} \\
+ & 2\left[\frac{m}{(\Delta x)^2\ (\Delta t)^2} + \frac{\beta\rho}{A\ (\Delta t)^2}\right] u_i^{n-1} - \frac{m}{(\Delta x)^2\ (\Delta t)^2}\left[u_{i+1}^{n-1} + u_{i-1}^{n-1}\right] \\
+ & \frac{k}{(\Delta x)^4}\left[u_{i+2}^{n} - 4u_{i+1}^{n} + 6u_i^{n} - 4u_{i-1}^{n} + u_{i-2}^{n}\right].
\end{aligned} \tag{11.20}$$

Let v be the exact solution of the partial differential equation with "frozen" coefficients. We then have

$$\begin{aligned}
& \left[\frac{m}{(\Delta x)^2\ (\Delta t)^2} + \frac{r}{(\Delta x)^2\ \Delta t}\right] v_{i+1}^{n+1} \\
- & \left[2\frac{m}{(\Delta x)^2\ (\Delta t)^2} + 2\frac{r}{(\Delta x)^2\ \Delta t} + 2\frac{\beta\rho}{A\ (\Delta t)^2}\right] v_i^{n+1} \\
+ & \left[\frac{m}{(\Delta x)^2\ (\Delta t)^2} + \frac{r}{(\Delta x)^2\ \Delta t}\right] v_{i-1}^{n+1} \\
= & \left[2\frac{m}{(\Delta x)^2\ (\Delta t)^2} + \frac{r}{(\Delta x)^2\ \Delta t} - \frac{s}{(\Delta x)^2}\right] v_{i+1}^{n} \\
- & 2\left[2\frac{m}{(\Delta x)^2\ (\Delta t)^2} + \frac{r}{(\Delta x)^2\ \Delta t} - \frac{s}{(\Delta x)^2} + 2\frac{\beta\rho}{A\ (\Delta t)^2}\right] v_i^{n} \\
+ & \left[2\frac{m}{(\Delta x)^2\ (\Delta t)^2} + \frac{r}{(\Delta x)^2\ \Delta t} - \frac{s}{(\Delta x)^2}\right] v_{i-1}^{n} \\
+ & 2\left[\frac{m}{(\Delta x)^2\ (\Delta t)^2} + \frac{\beta\rho}{A\ (\Delta t)^2}\right] v_i^{n-1} - \frac{m}{(\Delta x)^2\ (\Delta t)^2}\left[v_{i+1}^{n-1} + v_{i-1}^{n-1}\right] \\
+ & \frac{k}{(\Delta x)^4}\left[v_{i+2}^{n} - 4v_{i+1}^{n} + 6v_i^{n} - 4v_{i-1}^{n} + v_{i-2}^{n}\right] \\
+ & f\left[O((\Delta x)^2), O(\Delta t), O((\Delta t)^2)\right],
\end{aligned} \tag{11.21}$$

where $f\left[O((\Delta x)^2), O(\Delta t), O((\Delta t)^2)\right]$ is the truncation error of the discretized equation.

Denote by $E_i^n = u_i^n - v_i^n$ the error at spatial position i along the BM length at time index n. Then by subtracting Eq. 11.21 from Eq. 11.20 and dropping the term which corresponds to the truncation error, the following relation is obtained:

$$\begin{aligned}
& \left[\frac{m}{(\Delta x)^2\ (\Delta t)^2} + \frac{r}{(\Delta x)^2\ \Delta t}\right] E_{i+1}^{n+1} \\
- & \left[2\frac{m}{(\Delta x)^2\ (\Delta t)^2} + 2\frac{r}{(\Delta x)^2\ \Delta t} + 2\frac{\beta\rho}{A\ (\Delta t)^2}\right] E_i^{n+1} \\
+ & \left[\frac{m}{(\Delta x)^2\ (\Delta t)^2} + \frac{r}{(\Delta x)^2\ \Delta t}\right] E_{i-1}^{n+1} \\
= & \left[2\frac{m}{(\Delta x)^2\ (\Delta t)^2} + \frac{r}{(\Delta x)^2\ \Delta t} - \frac{s}{(\Delta x)^2}\right] E_{i+1}^{n} \\
- & 2\left[2\frac{m}{(\Delta x)^2\ (\Delta t)^2} + \frac{r}{(\Delta x)^2\ \Delta t} - \frac{s}{(\Delta x)^2} + 2\frac{\beta\rho}{A\ (\Delta t)^2}\right] E_i^{n} \\
+ & \left[2\frac{m}{(\Delta x)^2\ (\Delta t)^2} + \frac{r}{(\Delta x)^2\ \Delta t} - \frac{s}{(\Delta x)^2}\right] E_{i-1}^{n} \\
+ & 2\left[\frac{m}{(\Delta x)^2\ (\Delta t)^2} + \frac{\beta\rho}{A\ (\Delta t)^2}\right] E_i^{n-1} - \frac{m}{(\Delta x)^2\ (\Delta t)^2}\left[E_{i+1}^{n-1} + E_{i-1}^{n-1}\right] \\
+ & \frac{k}{(\Delta x)^4}\left[E_{i+2}^{n} - 4E_{i+1}^{n} + 6E_i^{n} - 4E_{i-1}^{n} + E_{i-2}^{n}\right].
\end{aligned} \tag{11.22}$$

Now we ask the questions: Suppose an error is introduced at time index n, how does the error propagate to time indices $n+1, n+2, n+3$, etc,...? Does the error decay or grow? As the purpose of the analysis is to study the behavior of the error as time progresses, the von Neumann method consists of expressing the error at any time index as a Fourier series.

Let the error, as a single term in the Fourier series, be expressed as

$$E_i^n = \alpha^n \; e^{ji\xi}, \tag{11.23}$$

then in order for $\mid E_i^n \mid \to 0$ as $n \to \infty$, α should be such that $\mid \alpha \mid < 1$. Substituting Eq. 11.23 into Eq. 11.22 gives

$$\begin{aligned}
& \left[\frac{m}{(\Delta x)^2 \; (\Delta t)^2} + \frac{r}{(\Delta x)^2 \; \Delta t}\right] \alpha^{n+1} \; e^{j(i+1)\xi} \\
- & \; 2\left[\frac{m}{(\Delta x)^2 \; (\Delta t)^2} + \frac{r}{(\Delta x)^2 \; \Delta t} + \frac{\beta\rho}{A \; (\Delta t)^2}\right] \alpha^{n+1} \; e^{ji\xi} \\
+ & \left[\frac{m}{(\Delta x)^2 \; (\Delta t)^2} + \frac{r}{(\Delta x)^2 \; \Delta t}\right] \alpha^{n+1} \; e^{j(i-1)\xi} \\
= & \left[2\frac{m}{(\Delta x)^2 \; (\Delta t)^2} + \frac{r}{(\Delta x)^2 \; \Delta t} - \frac{s}{(\Delta x)^2}\right] \alpha^{n} \; e^{j(i+1)\xi} \\
- & \; 2\left[2\frac{m}{(\Delta x)^2 \; (\Delta t)^2} + \frac{r}{(\Delta x)^2 \; \Delta t} - \frac{s}{(\Delta x)^2} + 2\frac{\beta\rho}{A \; (\Delta t)^2}\right] \alpha^{n} \; e^{ji\xi} \\
+ & \left[2\frac{m}{(\Delta x)^2 \; (\Delta t)^2} + \frac{r}{(\Delta x)^2 \; \Delta t} - \frac{s}{(\Delta x)^2}\right] \alpha^{n} \; e^{j(i-1)\xi} \\
+ & \; 2\left[\frac{m}{(\Delta x)^2 \; (\Delta t)^2} + \frac{\beta\rho}{A \; (\Delta t)^2}\right] \alpha^{n-1} \; e^{ji\xi} \\
- & \; \frac{m}{(\Delta x)^2 \; (\Delta t)^2}\left[\alpha^{n-1} \; e^{j(i+1)\xi} + \alpha^{n-1} \; e^{j(i-1)\xi}\right] \\
+ & \; \frac{k}{(\Delta x)^4}\left[\alpha^n\left(e^{j(i+2)\xi} - 4e^{j(i+1)\xi} + 6e^{ji\xi} - 4e^{j(i-1)\xi} + e^{j(i-2)\xi}\right)\right].
\end{aligned} \tag{11.24}$$

By dividing both sides of the equation by $\alpha^n \; e^{ji\xi}$ and using $e^{j\xi} = \cos\xi + j\sin\xi$, we have

$$\begin{aligned}
& \alpha\left[\frac{m}{(\Delta x)^2 \; (\Delta t)^2} + \frac{r}{(\Delta x)^2 \; \Delta t}\right] \cos\xi \\
- & \left[\frac{m}{(\Delta x)^2 \; (\Delta t)^2} + \frac{r}{(\Delta x)^2 \; \Delta t} + \frac{\beta\rho}{A \; (\Delta t)^2}\right] \alpha \\
= & \left[2\frac{m}{(\Delta x)^2 \; (\Delta t)^2} + \frac{r}{(\Delta x)^2 \; \Delta t} - \frac{s}{(\Delta x)^2}\right] \cos\xi \\
- & \left[2\frac{m}{(\Delta x)^2 \; (\Delta t)^2} + \frac{r}{(\Delta x)^2 \; \Delta t} - \frac{s}{(\Delta x)^2} + 2\frac{\beta\rho}{A \; (\Delta t)^2}\right] \\
+ & \left[\frac{m}{(\Delta x)^2 \; (\Delta t)^2} + \frac{\beta\rho}{A \; (\Delta t)^2} - \frac{m}{(\Delta x)^2 \; (\Delta t)^2}\cos\xi\right] \alpha^{-1} \\
+ & \; \frac{k}{(\Delta x)^4}\left[\cos 2\xi - 4\cos\xi + 3\right].
\end{aligned} \tag{11.25}$$

Introducing $\theta = \cos\xi \quad (-1 < \theta < 1)$ and noting that the coefficient of $\frac{k}{(\Delta x)^4}$ in Eq. 11.25 can be written as

$$\cos 2\xi - 4\cos\xi + 3 = 2\cos^2\xi - 1 - 4\cos\xi + 3 = 2(\theta - 1)^2,$$

we re-arrange Eq. 11.25 to

$$a\alpha^2 + b\alpha + c = 0, \tag{11.26}$$

where

$$\begin{aligned}
a &= \left(\frac{m}{(\Delta x)^2\ (\Delta t)^2} + \frac{r}{(\Delta x)^2\ \Delta t}\right)\theta - \left(\frac{m}{(\Delta x)^2\ (\Delta t)^2} + \frac{r}{(\Delta x)^2\ \Delta t} + \frac{\beta\rho}{A\ (\Delta t)^2}\right), \\
b &= -\left(2\frac{m}{(\Delta x)^2\ (\Delta t)^2} + \frac{r}{(\Delta x)^2\ \Delta t} - \frac{s}{(\Delta x)^2}\right)\theta \\
&+ \left(2\frac{m}{(\Delta x)^2\ (\Delta t)^2} + \frac{r}{(\Delta x)^2\ \Delta t} - \frac{s}{(\Delta x)^2} + 2\frac{\beta\rho}{A\ (\Delta t)^2}\right) - \left(2\frac{k}{(\Delta x)^4}(\theta-1)^2\right), \\
c &= \frac{m}{(\Delta x)^2\ (\Delta t)^2}\theta - \left(\frac{m}{(\Delta x)^2\ (\Delta t)^2} + \frac{\beta\rho}{A\ (\Delta t)^2}\right).
\end{aligned} \tag{11.27}$$

Note that since $-1 < \theta < 1$, a is always negative. The von Neumann stability condition hence becomes that both roots of Eq. 11.26, α_1 and α_2, satisfy $|\alpha_1| < 1$ and $|\alpha_2| < 1$.

Depending upon the sign of the discriminant ($b^2 - 4ac$, denoted by Δ) of Eq. 11.26, the roots can be either real or complex. These two cases have to be considered separately in the von Neumann analysis, which is pursued below.

Case where $\Delta < 0$

When $\mathbf{\Delta} < 0$ two complex roots are obtained from Eq. 11.26. In this case, the von Neumann analysis consists of using Fourier series with complex coefficients. The roots of Eq. 11.26 are:

$$\alpha_1 = \frac{-b + j\delta}{2a}, \tag{11.28}$$

$$\alpha_2 = \frac{-b - j\delta}{2a}, \tag{11.29}$$

where $\delta = \sqrt{-\mathbf{\Delta}}$. Because the roots are complex conjugate of each other, the condition for $|\alpha_1| < 1$ and $|\alpha_2| < 1$ is equivalent to

$$\frac{b^2}{4a^2} + \frac{\delta^2}{4a^2} < 1, \tag{11.30}$$

or

$$4ac < 4a^2 \tag{11.31}$$

after substituting δ. Since the coefficient a (defined in Eq. 11.27) is always negative, this stability condition 11.31 is reduced to just $c - a > 0$. Now, according to the definitions in Eq. 11.27, $c - a$ is just

$$\frac{r}{(\Delta x)^2 \Delta t}(1 - \theta), \tag{11.32}$$

which is always positive since $-1 < \theta < 1$ and $r, (\Delta x)^2, and \Delta t$ are all positive. We thus conclude that when the discriminant is negative, the von Neumann stability condition is automatically guaranteed for all possible values of θ, regardless of mesh sizes Δt and Δx.

Case where $\Delta > 0$

When $\mathbf{\Delta} > 0$, two real roots are associated with Eq. 11.26:

$$\alpha_1 = \frac{-b + \sqrt{\mathbf{\Delta}}}{2a}, \tag{11.33}$$

$$\alpha_2 = \frac{-b - \sqrt{\mathbf{\Delta}}}{2a}. \tag{11.34}$$

For this case, one part of von Neumann stability condition, $\mid \alpha_1 \mid < 1$, is equivalent to

$$\sqrt{\mathbf{\Delta}} > 2a + b, \tag{11.35}$$

and

$$\sqrt{\mathbf{\Delta}} < -2a + b; \tag{11.36}$$

and the other part of the stability condition, $\mid \alpha_2 \mid < 1$, is equivalent to

$$\sqrt{\mathbf{\Delta}} < -(2a + b), \tag{11.37}$$

and

$$\sqrt{\mathbf{\Delta}} > -(-2a + b). \tag{11.38}$$

(Note that stability requires that these four conditions be all simultaneously satisfied.)

Conditions 11.35 and 11.37 can be expressed by the single condition: $\mathbf{\Delta} < (2a+b)^2$. This, after substituting $\mathbf{\Delta}$, leads to

$$c < -(a + b) \qquad (a < 0). \tag{11.39}$$

Substituting back the definitions of a, b, and c above, we have an equivalence of the conditions 11.35 and 11.37:

$$\frac{s}{(\Delta x)^2}(\theta - 1) - 2\frac{k}{(\Delta x)^4}(\theta - 1)^2 < 0. \tag{11.40}$$

This is always satisfied, independent of mesh sizes Δt and Δx (note that $-1 < \theta < 1$). On the other hand, conditions 11.36 and 11.38 can be compressed to $\mathbf{\Delta} < (-2a+b)^2$, or

$$c < b - a. \tag{11.41}$$

Again, using the definitions of a, b, and c (Eq. 11.27), we obtain an equivalence of the conditions 11.36 and 11.38:

$$\begin{aligned} &\left(4\frac{m}{(\Delta x)^2\ (\Delta t)^2} + 2\frac{r}{(\Delta x)^2\ \Delta t} - \frac{s}{(\Delta x)^2}\right)\theta \\ - &\left(4\frac{m}{(\Delta x)^2\ (\Delta t)^2} + 2\frac{r}{(\Delta x)^2\ \Delta t} - \frac{s}{(\Delta x)^2} + 4\frac{\beta\rho}{A\ (\Delta t)^2}\right) \\ + &\left(2\frac{k}{(\Delta x)^4}(\theta - 1)^2\right) < 0. \end{aligned} \tag{11.42}$$

Multiplying by $(\Delta x)^2(\Delta t)^2$, rearranging and grouping, we simplify the stability condition of Eq. 11.42 to

$$\begin{aligned} & \left(s \times (\theta - 1) - \frac{2k \times (\theta - 1)^2}{(\Delta x)^2}\right)(\Delta t)^2 \\ - \quad & 2r \times (\theta - 1)\,\Delta t + 4(\Delta x)^2 \left(\frac{\beta\rho}{A} - \frac{m \times (\theta - 1)}{(\Delta x)^2}\right) \; > \; 0. \end{aligned} \tag{11.43}$$

The above analysis shows that condition 11.43 is the only one required for stability of solutions to the finite-difference equation.

11.4.2 Application of the stability analysis

In this section, we apply the stability condition 11.43 to select an optimal sampling interval Δt, given Δx (determined by the desired frequency resolution in the model), which allows for maximum computation efficiency for the model solution. It is optimal in the sense that any coarser sampling would likely produce unstable model outputs at high-CF regions, while any finer sampling would incur an unnecessarily large amount of computation for obtaining the model outputs. Ideally, Δt and Δx should be chosen as large as possible, but subject to the stability condition for the numerical solution.

In the established stability condition 11.43, θ is a free parameter whose value is bounded by one, and s, r, k, and m are BM stiffness, damping, stiffness coupling, and mass parameters, respectively, which were defined in Section 2. Since the above model parameters are functions of the spatial variable x, it is necessary to select very conservative values for Δt, given Δx, to ensure uniform stability of the numerical solution of the model.

To quantify such an application, we note first that the coefficient of Δt^2 in inequality 11.43,

$$A = s \times (\theta - 1) - \frac{2k \times (\theta - 1)^2}{(\Delta x)^2}, \tag{11.44}$$

is negative (since $s > 0, k > 0, -1 < \theta < 1$). Then we view the left-hand-side quantity in inequality 11.43,

$$\begin{aligned} f(\Delta t) \quad = \quad & \left(s \times (\theta - 1) - \frac{2k \times (\theta - 1)^2}{(\Delta x)^2}\right)(\Delta t)^2 \\ - \quad & 2r \times (\theta - 1)\,\Delta t + 4(\Delta x)^2 \left(\frac{\beta\rho}{A} - \frac{m \times (\theta - 1)}{(\Delta x)^2}\right), \end{aligned} \tag{11.45}$$

as a quadratic form of Δt, and denote the roots of $f(\Delta t) = 0$ by Δt_1 and Δt_2, respectively (with $\Delta t_1 \leq \Delta t_2$; both roots have been found to be always real). Standard algebra asserts that for $\Delta t_1 < \Delta t < \Delta t_2$, $A \times f(\Delta t) < 0$, or $f(\Delta t) > 0$ (due to the negativity of A). We have found empirically that $\Delta t_1 < 0$ and $\Delta t_2 > 0$ for all the model parameters. Hence, given values of Δx, θ, and the model parameters (as functions of BM position x), the stability condition expressed as the inequality 11.43 is equivalent to $0 < \Delta t < \Delta t_2$, where Δt_2 is a function of Δx, θ, and x. Detailed evaluation of such a function for the BM model has been presented in [Deng 93a] and will not be described here.

Based on the stability analysis above, in the time-domain numerical solution of the BM model, the input acoustic waveform is interpolated to a sampling rate as high as 25 kHz, in order to ensure uniform stability of model outputs and to avoid the accumulation of numerical errors. The output of the BM model is the temporal responses of 178 equally-spaced points (channels) along the BM spatial dimension x in our current simulation. The center frequencies (CFs) corresponding to the simulated points vary nearly logarithmically from most-basal channel 1 at CF = 8 kHz to most-apical channel 178 at CF = 100 Hz.

11.5 Computational Models for Inner Hair Cells and for Synapses to Auditory Nerve Fibers

To complete the cochlear model, it is necessary to connect the BM model to the model of inner hair cell (IHC) transducers and further to the auditory nerve (AN) fibers.

11.5.1 The inner hair cell model

The IHC model provides a functional role of a transducer from motions of the BM to the cell membrane receptor potential as input to the IHC-AN synapse model, where the receptor potential is a function of the angle of displacement of the stereocilia. Since the stereocilia of the IHCs are not attached to the tectorial membrane, the displacement angle is a function of the velocity of the fluid and therefore of the BM.

It appears that the signal processing of the auditory pathway does not strongly depend on the exact shape of the IHC transfer function [Wei 85]. Therefore, in our current study, we use a simple hyperbolic tangent function to describe the static IHC transfer function:

$$\nu_{IHC,k} = K \cdot \left[\tanh\left(\frac{0.707\nu_{input,k}}{P_D} - P_0\right) + \tanh P_0,\right],$$

where $K = \nu_{max}/(1 + \tanh P_0)$, $\nu_{input,k}$ is the output of the BM model, and $\nu_{IHC,k}$ is the output of the IHC model. The constants P_0 and P_D are determined empirically to match some published IHC data. The constant P_D is set to saturate the output at moderate input power levels.

11.5.2 The synapse model

The synapse model described here is adopted from the available literature [West 88]. The model uses "pools" of neurotransmitters which are separated by membranes of varying permeability. The voltage generated from the IHCs alters the permeabilities of the membranes and the volumes of the neurotransmitters in the pools. Three pools are simulated in the model: an *immediate* store, a *local* store, and a *global* store. The output of the synapse model is the spike rate, which is the product of the permeability and the concentration of the immediate store.

The model operates as follows: At rest, there is a base-spiking rate generated that reflects the spontaneous rate of the nerve fiber. When the stimulus is first exerted from the IHC, neurotransmitters are supplied from the immediate pool. However, the neurotransmitters from the immediate pool are soon exhausted. After this point, the neurotransmitters must be obtained from the local pool and then drawn through the immediate pool. In this way, more effort is required, after a short period of the stimulus, to draw neurotransmitters as they must pass through two membranes. When the stimulus continues on, the local pool of the neurotransmitters will be exhausted and more of them must be obtained from the global pool. This process requires even more effort, as the neurotransmitters must be drawn through yet another membrane.

Through the above process, the *adaptation* function of the synapse is implemented. Adaptation is the tendency of the synapse to respond less strongly to a sustained excitation than the onset of an excitation; it is similar to the function of a high-pass filter on the input. The process can be modeled by simple diffusion equations, where the rate of diffusion through the various membranes is equal to the permeability of the membrane multiplied by the difference in concentration across the membrane:

$$V\frac{\partial c}{\partial t} = -p\Delta c, \tag{11.46}$$

where V is the volume of neurotransmitter, c is the concentration of the neurotransmitter, and p is the permeability of the membrane. All the above qualities are functions of time. The dynamics of the transmitter release in the synapse are modeled by the coupled differential equations

$$V_I\frac{\partial c_I}{\partial t} = -p_I c_I + p_L(c_L - c_I) \tag{11.47}$$

$$V_L\frac{\partial c_L}{\partial t} = -p_L(c_L - c_I) + p_G(c_G - c_L), \tag{11.48}$$

where the subscript I denotes immediate store, L denotes local store, and G denotes global store. The global store concentration c_G is held constant.

The permeabilities of the membranes are derived from the IHC voltage ν_{IHC} by a simple linear function, and are then half-wave rectified; that is, for each store, we have

$$p'(t) = \frac{p_{MAX} - p_0}{\nu_{MAX}}\nu_{IHC}(t) + p_0,$$

$$p(t) = \begin{cases} p'(t) & p'(t) \geq 0 \\ 0 & p'(t) < 0 \end{cases},$$

where p_{MAX}, the maximum concentration, and p_0, the minimum concentration, are constants. The maxima are different for each pool, but the minima are the same. Similarly, we determine the volumes of each pool by

$$V(t) = (V_{MAX} - V_0)\nu'(t) + V_0,$$

where

$$\nu'(t) = \begin{cases} \nu_{IHC}(t) & \nu_{IHC}(t) \geq 0 \\ 0 & \nu_{IHC}(t) < 0 \end{cases}.$$

In this case as well, the maximum volumes are different for immediate and local pools, while the minima are the same. The volume of the global pool is infinite.

In our computer implementation, the concentration equations 11.47 and 11.48 were approximated in discrete time by

$$c_{I,k+1} = c_{I,k} + \frac{T}{V_{I,k}}\left[-p_{I,k}c_{I,k} + p_{L,k}(c_{L,k} - c_{I,k})\right]$$

$$c_{L,k+1} = c_{L,k} + \frac{T}{V_{L,k}}\left[-p_{L,k}(c_{L,k} - c_{I,k}) + p_{G,k}(c_G - c_{L,k})\right],$$

where T is the sampling interval. The output of the model is an instantaneous spiking rate, given by

$$s_k = p_{I,k}c_{I,k},$$

where s_k is in spikes/s. Support for this type of model over other models that involve automatic gain control (AGC) has been provided by numerous studies (e.g., [Smit 82]). As a summary, we list in Table 11.1 the parameters of the IHC model and the synapsis model.

p_D, p_0	Constants in IHC nonlinearity, $1e-3$ and .462, respectively
nu_{MAX}	Saturation value in IHC nonlinearity, $30e-3$ Volts
p_{IMAX}, p_{I0}	Maximum and resting permeabilities for the immediate storage 1.15 and 0.015, respectively
p_{LMAX}, p_{L0}	Maximum and resting permeabilities for the local storage, 0.1 and 0.015, respectively
p_{GMAX}, p_{G0}	Maximum and resting permeabilities for the global storage, 0.08 and 0.015, respectively
V_{IMAX}, V_{I0}	Maximum and resting volumes for the immediate storage, $3e-4$ and $1e-4$, respectively
V_{LMAX}, V_{L0}	Maximum and resting volumes for the local storage, $3e-3$ and $1e-3$, respectively
R_{spon},	Spontaneous spike rate (spikes/sec) for the synapse model, from 15 for low-rate to 70 for high-rate

Table 11.1: Parameter set of a cochlear model.

11.6 Interval-Based Speech Feature Extraction from the Cochlear Model Outputs

The output of the cochlear model described in the above sections is the simulated instantaneous firing rate (IFR) of the AN fibers. These simulation results can be used to extract the interval statistics, which are believed to be an important aspect of the robust features in terms of the temporal representation of the input acoustic signal in the peripheral auditory system. The interval statistics are constructed in the form of an inter-peak interval histogram (IPIH), which will be described in this section. On the other hand, the same form of the IPIH can be built from the available population data on the IFRs recorded from AN fibers using the identical acoustic stimuli. By comparing the IPIH based on the physiological AN firing data with that generated from the simulated IFR data using a range of values for the model parameters, it is possible to identify the key mechanisms embedded in the model and the model parameters responsible for matching the neural IPIH data.

11.6.1 Inter-peak interval histogram construction

The IPIH construction method described here is an improved version of the one described in [Seck 90]. The method is used to obtain robust interval statistics extracted from the IFR estimates occurring in either the actual or the simulated AN fibers.

First, the intervals between peaks in the IFRs during short segments of the response are measured and counted. Since the IFR waveforms are not sufficiently smooth, autocorrelation of 10-ms segments of the IFR data is used to identify the peaks. The square root of the autocorrelation function is smoothed with a Hamming window. This process is repeated every 3 ms to track the time-varying characteristics of the IFRs. The signed zero crossings of the first differences of the root autocorrelation are utilized to estimate the location of peaks in the IFR waveforms. The result of the above process is an interpeak interval signal for each IFR waveform. A histogram of the intervals occurring between the adjacent peaks within the first 5 ms of the root autocorrelation across all the simulated channels is constructed to form the IPIH.

In the IPIH construction, the amplitude of the peak at the start of the interval has been used as the weight to update the interval bin in the histogram. This implies that the IPIH formed has both synchrony and rate information encoded therein. For example, for a vowel stimulus there are high-amplitude resonances in the IFR waveforms around formant frequencies. As a result, the counts in the IPIH around the bins corresponding to the formant frequencies are much higher than the counts around other bins, leading to a spike-like IPIH. This weighting procedure prevents the IFR waveform at frequencies far from the formant frequencies, which generally have a low-amplitude oscillation near their CFs, from contributing significantly to the final IPIH. Therefore, after the IPIH construction, there is no need to explicitly add the rate information to the histograms. Further, in order to prevent the IFRs that belong to high-frequency channels from dominating the final shape of the IPIH, only the first three intervals of each root autocorrelation are used in generating each individual histogram. This implies that the length of the analysis window for each channel of the IFR waveform is roughly proportional to 1/CF. Since channels of the BM model are almost logarithmically spaced along the BM, the window length also varies logarithmically across the channels, having its minimum length at the high-frequency channels (channel 1) and the maximum at the last channel. To obtain an estimate of the spectrum of a speech segment, the IPIH data is averaged over the CF parameter for each segment. The aggregate IPIH data are used as a feature vector for analysis and modeling of the speech signals.

11.6.2 Matching neural and modeled IPIHs for tuning BM-model's parameters

As discussed before, one judicious approach to optimizing the parameters of the model is to seek a best match between the model output and the representative neural data available. Direct comparison between the neural firing pre-stimulus histograms and the model output poses difficulties arising from lack of appropriate measures for goodness-of-match. The IPIH summarizes highly compressed information of the original firing pattern while capturing key aspects of the temporal response characteristics. In the following we will use goodness-of-match between IPIHs constructed from the neural data and from the simulated IFRs as a criterion to tune some key parameters of the BM model and to examine the effect of these parameters on the simulated IPIH.

The stimuli in this matching experiment include synthetic syllables /da/ and /ba/ at 69 dB sound pressure level (SPL). The population AN fiber firings for these stimuli are recorded in a computer readable form. The formant trajectories of the two syllables are shown in Figs. 11.1a and b, respectively. The IPIHs constructed from the population AN data in response to the synthetic /da/ and /ba/ at 69 dB SPL are shown in Figs. 11.2a and b, respectively. Given these target neural IPIHs to be matched, the identical IPIH construction method is applied to the simulated IFRs generated from the cochlear model described in the above sections.

To examine the effect of the parameters in linear and nonlinear damping terms in the BM model described by Eqs. 11.7 and 11.8, a controlled set of experiments were conducted to demonstrate that the match between neural and model IPIHs is sensitive to

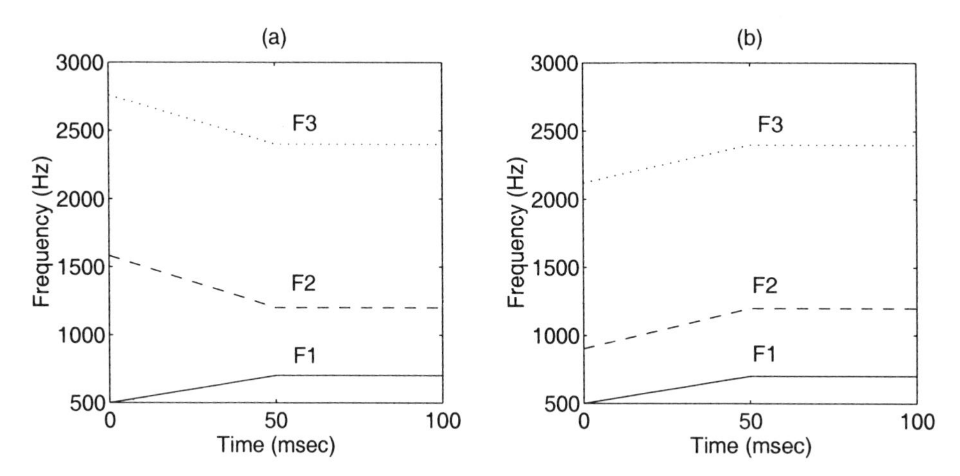

Figure 11.1: Formant trajectories for (a) synthetic /da/ and (b) synthetic /ba/ (after Sheikhzadeh and Deng [She 99], @Elsevier).

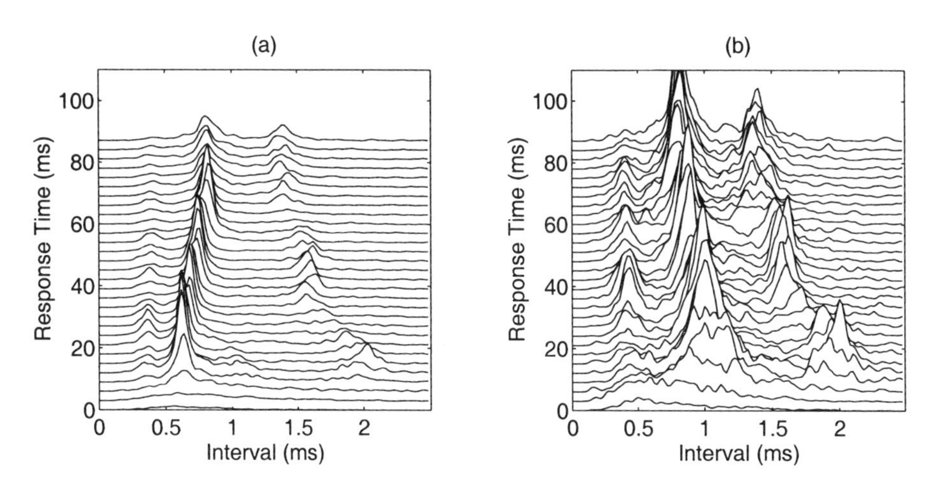

Figure 11.2: IPIH constructed from physiologically recorded neurograms in population AN fibers using 69 dB SPL (a) synthetic /da/ and (b) synthetic /ba/.

the parameters of the BM model and that the only certain parameter values can provide an optimal match. In these experiments, a low SPL for sounds is used to minimize the effects of the BM nonlinearity. These results offered the approximate range of the parameters according to goodness-of-fit to the desirable properties of the tuning curves. The BM model parameters are then fine tuned via goodness-of-fit to the neural IPIHs.

Tuning the linear component of the damping parameters

Figs. 11.3a-c are the IPIHs obtained using a low-SPL (49 dB) /da/ stimulus as input to the BM model, with three widely different damping parameters used in the model. Fig. 11.3a shows the IPIH with the parameters a, b, c, and d in Eq. 11.8 set to the nominal values of 0.4, 5, 20, and 15, respectively. These nominal values were obtained according to the optimal matching criterion via extensive experiments on fitting the transfer functions produced by the model to neural tuning curves. Comparing Fig. 11.3a to Fig. 11.2a from the neural IPIH, it is evident that all three formant tracks are well represented in our simulated IPIH. In contrast, when the damping parameters are artificially increased by a factor of four (i.e., set $a = 1.6, b = 20$), not only the resulting simulated BM filter transfer functions become artificially broadened ("wide-band"), but also the associated IPIH as shown in Fig. 11.3b becomes grossly mismatched with the neural IPIH (compared to Fig. 11.2a). The mismatch can be described as follows. First, the F3 formant track in Fig. 11.3b is lost. Second, the F2 track becomes artificially widened and a spurious track is inserted. These two aspects of mismatch can be understood because when the BM filters are widened by fabrication, the filters become able to pick up components from the low-frequency formants but at the same time lose the ability to resolve the high-frequency tracks.

Opposite types of mismatch were observed when the damping values are reduced by a factor of four from the nominal values (i.e., set $a = 0.1, b = 1.25$). Now the BM filters become artificially narrowed ("narrow-band"). Fig. 11.3c shows the IPIH constructed for the narrow-band BM model using the same low-SPL /da/ stimulus. As expected from the narrow-band nature of the BM filters, the associated IPIH cannot represent smooth formant transitions, as suggested by Fig. 11.3c.

The same set of experiments described above were conducted using a different stimulus /ba/ also at low SPL, and the same conclusion was reached. The three IPIHs constructed using nominal, broad-band, and narrow-band damping values are shown in Figs. 11.4a, b, and c, respectively.

Tuning the nonlinear component of the damping parameters

Tuning parameter γ in Eq. 11.7, which controls the strength of nonlinear component of the overall damping in the BM model, requires use of high-SPL stimuli. Since we have available to us only the neural population response data collected using stimuli /da/ and /ba/ at 69 dB SPL, we are limited within this context to claim optimality of the obtained parameter γ. Similar simulation experiments to those described in the previous subsection were conducted using the two speech tokens at 69 dB SPL, where the parameter γ was made to vary over a wide range. In our experiments, we fixed the linear components of the model parameter obtained from the low-SPL stimuli as described previously and varied the γ parameter to aim at the closest match of the simulated IPIH to its neural counterpart shown in Fig. 11.2. This process produced the best IPIH matching result when the γ value is set at 3×10^7. Figs. 11.5a and b show the simulated IPIHs for the 69 dB SPL /da/ and /ba/ stimuli, respectively. Comparing

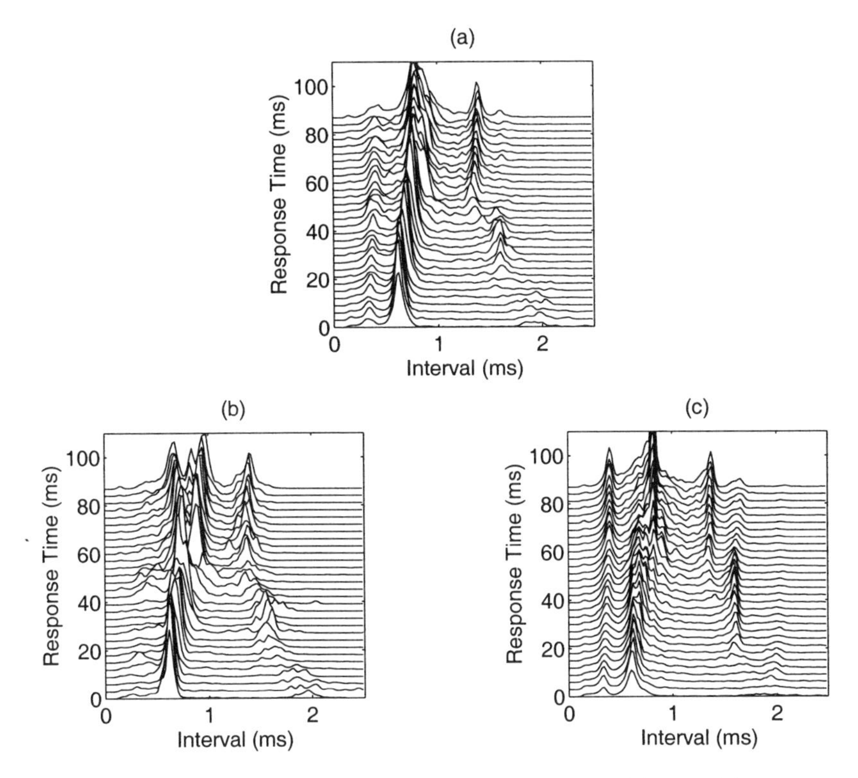

Figure 11.3: IPIH constructed from the IFR outputs of the cochlear model in response to synthetic /da/ at 49 dB SPL using (a) the BM model with the value of the linear damping component optimized via tuning-curve fitting, (b) the wideband BM model, and (c) the narrow-band BM model.

them with the corresponding neural IPIHs (Figs. 11.5a and b), we note that all detailed characteristics of the formant tracks in the neural IPIHs are duplicated.

11.7 Interval-Histogram Representation for the Speech Sound in Quiet and in Noise

In this section, we provide examples to illustrate how various classes of speech sounds, both in clean and in noisy environments, are represented in the form of IPIH discussed up to now. The example speech materials are the utterances of words *glass, found, heels,* and *semi,* all excised from a continuously spoken TIMIT sentence (male speaker). (TIMIT is a widely-used acoustic-phonetic continuous-speech corpus containing a total of 6300 sentences spoken by 630 male and female speakers from eight major dialect regions of the United States.) These words are chosen to be representative — they cover all major manner classes of American English: vowels (/ae/, /i/), diphthongs (/aw/, /ay/), liquid (/l/), nasals (/m/, /n/), fricatives (/s/, /z/, /f/), stop (/g/), and aspiration (/h/).

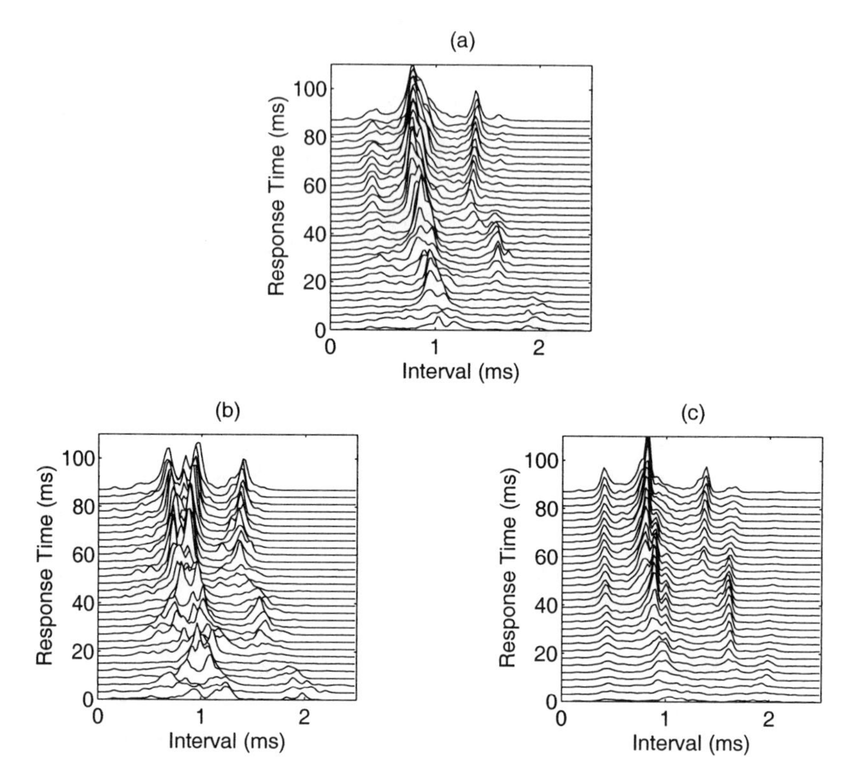

Figure 11.4: IPIH constructed from the IFR outputs of the cochlear model in response to synthetic /ba/ at 49 dB SPL using (a) the BM model with the value of the linear damping component optimized via tuning-curve fitting, (b) the wideband BM model, and (c) the narrow-band BM model.

11.7.1 Inter-peak interval histograms for clean speech

Fig. 11.6a is the IPIH constructed from the cochlear model output in response to a clean utterance *glass* at 69 dB SPL. A detailed view of a portion of the IPIH of Fig. 11.6a for the constituent vowel /ae/, with the vowel boundaries taken from the TIMIT segment file, is shown in Fig. 11.6b, and that for the constituent strident fricative /s/ is shown in Fig. 11.6c.

The main spectral component of the /g/ burst is located at around 1200 Hz, which has been clearly represented in the IPIH (Fig. 11.6a, near response time 20-30 ms) as the peak centered around 750 ms along the interpeak interval axis. Some relatively low spectral amplitudes in the /g/ burst that spread from 3500 Hz upwards are largely absent from the IPIH representation. The liquid /l/, characterized acoustically by relatively low F1 and F2 and very high F3 and by slow formant transitions, is also clearly identified in its IPIH representation (except that F3, which is seen acoustically above 6000 Hz only, is necessarily absent since our cochlear model has been run up to 5000 Hz). Particularly conspicuous in the IPIH representation (Fig. 11.6a) is the very slow F2 transition starting from interpeak interval 1 ms at time 65 ms and ending at interpeak interval 0.6 ms at

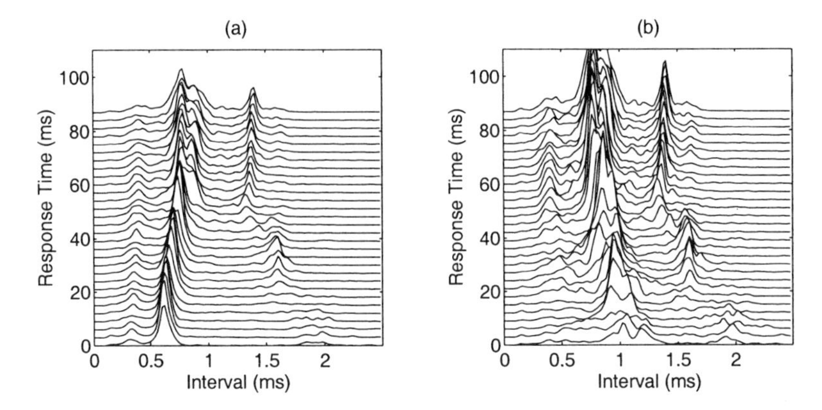

Figure 11.5: IPIH constructed from the IFR outputs of the cochlear model in response to 69 dB SPL (a) synthetic /da/ (b) synthetic /ba/.

time 150 ms. For the vocalic segment /ae/, F1, F2, and their dynamics show the most prominent peak trajectories in the IPIH (Figs. 11.6a and b). In contrast, F3 and F4 are represented as low-amplitude, rather noisy peaks in the IPIH. Finally, high-frequency fricative /s/ is represented in the IPIH as strong peaks in the low-interpeak-interval region (Fig. 11.6c).

The IPIH constructed from the model output in response to another utterance, *found* (at 69 dB SPL), is shown in Fig. 11.7a, with an enlarged portion for the /f/ segment shown in Fig. 11.7b and that for /aw/ in Fig. 11.7c. Again, all the major acoustic properties associated with the various segments of the word can be identified from the IPIH representation. For example, the relatively low amplitude of the /f/ spectrum, spreading from 600 Hz upwards, is represented in the IPIH as low-amplitude histogram peaks (in comparison to the peaks associated with the vocal segment starting from 125 ms in Fig. 11.7a); note these peaks are all located below interpeak-interval 1.7 ms in the IPIH (Fig. 11.7b). Also, the entire F2 transition of /aw/, which moves toward and ends near the F2-locus (1800 Hz) associated with the postvocalic alveolar consonant /n/, can be most clearly identified in the corresponding portion of the IPIH (Fig. 11.7c) with the F2 transition ending at precisely 550 (interpeak-interval 0.55 ms). Finally, F3 as the major spectral peak associated with the nasal segment /n/ in the utterance (around 2700 Hz from spectrogram) can be identified as a somewhat more prominent peak than other IPIH peaks associated with the lower-frequency formants.

The IPIHs constructed for utterances *heels* and *semi* are shown in Figs. 11.8a and b, respectively. The most prominent acoustic characteristic of these utterances is the extremely wide range for the formant transitions in the vocalic segments. For /i/ in *heels*, F2 moves drastically from near 2100 Hz down near 1300 Hz (F2 of the postvocalic /l/); this acoustic transition is reflected in the corresponding peak movement in the IPIH from about 0.48-ms interpeak interval (starting at 60 ms) to the interval of 0.75 ms (ending at around 200 ms). Similarly, the rising F1 transition in acoustics is represented as the falling IPIH peak. For /ay/ in *semi*, the rising F2 from 1200 Hz to 2000 Hz is transcribed to the falling IPIH peak from around 0.85 ms to 0.5 ms.

We have produced and analyzed the IPIHs for the words from several TIMIT sentences in much the same qualitative way as described above. From the analysis we find

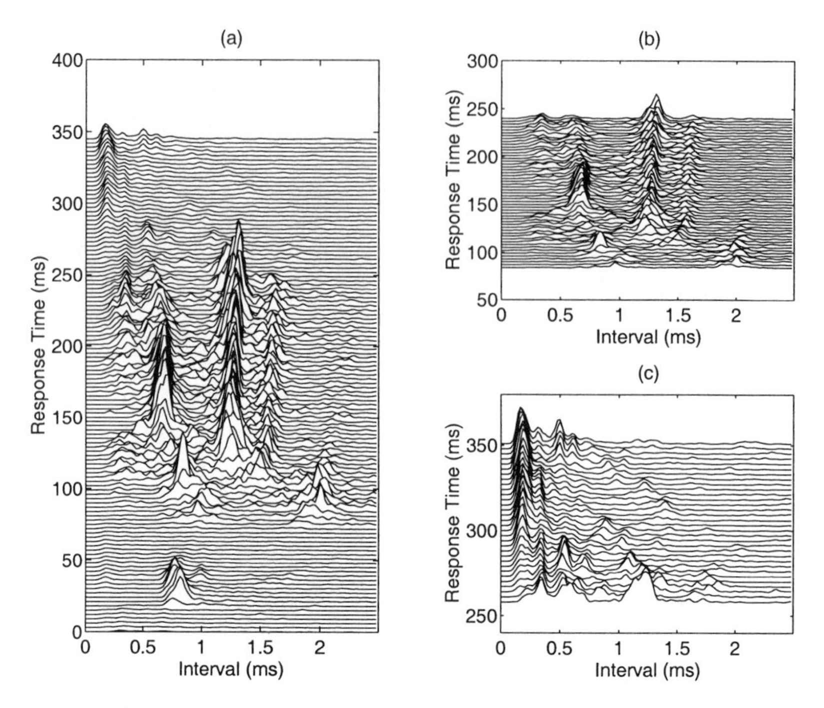

Figure 11.6: IPIH constructed from the cochlear model output in response to the clean utterance *glass* (excised from a TIMIT sentence). (a) IPIH for the whole word utterance; (b) Detailed view of the IPIH for the constituent vowel /ae/ (the vowel boundaries are taken from the TIMIT file); and (c) Detailed view of the IPIH for the constituent strident fricative /s/.

that the significant acoustic properties of all classes of American English sounds that can be identified from spectrogram reading can also be identified, albeit to a varying degree, from the corresponding IPIH.

11.7.2 Inter-peak interval histograms for noisy speech

To evaluate noise robustness of the speech representation in terms of the interval statistics collected from the auditory-nerve population, we performed the identical IPIH analysis for the speech sounds identical to the ones described above, except adding white Gaussian noise with 10-dB signal-to-noise-ratio (SNR) to the speech stimuli before running the auditory model. The SNR employed is the global SNR defined as

$$SNR = 10 \log \frac{\sum_{t=1}^{N} s^2(t)}{\sum_{t=1}^{N} n^2(t)},$$

where $s(t)$ and $n(t)$ represent time-waveforms for the speech utterance and the noise signal (both of length N samples), respectively. The resulting IPIHs for noisy utterances *glass*, *found*, *heels*, and *semi* are shown in Figs. 11.9a, b and 11.10a, b, respectively.

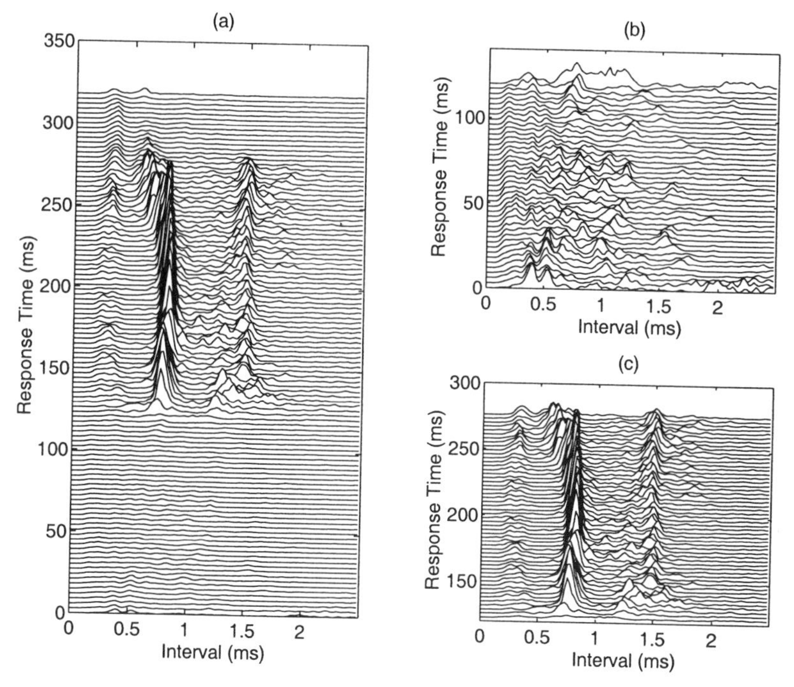

Figure 11.7: IPIH constructed from the cochlear model output in response to clean utterance *found* (excised from a TIMIT sentence). (a) IPIH for the whole word utterance; (b) Detailed view of the IPIH for the constituent weak fricative /f/, and (c) Detailed view of the IPIH for the constituent vowel /aw/.

Comparisons between these four IPIHs with the respective IPIHs for the clean version of the speech utterance (Figs. 11.6a, 11.7a, and 11.8a and b) demonstrate that the effects of the additive noise are mainly confined within the low-interpeak-interval (or high-frequency) region. Aside from some relatively minor distortions on the stop burst, the nasal murmur, and the weak fricative, the major characteristics in the IPIH representation described in the previous subsection for the clean speech have been preserved.

In contrast to the above IPIH-based temporal representation in the auditory domain, the differences in the acoustic (spectral) domain between the clean and noisy versions of the speech utterances are vast. To show this, we plot the conventional magnitude short-time Fourier transforms for the clean speech utterances *glass*, *found*, *heels*, and *semi* in Figs. 11.11a,b,c,d, and those for the corresponding noisy (10-dB SNR) speech utterances in Figs. 11.12a,b,c,d, respectively. Note that in each plot the x-axis represents the linear frequency which decreases from left to right, in a manner similar to the IPIH plots where the left side of the x-axis represents low interpeak interval corresponding to high frequency. (The conventional spectrographic displays tried for the noisy speech utterances show rather uniform blackness throughout the graphs with little differentiation.)

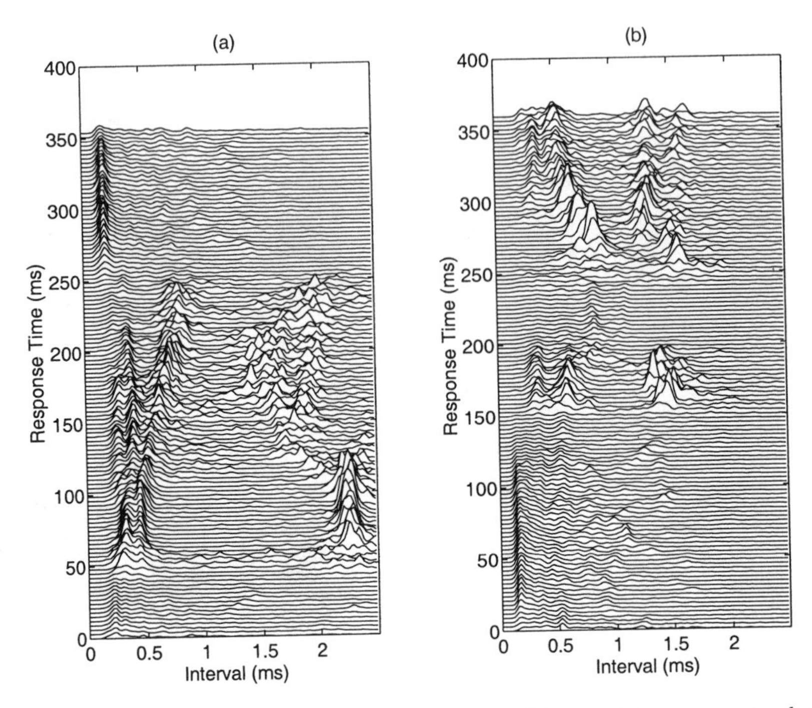

Figure 11.8: IPIHs constructed from the cochlear model output in response to clean utterances *heels* (a) and *semi* (b) (both excised from a TIMIT sentence).

11.8 Computational Models for Network Structures in the Auditory Pathway

The preceding sections were concerned with computational modeling of the early auditory processing stages peripheral to cochlear nuclei, as well as a computational analysis technique used to extract (interval-based) features from the model's outputs to speech stimuli. In this section, we will move one step up towards the central auditory nervous system. A computational, cochlear-nucleus-like network model for the study of speech encoding mechanisms associated with parts of the auditory system central to the auditory nerve will be described. This network model is based on physiological grounds, and is designed to gracefully interface with a cochlear model established earlier that incorporates a biophysically motivated dynamic nonlinearity in the basilar-membrane filtering function. This section is based on the study described in [She 95] aimed at addressing a long standing issue in auditory research, that is, how can the unique spatial-temporal patterns, shaped by the cochlea's nonlinear filtering, in the auditory nerve data in response to high-level speech sounds be transformed into a rate-place code in the auditory system in a physiologically plausible manner? Detailed neural mechanisms implemented in the network model to be described include neural inhibition, coincidence detection, short-term temporal integration of post synaptic potentials for action potential generation, and a conjectural temporal-to-rate conversion mechanism requiring the membrane time constant of a neuron to monotonically decrease with the neuron's CF. Model sim-

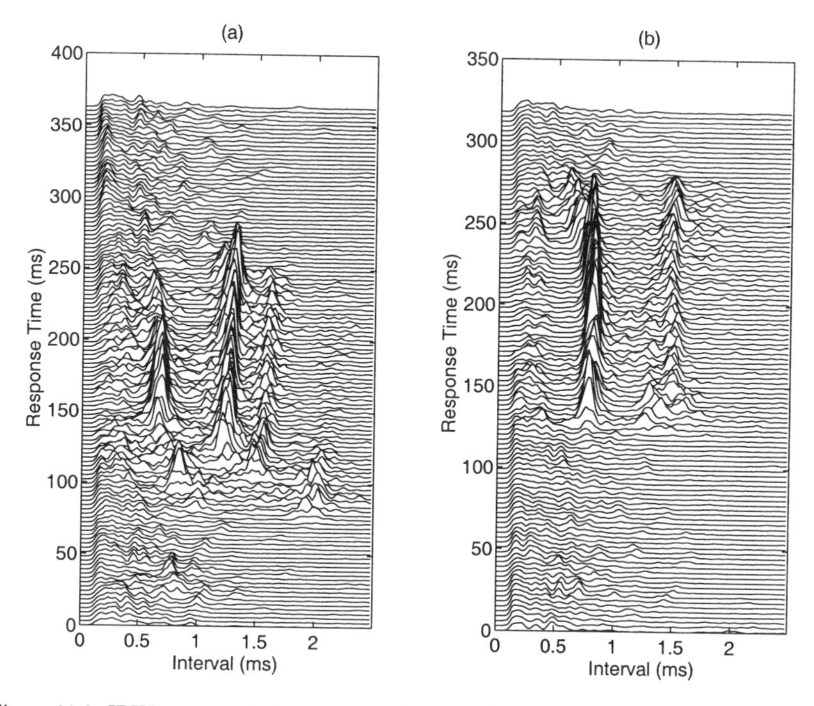

Figure 11.9: IPIHs constructed from the cochlear model output in response to utterances *glass* (a) and *found* (b), embedded in white noise with 10-dB SNR.

ulation experiments using both synthetic and natural speech utterances demonstrate that the auditory rate-place code constructed at the output of the network model is capable of reliable representation, with possible modification and/or enhancement, of the prominent spectral characteristics of the utterances exhibited in wideband spectrograms.

11.8.1 Introduction

In neuroanatomical terms, probably all auditory nerve (AN) fibers ascending towards higher levels of the auditory pathway terminate within the cochlear nucleus (CN) complex of the lower brainstem [Brug 92]. The CN is divided into a few major subdivisions mainly according to the type of cells contained therein. Each of the major subdivisions receives the tonotopic information from the entire AN array. This divergence of the AN fibers sets up the first stage in parallel processing of the acoustic information received by the ears [Brug 92, Irv 92]. It is believed that some subdivisions of the CN, the anteroventral cochlear nucleus (AVCN) in particular, are capable of faithfully transmitting the information contained in the AN firing pattern to higher auditory centers in the brain with little further processing, while other subdivisions, the dorsal cochlear nucleus (DCN) in particular, heavily transform both the temporal and the spectral aspects of the AN firing data [Osen 81, Brug 92, Rhod 92, Pick 88]. Concerning the nature of such an auditory transformation, of particular interest to both auditory theorists and experimentalists is the apparent loss of detailed temporal patterns of the AN data at

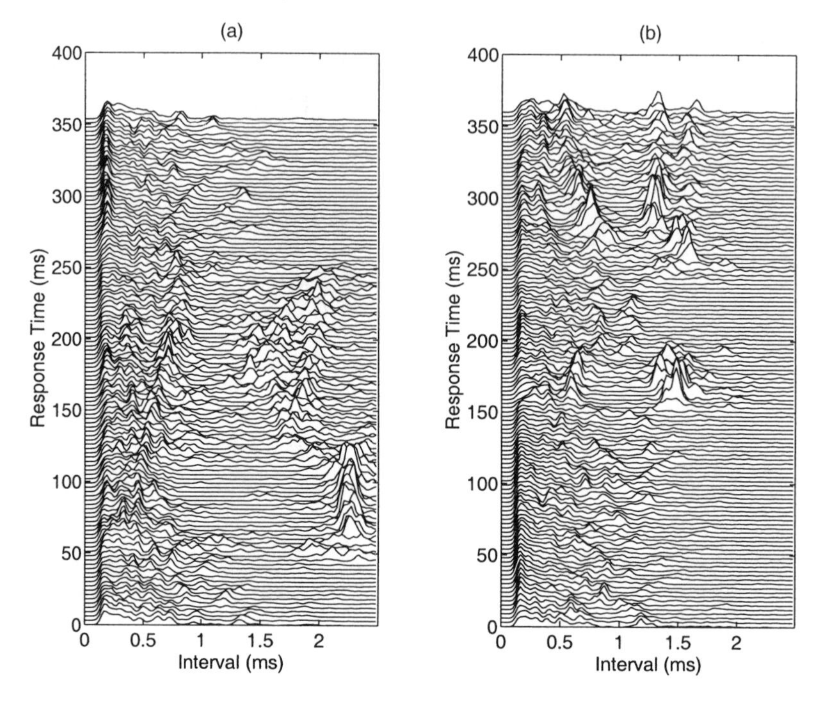

Figure 11.10: IPIHs constructed from the cochlear model output in response to utterances *heels* (a) and *semi*, embedded in white noise with 10-dB SNR.

the output of the DCN and below the inferior colliculus [Irv 92, Rhod 92]. This interest arises because the temporal response patterns of the AN fibers, particularly of those with center frequencies below 5 kHz, have been demonstrated to be important (although not the only) information carriers, at least for the vowel sounds [Young 79, Rhod 92]. A solution to this apparent conflict has been suggested, with neither supporting physiological evidence nor theoretical investigations, that the CN serves as a site for information conversion from a temporal code to a rate code [Rhod 92, Irv 92, Pick 88, Rhod 86].

Until recently, most of our knowledge about the encoding of sounds in the CN has been gained by the analysis of the responses of single neurons to relatively simple synthetic signals such as sinusoids, clicks, and wide-band noise [Rhod 92, Rhod 86, Rhod 85, Spir 91, Voi 88]. Natural sounds, speech and speech-like complex sounds in particular, differ significantly from the above simple sounds, both temporally and spectrally. Complexity of both the resonance structure and the dynamic properties of the speech signal is often responsible for perceptually important information contained in the signal. Nevertheless, the response patterns of the CN to speech-like complex sounds are difficult to infer from its responses to the spectrally simpler signals, due to various well known nonlinearities such as lateral suppression, inhibition, and rapid adaptation. There have been relatively few detailed studies on encoding of complex sounds in the CN. One related study is the one published in [Pont 89, Pont 91], where a relatively simple neural model for the DCN was developed based on conclusions drawn from the limited physiological data available on the CN (DCN in particular). Although the model was able to

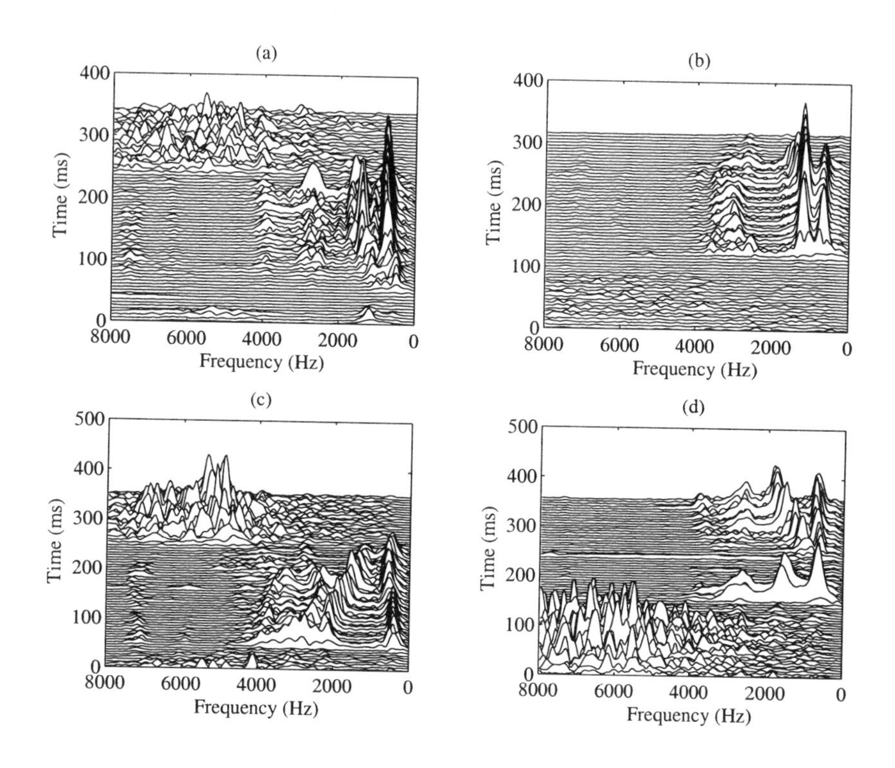

Figure 11.11: Magnitude short-time Fourier transforms for clean utterances *glass* (a), *found* (b), *heels* (c), and *semi* (d), respectively.

qualitatively match the neural responses of the DCN to noise and sinusoid stimuli, no attempt was made to match speech responses, and, due to the several well known types of auditory nonlinearities discussed above, the model cannot be expected to generalize and to produce correct neural responses to complex stimuli such as speech.

Because of the lack of modeling tools, as well as the lack of comprehensive physiological data, functional roles of the CN and higher centers in further processing the AN responses to speech and the possible mechanisms underlying this processing were not studied until recently. One study, based on a computational modeling approach, appeared recently in [She 95] and will be described in this section. In particular, we will describe how the speech-evoked AN data, which are extremely rich in their temporal structure, can be transformed into the rate-only response in parts of the higher auditory pathway, while retaining the spectral properties, both static and dynamic, of the input speech stimuli.

11.8.2 Modeling action potential generation in the auditory nerve

The portion of the auditory model up to the AN level has been described in the preceding sections of this chapter. The model output had been the IFR for each of the AN array, as analog signals as a function of time. In order to convert the IFR into neural

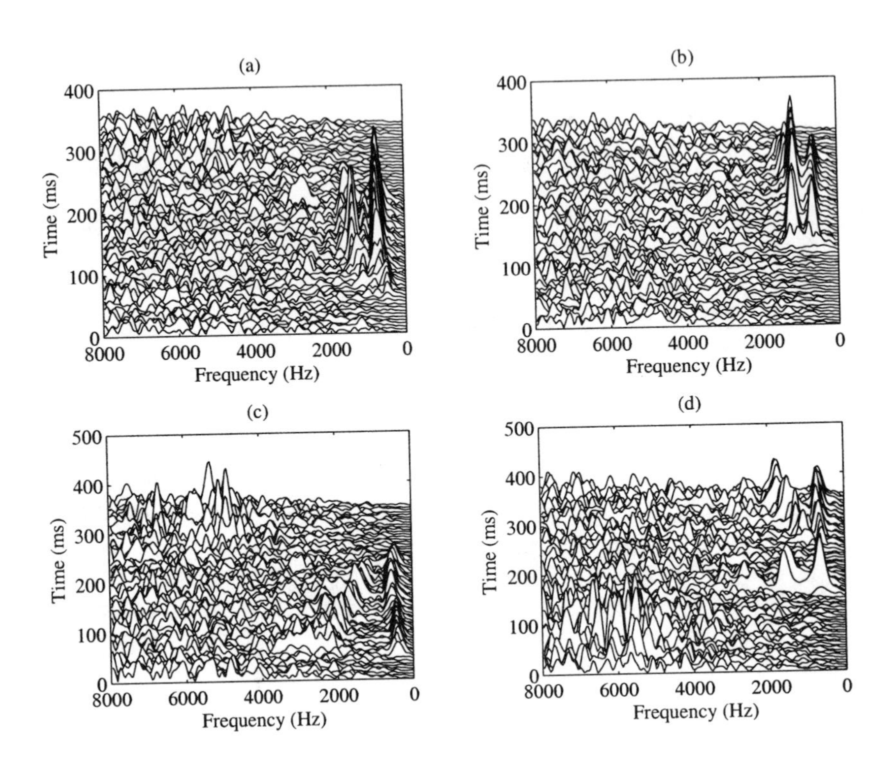

Figure 11.12: Magnitude short-time Fourier transforms for utterances *glass* (a), *found* (b), *heels* (c), and *semi* (d), respectively, all embedded in white noise with 10-dB SNR.

pulses required as the input to the CN-like neural network model, we need a model for generating action potential pulses, which is described now.

The model for action potential generation (or the spiking model, for short) produces a binary on/off spike sequence as the output, with the model input being the output of the IHC-AN synapse model. The spiking model implemented in [She 95] and described here is essentially a non-homogeneous Poisson discharge generator, but taking into account refractory effects. The model is similar to, but simpler than, those described in [Carn 93, West 88, Egge 85].

In the spiking model, the probability of AN firing is determined jointly by the output of the IHC-AN synapse model and by the refractory period. During the refractory period due to a previous spike, the AN firing is inhibited either completely or partially. For a very short interval (0.75 ms) after one spike, generation of a new spike is completely inhibited; this is called *absolute refractoriness.* After that interval, the refractoriness decays at an exponential rate. The effect of the refractoriness is also called the *discharge-history* effect, and is implemented in the spiking model as

$$H(d) = H_{max}\left[C_0 \exp\left(-\frac{d-d_{AR}}{S_0}\right) + C_1 \exp\left(-\frac{d-d_{AR}}{S_1}\right)\right], \tag{11.49}$$

where $d > d_{AR}$ is the time in milliseconds since the last discharge, $H_{max} = 100$ spikes/s is the maximum amount of firing rate reduction the refractoriness can cause, $d_{AR} = 0.75$ ms is the absolute refractoriness period, and C_0, C_1, S_0, and S_1 are empirically determined constants to fit the adaptation data of the AN firings. Note that Eq. 11.49 is valid only for $d > d_{AR}$; for $d < d_{AR}$, the firing rate is zero.

Given the discharge history function $H(d)$ by Eq. 11.49, the instantaneous spiking rate of the AN fiber is modified from that of the output of the IHC-AN synapse model (s_k) to

$$r_k = s_k - H(d). \tag{11.50}$$

Given the instantaneous spiking rate above, the firing decision is made by the firing probability $T \cdot r_k$, where T is the sampling interval. For each sampling interval, a random number q_k, uniformly distributed between 0 and 1, is produced by a standard random number generator. If $T \cdot r_k \geq q_k$, a spike is generated within that interval; otherwise no spike is generated. The spiking decision is in turn used to update the function $H(d)$ in Eq. 11.49 and the simulation for spike generation proceeds until after the input stimulus to the entire model terminates.

11.8.3 Neural-network models central to the auditory nerve

Here we first briefly review an earlier DCN model described in [Pont 91, Pont 89] (the P&D model henceforth). Although the P&D model was developed only for matching the DCN responses to simple, rather than complex, sounds, it nevertheless serves as a basis for which a more general auditory network model aimed at processing complex sounds has been developed [She 95].

The P&D model consists of three feed-forward neural layers, with each layer feeding the next and the AN fibers feeding all the layers. Although no accurate schematic for neural circuitry in the DCN exists, the limited anatomical data [Osen 81, Young 81, Manis 83], together with the physiological studies, constrain the P&D model to a reasonable extent.

The neuron model used in the P&D model is relatively simple. Each cell is characterized by the following parameters:

- Spontaneous discharge rate (spikes per second);
- Firing threshold (in arbitrary units);
- Propagation delay;
- Probability of cell firing when the membrane potential exceeds the firing threshold;
- Decay factor, which determines the amount by which the membrane potential drops over a fixed time interval when no input is given; and
- Reset factor, which determines the amount by which the membrane potential drops immediately after a spike is generated from the cell.

These parameters were manually adjusted in the P&D model for each Excitatory-Inhibitory Response Area (EIRA) cell type in order to match physiological data available for tone and noise stimuli.

Each of the three layers in the P&D model consists of neurons with different EIRA classifications. The first layer consisting of type III neurons receives excitatory inputs from the AN fibers. The second layer consisting of type II neurons receives excitatory inputs from the AN fibers and inhibitory inputs from the first layer (type III neurons). The final layer consisting of type IV neurons receives excitatory inputs from the AN fibers and inhibitory inputs from the second layer (type II neurons). The classification scheme used here has been discussed in detail in [Popp 92]. Basically, such a scheme has been based on the distinct receptive fields associated with the neurons. For example, type IV neurons contain mostly inhibitory regions but may also have a few excitatory areas with a prominent, low-threshold excitation region. Type II and III neurons, on the other hand, both contain lateral inhibitory sidebands, but they differ in spontaneous activity and in the response properties to wideband noise.

The three layers in the P&D model are constructed to account specifically for the response of the type IV neurons to tone and noise stimuli [Young 93]. The EIRA type IV cells are unresponsive to tone stimuli at high stimulus pressure levels; they have an inhibitory area at CF, and have excitatory areas bordering the inhibitory area to either side of the frequency. This suggests an inhibitory connection from the type II neurons, which have an excitatory area at CF. The type II neurons in the P&D model are made unresponsive to broad-band noise by a set of inhibitory connections to widely-spaced type III neurons, which respond to noise. In this way, the correct EIRA response pattern for each cell type can be built up from appropriate connections to other cell types. (Additional circuitry to form mild excitatory sidebands for the type IV cells by stellate cells in the PVCN, which was postulated in [Young 93], has not been incorporated into the P&D model.)

In the P&D model, the extent of the lateral connections between different layers are estimated by the extent of the innervation of the axons and dendrites of the neurons and by the physical dimensions of the DCN. The strength of the connections in the P&D model are determined empirically by extensive work on fitting available DCN physiological responses to simple sounds.

The P&D model outlined above is adopted as a basis for a more general auditory neural-network model developed in this study. We use a similar three-layer structure and interconnection strategy to those in the P&D model. Our model differs, however, from the P&D model in three key aspects. First, we use a much more accurate and realistic cochlear model incorporating crucial BM nonlinearity that we developed over the past several years (outlined in Section 2), as the early stage interfaced to the neural-network model. The BM nonlinearity in the peripheral model provides essential spatio-temporal codes for the speech input for further processing in the network model; the P&D model has been unable to process complex sounds in a meaningful way, due partly to the lack of a good peripheral interface. Second, we drastically change both the range and the strength of interconnections of neurons in the layered model from those in the P&D model. These changes are necessary for implementing a possible mechanism for converting a temporal speech coding scheme at the AN level to a rate coding scheme in the neural-network model (see Section 3.4 for detail). The issue of the nature of the coding schemes is of paramount significance for the study of auditory processing of complex sounds such as speech, but was not dealt with in the P&D model (which is largely concerned with fitting the model's responses to relatively simple sounds). Third, we introduce additional assumptions in the single neuron model used in the P&D model. These assumptions, in conjunction with the new neuronal interconnections, are also necessary for implementing the temporal-to-rate coding conversion mechanism.

Temporal vs. rate coding for complex sounds

An earlier landmark study by Sachs and Young [Sac 79] demonstrated that beyond an SPL of about 64 dB, the AN rate profiles reach a fully saturated range, at least for high spontaneous rate AN fibers. This saturation is due to a combination of saturation in receptor potentials of inner hair cells, of two-tone suppression in the basilar membrane and in AN, and of broadening of the tuning curves at higher intensities. Researchers have subsequently reassessed the earlier dismissal of rate-place encoding on the basis of the responses of AN fibers with low spontaneous rate ($<$ 10 spikes/sec)[Sac 86, Wins 86, Ains 96a]. The responses of these low spontaneous rate fibers do not show pronounced rate saturation. The lowest spontaneous rate group ($<$ 0.5 spikes/sec) constitutes a relatively small number of AN fibers (16 percent). For the CN responses to have a good rate-place encoding, this implies that the innervation pattern for the low spontaneous rate AN fibers is much more extensive than that of their high spontaneous rate counterpart [Rhod 92, Pick 88]. However, this has not yet been supported by physiological data.

In a concurrent study on the temporal patterns of AN responses to vowels, Young and Sachs [Young 79] observed that at the highest sound levels used in their study, significant temporal responses to the second and third formants remain in the AN data. They noted that as sound level is increased, the responses to the formants, particularly the first formant, increase near their places and spread to adjacent regions, particularly toward higher CFs [Young 79]. They used a method called Averaged Localized Synchronized Rate (ALSR) to extract the vowel formants at all sound levels. Fourier transforms of the firing patterns of ANs were used to measure the strength of phase locking to the individual frequency components of the stimulus. In quantifying the ALSR measure, phase locking at stimulus frequencies within half an octave of the fiber CF (an empirically determined range) was integrated.

A different mechanism has been proposed by Shamma [Sham 85b] for the representation of speech sounds in the temporal aspect of the AN firing pattern. The main motivation for this mechanism is the rapid phase shifts in the AN responses that occur near the CF locations corresponding to the formant frequencies. He attributed the shifts to an abrupt slow-down of the traveling wave components at or near the appropriate CF (this slow-down is closely related to the steep roll off of the high-frequency slopes of the cochlear filter) [Sham 85a, Sham 85b]. His analysis showed that the local phase shifts at the CF remain stable and local over a 70-dB SPL change of the input sound intensity. Based on this analysis, a lateral inhibitory network was proposed that extracts the spectral information from the AN responses using the cues contained in the rapid phase shifts across the AN fibers [Sham 85b].

Yet another temporal processing scheme has been proposed for the representation of speech sounds. Using a composite auditory model, Deng and Geisler used a cross-channel correlation algorithm to exploit the differences in the temporal patterns of discharge between fibers having slightly different CFs [Deng 87a]. In a related study, Deng and Kheirallah successfully used the proposed algorithm for dynamic formant tracking of noisy speech [Deng 93b].

A final form of the temporal processing scheme in the AN is the interval-based representation proposed and discussed in detail in [Seck 90, Ghit 88, Ghit 92, Ains 96b]. In these studies, the interval analysis gave rise to an inter-peak interval histogram or ensemble interval histogram. These histograms were demonstrated to be highly effective in extracting vowel formants, as well as the fundamental frequency, from the AN data at sound levels well into firing rate saturation in the AN.

Studies on the above various forms of temporal representation all conclude that the rate-saturated responses in the AN contain a sufficient amount of temporal information that enables faithful reconstruction of the resonance structure of complex sounds such as vowels. However, serious problems exist for the above proposed representations. With the ALSR representation, there is no evidence that the auditory system central to the AN is equipped with a machinery allowing for extraction of phase locking to frequencies only around each AN fiber's CF, while rejecting phase locking to other frequencies. In essence, the ALSR representation would require, unrealistically, the presence at the central end of each AN of a series of narrow bandpass filters centered at different CFs of the fibers. With the interval-based representation (inter-peak interval histogram or ensemble interval histogram), while providing superior formant-extraction performance to the ALSR representation [Seck 90], it, again, is not justifiable by physiological evidence. It ignores the place dependence of the AN responses and uses, instead, the aggregate interval-based histograms. Likewise, the problem with the lateral-inhibitory-network representation is that in the CN (at least in the DCN) there is no evidence for the existence of lateral inhibition required by the proposed lateral inhibitory network. Instead, evidence points to inhibitory connections between different layers of the DCN. The above problems associated with the temporal representations proposed in the earlier studies motivate our construction of the new model that is physiologically feasible and is capable of temporal-to-rate conversion essential for complex sound coding at higher levels in the auditory pathway.

Implementation of a temporal-to-rate conversion mechanism

The aim of our auditory neural-network model described here is to implement a physiologically plausible mechanism responsible for transforming the place-indexed temporal patterns of the AN responses into a place-indexed rate code at some level of the auditory pathway central to the AN. The model is partly motivated by the physiological studies of the DCN responses, which suggest the possibility of existence of a neural center for temporal-to-rate transformation. However, since our knowledge of the processing stages in the higher levels of the auditory system is limited, virtually any of the subcortical stages in the auditory system central to AN might be regarded as a potential site for such a transformation. The basic mechanism underlying the proposed temporal-to-rate conversion is coincidence detection, as qualitatively discussed in [Irv 92] and in [Rhod 92].

The coincidence detection mechanism can be implemented straightforwardly by a simple neuronal model. In the study of [Carn 92], a shot-noise threshold model, which is similar to the one proposed in [Colb 90], for AVCN cells is constructed. The study suggests that the two most important parameters determining the effectiveness of coincidence detection are 1) the amplitude of the individual inputs to the cell relative to the cell firing threshold, and 2) the time constant determining the duration of each input. Assuming that each cell receives input from fibers with a range of CFs, the individual input amplitudes should be less than the cell firing threshold so that only coincident inputs would drive the cell. On the other hand, the cell time constant parameter determines how close the two inputs should be in order to be regarded as coincident. The study of [Carn 92] found that the coincidence detection implemented by the simple shot-noise threshold model is sensitive to differences in spatiotemporal patterns across the population of AN responses for complex sounds containing different resonance structures.

The above finding motivates the construction of the neuron (cell) model with its outputs (firing rates) not only sensitive to differential resonance structures in stimuli but also capable of extracting, in conjunction with other components in the neural network model, the resonance structures of the stimuli. To this end, we assume firstly in our neuron model 1) that each cell receives inputs from fibers of a narrow range (determined empirically) of different CFs, and 2) that each cell performs a cross-correlation upon its inputs through the coincidence detection mechanism. These two assumptions are biologically justifiable for the neurons in the AVCN (e.g., [Carn 90, Carn 89, Carn 92, Carn 94]). In this regard, the first layer (type III layer) in our neural network model (and in the P&D model alike) can be generally treated as consisting of AVCN neurons. This then justifies use of the coincidence detection mechanism (which is basically a neural form of cross-correlation) in our neuron model. (The coincidence detection associated with our neuron model is also physiologically plausible for the brainstem neurons [Irv 92].)

In addition to the above two assumptions built into our neuron model, we introduce a further key property in the model: the cell decay factor or time constant has explicit dependency on the cell's tonotopical place (or equivalently on its center frequency). The need for this property arises from the nature of the place-specific temporal information contained in the AN responses to complex sounds that we discussed earlier. Let us now illustrate how this assumption for place dependency of the cell decay factor will work together with the coincidence detection mechanism to turn a temporal code to a rate code, and we will at the same time derive a specific form of this dependency accordingly.

Denote by A the amplitude of the excitatory post-synaptic potential (EPSP) as an AN input to a cell in the neural network and let the cell's threshold be TH. To allow the coincidence detection mechanism to operate, the amplitude of EPSP must be subthreshold, or $\frac{A}{TH} < 1$. Given a cell with center frequency CF, we adjust this CF-dependent (i.e., place dependent) cell decay factor $d(CF)$ such that two consecutive inputs received within the interval $1/CF$ would lead to a cell firing (shown in Fig. 11.13a for $\frac{A}{TH} > 0.5$). In contrast, when the interval T between two consecutive inputs is greater than $1/CF$, no cell firing would be triggered in our model construction (shown in Fig. 11.13b).

Let us formally define the cell decay factor to be the multiplier determining the reduction of the EPSP amplitude after an elapse of each short sampling period T_s before a new firing occurs. The condition for near-miss cell firing is then

$$A \times d(CF)^{\frac{T}{T_s}} + A = TH, \tag{11.51}$$

or using $T = 1/CF$, we arrive at the specific form of the place (CF)-dependent cell decay factor:

$$d(CF) = \left(\frac{TH}{A} - 1\right)^{(T_s \times CF)} \tag{11.52}$$

We offer the following interpretation for Eq. 11.52 and discuss its functional significance. Following the quantitative relation expressed in Eq. 11.52 as implemented in our model, the cell decay factor for a high-CF cell would be greater than that for a low-CF cell. The large cell decay factor leads to fast attenuation of the EPSP, making long-interval inputs from AN fibers much less likely to trigger the cell's firing than short-interval inputs. That is, the place dependence of the cell decay factor according to Eq. 11.52 blocks the lower-frequency contents in the input complex signal from influencing firing rates of higher-CF cells. On the other hand, interference of higher-frequency contents in the input complex signal with the rate response of a lower-CF cell

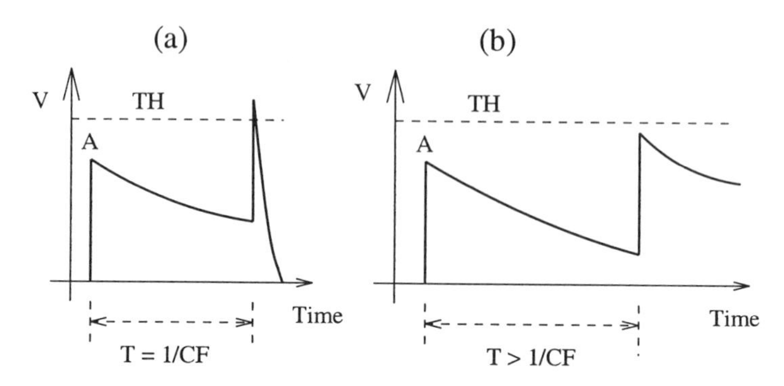

Figure 11.13: Cell membrane potentials (V) versus time. TH denotes the firing threshold. (a) Two spikes $T < 1/CF$ apart trigger a firing. (b) Two spikes $T > 1/CF$ apart do not trigger a firing (after Sheikhzadeh and Deng [She 99], @Elsevier).

is already eliminated at the level of the AN firing pattern; this has been due to a combination of a sharp BM filter cutoff on the high-frequency side and of BM nonlinearity ([Deng 87a, Deng 93b]), without recourse to any neuronal mechanisms. Therefore, the place-indexed temporal firing pattern present at the AN level can potentially be transformed to the tonotopically organized rate profile at the neural-network level without suffering from interference among various frequency regions. The key to the elimination of the interference (from low to high frequencies) is the explicit tonotopical (place) dependence of the cell decay factor according to Eq. 11.52.

While we have not been able to find explicit physiological evidence for tonotopically arranged cell decay factors for CN units, this general idea seems reasonable because of the strong tonotopic arrangement in the CN structure. The proposed place dependence of the cell decay factor might also be located in other subcortical stages of the auditory system. One potential candidate is the inferior colliculus. It has been shown that the sensitivity of the cells in the inferior colliculus to interaural time differences is tonotopically organized [Irv 92]. A coincidence detector with the tonotopically arranged cell decay factor would be able to show such a sensitivity. The place dependence of the cell decay factor we are proposing could be effectively implemented by similar cells in the inferior colliculus acting on monaural inputs. As will be seen from our simulation results (Section 4), our proposed mechanism is effective in transforming the temporal (interval) information to a rate-place code in a physiologically plausible way and does not involve complex or heuristic operations such as the ones required in constructing ALSR or ensemble interval histogram.

Neuronal interconnection and related parameters

The temporal-to-rate conversion mechanism via coincidence detection and place dependence of the cell decay factor described in the preceding subsection must be put into a complete network model with inputs from population responses of AN fibers in order to evaluate its performance for realistic signals. In line with speculations that a temporal-to-rate conversion mechanism might exist in DCN [Rhod 86, Rhod 92], we adopt the basic three-layer structure of the P&D DCN model but modify the model parameters according to the principles discussed in the preceding subsections. Due to the extensive modifications, the resulting neural-network model should no longer be regarded as a CN model.

We have run the original P&D DCN model for speech stimuli and observed (not reported here) that while the model matches the physiological responses to tone and noise stimuli well, it gives senseless results for speech and that the results for complex stimuli cannot be inferred from those of simple stimuli (due apparently to the nonlinear nature of the model). Nevertheless, we choose to keep many of the original P&D model parameters and the general interconnection strategy as long as they are not in contradiction with the proposed temporal-to-rate conversion mechanism. To quantitatively determine the model parameters as a starting point to construct the model, we use two synthetic stimuli (/da/ and /ba/), both at 69 dB SPL, and adjust the model parameters so that the temporal-to-rate conversion mechanism is reasonably effective at the output of each of the three layers. (Simulation results with use of one of these speech stimuli will be shown in Section 4.) The three-layer neural network model we construct and use in this study is shown in Fig. 11.14. The functional role of Layer 1, as designed, is to transform interval information in the AN responses to a rate code not through simple integration (as is the case in the P&D model) but by way of coincidence detection operative within a place-dependent time interval. The functional role of Layer 2 is to enhance the crude rate code represented in Layer 1 via a combination of neural inhibition and further coincidence detection. Finally, Layer 3, upon receiving inhibitory connections from Layer 2, inverts the rate-place code sharpened therein and leads to a reliable, temporal-based complex sound representation in the AN data.

While physiological evidence clearly points to the interconnections between type II and type IV cells, there is little evidence for the type III and type II cell interconnections [Pont 91]. It was suggested that in light of inhibitory interactions between VCN and DCN, type III cells might be replaced by a model of AVCN [Pont 91]. However, in a more recent study, evidence has been found of monosynaptic inhibitory connections between type II and type III units [Dav 94]. In our model shown in Fig. 11.14, we significantly modify the parameters of the first and second layers from those of the P&D DCN model. These modifications are necessary for the model to effectively couple coincidence detection with tonotopically arranged cell decay factors.

Details of the inter-layer neuronal interconnections and model parameter selection in our three-layer model (Fig. 11.14) are provided below. Layer 1 receives excitatory (denoted by solid lines) inputs from AN fibers. In the P&D model the cells in this layer had a cell firing threshold of 100 and an innervation width of 11 AN inputs, each having a connection weight of 100. Obviously, this arrangement forbids coincidence detection from occurring as the presence of a spike from any of the AN inputs will trigger a firing in a Layer-1 cell. In our model, we keep the firing threshold at 100 but change the connection weight from 100 to 90 for each of the AN inputs. To reduce smearing effects, the innervation width of the AN inputs to a Layer-1 cell is reduced from 11 in the P&D model to five. Finally, the decay factors are changed from a uniform value in the P&D model to a place-dependent function according to Eq. 11.52.

Layer 2 in our model receives inhibitory (denoted by dashed lines) inputs from Layer 1 and excitatory (solid lines) inputs from AN inputs. Again, we use coincidence detection for both excitatory and inhibitory inputs; i.e., a single excitatory input cannot trigger firing and a single inhibitory input cannot inhibit the cell. Eq. 11.52 is used for setting values of the cell decay factors. In the P&D model, seven AN inputs, each with a connection weight 170, innervate a Layer-2 cell having a firing threshold 450. The P&D model used only two inhibitory connections from Layer 1 to each of the Layer-2 cells with

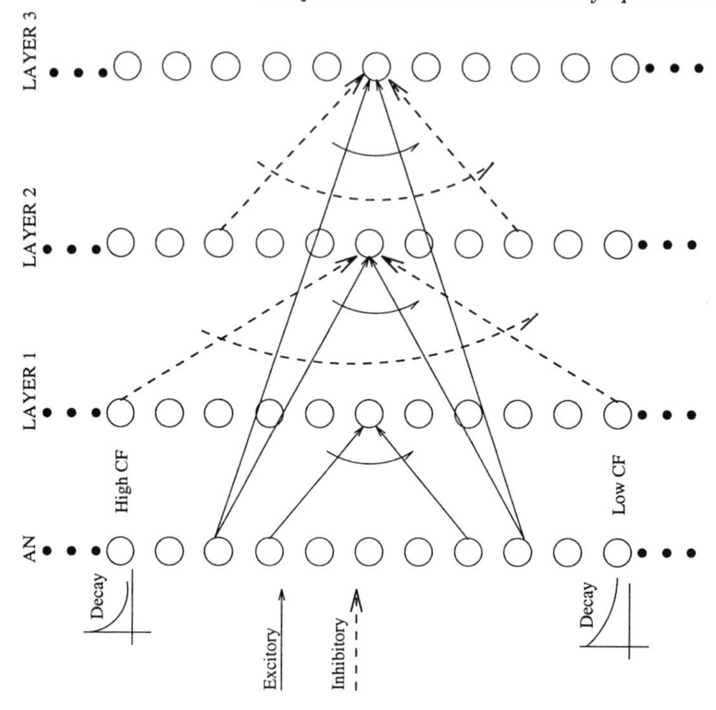

Figure 11.14: Block diagram of the layered auditory neural network model implemented in this study. Only 11 out of a total of 150 cells are shown for each layer, for clarity purposes. Interconnections are shown only for one central cell. Solid-line connections are excitatory and dashed-line connections are inhibitory. See text for details (after Sheikhzadeh and Deng [She 99], @Elsevier).

a fixed connection weight -140. The two Layer-1 cells providing inhibitory inputs were each 20 rows offset relative to the innervating Layer-2 cell's tonotopical location. Our new model maintains the firing threshold 450 and the AN innervation width seven. The connection weight from ANs to Layer-2 cells is increased to 250 (empirically determined according to the quality of the simulation results). Further, the inhibitory connections from Layer 1 to Layer 2 is increased from two to 11, each with slightly weaker connection weight (-100) than that in the P&D model. These new adjusted input connection parameters provide desirable inhibitory effects for the response peaks in the rate-place code in Layer 1. This not only makes the response peaks (local maxima) of the Layer-2 rate-place code appear as valleys (local minima) in the corresponding Layer-1 code, but also, as a result of coincidence detection, makes valleys (local minima) of the Layer-2 code sharper than the corresponding peaks in the Layer-1 code. Coincidence detection plays a key role here. The closer (in time) two spikes are in Layer-1 cells, the higher would be the possibility of inhibiting a cell firing in Layer 2. Now the cell decay factors we set according to Eq. 11.52 impose a place-dependent time window for the cells in Layer 2 in accepting inhibitory/excitatory input. After one excitatory input (from AN) has arrived, if the inhibitory inputs coincide within the time window, they can inhibit the cell and prevent a coincidence of the excitatory inputs from firing the Layer-2 cell.

On the other hand, if the inhibitory inputs do not coincide within the time window, the excitatory inputs would more likely trigger the cell firing. The spikes generated almost synchronously from tonotopically close cells of Layer 1 have greater inhibitory effects than tonotopically less close ones. Further, the inhibitory effects are especially evident for low-frequency cells that have longer time constants.

For Layer 3 of our neural-network model, we adopt essentially the same Layer-3 single neuron model used in the P&D model. The cell thresholds and input weights are modified slightly so that the cell does not act as a coincidence detector. Each Layer-3 cell receives seven excitatory inputs from ANs, each having a connection weight 500. It simultaneously receives seven inhibitory inputs with a connection weight −300 from Layer-2 cells with the cell threshold set at 400. The modification we have made to Layer 3 of the P&D model is to limit the range of inhibitory connections from Layer 2 to Layer 3. We reduce the number of tonotopically arranged connections from 53 in the P&D model to 7. This drastic reduction is motivated by our empirical observation that the innervation (to a single cell) of about 35% (53/150) of the total number of cells in each layer of the P&D model often undesirably smears the spectral information. The large range of innervation in the P&D model also appears to be inconsistent with the conclusions drawn from multi-unit DCN studies on the limited spread of inhibitory connections between type II and type IV cells orthogonal to the isofrequency sheets [Voi 90, Voi 88, Young 82]. In addition to the above modification of the innervation range, we also reduce the interconnection weights from −5000 in the P&D model to −300. This reduction of inhibitory effects makes our model consistent with the observation [Young 93] that the DCN responses to broad-band stimuli are excitatory rather than inhibitory responses. (With strong inhibitory weights −5000 as in the P&D model, no net excitatory responses would be produced.)

11.8.4 Model simulation with speech inputs

Some simulation results are shown now for the overall auditory model consisting of the cochlear model interfaced to the neural-network model using speech waveforms as the inputs. All the speech inputs presented to the model are set at such a high sound level that the AN firing rates from the model are fully saturated. This choice is made for the purpose of demonstrating the effectiveness of the proposed temporal-to-rate conversion mechanism at the supra-AN level in representing the spectral content. (At lower sound levels, the rate-place code of AN would show clear spectral peaks and, as a result, there would also be clear peaks in the rate-place code of the neural model output.)

In the model simulation experiments described here, each simulation run is repeated 200 times in order to obtain statistically reliable results. All the peri-stimulus time histograms (PSTHs) shown in this section from computer simulation of the model are constructed using a bin width 0.1 ms. To obtain a short-time rate-place code, a window length of 10 ms is used to calculate the average firing rate for each of the 150 different places (units) at various levels in the model, and we shift the window by 3 ms to update the firing rate information. Since each of the 150 places is associated in the mel scale with the center frequency of the place, this short-time rate-place code takes a similar form to the representation of the short-time mel-frequency spectrum.

The voicing portion of a synthetic /da/ syllable from the Klatt synthesizer, including conspicuous formant transitions from the consonantal /d/ place of articulation to the steady state of the vowel /a/, is used as the first testing signal in our model simulation experiments. The formant tracks of the syllable were shown in Fig. 11.1a. The waveform of this synthetic syllable, at 69 dB SPL, is used as the stimulus to the auditory model peripheral to the cochlear nucleus just described.

The short-time rate-place code is constructed directly from the above AN population responses. This rate-place code is shown in Fig. 11.15 with the temporal location of the windows displayed as the vertical axis. As Fig. 11.13 shows, the AN firing rate is fully saturated and no clear spectral peaks appear in the rate-place code.

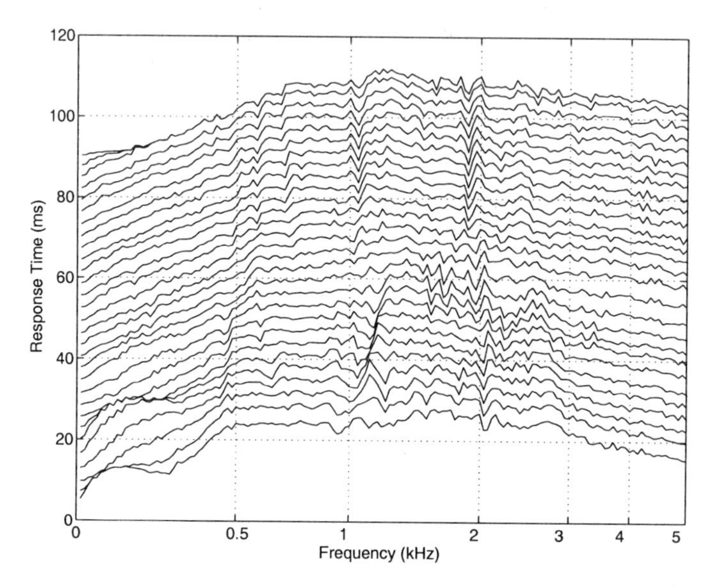

Figure 11.15: Short-time rate-place profile constructed from the AN population responses as the output of the cochlear model for a synthetic /da/ syllable at 69 dB SPL (after Sheikhzadeh and Deng [She 99], @Elsevier).

However, as discussed in earlier sections, the PSTHs of AN responses contain all the relevant spectral information about acoustic stimuli in the temporal activities of different AN bands. Specific to the stimulus of Fig. 11.1a, the temporal responses of ANs with CFs between 450 to 1200 Hz are dominated by intervals corresponding to the period of the first formant (F1). The places with CFs from 1200 to 2000 Hz are dominated by the second formant (F2) and the places with CFs between 2000 to 3000 Hz mostly contain F3 information. Above 3000 Hz, temporal responses tend to contain a mix of all three formants. Here, we observe place-indexed temporal information while the AN rate is fully saturated.

Exactly how such place-indexed temporal information in the AN population responses is converted to the rate-place code in the higher level of the auditory pathway is demonstrated in the auditory neural-network model simulation results presented below. Fig. 11.16 shows the rate-place code, constructed for the synthetic /da/ stimulus of Fig. 11.1a, for the first layer of the three-layer auditory neural-network model described in Section 3. In contrast to the uniformly flat rate-place code at the AN level where all the spectral information about the stimulus is lost, trajectories of small peaks appear in the places corresponding to the stimulus formants (F1 and F2, but not F3) in the rate-place code shown in Fig. 11.16. This (weak) spectral representation of the stimulus results from the first stage of the temporal-to-rate conversion mechanism discussed in

Section 3.4.

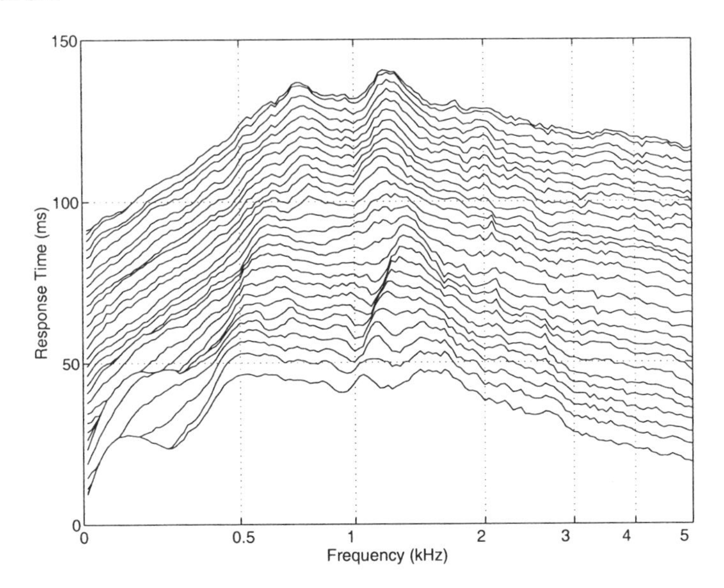

Figure 11.16: Short-time rate-place profile constructed from the responses of the first layer of the auditory neural-network model in response to a synthetic /da/ syllable at 69 dB SPL (after Sheikhzadeh and Deng [She 99], @Elsevier).

Fig. 11.17 shows the rate-place code for the second layer of the auditory neural-network model. Three clear trajectories of valleys manifest themselves at the places corresponding to transitions of formants F1, F2, and F3, respectively. Note that these valleys are much sharper and more distinct than the peaks of the first-layer rate-place code of Fig. 11.16. This stronger spectral representation of the stimulus results from the lateral inhibitory mechanism present in the model.

Finally, Fig. 11.18 shows the rate-place code for the third (final) layer of the model. Three clear trajectories of the peaks in the figure can be accurately mapped to the three formant tracks of the stimulus /da/ of Fig. 11.1a. This modeling result illustrates a close rate-place representation of the time-varying spectral information in the acoustic input signal at the output of the auditory neural-network model. This is so despite the high intensity of the input sound that saturates all AN firing rates in the peripheral model. We emphasize here that all the neurons in the neural-network model are biologically motivated and are unable to directly relay the temporal code associated with the input AN units. The close rate-place representation as shown in Fig. 11.18 results from coincidence detection, layered network interconnection, and lateral inhibition, the biologically motivated mechanisms that have been integratively incorporated into our computational model described earlier.

11.8.5 Discussion

The main points in the modeling study presented in this section can be summarized as follows. First, a computational auditory neural-network model can be constructed as a

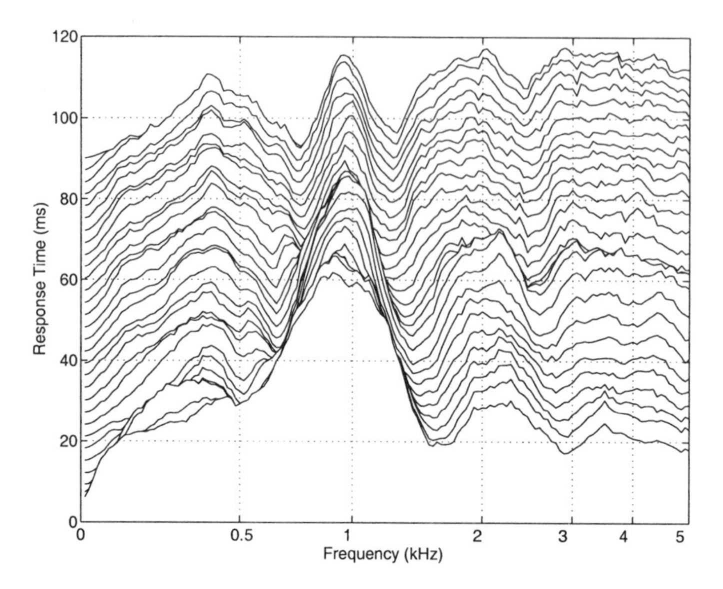

Figure 11.17: Short-time rate-place profile constructed from the responses of the second layer of the auditory neural-network model in response to a synthetic /da/ syllable at 69 dB SPL (after Sheikhzadeh and Deng [She 99], @Elsevier).

functional approximation to the complex-sound processing mechanisms associated with parts of the auditory system central to the AN. Second, armed with such a computational model, together with a cochlear model as the peripheral interface, a long standing issue in auditory research can be adequately addressed as to how the spatial-temporal patterns of the AN data in response to speech sounds can be transformed into the rate-only response (expressed as a rate-place code) in the auditory system while preserving or enhancing prominent spectral properties of the stimuli. Specifically, a temporal-to-rate conversion mechanism is discussed that requires the time constant parameter of a cell in the network model to follow a monotonically decreasing function of the CF of the cell. An explicit form of such a function has been given in Eq. 11.52. Third, in the model simulation experiments using a speech waveform as the input, it was shown that the proposed temporal-to-state conversion mechanism, based on coincidence detection, can be further enhanced by lateral inhibition implemented in the auditory network model. Each of the three layers in the network model plays a distinctive role to arrive at the robust rate-place code at the final layer of the model. Finally, the model simulation experiments using both speech utterances as inputs demonstrated that the auditory rate-place code reliably represents (with modification and/or enhancement) the prominent spectral characteristics of the utterances identifiable from the respective wideband spectrograms. These spectral characteristics are time varying in general, and the effectiveness of the rate-place code has been shown in [She 99] to be universal across all classes of speech sounds including vowels, liquids, fricatives, nasals, and stops that were chosen carefully from TIMIT as the test examples.

One distinguishing characteristic of the approach to auditory speech processing presented in this chapter is its focus on accurate, physically-based modeling. In our cochlear

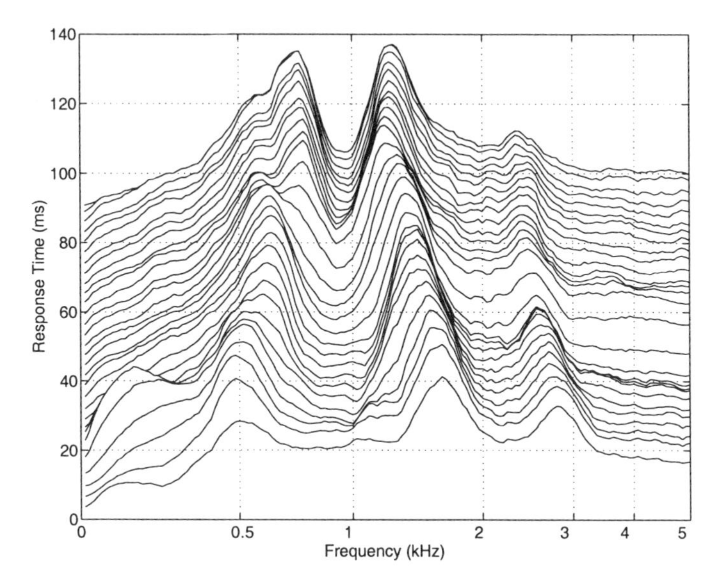

Figure 11.18: Short-time rate-place profile constructed from the responses of the third (final) layer of the auditory neural-network model in response to synthetic /da/ syllable at 69 dB SPL (after Sheikhzadeh and Deng [She 99], @Elsevier).

model, biophysical mechanisms underlying the BM vibration have been implemented, giving rise naturally and accurately to the dynamic nonlinearity in the BM filter function. This has been in contrast to many other cochlear models which functionally treated the BM vibration as simple linear digital filters. Such a physical-modeling approach has been extended to the auditory neural-network model, as presented in this section. Specifically, a natural interface between the cochlear model and the network model has been developed via the use of a spiking generator producing random sequences of AN action potentials as inputs to the network model. The use of action potentials as information carriers enables accurate implementation of the auditory mechanisms such as lateral inhibition, coincidence detection, temporal integration of synaptic potentials for new spike generation, and the temporal-to-state conversion mechanism discussed in this section. All the assumptions made in implementing these auditory mechanisms seem to be physiologically plausible. The response of each layer in the model is expressed in terms of the tonotopically arranged discharge rate profile with no requirements for the neurons to encode firing-interval information. Modeling of details of the auditory mechanisms endows the model with the capability of processing complex sounds in a way that may approximate the true biological auditory functionality. This aspect of the model also clearly distinguishes it from the P&D model, which can deal with only a limited number of simple sounds.

Another prominent characteristic of the auditory models discussed in this section and other sections in this chapter is the time-domain approach to the model solution by computer algorithms. This allows the various forms of nonlinearities inherent in the model, both the components peripheral and central to the AN, to be implemented precisely with little need for approximation. Further, this time-domain approach enables the model to process *arbitrary* input waveforms and hence allows us to reach conclusions of a most general nature. The main drawbacks of this time-domain approach are the implementation complexity and high computation requirement. These drawbacks have limited the model to only small-scale speech processing applications at present, but such a limitation should be removed in the near future with rapidly developing computing technology.

The computational models for auditory speech processing presented in this and other sections give one type of knowledge source for improving machine recognition of speech. We believe the key to *ultimate* success of machine speech recognition technology (i.e., achieving human speech recognition performance) is to move away from modeling surface phenomena of speech, and to move towards establishing a faithful computational model for the closed-loop human speech communication process. Auditory processing of speech is one integrated component crucial for the closed loop and its role is obvious. One specific use of the auditory models is to define the "goals" or targets, as a mixture of articulatory, acoustic, and auditory properties expressed in statistical terms, of speech production. In this way, the computational models for auditory speech processing discussed in this chapter and the computational models for speech production presented in Chapter 10 are naturally integrated within a general framework of modeling the entire closed-loop human speech communication process.

11.9 Summary

This chapter has been focused on computational modeling for auditory processing of speech, which constitutes another major area of computational phonetics, after computational modeling for speech production. The chapter covered mostly the peripheral auditory processing of speech, and has not covered areas of research related more to the central auditory processing (e.g., auditory scene analysis, auditory streaming, etc.).

In this chapter, a computational model with detailed mathematical formulation, based on a partial differential equation for one-dimensional nonlinear basilar membrane mechanics, was presented first. Details of the simulation steps, based on finite difference solutions to the partial differential equation, for the model's output for an arbitrary acoustic stimulus including a speech waveform were described. Both the frequency-domain and time-domain solutions were covered in detail. In addition, stability analysis for the time-domain solution was carried out, which contributed to the implementational efficiency of the solution by allowing for optimal choices of the time and spatial discretization steps with minimal loss of simulation accuracy.

After detailed presentations on the basilar membrane modeling, we then described computational models for inner hair cells and for the synapses from the inner hair cells to the auditory nerve fibers. The output of the synapse model as the final stage of the cochlear model was an array of instantaneous firing rates for each of the fibers. We provided two separate types of analysis exploiting this array of signals as the cochlear model's output. First, temporal interval statistics were computed over this array of signals. The results were subsequently used as a frame-based feature for the speech waveform. It was shown that this type of feature can effectively exploit detailed temporal information provided by the cochlear model, and it preserves and enhances much of the acoustic-phonetic properties of the original speech input under noisy environments. Details of the inter-peak interval histogram construction were provided, with examples given for the speech sounds both in quiet and in noise.

In the second type of analysis, the instantaneous firing rate in the auditory nerve was transformed into the action potential spike sequence that occurs randomly in time. The action potential spike sequence served as the input to a layered neural network model for the cochlear-nucleus-like structure. The computational models for such layered neural networks were presented, where the neural mechanisms of lateral inhibition and coincidence detection were used to account for the temporal-to-rate code conversion in the higher levels of the auditory system. Some speech examples were provided as the input to the cochlear model first and subsequently to the network model for the auditory pathway. Simulation results were provided to demonstrate the capability of the model to preserve time-varying spectral prominence information using the rate-place profile.

Part IV

SPEECH TECHNOLOGY IN SELECTED AREAS

Chapter 12

Speech Recognition

The analytical (mathematical) background, scientific (linguistic) perspective, and the computational aspect of the speech process that have been covered in the previous three parts (11 chapters) of this book have culminated in this final Part IV: Applications in speech technology. Due to space limitation, it is not possible to cover all areas of speech technology. In the final part of this book, we intend to cover only three selected areas of speech technology — speech recognition, speech enhancement, and speech synthesis — which best illustrate the applications of the dynamic modeling and optimization principles that have formed the basis for describing the speech process in the earlier chapters. Part IV starts with **automatic speech recognition** (ASR), also called machine speech recognition or **speech to text**, in this chapter. ASR is the most important area of speech technology, and it encompasses the widest range of concepts, principles, and theories about the speech process to which we have devoted all the previous chapters on its analytical, scientific, and computational backgrounds.

12.1 Introduction

ASR is the process of converting an acoustic signal of speech, captured by a microphone (or a microphone array), a telephone, or other transducers, to a text sequence typically in terms of a sequence of words. Depending on the ASR applications, these recognized words may be the final ASR results; examples of such applications are command and control, data entry, and document preparation or dictation. The recognized words can also serve as the input to further linguistic processing in order to achieve **speech understanding**, the process by which a computer maps an acoustic speech signal to some form of abstract meaning of the speech.

ASR has emerged as a promising area for applications such as dictation, telephone voice response system, database access, human-computer interactions, hands-free applications (such as car phones or voice-enabled PDAs), and web enabling via voice. Speech is the most direct and intuitive form of human communication; successful ASR thus can enhance the ease, speed, and effectiveness with which humans can direct machines to accomplish desired tasks. ASR has become an established research area, and has already created many successful products in the market place.

ASR can also be viewed as defining applications for artificial intelligence in computer science. Many important issues in ASR, such as feature extraction from auditory data, preservation of auditory perceptual invariance, integration of information across time and frequency, and exploration of distributed representations, are all in common with the key problems in artificial intelligence. It is a great scientific and technological challenge to understand how humans recognize speech and to simulate this ability in computers, a common goal in both ASR and artificial intelligence.

12.1.1 The speech recognition problem

The problem of speech recognition can be viewed in terms of the decoding problem in a source-channel representation. In this representation, the speaker consists of

- the information source specifying the intended word sequence $\mathbf{W}^* = w_1, w_2, ..., w_n$; and

- the speech production system as the noisy communication channel, whose output is the speech signal.

The recognizer is aimed at determining (decoding) the intended word sequence $\mathbf{W}^*$ produced by the speaker with as few errors as possible. It consists of

- the acoustic pre-processor, which extracts useful features from the speech signal and transforms it into a sequence of observation vectors $\mathbf{O} = \mathbf{o}_1, \mathbf{o}_2, ..., \mathbf{o}_T$; and

- the word decoder, which uses the observation vector sequence and the statistical knowledge about the language to produce the optimal output word sequence $\hat{\mathbf{W}}^* = \hat{w}_1, \hat{w}_2, ..., \hat{w}_n$.

The way a speech recognizer works involves several steps. First, speakers can talk directly into the microphone for (almost) real-time speech acquisition. They can also record speech on a handheld recordable device or PDA to be processed later, or use a telephone over a network to a server that runs the speech processing software. The next step is prefiltering and echo cancellation.

Speech consists of sound waves. Once they have been captured, these audio signals are converted from analog to digital representations and filtered to eliminate background noise from the signal. Then the speech audio signals are compressed. This reduces the bandwidth and storage needed to process the data and transport it through the computer. The following step is feature extraction. To further reduce the bandwidth requirement, the resulting audio signal is divided into shorter frames, which are processed to produce a series of feature vectors that are compared, in the following pattern matching step, to a pre-defined database of sound representations to determine the probability that the user uttered a particular sound. This comparison is based typically on a set of Hidden Markov Models (HMMs). The HMM, as has been covered in earlier chapters, is the most frequently used method of speech pattern comparison and recognition at present. In contrast to earlier template matching approaches, the HMM uses statistical modeling and libraries of word and grammar rules to select the highest probability outcome from a sequence of observation feature vectors. Pattern matching based on the HMM uses also a vocabulary database and grammar. This vocabulary and language model data depend on the language being spoken and the application. These help the HMM identify words and choose the most likely speech pattern. Finally, the output of a speech recognizer can be a file of text (for a dictation application) or an executed command (for a command and control application).

12.1.2 ASR system specifications

ASR systems can be characterized by a number of parameters. These parameters specify the capability of ASR systems and include:

- speaking mode (isolated words vs. continuous speech),
- speaking style (read speech with formal style vs. spontaneous and conversational speech with casual style),
- speaking situation (human-to-machine speech vs. human-to-human speech),
- speaker dependency (speaker dependent, speaker adaptive, vs. speaker independent),
- vocabulary size (small vocabulary of fewer than 20 words, medium vocabulary of fewer than 20,000 words, vs. large vocabulary of more than 20,000 words),
- language model (N-gram statistical model, finite state model, context free model, vs. context sensitive model),
- **perplexity** or **branching factor** (small perplexity of less than 10 vs. high perplexity of larger than 10)
- environment condition (high signal-to-noise-ratio of greater than 30 dB, medium SNR, vs. low SNR of lower than 10 dB),
- transducer (close-talking microphone, far-field microphone, microphone array, vs. telephone).

Some of the above parameters are briefly explained below. An isolated-word ASR system requires that the speaker pause briefly between words, and a continuous-speech recognition system does not.

Spontaneous or extemporaneously generated speech, which can happen either in human-machine communication (e.g., dictation of free-style e-mail) or in human-human communication (e.g., casual conversation among friends or in meetings), typically contains disfluencies. This casual style of speech is much more difficult for ASR than speech read formally from a script.

A speaker dependent system is developed to operate for a single speaker. A speaker independent system is developed to operate for any speaker of a particular type (e.g., American English). These systems are the most difficult to develop and most expensive, and accuracy is lower than with speaker dependent systems. However they are more flexible. A speaker adaptive system is developed to adapt its operation to the characteristics of new speakers. Its difficulty lies somewhere between speaker independent and speaker dependent systems.

Speaker dependent or adaptive ASR systems require speaker enrollment; that is, a user must provide samples of his or her speech before using the systems. These systems are usually easier to develop, cheaper to buy and more accurate, but not as flexible as speaker independent systems. Speaker dependent systems require much more enrollment data than speaker adaptive systems.

Recognition performance is generally lower when vocabularies are large, partly because a large vocabulary contains many similar-sounding words. Also, the computational complexity associated with a large vocabulary speech recognizer generally requires pruning in lexical access, which often reduces the recognition accuracy.

The language model, which is often expressed in terms of artificial grammars, is used to restrict the combination of words in the recognized text when speech is produced in a sequence of words. The most popular language model, used typically for large vocabulary ASR systems, is the N-gram statistical model (to be discussed later in detail in this chapter). The simplest linguistically motivated language model can be specified as a finite-state network, where the permissible words following each word are given explicitly. More general language models approximate natural language and are specified in terms of context-free or context-sensitive grammars.

One popular measure of the difficulty of the ASR task, combining the vocabulary size and the language model, is perplexity. Perplexity is loosely defined as the geometric mean of the number of words that can follow a word after the language model has been applied. For this reason, perplexity is also called the branching factor.

12.1.3 Dimensions of difficulty

ASR is a difficult problem due to the many sources of variability associated with the signal. The difficulty can be characterized along several dimensions. First, the acoustic realizations of phonemes, the basic contrastive sound units of which words are composed, are strongly dependent on the context in which they appear. The contextual variations can be rather dramatic at word boundaries, adding special difficulties to continuous speech recognition. The kind of contextual variations that affect the speech acoustics are usually very difficult to characterize at the phoneme level. The task becomes easier and more systematic if one adopts the phonological feature or articulatory gesture systems as we argued in Chapters 8 and 9 of this book.

Second, given the same phonetic context, speech variabilities can result from changes in the speaker's physical and emotional state, speaking style, speaking rate, or voice quality. For speech with a casual style, certain forms of phonological reorganization occur, which often cause some phonological features to be added, deleted, or modified so that phones may be added, deleted, or modified accordingly. In addition, phonetic reduction such as articulatory target undershooting also drastically changes speech articulation and acoustics from those in speech with a more formal, read style. The results of phonetic reduction are increased "static" confusability; that is, if one does not take account of the dynamic patterns caused by phonetic reduction, the static acoustic or articulatory properties of two otherwise distinct sounds often become indistinguishable. Speech with a casual style in general is associated with speech with a relatively fast speaking rate. However, there are mechanisms other than a casual speaking style that result in an increased speaking rate. Also importantly, due to the continuity constraint in speech articulation which is largely responsible both for contextual variations (coarticulation) and for phonetic reduction, there are intimate relationships among a number of phonetic factors with major influences on the difficulty of ASR. These factors include phonetic context, speaking rate, and speaking style.

Third, the same linguistic text consisting of the same word sequence, uttered by two speakers with the same speaking style and rate, can have drastically different speech signals. This difference is called across-speaker variabilities, and it results from differences in the respective speakers' sociolinguistic background, dialect, and in their vocal tract size and shape, etc. A crude division of all speakers gives rise to three major groups: 1) male adult speakers; 2) female adult speakers; and 3) children speakers. Among the three, children's speech is the most difficult for ASR.

Fourth, ignoring all the above sources of acoustic variabilities, changes in the acoustic environment (noise) where the speech recognizer is deployed and in the position and characteristics of the transducer (channel) also create additional variabilities in speech. One major challenge of ASR system design is to make the system robust to background noise and channel distortion. This noise and distortion can take the form of any additive environmental noise including a very difficult case of competing speaker(s), and of any convolutive transmission channel distortion including a very difficult case of acoustic reverberation with a long filter response. The noise may also include sounds created by the speaker himself, such as breath noise, lip smacks, clicks, coughs, etc.

12.1.4 Evaluation measures for speech recognizers

A speech recognizer is evaluated generally by the number of errors it suffers in predicting the correct word sequence hypotheses $\mathbf{W}^* = w_1, w_2, ..., w_n$. The sentence error rate (SER) can be computed in a test set of S sentences by

$$\text{SER} = \frac{1}{S}\sum_{s=1}^{S} \delta(\mathbf{W}_s^* \neq \hat{\mathbf{W}}_s),$$

where $\delta(\mathbf{W}_s^* \neq \hat{\mathbf{W}}_s)$ is the Kronecker delta function, which is one when $\mathbf{W}_s^* \neq \hat{\mathbf{W}}_s$ and zero otherwise.

A more common evaluation measure is the word error rate (WER). Let $M(\mathbf{W})$ be the total number of words in the word sequence $\mathbf{W}$, and let $E(\mathbf{W}_s^*, \hat{\mathbf{W}}_s)$ be the minimum number of word changes (including insertions, deletions, and substitutions) required to transform string $\mathbf{W}_s^*$ to string $\hat{\mathbf{W}}_s$. Then, the WER is defined by

Definition of the word error rate for speech recognizers:

$$\text{WER} = \frac{\sum_{s=1}^{S} E(\mathbf{W}_s^*, \hat{\mathbf{W}}_s)}{\sum_{s=1}^{S} M(\mathbf{W}_s^*)},$$

where the numerator is the minimum number of word changes between the correct and the recognized word sequences, and the denominator is the total number of words in the correct word sequence.

12.2 Mathematical Formulation of Speech Recognition

We now formalize the problem of speech recognition, introduced in the preceding section, based on the source-channel representation. This formalization, which we will describe below, provides a powerful statistical framework that underlies much of modern speech recognition research and system development.

12.2.1 A fundamental equation

Let the observation speech vector sequence be $\mathbf{O} = \mathbf{o}_1, \mathbf{o}_2, ..., \mathbf{o}_T$, which can be continuous-valued acoustic vectors [Young 95], or discrete-valued vector-quantized codes [Lev 83], or any other type of general acoustic measurements. Let $\mathbf{W}^* = w_1, w_2, ..., w_n$ be the sequence of words intended by the speaker who produces the acoustic record $\mathbf{O}$ above. The goal of a speech recognizer is to "guess" the most likely word sequence $\hat{\mathbf{W}}$ given the acoustic data $\mathbf{O}$. Bayes' decision theory, which we outlined in Chapter 6, provides a minimum Bayes risk solution to the above "guessing game," and the minimum Bayes risk can be made equivalent to minimum probability of error if the risk is assigned values of one or zero for incorrect and correct guesses, respectively. According to Bayesian decision theory, speech recognition can be formulated as a MAP decision rule (cf. Section 6.1 in Chapter 6), or equivalently as a top-down search problem over the allowable word sequences $\mathbf{W}$ based on the posterior probability $P(\mathbf{W}|\mathbf{O})$:

MAP decision rule for automatic speech recognition:

$$\begin{aligned}\hat{\mathbf{W}} &= \arg\max_{\mathbf{W}} P(\mathbf{W}|\mathbf{O}) \\ &= \arg\max_{\mathbf{W}} \frac{P(\mathbf{O}|\mathbf{W})P(\mathbf{W})}{P(\mathbf{O})} \\ &= \arg\max_{\mathbf{W}} P(\mathbf{O}|\mathbf{W})P(\mathbf{W}), \end{aligned} \tag{12.1}$$

where $P(\mathbf{W})$ is the prior probability that the speaker utters $\mathbf{W}$, which is independent of the acoustic data and is determined by the language model, and $P(\mathbf{O}|\mathbf{W})$ is the probability that the speaker produces (or the microphone of the speech recognizer receives) the acoustic data $\mathbf{O}$ if $\mathbf{W}$ is the intended word sequence by the speaker.

The denominator $P(\mathbf{O})$ in Eq. 12.1 above can be neglected as shown because the acoustic observation $\mathbf{O}$ is the same for all competing hypotheses $\mathbf{W}$.

Eq. 12.1 is sometimes called the fundamental equation of speech recognition because it most concisely summarizes the three basic elements of statistical ASR, which we outline below.

12.2.2 Acoustic model, language model, and sequential optimization

According to Eq. 12.1, after speech pre-processing which produces the acoustic vector sequence $\mathbf{O}$, the ASR problem can be reduced to the following three issues to be discussed in more detail in the remaining portion of this chapter:

- Acoustic modeling for speech production from the word sequence to acoustic vectors — how to accurately characterize and efficiently compute the probability $P(\mathbf{O}|\mathbf{W})$;
- Language modeling — how to accurately characterize and compute the probability $P(\mathbf{W})$;
- Search (sequential optimization) for the word sequence $\mathbf{W}$ — the optimization operation $\arg\max_{\mathbf{W}}$ in Eq. 12.1 to sequentially find the word sequence corresponding to the optimal value of the posterior probability $P(\mathbf{W}|\mathbf{O})$.

Eq. 12.1 is essentially a re-formulation of analysis-by-synthesis, cast in a consistent probabilistic framework: the synthesis phase is embedded in the assumption that acoustic data $\mathbf{O}$ is produced from the word sequence $\mathbf{W}$ (hence the necessity and possibility to evaluate the "production" probability $P(\mathbf{O}|\mathbf{W})$, and the analysis phase involves finding the solution $\hat{\mathbf{W}}$ which best matches (in a maximum-a-posteriori or MAP sense) the outcome of the production. From this analysis-by-synthesis interpretation of Eq. 12.1, we will see that good speech production theories, when meeting the computational requirements implied in Eq. 12.1, should have much to contribute to advancing ASR technology. Many of the computational models of speech production, which we discussed in Chapter 10, have been developed to aim at this goal.

Note that unless whole word models are used (which is suitable only for small vocabulary ASR), the acoustic problem above that involves $P(\mathbf{O}|\mathbf{W})$ typically consists of two separate sub-problems. The first one is to describe how a word and a word sequence can be represented in terms of sub-word units, a necessary step for large vocabulary ASR. This sub-problem is often referred to as that of **pronunciation modeling**. The sub-word units can be any of the units described in Chapters 8 and 9, including syllables, phonemes, allophones, phonological features, and articulatory gestures. The second sub-problem in acoustic modeling involves the mapping from each sub-word speech unit to acoustic observation vector sequences. We will discuss both of these sub-problems in acoustic modeling in some detail later in this chapter.

12.2.3 Differentially weighting acoustic and language models

The probabilities in the fundamental equation Eq. 12.1 provided by acoustic and language models are necessarily not exact. For example, the assumption of short-span temporal correlation typically made in the acoustic and language models, which will be discussed in more detail later in this chapter, assigns non-zero probabilities to sequences that could never occur. This causes the models to reserve only a fraction, not all, of the probability mass for the observed sequences. In addition, the overall ranges of the probability values computed from the acoustic and language models are typically quite different. This is because the acoustic model (such as HMMs) usually assigns a probability to sequences of high-dimensional, continuously-valued random vectors, while the language model assigns a probability to sequences of discrete random variables (i.e., words). Further, the sequences evaluated by the two models also have very different temporal densities. For the acoustic model, the density is about 100 frames per second or about 1000 frames per sentence, and for the language model, the density is as low as about 10 words per sentence.

The differences discussed above between the acoustic and language models suggests that the probabilities computed and combined in Eq. 12.1 should be weighted differently. That is, the acoustic and language models should be balanced.

One very easy and commonly used way for striking such balance is to use a fixed weighting factor, called the language model weight γ, to scale the language model vs. the acoustic model. This modifies posterior probability to

$$P(\mathbf{W}|\mathbf{O}) = \frac{P(\mathbf{O}|\mathbf{W})P^{\gamma}(\mathbf{W})}{P(\mathbf{O})},$$

where

$$P(\mathbf{O}) = \sum_{\mathbf{W}'} P(\mathbf{O}|\mathbf{W}')P^{\gamma}(\mathbf{W}').$$

The fundamental equation of Eq. 12.1 is subsequently modified into

$$\begin{aligned}\hat{\mathbf{W}} &= \arg\max_{\mathbf{W}} P(\mathbf{O}|\mathbf{W})P^{\gamma}(\mathbf{W}), \\ &= \arg\max_{\mathbf{W}} P^{\frac{1}{\gamma}}(\mathbf{O}|\mathbf{W})P(\mathbf{W}). \end{aligned} \tag{12.2}$$

12.2.4 Word insertion penalty factor

In addition to the language model weight, another commonly used weighting factor is called the **word insertion penalty**, which is used to compensate for the known weakness of the language model to provide an accurate probability for the sentence length in terms of the number of words. If we assume that the sentence length $|\mathbf{W}|$ has a discrete exponential distribution:

$$P(|\mathbf{W}| = w) = \frac{1}{C}\exp(\rho\, w),$$

then the posterior probability becomes

$$P(\mathbf{W}|\mathbf{O}) = \frac{P(\mathbf{O}|\mathbf{W})P^{\gamma}(\mathbf{W})\exp(\rho\,|\mathbf{W}|)}{C\,P(\mathbf{O})},$$

where

$$P(\mathbf{O}) = \sum_{\mathbf{W}'} P(\mathbf{O}|\mathbf{W}')P^{\gamma}(\mathbf{W}')\exp(\rho|\mathbf{W}'|)/C.$$

Eq. 12.2 is then subsequently modified to be

$$\begin{aligned}\hat{\mathbf{W}} &= \arg\max_{\mathbf{W}} P(\mathbf{O}|\mathbf{W})P^{\gamma}(\mathbf{W})\exp(\rho\,|\mathbf{W}|),\\ &= \arg\max_{\mathbf{W}} \log(P(\mathbf{O}|\mathbf{W})) + \gamma\log(P(\mathbf{W})) + \rho\,|\mathbf{W}|. \qquad (12.3)\end{aligned}$$

12.3 Acoustic Pre-Processor

12.3.1 What is acoustic pre-processing

Acoustic modeling as of the three basic issues in ASR discussed in the preceding section requires that the speech waveform $s(t)$ be processed to produce a sequence of feature vectors $\mathbf{O} = \mathbf{o}_1, \mathbf{o}_2, ..., \mathbf{o}_T$. This operation is called the acoustic pre-processor, or front-end processor or feature extractor, of ASR systems, since it generally does not depend on any phonetic class of speech. Many such pre-processing or analysis techniques have been covered in Chapter 2. In this section, we will only discuss some practical aspects of the commonly used acoustic pre-processors, and some special considerations of the speech representation in terms of the feature vectors appropriate for acoustic modeling in ASR.

In most ASR systems, the speech waveform $s(t)$ is typically sampled at a rate between 6.6 kHz (for telephone speech) and 16-20 kHz (for wideband speech), before being further processed to produce a sequence of feature vectors. The feature vectors typically have between 8 and 20 dimensions, and are usually computed every 10 ms with an (overlapped) analysis window of about 25 ms. These "static feature vectors" are often taken time differentials to form a set of "dynamic feature vectors". The first-order dynamic features are often called the delta parameters, and the second-order ones called the delta-delta parameters. The combined static and dynamic feature vectors are then fed to the acoustic model component of the recognizer, which estimates the probability that the portion of the waveform corresponds to a particular phonetic class. The feature representation of the speech waveform and the probability estimation may interact with each other [Rathi 97, Gales 02]. That is, what one sees as part of the feature representation another may see as part of the acoustic probability estimation or modeling process. In practice, however, if a processing operation is applied to all classes of speech, it is feature extraction. However, if the operation is applied differently to different phonetic hypotheses, then it becomes part of the acoustic modeling process.

Feature extraction aims to preserve the information needed to determine the phonetic class of a portion of the speech waveform while being invariant to other extraneous factors such as speaker differences, speaking rate differences, acoustic environments, and paralinguistic factors including the emotional state of the speaker. Given these properties, it is also desirable to make the extracted features as compact as possible.

12.3.2 Some common acoustic pre-processors

Most of the acoustic pre-processors in ASR systems concentrate primarily on properties of the speech signal attributable to the shape of the vocal tract rather than to the source excitation, whether the latter is generated at a supra-laryngeal constriction or at the larynx. While the extracted features are sensitive to whether the vocal folds are vibrating or not (the voiced/unvoiced distinction), they try to suppress the effects due to the fundamental frequency of vibration because of its strong dependency on speakers.

In most of the acoustic pre-processors in current ASR systems, the speech waveform is processed by a simple pre-emphasis filter giving a 6-dB/octave increase in gain over most of its range to make the average speech spectrum roughly flat. Then the short-term power spectrum is computed by taking successive overlapping portions of the preemphasized waveform (typically 25 ms long), tapering both ends with a bell-shaped window function, and applying a discrete Fourier transform. The resulting power spectrum has undesirable harmonic fine structure at multiples of the fundamental frequency. This is reduced by grouping neighboring sets of components together to form about 20 to 24 frequency bands. These filter bands are often made successively broader with increasing frequency above 1 kHz, usually according to the Mel frequency scale reflecting the frequency resolution of the human ear. This filter bank output can be equivalently obtained by computing the energy in the filter bands directly using a bank of digital filters.

Use of the power spectrum means that the short-term phase structure in the speech signal is ignored. This choice is made because our ears are largely insensitive to phase effects. The power spectrum associated with each filter bank is almost always represented on a log scale in a way that is consistent with the compressive nature of human auditory processing discussed in earlier chapters. This gives rise to the log channel (filter bank) energies. One advantage this provides is that the shape of the log channel energy profile is preserved when the gain applied to a signal varies. The channel energies in a log scale in this case are simply shifted up or down. Further, any linear filtering with convolutional effects on the waveform and any multiplicative effects on the linear power spectrum simply become additive constants on the log channel energy.

In typical acoustic pre-processors, the log channel energies are subject further to the cosine transformation, which is a version of the Fourier transform using only cosine basis functions and converts the set of log channel energies to a set of cepstral coefficients. Since the Mel frequency scale is used, these cepstral coefficients are also called **Mel-Frequency Cepstral Coefficients (MFCCs)**. One main purpose of the cosine transform is to create largely uncorrelated feature components. Since the shape of the spectrum imposed by the vocal tract is smooth, energy levels in adjacent bands tend to be correlated. Removing the correlation allows the number of parameters to be reduced while preserving the useful information. It also makes it easier to compute reasonably accurate probability estimates in a subsequent statistical matching process. Compared with the number of bands, typically only about half as many of these cepstral coefficients need be kept. The zero-th order MFCC (C_0) measures the overall loudness level of the speech signal. The first-order MFCC (C_1) measures the balance between the upper and lower halves of the log channel energies. The higher order coefficients are concerned with increasingly finer features in the log channel energy.

Since the vocal tract can be regarded as a lossless unbranched acoustic tube with plane-wave sound propagation, its effect on the excitation signal is that of a series of resonances. Hence, the vocal tract can be modeled approximately as an all-pole filter. The linear predictive coding (LPC) technique or autoregressive modeling, which was covered in Chapter 2, fits the parameters of an all-pole filter to the speech spectrum. This LPC-based technique provides a popular alternative method to cepstral coefficients discussed above.

However, the conventional LPC analysis has problems with certain signal degradations. The analysis has an inherently linear scale in frequency, and it is not very convenient to produce the mel-scale version of the LPC. Perceptual Linear Prediction (PLP) overcomes these difficulties. It combines the LPC and filter-bank approaches by fitting an all-pole model to the set of energies or loudness levels produced by a perceptually motivated filter bank, and then computes the cepstrum from the model parameters [Herm 90].

It is a common practice to augment information of the MFCC, LPC, or PLP with information on its rate of change over time. The simplest way to obtain this dynamic information would be to take the difference between consecutive frames. This gives rise to the so called **delta or dynamic parameters**. However, this simple method turns out to be too sensitive to random interframe variations. Consequently, linear trends are usually estimated over sequences of five or seven frames [Fur 86]. Many systems go further by estimating acceleration (delta-delta) features as well as linear rates of change (delta features) [Young 98]. These second-order dynamic parameters need even longer sequences of frames for reliable estimation.

It can be shown that any convolutional distortion affecting the shape or overall level of the spectrum would be associated with constant offsets in the log spectrum and cepstrum. Therefore, cepstral mean normalization can effectively compensate for such distortion. Note that the dynamic features are intrinsically immune to such constant normalization effects.

When first-order dynamic parameters are passed through a low-pass filter, something close to the original static parameters are recovered except that constant and very slowly varying features are reduced to zero, thus giving independence from constant or slowly varying channel characteristics. This technique is called RASTA (relative spectrum) and has been widely used in acoustic pre-processors.

As an alternative to the cosine transform approach to decorrelate the feature components, the principal components analysis (PCA) technique provides a more powerful transformation that can completely remove linear dependencies between sets of variables. This method can be used to decorrelate not just sets of energy levels across filter bank outputs but also combinations of parameter sets such as dynamic and static features, LPC, and PLP parameters. A double application of PCA with a weighting operation, known as linear discriminant analysis (LDA), can take into account the discriminative information needed to distinguish between speech sounds to generate a set of parameters.

The ultimate challenge in speech feature extraction is to close the gap between the superior performance of human listeners and automatic recognizers. This superiority is most remarkable when there is very little speech data to adapt to the voice of any specific speaker and when the acoustic environments are difficult. The fact that the superior performance of human recognition of speech persists even when nonsense words are used shows the importance of the acoustic-phonetic information for speech perception/recognition and that the superior performance cannot be explained by superior language modeling alone. Therefore, there is still much that remains to be done in improving acoustic pre-processors for more effective acoustic modeling in ASR.

12.4 Use of HMMs in Acoustic Modeling

ASR technology saw a significant breakthrough with the introduction of Hidden Markov Models (HMMs) and related data-driven statistical techniques in the 1970's and 1980's [Baker 75, Red 84, Rab 83, Jel 76, Bahl 83, Rab 89, Rab 93]. The main role of the HMMs has been their use as the acoustic model for computing the acoustic score of $P(\mathbf{O}|\mathbf{W})$ in fundamental equation Eq. 12.1. HMMs are still the dominant technology in today's ASR [Juang 00, Lee 96]. In earlier chapters, we already introduced the basic concept and parametric characterization of the HMM and its generalization as the trended HMM (Sections 3.5 and 4.2). We also discussed how the HMM (and the trended HMM) can be viewed as computational production models for the acoustic dynamics of speech, and presented a learning algorithm for estimating their parameters (Section 10.2). Based on the above background knowledge, in this section we will discuss various aspects of the use of the HMMs as the acoustic model specifically for speech recognition.

12.4.1 HMMs in ASR applications

In the currently most popular framework of ASR using fundamental equation 12.1, an inventory of elementary HMMs corresponding to basic linguistic units of speech (i.e., phonemes) is used to build a larger HMM for a word. A sequence of acoustic feature vectors extracted from the input speech waveform is seen as a realization of a concatenation of elementary processes described by the HMMs. As we discussed in Section 3.5, an HMM is a composition of two stochastic processes, a hidden Markov chain, which accounts for temporal variability, and an observable process, which accounts for spectral variability. This combination has proven to be rather powerful to encompass the most important sources of speech variability, while at the same time be sufficiently flexible to allow for efficient construction of practical ASR systems.

In ASR applications, two formal assumptions are usually made in constructing the HMMs. First, the first-order Markov hypothesis states that the past history has no influence on the chain's future evolution if the present is specified. Second, the conditional IID hypothesis says that given the state sequence, the observation sequence is an IID process. Thus, neither the Markov chain evolution nor past observations influence the present observation if the last chain transition is specified.

The HMMs used in ASR can be classified according to the nature of the output distribution functions. If the distributions are discrete, i.e., defined on finite spaces, then we have discrete HMMs. In this case, observations are vectors of symbols in a finite alphabet of a fixed number of different elements usually constructed via a process called vector quantization (VQ). For each one of the feature vector components, a discrete density is defined, and the distribution is obtained by multiplying the probabilities of each component. Notice that this definition assumes that the different components are independent.

On the other hand, if the HMM's output distribution is defined on a continuous observation space, then we have the continuous HMMs. In this case, strong restrictions have to be imposed on the functional form of the distributions, in order to have a manageable number of statistical parameters to estimate. The most popular approach is to characterize the model transitions with mixtures of base densities having a simple parametric form. The most typical example is the mixture Gaussian HMM as discussed in

Section 3.5, which can be parameterized by the mean vector and the covariance matrix for each mixture component and for each HMM state. To model complex distributions in this way, a rather large number of base densities have to be used in every mixture component. This may require a very large training corpus of data for the estimation of the distribution parameters. Problems arising when the available corpus is not large enough can be alleviated by sharing distributions among transitions of different models. In semi-continuous HMMs [Huang 90, Huang 90, Bell 89], for example, all mixtures are expressed in terms of a common set of base densities. Different mixture components are characterized only by different weights. A common generalization of semi-continuous HMM modeling consists of interpreting the input vector as composed of several components, each of which is associated with a different set of base distributions. The components are assumed to be statistically independent, hence the distributions associated with model transitions are products of the component density functions.

12.4.2 Relationships between HMM states and speech units

When an HMM arranged in a left-to-right manner is used to represent a word, each state corresponds to an acoustically stable region (i.e., a continuous set of frames with "stationary" observation vectors) of the word. If the number of states in the HMM is greater than the number of stable acoustic regions, then two or more states may join to represent one such region. Conversely, when the number of states is fewer than the number of stable regions, one state may absorb two or more different regions, leading to a large variance in the output distribution of this state. In the latter case, the HMM state uses its stationary distribution to fit the nonstationary data.

The use of a whole word as the speech unit for the HMM representation is only sensible for small vocabulary speech recognizers (such as digit or alphabet recognizers). For medium or large vocabulary recognizers or flexible-vocabulary recognizers, sub-word speech units must be used because otherwise it would not be possible to reliably train all the word models. For most sub-word based recognizers, words are usually represented by networks of phonemes. Each path in a word network represents a phonetic pronunciation of the word. In this case, an HMM is used to represent a phoneme and the HMM for a word is constructed by forming a network of the (small) phonemic HMMs. For a phonemic HMM, each state then corresponds to an acoustically stable region of the phoneme. The number of such regions is typically quite small, and hence popular phonemic HMMs usually have a small number of states (e.g., three states).

The main drawback of the phonemic HMMs is the large variances in the output distributions due to the fact that the same phoneme can have different acoustic realizations if pronounced in different contexts. This drawback has been reasonably well overcome by the introduction of allophonic HMMs. Allophones of a phoneme, as we discussed in Section 8.2, are the physical realizations of that phoneme in different phonetic contexts. The decision as to how many allophonic HMMs should be constructed in an ASR system for a given phoneme depends on many factors, e.g., the availability of enough training data to estimate the model parameters.

12.4.3 Construction of context-dependent HMMs

Given sufficiently detailed contexts, an allophonic or context-dependent phonetic HMM can be made to match the acoustic data for the same allophone quite closely. In principle, an allophone should be considered for every different word in which a phoneme appears. If the vocabulary is large, however, it is unlikely that there are enough data to train all these allophonic HMMs, so models for allophones are considered at a different level of detail; the level from high to low may include syllable, triphone, diphone, or context independent phoneme. The HMM's output distributions for an allophone having a certain degree of generality can be obtained by mixing the distributions of more detailed allophone models. The loss in specificity is compensated by a more robust estimation of the statistical parameters due to the increase of the ratio between training data and the number of free parameters that needs to be estimated.

A more principled approach to constructing allophonic HMMs consists of choosing allophones by clustering possible contexts. This choice can be made automatically and optimally with Classification and Regression Trees (CART). A CART is a binary tree having a phoneme at the root and a linguistic question about the context associated with each node in the tree. The questions are of the type: "Is the left phoneme a vowel?" For each possible answer (YES or NO) there is a link to another node with which other questions are associated. Algorithms have been developed and successfully used for growing and pruning CARTs based on automatically assigning questions to a node from a manually determined pool of questions. The leaves of the tree are labeled by an allophone symbol.

In addition to clustering allophonic HMMs to improve their reliability, output distributions at the HMM state level can be clustered (i.e., be made the same or tied). For example, the distributions for the central portion of the allophones of a given phoneme can be tied, reflecting the fact that they represent the stable (context-independent) physical realization of the central part of the phoneme, uttered with a more or less stationary configuration of the vocal tract. In general, all the models can be built by sharing distributions taken from a pool of a few thousand cluster distributions called **senones**.

12.4.4 Some advantages of the HMM formulation for ASR

The statistical formulation of acoustic modeling based on HMMs for ASR offers certain advantages over other approaches. First, the HMM has a simple and uniform structure. For example, all triphones can have the same left-to-right three-state topology, differing only in separate parameter values rather than in the HMM structure. This makes it easy to implement the recognizer.

Second, the HMM covers both the temporal and spectral variabilities in a rather flexible manner. While it does not model the acoustic dynamics of speech accurately, in many ASR tasks it may be sufficient to separate different classes of speech using such an inaccurate model.

Third, not only an allophone, a phoneme, a syllable, or a word can be represented by an HMM as we just described, an entire sentence can likewise be represented by one large composite HMM including the language model. In constructing such a large HMM, the transitions between words are made to be ordinary HMM transitions between the final state of the previous word and the initial state of the next word. The sub-HMMs for the constituent words are simply constructed according to the pronunciation dictionary and the transitions between phones are also ordinary HMM transitions between the final state of the previous phone and the initial state of the next phone.

Fourth, as a result of one single composite HMM representation for an entire sentence, the search for the optimal word sequence shown in Eq. 12.1 is simply a search for the best path through the trellis of this composite HMM. Highly efficient dynamic programming based algorithms, the Viterbi algorithm and its variants, exist for this purpose, which we already discussed in Section 5.3.

Fifth, highly efficient algorithms also exist for determining the HMM parameters directly from speech data, as we have touched on in Section 10.2. These algorithms are special cases of the EM algorithm, based on the maximum likelihood principle discussed in Section 5.6.

Finally, the same HMM formulation can be applied not only to English but equally well to other languages. The main requirement when applying the HMM formulation to a new language for its ASR is to make available the pronunciation dictionary of the new language.

12.5 Use of Higher-Order Statistical Models in Acoustic Modeling

12.5.1 Why higher-order models are needed

Some advantages of the HMM formulation that were just discussed have associated weaknesses in other aspects of ASR. Some of these weaknesses were discussed in Section 10.2 when we viewed and analyzed the HMM as a computational generative model for the dynamics of speech acoustics. Essentially, some simplifying assumptions made in the HMM formulation discard many key temporal properties of speech that we discussed in earlier chapters of this book. For example, the HMM framework assumes that the speech feature vector sequences are produced by a piecewise stationary process with instantaneous transitions between stationary HMM states. It also assumes that the probability of a feature vector associated with an HMM state depends only on that vector and the state. It thus discards the smooth, dynamic constraints of the speech production system. Further, the straightforward left-to-right sequential (i.e., linear) connection of phonetic HMMs to form a large HMM for a word or a sentence, in a manner commonly referred to as "beads on a string," often provides poor phonological generalization and results in a large number of context-dependent allophonic HMMs.

Speech recognition technology has advanced considerably over the past two decades (cf. [Lee 96]). The advances can be attributed to the use of a consistent statistical paradigm empowered by increasing quantities of speech data corpora, as well as by powerful algorithms developed for model learning from the data (cf. [Jel 76, Juang 00]). Up until recently, the technology has primarily been founded on the principle of statistical "ignorance" modeling, where generally unstructured speech models such as HMMs learn many parameters from massive amounts of directly observable speech acoustic data, with little concern paid to the mechanisms of the data generation. Such a blind, data-driven approach, which may be characterized as "missing science," may be fundamentally limited by its generalization capabilities. This approach has experienced greater and greater difficulties as the speech recognition tasks have become less and less constrained, while increasingly difficult, natural human-to-human speech related spontaneous ASR problems are attacked. These problems are often characterized by a wide range of mismatch factors, including inter-speaker, environmental, contextual, and speaking style (intra-speaker) variability, as well as their interactions, between the averaged training and instantaneous testing conditions. Solutions to such problems demand effective, dynamic generalization abilities to be built into the recognizer. These abilities may effectively be acquired from structural modeling of speech data generation mechanisms. For example, human speech production is such that faster speech naturally creates a greater degree of contextual variation, in a highly predictable manner, than speech spoken at a normal rate. However, the conventional, data-driven approach based on HMMs must take phonetic context and speaking rate as two separate factors (one discrete and the other continuous) to learn the acoustic variations. This leads to an undesirable explosion in the size of the conditioning factors and thus of recognizer parameters. Since the true nature of the acoustic variations is captured only in an inefficient way, the conventional HMM-based approach makes it difficult to achieve high-performance for unconstrained, natural speech recognition.

Advancement from the conventional HMM-based approach to overcome its weaknesses above can be made by establishing higher-order statistical models of speech that capture the internal structure of the speech dynamics. This new paradigm for ASR can and should still be adhered to the same Bayesian-decision theoretic pattern recognition framework as used by current HMM-based speech recognition technology. However, differing from the "missing science" approach, it makes extensive use of scientific knowledge about both the symbolic nature and dynamic behavior of human speech production and communication process. It also requires a new mathematical basis that generalizes substantially from the HMM, and also from the stochastic-segment modeling framework which we first describe below as some primitive forms of the higher-order statistical models for speech acoustics.

12.5.2 Stochastic segment models for speech acoustics

Stochastic segment models are a broad class of statistical models that generalize from the HMM and that intend to overcome some shortcomings of the HMM such as the conditional IID assumption and its consequences. The generalization is in the following sense: In an HMM, one frame of speech acoustics is generated by visiting each HMM state, while a variable-length sequence of speech frames is generated by visiting each "state" of a stochastic segment model, where that state is associated with a random sequence length.

Like an HMM, a stochastic segment model can be viewed as a generative process for the observation data sequences. It is intended to model the speech feature trajectories and temporal correlations that have been inadequately modeled by an HMM. As higher order models than HMMs, stochastic segment models tend to require more computation. However, it has been accepted that, with the increase in computational power and with the advance in decoding algorithms, they are viable for constructing ASR systems. A comprehensive review of stochastic segment models can be found in [Ost 96], and an early specific type of such models can be found in [Deng 91c, Deng 92b].

One excellent way to understand the various types of stochastic segment models and their relationships is to establish a hierarchy showing how the HMM is generalized by gradually relaxing its assumptions. Under the root node of a conventional, stationary-state HMM in this hierarchy, there are two main classes of its extended or generalized models. Each of these classes further contains several subclasses of models. This hierarchy can be described as follows:

- Nonstationary-state HMMs: This model class has also been called the trended HMM, constrained mean trajectory model, segmental HMM, or stochastic trajectory model, etc., with slight variations in technical detail according to whether the parameters defining the trend functions are random or not. This type of model has been described in detail in various parts of Chapters 3, 4, and 10 earlier in this book. The sample paths of the model are piecewise, explicitly defined stochastic trajectories (e.g. [Deng 91c, Deng 92b, Gish 93, Ghit 93a]).

- Polynomial trended HMM: The trend or trajectory function associated with each HMM state is a polynomial function of time frames.
  * Observable trend functions: Original versions of the trended HMM [Deng 91c, Deng 92b, Gish 93, Deng 94c] fall into this category where there is no uncertainty about the polynomial coefficients.
  * Hidden trend functions: The trend functions are "hidden" due to parameter uncertainty. The polynomial coefficients are random variables rather than deterministic parameters.
    · Discrete distributed polynomial coefficients: This gives rise to the mixture polynomial trended HMM [Deng 97d, Hon 00]
    · Continuous distributed polynomial coefficients: This is sometimes also called the segmental HMM, where the earlier versions have a polynomial order of zero [Holm 95, Gales 93] and more recent versions have an order of one [Holm 99] or two [Rathi 01].
- Exponential trended HMM: The trend function associated with each HMM state is a polynomial function of time frames [Deng 98].
- Non-parametric trended HMM: The trend function associated with each HMM state is not determined by any parametric form, but by the training data after performing dynamic time warping [Ghit 93a].

• Multi-region dynamic models: The sample paths of these models are piecewise, recursively defined stochastic trajectories (e.g., [Diga 93]). (Some intimate relations and implementational differences between the explicitly defined and recursively defined time series models have been discussed in [Kohn 88].) We have also discussed various aspects of this type of model in part of Chapters 4 and 10.

  - Autoregressive HMM: The trajectory function associated with each HMM state is defined by linear prediction or (recursively defined) autoregression function for the acoustic feature vectors [Ken 90, Deng 95a]. When the autoregression function is defined on the speech wave samples instead of on the feature vectors, we have the hidden filter model [Pori 88, She 94a]. The parameters for the autoregression are deterministic, and hence the trajectories defined in this model are not hidden. This type of model is also called the conditionally Gaussian model in [Ost 96].
  - Linear dynamic model: The models for the acoustic dynamics of speech described in [Diga 93, Deng 97e] uses not only the autoregression function to recursively define the trajectories, but also a noisy observation function to produce the observed acoustic data. Therefore, the trajectories are hidden by the observation noise. The actual effect of autoregression is to smooth the observed acoustic feature vectors of speech.

In addition to the many kinds of stochastic segment models with their classification above, there are other kinds of models discussed in [Ost 96]. They include: 1) nonlinear models for the acoustic dynamics of speech based on predictive neural networks; 2) models using segmental acoustic features in contrast to frame-based acoustic features; and 3) segmental models that use posterior distributions. We will not discuss these models here, but refer the readers to the survey paper of [Ost 96], where detailed discussions have been provided.

12.5.3 Super-segmental, hidden dynamic models

The many kinds of stochastic segment models just described generalize the conventional HMM by generating a variable-length sequence of speech frames in each state, rather than generating individual frames in a conditional IID manner (independent and identical distribution given the HMM state sequence). That is, these generalized models aimed mainly at overcoming the HMM's assumption of conditional, local IID. Yet the inconsistency between the HMM assumptions and the properties of the true speech process goes far beyond this local IID limitation. While in stochastic segment models the speech frames assigned to one segment have been made to be temporally correlated and/or time varying (hence not identically distributed any longer), the lengths of such segments are typically short, with the length of a phone duration or shorter. Long-term correlation among the phonetic units in a sentence has not been captured.

A more advanced class of speech models, which we call the **super-segmental models, hidden dynamic models**, or **structured speech models**, explicitly captures the long-term correlation among the phonetic units by imposing continuity constraints on the hidden dynamic variables internal to the acoustic observation data. Therefore, the concept of segment in the stochastic segment model is extended from the phone-sized units to the entire sentence, hence super-segmental. More delicate structures are needed in designing such models. Because of the need to introduce the hidden dynamic variables on which the continuity constraints are imposed, these models are also called the hidden dynamic models. Some typical examples of the super-segmental or hidden dynamic models have been presented in detail in Sections 10.3 and 10.4, when we discussed computational models for speech production. We now provide brief coverage of these and other kinds of hidden dynamic models designed to generalize the HMM and the stochastic segment models.

Many of the hidden dynamic models in the literature have been based on recursively defined trajectory models, generalizing from the earlier linear models aimed at dynamic modeling at the acoustic level [Diga 93, Ost 96, Deng 97e] to nonlinear models that take into account some articulatory mechanisms of speech production. Some proposals and empirical methods for modeling pseudo-articulatory dynamics or abstract hidden dynamics for the purpose of speech recognition can be found in [Bakis 91, Bakis 92, Blac 95, Gao 00, Rich 99, Pic 99], where the dynamics of a set of pseudo-articulators is realized either by IIR filtering from sequentially arranged, phoneme-specific target positions or by applying trajectory-smoothness constraints. In such work, due to the simple nature of the use of the pseudo-articulators, one very important property of the human speech production — compensatory articulation that requires modeling correlations among target positions of a set of articulators — could not be taken into account.

To incorporate crucial properties in human articulatory dynamics — including compensatory articulation, target behavior, and relatively constrained dynamics due to biomechanical properties of the articulatory organs — in a statistical generative model of speech, it appears critical to use essential properties of true, multidimensional articulators, rather than the pseudo-articulators. Because much of the acoustic variation observed in speech that makes speech recognition difficult can be attributed to articulatory phenomena, and because articulation is one key component in the closed-loop human speech communication chain, it is reasonable to expect that developing a faithful and explicit articulatory dynamic model and incorporating it into a comprehensive generative model of speech that serves to define the statistical structure of automatic speech recognizers will contribute to bridging the performance gap between human and machine speech recognition.

The hidden dynamic model proposed in [Deng 97c] explicitly captures compensatory articulation by modeling component-correlated articulatory targets. The model uses such stochastic targets defined for separate phonological units of speech in the utterance to smooth the multidimensional articulatory variables. The articulatory-to-acoustic mapping is then used to generate the acoustic observation vector sequences. In this model, as well as in other types of hidden dynamic models, the model parameters that control the hidden dynamics (sometimes called the "time constant" of the system matrix) and the parameters responsible for the mapping from the hidden variables to the observation variables are deterministic ones. These deterministic parameters have been generalized to random parameters in [Deng 99c] to provide further flexibility of the model. Such random parameters can be time invariant or be slowly time varying.

There is a substantial body of work done recently on a special version of hidden dynamic models where vocal tract resonance (VTR) is used to represent the target-directed hidden dynamics flowing from one speech unit to another [Deng 99a, Deng 00c, Ma 00, Pic 99]. The motivation of using the VTR hidden dynamics has been discussed in detail in Section 10.3.3. Importantly, the VTR has target-directed dynamic properties similar to the articulatory variables, but has a much lower dimension and a more intuitive connection to the easily inspected formant frequencies, thus facilitating model parameter initialization and reducing model implementation complexity. In the next section, as a case study, we will provide some details of such a hidden dynamic model and a summary of the speech recognition results.

12.5.4 Higher-order pronunciation models

As we discussed earlier in this section, a pronunciation model in acoustic modeling concerns how a word and a word sequence may be represented in terms of sub-word units. Pronunciation modeling is a necessary step for large vocabulary ASR because it is not possible to train all possible HMMs at the word level.

The pronunciation modeling in past ASR history has been based predominantly on the notion that a word is composed of a sequence of phonemes, which has been characterized as "beads on a string" and is fundamental to the traditional linear phonology discussed in Section 8.2. Considering the considerable progress in phonology as we outlined in Sections 8.3 to 8.7, the linear representation of words in the lexicon used in current-generation ASR has been extremely primitive. This simple "beads on a string" pronunciation or phonological model, however, has been quite satisfactory for read-style or scripted material (such as the *Wall Street Journal* corpus). This success is mainly due to the use of a large number of allophonic HMMs, enumerated based on the left and right phonetic contexts for each phone in clearly enunciated utterances.

For informal, casual style speech, which is typically found in spontaneous, human-to-human conversational speech (such as the Switchboard corpus consisting of telephone dialogues), the traditional phoneme-based pronunciation models have had very little success. Sophisticated models of this kind provide very little improvement on recognition performance, despite the use of massive amounts of training speech data with a similar informal speaking style. A very insightful analysis of the underlying reasons has been offered in the study of [McA 98], where it is demonstrated that, within the context of a fabricated-data experiment, ASR performance can be drastically improved when lexical representations match the phonetic labels. This suggests that a poor phone-based pronunciation model is one major reason for poor ASR performance (for the Switchboard corpus). An obvious problem with the phone-based pronunciation model is its inability to capture the pronunciation variation associated with casual speech. For example, due to the diversity of English syllabic forms (Sections 8.7 and 9.4), phone-based models have difficulties in capturing the contextual effects conditioned at this level of linguistic organization. More specifically, in the phone-based lexical representation, the possible ways of accounting for pronunciation variation are strongly limited: phone-level substitution, deletion, and insertion. This raises the question: whether moving away from such a limitation by adopting higher-order, nonlinear phonological models is capable of overcoming the current ASR performance limit for spontaneous speech.

In [Ost 00], a number of linguistically motivated approaches to constructing the higher-order pronunciation models have been reviewed. This includes the early work of [Deng 92d], where a set of parallel, discrete, multi-valued phonetic feature streams are used to provide constrained, asynchronous feature alignment in the same spirit as autosegmental and articulatory phonology (Chapter 9). The review also analyzes the work of [Kirc 96, King 98, Kirc 98], where a more flexible structure is proposed that treats the different features as independent streams synchronized at the syllable level. Use of syllables to re-synchronize the otherwise unconstrained feature alignment decouples the features from phones. This leads to models that generalize better across languages than models relying on phone-level alignment constraints as proposed in [Deng 92d].

The parallel feature model of [Deng 92d] has been limited to contextual dependency of adjacent phones, and is thus no more powerful than triphones if sufficient training data are made available. This limitation has been successfully overcome by a new paradigm based on the feature overlapping mechanism. The resulting higher order model is called the (multi-tiered) overlapping-feature model, and was presented in Chapter 9 of this book and in numerous publications [Deng 94e, Deng 96, Deng 97a, Deng 98, Sun 01, Deng 99d]. In particular, the syllable structure and other linguistic factors have been effectively used to control the feature overlaps, and speech recognition experiments have demonstrated the effectiveness of this higher order model compared with triphone-like linear models.

Further research on higher-order pronunciation models is needed to tightly integrate automatic learning of the multi-tiered pronunciation model and the remaining components of the phonetic model. Initial work on automatic learning of articulatory-feature overlapping patterns in the phonological model was reported in [Sun 00], based on a limited amount of articulatory data available with a fixed speaking style. Richer sets of articulatory data, with a wide range of speaking styles, will make the model derived from such automatic learning more effective. Other work, with promising results, on automatic learning of multi-tiered representations of speech events can be found in [Nock 00].

12.6 Case Study I: Speech Recognition Using a Hidden Dynamic Model

In this section, we provide a case study of one version of the super-segmental model that captures long-term correlation across speech units at the sentence level via the smoothness constraint in the VTR hidden dynamics as outlined in the previous section. This study illustrates the kind of process needed to be taken to formulate, learn, and evaluate hidden dynamic models. This case study is based on the work published in [Deng 00c].

12.6.1 Model overview

The hidden dynamic model described in this section is a super-segmental statistical coarticulatory model, a drastic departure from the conventional HMM-based approach to speech recognition. In the conventional approach, the variability in observed speech acoustics is accounted for by a large number of Gaussian distributions, each of which may be indexed by a discrete, "context" factor. The discrete nature of encoding the contextual (or coarticulatory) effect on speech variability leads to explosive growth of free parameters in the recognizers, and when the true source of the variability originates from causes of a continuous nature (such as in spontaneous speech), this approach necessarily breaks down. In contrast, the new approach focuses directly on the continuous nature of speech coarticulation and speech variability in spontaneous speech. In particular, the phonetic reduction phenomenon is explicitly modeled by a statistical dynamic system in the domain of VTR that is internal to, or hidden from, the observable speech acoustics. In this hidden dynamic model, the system matrix (encompassing the concept of time constants) is structured and constrained to ensure the asymptotic behavior in the VTR dynamics within each speech segment. Across speech segments in a speech utterance, a smoothness or continuity constraint is imposed on the VTR variables. The main consequence of this constraint is that the interacting factors of phonetic context, speaking rate and segment duration at any local temporal region are in combination exerting their influences on the VTR variable values (and hence the acoustic observations as noisy nonlinear functions of the VTR values) anywhere in the utterance. This gives rise to the property of long-term context dependence in the model without requiring use of context dependent speech units.

Some background work that leads to the development of this particular version of the model (i.e., with use of VTRs as the partially hidden dynamic states) has been extensive studies of spontaneous speech spectrograms and of associated speech production mechanisms. The spectrographic studies on spontaneous speech have highlighted the critical roles of smooth, goal-directed formant transitions (in vocalic segments, including vowels, glides, and liquids) in carrying underlying phonetic information in the adjacent consonantal and vocalic segments. The smoothness in formant movements (for vocalic sounds) and in VTR movements (for practically all speech sounds) reflects the dynamic behavior of the articulatory structure in speech production. The smoothness is not only confined within phonetic units but also across them. This cross-unit smoothness or continuity in the VTR domain becomes apparent after one learns to identify, by extrapolation, the "hidden" VTRs associated with most consonants, where the VTRs in spectrograms are either masked or distorted by spectral zeros, wide formant bandwidths, or by acoustic turbulences. The properties of the dynamic behavior change in a systematic manner as a function of speaking style and speaking rate, and the contextual variations of phonetic units are linked with the speaking style and rate variations in a highly predictable way.

One key characteristic of this model is the elimination of the need to enumerate contextual factors such as triphones. The contextual variations are automatically built into the goal-directed, globally smooth dynamic "state" equations governing the VTR movements during speech utterances. Moreover, the contextual variations are integrated into speaking rate variations that are controlled by a small number of shared dynamic model parameters. The sharing is based on physical principles of speech production.

To provide an overview, the version of the hidden dynamic model presented in this section consists of two separate components. They accommodate separate sources of speech variability. The first component has a smooth dynamic property, and is linear but nonstationary. The nonstationarity is described by left-to-right regimes corresponding to sequentially organized phonological units such as context-independent phones. Handling nonstationarity in this way is very close to the conventional HMMs; however, for each state (discrete as in the HMM), rather than having an IID process, the new model has a phonetic-goal-directed linear dynamic process with the physically meaningful entity of continuous state variables. Equipped with the physical meaning of the state variables (i.e., VTRs in the current version of the model), variability due to phonetic contexts and to speaking styles is naturally represented in the model structure with duration-dependent physical variables and with global temporal continuity of these variables. This contrasts with the conventional HMM approach where the variability is accounted for in a largely unstructured manner by accumulating an ever increasing model size in terms of the number of Gaussian mixture components. The second component, the observation model, is static and nonlinear. This lower-level component in the speech generation chain handles other types of variability including spectral tilts, formant bandwidths, relative formant magnitudes, frication spectra, and voice source differences. The two components combined form a nonstationary, nonlinear dynamic system whose structure and properties are well understood in terms of the general process of human speech production.

12.6.2 Model formulation

The mathematical formulation of the hidden dynamic model discussed in this section is a constrained and simplified nonlinear dynamic system. The basics of the nonlinear dynamic system has been covered in Section 4.5, and its use in computational models of speech production has been covered in Section 10.3. The specific dynamic system model here also consists of: 1) state equation, and 2) observation equation, which are described below.

State equation

A noisy, causal, and linear first-order "state" equation is used to describe the three-dimensional (F1, F2, and F3) VTR dynamics according to

$$\mathbf{z}(k+1) = \mathbf{\Phi}^j \mathbf{z}(k) + (\mathbf{I} - \mathbf{\Phi}^j)\mathbf{t}^j + \mathbf{w}_d(k), \qquad j = 1, 2, ..., J_P \tag{12.4}$$

where $\mathbf{z}(k)$ is the three-dimensional "state" vector at discrete time step k, $\mathbf{\Phi}^j$ and $\mathbf{t}^j$ are the system matrix and goal (or target) vector associated with the (sequentially or segmentally arranged) dynamic regime j that is related to the initiation of dynamic patterns in phone j, and J_P is the total number of phones in a speech utterance. Both $\mathbf{\Phi}^j$ and $\mathbf{t}^j$ are a function of time k via their dependence on dynamic regime j, but the time switching points are not synchronous with the phone boundaries. (In this case study, we define the phone boundary as the time point when the phonetic feature of manner of articulation switches from one phone to its next adjacent phone. The dynamic regime often starts ahead of the phone boundary in order to initiate the dynamic patterns of the new phone. This is sometimes called "look-ahead" or anticipatory coarticulation.) The time scale for evolution of dynamic regime j is significantly larger than that for time frame k. In Eq. 12.4, $\mathbf{W}_d(k)$ is the discrete-time state noise, modeled by an IID, zero-mean, Gaussian process with covariance matrix $\mathbf{Q}$.

The special structure in the state equation, which is linear in the state vector $\mathbf{z}(k)$ but nonlinear with respect to its parameters $\mathbf{\Phi}^j$ and $\mathbf{t}^j$, in Eq. 12.4 gives rise to two significant properties of the VTRs modeled by the state vector $\mathbf{z}(k)$. The first property is local smoothness; i.e., the state vector $\mathbf{z}(k)$ is smooth within the dynamic regime associated with each phone. The second, attractor or saturation property is related to the target-directed, temporally asymptotic behavior in $\mathbf{z}(k)$. This target-directed behavior of the dynamics described by Eq. 12.4 can be seen by setting $k \to \infty$, which forces the system to enter the local, asymptotic region where $\mathbf{z}(k+1) \approx \mathbf{z}(k)$. With the assumption of mild levels of noise $\mathbf{W}_d(k)$, Eq. 12.4 then directly gives the target-directed behavior in $\mathbf{z}(k)$: $\mathbf{z}(k) \to \mathbf{t}^j$.

An additional significant property of the state equation is the left-to-right structure in Eq. 12.4 for $j = 1, 2, ..., J_P$ and the related global-smoothness characteristics. That is, the local smoothness in state vector $\mathbf{z}(k)$ is extended across each pair of adjacent dynamic regimes, making $\mathbf{z}(k)$ continuous or smooth across an entire utterance. This continuity constraint is implemented in the current model by forcing the state vector $\mathbf{z}(k)$ at the end of dynamic regime j to be identical to the initial state vector for dynamic regime $j+1$. That is, the Kalman filter that implements optimal state estimation for dynamic regime $j+1$ is initialized by the $\mathbf{z}(k)$ value computed at the end of dynamic regime j.

Observation equation

The observation equation in the dynamic system model developed is nonlinear, noisy, and static, and is described by

$$\mathbf{o}(k) = h^{(r)}[\mathbf{z}(k)] + \mathbf{v}(k),$$

where the acoustic observation $\mathbf{o}(k)$ consists of MFCC measurements computed from a conventional speech preprocessor, $\mathbf{v}(k)$ is the additive observation noise modeled by an IID, zero-mean, Gaussian process with covariance matrix $\mathbf{R}$, intended to capture residual errors in the nonlinear mapping from $\mathbf{z}(k)$ to $\mathbf{o}(k)$. The multivariate nonlinear mapping, $h^{(r)}[\mathbf{z}(k)]$, is implemented by multiple switching MLPs (Multi-Layer Perceptrons), with each MLP associated with a distinct manner (r) of articulation of a phone. A total of ten MLPs (i.e., $r = 1, 2, ..., 10$) are used in the experiments reported in this chapter.

The nonlinearity is necessary because the physical mapping from VTR frequencies $(\mathbf{z}(k))$ to MFCCs $(\mathbf{o}(k))$ is highly nonlinear in nature. The noise used in the model Eq. 12.6.2 captures the effects of VTR bandwidths (i.e., formant bandwidths for vocalic sounds) and relative VTR amplitudes on the MFCC values. These effects are secondary to the VTR frequencies but they nevertheless contribute to the variability of MFCCs. Such secondary effects are quantified by the determinant of matrix $\mathbf{R}$, which, in combination with the relative size of the state noise covariance matrix $\mathbf{Q}$, plays important roles in determining relative amounts of state prediction and state update in the state estimation procedure.

In implementing the nonlinear function $h[\mathbf{z}(k)]$ (omitting index r for clarity henceforth) in Eq. 12.6.2, we used an MLP network of three linear input units ($\mathbf{z}(k)$ of F1, F2, and F3), of 100 nonlinear hidden units, and of 12 linear output units ($\mathbf{o}(k)$ of MFCC1-12). Denoting the MLP weights from input to hidden units as w_{jl}, and the MLP weights from hidden to output units as W_{ij}, we have

$$h_i(\mathbf{z}) = \sum_j W_{ij} \cdot g_j(\sum_l w_{jl} \cdot \mathbf{z}_l), \qquad (12.5)$$

where $i = 1, 2, ..., 12$ is the index of output units (i.e., component index of observation vector $\mathbf{o}_k$), $j = 1, 2, ..., 100$ is the index of hidden units, and $l = 1, 2, 3$ is the index of input units. In Eq. 12.5, the hidden units' activation function is the standard sigmoid function

$$g(x) = \frac{1}{1+\exp(-x)},$$

with its derivative

$$g'(x) = g(x)(1 - g(x)).$$

The Jacobian matrix for Eq. 12.5, which will be needed for the Extended Kalman Filter (EKF; see Section 5.8), can be computed in an analytical form:

$$\mathbf{H}_z(\mathbf{z}) \equiv \frac{d}{dZ}h(\mathbf{z}) = [H_{il}(\mathbf{z})] = \begin{pmatrix} \frac{\partial h_1}{\partial \mathbf{z}_1} & \frac{\partial h_1}{\partial \mathbf{z}_2} & \frac{\partial h_1}{\partial \mathbf{z}_3} \\ \frac{\partial h_2}{\partial \mathbf{z}_1} & \frac{\partial h_2}{\partial \mathbf{z}_2} & \frac{\partial h_2}{\partial \mathbf{z}_3} \\ \vdots & \vdots & \vdots \\ \frac{\partial h_{12}}{\partial \mathbf{z}_1} & \frac{\partial h_{12}}{\partial \mathbf{z}_2} & \frac{\partial h_{12}}{\partial \mathbf{z}_3} \end{pmatrix},$$

where

$$H_{il}(\mathbf{z}) = \sum_j \mathbf{W}_{ij} g[\sum_l w_{jl} g(z_l)][1 - g(\sum_l w_{jl} g(z_l))]w_{jl}.$$

12.6.3 Learning model parameters

The learning or parameter estimation method for the new speech model is based on the generalized EM algorithm. The EM algorithm is a two-step iterative scheme for maximum likelihood parameter estimation. Each iteration of the algorithm involves two separate steps, called the Expectation step (E step) and the Maximization step (M step), respectively. A formal introduction of the EM algorithm appeared in [Demp 77]. Examples of using the EM algorithm in speech recognition can be found in [Deng 91c, Deng 92b, Deng 93c, Diga 93]. The algorithm guarantees an increase (or strictly speaking, non-decrease) of the likelihood upon each iteration of the algorithm and guarantees convergence of the iteration to a stationary point for an exponential family. Use of local optimization, rather than the global optimization, in the M step of the algorithm gives rise to the generalized EM algorithm.

To derive the EM algorithm for the new model, we first use the IID noise assumption for $\mathbf{w}_d(k)$ and $\mathbf{v}(k)$ in Eqs. 12.4 and 12.6.2 so as to express the log-likelihood for acoustic observation sequence $\mathbf{o} = [\mathbf{o}(1), \mathbf{o}(2), \dots, \mathbf{o}(N)]$ and hidden task-variable sequence $\mathbf{z} = [\mathbf{z}(1), \mathbf{z}(2), \dots, \mathbf{z}(N)]$ as

$$\begin{aligned}\log L(\mathbf{z}, \mathbf{o}, \Theta) \quad = \quad & -\frac{1}{2}\sum_{k=0}^{N-1}\{\log|\mathbf{Q}| + [\mathbf{z}(k+1) - \mathbf{\Phi}\mathbf{z}(k) - (\mathbf{I} - \mathbf{\Phi})\mathbf{t}]^{\mathrm{Tr}} \\ & \qquad \mathbf{Q}^{-1}[\mathbf{z}(k+1) - \mathbf{\Phi}\mathbf{z}(k) - (\mathbf{I} - \mathbf{\Phi})\mathbf{t}]\} \\ & -\frac{1}{2}\sum_{k=1}^{N}\{\log|\mathbf{R}| + [\mathbf{o}(k) - h(\mathbf{z}(k))]^{\mathrm{Tr}}\mathbf{R}^{-1}[\mathbf{o}(k) - h(\mathbf{z}(k))]\} + \text{constant},\end{aligned}$$

where the model parameters Θ to be learned include those in the state equation Eq. 12.4 and those in the MLP nonlinear mapping functions Eq. 12.6.2: $\Theta = \{\mathbf{t}, \mathbf{\Phi}, \mathbf{W}_{ij}, w_{jl}, i = 1, 2, \dots, I; j = 1, 2, \dots, J; l = 1, 2, \dots, L\}$. (To simplify the algorithm description without loss of generality, estimation of additional model parameters of covariance matrices $\mathbf{Q}, \mathbf{R}$ for state and observation noises will not be addressed in this chapter. Also, the dynamic-regime index on parameters $\mathbf{t}, \mathbf{\Phi}$ and the phone-class index on parameters $\mathbf{W}_{ij}, w_{jl}$ are dropped because supervised learning is used. (In the current model implementation, $I = 3, J = 100, L = 12$.)

The E step

The E step of the EM algorithm involves computation of the following conditional expectation (together with a set of related sufficient statistics needed to complete evaluation of the conditional expectation):

$$\begin{aligned}\mathbf{Q}(\mathbf{z},\mathbf{o},\boldsymbol{\theta}) &= E\{\log L(\mathbf{z},\mathbf{o},\boldsymbol{\theta})|\mathbf{o},\boldsymbol{\theta}\}\\ &= -\frac{N}{2}\log|\mathbf{Q}| - \frac{N}{2}\log|\mathbf{R}|\\ &- \frac{1}{2}\sum_{k=0}^{N-1} E[\mathbf{e}_{k1}^{\mathrm{Tr}}\mathbf{Q}^{-1}\mathbf{e}_{k1}|\mathbf{o},\boldsymbol{\theta}] - \frac{1}{2}\sum_{k=1}^{N} E[\mathbf{e}_{k2}^{\mathrm{Tr}}\mathbf{R}^{-1}\mathbf{e}_{k2}|\mathbf{o},\boldsymbol{\theta}]\end{aligned}$$

where $\mathbf{e}_{k1} = \mathbf{z}(k+1) - \mathbf{\Phi}\mathbf{z}(k) - (\mathbf{I} - \mathbf{\Phi})\mathbf{t}$ and $\mathbf{e}_{k2} = \mathbf{o}(k) - h(\mathbf{z}(k))$, and E denotes conditional expectation on observation vectors $\mathbf{o}$.

This can be simplified by algebraic manipulation to

$$\begin{aligned}Q(\mathbf{z},\mathbf{o},\boldsymbol{\theta}) &= \underbrace{-\frac{N}{2}\log\{|\frac{1}{N}\sum_{k=0}^{N-1} E[\mathbf{e}_{k1}\mathbf{e}_{k1}^{\mathrm{Tr}}|\mathbf{o},\boldsymbol{\theta}]|\}}_{Q_1(\mathbf{z},\mathbf{o},\mathbf{\Phi},\mathbf{t})}\\ & \underbrace{-\frac{N}{2}\log\{|\frac{1}{N}\sum_{k=1}^{N} E[\mathbf{e}_{k2}\mathbf{e}_{k2}^{\mathrm{Tr}}|\mathbf{o},\boldsymbol{\theta}]|\}}_{Q_2(\mathbf{z},\mathbf{o},\mathbf{W}_{ij},w_{jl})} + \mathrm{C}.\end{aligned} \qquad (12.6)$$

Note that the state-equation's parameters $(\mathbf{\Phi}, \mathbf{t})$ are contained in Q_1 only, and the MLP weight parameters (W_{ij}, w_{jl}) in the observation equation are contained in Q_2 only. These two sets of parameters can then be optimized independently in the subsequent M step to be detailed in the next section.

The M step

The M step of the EM algorithm aims at optimizing the Q function in Eq. 12.6 with respect to model parameters $\boldsymbol{\theta}=\{\mathbf{t}, \mathbf{\Phi}, W_{ij}, w_{jl}\}$. For the model at hand, it seeks solutions for

$$\frac{\partial Q_1}{\partial \mathbf{\Phi}} \propto \sum_{k=0}^{N-1} E[\frac{\partial}{\partial \mathbf{\Phi}}\{[\mathbf{z}(k+1) - \mathbf{\Phi}\mathbf{z}(k) - (\mathbf{I} - \mathbf{\Phi})\mathbf{t}]^2\}|\mathbf{o},\boldsymbol{\theta}] = 0 \qquad (12.7)$$

$$\frac{\partial Q_1}{\partial \mathbf{t}} \propto \sum_{k=0}^{N-1} E[\frac{\partial}{\partial \mathbf{t}}\{[\mathbf{z}(k+1) - \mathbf{\Phi}\mathbf{z}(k) - (\mathbf{I} - \mathbf{\Phi})\mathbf{t}]^2\}|\mathbf{o},\boldsymbol{\theta}] = 0 \qquad (12.8)$$

$$\frac{\partial Q_2}{\partial \mathbf{W}_{ij}} \propto \sum_{k=1}^{N} E[\frac{\partial}{\partial \mathbf{W}_{ij}}\{[\mathbf{o}(k) - h(\mathbf{z}(k))]^2\}|\mathbf{o},\boldsymbol{\theta}] = 0 \qquad (12.9)$$

$$\frac{\partial Q_2}{\partial w_{jl}} \propto \sum_{k=1}^{N} E[\frac{\partial}{\partial w_{jl}}\{[\mathbf{o}(k) - h(\mathbf{z}(k))]^2\}|\mathbf{o},\boldsymbol{\theta}] = 0. \qquad (12.10)$$

Eq. 12.7 is a third-order nonlinear algebraic equation (in $\mathbf{\Phi}$ and $\mathbf{t}$), of the following form after some algebraic and matrix-calculus manipulation:

$$N\mathbf{\Phi}\mathbf{t}\mathbf{t}^{\mathrm{Tr}} - \mathbf{\Phi}\mathbf{t}\mathbf{A}^{\mathrm{Tr}} - \mathbf{\Phi}\mathbf{A}\mathbf{t}^{\mathrm{Tr}} - N\mathbf{t}\mathbf{t}^{\mathrm{Tr}} - \mathbf{t}\mathbf{A}^{\mathrm{Tr}} + \mathbf{B}\mathbf{t}^{\mathrm{Tr}} + \mathbf{\Phi}\mathbf{C} - \mathbf{D} = 0, \quad (12.11)$$

where

$$\mathbf{A} = \sum_{k=0}^{N-1} E[\mathbf{z}(k)|\mathbf{o},\boldsymbol{\theta}], \qquad \mathbf{B} = \sum_{k=0}^{N-1} E[\mathbf{z}(k+1)|\mathbf{o},\boldsymbol{\theta}],$$

$$\mathbf{C} = \sum_{k=0}^{N-1} E[\mathbf{z}(k)\mathbf{z}(k)^{\mathrm{Tr}}|\mathbf{o},\boldsymbol{\theta}], \quad \mathbf{D} = \sum_{k=0}^{N-1} E[\mathbf{z}(k+1)\mathbf{z}(k)^{\mathrm{Tr}}|\mathbf{o},\boldsymbol{\theta}].$$

Eq. 12.8 is another third-order nonlinear algebraic equation (in $\mathbf{\Phi}$ and $\mathbf{t}$) of the form:

$$N\mathbf{\Phi}^{\mathrm{Tr}}\mathbf{\Phi}\mathbf{t} - \mathbf{\Phi}^{\mathrm{Tr}}\mathbf{\Phi}\mathbf{A} - N\mathbf{\Phi}^{\mathrm{Tr}}\mathbf{t} - N\mathbf{\Phi}\mathbf{t} + \mathbf{\Phi}^{\mathrm{Tr}}\mathbf{B} + \mathbf{\Phi}\mathbf{A} + N\mathbf{t} - \mathbf{B} = 0. \quad (12.12)$$

The coefficients in Eqs. 12.11 and 12.12, A, B, C, and D, constitute the sufficient statistics, which can be obtained by the standard technique of EKF (see Section 5.8).

Solutions to Eqs. 12.9 and 12.10 for finding (W_{ij}, w_{jl}) to maximize Q_2 in Eq. 12.6 have to rely on approximation due to the complexity in the nonlinear function $h(\mathbf{z})$. The approximation involves first finding estimates of hidden variables $\mathbf{z}(k)$, $\mathbf{z}(k|k)$, via the EKF algorithm. Given such estimates, the conditional expectations in Eqs. 12.9 and 12.10 are approximated to give

$$\frac{\partial Q_2}{\partial \mathbf{W}_{ij}} \propto \sum_{k=1}^{N} [\mathbf{o}(k) - h(\mathbf{z}(k|k))]^{\mathrm{Tr}} \frac{\partial h(\mathbf{z}(k|k))}{\partial \mathbf{W}_{ij}} \quad (12.13)$$

$$\frac{\partial Q_2}{\partial w_{jl}} \propto \sum_{k=1}^{N} [\mathbf{o}(k) - h(\mathbf{z}(k|k))]^{\mathrm{Tr}} \frac{\partial h(\mathbf{z}(k|k))}{\partial w_{jl}}. \quad (12.14)$$

If the estimated state variable, $\mathbf{z}(k|k)$, is treated as the input to the MLP neural network defined in Eq. 12.5, and the observation, $\mathbf{o}(k)$, as the output of the MLP, then the gradients expressed in Eqs. 12.13 and 12.14 are exactly the same as those in the backpropagation algorithm [Bishop 97]. Therefore, the backpropagation algorithm is used to provide the estimates to the W_{ij} and w_{jl} parameters. The local-optimum property of the backpropagation algorithm in this M step makes the learning algorithm described in this section a generalized EM. The approximation used to obtain the gradients in Eqs. 12.13 and 12.14 makes the learning algorithm a pseudo-EM.

Extended Kalman filter for finding sufficient statistics

We have observed that, in the E step derivation of the EM algorithm shown in this section, the objective functions Q_1 and Q_2 in Eq. 12.6 contain a set of conditional expectations as sufficient statistics. These conditional expectations, A, B, C, and D in Eqs. 12.11 and 12.12, need to be computed during the M step of the EM algorithm. The Extended Kalman Filter or EKF algorithm provides a solution to finding these sufficient statistics. Also, as shown in Section 3.1.2, the EKF algorithm is also needed to approximate the gradients in Eqs. 12.9 and 12.10, before the M-step can be formulated as the backpropagation algorithm and can be carried out straightforwardly.

The EKF algorithm gives an approximate minimum-mean-square estimate to the state of a general nonlinear dynamic system. Our speech model discussed in Section

12.7.2 uses a special structure within the general class of the nonlinear dynamic system models. Given such a structure, the EKF algorithm developed is described here in a standard predictor-corrector format [Mend 95].

Denoting by $\hat{\mathbf{z}}(k|k)$ the EKF state estimate and by $\hat{\mathbf{z}}(k+1|k)$ the one-step EKF state prediction, the prediction equation for the special structure of our model has the form

$$\hat{\mathbf{z}}(k+1|k) = \mathbf{\Phi}\hat{\mathbf{z}}(k|k) + (\mathbf{I} - \mathbf{\Phi})\mathbf{t}.$$

The physical interpretation of Eq. 12.6.3 applied to our speech model is that the one-step EKF state predictor based on the current EKF state estimate will always move towards the target vector $\mathbf{t}$ for a given system matrix $\mathbf{\Phi}$. Such desirable dynamics comes directly from state equation Eq. 12.4, and it is, in fact, in exactly the same form as the noise-free model state equation.

Denote by $\mathbf{H}_z(\hat{\mathbf{z}}(k+1|k))$ the Jacobian matrix, as defined in Eq. 12.6.2, at the point of $\hat{\mathbf{z}}(k+1|k)$ for the MLP observation equation in our speech model, and denote by $h(\hat{\mathbf{z}}(k+1|k))$ the MLP output for the input $\hat{\mathbf{z}}(k+1|k)$. Then the EKF corrector (or filter) equation applied to our speech model is

$$\hat{\mathbf{z}}(k+1|k+1) = \hat{\mathbf{z}}(k+1|k) + \mathbf{K}(k+1)\{\mathbf{o}(k+1) - h(\hat{\mathbf{z}}(k+1|k))\},$$

where $\mathbf{K}(k+1)$ is the filter gain computed recursively according to:

$$\begin{aligned} \mathbf{K}(k+1) &= P(k+1|k)\mathbf{H}_z[\hat{\mathbf{z}}(k+1|k)] \\ &\quad \{\mathbf{H}_z[\hat{\mathbf{z}}(k+1|k)]P(k+1|k)\mathbf{H}_z^{\mathrm{Tr}}[\hat{\mathbf{z}}(k+1|k)] + \mathbf{R}(k+1)\}^{-1} \\ P(k+1|k) &= \mathbf{\Phi}P(k|k)\mathbf{\Phi}^{\mathrm{Tr}} + \mathbf{Q}(k+1) \\ P(k+1|k+1) &= \{\mathbf{I} - \mathbf{K}(k+1)\mathbf{H}_z[\hat{\mathbf{z}}(k+1|k)]\}P(k+1|k) \end{aligned} \tag{12.15}$$

In the above, $P(k+1|k)$ is the prediction error covariance and $P(k+1|k+1)$ is the filtering error covariance.

The physical interpretation of Eq. 12.6.3 as applied to the hidden dynamic model is that the amount of correction to the state predictor obtained from Eq. 12.6.3 is directly proportional to the accuracy with which the MLP is used to model the relationship between the state $\mathbf{z}(k)$ or VTRs and the observation $\mathbf{o}(k)$ or MFCCs. Such a matching error in the acoustic domain (called innovation in estimation theory) is magnified by the time-varying filter gain $\mathbf{K}(k+1)$, which is dependent on the balance of the covariances of the two noises $\mathbf{Q}$ and $\mathbf{R}$, and on the local Jacobian matrix which measures the sensitivity of the nonlinear function $h(\mathbf{z})$ represented by the MLP.

In using the EM algorithm to learn the model parameters, we require that all the conditional expectations (sufficient statistics) for the coefficients A, B, C and D in Eq. 12.12 be reasonably accurately evaluated. This can be accomplished, once the EKF's outputs become available, according to

$$\begin{aligned} E[\mathbf{z}(k)|\mathbf{o}] &= \hat{\mathbf{z}}(k|N) \approx \hat{\mathbf{z}}(k|k), \\ E[\mathbf{z}(k+1)|\mathbf{o}] &= \hat{\mathbf{z}}(k+1|N) \approx \hat{\mathbf{z}}(k+1|k+1), \\ E[\mathbf{z}(k)\mathbf{z}^{\mathrm{Tr}}(k)|\mathbf{o}] &= \mathbf{P}(k|N) + \hat{\mathbf{z}}(k|N)\hat{\mathbf{z}}^{\mathrm{Tr}}(k|N) \approx \mathbf{P}(k|k) + \hat{\mathbf{z}}(k|k)\hat{\mathbf{z}}^{\mathrm{Tr}}(k|k), \\ E[\mathbf{z}(k+1)\mathbf{z}(k)^{\mathrm{Tr}}|\mathbf{o}] &= \mathbf{P}(k+1,k|N) + \hat{\mathbf{z}}(k+1|N)\hat{\mathbf{z}}^{\mathrm{Tr}}(k|N) \\ &\approx \mathbf{P}(k+1,k|k+1) + \hat{\mathbf{z}}(k+1|k+1)\hat{\mathbf{z}}^{\mathrm{Tr}}(k|k). \end{aligned}$$

All the quantities on the right-hand sides above are computed directly from the EKF recursion, except for the quantity $\mathbf{P}(k+1, k|k+1)$, which is computed separately according to

$$\mathbf{P}(k+1, k|k+1) = [\mathbf{I} - \mathbf{K}(k+1)\mathbf{H}_z(\hat{\mathbf{z}}(k+1|k+1))]\mathbf{\Phi}\mathbf{P}(k|k).$$

12.6.4 Likelihood-scoring algorithm

In addition to the use of the EKF algorithm in the model learning as discussed so far, it is also needed in the likelihood-scoring algorithm, which we discuss now.

Using the basic estimation theory for dynamic systems (cf. Theorem 25-1 in [Mend 95]; see also in [Diga 93]), the log-likelihood scoring function for our speech model can be computed from the approximate innovation sequence $\tilde{\mathbf{o}}(k|k-1)$ according to

$$\begin{aligned} \log L(\mathbf{o}|\boldsymbol{\theta}) &= -\frac{1}{2}\sum_{k=1}^{N}\{\log|P_{\tilde{\mathbf{o}}\tilde{\mathbf{o}}}(k|k-1)| \\ &+ \tilde{\mathbf{o}}^{\mathrm{Tr}}(k|k-1)P_{\tilde{\mathbf{o}}\tilde{\mathbf{o}}}^{-1}(k|k-1)\tilde{\mathbf{o}}(k|k-1)\} + \text{Constant}, \end{aligned} \tag{12.16}$$

where the approximate innovation sequence

$$\tilde{\mathbf{o}}(k|k-1) = \mathbf{o}(k) - h(\hat{\mathbf{z}}(k|k-1)), \quad k = 1, 2, ..., N$$

is computed from the EKF recursion, and $P_{\tilde{\mathbf{o}}\tilde{\mathbf{o}}}$ is the covariance matrix of the approximate innovation sequence:

$$P_{\tilde{\mathbf{o}}\tilde{\mathbf{o}}}(k|k-1) = \mathbf{H}_z(\hat{\mathbf{z}}(k|k-1))P(k|k-1)\mathbf{H}_z^{\mathrm{Tr}}(\hat{\mathbf{z}}(k|k-1))^{\mathrm{Tr}} + \mathbf{R},$$

which is also computed from the EKF recursion.

For a speech utterance which consists of a sequence of phones with the dynamic regimes given, the log-likelihood scoring functions for each phone in the sequence as defined in Eq. 12.16 are summed to give the total log-likelihood score for the entire speech utterance.

12.6.5 Experiments on spontaneous speech recognition

Experimental paradigm and design parameters

In the experiments described in this section, an N-best list rescoring paradigm, according to the scoring algorithm Eq. 12.16, is used to evaluate the new recognizer based on the VTR-based coarticulatory model on the Switchboard spontaneous speech data. The N-best list of word transcription hypotheses and their phone-level segmentation (i.e., alignment) are obtained from a conventional triphone-based HMM, which also serves as the benchmark to gauge the recognizer performance improvement via use of the new speech model. The reasons for using the limited N-best rescoring paradigm in the current experiments are mainly computational ones.

The benchmark HMM system used in the experiments has word-internal triphones clustered by a decision tree, with a bigram language model. The total number of the parameters in this benchmark HMM system is approximately 3,276,000. This is the product of: 1) 39, which is the MFCC feature vector dimension; 2) 12, which is the number of Gaussian mixtures for each HMM state; 3) 2, which includes Gaussian means and diagonal covariance matrices in each mixture component; and 4) 3,500, which is the total number of the distinct HMM states clustered by the decision tree.

In contrast, the total number of parameters in the new recognizer is considerably smaller. The total 15,252 parameters in the recognizer consists of those from target parameters $42 \times 3 = 126$, those from diagonal dynamic system matrices $42 \times 3 = 126$, and those from MLP parameters $10 \times 100 \times (12 + 3) = 15,000$. These numbers are elaborated below.

First, a total of 42 distinct phone-like symbols are chosen, each of which is intended to be associated with a distinct three-dimensional (F1, F2, and F3) vector-valued target ($\mathbf{t}^j$) in the VTR domain. The phone-like symbol inventory and the VTR target values used to initialize the model parameter learning are based on the Klatt synthesizer setup. The values are slightly adjusted by examining some spectrograms of the Switchboard training data. Among the 42 phone-like symbols, 34 are context-independent. The remaining eight are context-dependent because their target VTRs are affected by the anticipatory tongue position associated with the following phone.

The next set of model parameters is the elements in the 42 distinct diagonal dynamic system matrices ($\mathbf{\Phi}^j$). Before the training, they are initialized based on the consideration that the articulators responsible for producing different phones have different intrinsic movement rates. This difference roughly translates to the difference in the VTR movement rates across the varying phones. For example, the VTR transitions for labial consonants (/b/, /m/, /p/) marked by "Lips" features are significant faster than those for alveolar consonants (/d/, /t/, /n/) marked by "Tongue-Blade" features. The VTR transitions for both labial and alveolar consonants are faster than those for velar consonants (marked by "Tongue Dorsum" features) and than those of vowels marked also by the "Tongue Dorsum" features.

The final set of model parameters in the recognizer are the MLP weights, $\mathbf{W}_{ij}$ and w_{jl}, responsible for the VTR-to-MFCC mapping. Unlike the target and system matrix parameters, which are phone dependent, we tie the MLPs approximately according to the distinct classes of manner of articulation (and voicing). Such tying reduces the MLP noise resulting from otherwise too many independently trained MLPs. On the other hand, by not tying all phones into one single MLP, we also ensure effective discrimination of phones using differential nonlinear mapping (from the smoothed physical VTR state variables to the MFCCs) even if the VTR targets are identical for different phones (a few phones have nearly identical VTR targets). The ten classes resulting from the tying and used in the current recognizer implementation are:

1) aw, ay, ey, ow, oy, aa, ae, ah, ao, ax, ih, iy, uh, uw, er, eh, el;
2) l, w, r, y;
3) f, th, sh;
4) s, ch;
5) v, dh, zh;
6) z, jh;
7) p, t, k;
8) b, d, g;
9) m, n, ng, en;
10) sil, sp.

In the above tying scheme, all vowels are tied using one MLP, because vowel distinction is based exclusively on different target values in the VTR domain. /s/ and /sh/ are associated with separate MLPs, because their target VTR values (not observable in the acoustic domain because of concurrent zeros and large VTR bandwidths) are similar to each other (this can be seen in terms of their similar ways in attracting the VTR (formant) transitions from the adjacent phones.) and hence their distinction will be based

	Benchmark HMM	VTR (HMM align)	VTR (hand align)
% Reference-at-Top	37.0%	38.8%	50.0%
Ave. Word Error Rate	39.2%	30.4%	22.8%

Table 12.1: Performance comparison of three recognizers for 18 utterances with the same speaker in training and testing (ref+5) (after Deng and Ma [Deng 00c], @AIP).

mainly on the different VTR-to-MFCC mappings. In this case, the acoustic difference between these two phones in terms of the greater amount of energy at lower frequency for /sh/ than for /s/ is captured by different MLP weights (which are trained automatically), rather than by differential VTR target values since the behavior of attracting adjacent phones' VTR transitions is similar between /s/ and /sh/.

For each of the ten distinct MLPs, we use 100 (nonlinear) hidden units, three (linear) input units, and 12 (linear) output units. This gives a total of $10 \times 100 \times (3 + 12)$ MLP weight parameters.

Experiments on small-scale N-best rescoring

In this set of experiments, the VTR-based model is trained with the design parameters outlined above using speech data from a single male speaker in the Switchboard data. A total of 30 minutes of the data are used which consist of several telephone conversations. Due to the use of only a single speaker, normalization problems are avoided for both the VTR targets and for the MFCC observations.

18 utterances (sentences) in one conversation are randomly selected as the test data from the same speaker that are disjoint from the training set. For these 18 utterances, an N-best list with $N = 5$ is generated, together with the phone alignments for each of the five-best hypotheses, by the benchmark HMM system. We then add the reference (correct) hypothesis together with its phone alignments to this list, making a total of six ("ref+5") hypotheses to be rescored by the VTR recognizer.

Under the identical conditions set out above, these 18 utterances are scored using the following three recognizers with the language model removed: 1) benchmark triphone HMM; 2) VTR model using automatically computed phone alignments (which determine the VTR dynamic regimes for each constituent phone) by the HMM for all the six hypotheses; and 3) VTR model using manually determined "true" dynamic regimes for the reference hypothesis according to spectrogram reading. A performance comparison of these three recognizers is shown in Table 12.1. Two performance measures are used in this comparison. First, among the 18 test utterances we examine the percentage when the correct, reference hypothesis scores higher than all the remaining five hypotheses. Second, we directly compute the word error rate (WER). The new, VTR-based recognizers are consistently better than the benchmark HMM, especially when the "true" dynamic regimes are provided and in this case the performance is considerably better.

We conduct a similar experiment to the above, using the same recognizers trained from a single male but choosing a separate male speaker's 10 utterances as the test data. Again, as shown in Table 12.2, the VTR-based recognizer with the "true" dynamic regimes gives significantly better performance than the others.

These experiments demonstrate superior performance of the VTR-based coarticulatory model when exposed to the reference transcriptions. They also highlight the

	Benchmark HMM	VTR (HMM align)	VTR (hand align)
% Reference-at-Top	30.0%	40.0%	50.0%
Ave. Word Error Rate	27.0%	25.7%	9.2%

Table 12.2: Performance comparison of three recognizers for 10 utterances with separate speakers in training and testing (ref+5) (after Deng and Ma [Deng 00c], @AIP).

	Benchmark HMM	VTR (HMM align)	Chance
Ave. WER (ref+5)	44.8 %	32.3%	45.0%
Ave. WER (ref+100)	56.1 %	50.2%	59.6%

Table 12.3: Performance comparison of two recognizers for a total of 1,241 test utterances when the recognizers are exposed to the reference transcription (ref+5 and ref+100) (after Deng and Ma [Deng 00c], @AIP).

importance of providing the true or optimal dynamic regimes to the model. Automatic searching for the optimal dynamic regimes is a gigantic computational problem and has not been addressed by the work reported in this chapter.

Experiments on large-scale N-best rescoring

In the large-scale experiments, the same recognizers trained from a single male speaker are kept but the size of the test set is significantly increased. All the male speakers from the test set are selected, resulting in a total of 23 male speakers comprising 24 conversation sides, 1241 utterances (sentences), 9970 words, and 50 minutes of speech as the test data. Because of the large test set and because of a lack of an efficient method to automate the optimization of the VTR-model dynamic regimes, we report in this section only the performance comparison between the benchmark HMM recognizer and the VTR recognizer with dynamic regimes suboptimally derived from the HMM phone alignments. In Table 12.3, we provide the performance comparison for the "ref+5" mode, and for the additional "ref+100" mode where the N-best list contains 100 hypotheses. In Table 12.3, we also add the "Chance" performance, which is used to calibrate the recognizers' performance. The chance WER is computed by ensemble averaging the WERs obtained by having a recognizer randomly choosing one hypothesis from the six possible ones (for the ref+5 mode or $N = 5$) or from the 101 possible ones (for the ref+100 mode or $N = 100$). For both the $N = 5$ and $N = 100$ cases, the VTR recognizer performs significantly better than the benchmark HMM recognizer, which is slightly better than the chance performance.

More detailed results of the above experiment for the VTR recognizer are shown in Table 12.4, where the average WER is shown as a function of N in the N-best list.

The same experiment as shown in Table 12.3 was conducted except no reference transcription was added to the N-best list. The results are shown in Table 12.5, with $N = 5$ and $N = 100$ in the N-best list, respectively. In both cases, the VTR recognizer performs nearly the same as the chance one, both slightly worse than the benchmark HMM

N	1	2	3	4	7	10	20
WER%	20.5	26.3	29.3	31.2	34.5	36.1	40.6
N	30	40	50	60	70	80	90
WER	43.3	44.6	46.4	47.7	48.5	49.5	50.1

Table 12.4: VTR recognizer's average WER% as a function of N in the N-best list (ref+N) (after Deng and Ma [Deng 00c], @AIP).

	Benchmark HMM	VTR (HMM align)	Chance
Ave. WER (5-best)	52.6 %	54.1%	54.0%
Ave. WER (100-best)	58.9 %	60.1%	60.2%

Table 12.5: Performance comparison of two recognizers for a total of 1,241 utterances when the recognizers are not exposed to the reference transcription (5-best and 100-best) (after Deng and Ma [Deng 00c], @AIP).

recognizer. This contrasts sharply with the superior performance of the VTR recognizer when it is exposed to the reference transcription shown in Tables 12.1-12.3. A reasonable explanation is that the long-span context-dependence property of the VTR model naturally endows the model with the capability to "lock-in" to the correct transcription and it at the same time increases the tendency for the model to "break-away" from partially correct transcriptions due to the influence of wrong contexts. Since nearly all the hypotheses in the N-best list contain a large proportion of incorrect words, they affect the matching of the model to the remaining correct words in the hypotheses through a context-dependence mechanism much stronger than the conventional triphone HMM.

12.7 Case Study II: Speech Recognition Using HMMs Structured by Locus Equations

In this section, we provide the second case study on a special type of HMM whose parameters are constrained to reflect some well understood acoustic-phonetic patterns of speech known as locus equations. The purpose of presenting this case study is to show how a basic understanding of the speech process can lead to better statistical models of speech within the HMM framework, and what the steps are involved to develop such models. This case study is based on the work published in [Deng 94a].

12.7.1 Model overview

It has been demonstrated that there is a great deal of rich phonetically discriminative information contained in the speech signal in the form of spectrogram without the need of higher-level knowledge (e.g., [Cole 90, Zue 91]). One particularly useful piece of such information which has been frequently exploited in spectrogram reading is the context (place of articulation) dependent trends in formant transitions associated with vocalic segments [Zue 91]. However, phonetic hypotheses made by human spectrogram readers have been largely qualitative in nature; in particular, no satisfactory analytical methods have been devised in the past to explicitly take advantage of the context-dependent formant transition information for the purpose of phonetic classification by computers.

The aim of this case study is to develop a quantitative model for the effect of formant transitions on the context dependence of phonetic units of speech. The model described here has been placed within the computational framework of HMM. This allows one to express qualitative reasoning, with varying degrees of uncertainty, used in spectrogram reading by means of a rigorous statistical language, leading to a model amenable to mathematical optimization.

The theoretical guidance for the development of the model described here is the concept of the acoustic "loci" and of the locus equations established from speech science. One early attempt which capitalizes on the classical Haskins "locus" concept [Dela 55] to impose constraints on the HMM state parameters was reported in [Deng 92e]. The very small improvement in the speech recognition rate reported in that study was disappointing, on the one hand, given the significant role of the acoustic transition on place classification demonstrated in both human speech perception and speech spectrogram reading experiments. On the other hand, such results point to possible inadequacy of the classic Haskins "locus" concept insofar as the concept was developed largely from speech synthesis experiments rather than from actual acoustic measurements.

Comprehensive acoustic measurements aimed at quantifying formant transition patterns in relation to vowel formant values have been available over the last twenty years [Sus 91, Sus 97, Near 87, Kew 82]. Studies reported in [Sus 97] provided negative support to the classic "locus" idea — the "locus" was only marginally present for labials and the palatal "locus" values overlapped significantly with the alveolar "locus" values. In [Lind 63, Near 87, Sus 91], the classic "locus" concept is generalized to that of "locus equations", which describe linear relationships between the formants at acoustic onset of the voiced stop consonant-vowel (CV) syllables and those at the midvowel nuclei. It is shown in [Sus 91] with a large quantity of acoustic data that the locus-equation slopes and intercepts are sufficiently separate for different places of articulation regardless of the vowel context. This conclusion has later been extended from stop CVs to other manner classes including nasals, fricatives, approximants, and unaspirated voiceless stops.

In light of the evidence in support of the relational invariant acoustic property expressed concisely by the locus equation and in light of the universally recognized role of formant transitions in auditory and visual (spectrogram reading) decoding of consonantal places of articulation, in this case study one asks the question as to whether speech recognition algorithms programmed in a computer can make effective use of such a role. Presented in this section is the development and evaluation of an HMM-based speech recognition system (called Locus-HMM for short) which capitalizes on the conceptualization of the locus equations as a basis for parametric modeling of phonetic contexts. The Locus-HMM is capable of abstracting the relational invariance (invariant with respect to the consonantal place; relational with respect to the vowel context) expressed in terms of slope and intercept parameters in the locus equation. This capability allows the model to generalize consonantal place features from a small amount of training data where the contextual information is only sparsely represented.

12.7.2 Model formulation

After motivating the use of locus equations as a quantitative measure of relationships between formant frequencies at the acoustic onset of CV syllables and those at the midvowel nuclei, the Locus-HMM is developed to incorporate this relationship as a constraint into the HMM structure. One can anticipate that use of this additional piece of information can lead to a more accurate and realistic model for speech acoustics than the one free from the locus-equation constraint (as in the conventional HMM).

To formulate the Locus-HMM, we consider a $2Q+1$-state continuous-density HMM, shown in Fig. 12.1, which represents the acoustic signal of a vocalic segment in a CVC environment. The first Q states represent the CV acoustic transition induced from the left consonantal context, followed by the Q-th state representing the acoustics of a possible vowel steady-state portion (which may be skipped for reduced and short vowels). The remaining Q states finish up the acoustic transition of the VC region.

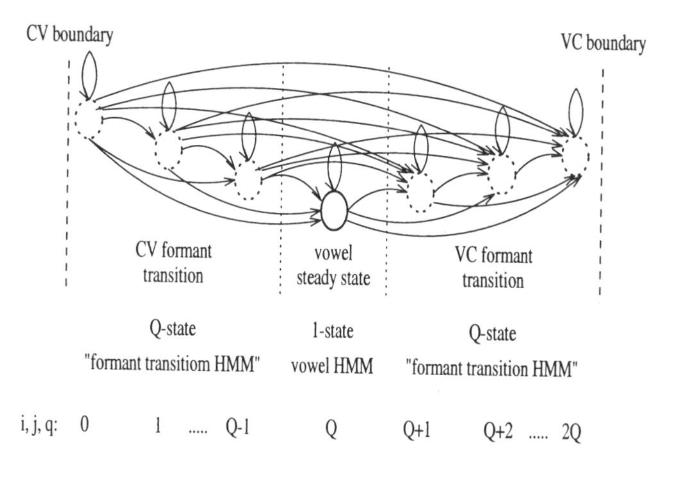

Figure 12.1: State diagram of the Locus-HMM with a total of $2Q+1$ states (after Deng and Braam [Deng 94a], @AIP).

The essence of the Locus-HMM, unlike the conventional HMM, is that the parameters in the Gaussian distributions associated with each state in the HMM shown in Fig. 12.1 are no longer independent of each other. Instead, the constraint imposed by the locus equation significantly reduces the degree of freedom in the possible parameter variations in the HMM. The locus equation, as supported by the evidence reported in [Near 87, Sus 91, Sus 97], states simply (in terms of our current HMM terminology) that the Gaussian mean of either the first state (μ_0) or the last state (μ_{2Q}) and the Gaussian mean of the Q-th state (μ_v) follow a linear relationship:

$$CV: \quad \mu_0 = m_c * \mu_v + b_c \tag{12.17}$$

$$VC: \quad \mu_{2Q} = m_{c'} * \mu_v + b_{c'} \tag{12.18}$$

where v is the label for a vowel, and c or c' refers to the left or right consonantal context. (Note that the first state in the Locus-HMM represents the acoustics near the CV syllable onset, and the last state represents the acoustics near the VC syllable offset.) m and b are the slope and intercept parameters of the locus equation specific to a consonant.

A further set of constraints are imposed in the Locus-HMM which are based on the assumption that the acoustic transitions are linear. This assumption translates to the linear interpolation among the Gaussian means of the transitional states in the Locus-

HMM:

$$\mu_q = \begin{cases} (1 - \frac{q}{Q})\mu_{cv} + \frac{q}{Q}\mu_v, & 0 \le q \le Q-1 \\ \mu_v, & q = Q \\ (1 - \frac{2Q-q}{Q})\mu_{vc} + \frac{2Q-q}{Q}\mu_v, & Q+1 \le q \le 2Q. \end{cases} \tag{12.19}$$

Integrating the locus-equation constraints Eqs. 12.17 and 12.18 into the interpolation constraint Eq. 12.19 gives us the following formulae governing the Gaussian mean of any state q in the Locus-HMM in terms of the model parameters m_c, b_c, and μ_v:

$$\mu_q = \begin{cases} (1 - \frac{q}{Q})m_c\mu_v + \frac{q}{Q}\mu_v + (1 - \frac{q}{Q})b_c, & 0 \le q \le Q-1 \\ \mu_v, & q = Q \\ (1 - \frac{2Q-q}{Q})m_c\mu_v + \frac{2Q-q}{Q}\mu_v + (1 - \frac{2Q-q}{Q})b_c, & Q+1 \le q \le 2Q. \end{cases} \tag{12.20}$$

In summary, the Locus-HMM formally consists of the following set of parameters:

$$\mathbf{\Phi} = \{\mu_v, \sigma_v, m_c, b_c, a_{ij}^v, \sigma_q^c\},$$

where

- μ_v is the mean in the Gaussian distribution associated with the HMM state describing mid-vowel acoustics for vowel $v, 1 \le v \le V$
- σ_v is the standard deviation in the Gaussian distribution associated with the HMM state describing mid-vowel acoustics for vowel $v, 1 \le v \le V$
- m_c is the locus equation slope specific to consonant $c, 1 \le c \le C$
- b_c is the locus equation intercept specific for consonant $c, 1 \le c \le C$
- a_{ij}^v is the transition probability of the HMM from state i to state j specific for each vowel $v, 1 \le v \le V$
- σ_q^c is the standard deviation in the Gaussian distribution associated with state q in the formant transition region for consonant c, $1 \le c \le C$

and V and C are the total number of distinct vowels and consonants in the vocabulary, respectively.

The transition probabilities $\{a_{ij}^v\}$ listed above are similar to those of the conventional HMM, in that they are constrained so that the Markov chain moves from left to right only. The transition probability matrix in our model is made specific to each vowel (a_{ij}^v is a function of vowel v). The parameter, σ_q^c, measures variation of the acoustic data associated with the state q of the Locus-HMM specific to each consonantal context c.

12.7.3 Learning locus-HMM parameters

The current success of conventional HMM-based speech recognition has been attributed to the automatic and effective methods for parameter learning [Rab 89]. In this case study, these learning methods are extended to the Locus-HMM, allowing the locus-equation parameters as well as other HMM parameters to be automatically determined from a small amount of training data in an optimal manner.

A Locus-HMM is fully specified with the parameter set: $\{\mu_v, \sigma_v, m_c, b_c, a_{ij}^v, \sigma_q^c\}$. We now first present a method for producing good initial estimates for these parameters, followed by detailed derivation of the re-estimation formulae based upon the EM algorithm, which improves upon these initial estimates. For simplicity of exposition, we assume that the observation vectors to be modeled are one-dimensional. For the multidimensional observation case, the derivation can be extended in a straightforward manner.

Initial parameter estimation

The success of the iterative EM algorithm depends largely on the quality of the initial model parameters. Arriving at good initial estimates for the various model parameters is generally an *ad hoc* process. For the Locus-HMM described in this case study, the model initialization methods are described below.

1. a_{ij}^v

The transition probabilities of our left-to-right Locus-HMM are initialized to be equally probable over the possible state transitions:

$$a_{ij}^v = \frac{1}{2Q+1-(i-1)}, \qquad 1 \le i \le 2Q+1, \quad i \le j \le 2Q+1, \quad 1 \le v \le V$$

2. μ_v, σ_v, **and** σ_q^c

Initial estimates for μ_v, σ_v, and σ_q^c are obtained using the procedure outlined below:

1. Determine the total number of states in the Locus-HMM. For CVC utterances, such a model will include two Q-state transitional regions and a 1-state mid-vowel region, with a total of $2Q+1$ states.
2. Allocate a pair of accumulators ($E_v[O_t]$, $E_v[O_t^2]$) for each vowel in the vocabulary, and Q pairs of accumulators ($E_{c,q}[O_t]$, $E_{c,q}[O_t^2]$) for each consonant in the vocabulary.
3. For each CVC token in the training data:
   (a) Determine the number of frames in the token.
   (b) According the phoneme sequence that makes up the token, generate a temporary HMM with each state i containing two temporary accumulators ($E_i[O_t]$, $E_i[O_t^2]$).
   (c) Assign an equal number of observation vectors to each state in the temporary HMM, with the "earliest" observation vectors assigned to the leftmost states and the "latest" observation vectors assigned to the rightmost states.
   (d) For each state i in the HMM, accumulate $E_i[O_t]$ and $E_i[O_t^2]$.
   (e) Add the results of (d) to the appropriate global accumulators.
4. After cumulative approximations of $E[O_t]$ and $E[O_t^2]$ for every consonant and vowel are obtained in step 3, estimate μ_v, σ_v, and σ_q^c according to

$$\begin{aligned}
\mu_v &= E_v[O_t], & 1 \le v \le V \\
\sigma_v &= \sqrt{E_v[O_t^2] - E_v[O_t]^2}, & 1 \le v \le V \\
\sigma_q^c &= \sqrt{E_{c,q}[O_t^2] - E_{c,q}[O_t]^2}, & 1 \le c \le C, \quad 1 \le q \le Q.
\end{aligned}$$

3. m_c and b_c

One appropriate method for initialization of the slope and intercept parameters would be to take laborious measurements on the acoustic data at syllable onsets and mid-vowel points and then perform linear regression. However, for the results reported in this section, the following simple initialization has been used for efficiency purposes:

$$m_c = 1.0, \qquad 1 \le c \le C$$
$$b_c = 0.0, \qquad 1 \le c \le C.$$

The EM algorithm for parameter re-estimation

Each iteration of the EM algorithm consists of two steps, the E (expectation) step and the M (maximization) step. In the E step, we need to evaluate the following conditional expectation:

$$Q(\mathbf{\Phi}|\mathbf{\Phi}_0) = E[\log P(O_1^T, \mathcal{S}|\mathbf{\Phi})|O_1^T, \mathbf{\Phi}_0], \tag{12.21}$$

where the expectation is taken over the "hidden" state sequence $\mathcal{S}$, $\mathbf{\Phi}$ refers to the complete set of the Locus-HMM parameters which are to be re-estimated during the current iteration, $\mathbf{\Phi}_0$ refers to the set of the Locus-HMM parameters that were estimated during the previous iteration (O_1^T denotes the acoustic observation sequences, such as formant frequencies or cepstral coefficients, of length T).

1. The E step

The Q function in Eq. 12.21 can be rewritten in terms of the expectation over state sequences $\mathcal{S}$ in the form of a weighted sum

$$Q(\mathbf{\Phi}|\mathbf{\Phi}_0) = \sum_{tokens} \sum_{\mathcal{S}} P(\mathcal{S}|O_1^T, \mathbf{\Phi}_0) \log P(O_1^T, \mathcal{S}|\mathbf{\Phi}). \tag{12.22}$$

Through a set of well established procedures described in Sections 5.6 and 10.2, in the E step, we can simplify Eq. 12.22 for the Locus-HMM into a form:

$$Q(\mathbf{\Phi}|\mathbf{\Phi}_0) = Q_1(\mathbf{\Phi}|\mathbf{\Phi}_0) + Q_2(\mathbf{\Phi}|\mathbf{\Phi}_0), \tag{12.23}$$

where

$$\begin{aligned} Q_1 &= \sum_{tok} \sum_{i=0}^{2Q} \sum_{j=0}^{2Q} \sum_{t=1}^{T-1} \xi_t(i,j) \log\ a_{ij}; \\ Q_2 &= \sum_{tok} \sum_{q=0}^{2Q} \sum_{t=1}^{T} \gamma_t(q)[-\frac{1}{2}\log(2\pi) - \log(\sigma_q) - \frac{1}{2}\frac{(O_t - \mu_q)^2}{\sigma_q^2}]. \end{aligned}$$

In Eq. 12.24, the weights $\gamma_t(q)$ and $\xi_t(i,j)$ can be calculated as part of the standard forward-backward algorithm.

2. The M step

Re-estimation of the Locus-HMM parameters involves maximization of the two quantities $Q_1(\mathbf{\Phi}|\mathbf{\Phi}_0)$ and $Q_2(\mathbf{\Phi}|\mathbf{\Phi}_0)$ separately. Throughout this section, one slight notation change will be made for clarity: the summation $\sum_{tok}$ will be replaced with the double summation $\sum_{c=1}^{C} \sum_{v=1}^{V}$, where C and V are the total number of consonants and vowels making up the training tokens. This change is made because all the model parameters are a function of either a consonant or a vowel, but not both — a requirement that is key to the parsimony of our context-dependent Locus-HMM. The maximization procedure for $Q_1(\mathbf{\Phi}|\mathbf{\Phi}_0)$ is identical to that of the conventional HMM and will not be reproduced here. The final re-estimation formula for the transition probabilities is

$$a_{ij}^v = \frac{\sum_{c=1}^{C}\sum_{t=1}^{T-1}\xi_t(i,j)}{\sum_{c=1}^{C}\sum_{t=1}^{T-1}\gamma_t(i)}, \quad 0 \le i \le 2Q, \quad i \le j \le 2Q, \quad 1 \le v \le V. \tag{12.24}$$

Re-estimation of the remaining model parameters via maximization of $Q_2(\mathbf{\Phi}|\mathbf{\Phi}_0)$ is described in detail below. Referring to the constraint expressed in Eq. 12.20, we note that the constraint has been imposed only on μ_v, m_c, and b_c. The parameters σ_v and σ_q^c are free of the constraint and hence can be re-estimated using the conventional formulae [Baum 72]:

$$\sigma_v = \sqrt{\frac{\sum_{c=1}^{C}\sum_{t=1}^{T}\gamma_t(Q)(O_t - \mu_v)^2}{\sum_{c=1}^{C}\sum_{t=1}^{T}\gamma_t(Q)}}, \quad 1 \le v \le V, \tag{12.25}$$

and

$$\sigma_q^c = \sqrt{\frac{\sum_{v=1}^{V}\sum_{t=1}^{T}\gamma_t(q)(O_t - \mu_q)^2}{\sum_{c=1}^{C}\sum_{t=1}^{T}\gamma_t(q)}}, \quad 0 \le q \le 2Q, \quad q \ne Q, \quad 1 \le c \le C. \tag{12.26}$$

To pursue re-estimation for parameters μ_v, m_c, and b_c, we first reduce the objective function in Eq. 12.24 to an equivalent form:

$$Q_e(\mathbf{\Phi}|\mathbf{\Phi}_0) = \sum_{c=1}^{C}\sum_{v=1}^{V}\sum_{q=0}^{2Q}\sum_{t=1}^{T}\gamma_t(q)(O_t - \mu_q)^2, \tag{12.27}$$

where μ_q is as defined in Eq. 12.20. Minimization of Q_e in Eq. 12.27 requires solution of the joint system of equations:

$$\frac{\partial Q_e}{\partial \mu_v} = 0, \quad \frac{\partial Q_e}{\partial m_c} = 0, \quad \frac{\partial Q_e}{\partial b_c} = 0, \qquad 1 \le c \le C, \quad 1 \le v \le V. \tag{12.28}$$

Carrying out the differentiations and substituting the constraint from Eq. 12.20 yields

$$\begin{aligned}
\frac{\partial Q_e}{\partial \mu_v} &= \sum_{c=1}^{C}\sum_{t=1}^{T}\sum_{q=0}^{Q-1}\gamma_t(q)(O_t - (1-\frac{q}{Q})m_c\mu_v - \frac{q}{Q}\mu_v - (1-\frac{q}{Q})b_c)((1-\frac{q}{Q})m_c + \frac{q}{Q}) \\
&+ \sum_{c=1}^{C}\sum_{t=1}^{T}\gamma_t(Q)(O_t - \mu_v) \\
&+ \sum_{c=1}^{C}\sum_{t=1}^{T}\sum_{q=Q+1}^{2Q}\gamma_t(q)(O_t - (1-\frac{2Q-q}{Q})m_c\mu_v - \frac{2Q-q}{Q}\mu_v - (1-\frac{2Q-q}{Q})b_c) \\
&\cdot ((1-\frac{2Q-q}{Q})m_c + \frac{2Q-q}{Q}) = 0, \qquad 1 \le v \le V
\end{aligned} \tag{12.29}$$

$$\begin{aligned}
\frac{\partial Q_e}{\partial m_c} &= \sum_{v=1}^{V}\sum_{t=1}^{T}\sum_{q=0}^{Q-1}\gamma_t(q)(O_t-(1-\frac{q}{Q})m_c\mu_v-\frac{q}{Q}\mu_v-(1-\frac{q}{Q})b_c)(1-\frac{q}{Q})\mu_v \\
&+ \sum_{v=1}^{V}\sum_{t=1}^{T}\sum_{q=Q+1}^{2Q}\gamma_t(q)(O_t-(1-\frac{2Q-q}{Q})m_c\mu_v-\frac{2Q-q}{Q}\mu_v-(1-\frac{2Q-q}{Q})b_c) \\
&\quad\cdot(1-\frac{2Q-q}{Q})\mu_v=0, \qquad 1\le c\le C \qquad (12.30)
\end{aligned}$$

$$\begin{aligned}
\frac{\partial Q_e}{\partial b_c} &= \sum_{v=1}^{V}\sum_{t=1}^{T}\sum_{q=0}^{Q-1}\gamma_t(q)[O_t-(1-\frac{q}{Q})m_c\mu_v-\frac{q}{Q}\mu_v-(1-\frac{q}{Q})b_c](1-\frac{q}{Q}) \\
&+ \sum_{v=1}^{V}\sum_{t=1}^{T}\sum_{q=Q+1}^{2Q}\gamma_t(q)[O_t-(1-\frac{2Q-q}{Q})m_c\mu_v-\frac{2Q-q}{Q}\mu_v- \\
&\quad(1-\frac{2Q-q}{Q})b_c]\cdot(1-\frac{2Q-q}{Q})=0, \quad 1\le c\le C. \qquad (12.31)
\end{aligned}$$

In theory, Eqs. 12.29, 12.30, and 12.31 can be solved simultaneously to produce the re-estimated values of μ_v, m_c, and b_c that we seek; it is a matter of solving $(2C+V)$ equations in $(2C+V)$ unknowns. However, in practice the system of equations is nonlinear and would generally require complex numerical procedures to solve.

To turn the above nonlinear problem into a linear problem while keeping theoretical rigor, we invoke the Generalized EM algorithm [Demp 77, Wu 83]. (The difference between the conventional EM algorithm and the Generalized EM algorithm lies in the M step: for the former, a global maximization involving all of the model parameters is required, whereas for the latter only a partial optimization — fixing a subset of parameters and optimizing the remaining — is needed.)

Using the Generalized EM algorithm, a linear set of equations for re-estimating μ_v, m_c, and b_c can be obtained by dividing the re-estimation process into a three-iteration cycle. On the first iteration of the cycle, the variance parameters σ_v and σ_q^c are re-estimated using equations Eq. 12.25 and Eq. 12.26 while the μ_v, m_c, and b_c parameters are held fixed at their current values $\overline{\mu_v}$, $\overline{m_c}$, and $\overline{b_c}$; this allows us to use the objective function Q_e in Eq. 12.27 for re-estimating the remaining parameters in the following iterations. On the second iteration of the cycle, the σ_v, σ_q^c, and μ_v parameters are fixed at their current values $\overline{\sigma_v}$, $\overline{\sigma_q^c}$, and $\overline{\mu_v}$, and Eqs. 12.30 and 12.31 are jointly solved for m_c and b_c; the resulting $2C$ equations in $2C$ unknowns are easily solved using linear methods. Similarly, on the third iteration of the cycle, the σ_v, σ_q^c, m_c, and b_c parameters are fixed at their current values $\overline{\sigma_v}$, $\overline{\sigma_q^c}$, $\overline{m_c}$, and $\overline{b_c}$, and Eq. 12.29 is solved for μ_v. This results in a system of V equations in V unknowns, which can also be solved using linear methods. In this way, linear re-estimation formulae for μ_v, m_c, and b_c can be obtained.

We now proceed to derive the re-estimation formulae for the parameters m_c and b_c. Setting $\mu_v=\overline{\mu_v}$ in Eq. 12.30 and rearranging terms yields:

$$
\begin{aligned}
& \sum_{v=1}^{V}\sum_{t=1}^{T}\sum_{q=0}^{Q-1}\gamma_t(q)(1-\frac{q}{Q})\overline{\mu_v}(O_t-\frac{q}{Q}\overline{\mu_v}) \\
+ & \sum_{v=1}^{V}\sum_{t=1}^{T}\sum_{q=Q+1}^{2Q}\gamma_t(q)(1-\frac{2Q-q}{Q})\overline{\mu_v}(O_t-(\frac{2Q-q}{Q})\overline{\mu_v}) \\
= & \sum_{v=1}^{V}\sum_{t=1}^{T}\sum_{q=0}^{Q-1}\gamma_t(q)(1-\frac{q}{Q})^2\overline{\mu_v}^2 m_c+\sum_{v=1}^{V}\sum_{t=1}^{T}\sum_{q=Q+1}^{2Q}\gamma_t(q)(1-\frac{2Q-q}{Q})^2\overline{\mu_v}^2 m_c \\
+ & \sum_{v=1}^{V}\sum_{t=1}^{T}\sum_{q=0}^{Q-1}\gamma_t(q)(1-\frac{q}{Q})^2\overline{\mu_v}b_c \\
+ & \sum_{v=1}^{V}\sum_{t=1}^{T}\sum_{q=Q+1}^{2Q}\gamma_t(q)(1-\frac{2Q-q}{Q})^2\overline{\mu_v}b_c, \qquad 1\le c\le C. \qquad (12.32)
\end{aligned}
$$

Define:

$$
\begin{aligned}
D_c &= \sum_{v=1}^{V}\sum_{t=1}^{T}\sum_{q=0}^{Q-1}\gamma_t(q)(1-\frac{q}{Q})\overline{\mu_v}(O_t-\frac{q}{Q}\overline{\mu_v}) \\
&+ \sum_{v=1}^{V}\sum_{t=1}^{T}\sum_{q=Q+1}^{2Q}\gamma_t(q)(1-\frac{2Q-q}{Q})\overline{\mu_v}(O_t-(\frac{2Q-q}{Q})\overline{\mu_v}) \\
E_c &= \sum_{v=1}^{V}\sum_{t=1}^{T}\sum_{q=0}^{Q-1}\gamma_t(q)(1-\frac{q}{Q})^2\overline{\mu_v}^2+\sum_{v=1}^{V}\sum_{t=1}^{T}\sum_{q=Q+1}^{2Q}\gamma_t(q)(1-\frac{2Q-q}{Q})^2\overline{\mu_v}^2 \\
F_c &= \sum_{v=1}^{V}\sum_{t=1}^{T}\sum_{q=0}^{Q-1}\gamma_t(q)(1-\frac{q}{Q})^2\overline{\mu_v}+\sum_{v=1}^{V}\sum_{t=1}^{T}\sum_{q=Q+1}^{2Q}\gamma_t(q)(1-\frac{2Q-q}{Q})^2\overline{\mu_v},
\end{aligned}
$$

then Eq. 12.32 can be rewritten as

$$
D_c = E_c m_c + F_c b_c, \qquad 1\le c\le C. \qquad (12.33)
$$

Following the same procedure for Eq. 12.31, we obtain:

$$
\begin{aligned}
& \sum_{v=1}^{V}\sum_{t=1}^{T}\sum_{q=0}^{Q-1}\gamma_t(q)(1-\frac{q}{Q})(O_t-\frac{q}{Q}\overline{\mu_v}) \\
+ & \sum_{v=1}^{V}\sum_{t=1}^{T}\sum_{q=Q+1}^{2Q}\gamma_t(q)(1-\frac{2Q-q}{Q})(O_t-(\frac{2Q-q}{Q})\overline{\mu_v})
\end{aligned}
\qquad (12.34)
$$

$$= \sum_{v=1}^{V}\sum_{t=1}^{T}\sum_{q=0}^{Q-1}\gamma_t(q)(1-\frac{q}{Q})^2\overline{\mu_v}m_c + \sum_{v=1}^{V}\sum_{t=1}^{T}\sum_{q=Q+1}^{2Q}\gamma_t(q)(1-\frac{2Q-q}{Q})^2\overline{\mu_v}m_c$$
$$+ \sum_{v=1}^{V}\sum_{t=1}^{T}\sum_{q=0}^{Q-1}\gamma_t(q)(1-\frac{q}{Q})^2 b_c$$
$$+ \sum_{v=1}^{V}\sum_{t=1}^{T}\sum_{q=Q+1}^{2Q}\gamma_t(q)(1-\frac{2Q-q}{Q})^2 b_c, \qquad 1 \le c \le C. \tag{12.35}$$

Further define:

$$G_c = \sum_{v=1}^{V}\sum_{t=1}^{T}\sum_{q=0}^{Q-1}\gamma_t(q)(1-\frac{q}{Q})(O_t - \frac{q}{Q}\overline{\mu_v}) + \sum_{v=1}^{V}\sum_{t=1}^{T}\sum_{q=Q+1}^{2Q}\gamma_t(q)(1-\frac{2Q-q}{Q})(O_t - (\frac{2Q-q}{Q})\overline{\mu_v})$$

$$H_c = \sum_{v=1}^{V}\sum_{t=1}^{T}\sum_{q=0}^{Q-1}\gamma_t(q)(1-\frac{q}{Q})^2 + \sum_{v=1}^{V}\sum_{t=1}^{T}\sum_{q=Q+1}^{2Q}\gamma_t(q)(1-\frac{2Q-q}{Q})^2.$$

We then rewrite Eq. 12.35 as:

$$G_c = F_c m_c + H_c b_c, \qquad 1 \le c \le C. \tag{12.36}$$

We now have $2C$ linear equations (Eqs. 12.33 and 12.36) for $2C$ unknowns ($m_c, b_c, 1 \le c \le C$). The solution of these unknowns has the following closed form:

$$b_c = \frac{D_c F_c - E_c G_c}{F_c^2 - E_c H_c}, \qquad 1 \le c \le C \tag{12.37}$$

and

$$m_c = \frac{D_c - F_c b_c}{E_c}, \qquad 1 \le c \le C. \tag{12.38}$$

The final parameters that need to be re-estimated are the mid-vowel Gaussian means, μ_v. This is accomplished by substituting $m_c = \overline{m_c}$, $b_c = \overline{b_c}$, and solving Eq. 12.29 for μ_v. Rearranging terms in Eq. 12.38 after the above substitution yields:

$$\sum_{c=1}^{C}\sum_{t=1}^{T}\sum_{q=0}^{Q-1}\gamma_t(q)((1-\frac{q}{Q})\overline{m_c} + \frac{q}{Q})(O_t - (1-\frac{q}{Q})\overline{b_c}) + \sum_{c=1}^{C}\sum_{t=1}^{T}\gamma_t(Q)O_t$$
$$+ \sum_{c=1}^{C}\sum_{t=1}^{T}\sum_{q=Q+1}^{2Q}\gamma_t(q)((1-(\frac{2Q-q}{Q})\overline{m_c} + \frac{2Q-q}{Q})(O_t - (1-\frac{2Q-q}{Q})\overline{b_c})$$
$$= \sum_{c=1}^{C}\sum_{t=1}^{T}\sum_{q=0}^{Q-1}\gamma_t(q)((1-\frac{q}{Q})\overline{m_c} + \frac{q}{Q})^2\mu_v + \sum_{c=1}^{C}\sum_{t=1}^{T}\gamma_t(Q)\mu_v$$
$$+ \sum_{c=1}^{C}\sum_{t=1}^{T}\sum_{q=Q+1}^{2Q}\gamma_t(q)((1-\frac{2Q-q}{Q})\overline{m_c} + \frac{2Q-q}{Q})^2\mu_v, \quad 1 \le v \le V. \tag{12.39}$$

Place of Articulation	Consonant	Slope	Intercept (Hz)
Labial	/b/	0.94	-210
	/p/	0.88	-109
Alveolar	/d/	0.10	1558
	/t/	0.14	1527
Velar	/g/	0.89	530
	/k/	3.64	-3654
Palatal	/g/	1.08	190
	/k/	0.71	764

Table 12.6: Locus equation parameters learned by the algorithm of Section II using whole F2 trajectories with no detection of syllable onsets and mid-vowel points (after Deng and Braam [Deng 94a], @AIP).

Finally, set

$$\begin{aligned} num_v &= \sum_{c=1}^{C}\sum_{t=1}^{T}\sum_{q=0}^{Q-1}\gamma_t(q)((1-\frac{q}{Q})\overline{m_c}+\frac{q}{Q})(O_t-(1-\frac{q}{Q})\overline{b_c})+\sum_{c=1}^{C}\sum_{t=1}^{T}\gamma_t(Q)O_t \\ &+ \sum_{c=1}^{C}\sum_{t=1}^{T}\sum_{q=Q+1}^{2Q}\gamma_t(q)((1-\frac{2Q-q}{Q})\overline{m_c}+\frac{2Q-q}{Q})(O_t-(1-\frac{2Q-q}{Q})\overline{b_c}) \\ denom_v &= \sum_{c=1}^{C}\sum_{t=1}^{T}\sum_{q=0}^{Q-1}\gamma_t(q)((1-\frac{q}{Q})\overline{m_c}+\frac{q}{Q})^2+\sum_{c=1}^{C}\sum_{t=1}^{T}\gamma_t(Q) \\ &+ \sum_{c=1}^{C}\sum_{t=1}^{T}\sum_{q=Q+1}^{2Q}\gamma_t(q)((1-\frac{2Q-q}{Q})\overline{m_c}+\frac{2Q-q}{Q})^2, \end{aligned}$$

then the re-estimation formula from Eq. 12.39 for μ_v becomes

$$\mu_v = \frac{num_v}{denom_v}, \qquad 1 \leq v \leq V. \tag{12.40}$$

Numerical results of the learning

The speech data examined consists of a total of 147 vowels in the CVC (C = one of the six stop consonants) context extracted from sentences of continuously spoken utterances in TIMIT with all the vowel boundaries given. The observation sequence is the trajectory of manually corrected and labeled formant frequencies sampled every 3 ms. Once the formant trajectories are made available, the learning algorithm described above can be applied to determine the "optimal" slope and intercept parameters for each consonant c (in conjunction with all other Locus-HMM parameters). In Table 12.7.3, the results are listed for the algorithmically learned slope and intercept parameters for F2 specific to each stop consonant (for /k/ and /g/, separate palatal and velar allophones are used).

12.7.4 Phonetic classification experiments

Data preparation

The speech data employed to evaluate the Locus-HMM in a set of phonetic classification experiments are extracted from the TIMIT corpus. The scope of this study is limited to a subset of the TIMIT sentences; in particular, speech data of only male speakers from three of the eight dialect regions (a total of 145 speakers) are used.

The acoustic data used in the experiments are from the vowels conditioned on the CVC context (C = stop consonant), confined generally within the boundaries provided by TIMIT. The stop consonant context is divided into eight classes:

$\{/b/\}, \{/d/\}, \{palatal/g/\}, \{velar/g/\}, \{/p/\}, \{/t/\}, \{palatal/k/\},$
$\{velar/k/\}.$

The vowel classes used for classification are defined as:

$\{/uw//ux/\}, \{/uh/\}, \{/ah//ax//ax-h/\}, \{/aa//ao/\}\{/ae/\}, \{/eh/\},$
$\{/ih//ix/\}, \{/ey/\}, \{/iy/\}, \{/ay/\}, \{/ow/\}, \{/aw/\}, \{/er//axr/\}.$

In the above, each set of braces {} is treated as a phonetically distinct entity; multiple phonemes within a single set of braces are "folded" together to serve as a single entity. Thus, the vocabulary consists of eight distinct stop allophones and thirteen distinct vowel classes, all constrained to occur in the CVC context.

In order to form observation sequences $O_1, O_2, ..., O_T$ suitable for our HMM-based classifiers, the original speech waveforms were preprocessed in two ways. First, the conventional mel-frequency Cepstrum coefficients (MFCC) are produced (they were found to not obey the constraints provided by the locus equations). Second, formant frequencies from the speech waveforms are accurately extracted manually. (For F2, it is expected to see the effectiveness of the Locus-HMM.)

Use of the MFCC preprocessor is intended to serve two purposes: to provide the baseline classification rate and to disprove effectiveness of the Locus-HMM used with the MFCC preprocessor. The static and dynamic MFCCs were computed in a standard way, resulting in a 15-dimensional vector per 10 ms frame. A total of 900 vowel tokens conditioned on CVC contexts were used as training data and 138 disjoint vowel tokens as test data for the experiments involving MFCCs as acoustic observations.

The true gain in Locus-HMM is expected when a formant-based preprocessor is deployed. To ensure accuracy, all the trajectories of formants F1, F2, and F3 for all the vowel utterances conditioned on CVC contexts were manually corrected. Formants were tracked at 3 ms intervals with a frequency resolution of 13 Hz. In total, formant extraction was performed for 285 CVC-conditioned vowel tokens, with 147 of these tokens serving as training data and the remaining 138 tokens as test data.

Results with formant preprocessor

In this set of experiments, the classification rate achieved using the Locus-HMM was compared to the baseline of a conventional HMM incorporating no locus-equation constraints. The training algorithm for the Locus-HMM has been described earlier and that for the conventional HMM is the standard Baum-Welch algorithm [Rab 89]. The classification decision rule is based on the maximum likelihood criterion for both the

Locus-HMM and the conventional HMM. The training and test data used consisted of formant trajectories for formants F1, F2, and/or F3. All seven possible permutations of {F1, F2, F3}, resulting in one, two, or three dimensional acoustic observation vectors as input to the HMM classifiers, were examined.

In Table 12.7.4, the classification rate, measured as the percentage of the correct class candidate scored with the highest likelihood among all 13 competitive class candidates, is shown as a function of the number of the states and of the way F1, F2, and F3 are combined. On the left three columns are listed the classification rates for the Locus-HMM, contrasting the classification rates achieved with the conventional HMM listed in the right three columns.

	Locus-HMM			Baseline		
Formants	5 states	9 states	17 states	5 states	9 states	17 states
F1	28.3	26.8	30.4	24.6	22.5	22.5
F2	35.5	34.9	34.2	31.2	27.0	29.7
F3	13.5	13.5	13.6	20.2	21.2	15.1
F1 + F2	45.7	42.8	45.7	42.0	37.0	37.7
F1 + F3	21.7	27.5	34.1	23.2	27.5	25.4
F2 + F3	27.1	29.9	28.2	35.2	36.0	34.2
F1+F2+F3	42.8	41.3	45.7	41.3	41.3	39.1

Table 12.7: Locus-HMM (left three columns) vs. conventional HMM (right three columns) with formant preprocessor: Classification rate as a function of configuration of formant vectors and of number of HMM states (after Deng and Braam [Deng 94a], @AIP).

The first row of Table 12.7.4 shows comparative classification rates between the Locus-HMM and conventional context-independent HMM (baseline) using one-dimensional F1 trajectories as the acoustic observations. The Locus-HMM consistently outperforms the baseline HMM. The performance improvement by the Locus-HMM using F2 trajectories as shown in the second row of Table 12.7.4, consistent again over all numbers of the HMM states, is more significant given the overall higher classification rates. Use of F3 trajectories produces very poor performance, both for the baseline HMM and for the Locus-HMM, as indicated in the third row of Table 12.7.4. The remaining experiments involve use of various combinations of F1, F2, and F3 as acoustic observations to the HMMs. The results shown in the fourth row of Table 12.7.4 using F1 and F2 as two-dimensional vectors demonstrate not only significantly higher classification rates than earlier one-dimensional formant experiments, but also superior performance achieved by the Locus-HMM in comparison with the baseline HMM. Other configurations of formants show similar results (except for the F1 + F3 combination).

The consistent conclusion drawn from this set of experiments is that when F2 is used as the acoustic observation, either on its own or as a component of the observation vector, the Locus-HMM consistently produces superior classification rates to the baseline HMM. This is gratifying in light of our original motivation for employing the locus equations and of our finding described earlier that the locus equations for F2 hold well and are distinctive across the different places of articulation.

Results with MFCC preprocessor

Despite the positive results obtained with use of the formant preprocessor just described, the overall classification rate has not been sufficiently high to warrant its practical application (45.7% maximum with the Locus-HMM, which has been lower than the result achievable with the state-of-the-art MFCC preprocessor.) As such, the following question can be asked: is there a mechanism by which use of the Locus-HMM can enhance the overall best classification rate? To answer this question, two sets of experiments were designed and reported in this and in the following subsections.

Instead of using formant data as acoustic observations, the set of experiments reported in this subsection involved use of MFCCs only. The training data consisted of four subsets of 900 vowel tokens in the CVC environment. The test data consisted of the same 138 vowel tokens used in the experiments described in the previous subsection.

There is no theoretical reason why the Locus-HMM should outperform the conventional HMM when MFCCs rather than formants are used. The purpose of our experiments is to determine how the Locus-HMM works "in practice" in reference to the conventional HMM and to investigate the effects of varying amounts of training data on the relative performance of the two models.

In Table 12.7.4, experimental results were summarized, showing the classification rates achievable by the Locus-HMM (left columns) and by the conventional HMM (right columns) as a function of the size of training data and of the number of HMM states. Listed in the first row are the performance figures of the two models when the training data is in plentiful supply: 900 tokens. For all different number of the HMM states, the conventional HMM consistently outperforms the Locus-HMM, suggesting that the relatively unrealistic locus-equation constraints imposed on the Locus-HMM describing the MFCC data are undesirable when there is a sufficient amount of training data available.

	Locus-HMM			Baseline		
MFCC's	5 states	9 states	17 states	5 states	9 states	17 states
900 training tokens	57.3	59.4	58.7	63.8	64.6	70.3
450 training tokens	51.5	53.6	55.8	60.9	58.0	58.0
225 training tokens	55.1	54.4	56.5	60.9	58.4	57.9
113 training tokens	52.2	52.9	49.3	47.8	47.8	44.9

Table 12.8: Locus-HMM (left columns) vs. conventional HMM (right columns) with MFCC preprocessor: Classification rate as a function of size of training data and of number of HMM states (after Deng and Braam [Deng 94a], @AIP).

For the second row of Table 12.7.4, the amount of available training data is reduced to one half of the previous amount, and the effect of this change on the performance of the two models is apparent. The conventional HMM experiences a sharp drop in performance, most notably where the number of states is large. The Locus-HMM also suffers a deterioration in performance, but it is much less severe. The large performance gap between the two models is closed considerably, down from over 10% in the previous experiment to 5% here. This highlights a possible advantage of the Locus-HMM: the Locus-HMM requires fewer parameters than a similarly sized HMM with no constraints among the model parameters, and hence it tends to be less affected by a reduction in the size of the training data.

The amount of training data is halved again, down to 225 tokens, with the performance figures shown in the third row of Table 12.7.4. The performance gap between the conventional HMM and the Locus-HMM continues to shrink.

Finally, the amount of training data is further halved to a total of 113 tokens with the performance figures listed in the fourth row of Table 12.7.4. The performance of the conventional HMM drops significantly from the 225-training-token case, and it becomes inferior to the performance of the Locus-HMM, which suffers only a small performance decrease due to reduction of training data. For the both models, the performance is lowered as the number of the HMM states increases. The superiority of the Locus-HMM in this case is interesting: it indicates that when there is very limited training data, the strength of the reduced parameter size associated with the Locus-HMM can outweigh the disadvantage of its possible inherent unsuitability for use with the MFCC preprocessor.

Results using both formant and cepstrum data with combined scores

The previous two sets of experiments have demonstrated that the Locus-HMM is superior to the conventional HMM for modeling formant (F2) trajectories, and the reverse is true in general for modeling MFCC trajectories. The phonetic classification results from our experiments indicate that it is not possible to achieve high classification rates using formant data alone; the best model utilizing all three formants could only muster 46% correct classification. The MFCC processor, however, performed appreciably better — 70% correct classification with a large amount of training data. The purpose of the experiments described here is to determine whether it is possible to "combine" the classification scores of the best MFCC-based and the formant-based HMMs in order to achieve a better overall classification rate than that of either type of the HMM classifiers individually.

Theoretically, it should be possible to improve the overall classification rate by combining the scores obtained by the separate classifiers, provided that:

1. Each of the classifiers produces sufficiently disjoint classification errors;
2. For the correctly classified tokens, both classifiers do the right job by a wide margin;
3. For the incorrectly classified tokens, both classifiers make mistakes by a narrow margin only.

It was found in the experiments that these three conditions hold well for the two classifiers which were individually evaluated as being the "best": the conventional HMM using the MFCC preprocessor trained with 900 vowel tokens (first row, right columns of Table 12.7.4), and the Locus-HMM using the formant preprocessor including F2 as a component (left columns of Table 12.7.4). The new classification rates, obtained by simply combining the scores of all the vowel classes that were already in place from the experiments described earlier, are listed in Table 12.7.4.

Shown in the first row of Table 12.7.4 are the new classification rates obtained via combinations of the classification results from the conventional HMM (MFCC, 900 training tokens) and from the Locus-HMM (F2 only). The combined results (left columns) are slightly but consistently superior to the best results based on MFCCs alone. As better and better versions of the Locus-HMM are used via incorporation of more and more formant measurements (F1+F2 in the second row and F1+F2+F3 in the third row), the superiority of the combined results becomes more significant over the baseline results. The best overall vowel classification rate achieved across all sets of the experiments is

74.7% when 1) formants F1+F2+F3 are used in the Locus-HMM and MFCCs used in the conventional HMM; and 2) at the same time, the results of both are combined.

	Combined Classifier			Best MFCC Classifier		
Formants + MFCC's	5 sts	9 sts	17 sts	5 sts	9 sts	17 sts
F2 + MFCC's	64.5	65.2	73.2	63.8	64.6	70.3
F1+F2 + MFCC's	66.7	68.1	73.9	63.8	64.6	70.3
F1+F2+F3 + MFCC's	65.2	69.6	74.7	63.8	64.6	70.3

Table 12.9: Combined classifier (left columns) vs. fixed best MFCC classifier (right columns): Classification rate as a function of ways of combining formant-based classification scores (using Locus-HMM) with the MFCC-based classification scores (conventional HMM with 900 training tokens) and of number of HMM states (after Deng and Braam [Deng 94a], @AIP).

12.8 Robustness of Acoustic Modeling and Recognizer Design

12.8.1 Introduction

One critical issue of acoustic modeling for speech recognition is robustness. Robustness here refers to the desire to maintain good recognition accuracy when the quality of the input speech is degraded by acoustic environment distortion (e.g., caused by additive noise and/or channel distortion). The robustness issue also arises when the articulatory, and phonetic, or acoustic (in addition to noise and linear channel distortion) characteristics of speech in the training and testing conditions are mismatched; for example, different speakers, or different speaking rates and styles in training and in testing. A speech recognizer that is relatively immune to these differences in training and testing and to the acoustic environment distortion is called a robust recognizer. More specifically, a speech recognizer immune to speaker variability is called a speaker-robust (speaker-adaptive or speaker-normalized) recognizer. A speech recognizer immune to acoustic environment variability is called a **environment-robust** (noise-robust or channel-robust) recognizer, and a speech recognizer immune to speaking style variability is called a **speaking-style** robust recognizer.

Major obstacles to environment robust recognition include acoustical degradations produced by additive noise (background noise such as babbles, car engine noise, door slam, etc.), the effects of linear filtering (such as frequency responses of microphone, reverberation, etc.), nonlinearities in transduction or transmission (such as speech codec, telephone and cellular phone distortion). A typical example of speaking-style robust recognition is that which overcomes diminished accuracy caused by changes in articulation produced by the presence of high-intensity noise sources, known as the Lombard effect.

Speaker variability has a rather different nature from environment variability. It may produce variations in dialect, speech rate, and different ways of implementing co-articulation. As a result, ASR systems that are designed to be speaker independent often exhibit dramatic degradations in recognition accuracy when training and testing conditions differ. However, techniques for adapting the HMMs to the new speaker and to the new environment have been rather similar. This has been caused mainly by a lack of simplistic mathematical description of speaker differences in the domain of the speech features (such as MFCCs). In the next subsections, some techniques that are common for speaker and environment robustness are presented.

Due to intense research in recent years, ASR systems have become much more robust with respect to both speaker variability and acoustical variability. In addition to achieving speaker robustness, many current systems can also automatically compensate for modest amounts of acoustical degradation caused by the effects of unknown noise and unknown linear filtering. As ASR systems are being transferred to real world applications, the need for greater robustness in recognition technology is becoming increasingly apparent. However, the performance of even the state-of-the art systems tends to deteriorate severely when the signal-to-noise ratio (SNR) is low, when the noise contains speech from other talkers (co-channel competing speech), when reverberation is present which causes inter-frame interference due to the long impulse response, or when the speaker's native language is not the one with which the system was trained.

The general approaches to robust speech recognition can be broadly classified into three categories. First, the speech signal at the sample level in the test condition is enhanced (or normalized) to combat the unwanted variability. This is called signal-space robust processing. Second, the speech feature (vector) is enhanced (or normalized) in the test condition so as to match the pre-trained models (HMMs) better. This is called feature-space robust processing. These two approaches may include processing the training data in the same way, followed by retraining the models. Third, the model parameters are adapted or normalized so as to match the unprocessed test speech features. This is called model-space robust processing. We will discuss the first two approaches in detail in the next chapter called speech enhancement as part of signal processing. In the next subsection, we will focus on the third approach, which is commonly considered as the problem of acoustic modeling rather than signal processing and is thus presented in this chapter instead of in the next chapter.

12.8.2 Model-space robustness by adaptation

MAP adaptation

One major technique designed to achieve model-space robustness, which has been widely used for transforming HMM parameters, is based on **maximum a posteriori (MAP)** estimation as introduced in Section 5.7. The MAP estimates of the HMM parameters have been successfully applied in speaker adaptation, where a newly enrolled speaker typically provides only a very small amount of speech data for use in updating the HMM parameters specific to the speaker (cf. [Lee 91, Gauv 94]). The same technique can also be applied to acoustic environment adaptation. The MAP estimation framework provides a natural way of incorporating prior information of the model parameters in the model training and adaptation process. Use of the prior information is particularly important for dealing with the difficulties arising from sparse training data, out of which the classical ML approach usually gives poor estimates of model parameters. This MAP approach has been shown to be reasonably effective for speaker adaptation in a number of speech recognition tasks where the time-invariant (conditioned on the HMM state) Gaussian densities are adapted to sparse training data uttered by new speakers. In such speaker adaptation applications, the prior distribution used in the MAP adaptation technique is generally derived from the speaker independent HMMs.

In Section 12.9, we will present a detailed case study on the development and evaluation of the MAP approach to adapting the parameters of the trended HMM. The derived MAP estimates are shown to reduce to the special case of the MAP estimates for the conventional HMM as were developed earlier in [Lee 91, Gauv 94, Lee 96a].

MLLR adaptation

A second major technique for model-space robustness, both for speaker adaptation and for acoustic environment adaptation, is called **maximum likelihood linear regression (MLLR)** estimation developed originally in [Legg 95]. This is essentially a generalization of least square estimation of the linear model (Section 5.5), or equivalently ML estimation (Section 5.6) for the Gaussian residual, from the fully observed case to the more complex case where discrete hidden variables (HMM states) exist. In a typical use of the MLLR, a set of linear (strictly affine) regression functions are defined to map both means and covariance matrices of HMMs from the speaker-independent or environment-independent models to speaker or environment adaptive models, based on the criterion of maximum likelihood for the adaptation data. This criterion is consistent with the HMM training (Baum-Welch algorithm) but it keeps the total number of free transformation parameters under control. Due to the hidden states in the HMM, the MLLR requires the EM algorithm. In the E-step, the auxiliary function Q is computed as a function of the affine transformation matrices, which constitute the MLLR parameters. In the M-step, the derivative of Q with respect to the MLLR parameters is set to zero, which results in a set of linear system of equations. The solution to the linear system of equations gives the ML estimate of the MLLR parameters. See [Legg 95] and [Huang 01] for detailed derivations.

While the solution to the MLLR parameters can be made separate for each of the HMM states, this becomes impractical because of the requirement for a large amount of adaptation data. To overcome this problem, the MLLR parameters can be tied across all Gaussians in all HMM states, making the HMM parameter transformation carried out by only one matrix. Between these two extremes, there are two other levels of granularity for MLLR parameter tying. First, tying by HMMs where all Gaussian mixture components associated with the same HMM are transformed by the same MLLR transformation matrix. Second, tying by states, where all mixture components in the same HMM state are updated by the same MLLR transformation matrix. Any specific way of MLLR parameter tying defines a set of regression classes.

In general, the more the mixture components are tied in the MLLR, the fewer parameters need to be estimated, but unfortunately, the greater approximation needs to be made. For a given amount of adaptation data, there is an optimal number of regression classes. A common rule of thumb for determining the requirement of the adaptation data is roughly two to four utterances for each regression class.

Compared with the MAP approach to speech recognizer adaptation, the MLLR approach requires a smaller amount of adaptation data since it can more effectively and flexibly tie the linear regression parameters across HMM states. The flexibility is provided by constructing a variable number of regression classes so that each class can be guaranteed to be associated with a fixed number of utterances in the adaptation data.

Speaking rate adaptation

Both the MAP and MLLR adaptation techniques just outlined have been successfully used to combat against speaker variability and acoustic environment variability. In principle, they can also be used to handle the variability arising from speaking rate differences between the training data and the test data. However, in casual speech, the speaking rate variation is typically very fast. Even within each conversational utterance, the speaking rate can change significantly from one portion of the utterance to another. This is in stark contrast to speaker or acoustic environment variability where it is very unlikely that within each utterance the speaker or the acoustic environment can vary.

The implication is that a successful speaking rate adaptation technique must be instantaneous and rapid. That is, a very small amount or virtually no adaptation data may be available. This makes the MAP and MLLR adaptation techniques difficult to apply to speaking rate adaptation in practice. It appears that in order to successfully achieve rapid adaptation to speaking rate in conversational speech, some structural properties of speech need to be exploited. The hidden dynamic models that we presented in Sections 10.3-10.5 represent such structural properties. Use of this type of speech models can conceivably provide more effective solutions to rapid speaking rate adaptation than existing adaptation techniques such as MAP and MLLR.

Quasi-Bayesian approach to dynamic adaptation

One interesting extension of the MAP adaptation technique is sequential parameter estimation, also called on-line dynamic adaptive learning, based on the approximate recursive Bayes estimate and addressing the problem of model-space robustness [Huo 97]. The essence of this technique is to dynamically refresh the hyperparameters in the prior distribution used in the MAP estimation (or other kinds of Bayesian estimation). Dynamic refreshing of the hyperparameters is important for the adaptation application where the model parameters change rapidly. For rapid speaker adaptation, the hyperparameter refreshing can be carried out at the end of each utterance. For rapid acoustic environment adaptation where the interference noise is highly nonstationary, the hyperparameter refreshing can occur even more frequently.

The mathematical basis for the sequential quasi-Bayesian adaptation is the sequential Bayes rule described in Section 3.2.3. It states that the posterior probability (of model parameters) conditioned on the observation sequence up until now can be used effectively as the prior for computing the posterior probability conditioned on the observation sequence including the future vectors. In general these posterior probabilities are not Gaussian, which creates computational difficulties. One solution is to use the so called **Laplace approximation** to replace the non-Gaussian posterior probabilities with a simple distribution such as a Gaussian. This gives rise to the quasi-Bayes procedure as presented in detail in [Huo 97]. By choosing the initial prior distribution to be the conjugate family of the complete-data distribution, the hyperparameters in the prior can be determined by matching appropriate terms in the relevant distribution forms or by taking the mode in the distribution.

12.8.3 Adaptive training

The various adaptation techniques described above for model-space robustness transform the HMM parameters so that the new HMMs match the statistical characteristics of the adaptation data or the unseen test data better than the HMMs before the parameter transformation took place. However, if the test data are severely distorted or are from a poorly articulating speaker, even under the ideal matched condition (i.e., the model transformation does a perfect job), the recognizer's performance will still be low.

One class of new techniques aimed at robustness of acoustic modeling and recognizer design has the potential to overcome this difficulty by a procedure known as **adaptive**

training. When this procedure is applied to handle speaker variability, it is called **speaker adaptive training (SAT)** [Ana 96] and it is called **noise adaptive training (NAT)** when corrupting noise is the main difficulty to overcome [Deng 00b]. The essence of the adaptive training is to create an underlying generic or compact model which, through possible transformation, can effectively compensate for extraneous acoustic variations (speaker variation, environment variation, etc.) not relevant to the speech discrimination task.

The adaptive training can be formulated rigorously in a Bayesian learning framework. In Section 12.10, we will present one version of this framework in a detailed case study (IV) and some speech recognition results.

12.9 Case Study III: MAP Approach to Speaker Adaptation Using Trended HMMs

In this case study, we describe a MAP approach to speaker adaptation with the use of the trended HMM. The purpose of this presentation is to show a concrete, trended-HMM example of a model-space robust transformation via the MAP technique. As we introduced in Section 4.2, the trended HMM is a generalized version of the Gaussian HMM where the Gaussian means in each HMM state are characterized by time-varying polynomial trend functions of the state sojourn time. Assuming uncorrelatedness among the polynomial coefficients in the trend functions, analytical results are obtained for the MAP estimates of the parameters including time-varying means and time-invariant precisions (inverse of the covariance matrices). A speech recognizer based on these results is implemented in speaker adaptation experiments using the TI46 corpora (Section 12.9.2). The experimental evaluation demonstrates that the trended HMM, with use of either the linear or the quadratic polynomial trend function, consistently outperforms the conventional, stationary-state HMM. The evaluation also shows that the unadapted, speaker-independent models are outperformed by the models adapted by the MAP procedure with as few as a single adaptation token (under supervision). Further, adaptation of polynomial coefficients alone is shown to be better than adapting both polynomial coefficients and precision matrices when fewer than four adaptation tokens are used, while the reverse is found with a greater number of adaptation tokens. This section is based on the work published in [Rathi 01].

The formulation of the (polynomial) trended HMM as discussed earlier in this book was proposed as a more accurate model for the description of the speech trajectories than the conventional, stationary-state HMM, and has been successfully used in some speech recognition tasks. The parameters of the trended HMM, especially those governing the state-dependent time-varying Gaussian means (such as the polynomial coefficients), can be trained by the Baum-Welch algorithm based on the observation-ML criterion [Deng 92b]. In ML training, the amount of speech data required for estimating model parameters is usually large; this is more so for the trended HMM than for the conventional HMM due to the additional polynomial coefficients to be trained. Therefore, for speaker adaptation applications where the amount of speech data for the new speaker is limited, new, robust training methods need to be devised for the trended HMM.

In this case study, the MAP approach is used to reduce the data requirement. The motivation for the use of the trended HMM for speaker adaptation is its higher accuracy for the dynamic patterns of speech associated with distinct speech classes, which is expected to lead to greater discriminability, at least theoretically, among these classes. However, in speech adaptation applications where the sparse training data problem usually presents practical difficulties with regard to parameter-estimation reliability, the theoretical advantage of using a more accurate speech model would be undermined if the estimation reliability cannot be assured. This calls for a greater demand to develop robust parameter-estimation and adaptation techniques for the trended HMM than for the conventional HMM, since in general the former contains a greater number of model parameters. An additional motivation for developing adaptation techniques for the trended HMM is that they would enable adaptation of the entire trajectory using all adaptation data within each state jointly. In contrast, the adaptation for the conventional HMM usually makes the effects frame by frame.

12.9.1 Derivation of MAP estimates for the trended HMM

To simplify the presentation of the approach, the data features are assumed to form a scalar-valued observation data sequence of length T: $o_t,\ t = 1, 2, \cdots, T$. Consider the trended HMM [Deng 94c] given by

$$o_t = \mathbf{x}_t^{\mathrm{Tr}}\mathbf{b}_i + n_t(\sigma_i^2), \tag{12.41}$$

where $n_t \sim \mathcal{N}(0, \sigma_i^2)$ is an IID scalar residual white Gaussian noise sequence, and $\mathbf{b}_i = [b_i(0)\ b_i(1)\ \cdots\ b_i(P)]^{\mathrm{Tr}}$ is a $(P+1)\times 1$ vector of state-dependent polynomial regression coefficients (i.e., P is the polynomial order), $\mathbf{x}_t = [(t-\tau_i)^0\ (t-\tau_i)^1\ \cdots\ (t-\tau_i)^P]^{\mathrm{Tr}}$ is a $(P+1)\times 1$ vector of exogenous explanatory variables with $(t-\tau_i)$ representing the sojourn time in state i. The MAP formulation requires a joint prior distribution for both $\mathbf{b}_i$ and σ_i^2, which are treated as random variables in the Bayesian analysis.

Let the prior information about $\mathbf{b}_i$ be conditioned on the value for σ_i^{-2} and be represented by a Gaussian random variable $\mathcal{N}(\mathbf{b}_i; \mathbf{\Phi}_i, \sigma_i^2\mathbf{M}_i)$. Its probability density function is then given by

$$f(\mathbf{b}_i|\sigma_i^{-2}) = (2\pi\sigma_i^2)^{-\frac{1}{2}}|\mathbf{M}_i|^{-\frac{1}{2}}\exp\left[-0.5\sigma_i^{-2}(\mathbf{b}_i-\mathbf{\Phi}_i)^{\mathrm{Tr}}\mathbf{M}_i^{-1}(\mathbf{b}_i-\mathbf{\Phi}_i)\right]. \tag{12.42}$$

Thus, prior to the observation of data samples, the best guess to the value of $\mathbf{b}_i$ is represented by the $(P+1)\times 1$ vector $\mathbf{\Phi}_i$, and the confidence in this guess is summarized by the $(P+1)\times(P+1)$ matrix $\sigma_i^2\mathbf{M}_i$; a lower degree of the confidence is represented by a larger (diagonal) element of $\mathbf{M}_i$. Knowledge about the exogenous variable $\mathbf{x}_t$ is presumed to have no effects on the prior distribution; hence Eq. 12.42 also describes the density $f(\mathbf{b}_i|\sigma_i^{-2}, \mathbf{x}_t)$. Following [Lee 91], it is convenient to describe the prior distribution not in terms of the variance σ_i^2 but rather in terms of the reciprocal of the variance, σ_i^{-2}, which is known as the **precision**. The prior distribution for precision σ_i^{-2} can be conveniently provided by the gamma distribution:

$$f(\sigma_i^{-2}) = \frac{q_i^{p_i}}{\Gamma(p_i)}\sigma_i^{-2(p_i-1)}\exp\left(-q_i\sigma_i^{-2}\right), \tag{12.43}$$

where $p_i > 0$ and $q_i > 0$ are parameters that describe the prior information. Thus, $f(\mathbf{b}_i, \sigma_i^{-2})$, the joint prior density for $\mathbf{b}_i$ and σ_i^{-2}, is given by the product of Eq. 12.42 and Eq. 12.43, or a normal-gamma distribution. The choice of such a prior density is made because the normal-gamma density is the conjugate density of the normal distribution, a fact that is essential for the analytical derivation of the MAP estimates.

Denote by $\boldsymbol{\theta}$ all the random parameters of the trended HMM: $\boldsymbol{\theta} = (\mathbf{b}_i, \sigma_i^2)$. According to the EM algorithm applied to the random-parameter case, the MAP estimates for the model parameters can be obtained according to

$$\hat{\boldsymbol{\theta}}_{MAP} = \arg\max_{\boldsymbol{\theta}} \left[\mathcal{Q}(\boldsymbol{\theta}|\boldsymbol{\theta}_0) + \sum_{i=1}^{N} \log f(\mathbf{b}_i, \sigma_i^{-2}) \right], \tag{12.44}$$

where the first term, the conditional expectation (E step in the EM algorithm) involving the log-likelihood function for the observation data, was derived previously in [Deng 92b] and is rewritten as

$$\mathcal{Q}(\boldsymbol{\theta}|\boldsymbol{\theta}_0) = \sum_{t=1}^{T}\sum_{i=1}^{N} \gamma_t(i) \left[\frac{1}{2}\log(\sigma_i^{-2}) - \frac{\sigma_i^{-2}}{2}(o_t - \mathbf{x}_t^{\mathrm{Tr}}\mathbf{b}_i)^2 \right]. \tag{12.45}$$

In making the Viterbi approximation to the total log-likelihood function, the quantity $\gamma_t(i)$ in Eq. 12.45 is either to be one, if the model generates o_t in state i at time t, or to be zero otherwise.

Maximization of Eq. 12.44 (which gives the MAP estimates) is accomplished by the M step of the EM algorithm. In order to proceed, all the optimization-independent terms and factors in the argument of Eq. 12.44 were removed. This gives an equivalent objective function of

$$\begin{aligned}
\hat{\mathcal{Q}}(\boldsymbol{\theta}|\boldsymbol{\theta}_0) &= \mathcal{Q}(\boldsymbol{\theta}|\boldsymbol{\theta}_0) + \sum_{i=1}^{N} \log f(\mathbf{b}_i, \sigma_i^{-2}) \\
&= \sum_{t=1}^{T}\sum_{i=1}^{N} \gamma_t(i) \left[\frac{1}{2}\log(r_i) - \frac{r_i}{2}(o_t - \mathbf{x}_t^{\mathrm{Tr}}\mathbf{b}_i)^2 \right] \\
&\quad + \sum_{i=1}^{N} \left[\frac{1}{2}\log(r_i) - \frac{1}{2}\log(|\mathbf{M}_i|) - \frac{r_i}{2}[\mathbf{b}_i - \boldsymbol{\Phi}_i]^{\mathrm{Tr}}\mathbf{M}_i^{-1}[\mathbf{b}_i - \boldsymbol{\Phi}_i] \right] \\
&\quad + \sum_{i=1}^{N} \left[p_i \log(q_i) + (p_i - 1)\log(r_i) - q_i r_i - \log(\Gamma(p_i)) \right],
\end{aligned} \tag{12.46}$$

where $r_i = \sigma_i^{-2}$ is the precision [Gauv 94].

The re-estimation formulas for the polynomial coefficients are obtained by solving

$$\frac{\partial \hat{\mathcal{Q}}(\boldsymbol{\theta}|\boldsymbol{\theta}_0)}{\partial \mathbf{b}_i} = 0,$$

which, after use of Eq. 12.46, becomes

$$\sum_{t=1}^{T} \gamma_t(i) \frac{\partial}{\partial \mathbf{b}_i} \left[-\frac{r_i}{2}(o_t - \mathbf{x}_t^{\mathrm{Tr}}\mathbf{b}_i)^2 \right] + \frac{\partial}{\partial \mathbf{b}_i} \left[-\frac{r_i}{2}[\mathbf{b}_i - \boldsymbol{\Phi}_i]^{\mathrm{Tr}}\mathbf{M}_i^{-1}[\mathbf{b}_i - \boldsymbol{\Phi}_i] \right] = 0,$$

$$\text{or} \qquad \sum_{t=1}^{T} \gamma_t(i)\mathbf{x}_t(o_t - \mathbf{x}_t^{\mathrm{Tr}}\mathbf{b}_i) - \mathbf{M}_i^{-1}[\mathbf{b}_i - \boldsymbol{\Phi}_i] = 0. \tag{12.47}$$

After rearranging the known terms on one side and the unknown terms on the other side, the above equation can be rewritten as:

$$\left[\sum_{t=1}^{T} \gamma_t(i)\mathbf{x}_t\mathbf{x}_t^{\mathrm{Tr}} + \mathbf{M}_i^{-1}\right] \mathbf{b}_i = \left[\sum_{t=1}^{T} \gamma_t(i)\mathbf{x}_t o_t + \mathbf{M}_i^{-1}\mathbf{\Phi}_i\right],$$

$$\text{or} \quad \hat{\mathbf{b}}_i = \underbrace{\left[\sum_{t=1}^{T} \gamma_t(i)\mathbf{x}_t\mathbf{x}_t^{\mathrm{Tr}} + \mathbf{M}_i^{-1}\right]^{-1}}_{\hat{\mathbf{M}}_i} \left[\sum_{t=1}^{T} \gamma_t(i)\mathbf{x}_t o_t + \mathbf{M}_i^{-1}\mathbf{\Phi}_i\right]. \tag{12.48}$$

Similarly, the re-estimation formulas for the variances are obtained by solving

$$\frac{\partial \hat{Q}(\boldsymbol{\theta}|\boldsymbol{\theta}_0)}{\partial r_i} = 0,$$

which, after use of Eq. 12.46 again, becomes

$$\begin{aligned}
&\frac{\partial}{\partial r_i} \sum_{t=1}^{T} \gamma_t(i) \left[\frac{1}{2}\log(r_i) - \frac{r_i}{2}(o_t - \mathbf{x}_t^{\mathrm{Tr}}\mathbf{b}_i)^2\right] + \\
&\frac{\partial}{\partial r_i} \left[\frac{1}{2}\log(r_i) - \frac{r_i}{2}[\mathbf{b}_i - \mathbf{\Phi}_i]^{\mathrm{Tr}}\mathbf{M}_i^{-1}[\mathbf{b}_i - \mathbf{\Phi}_i] + (p_i - 1)\log(r_i) - q_i r_i\right] = 0, \\
\text{or} \quad &\sum_{t=1}^{T} \gamma_t(i) \left[\frac{1}{2}r_i^{-1} - \frac{1}{2}(o_t - \mathbf{x}_t^{\mathrm{Tr}}\mathbf{b}_i)^2\right] + \frac{1}{2}r_i^{-1} + (p_i - 1)r_i^{-1} - \\
&\quad q_i - \frac{1}{2}[\mathbf{b}_i - \mathbf{\Phi}_i]^{\mathrm{Tr}}\mathbf{M}_i^{-1}[\mathbf{b}_i - \mathbf{\Phi}_i] = 0.
\end{aligned} \tag{12.49}$$

Again, after rearranging the known and unknown terms, the above equation can be rewritten as

$$\begin{aligned}
r_i^{-1}\left[\sum_{t=1}^{T} \frac{1}{2}\gamma_t(i) + \frac{1}{2} + (p_i - 1)\right] &= q_i + \frac{1}{2}\sum_{t=1}^{T} \gamma_t(i)(o_t - \mathbf{x}_t^{\mathrm{Tr}}\mathbf{b}_i)^2 \\
&+ \frac{1}{2}[\mathbf{b}_i - \mathbf{\Phi}_i]^{\mathrm{Tr}}\mathbf{M}_i^{-1}[\mathbf{b}_i - \mathbf{\Phi}_i],
\end{aligned} \tag{12.50}$$

or

$$\begin{aligned}
\hat{r}_i^{-1} &= \frac{q_i + \frac{1}{2}\sum_{t=1}^{T} \gamma_t(i)(o_t - \mathbf{x}_t^{\mathrm{Tr}}\mathbf{b}_i)^2 + \frac{1}{2}[\mathbf{b}_i - \mathbf{\Phi}_i]^{\mathrm{Tr}}\mathbf{M}_i^{-1}[\mathbf{b}_i - \mathbf{\Phi}_i]}{\frac{1}{2}\sum_{t=1}^{T} \gamma_t(i) + \frac{1}{2} + (p_i - 1)} \\
&= \frac{\overbrace{2q_i + \sum_{t=1}^{T} \gamma_t(i)(o_t - \mathbf{x}_t^{\mathrm{Tr}}\mathbf{b}_i)^2 + [\mathbf{b}_i - \mathbf{\Phi}_i]^{\mathrm{Tr}}\mathbf{M}_i^{-1}[\mathbf{b}_i - \mathbf{\Phi}_i]}^{\hat{q}_i}}{\underbrace{\sum_{t=1}^{T} \gamma_t(i) + 2p_i - 1}_{\hat{p}_i}}.
\end{aligned} \tag{12.51}$$

In the above Eq. 12.51, we denote the numerator and the denominator of the precision estimate by $\hat{q}_i$ and $\hat{p}_i$, respectively, and denote the variance estimate by $\hat{r}_i^{-1} = \hat{\sigma}_i^2$.

We now summarize the final results obtained above for the MAP estimates of $\mathbf{b}_i$, σ_i^2, and of their prior parameters as follows:

$$
\begin{aligned}
\hat{\mathbf{M}}_i &= \left[\sum_{t=1}^{T} \gamma_t(i)\mathbf{x}_t\mathbf{x}_t^{\mathrm{Tr}} + \mathbf{M}_i^{-1}\right]^{-1}, && (12.52)\\
\hat{\mathbf{\Phi}}_i &= \hat{\mathbf{M}}_i\left[\sum_{t=1}^{T} \gamma_t(i)\mathbf{x}_t o_t + \mathbf{M}_i^{-1}\mathbf{\Phi}_i\right], && (12.53)\\
\hat{q}_i &= 2q_i + \sum_{t=1}^{T} \gamma_t(i)(o_t - \mathbf{x}_t^{\mathrm{Tr}}\hat{\mathbf{b}}_i)^2 + (\hat{\mathbf{b}}_i - \mathbf{\Phi}_i)^{\mathrm{Tr}}\mathbf{M}_i^{-1}(\hat{\mathbf{b}}_i - \mathbf{\Phi}_i), && (12.54)\\
\hat{p}_i &= \sum_{t=1}^{T} \gamma_t(i) + 2p_i - 1, && (12.55)\\
\hat{\mathbf{b}}_i &= \hat{\mathbf{\Phi}}_i, && (12.56)\\
\hat{\sigma}_i^2 &= \frac{\hat{q}_i}{\hat{p}_i} && (12.57)
\end{aligned}
$$

Note that the MAP estimates of the (hyper-) parameters, $\hat{\mathbf{\Phi}}_i, \hat{\mathbf{M}}_i, \hat{q}_i$, and $\hat{p}_i$, for the prior distributions given by Eqs. 12.52-12.55 are actually obtained only in some intermediate stages of the M step in the EM algorithm. The objective of the M step is to estimate the actual parameters (i.e., not hyper-parameters) of the trended HMM — $\hat{\mathbf{b}}_i$ and $\hat{\sigma}_i^2$. In the implementation of the MAP-based speaker-adaptive speech recognition system, Eqs. 12.52–12.57 are programmed as such, which constitutes a full iteration of the MAP adaptive training procedure.

The difference between the ML estimation procedure and the MAP estimation procedure lies in the assumption made by the MAP procedure of a non-flat prior distribution of the parameters to be estimated. By using a flat or diffuse prior information, represented as $p_i = \frac{1}{2}$, $q_i = 0$ and $\mathbf{M}_i^{-1} = 0$, the MAP estimates for $\mathbf{b}_i$ and σ_i^2 become:

$$
\begin{aligned}
\hat{\mathbf{b}}_i &= \left[\sum_{t=1}^{T} \gamma_t(i)\mathbf{x}_t\mathbf{x}_t^{\mathrm{Tr}}\right]^{-1}\left[\sum_{t=1}^{T} \gamma_t(i)\mathbf{x}_t o_t\right], \quad \text{and} && (12.58)\\
\hat{\sigma}_i^2 &= \frac{\sum_{t=1}^{T} \gamma_t(i)\left[o_t - \mathbf{x}_t^{\mathrm{Tr}}\mathbf{b}_i\right]^2}{\sum_{t=1}^{T} \gamma_t(i)}. && (12.59)
\end{aligned}
$$

The above formulas for $\hat{\mathbf{b}}_i$ and $\hat{\sigma}_i^2$ would become identical to the ML estimates derived in [Deng 92b], which we presented in Section 10.2, for the total likelihood function case ($\gamma_t(i)$ values are between zero and one). The reduction of the MAP estimates to the ML estimates with flat priors serves to provide a "sanity-check" for the correctness of the analytical results derived above for the MAP estimates. If there are no samples available (i.e., the terms involving o_t are set to zero) for adaptation, then Eq. 12.48 and Eq. 12.51 would become:

$$
\begin{aligned}
\hat{\mathbf{b}}_i &= \mathbf{\Phi}_i && (12.60)\\
\hat{\sigma}_i^2 &= \frac{q_i}{p_i}, && (12.61)
\end{aligned}
$$

which are simply from the prior information. Hence the MAP estimates can be interpreted as a weighted average of the corresponding prior information and of the sample data. According to the reestimation formulas summarized in Eqs. 12.52-12.56, the weights are computed iteratively based on a combination of the prior speaker-independent model's parameters and of the new-speaker's observation data in a non-linear fashion.

It is also interesting and reassuring to note that the MAP estimation formulas for the trended HMM naturally revert to the familiar MAP estimation formulas for the conventional HMM [Gauv 94] when $P = 0$ was set. This can be seen by noting $\mathbf{x}_t = \mathbf{x}_t^{\mathrm{Tr}} = \mathbf{1}$ when $P = 0$.

12.9.2 Speaker adaptation experiments

Task and speech corpus

The task of speaker adaptation experiments conducted to evaluate the MAP adaptation approach described in Section II is recognition of 26 letters (words) in the English alphabet contained in the TI46 isolated-word corpus. The corpus contains speech materials from a total of 16 speakers, eight males and eight females. The speaker-independent (SI) training set in the experiments reported in this section consists of 26 spoken tokens for each word from each of six male and six female speakers. For the remaining four speakers (two males and two females), up to ten tokens of each word are used as the adaptation training data, and the remaining 16 tokens are used as the speaker-dependent test data.

Preprocessor and baseline systems

The same preprocessor in all types of the speech recognizers used in the experiments computes a vector of 26 elements consisting of 13 Mel-frequency cepstral coefficients (MFCCs) and 13 delta MFCCs for every 10-msec frame of speech. In computing the MFCCs, 25 Mel-spaced, triangular band pass filters are simulated; that is, the filters are spaced linearly from 0 to 1 kHz and exponentially from 1 kHz to 8.86 kHz, with the adjacent filters overlapped in the frequency range by 50%. Log FFT power spectral points are combined using a weighted sum to simulate the output of the triangular filter. The MFCCs are then computed according to

$$C_p = \sum_{r=1}^{25} S_r \cos\left(\frac{\pi p}{25}[r - 0.5]\right), \quad 0 \leq p \leq 12$$

where S_r is the log-power-spectral output of the rth mel-filter. The delta MFCCs are constructed by taking the difference between two frames forward and two frames backward from the current frame of the MFCCs. This temporal dynamic window length of 50 ms is found in our recognizers to be optimal in capturing the slope of the spectral envelope, i.e., the transitional information, and is hence used in the experiments. The augmented vectors consisting of the MFCCs and the delta MFCCs are provided as the data or observation input to speech recognizers (both trended and conventional HMMs) for every frame of the speech material.

Each word (English letter) is represented by a single left-to-right, three-state HMM (no skips allowed), an identical structure for the trended and the conventional HMMs. The covariance matrices for all the states of all the models are assumed to be diagonal, and are separate from each other (i.e., not tied). All transition probabilities are uniformly set to 0.5 (all transitions from a state are considered equally likely) and are not learned during the training process. Two types of baseline systems are used, for comparison purposes, to evaluate the effectiveness of the MAP adaptation approach. The first type is based on the speaker-dependent (SD) models, trained from the adaptation data (one to ten tokens for each word with a fixed speaker) using five iterations of the modified Viterbi algorithm [Deng 94c] with single mixture for each state in the HMMs. The same training method and program (cf. [Deng 94c]) is used for the trended HMM and for the conventional HMM except that in the latter the polynomial order is simply set to zero. The second type of baseline is based on the speaker-independent (SI) models, also with a single mixture distribution for each state in the HMMs (both trended and conventional). The parameters (polynomial coefficients in the trended HMM and Gaussian mean vector in the conventional HMM) of the single-mixture distribution are determined by combining the corresponding parameters of the mixture components, which are trained by the usual ML method using an ample amount of SI training data, as follows. The polynomial coefficients (or Gaussian mean vector) in each HMM state are taken to be a weighted average of the ML-trained polynomial coefficients (or Gaussian mean vectors) in the corresponding state. The combination formula is

$$\mathbf{b}_i(p) = \sum_{m=1}^{M} w_m \mathbf{b}_{im}(p), \quad p = 0, 1, \cdots, P, \tag{12.62}$$

where M is the number of mixture components used in each HMM state in the ML-trained SI models, w_m are the ML-trained mixture weight parameters, and $\mathbf{b}_{im}(p)$ are polynomial coefficients (in state i and mixture m) in the trended function. The variance of the single-mixture distribution is also determined by a weighted average of the ML-trained variances in the multiple mixture distribution using the SI training data according to

$$\sigma_i^2 = \sum_{m=1}^{M} w_m \sigma_{im}^2, \tag{12.63}$$

where σ_{im}^2 is the variance of the mth mixture component in state i.

Initialization and update of prior parameters

The initial prior-density parameters for the single-mixture HMMs are determined from the parameters of the regular trended or conventional HMMs (with multiple mixtures) that are trained by the ML method based on a large amount of SI training data. These initial parameters are computed according to

$$p_i = \frac{1}{\sum_{m=1}^{M} w_m \sigma_{im}^2}, \tag{12.64}$$

$$q_i = 1.0, \tag{12.65}$$

$$\mathbf{\Phi}_i(p) = \sum_{m=1}^{M} w_m \mathbf{b}_{im}(p), \tag{12.66}$$

$$\mathbf{M}_i(p) = \frac{1}{p_i \sum_{m=1}^{M} w_m (\mathbf{b}_{im}(p) - \mathbf{\Phi}_i(p))^2} \tag{12.67}$$

where $\mathbf{M}_i(p)$ is the pth element of the diagonal correlation matrix $\mathbf{M}_i$. After the initialization, the prior parameters are then updated over iterations of the batch MAP

Number of Adaptation Tokens	Polynomial Order P = 0 (SI: 70.0%)		
	SD	SA1	SA2
1	58.4%	79.0%	74.4%
2	71.2%	82.3 %	80.4%
3	77.7%	83.8%	83.0%
4	82.7%	84.8%	84.1%
5	85.4%	84.9%	85.8%
6	86.7%	86.6%	86.2%
7	87.6%	87.5%	87.1%
8	88.0%	88.5%	87.6%
9	87.9%	88.6%	88.9%
10	88.3%	88.7%	89.7%

Table 12.10: Summary of speaker adaptation results for conventional (flat-trended) HMM (P=0) (after Rathinavelu and Deng [Rathi 01], ©IEEE).

algorithm according to Eqs. 12.52-12.55. In the batch mode of MAP estimation, the prior parameters are updated only after processing all tokens for each iteration, in contrast with sequential adaptation where the prior parameters are adjusted at the end of processing each token. The sequential adaptation procedure will not be addressed here. In all experiments reported below, a total of five batch adaptation iterations were performed.

Performance evaluation

Speech recognition rates (correct percentage) on the task just described, averaged over two males and two females, are summarized in Table 12.10, Table 12.11, and Table 12.12 using the conventional HMMs (flat-trended with $P = 0$), the linearly trended HMMs, and the quadratic trended HMMs, respectively. Four experimental setups have been used: 1) speaker-independent (SI) as the first baseline type; 2) speaker-dependent (SD) as the second baseline type; 3) speaker-adaptation with only polynomial coefficients in the trend functions adapted (SA1); and 4) speaker-adaptation with both the polynomial coefficients and precision matrices adapted (SA2).

The percent-correct results in Tables 12.10-12.12 are shown as a function of the number of word tokens used in training (for the SD results) or in adaptation (for the SA1 and SA2 results) from a new speaker. In the SI baseline experiments, quadratically-trended HMM outperforms both the flat-trended ($P = 0$) and linearly-trended HMMs by about 23% and 6%, respectively. Comparing the results of Tables 12.11-12.12 with those in Table 12.10, the greater effectiveness of the MAP training for the trended HMMs than for the conventional HMMs is demonstrated. For example, in the SA1 experiments, the error rate reduction of 26.8% is obtained when moving from the $P = 0$ (83.77%) model to the $P = 1$ (88.11%) model with three adaptation tokens. The best recognition rate of 92.1% is achieved when both linear polynomial coefficients and precision matrices are adapted using all ten tokens of adaptation data. It is also observed that the quadratically-trended HMMs somewhat outperform the constant and linearly-trended HMMs when fewer than four adaptation tokens are used. For example, with three

Number of Adaptation Tokens	Polynomial Order P=1		
	(SI: 75.0%)		
	SD	SA1	SA2
1	46.8%	84.1%	75.8%
2	74.6%	86.8%	84.1%
3	82.5%	88.1%	87.0%
4	85.6%	88.6%	89.0%
5	86.5%	88.8%	90.1%
6	88.5%	89.7%	91.1%
7	88.6%	90.1%	91.9%
8	89.5%	90.1%	90.9%
9	90.2%	90.4%	91.5%
10	91.1%	91.7%	92.1%

Table 12.11: Summary of speaker adaptation results for linearly-trended HMM (P=1) (after Rathinavelu and Deng [Rathi 01], ©IEEE).

Number of Adaptation Tokens	Polynomial Order P=2		
	(SI: 76.8%)		
	SD	SA1	SA2
1	46.8%	84.1%	79.9%
2	74.0%	87.3%	85.2%
3	82.3%	88.9%	88.0%
4	84.9%	88.7%	89.0%
5	86.4%	89.5%	89.8%
6	88.3%	90.1%	90.3%
7	89.1%	90.4%	90.7%
8	89.4%	90.5%	90.8%
9	89.8%	90.9%	91.3%
10	91.0%	91.4%	91.5%

Table 12.12: Summary of speaker adaptation results for quadratically-trended HMM (P=2) (after Rathinavelu and Deng [Rathi 01], ©IEEE).

adaptation tokens, the quadratically-trended HMMs give an error rate reduction of 7% when compared with the linearly-trended HMMs and an error rate reduction of 32% compared with conventional HMMs. This at first glance may seem unexpected because the quadratically-trended HMMs have more parameters than other types of HMMs and hence would be less effective with a smaller amount of training/adaptation data. However, use of prior information from the SI models counters such an expectation, as is shown in the results. This suggests the effectiveness of the MAP adaptation algorithm on the quadratically-trended HMMs for sparse adaptation data.

Examination of the results in Tables 12.10-12.12 also shows that the recognition rate drops gradually with a decreasing number of adaptation tokens for both SA1 and SA2 experiments, with a somewhat faster drop for SA2 than for SA1. In contrast, for the SD baseline experiments, the recognition rate drops rapidly when the training tokens reduce from ten to one; the drop is faster for the trended HMMs (with a greater number of model parameters) than for the conventional HMMs. It is further observed from the results in Tables 12.10-12.12 that the MAP estimates (both SA1 and SA2 adaptation experiments) approach the ML estimates (SD baseline experiments) in performance when the number of training tokens increases from one to ten. This is reassuring because under the asymptotic condition, the posterior density would be dominated by the sample data likelihood function in theory, which can be confirmed by an examination of the MAP estimates given by Eqs. 12.56-12.57 based on Eqs. 12.52-12.55 when $T \to \infty$.

One key concern for the benefit of the trended HMMs over the conventional HMM in terms of adaptation effectiveness is the greater number of model parameters of the former for a fixed number of HMM states. To address this concern, experimental results are presented, comparing the trended and conventional HMMs under the condition of approximately equalized number of model parameters. Table 12.13 shows the MAP speaker adaptation results for conventional HMMs (P=0) with the number of HMM states for each alphabet (English letters) from three (as shown in Table 12.10) to four, five, and six, with the proportionally increased number of model parameters. In this way, the model with six states (last three columns in Table 12.13) will have roughly the same number of model parameters as those of the linearly-trended HMM (Table 12.11). However, comparisons of the performance figures between Table 12.11 and Table 12.13 clearly indicate the consistently greater adaptation effectiveness of the trended HMM. This suggests some inherent superiority of using trended HMMs in speaker adaptation: the superiority does not come from the additional parameters they use. Rather, these additional parameters define a structure in the model expressed as the polynomial trajectory associated with each HMM state. During adaptation, each state-bound polynomial trajectory is adapted based on joint and optimal use of all adaptation frames automatically assigned to the state.

12.10 Case Study IV: Bayesian Adaptive Training for Compensating Acoustic Variability

In this final case study, a version of the Bayesian learning framework for adaptive training and recognition is presented. This framework is used as a robust compensation strategy to deal effectively with extraneous (unwanted) acoustic variations for (model-space) robust speech recognition. The strategy described here extends the earlier **speaker-adaptive training** (SAT) of [Ana 96], and uses HMM parameter transformations to normalize the extraneous variations in the training data according to a set of pre-defined conditions. A "compact" model and the associated prior PDF's of transformation pa-

No. of Adapt. Tokens	No. HMM states=4 (SI:66.5%)			No. HMM states =5 (SI: 70.2%)			No. HMM states =6 (SI:68.3%)		
	SD	SA1	SA2	SD	SA1	SA2	SD	SA1	SA2
1	57.2%	80.3%	70.4%	57.2%	79.5%	70.4%	57.2%	79.3%	70.8%
2	71.1%	83.2%	78.3%	71.1%	80.6%	77.7%	71.1%	83.3%	78.8%
3	78.2%	84.1%	80.5%	78.2%	82.5%	79.8%	78.2%	84.1%	80.2%
4	82.8%	84.9%	84.3%	82.7%	83.7%	83.9%	82.7%	84.4%	84.5%
5	85.2%	85.8%	85.6%	85.3%	84.0%	85.0%	85.3%	84.9%	85.4%
6	86.1%	86.8%	86.5%	86.1%	85.4%	85.5%	86.1%	85.2%	85.9%
7	87.3%	86.6%	86.1%	87.3%	85.1%	85.6%	87.3%	85.2%	86.9%
8	87.7%	86.5%	86.7%	87.7%	85.4%	86.7%	87.7%	85.0%	87.4%
9	87.9%	86.3%	86.5%	87.9%	85.5%	86.5%	87.8%	85.6%	86.9%
10	88.0%	86.9%	87.1%	88.0%	85.5%	86.6%	87.9%	85.5%	86.8%

Table 12.13: Comparison of speaker adaptation results for conventional HMMs (P=0) with a varying number of HMM states and of model parameters (after Rathinavelu and Deng [Rathi 01], ©IEEE).

rameters are estimated using the ML criterion. In the testing phase, the generic model and the prior PDF's are used to search for the unknown word sequence. The proposed strategy is evaluated on the Switchboard task, and is used to deal with three types of extraneous variations and mismatch in conversational speech recognition: pronunciation variations, inter-speaker variability and telephone handset mismatch. Experimental results show that moderate word error rate reduction is achieved in comparison with a well-trained baseline HMM system under identical experimental conditions. This case study has been based on the published work of [Jia 02].

12.10.1 Background

In the conventional statistical paradigm of ASR, statistical models are usually estimated based on a large amount of training data. Then the estimated models are used to recognize unknown utterances. The training data (also called the found data) are usually obtained under as many different conditions as possible for the purpose of properly representing all possible incoming speech data in future use. Even though the data collection conditions may differ due to a wide range of factors, the conventional paradigm treats all training data collected under different conditions in an identical manner by simply pooling them together. The model parameters are then determined from the pooled data set via parameter estimation techniques, e.g., maximum likelihood (ML) or discriminant training. The variations contained in the data come from many different sources, and some of these sources are germane to the recognition problem or task while others are extraneous. An apparent shortcoming of the above training paradigm is that the large amount of pooled training data not only includes the pertinent variability (such as phonetic distinction), but also involves many other extraneous variations that are irrelevant to our modeling or recognition purpose and should therefore be compensated for. We call those variations existing in the data that are not directly related to our modeling or recognition purpose as **extraneous variations**. Obviously, extraneous variation varies from problem to problem. For instance, in a typical case of speech recognition, it is important to model the phonetically relevant variation sources. All other variabilities are considered to be extraneous, including those arising from speaker, transducer, telephone channel, speaking style, speaking rate, pronunciation change, etc. On the other hand, for the speaker recognition problem, speaker variations become pertinent while other variations are extraneous. The extraneous variations have several realization levels. Here the extraneous variations at the acoustic level are considered only. All other issues related to phonetic or higher levels will be beyond the scope of this study.

In a conventional implementation of speech recognizers, one generally does not have an explicit mechanism to compensate for the extraneous variations in the training procedure. In particular, when we recognize spontaneous speech, where many types of extraneous variations abound, the performance of speech recognition can be significantly affected. In the training phase, due to the extraneous variations, the training data may diverge substantially from what is assumed in the model. This would make the estimated models diverge from the desired behavior. In the testing phase, the deviation due to the extraneous variations can also be viewed as a special kind of mismatch between the models and the testing data. We now describe a robust strategy to deal with the extraneous acoustic variations in the training phase in order to achieve a "generic" (or compact) model that better reflects the pertinent variations in the speech recognition tasks.

Recently, researchers began to realize the importance of compensating the extraneous variations in the training phase in order to improve the modeling capability of the models. The SAT technique proposed in [Ana 96] is an important step in the direction of a robust training strategy. In SAT, some linear regression transformations are often used to normalize inter-speaker variations in the speech data to construct a "compact" model. An iterative algorithm was proposed in [Ana 96] to estimate the parameters of both transformations and compact models in a sequential mode based on the ML criterion. The work reported in [Huo 99] shows another way to normalize "irrelevant variability" in the training phase for the purpose of learning a model structure (HMM state tying). More recently, in [Gales 98, Gales 00], another interesting robust training method was proposed for the same purpose as in the work reported here. In that work, several clusters are pre-defined and a canonical model is estimated for each cluster based on the ML criterion in the training phase. During testing, an interpolated model of all cluster-specific canonical models is constructed for each separate test utterance. The interpolating weights are estimated on-line from each current utterance.

In the case study described here, a new robust training strategy is presented to compensate for and to normalize the extraneous variations, with a solid Bayesian-theoretic foundation and with practical effectiveness. It differs from the previous work discussed above in its novel use of the *distribution* of the transformation parameters in a Bayesian framework. Briefly, each utterance in the training set is labeled with one of a set of pre-defined conditions, depending on the nature of the extraneous variation to be compensated, such as speaker ID, speaking style, pronunciation, transducer, transmission channel, etc. The data from different conditions are first normalized by using some appropriate transformations before they are pooled to estimate a "generic" (or compact) model. Meanwhile, a prior distribution of transformation parameters is also automatically estimated from the data to represent the knowledge of all possible transformations used across the various "conditions" in the training phase. In this way, the extraneous variation is adequately compensated for and the generic model will converge properly to represent the pertinent variations in question. In the testing/decoding phase, based on the generic model and the prior distribution of transformation parameters, a new search algorithm is used to decode the unknown input utterance according to Bayesian Predictive Classification (BPC) proposed in [Jia 99].

In order to obtain the "generic" acoustic models that can adequately describe phonetically relevant variation sources, the strategy presented here is used to normalize and/or compensate for several types of major extraneous variations in spontaneous speech recognition. Throughout this section, the Switchboard corpus is taken as the evaluation data set. There are several interesting aspects in this corpus for the evaluation of the robust training strategy. First, in the Switchboard task, the pronunciation variation in conversational speech is shown to be one major extraneous variation source hampering speech recognition. Thus, one can justifiably define the "condition" that characterizes the pronunciation variation, and in this case the proposed strategy can be employed to compensate for the pronunciation variation. Second, like SAT, the robust training strategy can be utilized to normalize the inter-speaker differences that also clearly exhibit themselves in the Switchboard corpus. Here, the "condition" is defined based on the speaker ID. Third, the robust training strategy is also used to normalize the extraneous variation related to the mismatches caused by different telephone handsets in the Switchboard corpus. Here, the information about the telephone number of each participant in both conversation sides in the Switchboard corpus is used to define the "condition." In order to facilitate the implementation, only a very simple transformation, with piecewise-linear functions, is used to normalize and/or compensate all of these three types of extraneous acoustic variations in the conversational telephony speech data of the Switchboard corpus.

12.10.2 Overview of the compensation strategy

Following the idea originally presented in [Ana 96], suppose we have a generic (or compact) mixture Gaussian HMM (Section 3.5) $\lambda_c = (\pi, \mathbf{A}, \boldsymbol{\theta})$ for each speech unit $\mathbf{W}$ that we desire to model, where π is the initial state distribution, $\mathbf{A} = \{a_{ij} \mid 1 \leq i, j \leq N\}$ is the transition matrix, and $\boldsymbol{\theta}$ is the parameter vector composed of mixture parameters $\boldsymbol{\theta}_i = \{w_{ik}, m_{ikd}, r_{ikd}\}_{k=1,2,\cdots,K; d=1,2,\cdots,D}$ for each state i, where K denotes the number of Gaussian mixture components in each state and D denotes the dimension of feature vectors. The state observation PDF is assumed to be a mixture of multivariate Gaussian PDF's with diagonal precision matrices:

$$p_i(o) = \sum_{k=1}^{K} w_{ik} \cdot \prod_{d=1}^{D} \sqrt{\frac{r_{ikd}}{2\pi}} \exp[-\frac{r_{ikd}}{2}(o_d - m_{ikd})^2]. \tag{12.68}$$

We denote all training data for $\mathbf{W}$ as $\mathbf{o} = \{\mathbf{o}^{(r)} \mid r = 1, 2, \cdots, R\}$, where $\mathbf{o}^{(r)}$ stands for those data collected under condition r and we have a total of R different conditions. The condition is defined according to the extraneous variations to be normalized, which will be explained for the specific examples in detail in the following subsections. Then, we aim to choose some proper transformations to normalize/compensate for the extraneous variations in speech signals. In other words, we need to choose a set of transformations $\{T_\eta^{(r)}(\cdot) \mid r = 1, 2, \cdots, R\}$, for the generic model λ_c. Each transformation $T_\eta^{(r)}(\cdot)$, which is parameterized by η, corresponds to a specific condition r so that for each condition r ($r = 1, 2, \cdots, R$) the transformed model $T_\eta^{(r)}(\lambda_c)$ gives a better description for the data $\mathbf{o}^{(r)}$ that are collected under this condition. The same SAT algorithm in [Ana 96] could be used to estimate the compact model λ_c and the corresponding transformations

$T_\eta^{(r)}(\cdot)$ according to the ML criterion. However, in the testing phase, it would not be appropriate to use the compact model λ_c to evaluate the testing data directly because λ_c would not match the original data due to the involved transformations. Furthermore, we do not know which transformation should be used for each testing utterance because we have no idea of which *condition* the test data come from. To solve this problem, Bayesian prediction can be used and has been implemented. The specific transformation parameters η for different "conditions" are viewed as some sampling outcomes from a prior PDF for the transformation parameters, denoted as $\rho(\eta)$. In the training stage, the prior $\rho(\eta)$ is simultaneously estimated to represent the knowledge of all transformations possibly used in the training stage. In the testing phase, the BPC algorithm helps to make an optimal decision given the information supplied by the prior $\rho(\eta)$.

Before deriving the robust training and testing algorithms, the functional form for the transformation $T_\eta^{(r)}(\cdot)$ and that for the prior PDF of transformation parameters $\rho(\eta)$ have to be carefully determined. First, for the transformation $T_\eta^{(r)}(\cdot)$, the requirements are: i) the transformation is sufficiently powerful to normalize the acoustic difference caused by extraneous variations; and ii) the transformation form is simple enough so that Bayesian prediction is tractable in the decoding phase. One possible choice is a piecewise linear transformation. In this case study, we choose the simplest transformation form, namely the bias vector plus the mean vector of an HMM:

$$m'_{ikd} = m_{ikd} + \beta_d \quad (d = 1, 2, \cdots, D), \tag{12.69}$$

where $\boldsymbol{\beta} = \{\beta_1, \beta_2, \cdots, \beta_d\}$ denotes the transformation parameters. We assume that all other HMM parameters remain unchanged. In principle, each transformation could be related or tied to any different segment of the speech signal. For simplicity, we assume, however, that each transformation is HMM state-dependent. That is, we use different transformations for different HMM states and the transformations of various states are tied based on the triphone state-tying in the entire HMM set. Second, to determine the prior PDF for the transformation parameters, $\rho(\boldsymbol{\beta})$, in order to have a simple form in the decoding stage, we choose the following prior PDF based on the concept of a natural conjugate prior [DeG 70]:

$$\rho(\boldsymbol{\beta}) = \prod_{d=1}^{D} \sqrt{\frac{\tau_d}{2\pi}} \exp[-\frac{\tau_d}{2}(\beta_d - \mu_d)^2], \tag{12.70}$$

where $\{\mu_d, \tau_d \mid d = 1, 2, \cdots, D\}$ are the hyperparameters.

12.10.3 Bayesian adaptive training algorithm

We now integrate the above robust adaptive training ideas into the conventional acoustic modeling method used in a large vocabulary speech recognition system; i.e., triphone model state tying based on phonetic decision trees [Young 94]. The algorithm that implements this strategy consists of the following steps:

1. Define a set of "conditions" according to the specific extraneous variations to be normalized or compensated. Specifically, we define a total of R different "conditions," and each is indexed by r $(1 \leq r \leq R)$.

2. Build a baseline system based on the conventional HMM approach.

3. Align all speech utterances in the whole training set to obtain the Viterbi segmentation for each utterance at the HMM's state level; then, label each frame of feature vectors with the "condition" r $(1 \leq r \leq R)$ where the feature vector belongs.

4. Tying states of all triphone models and parameter estimation: build a single decision-tree for each state of phone models, based on all data belonging to its corresponding triphone states. For each tied state of the tri-phone models (i.e., every leaf node in the decision tree):

   (a) According to the above alignment results, pool all labeled data together, $\mathbf{o} = \{\mathbf{o}^{(r)} \mid r = 1, 2, \cdots, R\}$, where $\mathbf{o}^{(r)}$ denotes all data labeled with condition r. Use the state distribution in the current leaf node as the initial estimate of the compact model λ_c for this tied state. Here, λ_c is a mixture Gaussian model, i.e., $\lambda_c = \{w_k, m_k, r_k \mid 1 \leq k \leq K\}$.

   (b) Given the current λ_c, estimate all R transformations $\{T_\eta^{(r)}(\cdot) \mid r = 1, 2, \cdots, R\}$ for each condition r based on the data $\mathbf{o}^{(r)} = \{o_t^{(r)} \mid 1 \leq t \leq T^{(r)}\}$ (See [Jia 02] for detailed derivation):

   For each dimension $d = 1, 2, \cdots, D$ (using $\beta^{(r)}[d] = 0$ as initialization),

$$\beta^{(r)}[d] = \frac{\sum_{t=1}^{T^{(r)}} \sum_{k=1}^{K} \xi_t^{(r)}(k) \cdot r_{kd} \cdot (o_{td}^{(r)} - m_{kd})}{\sum_{t=1}^{T^{(r)}} \sum_{k=1}^{K} \xi_t^{(r)}(k) \cdot r_{kd}}, \tag{12.71}$$

   where $\xi_t^{(r)}(k)$ denotes the posterior probability that the mixture component l_t takes the value of k, i.e.,

$$\xi_t^{(r)}(k) = \Pr(l_t = k \mid o_t^{(r)}, \boldsymbol{\beta}^{(r)}) = \frac{w_k \cdot \mathcal{N}(o_t^{(r)} \mid m_k + \boldsymbol{\beta}^{(r)},\ r_k)}{\sum_{k=1}^{K} w_k \cdot \mathcal{N}(o_t^{(r)} \mid m_k + \boldsymbol{\beta}^{(r)},\ r_k)} \tag{12.72}$$

   (c) Re-estimate the compact model λ_c (See [Jia 02] for detailed derivation):

   For $1 \leq k \leq K$ and $1 \leq d \leq D$

$$m_{kd} = \frac{\sum_{r=1}^{R} \sum_{t=1}^{T^{(r)}} \xi_t^{(r)}(k) \cdot (o_{td}^{(r)} - \beta^{(r)}[d])}{\sum_{r=1}^{R} \sum_{t=1}^{T^{(r)}} \xi_t^{(r)}(k)} \tag{12.73}$$

$$r_{kd} = \frac{\sum_{r=1}^{R} \sum_{t=1}^{T^{(r)}} \xi_t^{(r)}(k)}{\sum_{r=1}^{R} \sum_{t=1}^{T^{(r)}} \xi_t^{(r)}(k) \cdot (o_{td}^{(r)} - m_{kd} - \beta^{(r)}[d])^2} \tag{12.74}$$

$$w_k = \frac{\sum_{r=1}^{R} \sum_{t=1}^{T^{(r)}} \xi_t^{(r)}(k)}{\sum_{r=1}^{R} \sum_{t=1}^{T^{(r)}} \sum_{k=1}^{K} \xi_t^{(r)}(k)} \tag{12.75}$$

   (d) Go to step Eq. 4b unless some convergence conditions are met.

   (e) Estimate the hyperparameters $\{\mu_d, \tau_d \mid 1 \leq d \leq D\}$ of the prior PDF $\rho(\boldsymbol{\beta})$ for the current tied state:

   For $1 \leq d \leq D$,

$$\mu_d = \frac{\sum_{r=1}^{R}\sum_{t=1}^{T^{(r)}}\sum_{k=1}^{K} \xi_t^{(r)}(k)\cdot\beta^{(r)}[d]}{\sum_{r=1}^{R}\sum_{t=1}^{T^{(r)}}\sum_{k=1}^{K} \xi_t^{(r)}(k)} = \frac{\sum_{r=1}^{R} T^{(r)}\cdot\beta^{(r)}[d]}{\sum_{r=1}^{R} T^{(r)}} \quad (12.76)$$

$$\tau_d = \frac{\sum_{r=1}^{R}\sum_{t=1}^{T^{(r)}}\sum_{k=1}^{K} \xi_t^{(r)}(k)}{\sum_{r=1}^{R}\sum_{t=1}^{T^{(r)}}\sum_{k=1}^{K} \xi_t^{(r)}(k)\cdot(\beta^{(r)}[d]-\mu_d)^2} = \frac{\sum_{r=1}^{R} T^{(r)}}{\sum_{r=1}^{R} T^{(r)}\cdot(\beta^{(r)}[d]-\mu_d)^2}, \quad (12.77)$$

where $\{\mu_d, \tau_d \mid 1 \leq d \leq D\}$ are tied for all HMM states related to the current leaf node.

12.10.4 Robust decoding using Bayesian predictive classification

According to [Jia 99], the **Bayesian predictive classification** (BPC) decision rule makes a speech recognizer minimize the overall recognition error when the expectation is taken with respect to the model parameter uncertainty described by its prior PDF. Assuming that the functional form of the parameter transformation is exactly known and that all available information about the transformation parameters is completely contained in the prior PDF $\rho(\eta)$, given unknown observation $\mathbf{o}$, then we can express such an optimal recognition result $\hat{\mathbf{W}}$ as

$$\begin{aligned}\hat{\mathbf{W}} &= \arg\max_{\mathbf{W}} \Pr(\mathbf{W})\cdot\int_\eta \Pr(\mathbf{o} \mid T_\eta(\lambda_c), \mathbf{W})\cdot\rho(\eta)\,\mathrm{d}\eta \\ &= \arg\max_{\mathbf{W}} \Pr(\mathbf{W})\cdot\int_\eta \sum_{s,l} \Pr(\mathbf{o}, s, l, \mid T_\eta(\lambda_c), \mathbf{W})\cdot\rho(\eta)\,\mathrm{d}\eta \\ &\approx \arg\max_{\mathbf{W}} \Pr(\mathbf{W})\cdot\max_{s,l}\int_\eta \Pr(\mathbf{o}, s, l, \mid T_\eta(\lambda_c), \mathbf{W})\cdot\rho(\eta)\,\mathrm{d}\eta, \end{aligned} \quad (12.78)$$

where s, l denote the HMM state sequence and the Gaussian component label sequence, respectively. In the above equation, the Viterbi approximation [Jia 99] has been adopted to make the integral tractable. This approximate BPC is called Viterbi BPC or VBPC. Based on the work in [Jia 99], where prior PDF's are defined for all HMM parameters and the integral is taken with respect to the HMM parameters, we now introduce a new transformation-based structure constraint into the BPC method. Therefore, Eq. 12.78 can be thought as a kind of "constraint-based" VBPC.

In this improved BPC, the optimal HMM parameters for each testing utterance are assumed to follow some constraints, which are established by applying transformations into a known "generic" HMM λ_c. The transformations are known exactly except for a small set of parameters η treated as random variables. It is further assumed that our prior knowledge about the transformation parameters η is contained in a prior PDF $\rho(\eta)$. Under these assumptions, the optimal decision rule will be the "constraint-based" BPC shown in Eq. 12.78. When the number of the transformation parameters is much fewer than that of HMM ones, the "constraint-based" BPC makes it easier to determine the prior PDF.

The linear transformation as shown in Eq. 12.69 for $T_\eta(\cdot)$ is adopted, where each transformation is associated with an HMM state. We now present a frame-synchronous search algorithm to implement the above "constraint-based" BPC rule; the search algorithm was modified from the general BPC algorithm presented in [Jia 99]. According to Eq. 12.78, the value of the integral depends on the path in the HMM. This makes it difficult to derive a recursive algorithm to compute an accurate value of the integral. The solution to this problem adopted was to incorporate the calculation of the integral into the Viterbi search. For each time frame, the integration over the transformation parameters is computed for all active hypothesized partial paths. Then, for each node in the search network, all incoming partial paths are merged by selecting the one with the largest integral value. The selected path is propagated and the integral is recomputed according to the extended partial path. The search procedure is then repeated until the end of the utterance. In this way, it is possible to achieve a Viterbi approximation of the integral.

Given a test utterance $\mathbf{o} = (o_1, o_2, \cdots, o_T)$, the generic model λ_c, and the prior PDF $\rho(\boldsymbol{\beta})$ shown in Eq. 12.70 (with the hyperparameters estimated from Eqs. 12.76 and 12.77, the recursive search procedure for accomplishing the computation in Eq. 12.78 is described as follows:

$$\tilde{b}_i(o_t) = \arg \max_{1 \le k \le K} \int w_{ik} \cdot \mathcal{N}(o_t | m_{ik} + \boldsymbol{\beta}_i, r_{ik}) \cdot \rho(\boldsymbol{\beta}_i)\, \mathrm{d}\boldsymbol{\beta}_i$$

$$= \arg \max_{1 \le k \le K} \quad w_{ik} \cdot \prod_{d=1}^{D} \sqrt{\frac{\tau_d^{(i)} r_{ikd}}{2\pi(\tau_d^{(i)} + r_{ikd})}} \exp[-\frac{\tau_d^{(i)} r_{ikd}}{2(\tau_d^{(i)} + r_{ikd})}(o_{td} - m_{ikd} - \mu_d^{(i)})^2]$$

where $\boldsymbol{\beta}_i$ denotes the transformation parameters related to state i. Here, $\delta_t(i)$ denotes the partial predictive value based on the optimal partial path arriving at state i at the time instant t. The corresponding best partial path is represented by a chain of points starting from $\psi_t(i)$.

(1) Initialization (for $t = 1$)

$$\delta_1(i) = \pi_i \cdot \tilde{b}_i(o_1) \quad 1 \le i \le N \tag{12.79}$$

$$\psi_1(i) = 0 \quad 1 \le i \le N \tag{12.80}$$

(2) Recursion: for $2 \le t \le T$, $1 \le j \le N$, do

(2.1) Path-merging in state j:

$$\delta_t(j) = \max_{1 \le i \le N} [\, \delta_{t-1}(i) \cdot a_{ij}] \tag{12.81}$$

$$\psi_t(j) = \arg \max_{1 \le i \le N} [\, \delta_{t-1}(i) \cdot a_{ij}] \tag{12.82}$$

(2.2) Update the partial predictive value:

If the state j is involved for the first time in the computation of $\delta_t(j)$),then

$$\delta_t(j) = \delta_t(j) \times \tilde{b}_j(o_t); \tag{12.83}$$

else

$$\delta_t(j) = \delta_t(j) \times \frac{\tilde{b}_j(o_{j1}, o_{j2}, \cdots, o_{jL_j})}{\tilde{b}_j(o_{j1}, o_{j2}, \cdots, o_{j(L_j-1)})}, \tag{12.84}$$

where L_j is the accumulated number of feature vectors belonging to state j, based on the optimal partial path up to the time instant t; o_{j_i} denotes the ith vector in the state j; and $\tilde{b}_j(o_{j_1}, o_{j_2}, \cdots, o_{j_{L_j}})$ denotes the contribution of data $\{o_{j_1}, o_{j_2}, \cdots, o_{j_{L_j}}\}$, residing in state j, to the partial predictive value $\delta_t(j)$:

$$\tilde{b}_j(o_{j_1}, o_{j_2}, \cdots, o_{j_n}) = \int p(o_{j_1}, o_{j_2}, \cdots, o_{j_n} \mid m_{jk} + \boldsymbol{\beta}_j, r_{jk}) \cdot \rho(\boldsymbol{\beta}_j)\, \mathrm{d}\boldsymbol{\beta}_j. \tag{12.85}$$

(3) Termination

$$\tilde{p}(\mathbf{o}|\mathbf{W}) \approx \max_i \ \delta_T(i) \tag{12.86}$$

$$s_T^* = \arg\max_i \delta_T(i). \tag{12.87}$$

(4) Path (state sequence) Backtracking

$$s_t^* = \psi_{t+1}(s_{t+1}^*) \qquad t = T-1, T-2, \cdots, 1. \tag{12.88}$$

In Eq. 12.85, $\tilde{b}_j(o_{j_1}, o_{j_2}, \cdots, o_{j_n})$ can be approximated based on the "closest" mixture component label sequence corresponding to the data $\{o_{j_1}, o_{j_2}, \cdots, o_{j_n}\}$:

$$\begin{aligned} \tilde{b}_j(o_{j_1}, o_{j_2}, \cdots, o_{j_n}) &\approx \prod_{k=1}^{K} w_{jk}^{L'_k} \cdot \tilde{f}_{jk}(o_{l_1^k}, \cdots, o_{l_{L'_k}^k}) \\ &= \prod_{k=1}^{K} w_{jk}^{L'_k} \cdot \prod_{d=1}^{D} \tilde{f}_{jkd}(o_{l_1^k d}, \cdots, o_{l_{L'_k}^k d}) \end{aligned} \tag{12.89}$$

where $\{o_{j_1}, o_{j_2}, \cdots, o_{j_n}\}$ denote feature vectors belonging to state j in $\mathbf{o}$, among which $l_1^k \cdots l_{L'_k}^k$ denote labels of the vectors "closest" to the mixture component k of state j. Then we have

$$\tilde{f}_{jkd}(o_{1d}, o_{2d}, \cdots, o_{vd}) = \sqrt{(\frac{r_{jkd}}{2\pi})^v \frac{\tau_d^{(j)}}{v r_{jkd} + \tau_d^{(j)}}}$$
$$\exp[-\frac{v r_{jkd}}{2}(\overline{o_v^2} - \overline{o}_v^2)] \exp[-\frac{v r_{jkd} \tau_d^{(j)}}{2(v r_{jkd} + \tau_d^{(j)})}(\overline{o}_v - \mu_d^{(j)})^2],$$

where

$$\overline{o}_v = \frac{1}{v} \sum_{i=1}^{v} (o_{id} - m_{jkd}) \tag{12.90}$$

and

$$\overline{o_v^2} = \frac{1}{v} \sum_{i=1}^{v} (o_{id} - m_{jkd})^2. \tag{12.91}$$

In the above VBPC, a single best path is searched to compute the integral-based

	Sub	Del	Ins	WER
10-hr baseline	43.94	17.92	3.49	65.39
RobustPron-I	41.94	18.58	4.31	64.84
RobustPron-II	42.61	18.17	3.90	64.68

Table 12.14: Performance (in %) comparison with the 10-hr baseline system when the robust method is used to deal with pronunciation variations (after Jiang and Deng [Jia 02], ©IEEE).

Bayesian prediction instead of calculating the integral over all possible paths, as shown in Eqs. 12.78 and 12.89. As in [Jia 99], VBPC is found to generally lead to a rather good approximation because the contribution from the best path almost always dominates the entire Bayesian prediction.

12.10.5 Experiments on spontaneous speech recognition

The Bayesian adaptive training algorithm and the related BPC-based decoding algorithm have been applied to several types of extraneous acoustic variations in spontaneous speech recognition. In this case study, a fast evaluation set of the Switchboard corpus is used, which is approximately 10 hours in duration. It is called the "10-hr" set hereafter. In the following recognition experiments, a baseline system is built first from the 10-hr training data according to the conventional training method. Then, starting from the baseline system, the new training method is used to deal separately with three different types of the extraneous variations in the Switchboard task, i.e., the pronunciation variation, speaker difference and handset mismatch. The experimental setup and comparative results will be reported in this section, together with some discussions of the experimental results and of the computation complexity of the algorithm used.

The 10-hr baseline system

HTK v2.2 is used to implement the baseline system [Young 98], where a 39-dimension feature vector is used that consists of 12 MFCC's with log-energy, and corresponding delta and acceleration coefficients. Cepstral mean normalization is performed at the utterance level for both training and testing data. The acoustic models are 3-state, 5-mixture-per-state, word-internal triphone HMM's. The standard phonetic decision-tree method is used for state-tying. After the tying, the total number of all distinct tied-states is reduced to approximately 2000. The dictionary consists of all words (about 6474 words) occurring in the 10-hr training set. Multiple pronunciations are used for some words. The language model is the back-off bigram model trained only on the transcriptions of all utterances in the 10-hr set. The test set consists of 200 utterances (a total of 1948 words) randomly selected from the evaluation test set in WS96, which is disjointed from the 10-hr set.

The recognition performance of the baseline system with these 200 test utterances is shown in the first line of Table 12.10.5, i.e., with a 65% word error rate (WER). The performance is close to the best baseline results.

	Sub	Del	Ins	WER
10-hr baseline	43.94	17.92	3.49	65.39
RobustSpk	43.28	17.57	3.33	64.18

Table 12.15: Performance (in %) comparison with the 10-hr baseline system when the robust method is used to deal with speaker difference (after Jiang and Deng [Jia 02], ©IEEE).

	Sub	Del	Ins	WER
10-hr baseline	43.94	17.92	3.49	65.39
RobustHandset	42.94	18.20	3.44	64.58

Table 12.16: Performance (in %) comparison with the 10-hr baseline system when the robust method is used to deal with handset mismatch (after Jiang and Deng [Jia 02], ©IEEE).

Dealing with pronunciation variations

According to [Byrn 97, McA 98], in the Switchboard task, pronunciation variation in conversational speech is a major extraneous variation source hampering speech recognition performance. How to cope with pronunciation variations in conversational speech recognition has been studied by many researchers, e.g., [Byrn 97]. It is straightforward to incorporate multiple pronunciations in the search network for some words. However, this strategy also increases the perplexity of the search network and makes it more confusable. In this experiment, attempts were made to deal with the pronunciation variations by using our robust training strategy. In principle, the speech data that come from the same word may be treated as from different "conditions" if the word is pronounced differently. That is, the set of all "conditions" can be defined by all distinct pronunciations of all words in the vocabulary. In this way, the above robust training approach can be directly used to normalize for the acoustic variations caused by pronunciation differences.

One of the most important implementation issues here is how to define the speech conditions and partition the data into different conditions. It is crucial to have a good tradeoff between the number of conditions and the amount of data used for each condition. In order to obtain reliable estimation for the transformation parameters for each condition, it is important to ensure that there is enough data for each condition. In this case study, the baseline recognition system and a phoneme recognizer were used to automatically determine pronunciation of every word in the training data. The following two methods were explored in defining a *condition* for pronunciation variations:

(I) Each utterance in the training set is aligned with its text word by word, based on its transcription by using the baseline system to obtain the segmentation information for every single word. Then, phone recognition is performed on acoustic phone models in the baseline system for each word according to the above alignment boundary. The phoneme recognition results are viewed as the pronunciation of this word. However, this method usually causes too many pronunciations for each word. Thus, a very simple measure is used for the distance between two pronunciations, e.g., the number of different phonemes, to cluster all different pronunciations of each word into four classes or fewer. In training step (4a) of section 12.10.3, all data from the same word and the same pronunciation class are treated as coming from the same condition. This method is denoted as *RobustPhon-I* hereafter.

(II) Phoneme recognition is directly performed for each utterance in the training set to obtain the phoneme sequence for the sentence by using the baseline system, where the phoneme HMM's are used without any language model. In training step (3) described in section 12.10.3, each feature vector is labeled with the recognized phoneme where the vector belongs. When doing decision-tree state-tying in training step (4a), all data in this state that correspond to the same recognized phoneme are treated as coming from the same condition. This method is denoted as *RobustPhon-II* hereafter.

The *RobustPhon-I* and *RobustPhon-II* methods are implemented under the same experimental condition as that of the baseline system. From the comparative experimental results in Table 12.10.5, where **Sub**, **Del** and **Ins** denote the substitute, deletion, and insertion error rate, respectively, it is observed that the robust training method gives close to 1% reduction in word error rate (WER) over the baseline system. It is also noted that, for the 10-hr data set, the *RobustPhon-II* achieves somewhat better results because *RobustPhon-I* usually causes too many conditions and, in turn, too few training data for some conditions.

In both *RobustPhon-I* and *RobustPhon-II*, we can identify several factors that influence the final performance. The first factor is the poor phoneme recognition results when transcribing the Switchboard data by using the baseline system. The high error rate in phoneme recognition causes the conditions related to pronunciation variations not to be well defined. The second factor is that the functional form Eq. 12.69 for transformation may not be powerful enough to normalize for acoustic variations caused by the pronunciation changes.

Dealing with inter-speaker differences

Inter-speaker difference is another major source of extraneous variations for any speaker-independent speech recognition system. In this experiment, the robust training strategy discussed above is used to compensate/normalize for the inter-speaker difference in the Switchboard data. Here, each "condition" in training step (1) is related to each speaker in the training set (Section 12.10.3). Then, in training step (3), every feature vector in the entire training set is labeled with the speaker who utters the current sentence. In step (4a), for every tied state, all speech data which come from the same speaker are considered under the same condition. This method is denoted as *RobustSpk* hereafter. Recognition performance comparison of *RobustSpk* with the baseline system is shown in Table 12.10.5. From the results, we note that when the robust training strategy is used to normalize for the inter-speaker difference, a 1.2% WER reduction over the baseline system was achieved. We also see that the performance improvement here is somewhat larger than for both *RobustPhon-I* and *RobustPhon-II*. One possible reason is that the condition here is well-defined because the condition is decided solely by the speaker ID and is independent of the performance of the baseline system.

Dealing with handset mismatches

The Switchboard data consist of recorded telephone conversations among a set of registered participants. A participant would initiate a conversation by calling an automaton that would find another participant to receive the call. The automaton would note the telephone numbers used by both participants. We generally assume that when the phone numbers are the same, the handsets are also the same, though there may be exceptions. Based on the information of telephone numbers, the robust training strategy presented earlier in this section is similarly used to normalize the acoustic variations caused by handset mismatch. Here, each condition is related to one telephone number recorded in the training set. In training step (4a), all training data from the same telephone number are considered to be under the same condition. This method is denoted as *RobustHandset* hereafter. The comparative results in Table 12.10.5 show that the robust training method *RobustHandset* achieves nearly 1% WER reduction compared to the baseline system.

Discussion

Although moderate WER reduction for the Switchboard task has been observed from all the promising experimental results above, the performance improvement is smaller than what one could have expected. One possible reason is that the Switchboard task is an extremely difficult one, and the data contain many other types of variabilities that have not been addressed in this case study.

One important issue is the computational complexity of the new robust training approach. Compared with the conventional training method and SAT, the robust training algorithm here does not significantly increase the computational complexity. However, as discussed in [Jia 99], the decoding algorithm based on BPC demands much more computation or memory overload than the normal Viterbi search algorithm. As shown in [Jia 99, Jia 98], this usually does not cause any problem for small-vocabulary tasks. For the Switchboard task, where the search network is constructed from a vocabulary of several thousand words, a fast implementation version of the VBPC search algorithm usually requires a memory greater than 1,000 Megabytes. Although the fast version of the VBPC search has a similar running speed to the normal Viterbi search, memory requirements are not affordable for most current machines. Thus, it is very important to have a good programming design to achieve a good tradeoff between the speed and memory. Even so, in most cases, in order to have an acceptable speed of response, heavy pruning is necessary in the search algorithm.

From the experimental results presented here, over 1% (absolute) WER reduction is observed separately for each type of extraneous variations. It will be interesting to see whether one can have additive improvements when the method is used to jointly deal with all three types of variations. In doing so, however, one will face a serious problem of "sparse data" when jointly normalizing three types of variations. For example, the total number of "conditions" is only from several tens to several hundreds, as described in Subsections 12.10.5, 12.10.5 and 12.10.5. If we jointly deal with three types of variations, the total number of "conditions" will increase to around one million. The training data will not be enough for most conditions unless we have a good method to tie some conditions together.

12.11 Statistical Language Modeling

12.11.1 Introduction

Most of the sections in this chapter until now have focused on acoustic modeling for $P(\mathbf{O}|\mathbf{W})$, one of the three basic problems in ASR as exemplified in the fundamental equation of Eq. 12.1 described in Section 12.2. The third problem, also called search or decoding, is a sequential optimization problem for the word sequence $\mathbf{W}$ that maximizes the posterior probability $P(\mathbf{W}|\mathbf{O})$ as shown in Eq. 12.1. There is a vast literature on many solutions to this problem, which will not be covered in this book. For interested readers, we refer them to excellent references in [Ney 00] and in (Chapters 12 and 13) of the book [Huang 01].

This final section of the chapter is devoted to language modeling, the second basic problem of ASR according to Eq. 12.1. A statistical language model is a probability distribution $P(\mathbf{W})$ over all possible sentences $\mathbf{W} = w_1, w_2, ..., w_n$. It plays the role of the prior for ASR in the Bayesian framework exemplified by the fundamental equation Eq. 12.1. Most of the current language models decompose the probability of a sentence $\mathbf{W}$ into a product of conditional probabilities (discussed in Section 3.2) according to

$$P(\mathbf{W}) = P(w_1, w_2, ..., w_n) = \prod_{i=1}^{n} P(w_i|w_1, w_2, ..., w_{i-1}) = \prod_{i=1}^{n} P(w_i|h_i),$$

where w_i is the ith word in the sentence, and $h_i = w_1, w_2, ..., w_{i-1}$ denotes the history for word w_i.

Language modeling is critical to many applications that process human language with less than complete knowledge. In addition to the applications for ASR, it also finds important applications for machine translation, optical character recognition, handwriting recognition, and spelling and grammar correction (which will not be covered in this section). Formal theories of grammar so far have not proved successful to account adequately for actual natural usage of language. Generalization of formal grammars to include statistical distributions have also not proved to be superior in practice to simple Markov models (N-grams, to be discussed right after this subsection) estimated from larger amounts of data. However, with the advent of huge and increasing amounts of textual corpora, a breakthrough in language modeling can be expected when we successfully integrate linguistic knowledge with statistical learning techniques.

This section is confined to the applications of language modeling for ASR applications. As we discussed earlier, one popular measure of the difficulty of the ASR task and the quality of the language model is perplexity. Perplexity can be interpreted as the geometric average branching factor of the language according to the model. It is a function of both the language and the model, and it measures how good the model is: the better the model, the lower the perplexity (although many counter-examples exist). In the remaining portion of this section, we will describe several existing major types of language models.

12.11.2 N-gram language modeling

Definition

The N-grams are the state-of-the-art language model underlying much of the current ASR technology. Nearly all commercial ASR products use an N-gram language model in one form or another. Essentially, an N-gram is a Markov-chain approximation to the general, rigorous language model of Eq. 12.11.1 by ignoring elements of history longer than N words in the past:

$$P(w_i|h_i) \approx P(w_i|w_{i-N+1}, ..., w_{i-2}, w_{i-1}).$$

This then defines the N-gram language model as

Definition of the N-gram language model:

$$P(\mathbf{W}) = P(w_1, w_2, ..., w_n) = \prod_{i=1}^{n} P(w_i|w_{i-N+1}, ..., w_{i-2}, w_{i-1}). \tag{12.92}$$

As we recall from Section 3.4.3, this N-gram model is simply the Markov property (of order N) expressed as the conditional PDFs.

In the N-gram model, the value of N is chosen based on the balance between the training reliability (the higher the N, the less reliable in estimation for a given set of training data) and the model accuracy (the higher the N, the more accurate the model if adequately trained). The most common trigram model is obtained when setting $N = 3$ in Eq. 12.92, which is used with large training corpora (millions of words or more). The bigram model where $N = 2$ in Eq. 12.92 is frequently used with smaller training corpora. The monogram language model (where $N = 1$ or $P(w_i|h_i) \approx P(w_i)$, where h_i is the prior word history of the utterance) and the flat language model (or zero-gram model where $P(w_i|h_i) \approx$ constant) have very little or no power as the prior in constraining the word sequence in language.

Parameter estimation

It can be easily shown that the maximum likelihood (ML) estimate for the N-gram model parameters is a procedure of counting followed by division. Estimating the trigram and even the bigram (discrete) probabilities using this ML estimate, however, is a well known sparse estimation problem, even under the condition of very large corpora. Simply said, the conventional ML technique based on straightforward counting does not give reliable estimates for the N-gram model parameters.

To improve estimation reliability for the N-gram models, one needs to use smoothing techniques. Common smoothing techniques include backing off to lower-order N-grams, linear interpolation of N-grams with different orders, discounting the ML estimates, and variable-length N-grams.

In addition to smoothing, another common technique to overcome sparseness is word clustering. Suppose word w_j is clustered into class C_j. Then we can approximate the bigram probability by

$$P(w_i|w_{i-1}) \approx P(w_i|C_{i-1}),$$

and the trigram probability can be approximated by

$$P(w_i|w_{i-1}, w_{i-2}) \approx P(w_i|C_{i-1}, C_{i-2}),$$

or

$$P(w_i|w_{i-1}, w_{i-2}) \approx P(w_i|C_i)P(C_i|w_{i-1}, w_{i-2}),$$

or

$$P(w_i|w_{i-1}, w_{i-2}) \approx P(w_i|C_i)P(C_i|C_{i-1}, C_{i-2}).$$

The counting based on the classes on the right-hand sides of the above is always much more reliable than that based on the word on the left-hand sides. The techniques for deriving the clustering classes include manual clustering of semantic categories, manual clustering of syntactic categories (e.g., parts of speech), and automatic, iterative clustering based on information-theoretic criteria.

12.11.3 Decision-tree language modeling

Decision tree algorithms were developed and applied to language modeling in [Bahl 89]. A decision tree partitions the space of the histories h of word w by asking binary questions about the history at each of the internal nodes. The training data at each leaf are then used to construct a probability distribution $P(w|h)$ over the next word. To improve the reliability of the estimate, the leaf distribution is interpolated with the internal node distributions in the decision tree found along the path to the node. The decision trees are determined by "greedily" selecting the most informative question. Pruning and cross validation are applied also in finalizing the decision tree.

The main technical challenge in decision-tree language modeling is the huge space of word histories and the even greater space of possible questions. To overcome these difficulties, the classes of questions are strongly constrained and efficiently searched. Rapid optimal binary partitioning of the vocabulary is also developed.

12.11.4 Context-free grammar as a language model

A context-free grammar (CFG) is defined by a vocabulary (a set of words usually, or more formally, a collection of terminal symbols), a set of non-terminal symbol, and a set of production rules. It is a well-understood generative model for natural language. A CFG is generative in that sentences can be generated from it by starting with an initial non-terminal symbols, followed by repeated applications of the production rules. Each application of the rules transforms a non-terminal symbol into a sequence of terminal symbols (words) and non-terminal ones. The generative process continues until the generated sequence contains no non-terminal symbols. A CFG can be created from parsed and annotated corpora.

Compared with the N-gram model that largely ignores the linguistic structure, the CFG-based language model reflects such a structure. The structure is usually expressed as a set of rules that have a single atomic grammatical category on the left-hand side of each rule, and a sequence of atomic categories and words on the right-hand side. Examples of the rules are

```
S  -> NP VP   (Sentence consists of Noun Phrase followed by Verb Phrase)
NP -> Det N   (Noun Phrase consists of Determinant followed by Noun)
NP -> NP PP   (or NP consists of another NP followed by Preposition Phrase)
PP -> P NP    (PP consists of Preposition followed by Noun Phrase)
VP -> V NP    (Verb Phrase consists of Preposition followed by Noun Phrase)
N  -> word    (Noun consists of word)
...
```

One important characteristic of the rules such as the above is the ability to define the grammatical categories via mutual recursions. For example, in the third rule above, NP appears on both the left- and right-hand sides. This represents the gross surface structure of natural language expressions, and makes it suitable to define linguistically-based language models.

The use of the structure of natural language is important if the amount of training data is not sufficient to estimate all parameters of the N-gram language model. The structure captured by the CFG allows a generalization to be made from one set of data to another set. However, a major challenge of using a CFG as the language model for speech recognition is the **coverage problem** where the designed CFG (usually for a specific speech recognition task) may not validate all possible sentences uttered by speakers. Another major challenge is the difficulty of the search (decoding) component in a speech recognizer that looks for the optimal word sequence that is constrained by the grammar specified by the CFG. This latter challenge has been addressed by intensive efforts that efficiently incorporated CFGs into the search architecture in real-time speech recognizers [Moore 99].

Probabilistic CFG

A principled extension of the CFG discussed so far is the probabilistic CFG. This is a statistical natural language grammar that can be used as a statistical language model. Note that in the conventional CFG, the probability of a sentence is limited to one or zero only, depending on whether it follows the CFG rules or not. In contrast, in the probabilistic CFG used as a language model, the probability of a sentence is usually a continuous value, much like N-gram language models.

In the probabilistic CFG, each CFG rule is associated with a conditional probability for the right-hand-side symbol(s) given the left-hand-side one(s). As an example, consider the CFG rule:

```
NP -> NP PP,
```

which could be assigned a probability of 0.3. This would mean that given a phrase of category NP (noun phrase), there is a probability of 0.3 that it consists of another NP category followed by a category of preposition phrase (PP) (e.g., *a man with a telescope*). Let the top-level category of the grammar have a probability of one; then this uniquely defines a probability for each analysis tree defined by the CFG grammar rules. Such a probability for the tree is determined by multiplying all the conditional probabilities for the rules that are used by the analysis, and is used as the probability for the analyzed sentence as the language model score.

Thus, a generalization of the CFG gives rise to the probabilistic CFG, where a probabilistic distribution is assigned to the transitions from each non-terminal symbol. This induces a distribution over an entire set of sentences. The EM algorithm for ML estimation of these transition probabilities using annotated corpora has been developed, and is called the Inside-Outside algorithm. It is known, however, that the Inside-Outside algorithm suffers severely from the local optimum problem.

12.11.5 Maximum-entropy language modeling

The maximum-entropy (ME) model is also called the exponential model [Rosen 96]. It solves the problem of data fragmentation that is associated with all models discussed so far. Data fragmentation refers to the fact that more detailed modeling necessarily leads to less data in estimating new parameters. For example, as the decision tree grows, each leaf contains fewer and fewer data points for parameter estimation. Likewise, as the N in the N-gram model grows, fewer and fewer data are available for training the N-gram probabilities.

The problem of this kind of data fragmentation can be addressed by the ME or exponential model, which constrains the word-history conditional probability by the fixed form of

$$P(w|h) = \frac{1}{\mathbf{z}(h)} \exp\left[\sum_i \lambda_i f_i(h,w)\right],$$

where λ_i's are the model parameters, $\mathbf{z}(h)$ is a normalization term, and the features $f_i(h,w)$ are arbitrary functions of the word-history pair.

It can be shown that the ML estimate for this exponential model is consistent with the ME distribution, which is the distribution with the highest entropy among all distributions satisfying a constraint for the ML estimate of the parameters in Eq. 12.11.5. This constraint provides the strength of incorporating arbitrary knowledge sources while avoiding data fragmentation.

One major weakness of the ME model is the computational burden for optimization, which makes it sometimes infeasible. In addition, the choice of features included in the model of Eq. 12.11.5 remains a challenging problem.

12.11.6 Adaptive language modeling

Adaptive language modeling addresses the issue of dealing with non-homogeneous natural language sources. This non-homogeneity includes varying language topics, genres, and styles, as well as varying times when the language (or speech) is produced.

When the test data come from a source to which the language model has not been exposed during training, those test data are the only useful information for language model adaptation. This problem is called cross-domain adaptation. One very effective technique for solving this type of adaptation problem is to continuously update the history, h, of each test word, w, and to use it to create a run-time, dynamic N-gram probability called the cache probability $P_{cache}(w|h)$. The final adapted language model probability is computed by interpolating this cache probability with the static language model probability:

$$P_{adapted}(w|h) = \lambda P_{cache}(w|h) + (1-\lambda)P_{static}(w|h).$$

In the above, the interpolation weight λ is optimized separately on held-out data. The adapted language model based on Eq. 12.11.6 is called the **cache language model**, which has been described in detail in [Kuhn 90].

Sometimes a language model is developed where test data come from the same source as the training data but the training data may consist of many heterogeneous subsets of data. This task, called within-domain language model adaptation, is similar to the adaptive training discussed earlier for acoustic modeling. Estimating the language model for such within-domain adaptation can be accomplished by several steps: 1) Cluster the training data by the principal source of variability, such as topic. 2) Identify at run-time the topic of the portion of the test data for which the adapted model is to be established. 3) Isolate the subsets of the clustered training data that correspond to the same topic in the test set. 4) Estimate the specific language model parameters using these subsets of the training data. 5) Interpolate this specific language model with the general model trained using all heterogeneous training data.

The last step of interpolation is important because the specific model tends to be biased while being less variable. The opposite is true for the general language model. The interpolation gives an appropriate tradeoff between the two.

12.12 Summary

In this chapter, we introduced the most important area of speech technology, automatic speech recognition (ASR) or speech-to-text. We first presented the general problem statement for ASR in terms of the decoding problem in a source-channel representation. We then provided some general specifications for ASR systems, including speaking mode and style, speaker dependency, vocabulary size, environmental condition, transducer, language model, and perplexity. Several dimensions of difficulty in ASR were discussed, including context dependency, speaking rate and style variability, cross-speaker variability and accent, and noisy acoustic environments. Evaluation measures for ASR were also briefly outlined.

We then presented the mathematical formulation of ASR, and in particular, the fundamental equation of ASR based on Bayes' decision theory and the MAP decision rule. This fundamental equation breaks down the ASR problem into three sub-problems: acoustic modeling, language modeling, and sequential optimization or decoding. In the remaining sections of this chapter, acoustic modeling was the main focus of coverage, with the final section dedicated to language modeling. No special attempts were made to cover decoding.

In discussing acoustic modeling, we first described acoustic pre-processing. Some commonly used acoustic pre-processors were outlined, including MFCCs, PLP, RASTA, LPC parameters, and their temporal derivatives.

Following acoustic pre-processing, the use of HMMs and of higher-order (beyond HMMs) statistical models in acoustic modeling was discussed in some detail. The conventional HMMs used in most current ASR systems have some clear advantages, resulting from the possibility of using a single composite HMM representation for an entire sentence. Some weaknesses of the HMM were then outlined, prompting the need to use higher-order statistical models. Three classes of higher-order models were discussed. The first class is stochastic segment models, where hierarchical categorization of these models was provided. They include nonstationary-state HMMs, with sub-categories of the polynomial trended HMM, exponentially trended HMM, and non-parametric trended HMM. They also include multi-region (switching) dynamic models, with sub-categories of the autoregressive HMM (also called the hidden filter model) and the linear dynamic system model. The second class of higher-order models consists of hidden dynamic models, also called structured speech models or super-segmental models. The basic philosophy behind such structured speech models is the opportunity to use the continuity of the internal speech dynamics across speech units to capture the contextual effects, and to use the specially designed shapes of the hidden dynamics to capture phonetic reduction arising from fast or casual speech. Both of these aspects were discussed in more detail in Chapter 10. A detailed case study was presented to illustrate the implementation and preliminary evaluation of one particular version of the hidden dynamic model. The third class of higher-order models comprises nonlinear phonological or pronunciation models. The multi-tiered, overlapping-feature model that was discussed in detail in Chapter 9 is typical of such models.

Part of this chapter is devoted to the robustness issue in acoustic modeling and in recognizer design. We covered various adaptation techniques for model-space robustness, including MAP adaptation, MLLR adaptation, speaking rate adaptation, and dynamic adaptation using quasi-Bayesian approaches. We also briefly discussed adaptive training (such as speaker or noise adaptive training), where adaptation techniques were applied to the training data and re-training took place.

This chapter largely consists of four case studies on acoustic modeling in ASR. They are: 1) a VTR-based hidden dynamic model of speech that uses first-order temporal dynamics; 2) a special kind of HMM that is structured by the locus equation for a regular pattern in the formant movement dynamics; 3) speaker adaptation with the use of the polynomially trended HMM based on the MAP adaptation technique; and 4) adaptive training using the Bayesian technique, where the transformation parameters were treated as random vectors with appropriate statistical distributions. Learning algorithms and ASR experimental results were presented in detail for all these case studies.

Finally, we provided an overview of language modeling for ASR. The topics covered included N-gram models, decision-tree based models, context-free grammar and probabilistic context-free grammar models, maximum-entropy models, and adaptive models.

Chapter 13

Speech Enhancement

The second area of speech technology applications covered in the final Part IV of this book is speech enhancement. Speech enhancement refers to the restoration of clean speech, either in the form of speech waveforms for enhanced human perceptual listening or in the form of speech features for enhanced or robust speech recognition, starting from speech corrupted by distorting acoustic environments. This distortion may be due to additive ambient noise, linear or nonlinear channel (convolutional) distortion, or interfering speech. As in ASR, the dynamic modeling and optimization principles as main threads throughout this book are also clearly illustrated in speech enhancement applications, to which we devote this chapter.

13.1 Introduction

Speech communication is generally easy when speaker and listener are close to each other in a quiet acoustic environment. It becomes difficult and fatiguing, however, under conditions with ambient or background noise (including interfering speakers). The speech sounds are often masked by the noise, resulting in a reduction of speech discrimination. This often degrades both speech intelligibility and speech naturalness (quality). Another common form of speech distortion occurs when speech is transmitted across telephone/cell-phone channels, loudspeakers, or headphones. In this case, the quality of speech can be degraded during data capture by microphones, and during data transmission and reproduction. In order to restore or improve speech quality under these degrading conditions, speech enhancement algorithms have been developed to reduce background noise and to suppress channel or speaker interference.

The main objective of speech enhancement is to improve one or more perceptual aspects of speech, such as overall quality, intelligibility for human or machine recognizers, or degree of listener fatigue. In the presence of background noise, the human auditory system is capable of employing effective mechanisms to reduce the effect of noise on speech perception. Although such mechanisms are not well enough understood in our present state of knowledge to allow the design of speech enhancement systems based on auditory principles that would give comparable human speech enhancement capabilities, many practical methods for speech enhancement have already been developed. Digital signal processing (DSP) techniques for speech enhancement include spectral subtraction [Lim 78b, Boll 79, Lim 78a], adaptive filtering, and suppression of nonharmonic frequencies [Pars 76]. Some of these techniques either require a second microphone to provide the noise reference, or require that the characteristics of noise be relatively stationary.

Many of these requirements often cannot be met in most practical applications. Spectral subtraction, with no need for a second microphone and with the capability of handling noise nonstationarity to some extent, has been one of the more successful DSP methods. One major problem with this method is the annoying nonstationary "musical" [Boll 91] background noise associated with the enhanced speech; it also is unable to effectively cope with rapid variations in noise characteristics (e.g., simple noise amplitude variations). The basic advantage of this method is its implementation simplicity and relatively light computation requirements. We will describe spectral subtraction techniques in some detail in this chapter.

Enhancement methods which are based on stochastic models (such as the HMM or trended HMM covered in earlier chapters) are able to overcome weaknesses of the DSP techniques by modeling general properties of clean speech and noise. They accommodate the nonstationarity of speech with multiple states connected with transition probabilities of a Markov chain, or by explicitly modeling time-varying trajectories with each Markov state. Using multiple states in the HMM for noise enables the speech enhancement system to relax the assumption of noise stationarity also. We will also describe this different type of technique in the framework of Wiener filtering in some detail in this chapter.

13.2 Classification of Basic Techniques for Speech Enhancement

Before we describe any specific techniques, let us first provide some general classification of the many kinds of speech enhancement methods. Because of the importance of the speech enhancement applications, many techniques have been developed in the past three decades. Several different ways can be used to classify these techniques.

13.2.1 Classification by what and how information is used

First, the techniques can be grouped by the manner in which the speech is modeled as the prior information for the enhanced speech. For the techniques that exploit speech knowledge, they can be based on statistical models where a given mathematical optimization criterion (such as MAP or MMSE discussed in Chapter 5) is used, or based on perceptual aspects of the speech process that are less quantifiable. We in this chapter will mainly focus on those that use statistical models of speech with well-defined mathematical optimization criteria.

13.2.2 Classification by waveform or feature as the output

Second, speech enhancement techniques can also be classified depending on whether the enhanced speech is in the form of cleaned speech waveforms or in the form of improved speech features derived from speech corrupted by distorting acoustic environments. The former mainly serves the purpose of enhanced human perceptual listening. The latter mainly serves the purpose of robust speech recognition, also called feature-space ASR robustness. Based on different kinds of features as the output of speech enhancement algorithms, further subclassifications can be made. For example, we will devote the final section of this chapter to the various techniques developed for such feature-space ASR robustness.

13.2.3 Classification by single or multiple sensors

Third, depending on whether one or more microphones (sensors) are used, enhancement techniques can be divided into single-channel and multi-channel ones. Under the single-channel condition, characterization of noise is necessary, either by statistical modeling or by estimating noise features during the non-speech period and extrapolating them to the processing frames under an assumption of local stationarity. Under the multiple-channel condition, more information is available to achieve better performance than with single-channel techniques. The acoustic sound waves arrive at each microphone at slightly different times, which can be effectively exploited. Sometimes no acoustic barriers exist between the microphones, and hence the issue of cross talk must be addressed. In this chapter, we will focus only on single-channel speech enhancement methods.

13.2.4 Classification by the general approaches employed

Fourth, speech enhancement techniques can be categorized in terms of the general approaches taken and the assumptions made. When no prior information about clean speech is used, a very sensible approach to remove noise is to estimate it first during non-speech activity and then subtract it from adjacent noisy speech. This is typically done in the short-term (linear) spectral or power spectral domain, giving rise to the name **spectral subtraction**. When prior knowledge about clean speech and noise or their statistical properties is exploited, the (linear) **Wiener filtering** approach can be used to effectively re-synthesize minimal-error speech signals or features, following estimation of model parameters that characterize the clean speech. We note that if use of the prior knowledge about the clean speech is replaced by subtracting a noise estimate from noisy speech, then the Wiener filtering approach becomes equivalent to that of spectral subtraction.

While both spectral subtraction and Wiener filtering approaches operate on the linear spectral (or power spectral) domain, advantages can be gained by using the log-spectral or cepstral (i.e., linearly weighted log-spectral) domain. The advantages are most obvious when enhancement of speech features (rather than waveforms) is needed for feature-space ASR robustness. This is because nearly all front ends in ASR systems (cf. Chapter 13) employ log-domain features in one way or another, and the enhancement in the same domain provides the most direct enhanced input to the ASR systems. In the log domain, however, an additional difficulty arises because the simple linear distortion process (e.g., additive noise or convolutional distortion) now becomes more a complex nonlinear process. Mathematical characterization of this process is sometimes referred to as the acoustic distortion or environment model.

Another general type of approaches to speech enhancement is to exploit the harmonic structure of voiced speech, called the comb filtering approach. If the period of noisy voiced speech can be determined, then comb filters can be applied in the frequency domain to increase the Signal-to-Noise Ratio (SNR). Comb filtering assumes that the noise is additive and is short-term stationary (due to the delay in pitch detection). It is also naturally limited to voiced speech, and will not be applicable to unvoiced speech, speech segments with fast transitions, or voiced fricatives. Comb filtering multiplies, in the frequency domain, the observation signal by a sequence of delta functions whose interval is the fundamental frequency of the speech signal. This approach depends naturally upon accurate estimations of the period of the noisy speech, which must be tracked as it varies over time.

Lastly, the computational auditory scene analysis approach to speech enhancement provides a solution to the problem of extracting information concerning a sound source embedded in a noisy auditory background. The auditory background can be environmental noise, but also other speakers and music. One way of looking upon this problem is to consider the auditory scene as a mixture of several "auditory objects" and to design a pre-processor where components are separated and grouped object-by-object before identification. This approach is generally more complex than other methods, but perceptual data suggests that this is the strategy employed by human listeners. Most attention lies in the area of "grouping" the low-level auditory objects. The acoustic properties and auditory scene analysis models currently available suggest that onset-offset time, temporal dynamics of amplitude and frequency (e.g., pitch and formant trajectory), and spatial location are the most important features which govern auditory grouping decisions.

In the rest of this chapter, we will describe several common techniques of speech enhancement based mainly on this last (fourth) way of classification.

13.3 Spectral Subtraction

Spectral subtraction is a noise suppression technique used to reduce the effects of added noise in speech. It estimates the power of clean speech by explicitly subtracting the noise power from the noisy speech power. This of course assumes that the noise and speech are uncorrelated and additive in the time domain. Also, as spectral subtraction techniques necessitate estimation of noise during pauses, it is supposed that noise characteristics change slowly. However, because noise is estimated during speech pauses, this makes the method computationally efficient. Unfortunately, for these reasons, spectral subtraction is beset by a number of problems. First, because noise is estimated during pauses the performance of a spectral subtraction system relies upon a robust noise/speech classification system. If a misclassification occurs this may result in a misestimation of the noise model and thus a degradation of the speech estimate. Spectral subtraction may also result in negative power spectrum values, which are then reset to non-negative values. This results in residual noise known as musical noise. Finally subtraction techniques cannot be used in the logarithmic spectrum domain because noise becomes signal dependent. Spectral subtraction has been used for various kinds of speech enhancement applications, including feature-space ASR robustness. In a speech enhancement application it has been shown that, at 5 dB SNR, the quality of the speech signal is improved without decreasing intelligibility. However, at lower SNR speech this performance reduces rapidly. When used in ASR the trade-off between SNR improvement and spectral distortion is important, although various attempts have been made to reduce musical noise. One method proposed a scheme where the frame-by-frame randomness of the noise is measured. For a given frequency bin, the residual noise is suppressed by replacing its current value with a minimum value chosen from the adjacent analysis frame.

Continuous spectral subtraction techniques have also been applied to avoid the problem of speech boundary detection in noise. With this method a smoothed estimate of the long term spectrum is continuously calculated and subtracted from the system. However, it still requires the detection of occasional periods of non-speech activity to update its noise model.

To provide a mathematical description of the spectral subtraction technique, we write the spectrum of the noisy speech in terms of that of the clean speech and additive noise (the simplest acoustic distortion model):

$$y(t) = x(t) + n(t),$$

where the clean speech is denoted by $x(t)$, additive noise by $n(t)$, and noisy speech by $y(t)$; all are within the same short-time frame and are assumed to be stationary within that frame. The spectrum (DFT) of the noisy speech can also be written as

$$Y(\omega) = X(\omega) + N(\omega),$$

where $Y(\omega)$, $X(\omega)$, and $N(\omega)$ are the spectra of $y(t)$, $x(t)$, and $n(t)$, respectively.

In implementing spectral subtraction, FFT is performed on each frame of the noisy signal to estimate the spectrum of the noisy speech. The estimate of the noise spectrum is updated during periods of presumed non-speech activity. (An autocorrelation-based voicing and pitch detector can used for speech detection.) When no speech is detected, the signal is assumed to be noise and the magnitude of the noise spectral estimate, $\mid \hat{N}(\omega) \mid$, is updated as

$$\mid \hat{N}(\omega) \mid^2 = \Gamma_n \mid \hat{N}_{old}(\omega) \mid^2 + (1 - \Gamma_n) \mid Y(\omega) \mid^2,$$

where $\mid Y(\omega) \mid$ is the spectral magnitude of the current frame, Γ_n is a decay factor, and $\mid \hat{N}_{old}(\omega) \mid$ is the magnitude of noise spectral estimate before the update. The estimated noise spectral magnitude squared is then subtracted from the short-time squared spectral magnitude of the degraded speech estimated in the frequency domain. Enhanced speech is obtained by reconstructing the speech using the modified magnitude and the original (noisy) phase [Boll 79, Lim 78a]:

Basic operation in spectral subtraction:

$$X(\omega) = \left[\mid Y(\omega) \mid^2 - \mid \hat{N}(\omega) \mid^2\right]^{1/2} e^{j\Theta_y(\omega)}, \tag{13.1}$$

where $\Theta_y(\omega)$ is the phase of the original noisy speech.

We can interpret the above spectral subtraction in terms of Wiener filtering (discussed in the immediately following section), where the transfer function of the filter is

$$H(\omega) = \left(\frac{\mid Y(\omega) \mid^2 - \mid \hat{N}(\omega) \mid^2}{\mid Y(\omega) \mid^2} \right)^{1/2} \tag{13.2}$$

for each frame. Then, the formula in Eq. 13.1 can be calculated using the linear system operation:

$$X(\omega) = H(\omega) Y(\omega).$$

As we discussed earlier, spectral subtraction relies on the critical assumption that noise is stationary, and works well only when the noise spectrum is not changed significantly from the last frame from which the noise estimate was updated. In practice, due to the random nature of the noise and its inaccurate estimation and due to ignorance of their phase relationships, the difference between the noisy speech and noise power spectra $\mid Y(\omega) \mid^2 - \mid \hat{N}(\omega) \mid^2$ in Eqs. 13.1 and 13.2 may not always be positive. Therefore, spectral subtraction techniques usually set a threshold T in Eq. 13.1:

$$X(\omega) = \left[\max(\mid Y(\omega) \mid^2 - \mid \hat{N}(\omega) \mid^2, F)\right]^{1/2} e^{j\Theta_y(\omega)}, \tag{13.3}$$

where F can be either an absolute floor level or a small fraction of the power spectrum of the noisy speech frame $| Y(\omega) |^2$.

13.4 Wiener Filtering

While the Wiener filter Eq. 13.2 may use the run-time estimate of clean speech via spectral subtraction (as the numerator in Eq. 13.2) to perform speech enhancement, its real strength comes from exploiting prior knowledge of the clean speech. Such prior knowledge can be expressed as a Gaussian mixture model, HMM, or trended HMM pre-trained from a reasonable amount of clean speech data that captures all principal spectral shapes, and possibly their dynamics, for the clean speech. We now introduce the basics of Wiener filtering for speech enhancement.

According to the Wiener filtering results (as LMMSE estimator) discussed in Section 5.7.4, the Wiener filter is a linear filter that produces an optimal estimate of the clean speech signal in the MSE sense. In the frequency domain, the Wiener filter has the transfer function of

$$H(\omega) = \frac{| X(\omega) |^2}{| X(\omega) |^2 + | N(\omega) |^2}. \tag{13.4}$$

However, in practice, the Wiener filter Eq. 13.4 cannot be directly used to filter the input noisy speech because the power spectrum of the clean speech $| X(\omega) |^2$ is unknown and so is the noise power spectrum $| N(\omega) |^2$. Further, both clean speech and noise are only short-term stationary, and hence their power spectra often vary as a function of time frame t. Eq. 13.4 is usually approximated by using the estimated power spectra of clean speech and noise on a frame-by-frame basis:

Basic transfer function of the time-varying Wiener filter:

$$H(\omega, t) = \frac{| \hat{X}(\omega, t) |^2}{| \hat{X}(\omega, t) |^2 + | \hat{N}(\omega, t) |^2}. \tag{13.5}$$

Accurate estimation of the clean speech spectrum is the main challenge of the Wiener filtering approach to speech enhancement. As discussed earlier, if we were to use spectral subtraction to estimate it, then the Wiener filtering approach would become identical to spectral subtraction. There are two general ways to meet this challenge in the framework of Wiener filtering. First, some kind of models for the clean speech spectrum is employed. The models are not direct estimation of the clean speech spectrum, but rather its statistical description, which serves as the prior information. Examples of this statistical description are Gaussian mixture models or HMMs for speech. Second, iterations are used so that the initially poor estimate of the clean speech spectrum may be gradually improved, upon the repeated use of the Wiener filter. For example, an AR (LPC or all-pole) model (See Chapter 2) can be used as a simple speech model. The AR coefficients and gain parameters are estimated using the output of the Wiener filter (which gives reasonably clean speech), and the resulting AR model is used to give an estimate of the clean speech spectrum for the next iteration of the Wiener filter [Lim 78b].

Generalizations of the Wiener filter can be made to empirically improve its performance in speech enhancement (and in other areas of signal processing, such as image restoration). One generalization is to employ two free parameters: a noise-scaling factor

b and a power exponent a:

$$H(\omega,t) = \left(\frac{|\hat{X}(\omega,t)|^2}{|\hat{X}(\omega,t)|^2 + b\,|\hat{N}(\omega,t)|^2} \right)^a . \tag{13.6}$$

In the above, if we set $a = 1$ and $b = 1$, we obtain the standard Wiener filter. When $a = 0.2$ and $b = 1$, it becomes power spectral filtering.

13.5 Use of HMM as the Prior Model for Speech Enhancement

We now describe a particular approach to speech enhancement in the framework of Wiener filtering using the HMM as the prior speech model. This approach takes advantage of the capability of the HMM to segment a speech utterance (as well as noise) into quasi-stationary segments. This approach was originally developed and described in [Ephr 89, Ephr 92].

As we discussed in Chapter 12, the HMM has long been used for speech modeling with applications to speech recognition. Since the late 1980's, it has also been applied to speech enhancement. There are significant differences in applying HMMs for recognition and enhancement purposes. In speech recognition, a separate model for every speech unit (allophone, phoneme, or word) is trained. The trained model is intended to contain an ordered sequence of stochastic properties for utterances corresponding to that speech unit. Therefore, following the necessary temporal constraints of one-dimensional speech signals, the model must be left-to-right, i.e., transitions from a higher-indexed state to a lower-indexed state (e.g., from a later phone to an earlier phone) are generally prohibited. For a left-to-right model, if similar states (corresponding to similar signal properties) can happen in different time frames, they must be retained as different states even though they contain the same statistical information. An objective in speech recognition is to find models (for different speech units) with maximal separation, so that they give as different likelihoods for a single testing token as possible. This requires that the models best preserve the distinctive statistical properties of the training data.

The modeling problem in speech enhancement is rather different. Its objective is to average parameters from the speech signal and extract the general spectral characteristics of speech, regardless of the phoneme, word, or sentence pronounced. This is done to distinguish speech from noise, and not to distinguish different units of speech. Thus the structure of the speech model for enhancement should be different from that for speech recognition. Firstly, we desire to accommodate all speech characteristics in a single, compact model. Secondly, the model is not supposed to capture distinctive properties of speech within different utterances; rather, it is to capture the global characteristics of speech. Thirdly, the temporal order of the states in the model need not be constrained since there is a single, global model for speech and different state sequences for the same state ensemble can represent distinct utterances. As a result, the speech model for enhancement is structured to be ergodic; i.e., there is no special constraint on the transition probabilities of the HMM. This also makes the model less redundant since each distinct spectral shape of speech or noise needs to be represented only once in the model.

13.5.1 Training AR-HMMs for clean speech and for noise

The HMMs for clean speech and for noise discussed here for speech enhancement are ergodic mixture auto-regressive hidden Markov models (AR-HMM) [Juang 85]. These HMMs enable us to parametrically model the speech and noise spectral shapes. The output PDF of each mixture of the AR-HMMs is assumed to be a Gaussian AR process. The likelihood of a training data sequence given the HMM (for clean speech or for noise) parameters is expressed in terms of the transition probabilities, mixture weights, and conditional output PDF [Ephr 89]. For implementation efficiency, the output PDF is approximated by a sum of the products of the data and model autocorrelation coefficients [Juang 85].

The model parameter set for an AR-HMM with M states and L mixtures is defined as $\lambda_x = (\pi, a, c, h)$, where $\pi = \{\pi_\beta\}$ is the set of initial state probabilities, $a = \{a_{\alpha\beta}\}$ is the set of state transition probabilities, $c = \{c_{\gamma|\beta}\}$ is the set of mixture weights, and $h = \{h_{\gamma|\beta}\}$ with $h_{\gamma|\beta}$ being the AR parameter set of a zero-mean N_x-th order Gaussian AR output process corresponding to state and mixture pair $(\beta, \gamma), h_{\gamma|\beta} = \{h_{\gamma|\beta}(0), h_{\gamma|\beta}(1), \ldots, h_{\gamma|\beta}(N_x), \sigma^2_{\gamma|\beta}\}$, $h_{\gamma|\beta}(0) = 1$, $\sigma^2_{\gamma|\beta}$ being the variance (AR gain) for $\alpha, \beta = 1, \ldots, M$ and $\gamma = 1, \ldots, L$. Given a K-dimensional training data sequence $x = \{x_t\}, x_t \in R^K$, a maximum likelihood (ML) estimate of the parameter set λ_x is obtained and maximized through the EM re-estimation algorithm [Juang 85].

Since the EM-based algorithm generally does not guarantee global optimization of the objective function, it is important to devise a good initial model. Vector quantization (VQ) measures [Gray 80, Ger 92] have been used to effectively estimate the initial model parameters, and the popular algorithm of [Buzo 80] has been used to design the VQ codebook [Sam 98]. To obtain the initial estimate for the (π, a, c) parameters, the training data sequence is encoded according to the designed codebook. π, a, and c are then obtained by using the frequency counts associated with them.

13.5.2 The MAP enhancement technique

For *maximum a posteriori* (MAP) estimation of the clean speech signal [Ephr 89], the EM algorithm (see Section 5.6) can be employed to construct the speech enhancement algorithm. Let k denote the iteration index (initially set to zero). First, the weight sequence,

$$q(\beta, \gamma \mid x(k)) = \{q_t(\beta, \gamma \mid x(k)), \quad t = 0, \ldots, T\},$$

is computed for all possible states β, mixture components γ, and time frames t using the forward-backward algorithm; $q_t(\beta, \gamma \mid x(k))$ is the conditional probability of being in state β and choosing mixture component γ at time frame t, given an estimate of the clean speech $x(k)$:

$$q_t(\beta, \gamma \mid x(k)) = P(s_t = \beta, m_t = \gamma \mid x(k)).$$

Associated with each state and mixture-component pair (β, γ), there is a set of AR (LPC) coefficients that can be used in combination with the noise AR process to form a Wiener filter $H_{\gamma|\beta}(\theta)$. This Wiener filter has the form of Eq. 13.5, where the estimated power spectrum of clean speech is derived from the AR model from the specific state and mixture-component pair (β, γ) in the pre-trained AR-HMM [Lim 78b, Ephr 89, Sam 98].

A new estimate of the clean speech is then computed by filtering the noisy speech through a weighted sum of the Wiener filters, the weights being $q_t(\beta, \gamma \mid x(k))$ for each time frame t. This estimate of clean speech is then used to find a probability sequence $q(\beta, \gamma \mid x(k))$, thus supplying a new sequence of Wiener filters and another estimate of the clean speech. This iterative process continues until a preset convergence criterion is reached. In the first iteration, noisy speech is used as an estimate of the clean speech. For each time frame t, such enhancement can be carried out efficiently in the frequency domain according to [Ephr 89]

$$x_{t,\theta}(k+1) = \left[\sum_{\beta=1}^{M} \sum_{\gamma=1}^{L} q_t(\beta, \gamma \mid x(k)) H_{\gamma|\beta}^{-1}(\theta) \right]^{-1} y_{t,\theta}, \tag{13.7}$$

where x and y denote the clean signal and the noisy signal, respectively, and subscript θ indicates the frequency domain components. A block diagram of this MAP-based enhancement technique is shown in Fig. 13.1.

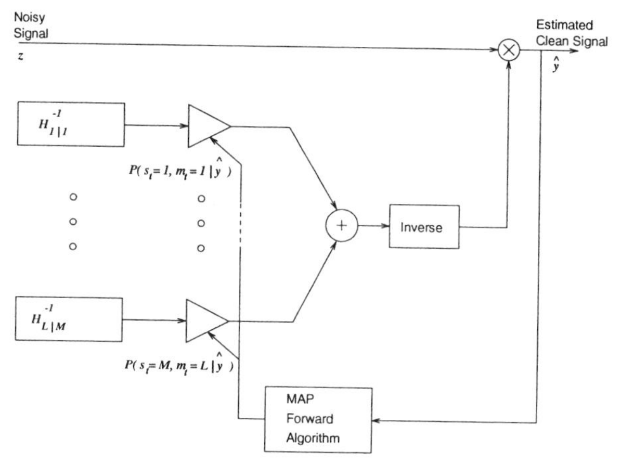

Figure 13.1: Block diagram for the MAP enhancement technique (after Sameti et al. [Sam 98], ©IEEE).

13.5.3 The approximate MAP enhancement technique

The approximate MAP enhancement technique originally presented in [Ephr 89] has its block diagram shown in Fig. 13.2. In the AMAP, a single state and mixture-component pair is assumed to dominate the sequence at each time frame, thus constraining the filter weights to be one for only one state and mixture pair, and zero for the others. Given an estimate of the clean speech signal, estimation of the most likely sequence of states and mixture components is carried out by applying the Viterbi algorithm using the path metric:

$$\begin{aligned} \ln \pi_\beta &+ \ln c_{\gamma|\beta} + \ln b(x_0(k) \mid m_0 = \gamma, s_0 = \beta) && \text{for } t = 0, \\ \ln a_{\alpha\beta} &+ \ln c_{\gamma|\beta} + \ln b(X_t(k) \mid m_t = \gamma, s_t = \beta) && \text{for } 1 \le t \le T, \end{aligned}$$

where $\alpha, \beta = 1, \ldots, M$, $\gamma = 1, \ldots, L$, and s_t and m_t are the state and mixture components at time frame t, respectively. At each t, a frame of noisy speech, y_t, is enhanced using the Wiener filter corresponding to the most probable state and mixture-component pair.

Note that both the MAP and AMAP enhancement algorithms are iterative since they use the enhanced speech as an estimate of the clean speech that is theoretically required by the formulations. Increased iterations make the enhanced speech closer to the clean speech. Also note that neither of these methods is capable of handling nonstationary noise, as the construction of the filter weights is based on the clean signal information only and ignores noise variation.

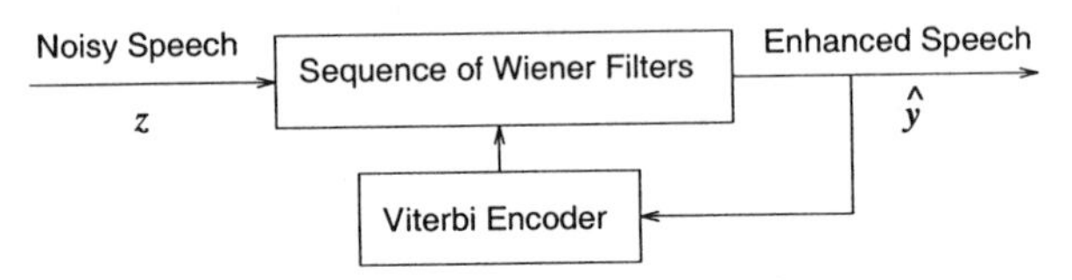

Figure 13.2: Block diagram for the approximate MAP enhancement technique (after Sameti et al. [Sam 98], ©IEEE).

13.5.4 The MMSE enhancement technique

Some detailed description of the MMSE (minimum mean square error) enhancement technique has been provided in [Ephr 92, Sam 98]. For this technique, a multiple state and multiple mixture-component noise model is employed to accommodate nonstationarity in noise. Fig. 13.3 shows a simplified block diagram of the MMSE enhancement technique. It is designed to optimize the mean square error as the objective function, a rather different one than in the MAP technique. According to the theorem for the MMSE computation (Section 5.7.1), the MMSE enhancement technique computes the following conditional expectation:

$$\hat{g}(x_t) = E\left\{g(x_t) \mid y_0^t\right\}, \tag{13.8}$$

where $g(\cdot)$ is a function on R^K, $E\{\cdot \mid \cdot\}$ denotes conditional expectation, and $y_0^t = \{y_0, \ldots, y_t\}$ is the noisy speech data from time zero to time t.

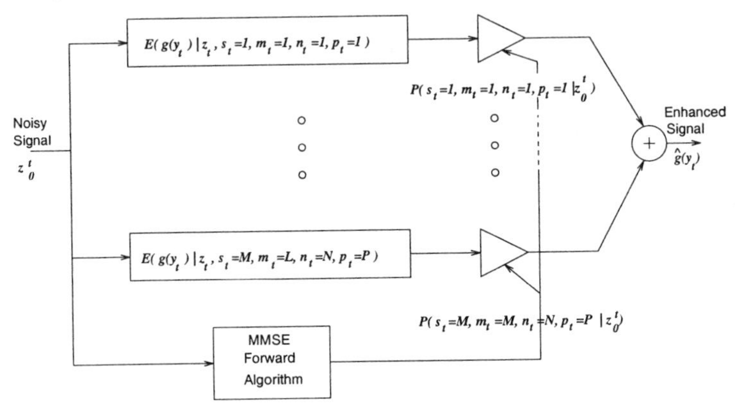

Figure 13.3: Block diagram for the MMSE enhancement technique (after Sameti et al. [Sam 98], ©IEEE).

Let n_t and p_t denote the state and mixture component of noise at time frame t, and let N and P denote the number of states and mixture components of the noise AR-HMM.

$\hat{g}(x_t)$ can then be calculated from

$$\hat{g}(x_t) = \sum_{\beta=1}^{M}\sum_{\gamma=1}^{L}\sum_{\xi=1}^{N}\sum_{\delta=1}^{P} q_t(\beta,\gamma,\xi,\delta \mid y_0^t) E\left\{g(x_t) \mid y_t, s_t=\beta, m_t=\gamma, n_t=\xi, p_t=\delta\right\} \quad (13.9)$$

$$(13.10)$$

where

$$\begin{aligned} q_t(\beta,\gamma,\xi,\delta \mid y_0^t) &= P(s_t=\beta, m_t=\gamma, n_t=\xi, p_t=\delta \mid y_0^t) \\ &= \frac{F_t(\beta,\gamma,\xi,\delta,y_0^t)}{\sum_{\beta=1}^{M}\sum_{\gamma=1}^{L}\sum_{\xi=1}^{N}\sum_{\delta=1}^{P} F_t(\beta,\gamma,\xi,\delta,y_0^t)} \end{aligned} \quad (13.11)$$

is the posterior joint probability of speech state β, mixture component γ, noise state ξ, and mixture-component δ at time t given the noisy signal y_0^t. In Eq. 13.11, we have

$$F_0(\beta,\gamma,\xi,\delta,y_0) = \pi_\beta \cdot c_{\gamma|\beta} \cdot \pi_\xi \cdot c_{\delta|\xi} \cdot b(y_0 \mid \beta,\gamma,\xi,\delta) \quad (13.12)$$

$$F_t(\beta,\gamma,\xi,\delta,y_0^t) = \sum_{\{s_0^{t-1}:s_t=\beta\}} \sum_{\{m_0^{t-1}:m_t=\gamma\}} \sum_{\{n_0^{t-1}:n_t=\xi\}} \sum_{\{p_0^{t-1}:p_t=\delta\}} \prod_{\tau=0}^{t} a_{s_{\tau-1}s_\tau} \cdot c_{m_\tau|s_\tau} \cdot a_{n_{\tau-1}n_\tau} \cdot c_{p_\tau|n_\tau} \cdot b(y_\tau \mid s_\tau, m_\tau, n_\tau, p_\tau) \quad (13.13)$$

for $t > 0$, where

$$b(y_t \mid s_t, m_t, n_t, p_t) = \frac{\exp\left\{-\frac{1}{2}y_t^{\mathrm{Tr}}(\Sigma_{m_t|s_t} + \Sigma_{p_t|n_t})^{-1}y_t\right\}}{(2\pi)^{K/2}\left[\det(\Sigma_{m_t|s_t} + \Sigma_{p_t|n_t})\right]^{1/2}} \quad (13.14)$$

is the conditional PDF of the noisy signal y_t given that the clean signal is in state s_t with mixture component m_t and the noise frame corresponds to state n_t and mixture component p_t. In Eq. 13.14, $\Sigma_{\gamma|\beta} = \sigma_{\gamma|\beta}^2(A_{\gamma|\beta}^{\mathrm{Tr}} A_{\gamma|\beta})^{-1}$ is the covariance matrix of the Gaussian output process associated with state β and mixture γ of the speech AR-HMM, $\sigma_{\gamma|\beta}^2$ is the variance of the innovation process of the AR source, and $A_{\gamma|\beta}$ is a $K \times K$ lower triangular Toeplitz matrix in which the first $N_x + 1$ elements of the first column constitute the coefficients of the AR process, $h_{\gamma|\beta}(\cdot)$. Similarly, $\Sigma_{\delta|\xi}$ is the covariance matrix of the Gaussian output process associated with state ξ and mixture component δ of the noise AR-HMM. Note that for the Gaussian HMMs that represent speech and noise, the noisy process y_0^t is also a Gaussian process.

Eq. 13.9 shows that the MMSE estimator of $g(x_t)$ given y_0^t is a weighted sum of the individual MMSE estimators of the output processes generated by the composite speech and noise AR-HMM states and mixture components, where the weights are the posterior probabilities that the individual estimators are the correct ones conditioned on the given noisy signal. The conditional expectations on the right-hand side of Eq. 13.9 are given by

$$E\left\{g(x_t) \mid y_t, s_t, m_t, n_t, p_t\right\} = \int g(x_t) p_{\lambda_x\lambda_v}(x_t \mid y_t, s_t, m_t, n_t, p_t) dx_t, \quad (13.15)$$

where $p_{\lambda_x\lambda_v}(x_t \mid y_t, s_t, m_t, n_t, p_t)$ denotes the conditional PDF of the clean signal x_t given state s_t with mixture component m_t at time t, and given the noisy signal y_t. The exact evaluation of Eq. 13.15 for a general nonlinear function $g(\cdot)$ is not trivial. For $g_1(x_t) = \{X_t(k), k = 0, \ldots, K-1\}$ where $X_t(k)$ is the k-th component of the DFT of

x_t, Eq. 13.15 can be shown to be the Wiener filter with the form of Eq. 13.5. Other functions that are useful in speech enhancement are

$$g_2(x_t) = \{\mid X_t(k) \mid, k = 0, \ldots, K-1\} \quad (13.16)$$
$$g_3(x_t) = \{\mid X_t(k) \mid^2, k = 0, \ldots, K-1\} \quad (13.17)$$
$$g_4(x_t) = \{\log \mid X_t(k) \mid, k = 0, \ldots, K-1\}. \quad (13.18)$$

Using the MMSE enhancement technique as shown in Figure 13.3, no iterations are necessary, unlike with the MAP counterpart. This shows superiority of the MMSE technique over the MAP one; the latter requires many iterations in order to achieve an acceptable result. The more significant superiority of the MMSE technique over the MAP counterpart, however, is its ability to deal with nonstationary noise due to its inherent capability to compute filter weights given the noisy signal instead of an estimate of the clean signal.

Eqs. 13.9–13.14 indicate that obtaining $\hat{g}(x_t)$ may be costly in terms of computational complexity. For each frame t, a large number of filter weights has to be calculated using expensive calculation of Eqs. 13.13 and 13.14. This makes the enhancement procedure time consuming. In the next section (Case Study I), we will describe some implementation detail that significantly reduces the computation.

13.5.5 Noise adaptation

There are usually a large number of diversified types of noise, with possibly time-varying spectral characteristics, in the acoustic environment in which speech enhancement systems are intended to be deployed. It is always an advantage for the enhancement system to have *a priori* knowledge about the nature of the noise. Enhancement methods that make assumptions about the noise type are deficient in terms of functionality under various corrupting noise types. The HMM-based enhancement techniques discussed so far in this section inherently rely on the type of training data for noise. As expected, such a system can handle only the type of noise that has been used for training the noise HMM. Therefore, data from various noise types should be used for training the noise HMM. This creates the problem of a large model size for the noise HMM, making the search space expand linearly with the number of noise types and making computation cost grow drastically. Furthermore, the unwanted large search space degrades the system performance by introducing possible new sources of error in computing posterior probabilities as the filter weights in the enhancement algorithms.

A noise adaptation algorithm has been described in [Sam 98] which enables the enhancement techniques to handle arbitrary types of corrupting noise, and limits the growth in computation complexity while the number of the noise types is increasing. This algorithm, with the block diagram shown in Fig. 13.4, carries out noise-model selection and adaptation of the variances (LPC gains) of the Gaussian AR processes associated with the noise HMMs. During intervals of non-speech activity, the Viterbi algorithm is performed on noise data using different noise models. By scaling the gain term in every HMM mixture component by a single factor and performing the Viterbi scoring, the model gain is coarsely optimized. The noise HMM generating the best score is selected and a fine scaling adjustment is carried out to adapt to the noise level using the Viterbi algorithm again. This procedure is based on the assumption that noise training sequences with similar characteristics, but varying levels, result in AR-HMMs differing only in the AR gains (not in spectral shapes). In order to avoid confusing unvoiced speech (mainly fricatives) with non-speech segments contaminated with noise, only segments more than 100 ms long are used for updating the noise model. Since few fricatives or other unvoiced phones last longer than 100 ms, the system will rarely confuse speech with pure noise intervals.

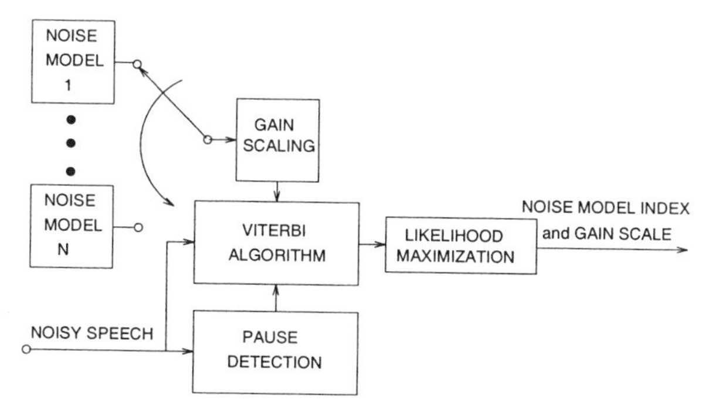

Figure 13.4: Block diagram of the noise adaptation technique (after Sameti et al. [Sam 98], ©IEEE).

Note that the MMSE enhancement technique does not require noise model updating as often as the spectral subtraction method, since it can handle noise nonstationarity within a specific noise type due to the use of the multiple state and mixture noise model. Noise model updating here is only to switch to the model representing a new noise type if required. Selection of different spectra and gains within a specific type of noise is carried out by the forward algorithm for each speech frame. A corrupting noise with continuously variable power can be easily handled by the MMSE method without the requirement to update the noise model type; in contrast, the spectral subtraction method fails to follow the continuous noise power variations. This method of noise model selection can successfully cope with noise level variations as well as different noise types, as long as the corrupting noise has been modeled during the training process. Further, the method keeps the noise model sufficiently compact so that excessive computation cost in enhancement is avoided. Assume a 3-state and 3-mixture-component HMM is required to model each noise type, and assume that five noise types are to be dealt with. Without the noise adaptation algorithm, 45 possible output distributions have to be searched to select a noise PDF. Using the noise adaptation algorithm, this search space is reduced to only nine output distributions at a time. The only extra computation is due to the selection of the appropriate noise model once every few seconds during the non-speech activity.

13.6 Case Study I: Implementation and Evaluation of HMM-Based MMSE Enhancement

This case study is extracted from the work published in [Sam 98], whose goal is to implement a real-time speech enhancement system based on the MMSE technique just described. The system was efficiently implemented, which reduced the computation requirement to that comparable to the conventional DSP method.

13.6.1 Double pruning the MMSE filter weights

Computation of the MMSE forward probability and filter weights, which constitutes a major computational load, is carried out according to Eq. 13.11. For speech and noise HMMs of sizes $M \times L$ and $N \times P$, these equations require the calculation of $M \times L \times N \times P$ filter weights and the same number of PDF values for each time frame t. Since most of these weights are negligible due to their extremely small values (orders as small as 10^{-200}), an efficient pruning method was devised and implemented to reduce the computation cost, as well as the memory requirement, of the system.

Eq. 13.13 can be rewritten in the following recursive form:

$$F_t(\beta, \gamma, \xi, \delta, y_0^t) = \sum_{\beta'=1}^{M} \sum_{\gamma'=1}^{L} \sum_{\xi'=1}^{N} \sum_{\delta'=1}^{P} F_{t-1}(\beta', \gamma', \xi', \delta', y_0^{t-1}) \cdot a_{\beta'\beta} \cdot a_{\xi'\xi} \cdot c_{\gamma|\beta} \cdot c_{\delta|\xi} \cdot b(y_t \mid \beta, \gamma, \xi, \delta), \tag{13.19}$$

where $b(\cdot)$ is calculated from Eq. 13.14. For pruning, $b(\cdot)$ values are first normalized by their maximum value. (This does not affect the filter weights since the forward probabilities appear both in the numerator and denominator in Eq. 13.11.) Then all the $b(\cdot)$ values less than an empirically determined certain threshold are deleted, and Eq. 13.19 is calculated only for the remaining $b(\cdot)$'s. A second pruning is performed for F_{t-1} and only the significant values of F_{t-1} are used to calculate Eq. 13.19. This double pruning method allows the computation cost of the enhancement process to be independent of the size of the speech and noise HMMs, since the number of saved filter weights does not directly depend on the model size. Without this pruning, the computation cost would increase proportionally with the speech and noise model sizes.

13.6.2 PDF approximation for noisy speech

Calculation of Eq. 13.14 is very costly because of the inversion of the $K \times K$ covariance matrix ($K = 256$ in the system of [Sam 98]) and multiplication of large matrices (for $K = 256$, computation cost is of the order $K^3 = 1.6 \times 10^7$). Since the summation of two AR processes is not necessarily an AR process, the assumption of a structured covariance matrix (in order for $\Sigma_{y_t} = \Sigma_{\gamma|\beta} + \Sigma_{\delta|\xi}$ to be decomposable into Toeplitz matrices comprised of AR coefficients of process y_t [Gray 93]) for noisy speech is generally invalid. To avoid the expensive calculation, an approximation method was devised for the inversion of the noisy covariance matrix. For any process y_t, the covariance matrix can be written in the form of [Kay 88]:

$$\Sigma_{y_t}^{-1} = CP^{-1}C^{\mathrm{Tr}}, \tag{13.20}$$

where C and P are upper triangular and diagonal $(K+1)\times(K+1)$ matrices, respectively, of the forms:

$$C = \begin{bmatrix} 1 & a_1(1) & a_2(2) & \cdots & a_K(K) \\ 0 & 1 & a_2(1) & \cdots & a_K(K-1) \\ \vdots & \vdots & \vdots & \ddots & \vdots \\ 0 & 0 & 0 & \cdots & 1 \end{bmatrix} \tag{13.21}$$

$$P = \mathrm{diag}(r_{yy}(0), E_1, E_2, \ldots, E_K), \tag{13.22}$$

where $a_i(j)$ is the ith coefficient of the jth order linear predictor for the process y_t, $r_{yy}(i)$ is the ith autocorrelation coefficient of the process y_t, and E_i is the squared prediction error for the ith-order linear predictor. The exponent term in Eq. 13.14 needs

$$D = y_t^{\mathrm{Tr}} \Sigma_{y_t}^{-1} y_t$$

to be calculated. From Eq. 13.20 we can write

$$D = y_t^{\mathrm{Tr}} C P^{-1} C^{\mathrm{Tr}} y_t = \left[y_t^{\mathrm{Tr}} C (P^{-1})^{\frac{1}{2}} \right] \left[(P^{-1})^{\frac{1}{2}} C^{\mathrm{Tr}} y_t \right];$$

so

$$D = \left[y_t^{\mathrm{Tr}} U^{\mathrm{Tr}} \right] [U y_t] = [U y_t]^{\mathrm{Tr}} [U y_t], \tag{13.23}$$

where $U = (P^{-1})^{\frac{1}{2}} C^{\mathrm{Tr}}$. Thus, the inversion of the $K \times K$ matrix is avoided, but the problem of multiplying large matrices still remains to be resolved. Note that $\det(\Sigma_{y_t}) = \det(P)$, with P being a diagonal matrix; so $\det(\Sigma_{y_t})$ is found by calculating the product of diagonal elements of matrix P. To resolve the second computation problem, an approximated U (instead of U) is calculated by considering the process y_t as an AR process of a higher order than the orders of either of the two processes x_t and v_t (the clean signal and noise). For an AR process of order p, for $j \geq p$ we have:

$$\begin{aligned} a_j(i) &= \begin{cases} a_p(i) & i = 1, 2, \ldots, p \\ 0 & i > p \end{cases} \\ E_j &= E_p. \end{aligned}$$

Therefore U will be a Toeplitz matrix after its pth row,

$$U = \begin{bmatrix} r_{yy}(0) & 0 & 0 & 0 & 0 & 0 & \cdots & 0 \\ a_1(1) & E_1^{-\frac{1}{2}} & 0 & 0 & 0 & 0 & \cdots & 0 \\ a_2(2) & a_2(1) & E_2^{-\frac{1}{2}} & 0 & 0 & 0 & \cdots & 0 \\ \vdots & \vdots & \vdots & \vdots & \vdots & \vdots & \vdots & \vdots \\ a_p(p) & \cdots & a_p(1) & E_p^{-\frac{1}{2}} & 0 & 0 & \cdots & 0 \\ 0 & a_p(p) & \cdots & a_p(1) & E_p^{-\frac{1}{2}} & 0 & \cdots & 0 \\ 0 & 0 & \ddots & \ddots & \ddots & \ddots & \ddots & \vdots \\ 0 & 0 & 0 & a_p(p) & \cdots & a_p(1) & E_p^{-\frac{1}{2}} & 0 \\ 0 & 0 & 0 & 0 & a_p(p) & \cdots & a_p(1) & E_p^{-\frac{1}{2}} \end{bmatrix}.$$

Thus, U can be separated into two parts: the first part comprised of the first p rows and the second part of the other $K - p$ rows. Multiplication of the first p rows is done easily due to the small value of p compared to K ($p = 14$ and $K = 256$ in our system). The second part of the matrix has a circular structure, and for implementation efficiency the output PDF can be approximated by the sum of the products of the autocorrelation coefficients of the data and of the model AR parameters [Juang 85] as follows.

For a zero-mean pth-order Gaussian AR output process with the AR parameter set of $a_p = \{a_p(0), a_p(1), \ldots, a_p(p)\}$, $a_p(0) = 1$, gain σ^2 and observation y_t with vector size

K, if $K \gg p$ then the output PDF can be approximated by

$$b(y_t) = \frac{\exp\{-\alpha/(2\sigma^2)\}}{(2\pi\sigma^2)^{K/2}}, \tag{13.24}$$

where α is defined as

$$\alpha = r_t(0)R_p(0) + 2\sum_{m=1}^{p} r_t(m)R_p(m). \tag{13.25}$$

The terms $r_t(m)$ and $R_p(m)$ are simply autocorrelation sequences defined as

$$\begin{aligned} r_t(m) &= \textstyle\sum_{n=0}^{K-m-1} y_t(n)y_t(n+m), \\ R_p(m) &= \textstyle\sum_{n=0}^{p-m-1} a_p(n)a_p(n+m). \end{aligned}$$

Using the method above, generation of covariance matrices of clean speech, Σ_{x_t}, and of noise, Σ_{v_t}, separately (for calculating Σ_{y_t}) is avoided. Instead, the autocorrelation coefficients of the clean speech and noise processes are calculated from their AR coefficients. Assuming additivity and independence of the noise and original speech signal, their autocorrelation coefficients are added to obtain the autocorrelation coefficients of the noisy speech. The Levinson-Durbin [Kay 88] recursion is performed on the calculated autocorrelation coefficients to find the AR coefficients of the noisy process, a_p, and the error prediction terms, E_i. The matrix U can then be calculated. However, most of the calculation to get D (from Eq. 13.23) is due to the lower (circulant) segment of U since $K >> p$. Moreover, this part of the calculation is further simplified by approximating D with α as shown in Eq. 13.25. In this way, the computation cost for calculating the noisy process PDF Eq. 13.14 is drastically reduced from the order of K^3 to the order of pK.

13.6.3 Overview of speech enhancement system and experiments

The speech data used in the evaluation experiments of [Sam 98] were selected from sentences in the TIMIT database. One hundred sentences spoken by 13 different speakers with a sampling rate of 16 kHz were used to train the clean speech model. Each frame of speech included 256 speech samples (the equivalent of 16 ms). No interframe overlap was used when training the speech model. In all the experiments, each HMM consisted of 5 states and 5 mixtures. The sentences used for enhancement tests were selected such that there were no common sentences or speakers between the training and enhancement sets. A 50% overlap between adjacent frames was used in the enhancement procedure.

A block diagram of the implemented MMSE enhancement system is shown in Fig. 13.5. Each frame of noisy speech is first preprocessed, to obtain its AR coefficients. The components inside the dashed lines in the figure implement the noise-model adaptation method described in detail in Section 13.5.5. Briefly, the noisy signal during long periods of non-speech activity is first fed into a Viterbi-like forward algorithm. Then, the likelihood for each pre-trained noise HMM is calculated and compared with likelihoods for the other noise HMMs and the model associated with the highest likelihood determines the selected noise model. Using the selected noise HMM parameters and the clean speech model, the preprocessed noisy speech is input to the MMSE forward algorithm, specified in Eq. 13.11, which generates the weights for the Wiener filters.

In the meantime, all Wiener filters for each combination of the state and mixture pairs in the speech and noise models are calculated. A single weighted filter is constructed for each frame of noisy speech using the calculated filter weights and the pre-trained Wiener

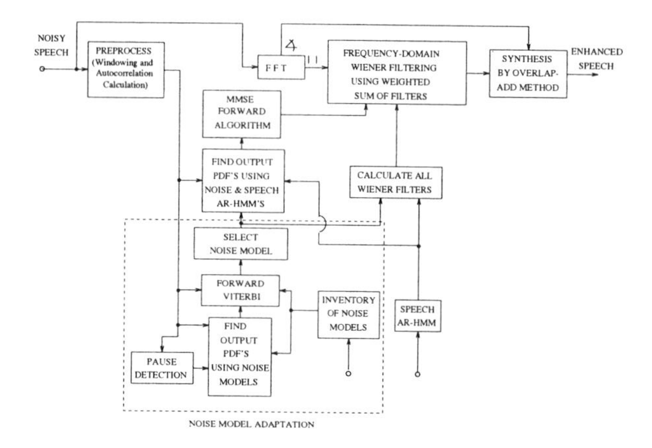

Figure 13.5: Block diagram of all components of the MMSE enhancement system (HMM as the prior model for clean speech) (after Sameti et al. [Sam 98], ©IEEE).

filters. The filtering of the noisy signal is carried out using the weighted filter. This generates the spectral magnitude of the enhanced speech signal. Using this magnitude and the noisy speech's phase information, an inverse FFT is performed to obtain the time-domain enhanced speech via the standard overlap-and-add method [Lim 88].

In speech enhancement experiments, three different types of noise were used: white noise, simulated helicopter noise (obtained by modulating the white noise with a 5-Hz sinusoidal waveform), and multi-talker low-pass noise that was recorded live at a cocktail party. Since the MAP and AMAP algorithms cannot cope with nonstationary noise, they were used only for the sentences corrupted with white noise. Spectral subtraction and the MMSE methods can handle noise nonstationarity, and hence their performances on all three types of noise were compared. For the MMSE enhancement, the noise HMMs contained three states and three mixtures. Noise models containing five states and five mixtures were also used in a few tests, and were found to not result in notable improvements over the models with three states and three mixtures.

13.6.4 Enhancement results using SNR as an evaluation measure

A global measure of SNR is used first in this case study as the objective evaluation criterion for speech enhancement, and is defined by

$$SNR = 10 \log \frac{\sum_{n=1}^{K} x^2(n)}{\sum_{n=1}^{K} [x(n) - \hat{x}(n)]^2},$$

where K is the frame-length, $x(n)$ is the clean speech signal and $\hat{x}(n)$ the enhanced speech signal. In the evaluation, input SNRs varied from 0 to 20 dB. Spectral subtraction and several types of HMM-based enhancement techniques described in this section were implemented and compared. Figs. 13.6, 13.7, and 13.8 show the output SNRs of these enhancement algorithms averaged over ten different test sentences corrupted by white noise, helicopter noise, and multi-talker noise, respectively.

As evident from these figures, the HMM-based techniques always outperform spectral subtraction. For the white noise case, the HMM-based systems have an advantage of at

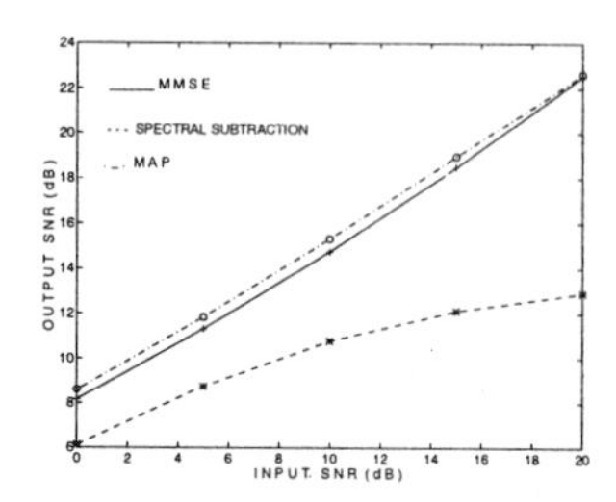

Figure 13.6: Comparison of MAP, MMSE and spectral subtraction systems for speech signals corrupted by white noise (after Sameti et al. [Sam 98], ©IEEE).

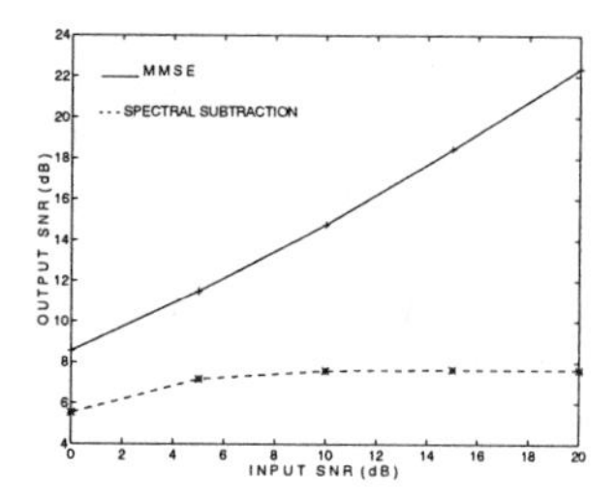

Figure 13.7: Comparison of MAP, MMSE and spectral subtraction systems for speech signals corrupted by helicopter noise (after Sameti et al. [Sam 98], ©IEEE).

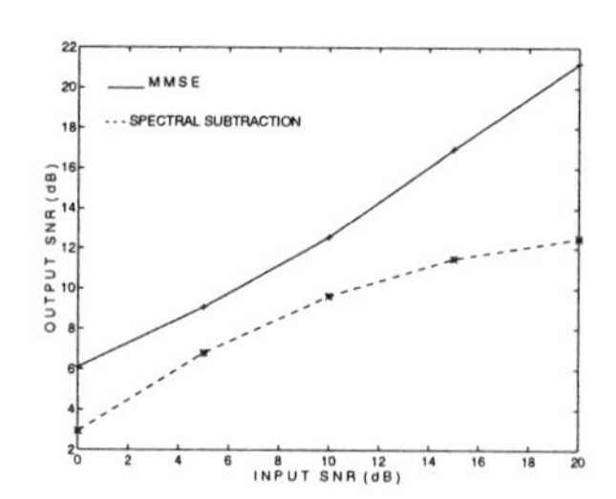

Figure 13.8: Comparison of MAP, MMSE and spectral subtraction systems for speech signals corrupted by multitalker noise (after Sameti et al. [Sam 98], ©IEEE).

least 2.5 dB SNR over the spectral subtraction system, and since the noise is stationary, the performances of the MMSE and MAP systems are similar to each other. For the two nonstationary-noise cases (Figs. 13.7 and 13.8), while the MMSE technique results in an almost linear input-output relation with respect to the SNR values, spectral subtraction tends to saturate in output SNR at high input SNRs and falls behind the MMSE technique by at least 2.5 dB even at low input SNRs. Spectral subtraction fails to handle noise nonstationarity that is as simple as the simulated, highly regular helicopter noise. In fact, for input SNRs of greater than about 10 dB, the spectral subtraction method degrades the signal such that the output SNR is lower than the input SNR. These results are consistent with the results of subjective evaluations presented next. In these cases, listeners prefer an unprocessed noisy sentence over an enhanced one using spectral subtraction.

13.6.5 Enhancement results using subjective evaluation

For spectral subtraction, it was found that the process of dynamic reduction of spectral energy often introduces an audible artifact, a "musical"-like signal-dependent interference. Since spectral subtraction raises the SNR without knowledge about speech characteristics, low-amplitude speech signals such as stops tend to be lost at input SNRs below 5 dB. This reduces the effectiveness of the algorithm in enhancing speech intelligibility. Under low input SNR conditions, the problem of musical noise bothered listeners extensively. Although the SNRs were improved in these cases, some listeners could not tolerate the musical noise. For tests using higher input SNR (10 dB and more), the noise reduction was not carried out efficiently and the musical noise was also generated, although not as strong as the low input SNR cases. In all cases, some listeners preferred the non-processed signal over the enhanced one.

On the other hand, since the HMM-based enhancement techniques use speech information already embedded in the trained model, their output intelligibility is expected to be better than for spectral subtraction, at a cost of higher system implementation complexity. This is particularly true for the MMSE enhancement strategy, since it is capable of coping with noise nonstationarities. The SNR results just presented have indirectly reflected this fact.

In the work of [Sam 98], mean opinion score (MOS) comparative evaluations were conducted for the MMSE enhancement technique and spectral subtraction. Both of the techniques were scored by five native English speakers using MOS. Fig. 13.9 shows the MOS results averaged over 10 test sentences contaminated by the three types of noise (denoted by W for white noise, H for simulated helicopter noise, and M for multi-talker noise), each at 0, 5, and 10 dB input SNR levels. The results show that the MMSE technique consistently outperforms spectral subtraction by one score on average. In general, the MOS results are strongly consistent withthe SNR objective evaluations.

13.7 Case Study II: Use of the Trended HMM for Speech Enhancement

In this second case study, which is based on the work published in [Sam 02], we extend the use of the stationary-state HMM (as the prior model for clean speech just presented) to the use of the trended HMM, for the same purpose of representing the spectral characteristics of clean speech for speech enhancement. In this new application, the trended HMM, which has been covered in numerous places earlier in this book, serves as the

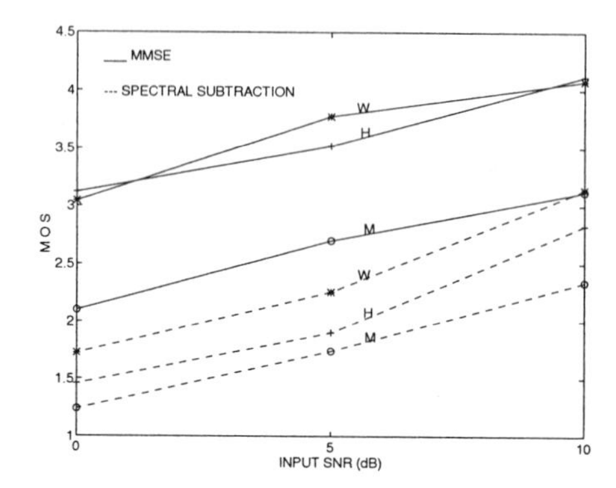

Figure 13.9: MOS results for MMSE enhancement vs. spectral subtraction, averaged over ten sentences evaluated by five listeners. W: white noise; H: helicopter noise; M: multitalker noise (after Sameti et al. [Sam 98], ©IEEE).

prior model for the clean speech statistics in deriving the MMSE estimator. The MMSE formulation is established where the trended HMM is used as the clean speech enhancement experiments are conducted, demonstrating superiority of the trended HMM over the conventional, stationary-state HMM in the enhancement performance for low SNRs. Detailed diagnostic analysis on the speech enhancement system's operation shows that the superiority arises from the ability of the trended HMM to fit the spectral trajectory of the signal embedded in noise more closely than the stationary-state HMM.

13.7.1 Formulation of the prior model

As discussed earlier in this book, the trended HMM was intended to overcome the weaknesses of the conventional HMM by permitting nonstationarity within the Markov states. In the trended HMM, the observation vector $\mathbf{x}_t$ is composed of a deterministic trend function plus the residual:

$$\mathbf{x}_t = \mathbf{f}_t + N_t,$$

where $\mathbf{f}_t$ is the deterministic trend function at time frame t, and N_t is the stationary residual; $\mathbf{x}_t$, $\mathbf{f}_t$, and N_t are all $K \times 1$ vectors. N_t is taken to be an IID, zero-mean Gaussian source. A trended HMM is completely characterized by the following parameter set: 1) Transition probabilities, $a = \{a_{ij}\}$, $i, j = 1, 2, \ldots, M$ of the Markov chain (no left-to-right constraints are imposed); 2) Mixture weights, $c = \{c_{m|i}\}$, $i = 1, 2, \ldots, M; m = 1, 2, \ldots, L$ of the Markov chain with a total of M states and L mixture components; 3) Parameters $\Theta = \{\Theta_{i,m}\}, i = 1, 2, \ldots, M; m = 1, 2, \ldots, L$ in the deterministic trend function $\mathbf{f}_t(\Theta_{i,m})$, dependent on state i and mixture component m in the Markov chain; and 4) Covariance matrices, $\Sigma = \{\Sigma_{i,m}\}, i = 1, 2, \ldots, M; m = 1, 2, \ldots, L$, of the zero-mean Gaussian IID residual $N_t(0, \Sigma_{i,m})$, which are also state and mixture-component dependent.

Given the above model parameters, the observation vector sequence $\mathbf{x}_t$ (indexed by (i, m)), $t = 0, 1, \ldots, T-1$ is generated from the model according to

$$\mathbf{x}_t^{(i,m)} = \mathbf{f}_t(\Theta_{i,m}) + N_t(0, \Sigma_{i,m}),$$

where state i and mixture component m at a given time frame t is determined by the evolution of the Markov chain characterized by a_{ij} and $c_{m|i}$.

The mixture version of the trended HMM has the same underlying Markov chain as in the conventional, stationary-state HMM. The time-varying means are expressed explicitly as polynomials of state-sojourn time. We thus have

$$\mathbf{x}_t^{(i,m)} = \sum_{r=0}^{R} \mathbf{B}_{i,m}(r)\mathbf{h}_r(t - \tau_i) + N_t(0, \Sigma_{i,m}), \tag{13.26}$$

where the first term is due to a polynomial regression function (order R) depending on the state (i) and mixture component (m), with $\mathbf{B}_{i,m}(r)$ being the polynomial coefficients, with τ_i registering the time when state i in the HMM is just entered before regression on time takes place. $\mathbf{h}_r(.)$ is an r-th order polynomial function. (To take account of all components of vector $\mathbf{x}_t$ of dimension K, we have a matrix with dimension $K \times (R+1)$ for all the polynomial coefficients.) In the model implementation, orthogonal polynomials are chosen for their superior stability properties in parameter estimation. The final term in Eq. 13.26 is the residual noise, which is assumed to be the output of an IID, zero-mean Gaussian source depending on states and mixture-components, with a time-invariant covariance matrix $\Sigma_{i,m}$. Note that in Eq. 13.26 only the covariance matrix $\Sigma_{i,m}$ and polynomial coefficients $\mathbf{B}_{i,m}(r)$ are considered as *true* model parameters; τ_i is merely an auxiliary parameter for the purpose of obtaining maximal accuracy in estimating $\mathbf{B}_{i,m}(r)$ (over all possible τ_i values).

13.7.2 Derivation of the MMSE estimator using the prior model

Assume the clean speech signal in a chosen feature domain $\mathbf{x} = \{\mathbf{x}_t, t = 0, \ldots, T-1\}$ is corrupted with independent additive noise $\mathbf{v} = \{\mathbf{v}_t, t = 0, \ldots, T-1\}$, resulting in the noisy speech signal $\mathbf{y} = \{\mathbf{y}_t = \mathbf{x}_t + \mathbf{v}_t, t = 0, \ldots, T-1\}$. Let $\mathbf{x}_0^t, \mathbf{v}_0^t$, and $\mathbf{y}_0^t$ denote the vectors of clean speech, noise, and noisy speech processes, respectively, from time frame 0 to t. Let λ_x, λ_v, and λ_y denote the model parameters for the clean speech, noise and noisy speech, respectively. Further, let $\hat{\mathbf{x}} = \{\hat{\mathbf{x}}_t, t = 0, \ldots, T-1\}$ denote the MMSE estimate of the clean speech sequence. For the mean square error,

$$\text{Error} = E[(\hat{\mathbf{x}}_t - \mathbf{x}_t)^{\text{Tr}}(\hat{\mathbf{x}}_t - \mathbf{x}_t)],$$

to be minimized, $\hat{\mathbf{x}}_t$ should be

$$\hat{\mathbf{x}}_t = E\{\mathbf{x}_t \mid \mathbf{y}_0^t\},$$

according to the well-established estimation theory, where $E\{\cdot \mid \cdot\}$ denotes conditional expectation. We are interested in estimating a function g of the clean speech. The optimal estimate is

$$\hat{g(\mathbf{x}_t)} = E\{g(\mathbf{x}_t) \mid \mathbf{y}_0^t\} = \int_{-\infty}^{\infty} g(\mathbf{x}_t) f_{\lambda_x|\lambda_y}(\mathbf{x}_t \mid \mathbf{y}_0^t) d\mathbf{x}_t, \tag{13.27}$$

where $f_{\lambda_x|\lambda_y}$ denotes the conditional probability density function defined according to the model parameters λ_x and λ_y. Using Bayes' rule we can write:

$$f_{\lambda_x|\lambda_y}(\mathbf{x}_t \mid \mathbf{y}_0^t) = \frac{f_{\lambda_y|\lambda_x}(\mathbf{y}_0^t \mid \mathbf{x}_t) f_{\lambda_x}(\mathbf{x}_t)}{f_{\lambda_y}(\mathbf{y}_0^t)}. \tag{13.28}$$

Substitution of Eq. 13.28 into Eq. 13.27 gives:

$$E\{g(\mathbf{x}_t) \mid \mathbf{y}_0^t\} = \frac{\int_{-\infty}^{\infty} g(\mathbf{x}_t) f_{\lambda_y|\lambda_x}(\mathbf{y}_0^t \mid \mathbf{x}_t) f_{\lambda_x}(\mathbf{x}_t) d\mathbf{x}_t}{f_{\lambda_y}(\mathbf{y}_0^t)}. \tag{13.29}$$

For two independent random variables $\mathbf{x}$ and $\mathbf{v}$, the PDF of their sum $\mathbf{y}$ equals the convolution of their respective PDF's. Hence,

$$\begin{aligned} f_{\lambda_y|\lambda_x}(\mathbf{y}_0^t \mid \mathbf{x}_t) &= \int_{-\infty}^{\infty} f_{\lambda_x}(\mathbf{x}_0^t \mid \mathbf{x}_t) f_{\lambda_v|\lambda_x}(\mathbf{y}_0^t - \mathbf{x}_0^t \mid \mathbf{x}_t) d\mathbf{x}_0^t \\ &= \int_{-\infty}^{\infty} f_{\lambda_x}(\mathbf{x}_0^{t-1}) f_{\lambda_v}(\mathbf{v}_0^t) d\mathbf{x}_0^t \\ &= f_{\lambda_v}(\mathbf{v}_t) \int_{-\infty}^{\infty} f_{\lambda_x}(\mathbf{x}_0^{t-1}) f_{\lambda_v}(\mathbf{v}_0^{t-1}) d\mathbf{x}_0^t \\ &= f_{\lambda_v}(\mathbf{v}_t) f_{\lambda_y}(\mathbf{y}_0^{t-1}). \end{aligned} \tag{13.30}$$

Note that in the above we assume $f_{\lambda_x}(\mathbf{x}_0^t \mid \mathbf{x}_t) = f_{\lambda_x}(\mathbf{x}_0^{t-1})$. This implies independence of vectors $\mathbf{x}_t$ for different time frames t (given the model parameters λ_x), as assumed in all types of HMMs, including the trended HMM. However, the key difference between the trended HMM and the conventional HMM is the different parameter sets λ_x; one defines smoothly-varying trajectories and the other does not. Given such parameterization, the $\mathbf{x}_t$ vectors necessarily exhibit the dynamic behavior in the trended HMM despite the independence assumption on the residual noise sequence (and hence an independence assumption on $\mathbf{x}_t$ given λ_x).

By substituting Eq. 13.30 into Eq. 13.29, we obtain:

$$E\{g(\mathbf{x}_t) \mid \mathbf{y}_0^t\} = \frac{\int_{-\infty}^{\infty} g(\mathbf{x}_t) f_{\lambda_v}(\mathbf{v}_t) f_{\lambda_y}(\mathbf{y}_0^{t-1}) f_{\lambda_x}(\mathbf{x}_t) d\mathbf{x}_t}{f_{\lambda_y}(\mathbf{y}_0^t)}. \tag{13.31}$$

Note that Eq. 13.31 above is a general result with no assumptions about any particular models used to describe the statistics of the speech and noise signals. In order to devise a practical speech enhancement algorithm, some modeling assumptions are needed. In this work, the trended HMM is assumed for the short-time DFT (magnitude) sequences of clean speech, and the stationary-state HMM is used for the short-time DFT sequences of noise. Given such modeling assumptions, the PDF's in Eq. 13.31 can be detailed as follows. $f_{\lambda_x}(\mathbf{x}_t)$ and $f_{\lambda_v}(\mathbf{v}_t)$ can be written, according to the law of total probability, as

$$f_{\lambda_x}(\mathbf{x}_t) = \sum_{\beta=1}^{M}\sum_{\gamma=1}^{L}\sum_{d=1}^{t} f_{\lambda_x}(\mathbf{x}_t \mid s_t = \beta, m_t = \gamma, d_t = d) \cdot f_{\lambda_x}(s_t = \beta, m_t = \gamma, d_t = d) \tag{13.32}$$

$$f_{\lambda_v}(\mathbf{v}_t) = \sum_{\xi=1}^{N}\sum_{\delta=1}^{P} f_{\lambda_v}(\mathbf{v}_t \mid n_t = \xi, p_t = \delta) \cdot f_{\lambda_v}(n_t = \xi, p_t = \delta), \tag{13.33}$$

where s_t and m_t denote the speech state and mixture component, respectively, at time t, d_t is the duration of state s_t from the time of entry into that state to time t, and n_t and p_t denote the noise state and mixture component, respectively, at time t. Similarly,

$f_{\lambda_y}(\mathbf{y}_0^t)$ can be written as:

$$\begin{aligned} f_{\lambda_y}(\mathbf{y}_0^t) &= \sum_{\mathcal{S}_0^t}\sum_{\mathcal{M}_0^t}\sum_{\mathcal{N}_0^t}\sum_{\mathcal{P}_0^t} f(\mathcal{S}_0^t, \mathcal{M}_0^t, \mathcal{N}_0^t, \mathcal{P}_0^t, \mathbf{y}_0^t) \\ &= \sum_{\mathcal{S}_0^t}\sum_{\mathcal{M}_0^t}\sum_{\mathcal{N}_0^t}\sum_{\mathcal{P}_0^t} f_s(\mathcal{S}_0^t) f_m(\mathcal{M}_0^t \mid \mathcal{S}_0^t) f_n(\mathcal{N}_0^t \mid \mathcal{M}_0^t, \mathcal{S}_0^t) f_p(\mathcal{P}_0^t \mid \mathcal{N}_0^t, \mathcal{M}_0^t, \mathcal{S}_0^t) \\ &\qquad f_{\lambda_y}(\mathbf{y}_0^t \mid \mathcal{P}_0^t, \mathcal{N}_0^t, \mathcal{M}_0^t, \mathcal{S}_0^t), \end{aligned} \tag{13.34}$$

where $f_s(\mathcal{S}_0^t)$ is the probability of the clean speech state sequence $\mathcal{S}_0^t$, $f_m(\mathcal{M}_0^t \mid \mathcal{S}_0^t)$ is the probability of the sequence of clean speech mixture-component sequence $\mathcal{M}_0^t$ given the state sequence of clean speech $\mathcal{S}_0^t$, $f_n(\mathcal{N}_0^t \mid \mathcal{M}_0^t, \mathcal{S}_0^t)$ (which, due to independence of the clean speech and noise sequences, equals $f_n(\mathcal{N}_0^t)$) is the probability of the noise state sequence, $f_p(\mathcal{P}_0^t \mid \mathcal{N}_0^t, \mathcal{M}_0^t, \mathcal{S}_0^t) = f_p(\mathcal{P}_0^t \mid \mathcal{N}_0^t)$ is the probability of the noise mixture-component sequence given the noise state sequence, and $f_{\lambda_y}(\mathbf{y}_0^t \mid \mathcal{P}_0^t, \mathcal{N}_0^t, \mathcal{M}_0^t, \mathcal{S}_0^t)$ is the PDF of the noisy speech output sequence $\mathbf{y}_0^t$ given $\{\mathcal{P}_0^t, \mathcal{N}_0^t, \mathcal{M}_0^t, \mathcal{S}_0^t\}$.

The components of the product on the right-hand side of Eq. 13.34 are

$$f_s(\mathcal{S}_0^t) = \prod_{\tau=0}^{t} a_{s_{\tau-1}s_\tau} \tag{13.35}$$

$$f_m(\mathcal{M}_0^t \mid \mathcal{S}_0^t) = \prod_{\tau=0}^{t} f_m(m_\tau \mid s_\tau) = \prod_{\tau=0}^{t} c_{m_\tau \mid s_\tau} \tag{13.36}$$

$$f_n(\mathcal{N}_0^t \mid \mathcal{M}_0^t, \mathcal{S}_0^t) = f_n(\mathcal{N}_0^t) = \prod_{\tau=0}^{t} a'_{n_{\tau-1}n_\tau} \tag{13.37}$$

$$\begin{aligned} f_p(\mathcal{P}_0^t \mid \mathcal{N}_0^t, \mathcal{M}_0^t, \mathcal{S}_0^t) &= f_p(\mathcal{P}_0^t \mid \mathcal{N}_0^t) \\ &= \prod_{\tau=0}^{t} f_p(p_\tau \mid n_\tau) = \prod_{\tau=0}^{t} c'_{p_\tau \mid n_\tau} \end{aligned} \tag{13.38}$$

$$\begin{aligned} f_{\lambda_y}(\mathbf{y}_0^t \mid \mathcal{P}_0^t, \mathcal{N}_0^t, \mathcal{M}_0^t, \mathcal{S}_0^t) &= \prod_{\tau=0}^{t} f_{\lambda_y}(\mathbf{y}_\tau \mid s_\tau, m_\tau, n_\tau, p_\tau, d_\tau) \\ &= \prod_{\tau=0}^{t} b(\mathbf{y}_\tau \mid s_\tau, m_\tau, n_\tau, p_\tau, d_\tau), \end{aligned} \tag{13.39}$$

where a_{ij} denotes the speech state transition probability from state i to state j and $c_{m|j}$ is the probability of choosing speech mixture component m given speech state j, a'_{ij} and $c'_{m|j}$ are similarly defined for the noise model, $b(\mathbf{y}_t \mid s_t, m_t, n_t, p_t, d_t)$ denotes the PDF of noisy observation $\mathbf{y}_t$ at time t given the quadruple of speech state, speech mixture component, noise state, and noise mixture component (s_t, m_t, n_t, p_t) and the duration of the speech state up to time t, d_t. Given (s_t, m_t, d_t), the clean speech PDF at time t is Gaussian and so is the noise PDF at time t given (n_t, p_t). Due to the independence of the clean speech and noise processes, the noisy speech PDF (Gaussian) can be written as

$$b(\mathbf{y}_t \mid s_t, m_t, n_t, p_t, d_t) = N_t[\Lambda(s_t, m_t, n_t, p_t, d_t), \Sigma_{s_t,m_t} + \Sigma_{n_t,p_t}], \tag{13.40}$$

where Σ_{s_t,m_t} denotes the covariance matrix of the clean speech for (s_t, m_t) and Σ_{n_t,p_t} denotes the noise covariance matrix for (n_t, p_t). In Eq. 13.40, the mean of the Gaussian,

$\Lambda(s_t, m_t, n_t, p_t, d_t)$, is defined as

$$\Lambda(s_t, m_t, n_t, p_t, d_t) = \left[\mathbf{y}_t - \sum_{r=0}^{R} \mathbf{B}_r(s_t, m_t, n_t, p_t)\mathbf{h}_r(d_t) \right], \tag{13.41}$$

where $\mathbf{h}_r(d_t)$ is the value of the Legendre orthogonal polynomial of order r for the duration d_t, and where $\mathbf{B}_r(s_t, m_t, n_t, p_t)$ is the r-th order trend polynomial coefficient of the noisy speech model. These polynomial coefficients are found from:

$$\begin{aligned} \mathbf{B}_0(s_t, m_t, n_t, p_t) &= \mathbf{B}_0(s_t, m_t) + \mu(n_t, p_t) && \text{and} \\ \mathbf{B}_r(s_t, m_t, n_t, p_t) &= \mathbf{B}_r(s_t, m_t) && \text{for} \quad r > 0. \end{aligned} \tag{13.42}$$

In Eq. 13.42, $\mathbf{B}_r(s_t, m_t)$ denotes the r-th order polynomial coefficients of the clean speech trended HMM for the state and mixture-component pair (s_t, m_t), and $\mu(n_t, p_t)$ is the mean of the noise standard HMM for the state and mixture-component pair (n_t, p_t).

Now we return to the computation of the MMSE estimate specified in Eq. 13.31. Use of Eqs. 13.34-13.39 results in

$$f_{\lambda_y}(\mathbf{y}_0^{t-1}) = \sum_{\mathcal{S}_0^{t-1}} \sum_{\mathcal{M}_0^{t-1}} \sum_{\mathcal{N}_0^{t-1}} \sum_{\mathcal{P}_0^{t-1}} \prod_{\tau=0}^{t-1} a_{s_{\tau-1}s_\tau} c_{m_\tau|s_\tau} a'_{n_{\tau-1}n_\tau} c'_{p_\tau|n_\tau} b(\mathbf{y}_\tau \mid s_\tau, m_\tau, n_\tau, p_\tau, d_\tau). \tag{13.43}$$

Eq. 13.43 can now be used to calculate $f_{\lambda_y}(\mathbf{y}_0^t)$ in the denominator of Eq. 13.31 by replacing $t-1$ with t.

In order to write the computation of the MMSE estimate in a clean form, we first define

$$G_t(\beta, \gamma, \xi, \delta, d, \mathbf{y}_0^t) = f_{\lambda_y}(\mathbf{y}_0^t \mid s_t = \beta, m_t = \gamma, n_t = \xi, p_t = \delta, d_t = d) \tag{13.44}$$

$$= \sum_{\{\mathcal{S}_0^t : s_t = \beta\}} \sum_{\{\mathcal{M}_0^t : m_t = \gamma\}} \sum_{\{\mathcal{N}_0^t : n_t = \xi\}} \sum_{\{\mathcal{P}_0^t : p_t = \delta\}} \prod_{\tau=0}^{t} a_{s_{\tau-1}s_\tau} \cdot c_{m_\tau|s_\tau} \cdot a'_{n_{\tau-1}n_\tau} \cdot c'_{p_\tau|n_\tau} \cdot b(\mathbf{y}_\tau \mid s_\tau, m_\tau, n_\tau, p_\tau, d_\tau).$$

The denominator of Eq. 13.31 can then be written as

$$f_{\lambda_y}(\mathbf{y}_0^t) = \sum_{\beta=1}^{M} \sum_{\gamma=1}^{L} \sum_{\xi=1}^{N} \sum_{\delta=1}^{P} \sum_{d=1}^{t} G_t(\beta, \gamma, \xi, \delta, d, \mathbf{y}_0^t).$$

Further, the product of the three PDF's in the integrand of the numerator in Eq. 13.31 can be written in a compact form derived from Eq. 13.32, Eq. 13.33, Eq. 13.43, and Eq. 13.45:

$$f_{\lambda_y}(\mathbf{y}_0^{t-1}) f_{\lambda_x}(\mathbf{x}_t) f_{\lambda_v}(\mathbf{v}_t) = \sum_{\beta=1}^{M} \sum_{\gamma=1}^{L} \sum_{\xi=1}^{N} \sum_{\delta=1}^{P} \sum_{d=1}^{t} G_t(\beta, \gamma, \xi, \delta, d, \mathbf{y}_0^t) f_{\lambda_x}(\mathbf{x}_t \mid s_t = \beta, m_t = \gamma, n_t = \xi, p_t = \delta, d_t = d) \tag{13.45}$$

On substitution of Eq. 13.43 and Eq. 13.45 in Eq. 13.31, we finally obtain the MMSE estimate

$$\begin{aligned}E\{g(\mathbf{x}_t) \mid \mathbf{y}_0^t\} &= \sum_{\beta=1}^{M}\sum_{\gamma=1}^{L}\sum_{\xi=1}^{N}\sum_{\delta=1}^{P}\sum_{d=1}^{t} \mathcal{W}_t(\beta,\gamma,\xi,\delta,d,\mathbf{y}_0^t)\cdot \\ &\quad \int_{-\infty}^{\infty} g(\mathbf{x}_t) f_{\lambda_x}(\mathbf{x}_t \mid s_t=\beta, m_t=\gamma, n_t=\xi, p_t=\delta, d_t=d)d\mathbf{x}_t \\ &= \sum_{\beta=1}^{M}\sum_{\gamma=1}^{L}\sum_{\xi=1}^{N}\sum_{\delta=1}^{P}\sum_{d=1}^{t} \mathcal{W}_t(\beta,\gamma,\xi,\delta,d,\mathbf{y}_0^t)\cdot \\ &\quad E\{g(\mathbf{x}_t) \mid s_t=\beta, m_t=\gamma, n_t=\xi, p_t=\delta, d_t=d\}, \end{aligned} \tag{13.46}$$

where the weights $\mathcal{W}_t(\beta,\gamma,\xi,\delta,d,\mathbf{y}_0^t)$ are defined as

$$\mathcal{W}_t(\beta,\gamma,\xi,\delta,d,\mathbf{y}_0^t) = \frac{G_t(\beta,\gamma,\xi,\delta,d,\mathbf{y}_0^t)}{\sum_{\beta=1}^{M}\sum_{\gamma=1}^{L}\sum_{\xi=1}^{N}\sum_{\delta=1}^{P}\sum_{d=1}^{t} G_t(\beta,\gamma,\xi,\delta,d,\mathbf{y}_0^t)} \tag{13.47}$$

$$\text{for} \quad 1 \le \beta \le M,\ 1 \le \gamma \le L,\ 1 \le \xi \le N,\ 1 \le \delta \le P,\ 1 \le d \le t.$$

A very interesting interpretation emerges from an examination of Eq. 13.46: the MMSE estimate of the speech signal can be expressed as a weighted average of the state- and mixture-component-conditioned signal expectations over all possible combinations of speech and noise HMM states and mixture components. The main burden for computing the MMSE estimate is now reduced to the calculation of these state- and mixture-component-conditioned expectations:

$$\begin{aligned}&E\{g(\mathbf{x}_t) \mid s_t=\beta, m_t=\gamma, n_t=\xi, p_t=\delta, d_t=d\} = \\ &\quad \int_{-\infty}^{\infty} g(\mathbf{x}_t) f_{\lambda_x}(\mathbf{x}_t \mid s_t=\beta, m_t=\gamma, n_t=\xi, p_t=\delta, d_t=d)d\mathbf{x}_t,\end{aligned} \tag{13.48}$$

which we address below.

To compute Eq. 13.48, function $g(\cdot)$ has to be specified first. For the stationary-state HMM (a special case of the trended HMM), the conditional expectation Eq. 13.48 has been evaluated by Ephraim [Ephr 90]. Different choices of the $g(\cdot)$ function incur varying computational costs. The least amount of computation happens with the choice of

$$g(\mathbf{x}_t) = \{X_t(k), k=0,\ldots,K-1\},$$

where $X_t(k)$ is the k-th DFT (magnitude) component of $\mathbf{x}_t$. In addition to the motivations discussed in detail in Section 2.2, this computational consideration gives another motivation for use of the DFT as the speech feature in this work; that is, we choose $g(\cdot)$ according to Eq. 13.7.2 in the computation of the MMSE estimate.

In determining the integral in Eq. 13.48, we generalize a result of [Ephr 90] from stationary-state HMMs to trended HMMs. The generalized result is that the linear estimate using the MMSE criterion for the k-th component of g (denoted by $g(k)$) given by

$$E\{g(k) \mid \mathbf{y}_t, s_t, m_t, n_t, d_t\} = \int X_t(k) f_{\lambda_y|\lambda_v}(X_t(k) \mid \mathbf{y}_t, s_t, m_t, n_t, d_t)dX_t(k) \tag{13.49}$$

is Gaussian distributed with mean $\widetilde{H}_{s_t,m_t,n_t,p_t,d_t}(k)Y_t(k)$. Here, $Y_t(k)$ is the k-th component of the DFT of $\mathbf{y}_t$, and $\widetilde{H}_{s_t,m_t,n_t,p_t,d_t}(k)$ is the k-th component of the DFT of the frequency-domain Wiener filter output (given state s_t, mixture component m_t, and duration d_t in the clean speech model, and given state n_t and mixture component p_t in the noise model). Note that this Wiener filter changes its transfer function over every single time frame because the transfer function is determined by the polynomial trend function in the trended HMM, which changes smoothly over time frames by design.

Returning to the computation of the sum in Eq. 13.45, which determines the weights in Eq. 13.47 contributing to the aggregate Wiener filter, we are able to establish an efficient recursive form. The recursive formula for the calculation is

$$\begin{aligned} G_t(\beta,\gamma,\xi,\delta,0,\mathbf{y}_0^t) &= \left[\sum_{\beta'=1}^{M}\sum_{\gamma'=1}^{L}\sum_{\xi'=1}^{N}\sum_{\delta'=1}^{P}\sum_{d'=0}^{t} G_{t-1}(\beta',\gamma',\xi',\delta',d',\mathbf{y}_0^{t-1})\cdot a_{\beta'\beta}\cdot a'_{\xi'\xi}\right]\cdot \\ &\quad c_{\gamma|\beta}\cdot c'_{\delta|\xi}\cdot b(\mathbf{y}_t\mid\beta,\gamma,\xi,\delta,0) \\ G_t(\beta,\gamma,\xi,\delta,d,\mathbf{y}_0^t) &= \left[\sum_{\xi'=1}^{N}\sum_{\delta'=1}^{P} G_{t-1}(\beta,\gamma,\xi',\delta',d-1,\mathbf{y}_0^{t-1})\cdot a_{\beta\beta}\cdot a'_{\xi'\xi}\right]\cdot \\ &\quad c_{\gamma|\beta}\cdot c'_{\delta|\xi}\cdot b(\mathbf{y}_t\mid\beta,\gamma,\xi,\delta,0), \qquad \text{for } 0<d\le t. \end{aligned} \tag{13.50}$$

Note that the above recursion has an additional dimension of state-sojourn time (d'), which results from the time-varying means (i.e., polynomial trend functions) in the trended HMM output distributions.

The computation in Eq. 13.46 and that in the related terms by recursive updating formula Eq. 13.50 form the core computation in the MMSE algorithm. Given the time-varying Wiener filters, the multiple nested summations in Eq. 13.46 give a total of $M\cdot L\cdot N\cdot P\cdot t$ summations for each frame (at time t). To compute all T frames, the total number of summations is $M\cdot L\cdot N\cdot P\cdot T^2/2$. Similarly, the total number of summations required in computing the quantities in Eq. 13.50 is $M\cdot L\cdot N\cdot P\cdot T^2/2$.

13.7.3 Implementation of the MMSE enhancement technique

A simplified block diagram of the trended HMM-based enhancement system using the DFT magnitude as speech features is shown in Fig. 13.10. In principle, all possible Wiener filters have to be computed according to the rigorous derivation of the MMSE technique just presented. The number of the possible Wiener filters using the trended HMM as the clean speech model is far more than that for the standard HMM case. Since the mean of the clean speech model for a given state and mixture-component pair is still a function of the additional parameter of state sojourn time, an additional dimension is added to the calculation of the MMSE forward algorithm for finding the filter weights. For each time frame t, a total of $t\times M\times L\times N\times P$ Wiener filters and their corresponding weights have to be calculated. The weights as shown in Eq. 13.47 are functions of the speech and mixture-component pairs of both clean speech and noise models and the clean speech duration up to time t. In theory, the possible number of duration values that a speech state may have at time t is t itself. Therefore, in terms of the computational load, compared with the standard HMM-based MMSE enhancement algorithm, the factor t is multiplied by the total number of possible Wiener filters.

In parallel with calculation of the Wiener filters, each frame of the time-domain noisy input speech is preprocessed and transformed to the feature domain (DFT magnitude). Then, for each frame of the noisy speech, the MMSE forward algorithm is performed. For this, the PDF of the noisy speech has to be calculated. The noisy speech PDF can

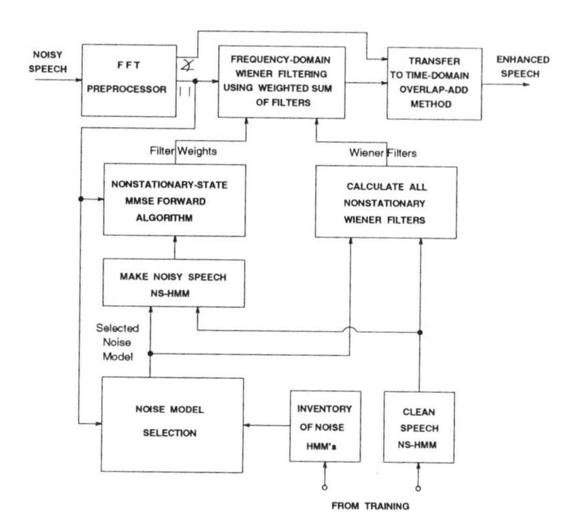

Figure 13.10: Block diagram of the MMSE enhancement technique using the trended HMM for the prior spectral model of clean speech (after Sameti and Deng [Sam 02], @Elsevier).

be specified with the trend polynomial coefficients (**B** matrix) and the covariance matrix. Since speech and noise are additive and independent, the polynomial coefficients of the noisy speech (**B** matrix) is found by adding up the corresponding elements of the **B** matrix of the clean speech model and the means of the noise model. For each time frame t, a total of $M \times L \times N \times P$ matrices of polynomial coefficients are calculated. These calculations are extremely simple compared to the calculations necessary for the construction of the noisy PDF in the AR-HMM structure, even after the approximations employed to reduce the computation cost [Sam 98]. By using diagonal covariance matrices for the trended HMMs, the problem of high computation cost for the inversion of the covariance matrix for calculation of the output likelihoods is also avoided.

Despite all the simplicity for finding the PDF of the noisy speech in the trended HMM framework, the encoding space is still significantly larger than that in the standard HMM case, due to the additional dimension of the speech state duration. This problem has been solved using the double pruning algorithm described in [Sam 98]. For the trended HMM the pruning algorithm is of great importance, because in addition to pruning the calculations for finding the filter weights, it is employed to reduce the number of Wiener filter transfer functions. In the standard HMM enhancement framework, it is practical to calculate all possible Wiener filters and save them before the actual enhancement procedure starts. However, in the trended HMM framework, due to the extensive number of possible Wiener filters, it is better to avoid such calculation. Therefore, in the trended HMM enhancement technique, the necessary Wiener filters for a specific time frame of noisy speech are calculated after the calculation of the filter weights for that specific time frame.

After calculating the PDF of the noisy speech, the extended MMSE forward algorithm is used to find the Wiener filter weights for each time frame. The weights are employed to generate a single weighted Wiener filter from the inventory of the previously calculated Wiener filters. The weighted filter is applied to the noisy frame of speech in the frequency domain. The filtered frequency-domain speech signal is transformed to the time domain using the original noisy signal phase, and the output enhanced signal is synthesized with the overlap-add method.

13.7.4 Approximate MMSE enhancement technique

The MMSE algorithm which employs a soft-decision method for building a weighted-sum filter can be approximated by using a single Wiener filter for each time frame. In this approximate MMSE (AMMSE) method, the most likely Wiener filter for each time frame is selected using a modified version of the extended Viterbi algorithm for the trended HMM, described in Section 5.3.3. This improves the computational efficiency of the enhancement algorithm. A block diagram of the AMMSE technique for the case of the DFT magnitude as the speech feature is shown in Fig. 13.11.

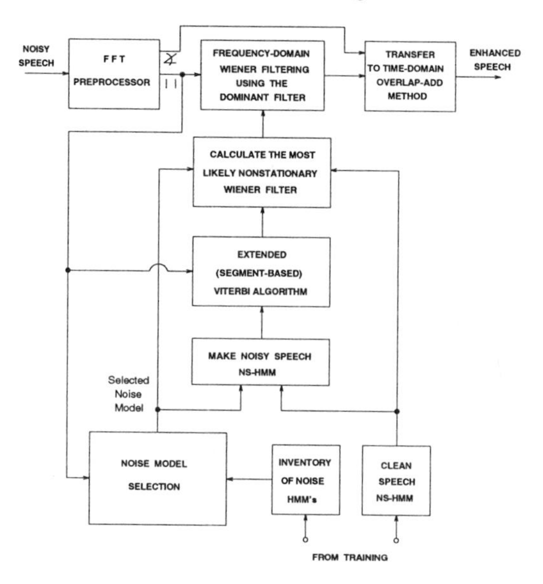

Figure 13.11: Block diagram of the Approximate MMSE enhancement technique (after Sameti and Deng [Sam 02], @Elsevier).

As in the MMSE technique, for the AMMSE the noisy signal is preprocessed and transformed to a DFT magnitude. Using the selected noise HMM and the clean speech's trended HMM, a new trended HMM is built for the noisy speech. Using this newly constructed noisy speech model, the extended Viterbi algorithm is performed on the preprocessed noisy speech data to provide the segmentation information for the noisy observation sequence. The extended Viterbi algorithm determines the most likely speech state, speech mixture-component, noise state, and noise mixture-component quadruple for each frame of the noisy speech signal. For each time frame, having the most likely state, mixture component, and state duration for the clean speech and the most likely state and mixture-component pair for the noise, a single Wiener filter is built from the model parameters. The noisy speech data belonging to that frame is filtered with the constructed Wiener filter. The resulting frequency-domain enhanced frame of speech is transformed into the time domain using the noisy speech phase and the overlap-add method.

In the AMMSE enhancement technique, the problem is to fit the clean speech data, given the noisy speech. More accurate data fitting is equivalent to a more appropriate Wiener filter and better noise reduction as a result. Therefore, this method can be used to analyze the enhancement system's behavior. The AMMSE is simpler than the MMSE technique in that it uses only one Wiener filter for each frame and avoids the costly calculation of the filter weights and the weighted-sum filter. However, the AMMSE algorithm relies on segmentation information, which is obtained by the extended Viterbi algorithm. The extended Viterbi algorithm for the multiple-state and mixture-component noise model would itself be computationally costly. The algorithm confronts a five-dimensional search space comprised of the noise and speech states and mixture components, in addition to the speech state duration. For a single-state noise, however, the AMMSE is more efficient than the MMSE enhancement technique.

The computation of the AMMSE algorithm mainly lies in that of the extended Viterbi algorithm. To summarize, the computation is quadratic in T (as opposed to the conventional Viterbi algorithm, which has computation linear in T). However, if the durational properties of speech are taken into account, the computation can be reduced to that linear in T, with a large constant equal to the maximum duration of the speech unit.

13.7.5 Diagnostic experiments

In the diagnostic experiments described here, some individual components of the enhancement technique are analyzed. The AMMSE technique is used as the enhancement algorithm since it enables one to analyze detailed data-fitting behavior in the enhancement process and to investigate the relation between the observed goodness-of-fit and the overall performance of the enhancement system.

In the AMMSE-based enhancement system, the extended Viterbi algorithm obtains the possible segmentation information for the desired *clean* speech signal using the available *noisy* speech signal. To achieve this goal, the noisy-speech-trended HMM is artificially built from the pre-trained clean speech model and the noise model.

An utterance comprised of a portion of a sentence from the TIMIT speech database was used for both training and testing. DFT magnitude vectors containing 129 components with a resolution of 62.5 Hz were used as the speech and noise features. Each frame of the noisy speech was taken to be 256 samples long, corresponding to 16 ms of data. The overlap between adjacent frames was set to 50%. Trended HMMs with four states and four mixture components and of orders 0, 1, and 2 were trained with the utterance. Noisy speech was generated by adding Gaussian white noise to the test utterance with zero-dB SNR. The AMMSE enhancement procedure was performed on the test utterance, and the output SNRs and total likelihoods for varying polynomial orders of the speech-trended HMM were then calculated. Data-fitting results were obtained during the segmentation stage, where the polynomial functions of time were concatenated sequentially, according to the selected HMM state/mixture sequence, to approximate the data trajectories. To establish the highest possible performance of the speech enhancement system, the clean speech spectra were used in the system to construct Wiener filters, instead of using the polynomials obtained from the output of the extended Viterbi algorithm. The SNR obtained via use of the clean speech information sets the upper limit of the system performance. The output SNRs and the associated likelihoods signaling goodness of data-fitting to the models are presented in Table 13.1. A direct correlation between the goodness of data fitting and the speech enhancement performance measured by the output SNRs is clearly demonstrated by these results. The results in Table 13.1 also show that as the polynomial order in the trended HMM is increased from zero to two, the performance of enhancement approaches the theoretical limit determined by using the clean speech spectra to derive the Wiener filters.

	Order 0	Order 1	Order 2	Clean Data
Output SNR (dB)	9.45	10.53	11.01	11.43
Log-Likelihood	-1.6196e+05	-1.5883e+05	-1.5769e+05	–

Table 13.1: The output SNRs and log-likelihoods from the diagnostic tests with different orders of the NS-HMM trend polynomial and the output SNR using clean speech information (after Sameti and Deng [Sam 02], @Elsevier).

To examine the detailed behavior of the comparative system performance exhibited in Table 13.1, we show data-fitting results for modeled DFT feature vectors (polynomials) fitted to real DFT feature vectors. For illustration purposes, only one representative DFT component (at 187.5 Hz) is shown in Fig. 13.12 and another representative DFT component (at 1687.5 Hz) is shown in Figure 13.13. The results in Figs. 13.12 and 13.13 and our observations on several other DFT components show that the DFT components that are less affected by noise were estimated more accurately than those affected more by noise. This implies that the method of calculating the noisy speech PDF and that of building a good noisy speech model are crucial in finding a good estimate of the clean speech signal. Note that the likelihood obtained from the extended Viterbi algorithm is that of the noisy speech observations given the artificially generated noisy speech model, and it is desirable that this likelihood be a reliable indication of that for the clean speech data given the clean speech model. The data-fitting observations shown in Figures 13.12 and 13.13 are supportive of a positive correlation between the two likelihoods.

The results in Table 13.1 discussed earlier showed that use of higher polynomial orders in the trended HMM yields both better likelihoods and higher SNRs than use of lower orders (in particular, order zero). The data-fitting results shown in Figures 13.12 and 13.13 provided some underlying reasons for this improvement: the trended HMM is doing its job of smoothly (except when HMM state transitions occur) following speech spectral variations over time, with a better job done using higher orders of polynomial trend functions. The consistency of the clean speech data-fitting results with the associated likelihoods indicates that the parameters of the noisy speech model are reasonably accurately estimated. Further, the consistency between goodness of data-fitting and the output SNRs indicates that the construction of the dominant Wiener filter and the procedure of frequency-domain filtering and reconstruction of time-domain speech signals have been done correctly.

It may be argued that the smooth trajectories across frames inherent in the trended HMM could also be realized by the Gaussian-mixture model if different mixture components could be automatically chosen to fit the smooth trajectories in the speech data. To examine such a possibility, we conduct speech enhancement experiments by constraining the number of HMM states to be one and obtaining a Gaussian-mixture model for clean speech. Keeping the same number of Gaussian components (16 in total), we obtained the output SNR which is worse than the HMM — 9.05 dB for the Gaussian-mixture model versus 9.45 dB for the conventional HMM and 11.01 for the trended HMM. This suggests that the Gaussian-mixture model appears unable to automatically choose the

a)

b)

c)

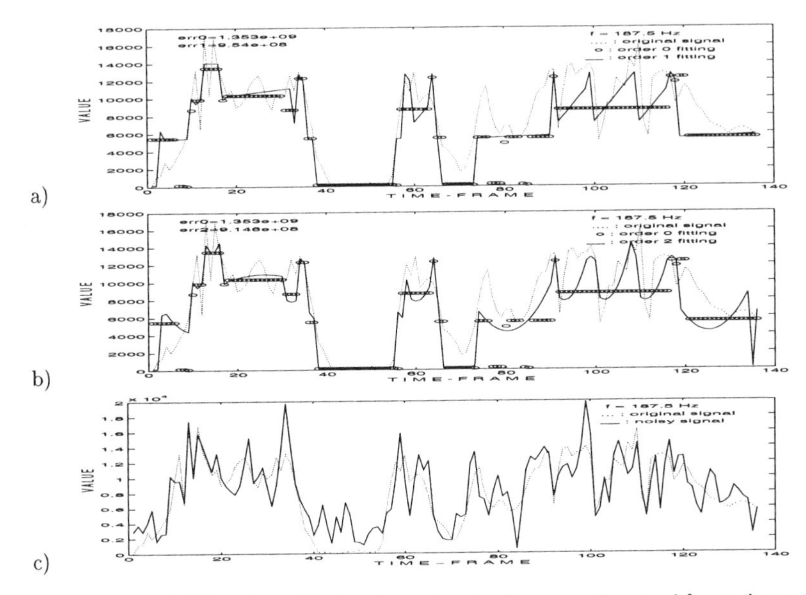

Figure 13.12: HMM fitting with the clean signal using noisy input with varying polynomial orders. DFT magnitude component 3, frequency=187.5 Hz: a) order 0 and order 1 fitting, b) order 0 and order 2 fitting, c) noisy input and clean signal. err0: total fitting error between the original signal and the order-0 model; err1: error between the original signal and the order-1 model; err2: error between the original signal and the order-2 model (after Sameti and Deng [Sam 02], @Elsevier).

"correct" mixture components given only the noisy speech data available. Since the trended HMM forces smooth transitions across frames in the model structure itself, it guarantees a natural fit between the data trajectory and the model component, thereby outperforming both the Gaussian-mixture model and the conventional HMM.

The next diagnostic experiment was motivated by the following reasoning. As explained in Section 13.7, the AMMSE speech enhancement method employs the extended Viterbi algorithm to obtain the segmentation information for the clean speech signal. Given a valid segmentation of the clean speech, proper estimation of the signal is possible using the clean speech model parameters. However, this estimation is not sufficient to generate high-quality enhanced speech. This is due to the fact that the estimated clean speech is the mean of a random process, and only a single realization of the random process gives rise to the enhanced speech. Since the noisy version of this realization is available, use of Wiener filtering is one way to obtain the clean speech signal. In other words, a Wiener filter is constructed using the acquired estimation of the clean speech and the noise spectrum, and frequency-domain filtering of the noisy speech spectrum is performed to obtain the enhanced speech. Now, to illustrate the role of the Wiener filtering in speech enhancement, we in this diagnostic experiment deliberately eliminated the Wiener filtering stage. In the implementation, we first estimated the means of the polynomial trajectories in the trended HMM (for the clean speech spectral sequence) using the extended Viterbi algorithm, and then straightforwardly transformed the estimated means into the time domain. (This transformation does not lose information because

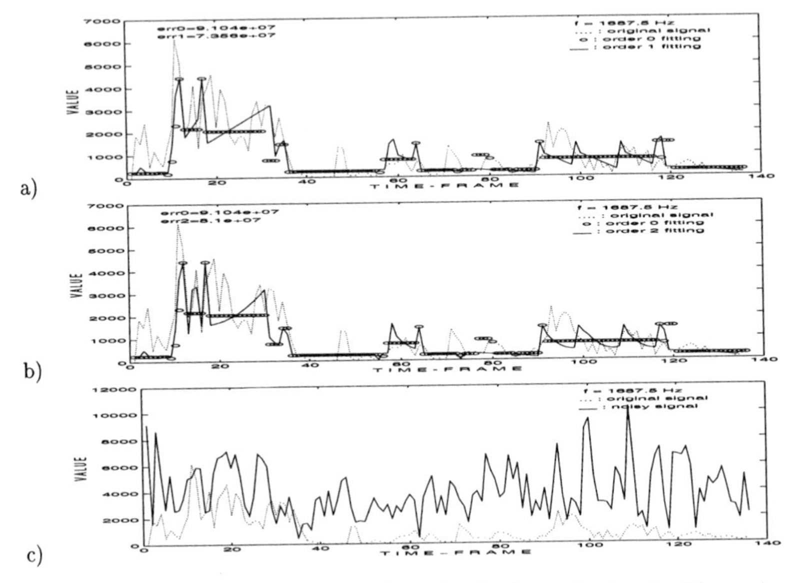

Figure 13.13: HMM fitting with the clean signal using noisy input with varying polynomial orders. DFT magnitude component 27, frequency=1687.5 Hz. a) order 0 and order 1 fitting, b) order 0 and order 2 fitting, c) noisy input and clean signal. err0: total fitting error between the original signal and the order-0 model; err1: error between the original signal and the order-1 model; err2: error between the original signal and the order-2 model (after Sameti and Deng [Sam 02], @Elsevier).

the DFT used is a fully reversible feature.) The SNRs of the reconstructed time-domain speech utterance were calculated, with the results shown in Table 13.2. Here, we observe that the quality of such reconstructed speech signals is significantly lower than that of their counterparts obtained with the use of Wiener filters. Nevertheless, we also observe from Table 13.2 that the absolute SNRs still improve as the polynomial order in the trended HMM used to fit the speech data increases from zero to two.

	Order 0	Order 1	Order 2
Output SNR (dB)	6.60	7.50	7.81

Table 13.2: The output SNRs from the diagnostic tests using only the estimated mean trajectories of the clean speech, with different orders of the NS-HMM trend polynomial (after Sameti and Deng [Sam 02], @Elsevier).

The main conclusion drawn from the diagnostic experiments presented in this subsection is that the superior speech enhancement performance (measured by the output SNRs) achieved by the trended HMM, compared to the conventional stationary-state HMM, derives from the superior ability of the trended HMM to fit the speech data. Therefore, the problem of speech enhancement is to a large extent equivalent to that of correct segmentation of, and accurate spectral-mean approximation (fitting) to, the "hidden" clean speech signal given the noisy speech data.

13.7.6 Speech waveform enhancement results

We now discuss some results of speech enhancement experiments in this case study using the trended HMM as the clean speech model. The MMSE enhancement algorithm described earlier was used throughout the experiments. This case study tested polynomial orders of 0, 1, 2, 3, and 4 in the trended HMM, and used two types of noise, white noise and simulated helicopter noise, as the additive noise to generate noisy speech data in all experiments. All experiments were carried out using 129 components of the DFT magnitude as the features for speech and noise data.

Experiments were run on arbitrarily chosen TIMIT sentences. The two types of additive noise, each having input SNRs ranging from 0 dB to 15 dB with increments of 5 dB, were used. The formal measure of the enhancement performance is the output SNR. The results of the experiments are shown in Tables 13.3 and 13.4, for the use of simulated white noise and simulated helicopter noise, respectively, in a typical sentence. In the tables, the output SNR as the performance measure is shown as a function of the input SNR and of the polynomial order in the trended HMM used in the MMSE enhancement algorithm.

Input SNR (dB)	Order 0	Order 1	Order 2	Order 3	Order 4
0	7.51	7.60	7.62	7.65	7.67
5	8.22	9.10	9.12	9.13	9.15
10	8.97	10.95	10.95	11.02	11.15
15	9.91	12.15	12.33	12.38	12.50

Table 13.3: The output SNRs for different input SNRs and different orders of NS-HMM. White additive noise is used (after Sameti and Deng [Sam 02], @Elsevier).

Input SNR (dB)	Order 0	Order 1	Order 2	Order 3	Order 4
0	7.98	8.15	8.21	8.22	8.30
5	8.70	9.62	9.71	9.81	9.88
10	9.15	10.92	10.95	11.12	11.20
15	9.81	12.20	12.32	12.38	12.55

Table 13.4: The output SNRs for different input SNRs and different orders of NS-HMM. Simulated helicopter noise is used (after Sameti and Deng [Sam 02], @Elsevier).

In analyzing the enhancement results, general consistency is observed across different speech utterances and across noise types. First, enhancement is equally effective for stationary and nonstationary noises. Second, enhancement is most effective for low input SNRs (0 dB and 5 dB). Third, advantages of the trended HMM (order greater than zero) over the stationary-state HMM (i.e., order-zero-trended HMM) were observed across all input SNRs and across both noise types. Fourth, the SNR improvements gained using the trended HMM with orders higher than one are marginal, but the superiority of the order-one trended HMM over the order-zero trended HMM is considerable. This again is consistent across all input SNRs and across both noise types. As reported in [Sam 02], some informal listening results indicate that the improvement from use of the order-zero trended HMM to use of the order-one trended HMM is perceptually detectable by listeners, but no perceptual differences are found with use of other varying polynomial orders in the trended HMM.

13.8 Use of Speech Feature Enhancement for Robust Speech Recognition

As we discussed earlier in this chapter, in addition to restoring clean speech waveforms for improved human perceptual listening, speech enhancement can also be applied to restore clean speech features for feature-space robust ASR via processing the distorted speech (waveforms or features) caused by acoustic environments. There are two special considerations for such speech feature enhancement in contrast to the speech waveform enhancement introduced earlier. First, since the purpose of speech feature enhancement is to improve ASR performance, the optimization criteria used for the enhancement need to take account of those for ASR system design. In particular, since Mel-scaled, log-domain speech features (such as cepstra or Mel log-channel energies) have proved to be more effective than linear-domain features (such as DFT magnitudes or speech waveforms) for ASR, it should be more desirable to formulate speech feature enhancement algorithms directly in the log domain than in the linear domain, as was discussed earlier in this chapter for speech waveform enhancement. Second, due to the log-domain formulation of speech feature enhancement, the mathematical description of the generation process of the distorted speech features (in the log domain) will be more complex than that for the linear domain. This generation process is often called the **acoustic distortion model** or environmental model.

The acoustic distortion model for the linear-domain enhancement algorithms is simply addition of signals (between clean speech and noise), as well as convolution (for waveforms) or multiplication (for DFTs) of signals. For the log-domain enhancement algorithms, the acoustic distortion model is much more complex and involves nonlinearity among the log-domain quantities of the distorted speech, clean speech, noise, and the channel impulse response function. Such a model will be presented in this section, following an introduction to the roles of speech enhancement in feature-space robust ASR. model Using the log-domain nonlinear acoustic distortion model, we will then present the use of speech prior models and related feature enhancement algorithms formulated in the same log-domain. The material of this section is based on the work published in [Deng 02].

13.8.1 Roles of speech enhancement in feature-space robust ASR

While dramatic progress has been made during the past decade in the various fields of speech recognition technology, it is widely accepted that the technology today is still far from the ultimate goal of seamless human-machine communication. One of the major problems that still remain unsolved in the current recognition technology is noise robustness (e.g., [Vii 01, Pea 01]). Two major classes of techniques for noise-robust speech recognition include: 1) the model-domain approach as we discussed in Chapter 12, where the speech models in the recognizer are modified or adapted to match the statistical properties of the unmodified noisy test speech; and 2) the feature-domain approach to be discussed in this chapter, where the noisy test speech (and possibly the "noisy" training speech as well) is modified or enhanced to move towards clean speech as much as possible. The work of [Deng 00b, Deng 01, Ace 00] demonstrated remarkably superior performance of the feature-domain approach over the model-domain one. When the training speech is corrupted intentionally, followed by feature enhancement (i.e., front-end denoising) and retraining of the HMM system, a higher performance is achieved than that under the matched noisy condition, which sets the limit of the model-domain approach.

Towards solving the noise robustness problem based on the feature-domain approach, a number of statistical techniques have been developed based on parametric modeling of nonlinear acoustic distortion models or on the use of other information [Deng 02, Ace 93, More 96a, More 96b, Ace 00, Deng 01, Frey 01, Kim 98, Seg 01, Deng 00b, Deng 01, Drop 01]. In this section, we will elaborate on a particular approach taken in the work of [Deng 02] which incorporates joint static and dynamic cepstral or log-channel energy features in the Bayesian framework for effective speech feature enhancement. The feature enhancement algorithm uses the full posterior information that integrates the predictive information from the nonlinear acoustic distortion model, the prior information based on the static clean speech cepstral distribution, and the prior information based on the frame-differential dynamic cepstral distribution. This algorithm can be viewed as a generalization or modification of feature-space spectral subtraction techniques. As discussed earlier in this chapter, spectral subtraction works by obtaining a noise estimate in the linear spectral domain and then subtracting that from the noisy observation in the same domain. The subtraction residual gives the spectral estimate for clean speech, but there are no mechanisms to reject the subtraction result even if it deviates substantially from the clean speech statistics (due, for instance, to a poor noise estimate). This is because there is no prior model or template for clean speech that can be used to verify the "reasonableness" of the spectral subtraction result. The new statistical technique presented in this section provides a formal framework to overcome this deficiency, and can be viewed as generalized, probabilistic spectral "subtraction" or noise removal. When all the signals are represented in the cepstral or log domain, which we believe is most effective for speech recognition applications, such generalized "noise removal" that takes into account the prior information is achieved via optimal statistical estimation using a nonlinear environment model. This process has been formalized as an efficient feature enhancement algorithm and will be described in detail in this section.

13.8.2 A statistical model for log-domain acoustic distortion

We now describe a statistical environment model for the log-spectral-domain acoustic distortion. This will allow us to compute the conditional likelihood for the noisy speech observation, in the same log-domain domain, given all relevant information (such as clean speech and noise).

Following the standard discrete-time, linear system model for the acoustic distortion in the time domain and in the frequency domain [Ace 93, More 96a], we have the following relationship among the noisy speech (y, Y), clean speech (x, X), additive noise (n, N), and channel impulse response h with corresponding transfer function H:

$$y[t] = x[t] * h[t] + n[t],$$

and

$$Y[k] = X[k]H[k] + N[k], \qquad (13.51)$$

respectively, where $*$ denotes circular convolution, and k is the frequency-bin index in the DFT for a fixed-length time window.

Power spectra of the noisy speech can then be obtained from the DFT in Eq. 13.51 by

$$|Y[k]|^2 = |X[k]|^2|H[k]|^2 + |N[k]|^2 + 2(X[k]H[k])N[k]\cos\theta_k,$$

where θ_k denotes the (random) angle between the two complex variables $N[k]$ and $(X[k]H[k])$.

We now apply a set of Mel-scale filters (L in total) to power spectra $|Y[k]|^2$ in the frequency domain, where the lth filter is characterized by the transfer function $W_k^{(l)} \geq 0$. This produces L channel (Mel-filter bank) energies of

$$\sum_k W_k^{(l)}|Y[k]|^2 = \sum_k W_k^{(l)}|X[k]|^2|H[k]|^2 + \sum_k W_k^{(l)}|N[k]|^2 + 2\sum_k W_k^{(l)}(X[k]H[k])N[k]\cos\theta_k, \tag{13.52}$$

with $l = 1, 2, ..., L$.

Denoting the various channel energies in Eq. 13.52 by

$$|\tilde{Y}^{(l)}|^2 = \sum_k W_k^{(l)}|Y[k]|^2, \quad |\tilde{X}^{(l)}|^2 = \sum_k W_k^{(l)}|X[k]|^2, \quad |\tilde{N}^{(l)}|^2 = \sum_k W_k^{(l)}|N[k]|^2,$$

and

$$|\tilde{H}^{(l)}|^2 = \frac{\sum_k W_k^{(l)}|X[k]|^2|H[k]|^2}{|\tilde{X}^{(l)}|^2} \approx \sum_k W_k^{(l)}|H[k]|^2,$$

we simplify Eq. 13.52 to

$$|\tilde{Y}^{(l)}|^2 = |\tilde{X}^{(l)}|^2|\tilde{H}^{(l)}|^2 + |\tilde{N}^{(l)}|^2 + 2\lambda^{(l)}|\tilde{X}^{(l)}||\tilde{H}^{(l)}||\tilde{N}^{(l)}|, \tag{13.53}$$

where we define

$$\lambda^{(l)} \equiv \frac{\sum_k W_k^{(l)}(\tilde{X}[k]\tilde{H}[k])\tilde{N}[k]\cos\theta_k}{|\tilde{X}^{(l)}||\tilde{H}^{(l)}||\tilde{N}^{(l)}|}.$$

$\lambda^{(l)}$ is a scalar, which can be shown to have its value between -1 and 1.

Define the log channel energy vectors:

$$\boldsymbol{y} = \begin{bmatrix} \log|\tilde{Y}^{(1)}|^2 \\ \log|\tilde{Y}^{(2)}|^2 \\ \cdots \\ \log|\tilde{Y}^{(l)}|^2 \\ \cdots \\ \log|\tilde{Y}^{(L)}|^2 \end{bmatrix}, \quad \boldsymbol{x} = \begin{bmatrix} \log|\tilde{X}^{(1)}|^2 \\ \log|\tilde{X}^{(2)}|^2 \\ \cdots \\ \log|\tilde{X}^{(l)}|^2 \\ \cdots \\ \log|\tilde{X}^{(L)}|^2 \end{bmatrix}, \quad \boldsymbol{n} = \begin{bmatrix} \log|\tilde{N}^{(1)}|^2 \\ \log|\tilde{N}^{(2)}|^2 \\ \cdots \\ \log|\tilde{N}^{(l)}|^2 \\ \cdots \\ \log|\tilde{N}^{(L)}|^2 \end{bmatrix}, \quad \mathbf{h} = \begin{bmatrix} \log|\tilde{H}^{(1)}|^2 \\ \log|\tilde{H}^{(2)}|^2 \\ \cdots \\ \log|\tilde{H}^{(l)}|^2 \\ \cdots \\ \log|\tilde{H}^{(L)}|^2 \end{bmatrix}, \tag{13.54}$$

and define the vector:

$$\boldsymbol{\lambda} = \begin{bmatrix} \lambda^{(1)} \\ \lambda^{(2)} \\ \cdots \\ \lambda^{(l)} \\ \cdots \\ \lambda^{(L)} \end{bmatrix}.$$

Eq. 13.53 can now be written as

$$e^{\boldsymbol{y}} = e^{\boldsymbol{x}} \bullet e^{\mathbf{h}} + e^{\boldsymbol{n}} + 2\,\boldsymbol{\lambda} \bullet e^{\boldsymbol{x}/2} \bullet e^{\mathbf{h}/2} \bullet e^{\boldsymbol{n}/2} = e^{\boldsymbol{x}+\mathbf{h}} + e^{\boldsymbol{n}} + 2\,\boldsymbol{\lambda} \bullet e^{(\boldsymbol{x}+\mathbf{h}+\boldsymbol{n})/2}, \tag{13.55}$$

where the $\bullet$ operation for two vectors denotes element-wise product, and each exponentiation of a vector above is also an element-wise operation.

To obtain the log channel energy for noisy speech, we apply the log operation on both sides of Eq. 13.55:

$$y = \log\left[e^{x+\mathbf{h}} \bullet (\mathbf{1} + e^{n-x-\mathbf{h}} + 2\ \boldsymbol{\lambda} \bullet e^{\frac{x+\mathbf{h}+n}{2}-x-\mathbf{h}})\right] = x+\mathbf{h}+\log[\mathbf{1}+e^{n-x-\mathbf{h}}+2\ \boldsymbol{\lambda}\bullet e^{\frac{n-x-\mathbf{h}}{2}}]$$

This can be further simplified to

$$\begin{aligned} y &= x + \mathbf{h} + \log\left[(\mathbf{1} + e^{n-\mathbf{h}-x}) \bullet [\mathbf{1} + 2\ \boldsymbol{\lambda} \bullet e^{\frac{n-x-\mathbf{h}}{2}} \bullet /(\mathbf{1} + e^{n-x-\mathbf{h}})]\right] \\ &= x + \mathbf{h} + \log(\mathbf{1} + e^{n-\mathbf{h}-x}) + \log[\mathbf{1} + 2\ \boldsymbol{\lambda} \bullet e^{\frac{n-x-\mathbf{h}}{2}} \bullet /(\mathbf{1} + e^{n-x-\mathbf{h}})] \\ &= x + \mathbf{h} + \log(\mathbf{1} + e^{n-\mathbf{h}-x}) + \log[\mathbf{1} + \boldsymbol{\lambda} \bullet / \cosh(\frac{n - x - \mathbf{h}}{2})], \end{aligned} \tag{13.56}$$

where "$\bullet/$" denotes element-wise vector division in the above. Further, assuming $\boldsymbol{\lambda} \ll \cosh(\frac{n-x-\mathbf{h}}{2})$ and using $\log(1+x) \approx x$ for $x \ll 1$, we obtain

A log-domain acoustic-distortion model for the effect of additive noise and linear channel distortion:

$$y \approx x + \mathbf{h} + \log(\mathbf{1} + e^{n-\mathbf{h}-x}) + \boldsymbol{\lambda} \bullet / \cosh(\frac{n - x - \mathbf{h}}{2}), \tag{13.57}$$

where y, x, n, and h are log-channel energies as defined in Eq.13.54.

Due to the generally small values of the last term in Eq. 13.56, this acoustic distortion model can be interpreted as a predictive mechanism for y, where the predictor is

$$\hat{y} = x + h + g(n - x - h),$$

in which

$$g(z) = \log\big(\mathbf{1} + e^{z}\big).$$

The small prediction residual in Eq. 13.56:

$$\mathbf{r} = \boldsymbol{\lambda} \bullet / \cosh(\frac{n - x - \mathbf{h}}{2}), \tag{13.58}$$

is complicated to evaluate and to model. It is therefore represented by an "ignorance" model as a Gaussian random vector. Using the stereo speech data consisting of about 10,000 digit sequences in the training set of the Aurora2 database and using Eq. 13.56, the study of [Deng 02] empirically verified that the average value of $\boldsymbol{\lambda} \bullet / \cosh(\frac{n-x-\mathbf{h}}{2})$ is very close to zero for each vector element. It was also empirically observed that the distribution of $\boldsymbol{\lambda}$ has Gaussian shapes (subject to the truncation above +1 and below −1). Therefore, as a reasonable choice, the zero mean vector can be reasonably used in the Gaussian distribution as an approximate model for the prediction residual.

The covariance matrix of the modeled Gaussian random vector for the prediction residual Eq. 13.58 is in principle a function of the (instantaneous) SNR. However, for reducing implementation complexity associated with the SNR dependency, one fixed diagonal covariance matrix of the residual noise can be estimated by pooling the training data with all available SNRs. Assuming a fixed, SNR-independent covariance matrix $\boldsymbol{\Psi}$, the statistical model for the acoustic environment is thus established as

$$y = \underbrace{x + h + g(n - x - h)}_{\hat{y}} + \mathbf{r}, \tag{13.59}$$

with $\mathbf{r} \sim \mathcal{N}(\mathbf{r}; \mathbf{0}, \Psi)$.

Further simplification can be made by taking account of additive noise only. The channel distortion can be handled via a separate process of cepstral mean normalization. This further simplifies the model of Eq. 13.59 into

$$y = \underbrace{x + g(n - x)}_{\hat{y}} + \mathbf{r}. \tag{13.60}$$

The Gaussian assumption for the residual $\mathbf{r}$ in the model of Eq. 13.60 allows straightforward computation of the likelihood for the noisy speech observation according to

$$p(y|x, n) = \mathcal{N}(y; x + g(n - x), \Psi) = \mathcal{N}(y; \hat{y}, \Psi). \tag{13.61}$$

This likelihood model is a key component in deriving the MMSE estimator for speech feature enhancement, which will be described shortly.

13.8.3 Use of prior models for clean speech and for noise

In addition to the acoustic distortion model for the likelihood evaluation of y, the MMSE estimation framework for speech feature enhancement also requires "prior" models for the statistical behavior of clean speech features and of noise features. Both the speech and noise features are known to be nonstationary. The mechanisms to capture the nonstationarity in the prior model for clean speech can be: 1) using dynamic features which take the local, time difference of static features; and 2) using multiple modes (mixtures) in the probability distribution, allowing the mode to switch freely at different times in an unstructured manner. In contrast, the most effective mechanism to accommodate nonstationarity in the prior model for noise is to directly represent the prior properties of the noise features in an explicitly time-indexed fashion.

Prior model for clean speech incorporating dynamic features

The prior model can be designed to take into account both the static and dynamic properties of clean speech, in the domain of log Mel-channel energy (or equivalently in the cepstral domain via a fixed, linear transformation). The simplest way of capturing the dynamic property is to use the frame-differential, or "delta" feature, defined by

$$\Delta x_t \equiv x_t - x_{t-1},$$

where a one-step, backward time (frame) difference is used.

The choice of the functional form of the probability distribution for both the static and delta features of clean speech is often motivated by simplicity in the enhancement algorithm implementation, as a mixture of multivariate Gaussians, where in each Gaussian component the static and delta features are assumed to be uncorrelated with each other. This gives the joint PDF:

$$p(\boldsymbol{x}_t, \Delta \boldsymbol{x}_t) = \sum_{m=1}^{M} c_m \mathcal{N}(\boldsymbol{x}_t; \boldsymbol{\mu}_m^{\boldsymbol{x}}, \boldsymbol{\Sigma}_m^{\boldsymbol{x}}) \mathcal{N}(\Delta \boldsymbol{x}_t; \boldsymbol{\mu}_m^{\Delta \boldsymbol{x}}, \boldsymbol{\Sigma}_m^{\Delta \boldsymbol{x}}). \tag{13.62}$$

A standard EM algorithm can be used to train the mean and covariance parameters $\boldsymbol{\mu}_m^{\boldsymbol{x}}, \boldsymbol{\Sigma}_m^{\boldsymbol{x}}, \boldsymbol{\mu}_m^{\Delta \boldsymbol{x}}$, and $\boldsymbol{\Sigma}_m^{\Delta \boldsymbol{x}}$ for the log-domain features. When the cepstral features are used, the mean vectors in the log Mel-channel energy domain can be obtained via the linear transform using the inverse cosine transformation matrix. The diagonal elements of the two covariance matrices in the log Mel-channel energy domain can be computed also from those in the cepstral domain, using the inverse cosine transformation matrix and its transpose.

Note that due to the inclusion of the delta feature in Eq. 13.62, the speech frame $\boldsymbol{x}_t$ is no longer independent of its previous frame. This allows the trajectory information of speech to be captured as part of the prior information. Compared with conventional approaches which exploit only the static features in the Gaussian mixture model, the additional information source provided by Eq. 13.62 is the new dynamic parameters $\boldsymbol{\mu}_m^{\Delta \boldsymbol{x}}$ and $\boldsymbol{\Sigma}_m^{\Delta \boldsymbol{x}}$, which cannot be inferred from the static parameters $\boldsymbol{\mu}_m^{\boldsymbol{x}}$ and $\boldsymbol{\Sigma}_m^{\boldsymbol{x}}$. These orthogonal sources of information permit more accurate characterization of the prior statistical properties of the clean-speech process.

Prior model for noise features using a time-varying, fixed-point estimate

In principle, in the Bayesian framework, it is desirable to provide a prior distribution for the noise parameter $\boldsymbol{n}$. In many cases, the noise distribution needs to be nonstationary or time-varying; that is, the noise distribution is a function of time frame t. When a limited amount of noisy speech training data is available, even assuming a simple Gaussian model for the noise feature with a time-varying mean and variance, accurate estimation of these parameters is still difficult. Simplification can be made where the noise feature is assumed to be deterministic, which can be tracked sequentially directly from the individual noisy test utterance. This is equivalent to assuming a nonstationary (degenerated) Gaussian model as the prior for noise, where the mean vector indexed separately for each time frame t is known and where the covariance matrix is fixed to be zero; i.e., the prior probability distribution for noise is reduced to a time-varying, vector-valued, delta function:

$$p(\boldsymbol{n}_t) = \delta(\boldsymbol{n}_t - \bar{\boldsymbol{n}}_t). \tag{13.63}$$

The above method of dealing with noise nonstationarity can be considered as a "non-parametric" technique, where the noise variable $\bar{\boldsymbol{n}}_t$ is explicitly indexed by each time t, rather than being drawn from a parametric distribution. This is in contrast to the use of the time-invariant mixture model as a parametric method for capturing speech nonstationarity as described earlier in this section. The speech nonstationarity is implicitly embedded via the possibility of mode (component) switching in the mixture model. This parametric technique is appropriate for modeling clean speech, since the training set used to estimate the parameters in the mixture model can often easily cover the acoustic space of the "hidden" clean speech responsible for generating the noisy speech observations. However, this same parametric technique would not be appropriate for modeling noise nonstationarity. This is because the noise types and levels are too numerous, and they are too difficult to predict in advance for training a mixture model with a full coverage of the acoustic space for the time-varying noise embedded in the test data. Adaptation to the test data is necessary, and the very small amount of the adaptation data in each changing test utterance makes parametric techniques ineffective.

13.8.4 Use of the MMSE estimator

Given the prior models for clean speech (Eq. 13.62) and for noise (Eq. 13.63), and given the likelihood model (Eq. 13.61), an application of Bayes' rule in principle would give the posterior probability for the clean speech conditioned on the noisy speech observations. This computation, however, is highly complex, since it would require very expensive nonlinear techniques, due to the use of the log-domain environment model. The computation can be made feasible in two ways. First, linearization on the nonlinear predictor, $\hat{\boldsymbol{y}}$, in the likelihood model Eq. 13.61 is made. The approximation accuracy can be improved via an iterative technique in nonlinear signal processing [Mend 95], which was previously successfully applied to speech enhancement in [Frey 01] and for spontaneous speech recognition in [Deng 00c]. Second, while computing the entire posterior probability would be desirable for an integrated system for signal processing and speech recognition, speech feature enhancement as front-end signal processing of primary concern to robust ASR does not require the complete posterior probability. A point estimate, such as the MMSE estimate which we present here, will often suffice for the desired application.

As introduced in Section 5.7, given the observation vector $\boldsymbol{y}$, the MMSE estimator $\hat{\boldsymbol{x}}$ for the random vector $\boldsymbol{x}$ is one that minimizes the MSE distortion measure of

$$MSE \equiv E[(\boldsymbol{x} - \hat{\boldsymbol{x}})^{\mathrm{Tr}}(\boldsymbol{x} - \hat{\boldsymbol{x}})],$$

or

$$\hat{\boldsymbol{x}} = \arg\min_{\hat{\boldsymbol{x}}} MSE = \arg\min_{\hat{\boldsymbol{x}}} E[(\boldsymbol{x} - \hat{\boldsymbol{x}})^{\mathrm{Tr}}(\boldsymbol{x} - \hat{\boldsymbol{x}})].$$

From the fundamental theorem for MMSE computation (see Section 5.7.1), the MMSE estimator is the expected value of the posterior probability $p(\boldsymbol{x}|\boldsymbol{y})$:

$$\hat{\boldsymbol{x}} = E[\boldsymbol{x}|\boldsymbol{y}] = \int \boldsymbol{x} p(\boldsymbol{x}|\boldsymbol{y}) d\boldsymbol{x}.$$

This becomes

$$\hat{\boldsymbol{x}} = \frac{\int \boldsymbol{x} p(\boldsymbol{y}|\boldsymbol{x}) p(\boldsymbol{x}) d\boldsymbol{x}}{p(\boldsymbol{y})} \tag{13.64}$$

after using Bayes' rule.

While the MMSE estimator is generally more difficult to derive than some other estimator for random parameters, such as the *maximum a posteriori* (MAP) estimator, we present the MMSE estimator for two reasons. First, in much of the past work on speech enhancement using HMMs, the MMSE estimator has in practice exhibited consistently superior enhancement performance compared to the (approximate) MAP estimator (cf. [Ephr 92, Sam 98]). The second reason is a theoretical one. Although the MMSE estimator is defined for the MSE distortion measure, its optimality also extends to a large class of other distortion measures (under only some mild assumptions). This property does not hold for the MAP estimator. Because the perceptually significant distortion measure for speech is unknown, the wide coverage of the distortion classes by the MMSE estimator with the same optimality is highly desirable.

13.8.5 MMSE estimator with prior speech model of static features

To facilitate the derivation of the MMSE estimator with a prior speech model for joint static and dynamic features, we in this subsection first derive the estimator with a prior speech model for static features only. The result will be extended to the desired case in the next subsection.

In this derivation, the prior model for clean speech is a simplified version of model Eq. 13.62, and frame index t is dropped since the model is independent of t:

$$p(\boldsymbol{x}) = \sum_{m=1}^{M} c_m \underbrace{\mathcal{N}(\boldsymbol{x}; \boldsymbol{\mu}_m^{\boldsymbol{x}}, \boldsymbol{\Sigma}_m^{\boldsymbol{x}})}_{p(\boldsymbol{x}|m)}. \tag{13.65}$$

The derivation starts from Eq. 13.64, from which we use

$$p(\boldsymbol{y}|\boldsymbol{x}) = \int p(\boldsymbol{n})p(\boldsymbol{y}|\boldsymbol{x},\boldsymbol{n})d\boldsymbol{n} \qquad \text{and} \qquad p(\boldsymbol{x}) = \sum_{m=1}^{M} c_m p(\boldsymbol{x}|m),$$

to obtain

$$\hat{\boldsymbol{x}} = \frac{\sum_{m=1}^{M} c_m \int\int \boldsymbol{x} p(\boldsymbol{n})p(\boldsymbol{x}|m)p(\boldsymbol{y}|\boldsymbol{x},\boldsymbol{n})d\boldsymbol{x}d\boldsymbol{n}}{p(\boldsymbol{y})}. \tag{13.66}$$

Using the deterministic prior noise model Eq. 13.63, Eq. 13.66 is simplified to

$$\hat{\boldsymbol{x}} = \frac{\sum_{m=1}^{M} c_m \int \boldsymbol{x} p(\boldsymbol{x}|m)p(\boldsymbol{y}|\boldsymbol{x},\bar{\boldsymbol{n}})d\boldsymbol{x}}{p(\boldsymbol{y})}. \tag{13.67}$$

Using the likelihood model Eq. 13.61, we now evaluate the integral in Eq. 13.67 as

$$\mathbf{I}_m = \int \boldsymbol{x} p(\boldsymbol{x}|m)p(\boldsymbol{y}|\boldsymbol{x},\bar{\boldsymbol{n}})d\boldsymbol{x} = \int \boldsymbol{x}\mathcal{N}(\boldsymbol{x}; \boldsymbol{\mu}_m^{\boldsymbol{x}}, \boldsymbol{\Sigma}_m^{\boldsymbol{x}})\mathcal{N}(\boldsymbol{y}; \boldsymbol{x} + \boldsymbol{g}(\bar{\boldsymbol{n}} - \boldsymbol{x}), \boldsymbol{\Psi})d\boldsymbol{x}, \tag{13.68}$$

where $\boldsymbol{y}$ and $\bar{\boldsymbol{n}}$ are treated as constants. This integral, unfortunately, does not have a closed-form result due to the nonlinear function of $\boldsymbol{x}$ in $\boldsymbol{g}(\bar{\boldsymbol{n}} - \boldsymbol{x})$. To overcome this, we linearize the nonlinearity using a truncated Taylor series. The first-order Taylor series has the form of

$$\boldsymbol{y} \approx \boldsymbol{x} + \boldsymbol{g}(\bar{\boldsymbol{n}} - \boldsymbol{x}_0) + \boldsymbol{G}(\boldsymbol{x}_0)(\boldsymbol{x} - \boldsymbol{x}_0) + \mathbf{r},$$

where $\boldsymbol{x}_0$ is the fixed expansion point, and $\boldsymbol{G}(\boldsymbol{x}_0)$ is the gradient of function $\boldsymbol{g}(\bullet)$ evaluated at $\boldsymbol{x}_0$. The zero-th order Taylor series expansion on $\boldsymbol{g}(\bullet)$ has a much simpler form:

$$\boldsymbol{y} \approx \boldsymbol{x} + \boldsymbol{g}(\bar{\boldsymbol{n}} - \boldsymbol{x}_0) + \mathbf{r}.$$

This approximation simplifies the likelihood model Eq. 13.61 to

$$p(\boldsymbol{y}|\boldsymbol{x},\boldsymbol{n}) = \mathcal{N}(\boldsymbol{y}; \underbrace{\boldsymbol{x} + \boldsymbol{g}(\bar{\boldsymbol{n}} - \boldsymbol{x}_0)}_{\hat{\boldsymbol{y}}}, \boldsymbol{\Psi}), \tag{13.69}$$

which will be used in the remaining derivation in this section.

Now, the integral of Eq. 13.68 becomes

$$\begin{aligned}\mathbf{I}_m &= \int \boldsymbol{x}\mathcal{N}(\boldsymbol{x};\boldsymbol{\mu}_m^{\boldsymbol{x}},\boldsymbol{\Sigma}_m^{\boldsymbol{x}})\mathcal{N}(\boldsymbol{y};\boldsymbol{x}+\boldsymbol{g}_0,\boldsymbol{\Psi})d\boldsymbol{x} \\ &\propto \int \boldsymbol{x}e^{-\frac{1}{2}[(\boldsymbol{x}-\boldsymbol{\mu}_m^{\boldsymbol{x}})^{\text{Tr}}(\boldsymbol{\Sigma}_m^{\boldsymbol{x}})^{-1}(\boldsymbol{x}-\boldsymbol{\mu}_m^{\boldsymbol{x}})+(\boldsymbol{y}-\boldsymbol{x}-\boldsymbol{g}_0)^{\text{Tr}}\boldsymbol{\Psi}^{-1}(\boldsymbol{y}-\boldsymbol{x}-\boldsymbol{g}_0)]}d\boldsymbol{x}. \qquad (13.70)\end{aligned}$$

where $\boldsymbol{g}_0 = \boldsymbol{g}(\bar{\boldsymbol{n}}-\boldsymbol{x}_0)$, which can be treated as a constant now. After fitting the exponent in Eq. 13.70 into a standard quadratic form in $\boldsymbol{x}$, and using $\int \boldsymbol{x}\mathcal{N}(\boldsymbol{x};\boldsymbol{\mu},\boldsymbol{\Sigma})d\boldsymbol{x} = \boldsymbol{\mu}$, a closed-form result is obtained as

$$\begin{aligned}\mathbf{I}_m &= (\boldsymbol{\Sigma}_m^{\boldsymbol{x}}+\boldsymbol{\Psi})^{-1}[\boldsymbol{\Psi}\boldsymbol{\mu}_m^{\boldsymbol{x}}+\boldsymbol{\Sigma}_m^{\boldsymbol{x}}(\boldsymbol{y}-\boldsymbol{g}_0)]\mathcal{N}(\boldsymbol{y};\boldsymbol{\mu}_m^{\boldsymbol{x}}+\boldsymbol{g}_0,\boldsymbol{\Sigma}_m^{\boldsymbol{x}}+\boldsymbol{\Psi}) \\ &= [\mathbf{W}_1(m)\boldsymbol{\mu}_m^{\boldsymbol{x}}+\mathbf{W}_2(m)(\boldsymbol{y}-\boldsymbol{g}_0)]N_m(\boldsymbol{y}), \qquad (13.71)\end{aligned}$$

where we introduced the weighting matrices $\mathbf{W}_1(m) = (\boldsymbol{\Sigma}_m^{\boldsymbol{x}}+\boldsymbol{\Psi})^{-1}\boldsymbol{\Psi}$ and $\mathbf{W}_2(m) = \mathbf{I}-\mathbf{W}_1(m)$, and where

$$N_m(\boldsymbol{y}) = \mathcal{N}(\boldsymbol{y};\boldsymbol{\mu}_m^{\boldsymbol{x}}+\boldsymbol{g}_0,\boldsymbol{\Sigma}_m^{\boldsymbol{x}}+\boldsymbol{\Psi})$$

can be easily shown to be the likelihood of observation $\boldsymbol{y}$ given the m-th component in the clean speech model and under the zero-th order approximation made in Eq. 13.69; i.e.,

$$p(\boldsymbol{y}|m) \approx N_m(\boldsymbol{y}).$$

The denominator in Eq. 13.67,

$$p(\boldsymbol{y}) = \sum_{m=1}^{M} c_m \int p(\boldsymbol{x}|m)p(\boldsymbol{y}|\boldsymbol{x},\bar{\boldsymbol{n}})d\boldsymbol{x},$$

can be easily computed in closed form also under the zero-th order approximation of Eq. 13.69:

$$p(\boldsymbol{y}) = \sum_{m=1}^{M} c_m p(\boldsymbol{y}|m) = \sum_{m=1}^{M} c_m N_m(\boldsymbol{y}). \qquad (13.72)$$

Now, substituting Eqs. 13.71 and 13.72 into Eq. 13.67, we obtain the final closed-form MMSE estimator:

$$\begin{aligned}\hat{\boldsymbol{x}} &= \frac{\sum_{m=1}^{M} c_m N_m(\boldsymbol{y})[\mathbf{W}_1(m)\boldsymbol{\mu}_m^{\boldsymbol{x}}+\mathbf{W}_2(m)(\boldsymbol{y}-\boldsymbol{g}_0)]}{\sum_{m=1}^{M} c_m N_m(\boldsymbol{y})} \\ &= \sum_{m=1}^{M} \gamma_m[\mathbf{W}_1(m)\boldsymbol{\mu}_m^{\boldsymbol{x}}+\mathbf{W}_2(m)(\boldsymbol{y}-\boldsymbol{g}_0)], \qquad (13.73)\end{aligned}$$

where

$$\gamma_m(\boldsymbol{y}) = \frac{c_m N_m(\boldsymbol{y})}{\sum_{k=1}^{M} c_k N_k(\boldsymbol{y})}$$

is the posterior probability $p(m|\boldsymbol{y})$ for the mixture component.

The MMSE estimator for clean speech in Eq. 13.73 has a clear interpretation. The component in the first term, $\boldsymbol{\mu}_m^{\boldsymbol{x}}$, is the prior mean vector in the clean speech model.

The component in the second term:

$$\boldsymbol{y} - \boldsymbol{g}_0 \approx (\boldsymbol{x} + \boldsymbol{g}_0 + \mathbf{r}) - \boldsymbol{g}_0 = \boldsymbol{x} + \mathbf{r}$$

is the true clean speech vector perturbed by a small zero-mean residual, and can be interpreted as the prediction of clean speech when no prior information on speech statistics is available. After the prior information is made available, each summand in the estimator of Eq. 13.73 is a weighted sum of these two terms (for each mixture component), where the weights are determined by the relative sizes of the variances, $\boldsymbol{\Psi}$ and $\boldsymbol{\Sigma}_m^{\boldsymbol{x}}$, in the likelihood prediction model and in the prior speech model, respectively. The final MMSE estimator is another outer-loop, weighted sum of this combined prediction with each weight being the posterior probability for each mixture component.

13.8.6 Estimation with prior speech model for joint static and dynamic features

We now derive the (conditional) MMSE estimator using a more complex prior speech model Eq. 13.62 with dynamic features, instead of model Eq. 13.65 with static features only. Given the estimated clean speech feature in the immediately past frame, $\hat{\boldsymbol{x}}_{t-1}$, the conditional MMSE estimator for the current frame t becomes

$$\hat{\boldsymbol{x}}_{t|t-1} \equiv E[\boldsymbol{x}|\boldsymbol{y}, \hat{\boldsymbol{x}}_{t-1}].$$

Following a similar derivation for Eq. 13.67, its counterpart result is

$$\begin{aligned}
\hat{\boldsymbol{x}}_{t|t-1} &= \frac{\sum_{m=1}^{M} c_m \int \boldsymbol{x}_t p(\boldsymbol{x}_t|m, \hat{\boldsymbol{x}}_{t-1}) p(\boldsymbol{y}_t|\boldsymbol{x}_t, \bar{\boldsymbol{n}}_t) d\boldsymbol{x}_t}{p(\boldsymbol{y}_t)} \\
&\approx \frac{\sum_{m=1}^{M} c_m \int \boldsymbol{x}_t p(\boldsymbol{x}_t|m, \boldsymbol{x}_{t-1}) p(\boldsymbol{y}_t|\boldsymbol{x}_t, \bar{\boldsymbol{n}}_t) d\boldsymbol{x}_t}{p(\boldsymbol{y}_t)}, \qquad (13.74)
\end{aligned}$$

where we used the approximation

$$\begin{aligned}
p(\boldsymbol{x}_t|m, \hat{\boldsymbol{x}}_{t-1}) &= \int p(\boldsymbol{x}_t, \boldsymbol{x}_{t-1}|m, \hat{\boldsymbol{x}}_{t-1}) d\boldsymbol{x}_{t-1} \\
&= \int p(\boldsymbol{x}_t|\boldsymbol{x}_{t-1}, m, \hat{\boldsymbol{x}}_{t-1}) p(\boldsymbol{x}_{t-1}|\hat{\boldsymbol{x}}_{t-1}) d\boldsymbol{x}_{t-1} \\
&= \int p(\boldsymbol{x}_t|\boldsymbol{x}_{t-1}, m) p(\boldsymbol{x}_{t-1}|\hat{\boldsymbol{x}}_{t-1}) d\boldsymbol{x}_{t-1} \\
&= \int p(\boldsymbol{x}_t|\boldsymbol{x}_{t-1}, m) \mathcal{N}(\boldsymbol{x}_{t-1}; \hat{\boldsymbol{x}}_{t-1}, \boldsymbol{\Sigma}^{\hat{\boldsymbol{x}}_{t-1}}) d\boldsymbol{x}_{t-1} \\
&\approx p(\boldsymbol{x}_t|m, \boldsymbol{x}_{t-1}). \qquad (13.75)
\end{aligned}$$

The approximation in the last step is due to the assumption of zero variance, $\boldsymbol{\Sigma}^{\hat{\boldsymbol{x}}_{t-1}} = \mathbf{0}$, in the presumed Gaussian for $p(\boldsymbol{x}_{t-1}|\hat{\boldsymbol{x}}_{t-1})$; that is, the MMSE estimator $\hat{\boldsymbol{x}}_{t-1}$ is expediently assumed to have incurred no error: $p(\boldsymbol{x}_{t-1}|\hat{\boldsymbol{x}}_{t-1}) = \delta(\boldsymbol{x}_{t-1} - \hat{\boldsymbol{x}}_{t-1})$.

To compute the integral in the above Eq. 13.74, we first evaluate the conditional prior of

$$\begin{aligned}
p(\boldsymbol{x}_t|m, \boldsymbol{x}_{t-1}) &\propto p(\boldsymbol{x}_t, \boldsymbol{x}_t - \boldsymbol{x}_{t-1}|m) = \mathcal{N}(\boldsymbol{x}_t; \boldsymbol{\mu}_m^{\boldsymbol{x}}, \boldsymbol{\Sigma}_m^{\boldsymbol{x}}) \mathcal{N}(\Delta \boldsymbol{x}_t; \boldsymbol{\mu}_m^{\Delta \boldsymbol{x}}, \boldsymbol{\Sigma}_m^{\Delta \boldsymbol{x}}) \\
&\propto e^{-\frac{1}{2}[(\boldsymbol{x}_t - \boldsymbol{\mu}_m^{\boldsymbol{x}})^{\mathrm{Tr}} (\boldsymbol{\Sigma}_m^{\boldsymbol{x}})^{-1} (\boldsymbol{x}_t - \boldsymbol{\mu}_m^{\boldsymbol{x}}) + (\boldsymbol{x}_t - \boldsymbol{x}_{t-1} - \boldsymbol{\mu}_m^{\Delta \boldsymbol{x}})^{\mathrm{Tr}} (\boldsymbol{\Sigma}_m^{\Delta \boldsymbol{x}})^{-1} (\boldsymbol{x}_t - \boldsymbol{x}_{t-1} - \boldsymbol{\mu}_m^{\Delta \boldsymbol{x}})} \qquad (13.76)
\end{aligned}$$

Fitting the exponent in Eq. 13.76 into the standard quadratic form in $\boldsymbol{x}_t$, we have

$$p(\boldsymbol{x}_t|m, \boldsymbol{x}_{t-1}) = \mathcal{N}(\boldsymbol{x}_t; \boldsymbol{\mu}_m, \boldsymbol{\Sigma}_m), \tag{13.77}$$

where

$$\boldsymbol{\mu}_m = (\boldsymbol{\Sigma}_m^{\boldsymbol{x}} + \boldsymbol{\Sigma}_m^{\Delta \boldsymbol{x}})^{-1}\boldsymbol{\Sigma}_m^{\Delta \boldsymbol{x}}\boldsymbol{\mu}_m^{\boldsymbol{x}} + (\boldsymbol{\Sigma}_m^{\boldsymbol{x}} + \boldsymbol{\Sigma}_m^{\Delta \boldsymbol{x}})^{-1}\boldsymbol{\Sigma}_m^{\boldsymbol{x}}(\boldsymbol{x}_{t-1} + \boldsymbol{\mu}_m^{\Delta \boldsymbol{x}}),$$

and

$$\boldsymbol{\Sigma}_m = (\boldsymbol{\Sigma}_m^{\boldsymbol{x}} + \boldsymbol{\Sigma}_m^{\Delta \boldsymbol{x}})^{-1}\boldsymbol{\Sigma}_m^{\boldsymbol{x}}\boldsymbol{\Sigma}_m^{\Delta \boldsymbol{x}}. \tag{13.78}$$

Using the same zero-th order approximation, Eq. 13.69, to the nonlinear function in the likelihood model, and substituting Eqs. 13.77-13.78 into Eq. 13.74, we obtain the final result for the conditional MMSE estimator:

$$\hat{\boldsymbol{x}}_{t|t-1} \approx \sum_{m=1}^{M} \gamma_m[\mathbf{V}_1(m)\boldsymbol{\mu}_m^{\boldsymbol{x}} + \mathbf{V}_2(m)(\hat{\boldsymbol{x}}_{t-1} + \boldsymbol{\mu}_m^{\Delta \boldsymbol{x}}) + \mathbf{V}_3(m)(\boldsymbol{y} - \boldsymbol{g}(\bar{\boldsymbol{n}} - \boldsymbol{x}_0))], \tag{13.79}$$

where

$$\mathbf{V}_1(m) = \underbrace{(\boldsymbol{\Sigma}_m^{\boldsymbol{x}} + \boldsymbol{\Psi})^{-1}\boldsymbol{\Psi}}_{\mathbf{W}_1}(\boldsymbol{\Sigma}_m^{\boldsymbol{x}} + \boldsymbol{\Sigma}_m^{\Delta \boldsymbol{x}})^{-1}\boldsymbol{\Sigma}_m^{\Delta \boldsymbol{x}},$$

$$\mathbf{V}_2(m) = \underbrace{(\boldsymbol{\Sigma}_m^{\boldsymbol{x}} + \boldsymbol{\Psi})^{-1}\boldsymbol{\Psi}}_{\mathbf{W}_1}(\boldsymbol{\Sigma}_m^{\boldsymbol{x}} + \boldsymbol{\Sigma}_m^{\Delta \boldsymbol{x}})^{-1}\boldsymbol{\Sigma}_m^{\boldsymbol{x}},$$

and

$$\mathbf{V}_3(m) = \mathbf{W}_2(m) = (\boldsymbol{\Sigma}_m^{\boldsymbol{x}} + \boldsymbol{\Psi})^{-1}\boldsymbol{\Sigma}_m^{\boldsymbol{x}}, \tag{13.80}$$

and where $\boldsymbol{x}_{t-1} \approx \hat{\boldsymbol{x}}_{t-1}$ is used. Note that

$$\mathbf{V}_1(m) + \mathbf{V}_2(m) + \mathbf{V}_3(m) = \mathbf{I} \qquad \forall\, m.$$

This also provides a clear interpretation of Eq. 13.79, generalizing from Eq. 13.73 given earlier; that is, each summand in Eq. 13.79, as a mixture-component (m)-specific contribution to the final estimator, is a weighted sum of three terms. The unweighted first two terms are derived from the static and dynamic elements in the prior clean speech model, respectively. The unweighted third term is derived from the predictive mechanism based on the linearized acoustic distortion model in the absence of any prior information.

Note also that under the limiting case where $\boldsymbol{\Sigma}_m^{\Delta \boldsymbol{x}} \to \infty$, we have

$$\mathbf{V}_1(m) \to \mathbf{W}_1(m), \quad \textit{and} \quad \mathbf{V}_2(m) \to \mathbf{0}.$$

Then the conditional MMSE estimator Eq. 13.79 reverts to the MMSE estimator Eq. 13.73 when no prior for dynamic speech features is exploited. This shows a desirable property of Eq. 13.79 since when $\boldsymbol{\Sigma}_m^{\Delta \boldsymbol{x}} \to \infty$ the effect of using the prior for dynamic features should indeed be diminishing to null. As the opposite limiting case, let $\boldsymbol{\Sigma}_m^{\Delta \boldsymbol{x}} \to \mathbf{0}$. We then have

$$\mathbf{V}_1(m) \to \mathbf{0}, \qquad \text{and} \quad \mathbf{V}_2(m) \to \mathbf{W}_1(m);$$

i.e., only the prior information for the dynamic speech features is used for speech feature enhancement.

13.8.7 Implementation issues

In this subsection, some key implementation issues are discussed for the speech feature enhancement algorithm Eq. 13.79, which was derived in the preceding subsection.

Initialization and iterative refinement of the Taylor series expansion point

While deriving the conditional MMSE estimator Eq. 13.79, as well as its limiting case Eq. 13.73, we left untouched the key issue of how to choose the Taylor series expansion point, $\boldsymbol{x}_0$, in approximating the likelihood model in Eq. 13.69. This crucial issue for the algorithm implementation is resolved in two ways. First, the following crude but reasonable estimate for the "clean" speech is used to initialize the Taylor series expansion point:

$$\boldsymbol{x}_0 = \arg\max_{\boldsymbol{\mu}_m^{\boldsymbol{x}}} p(\boldsymbol{y}|m)\underbrace{\boldsymbol{x} + \boldsymbol{g}(\bar{\boldsymbol{n}} - \boldsymbol{x}_0(m))}_{\hat{\boldsymbol{y}}}, \boldsymbol{\Psi}) \approx arg\max_{\boldsymbol{\mu}_m^{\boldsymbol{x}}} \mathcal{N}[\boldsymbol{y}; \boldsymbol{\mu}_m^{\boldsymbol{x}} + \boldsymbol{g}(\bar{\boldsymbol{n}} - \boldsymbol{\mu}_m^{\boldsymbol{x}}), \boldsymbol{\Psi}].$$

Second, this initial estimate is refined successively via iterations using the conditional MMSE estimator Eq. 13.79. This turns the algorithm Eq. 13.79 into an iteration, which will be formalized shortly.

The motivation for the use of iterations is a simple recognition that the accuracy of the truncated Taylor series approximation to a nonlinear function is determined largely by the accuracy of the expansion point to the true variable value of the function's argument (given the fixed expansion order), and that Eq. 13.79 is simply the "best" available estimate of that true variable value. Therefore, the successive refinement of this estimate should improve the Taylor series approximation accuracy and hence the new estimator's quality. The use of iterations is also motivated by its success in the work of [Frey 01], where a large number of Taylor series expansion points is used. This contrasts with the single-point expansion used in this work, which considerably cuts down the computational load in our algorithm.

The convergence property of our iterative algorithm has not been systematically explored. However, under the special case where $\mathbf{W}_1 = \mathbf{0}$ (using no prior information), Eq. 13.79 becomes the well known fixed-point iterative solution of solving for $\boldsymbol{x}$ in the nonlinear equation:

$$\boldsymbol{x} = \boldsymbol{y} - \boldsymbol{g}(\bar{\boldsymbol{n}} - \boldsymbol{x}) = \boldsymbol{y} - \log\left(1 + e^{\bar{\boldsymbol{n}} - \boldsymbol{x}}\right) = \mathbf{h}(\boldsymbol{x}). \tag{13.81}$$

The convergence property for this special case of the algorithm can be found in standard numerical analysis textbooks (e.g., [Math 99]). Since the gradient of the right hand side of Eq. 13.81 is always less than one:

$$\nabla \mathbf{h}(\mathbf{x}) = \mathbf{1} - \frac{\mathbf{1}}{\mathbf{1} + e^{\bar{\boldsymbol{n}} - \boldsymbol{x}}} < \mathbf{1},$$

the iterative solution to Eq. 13.81 is guaranteed to converge, as has been observed in our experimental work also.

Variance scaling

Another important issue that has been explored in the study of [Deng 02] in the algorithm implementation is the variance weighting aimed at balancing the contributions of the static and dynamic feature priors with the overall qualities of denoising and of speech recognition performance. It has been found that the use of the variances in the clean speech model estimated from the training data set alone does not give optimal performance. This suggests that the information provided by the static cepstral means and by the differential cepstral means in the clean speech model do not consistently complement each other in enhancing the accuracy of the model, based on the simple parametric form of PDF given by Eq. 13.62, in representing the true, underlying dynamics of clean speech features. (This has been a well-known problem in statistical modeling of speech-feature dynamics. A search for solutions to this problem has produced a number of advanced statistical models beyond the conventional HMM [Ost 96, Deng 92b, Deng 94c].)

An additional reason why the use of the variances in the clean speech model, which are estimated from the training data set, is not very desirable may be attributed to the approximation made in Eq. 13.75: $\mathbf{\Sigma}^{\hat{\boldsymbol{x}}} = \mathbf{0}$. Due to the necessarily imperfect estimator, this variance cannot be zero. Not accounting for such a variance is a source of the above problem.

Rather than providing a rigorous, expensive solution, a simple yet effective way of problem fixing has been provided in the study of [Deng 02]. The contributions of the static and dynamic feature priors are adjusted by empirically scaling the variance $\mathbf{\Sigma}_m^{\Delta \boldsymbol{x}}$ (including all diagonal elements) in Eq. 13.79. (This process is similar to the empirical weighting of the language model score, which has been adopted by almost all speech recognition systems.) This keeps the estimation algorithm intact while slightly changing the weights in Eq. 13.79 to:

$$\mathbf{V}_1(m,\rho) = (\mathbf{\Sigma}_m^{\boldsymbol{x}} + \mathbf{\Psi})^{-1}\mathbf{\Psi}(\mathbf{\Sigma}_m^{\boldsymbol{x}} + \rho\, \mathbf{\Sigma}_m^{\Delta \boldsymbol{x}})^{-1}(\rho\, \mathbf{\Sigma}_m^{\Delta \boldsymbol{x}}) \qquad \text{and} \tag{13.82}$$

$$\mathbf{V}_2(m,\rho) = (\mathbf{\Sigma}_m^{\boldsymbol{x}} + \mathbf{\Psi})^{-1}\mathbf{\Psi}(\mathbf{\Sigma}_m^{\boldsymbol{x}} + \rho\, \mathbf{\Sigma}_m^{\Delta \boldsymbol{x}})^{-1}\mathbf{\Sigma}_m^{\boldsymbol{x}}, \tag{13.83}$$

where ρ is the variance scaling factor. The effects of choosing different ρ will be studied and reported in Section 6.3.

Algorithm description

Summarizing the implementation considerations above, the complete execution steps for the speech feature enhancement algorithm are described now. First, train and fix all parameters in the clean speech model: $c_m, \boldsymbol{\mu}_m^{\boldsymbol{x}}, \boldsymbol{\mu}_m^{\Delta \boldsymbol{x}}, \mathbf{\Sigma}_m^{\boldsymbol{x}}$, and $\mathbf{\Sigma}_m^{\Delta \boldsymbol{x}}$. Then, compute the noise estimates, $\bar{\boldsymbol{n}}_t$, for all frames of all test data based on the sequential tracking algorithm. Further, precompute the weights $\mathbf{V}_1, \mathbf{V}_2$, and $\mathbf{V}_3$, which are dependent on only the known model parameters, according to Eqs. 13.82, 13.83 and 13.80.

Next, fix the total number, J, of intra-frame iterations. For each frame $t = 2, 3, ..., T$ in a noisy utterance $\boldsymbol{y}_t$, set iteration number $j = 1$, and initialize the clean speech estimator by

$$\hat{\boldsymbol{x}}_t^{(1)} = \arg\max_{\boldsymbol{\mu}_m^{\boldsymbol{x}}} \mathcal{N}[\boldsymbol{y}_t; \boldsymbol{\mu}_m^{\boldsymbol{x}} + \boldsymbol{g}(\bar{\boldsymbol{n}}_t - \boldsymbol{\mu}_m^{\boldsymbol{x}}), \mathbf{\Psi}]. \tag{13.84}$$

Then, execute the following steps sequentially over all time frames:

- Step 1: Compute

$$\gamma_t^{(j)}(m) = \frac{c_m \mathcal{N}(\boldsymbol{y}_t; \boldsymbol{\mu}_m^{\boldsymbol{x}} + \boldsymbol{g}^{(j)}, \mathbf{\Sigma}_m^{\boldsymbol{x}} + \mathbf{\Psi})}{\sum_{m=1}^{M} c_m \mathcal{N}(\boldsymbol{y}_t; \boldsymbol{\mu}_m^{\boldsymbol{x}} + \boldsymbol{g}^{(j)}, \mathbf{\Sigma}_m^{\boldsymbol{x}} + \mathbf{\Psi})},$$

  where $\boldsymbol{g}^{(j)} = \log\big(\mathbf{1} + e^{\bar{\boldsymbol{n}}_t - \hat{\boldsymbol{x}}_t^{(j)}}\big)$.

- Step 2: Update the estimator:

$$\begin{aligned}\hat{\boldsymbol{x}}_t^{(j+1)} &= \textstyle\sum_m \gamma_t^{(j)}(m)[\mathbf{V}_1(m,\rho)\boldsymbol{\mu}_m^{\boldsymbol{x}} + \mathbf{V}_2(m,\rho)\boldsymbol{\mu}_m^{\Delta\boldsymbol{x}}] + [\sum_m \gamma_t^{(j)}(m)\mathbf{V}_2(m,\rho)]\hat{\boldsymbol{x}}_{t-1}^{(J)} \\ &\quad + [\textstyle\sum_m \gamma_t^{(j)}(m)\mathbf{V}_3(m)](\boldsymbol{y} - \boldsymbol{g}(\bar{\boldsymbol{n}} - \hat{\boldsymbol{x}}_t^{(j)})).\end{aligned}$$

- Step 3: If $j < J$, increment j by 1, and continue the iteration by returning to Step 1. If $j = J$, then increment t by 1, and start the algorithm again by resetting $j = 1$ to process the next time frame until the end of the utterance $t = T$.

13.9 Summary

We have devoted this chapter to speech enhancement, another important area of speech technology. The general aim of speech enhancement is to restore clean speech from its distorted version, either in the form of speech waveforms or in the form of speech feature vectors after pre-processing. After a general introduction to speech enhancement, the chapter started by providing three ways of classifying the many kinds of speech enhancement techniques available in the literature. The techniques can be classified either by the kind of prior information used in the enhancement process, by the kind of output produced by the enhancement process, by the number of sensors (single or multiple) employed during the enhancement process, or by the kind of general approaches taken by the enhancement process.

The chapter then covered several of the most common general approaches to speech enhancement. These include spectral subtraction, Wiener filtering, and optimally weighted-sum Wiener filtering based on the MMSE enhancement technique and on the prior speech model of HMM. We also covered the iterative enhancement technique using the MAP criterion with the same HMM as the prior speech model. For the HMM-based approaches, we discussed the specific issue of noise nonstationarity. This issue was addressed by using the multiple-state HMM to represent the noise process, where the spectral prototypes associated with the HMM states form an inventory (dictionary) of the possible kinds of noises that may be encountered in noisy speech.

We then proceeded to provide the first case study on the implementation and evaluation of an HMM-based MMSE enhancement system. Some important practical issues were addressed, and an overview of the enhancement system was provided. Experimental results based on both the SNR and MOS subjective evaluation were presented.

In the second case study presented, the polynomially trended HMM was used as the prior speech model. This is an interesting generalization of the HMM-based approach discussed earlier. The result of the MMSE enhancement algorithm was shown to be a weighted sum of Wiener filters, where each filter contains one prototype clean speech spectral sequence represented by one of the polynomial trend functions in the trended HMM. Experimental results were presented that demonstrated the usefulness of generalizing from the HMM to the trended HMM in speech enhancement.

Finally in this chapter we switched from the focus of speech-waveform enhancement to that of speech-feature enhancement. This topic has gained increasing interest recently since it is directly relevant to (feature-space) robust speech recognition. We first presented a statistical acoustic distortion model in the log domain (e.g., log-spectra or cepstra). We then described the mixture of a Gaussian prior model for clean speech, both with the static and the dynamic features. Given the prior model and the acoustic distortion model, the MMSE estimate of clean speech features can be derived using Taylor series truncation. Some key implementation issues for MMSE enhancement were then addressed.

Chapter 14

Speech Synthesis

The third area of speech technology applications, covered in Part IV of this book, is speech synthesis. Strictly speaking, speech synthesis refers to the automatic generation of a speech signal, also called waveform synthesis, starting typically from a phonetic transcription with its associated prosody. It often uses previously analyzed digital speech data. Such waveform synthesis is usually a final module of a larger **text-to-speech (TTS)** system, which starts from a textual input (such as a word sequence). A TTS system first performs linguistic and text processing, including text normalization. It then uses "letter-to-sound" conversion to generate the phonetic transcription, which is fed into the speech synthesis module. A separate prosody or intonation generation module provides an additional input to the speech (waveform) synthesis module. Thus these modules act as a **front-end** processor to the speech synthesis. As in the previous two application areas just covered, i.e., speech recognition and enhancement, both dynamics and optimization play important roles in speech synthesis. In this chapter, we will emphasize such roles in covering basic techniques for waveform synthesis, as well as text processing and intonation generation that form integral parts of a full TTS system.

14.1 Introduction

Among all the challenges in speech processing, that of generating a good synthetic voice is perhaps the most intriguing to the layman. Only research specialists truly appreciate success in certain applications such as speech coding (only failures are apparent, when coders degrade speech quality), and most people are impatient with speech recognition systems, expecting perfect understanding by machines when they talk to the machines (as they are used to receiving such services from human operators). When people hear a computer talking, however, they react differently. One reaction is simply that the synthetic voice is just playback of a human, as it often is in simple applications, such as in basic telephone directory services or indeed in virtually all automatic answering machine systems. On the other hand, when humans hear a computerized voice pursuing a true dialogue with them or even when giving them more detailed information than just a rote message, they often still marvel at the computer's ability to perform artificial intelligence (even now, decades after the personal computer first made major inroads in people's lives).

The ability to impress people with automatic generation of speech from text is proportional to the challenge such synthesis faces in approaching true naturalness for a diverse and lengthy text. In a sense, speech synthesis faces both a simpler and more difficult challenge than ASR (automatic speech recognition). A basic method of just playing back previously-recorded (and coded) speech and denoting such a signal as synthetic speech is almost trivial in its concept and execution, assuming reasonably that one has mastered appropriate digitization, storage and delivery systems with today's technology. As today's answering machines have largely shown, such technology is indeed straightforward for many simple applications, where the spoken messages are small in number. For most such machines, one can easily ask a "good speaker" to record a few dozen (or more) messages to handle most user requirements. In the case of needing to synthesize longer digit strings, which can easily number in the many thousands, one exploits a standard synthesis technique of concatenating short sections of speech (from a small database, e.g., of ten single digits) to form longer numbers (e.g., 16-digit credit card numbers or 10-digit phone numbers). The resulting quality is less perfect than with simple playback, however, and thus arises the challenge for speech synthesis: to maintain speech quality as the task grows from uttering a few dozen phrases to that of synthesizing from any given text.

It is natural to compare the apparently inverse tasks of synthesizing and recognizing speech via computer. Since most humans do the parallel tasks of speaking and listening equally well, it is tempting to think that they are indeed inverse tasks, and that success in one endeavor would necessarily lead to corresponding success for the other task. When one looks at the ways humans produce and perceive speech, however, it is easy to note major differences in the way speech is formulated and received. The simple fact that all mammals have similar hearing mechanisms, while only humans talk, is enough to suggest major differences in the synthesis and recognition tasks. Clearly, the superior brain power of humans (among all the mammals) is a major factor here, but the obvious anatomical differences between the vocal tract and the ear are enough to suggest that one will not easily succeed at converting either system to work in the inverse direction.

It is easy to say that automatic synthesis is a simulation of the vocal tract, while corresponding recognition tries to model the ear. However, in designing a good speech synthesis system, one must account for the precision of the ear, in making sounds that can be discerned by the ear. Correspondingly, recognizers should not seek parameterizations of speech signals that go beyond the humans' ability to control their vocal tracts. Hence, the reality of designing speech synthesis and ASR is that the tasks are quite different, and success in one need not lead to success in the other. Both, however, should exploit knowledge about the human processes of speech communication, even though a computer simulation need not follow all such aspects. For example, an airplane need not flap its wings like a bird to fly, instead taking advantage of superior engine power, but it must follow the basic aerodynamics of wing design. Taking an example closer to artificial intelligence, a chess-playing program need not follow the thought processes of a grandmaster, instead taking advantage of superior memory and computation power than a human, but it must nonetheless understand the basic rules of the game in order to structure its search algorithm for success. It is in such parallels that we see the successes and failures of speech synthesis and ASR. ASR has succeeded in simple tasks such as understanding spoken telephone numbers, but not so far in more general tasks such as ordinary conversations. The reasons are largely due to a poor exploitation of structure in the speech signal in current high-powered ASR methods.

As for speech synthesis, success is sometimes inversely proportional to the size of the set of sentences to be spoken, i.e., the **vocabulary** of the synthesizer. When we readily exploit the computer's memory and power advantages by simply storing human-produced utterances for playback as needed in speech synthesis, the **synthetic voice** is a clear success. For many applications that require a much more diverse set of utterances, this simple approach is inadequate, and we must use a more general approach, which is the subject of this chapter.

The line separating simple and complex speech synthesis tasks is vague, since practical applications range unevenly from systems merely pronouncing numbers to those needing access to unlimited texts. For intermediate speech synthesis cases, the best system may be a hybrid between simple playback and more advanced synthesis techniques. For now, however, we will approach the speech synthesis task in its full generality, assuming that we wish to generate a synthetic voice of humanlike quality, uttering any given text (i.e., unlimited vocabulary).

Unlike in speech coding, where the input is a speech signal and the output can be measured in terms of a faithful reproduction (after the constraints of reduced bit rate and degraded communication channels are imposed), the speech synthesis input is usually a text. Thus, to evaluate speech synthesis performance, we often do not have an ideal utterance for comparison, which makes any objective measure very difficult. On the other hand, like much of low-bit-rate speech coding, speech synthesis often relies on human listeners for subjective evaluations. As in coding, the quality of the output synthetic speech can be measured in terms of intelligibility and naturalness. Speech synthesis has progressed sufficiently in recent years that many systems no longer are concerned with intelligibility (e.g., current synthetic speech can often be equal to human speech, in terms of understanding, although not all speech synthesis is so advanced). Speech synthesis for unlimited text, however, usually suffers from distortions that make it easy for humans to distinguish the lower-quality synthetic voice from that of humans. The distortions vary significantly depending on the synthesis method. As described below, the major challenge to improve unlimited-vocabulary TTS is to reduce these distortions while keeping practical costs low (memory, computation, human training, etc.).

As befits the theme of this book, major issues in meeting future speech synthesis challenges concern optimization and dynamics. Synthesis researchers seek ways to optimize the database information that is stored to allow speech synthesis and to optimize the concatenation method of forming the synthetic voice. The dynamic nature of speech is critical in improving speech synthesis quality, because it is largely in the transitions between sounds (at all levels ranging from phonemes to phrases) that the synthetic voice continues to manifest its unnaturalness. We will focus on these points in the discussions below.

14.2 Basic Approaches

Synthetic voices are made by concatenating units of sound that have been previously stored in a reference database [All 92] (Fig. 14.1). The contents of these units and methods of concatenation vary, but the principle of concatenation is universal for speech synthesis involving all but the briefest messages. Currently, there are two major approaches using either spectral units or waveform units. The traditional approach, for some decades now (and still often found in commercial products), has been to store spectral templates for individual short sound units, to fetch them as needed to produce an utterance, to adjust their values dynamically in time to simulate coarticulation, and finally to excite a dynamic digital filter (simulating the vocal tract) with an excitation (often modeling vocal cord behavior). Recently, use of actual speech waveforms has become increasingly popular, where stored waveforms of various sizes are fetched as needed, with adjustments made mostly at unit boundaries, but sometimes more generally throughout the

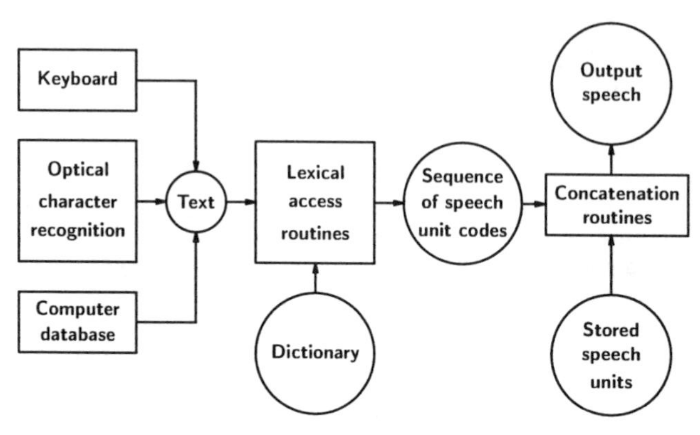

Figure 14.1: Steps in speech synthesis.

utterance.

With computer memory costs declining rapidly, storage efficiency is of lesser concern these days. Nonetheless, consider first a voice response system that minimizes memory needs by generating synthetic speech from sequences of very brief basic sounds, such as phonemes. This has great flexibility, since most languages have fewer than 40 such distinct linguistic sounds. The major disadvantage is that the acoustic features of these short concatenated sounds (typically having durations of 50-200 ms) must be significantly adjusted at their frequently occurring boundaries to avoid very discontinuous speech. The normal pronunciation of phonemes in a sentence varies greatly with its phonetic context (e.g., neighboring phonemes, intonation, and speaking rate). Thus small-unit synthesizers, while storage-efficient, are generally complicated and provide less natural output speech than systems employing larger units.

The need to smooth acoustic parameters of the concatenated sounds at their boundaries decreases as concatenated units increase in size, owing to the smaller number of boundaries in the synthesized speech. Smoothing is much simpler when the joined units have similar acoustics at the boundaries (e.g., with diphones, see immediately below). However, the effects of coarticulation often extend beyond adjacent phonemes; e.g., in the word 'strew,' the lips round in anticipation of the ensuing vowel during the initial three consonants /s/, /t/, and /r/, with corresponding acoustic effects that short diphone units cannot account for (unless numerous context-dependent diphones are stored, much like in automatic speech recognizers).

14.3 Choice of Units

For generality of vocabulary and flexibility, commercial synthesizers often concatenate small units. In addition to phone-sized units, words and diphones are popular units. Diphones are speech elements having the size of phones, but derived from phone sequences and divided in the middle of each phone. Thus each diphone preserves the transition between adjacent phonemes (e.g., for a two-vowel sequence /ia/, the corresponding diphone would be the second half of /i/ followed immediately by the first half of /a/). Memory needs increase significantly for diphones (as opposed to phones) since, assuming a language with N distinct phones (e.g., approximately 32 for English), there are N^2 diphones. In addition, since diphones correspond to dynamic entities, whereas phones are often steady-state units, each diphone typically stores more data per unit.

To quantify some of these issues, note that English has more than 50,000 common words. (Most native speakers only use perhaps 5000, but such word sets vary significantly from person to person.) The average word length is only several phones, and the typical word lasts less than a second. With efficient coding methods (see below) of about 4 kbits/s, a 200 Mbit storage is feasible using word units, with the possibility of significant reductions if universal vocabulary (i.e., unlimited TTS) is sacrificed. However, it is quite fatiguing for a single speaker to utter thousands of words in a uniform fashion. This human-factors issue is a significant problem for synthesizers storing large amounts of speech from individual speakers, whose voices are simulated directly as synthetic speech.

One way to try to reduce the amount of training speech for speech synthesis is to consider smaller units than words. For example, almost all English words can be generated through the use of several thousand **morphemes**, the basic meaningful elements of which words consist; e.g., 'antidisestablishmentarianism' has the root morpheme 'establish' as well as two prefixes and four suffixes. A morphemic approach has the advantage of preserving syntactic information, since suffixes often specify a word's role in a sentence (i.e., the need for a large full-word dictionary, where each entry has the word's pronunciation and part-of-speech, is less obvious when a reduced morpheme dictionary could supply the same results). This approach requires a word decomposition algorithm, which tries to strip affixes from each word in an input text. Since there are only a few dozen such affixes in most languages and since they often affect pronunciation (e.g., the third vowel in 'algebra' vs. 'algebraic'), this decomposition reduces memory needs (although some extra computation is needed).

The use of syllables as units could reduce size further, although syntactic information is lost. If the speech unit is smaller than a syllable, there is little incentive to restrict vocabulary. For larger units, vocabulary size can influence efficiency; however, with diphones or phones as units, the synthesizer vocabulary can be unlimited with little additional memory cost. For example, 2000 diphones suffice to synthesize almost all words, but a small, basic vocabulary of 1000 words would still likely require 1000 diphones.

14.4 Synthesis Methods

The degree to which human speech production is directly or indirectly simulated in speech synthesis procedures varies substantially. Theoretically, an articulatory approach in which muscle commands (or even thought processes) are employed in the initial stages of speech synthesis would be the closest simulation of the way humans speak. However, despite many attempts, articulatory speech synthesis has never come close to the quality of most commercial devices using either a spectral or waveform concatenation approach. Instead, a **terminal-analog** approach is usually followed, in which the speech waveform (i.e., the output of the vocal tract's 'terminus,' or mouth) is generated, either directly in the time domain or via its spectrum. Since such an output in natural speech is the result of an excitation convolved with a vocal tract response (both varying dynamically), speech synthesis usually tries to model both the excitation and filter phenomena. The spectral approach allows direct manipulation of these two major aspects of speech generation, while the waveform method necessarily does this indirectly.

14.4.1 Articulatory method for speech synthesis

While speech generation by modeling of vocal-tract movements has been less successful than terminal-analog synthesis, we discuss articulatory synthesis first because it is the most basic approach conceptually. Direct modeling of articulatory motion avoids most of the complicated modeling of acoustic-phonetics that many other speech synthesis methods must do. Most speech synthesis systems model the acoustic cues for the different phonemes that they try to simulate as complicated functions of phonetics, speaking rate, and intonation. In particular, the effects of coarticulation have to be determined by rule in terminal-analog systems, but good coarticulation modeling in articulatory speech synthesis is a direct outcome of good system models. We divide articulatory speech synthesis into three groups, according to their level of control in speech production: neuromotor, articulator, and vocal-tract shape [Matt 70].

Let's start at the highest cognitive level for speech production - the brain. In order to speak, a person conceives a message and the brain sends corresponding commands to appropriate vocal-tract muscles, which cause certain articulators to move and to change the shape of the vocal tract. Thus, at the most basic level, articulatory synthesis transforms a desired phoneme sequence (input to a speech synthesizer) into a set of **neuromotor** commands to the muscles, according to relevant models from electromyographic studies [Hiki 70]. Such a system also requires models to relate muscle commands to articulatory movements and to convert each set of articulator positions into vocal-tract shapes or **area functions**, sets of N cross-sectional areas A_m (for $m = 1, \ldots, N$) for N short sections constituting the vocal tract as in standard vocal-tract models with reflection coefficients. A temporal sequence of such area functions $A_m(n)$ (as a function of time frames n) can be used to specify a time-varying lattice filter whose reflection coefficients are specified by the ratios of adjacent areas. The resulting synthesizer is usually that of LPC synthesis; recall that in LPC the reflection coefficients are determined indirectly by acoustical analysis; in articulatory speech synthesis, the parameters are found directly from articulatory models of the vocal tract. (Once the A_m's are determined, other synthesis architectures are feasible; e.g., one could use the A_m's to determine formant frequencies and their bandwidths, for use in a formant synthesizer [Cok 76].)

The above neuromotor approach to synthesis is quite complex, as it requires relating each phoneme directly to a set of muscle commands. As such, it is perhaps the most complicated way to do speech synthesis (especially given our lack of knowledge of how the brain controls various muscles). The major advantage of articulatory speech synthesis is the natural way that coarticulation is handled directly from the muscle commands for individual articulators, but while the commands for individual speech gestures (e.g., lip closing) may well be similar for all phonemes that use that gesture, the relationships between commands and gestures are very complicated.

Instead of operating at the neuromotor level, we could bypass that level and attempt articulatory synthesis by specifying articulator positions for each phoneme. In this way, we avoid examining direct brain processes, but we must now explicitly account for vocal-tract behavior that is likely more efficiently described at the muscle level. Because experimental data are much more available for articulator motion than for muscle behavior, articulator synthesis has been used to generate speech more often than neuromotor synthesis. Typical attempts presume that approximately ten parameters are sufficient to properly describe articulatory behavior, typically one parameter to measure the degree of velum lowering, one parameter each for lip protrusion and closure, and two each (vertical and horizontal) for the tongue body and tip [Matt 70],[Cok 76], while other parameters may be jaw height, pharynx width, and larynx height [Merm 73],[Rub 81].

As in all synthesizers, a set of values (e.g., targets) for such parameters for articulator synthesis would be stored in a computer memory for each phoneme, and the synthesizer would construct a temporal sequence of parameter values by interpolation among the target values. As is the case in many forms of speech synthesis, smoothing the parameter trajectories would follow estimated coarticulation phenomena (e.g., various time constants for different articulators, such as fast movement allowed for the flexible tongue tip and slower motion for the jaw). Since many phonemes impose constraints on several articulators, each phoneme would be interpreted as having both critical target parameters and others of lesser importance. The important ones would be more resistant to coarticulation; e.g., for /p/, lip and velum closures are essential, while few constraints are imposed on the tongue.

While the source of information to design neuromotor synthesis is mostly limited to invasive electrical probes (e.g., electomyography), models of articulator motion can usually be derived from X-rays, ultrasound, or magnetic imaging. These would all be observed from motion of the vocal tract during natural utterances. Unfortunately, most such data gathering is costly, and often is limited to display only two dimensions, rather than the three dimensions of the vocal tract. While the clearest results in the past have come from X-rays, such data is now very limited (e.g., X-ray microbeams) available due to the danger of X-ray exposure. Furthermore, even assuming access to adequate vocal-tract data, the typical articulator models used for speech synthesis have been very crude in the past; i.e., the actual vocal-tract shape is only very roughly approximated by the simplified models using only about ten parameters. As a result, synthetic speech from all forms of articulatory synthesis have been based on generalizations from very limited data, and has yielded lower-quality speech than has terminal-analog synthesis.

A third approach to articulatory synthesis starts from vocal-tract shapes, rather than muscle or articulator behavior. This method bypasses modeling of both the muscles and articulators. Instead, phonemes are represented by area-function targets. This approach has greater potential to approach the quality of terminal-analog synthesis, since it starts from parametric representations that are fairly similar to those of spectral synthesis (e.g., vocal-tract areas are readily converted to digital filters, as in LPC synthesis). Such an area-function approach, however, loses much of the efficiency of the more compact neuromotor and articulator models and requires direct specification of all coarticulation, in terms much less elegant than the other articulatory methods. There may nonetheless be an advantage to manipulating coarticulation by vocal-tract areas rather than by spectral parameters (the latter is done in many terminal-analog synthesizers).

Appropriate rules for coarticulation in terms of vocal-tract shapes would seem to be less clear than for other articulatory synthesizers; e.g., for the latter, clear physical constraints impose simple guidelines for movements of muscles or articulators [Merm 73]. Vocal-tract-area synthesizers, however, would perhaps allow more accurate modeling of transient phonemes that have abrupt changes of area than other forms of synthesis. In addition, direct control of areas might allow better automatic generation of noise excitation at narrow constrictions, and allow their derivation directly from aerodynamic considerations.

14.4.2 Spectral method for speech synthesis

In the spectral speech synthesis procedure, a simplified spectral envelope is stored for individual sound units, usually very short units such as phonemes or diphones. An equally-simplified excitation, modeling (in natural speech) either glottal vibration or noise produced at a constriction in the vocal tract, is convolved with the filter described

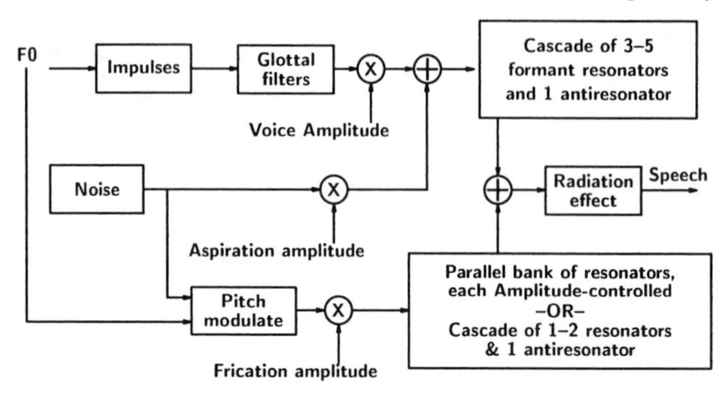

Figure 14.2: Simplified organization of a formant synthesizer.

by the spectral envelope, to yield the synthetic voice. Typically, the filter is specified in terms of resonance (formant) center frequencies and bandwidths, over a frequency range up to approximately 5 kHz (higher frequencies add little to the quality of synthetic speech, but can easily be added if desired for more pleasant broadband effects). Alternatively, an LPC spectrum can be used. In formant-based speech synthesis, usually the lowest four resonance center frequencies are allowed to vary, along with the lowest three bandwidths (dynamic variation in the higher-order parameters is usually omitted, as it has little perceptual effect) (Fig. 14.2). This spectral manipulation is supposed to model the coarticulation effects of the vocal tract as it changes shape to produce different speech sounds.

Formant synthesis

Traditional speech synthesis employs a digital filter to generate the output synthetic speech, in which the spectral response for each phonetic sequence is directly simulated by dynamic manipulation of the multiplier coefficients of the filter. One basic approach is that of **formant synthesis** [Klatt 87a], in which the filter has either a cascade or parallel structure of second-order digital components, each simulating one resonance in the vocal-tract model (with a few additional components to handle other effects). A cascade structure is most common because it approximates speech spectra well for vowels and allows easy global control of intensity using only one parameter. Fig. 14.3 displays one specific implementation that has been widely used, in part owing to the early publication of actual program code [Klatt 80]. This approach has both a cascade structure (for easy control of sonorants) and a parallel structure (for frication sounds).

Second-order digital resonators are regularly employed in formant synthesis because higher-order filters need extra bits (per coefficient) to achieve the same spectral accuracy [Opp 89]. It is only necessary to vary the lowest four formants; higher formants are often kept fixed, with little perceptual degradation. The synthesizer of Fig. 14.3 has 19 control parameters that may change for each 10-ms frame. While high-quality synthetic speech seems to require four dynamic formants, simpler synthesizers may leave F4 fixed and keep all formant bandwidths constant (except for F1, since F1's bandwidth is most noticeable for nasals versus vowels). One early synthesizer, for example, used constant bandwidths of 60, 100, and 120 Hz for F1-F3, respectively (with F1 bandwidth increasing to 150 Hz during nasals) [Rab 68].

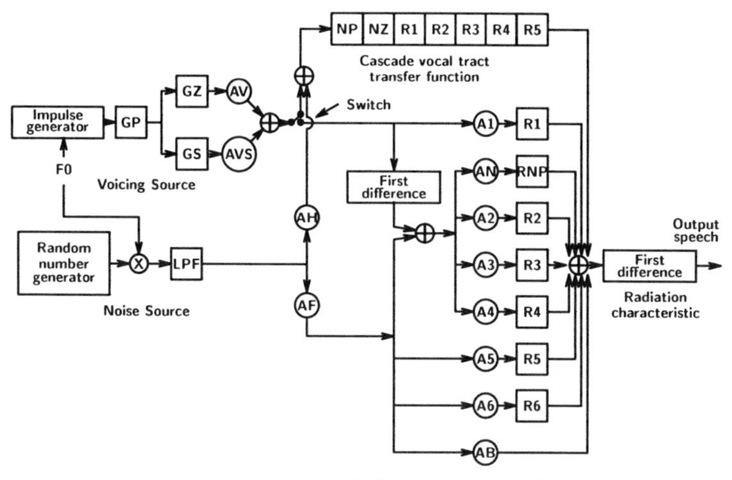

Figure 14.3: Block diagram of a cascade-parallel formant synthesizer. Digital resonators are indicated by the prefix *R* and amplitude controls by the prefix *A*. Each resonator R_n has associated resonant frequency and bandwidth control parameters (after Klatt [Klatt 80], @AIP).

Noise excitation source

Formant synthesizers generally allow more complex types of excitation, to exploit their more complex architecture (as compared to LPC synthesis). As in most spectrally-based synthesizers, the excitation is usually a periodic train of pulses to simulate voiced speech, and pseudo-random noise for unvoiced speech. Formant synthesizers allow the noise to be shaped periodically (at the F0 rate) to better model voiced frication. Whereas the glottal pulses for voiced excitation decrease in intensity with frequency at about -12 dB/octave, the noise excitation for unvoiced sounds is better modeled by a flat spectrum. A further factor is that the intensities of natural noise samples have an approximately Gaussian distribution. Noise samples for speech excitation are usually produced via a pseudo-random number generator, which yields a flat intensity spectrum with a uniform (non-Gaussian) distribution. If we sum several (e.g., 16) of such random numbers, an approximate Gaussian noise sample is obtained (the central limit theorem of probability says that adding independent identically-distributed random variables approaches a Gaussian) [Klatt 80].

The noise source for constrictions within the mouth cavity excites mostly the upper portion of the vocal tract (i.e., from the source to the lips). Owing to this short acoustic tube, most frication energy is at much higher frequency than for the vowels. As a result, very little obstruent energy is found below 2 kHz. The non-excitation of F1 (and usually F2 as well) means that simulation of obstruents is more readily accomplished with a parallel bank of resonators, where the amplitude for each filter resonance is separately controlled (Fig. 14.3). Switching between series and parallel filter structures for voiced and unvoiced sections of voice synthesis often leads to speech sounding as if two people were uttering the different sections in turn; i.e., it sounds like hissing noises superimposed on speech of vowels and sonorants, rather than an integrated single voice [Kerk 93]. Some hybrid forms of synthesis combining formant and waveform synthesis may yield better transitions between voiced and unvoiced sounds [Fries 94].

Parallel formant synthesis

As noted in the previous section, a parallel bank of filters is commonly used to generate obstruents. It could be used to make vowels as well, but then each formant amplitude must be individually specified. This amplitude control freedom is essential for obstruents, but leads to unnecessary algorithmic complexity for sonorants, while simplifying the overall hardware (i.e., no need for a cascade set of filters). A disadvantage of a parallel-only approach is that unintentional spectral zeros appear in such systems. This is easily seen by considering the simple case of combining two single-pole systems $1/(z-a)$ and $1/(z-b)$:

$$1/(z-a) + 1/(z-b) = (2z-a-b)/((z-a)(z-b)).$$

The resulting second-order system retains poles at the original locations a and b, but also has an unwanted zero at $(a+b)/2$, which would not appear with a cascade combination of the two subfilters.

Phase effects also need to be accounted for in parallel filters. Formants are usually alternated in sign when summed in parallel, (i.e., A1 − A2 + A3 − A4 . . .), because each formant frequency (modeled by a complex-conjugate pair of poles) is associated with a 180° phase shift in the vocal-tract transfer function.

Nasal synthesis via formants

Most sonorants have similar formant structure, but nasals are different because of the involvement of the nasal tract, which extends the vocal tract and provides an acoustic side-branch. The longer total tract length means that nasals should have one or two more formants than for other sonorants. The 21-cm length (vs. 17 cm) leads to about 15% smaller spacing between formants. Thus, for synthesis of 5-kHz bandwidth, one extra formant is needed (and two for wideband 10-kHz synthesis).

In addition to extra formants, nasals have spectral zeros, which other sonorants do not have. Thus a normal cascade of five second-order resonators suffices for most sonorants, but an extra resonator and an antiresonator are often included in the cascade to accommodate nasals. Additional zeros at higher frequencies in naturally-produced nasals are often ignored for synthesis purposes, as having little perceptual importance. The retained lowest zero and one extra pole (at a neighboring frequency) are often combined as a pole/zero pair, since the same second-order filter can have both feedforward and feedback coefficients. For non-nasal sounds, these coefficients are set so as to have the zero and pole cancel each other.

LPC synthesis

Formant synthesis requires a thorough manual development of coarticulation rules (i.e., a detailed algorithm) to specify the complex movement of resonances in different phonetic contexts. Owing to the difficulty of trying to model all phonetic variation in developing an expert system, speech synthesis designers have often preferred to use instead the automatic method of linear predictive coding (LPC) (discussed earlier in Chapter 2 of this book). In addition, the filter for LPC has a simpler structure than for formant synthesis, because all spectral properties of speech (except for intensity and periodicity, which present equal challenges to almost all speech synthesis methods (see below)) are included in the LPC coefficients. A lattice filter is typically used because the multiplier coefficients (modeling boundaries between adjacent cylindrical sections of a vocal tract) can be linearly interpolated between successive frames (for smoother speech) without yielding an unstable filter (neither LPC nor formant synthesis employs direct-form filters, due to their sensitivity to quantization noise and risk of instability). LPC however shares with waveform synthesis (see below) an inflexibility in not allowing simple transformations of the stored data to simulate different voices. Modifying formant frequencies (to create additional synthetic voices) can alter speaker-related aspects of a synthetic voice much more easily than is possible with simpler parametric methods.

At unit boundaries, several techniques are used to smooth changes. Use of more appropriate parameters, such as LSFs, can help, because they have easier physical interpretation than more traditional speech coding parameters such as LPC reflection coefficients. Simple linear and nonlinear interpolation methods have been tried with varying degrees of success, with the evaluation being done perceptually [Wout 01].

Excitation for spectral synthesis

The synthesis filters' excitation (whether LPC- or formant-based) is often a regularly-spaced sequence of impulses (for voiced speech) and pseudo-random noise (for unvoiced speech). This simulated vocal-tract excitation has traditionally been varied in amplitude (to simulate dynamic energy levels in normal speech) and in periodicity (the rate of vibration of the vocal cords, also called the fundamental frequency or F0). Over decades, studies of speech production have led to numerous models for these parameters (formants, amplitude, F0). Their careful, manual dynamic manipulation can yield excellent-quality synthetic speech, but rule-based approaches (i.e., automatic speech synthesis, and not manual synthesis) often fall short of producing natural-sounding speech. Inadequate modeling of dynamic variation in formant movements (due to still poorly understood aspects of coarticulation) often leads to artifacts in the synthetic speech, and even to perceptions that the synthesizer has a foreign accent. Similar comments can be made about excitation modeling, although even less is understood about the subtleties of how vocal cord behavior influences speech quality. It is well known that cord periodicity directly influences pitch perception (the relationship is monotonic, but nonlinear), but there remains the general problem of relating the linguistic content of a sentence to how the pitch should vary to sound appropriately natural. In addition, the common use of a very simple pulse model to excite the vocal-tract filter often leads to buzzy, mechanical-quality speech. One could summarize the current challenges for speech synthesis as better modeling of three aspects of speech generation: coarticulation, intonation, and excitation. Inferior sound quality is usually due to inadequate modeling of one or more of these three aspects of human speech production. We next discuss using speech waveforms directly for synthesis, a solution that tries to bypass the excitation problem.

14.4.3 Waveform methods for speech synthesis

Conceptually, the waveform method for speech synthesis is even simpler than spectral techniques. Brief portions of actual speech waveforms are chained together, one after the other as specified by the input text, with adjustments made for: 1) amplitude (scaling to energy levels appropriate for each unit in its broader sentential context), 2) period (modifying F0 from its original stored value), and 3) other, finer aspects of the stored unit. Normally, the third category would refer to coarticulation effects (e.g., in the spectral method, smoothing out formant changes when concatenating units representing different vocal tract shapes). However, the waveform method has no access to such appropriate physical models, so the corresponding adjustments made in waveform methods are less well motivated by speech production theory than is smoothing in spectral methods.

All these modifications have as an objective to transform the individual stored speech units to correspond more closely to what they would be in each phonetic context in ordinary speech. Since humans normally vary amplitude at linguistic levels that are broader than phonemes, intensity must be adjusted so that the synthetic speech does not sound as if there is a rapid, random variation of the volume control for the synthetic voice. While phonemic units may be stored with a suitable average amplitude, the level in normal speech varies significantly in complex ways with stress, speaking rate, syntactic word type, etc.

Such amplitude adjustments need to be made with spectral synthesis methods as well. Fundamental period (F0) adjustments, however, are quite different in the spectral and waveform approaches, since the spectral method explicitly controls the excitation period. In the waveform approach, which explicitly stores actual pitch periods, the lengths of individual periods must be directly changed to accommodate varying F0, according to the needs of the synthetic utterance (e.g., to cue syntactic and semantic phenomena). The simplest way to change F0 is to truncate each period, eliminating its final few samples, if period shortening is desired (i.e., to raise F0), or to extend each period, if F0 must fall. Extensions can be done by extrapolation of the final few samples, or via some other method interpolating with adjacent periods. (The even simpler method of just inserting zero-amplitude samples at the end of each period to be lengthened is sometimes used.) All these simple adjustments, however, are often detectable perceptually as inappropriate changes not found in natural speech, although the distortions may not be too annoying when changes are small. These F0 adjustments however pose a major problem for the waveform concatenation method when the necessary changes are large. Small changes involving the (often) weak final samples in each pitch period may be tolerated perceptually. Given that speakers can vary their F0 over a range as large as an octave in normal speech, and occasionally do so within single syllables (and not just while singing), it is important to be able to produce speech with appropriately dynamic F0.

Since it is easy to capture natural speech signals via an A/D converter and to use actual sampled waveforms to provide synthetic speech, one may ask what is the simplest approach for voice response systems. Compact disks store audio signals (including speech) with 16-bit sampling at 44,100 samples per second. Such high storage rates are rarely used for speech synthesis, due to their storage cost. A more common sampling rate is 10,000/s, which preserves acoustic frequencies up to almost 5 kHz (following the Nyquist theorem) and still allows very natural speech. The high-frequency aspects of speech are lost in such a case, but this has little impact on intelligibility. Furthermore, since many speech synthesis applications use telephone lines, which heavily attenuate frequencies outside the 300-3200 Hz range, sampling rates above 8 kHz are hard to justify for telephone speech.

Basic sampling via linear PCM (pulse code modulation) often uses 12 or 16 bits/sample, which consequently requires storage memory of 120–160 kbits/sec. Such high data rates are difficult to justify for any application except those employing small vocabularies. Even with advanced coding schemes (e.g., CELP at 4-10 kbits/sec), the straightforward playback approach is only acceptable for small-vocabulary speech synthesis. Notwithstanding rapidly decreasing memory costs, it remains impossible to store all the needed speech signals for general speech synthesis.

14.4.4 Phase mismatch

A serious problem for many types of concatenative synthesis is that of **phase mismatch**. When stored speech units are joined together, there must be sufficient acoustic continuity at the boundaries to convince the listener that the speech sounds natural. Since the stored units were extracted from different speech signals originally (except for large, sentential units, which are reproduced without concatenation), there is no inherent guarantee of continuity in either spectral amplitude or phase. Often the units are chosen and stored based on spectral amplitude criteria, which tend to reduce the continuity problems in that domain, at least for large, well-designed databases. However, spectral phase is a more difficult problem. One source of phase mismatch is simply the timing of pitch periods. Many systems concatenate units corresponding to supposed pitch periods, which often presume a synchronization with the glottal closure instants of the vocal cords, which initiate each major increase of amplitude in each pitch period. Small inaccuracies in choosing such precise sample points can lead (in the synthetic speech) to perception of a harshness or a garbled type of speech. A second source of mismatch is due to the phase of the vocal tract response, which theoretically should be the same for the identical vocal tract shapes that speech synthesis systems try to simulate at unit boundary joins. However, in practice, there are small changes in both amplitude and phase in the stored parameters.

Some systems use manually-labeled pitch periods (e.g., PSOLA), but this is increasingly difficult as speech synthesis databases increase in size, to exploit cheaper computer memory while reducing spectral variation at boundaries [Geo 97]. Recent work with a **harmonic-plus-noise** model has explored simple ways to minimize phase mismatch [Syrd 98].

Sinusoidal models have been explored for speech synthesis, exploiting the fact that natural voiced speech exhibits a smooth flow of harmonics uniformly spaced in frequency, all synchronized to remain as multiples of the dynamic F0 [Quat 92]. Such models often handle spectral amplitude changes by allowing individual harmonics to appear ('be born') or disappear ('die'), as the spectral envelope requires. One advantage of this approach is the reduced need for precise pitch period identification [O'Br 01].

14.5 Databases

The quality of synthetic speech often varies in proportion to the size of the speech units that are concatenated to yield the output speech signal. If the units correspond to full phrases or sentences, synthetic voices can approach the quality of natural speech. However, as part of the system development, each phrase must be uttered by one speaker with intonation appropriate for all target sentences. One can easily detect the inferior quality of some elementary synthesizers which simply concatenate words that were originally spoken separately.

Since speech synthesis systems are usually designed to handle all textual input (for a given language), most speech synthesis databases are stocked with a large number of small units (corresponding typically to phonemes, allophones or diphones). If all such possible units are represented in the database, the system can accommodate all requests to synthesize a text, even for texts containing new words (even many foreign words, although often with less appropriate pronunciation). Early synthesis systems efficiently used quite small inventories, e.g., simple phone- or allophone-based speech synthesis can function with fewer than 100 stored units, although such systems must do significant modifications to the stored units to simulate coarticulation. More recently, the trend has been to store thousands of units, all automatically extracted from training speech [Dutoit 97, Ding 97].

Database design has been a significant area of recent speech synthesis research. It had been earlier thought that perhaps a thousand or so diphone units might be a limit for speech synthesis systems, on the assumption that any individual speaker would find it very difficult to maintain uniformity while speaking more than a thousand units. Indeed, this issue of uniform pronunciation (in terms of clarity, effort, intonation, speaking rate, etc.) can seriously affect speech synthesis quality. One can, however, compensate for undesired intraspeaker variation by being selective in unit storage, i.e., record the chosen speaker over a long period of time and automatically select only the desirable units to retain for the database. This requires a mechanism to choose the appropriate units, according to some naturalness criterion. In earlier trials, the system designers would personally listen to all units and re-record those that did not sound right, following some subjective criterion (based on trying to achieve uniformity). As the databases increased in size, this became less feasible, and automatic methods are now preferred. However, it is not always clear which objective basis to use for retaining speech units.

A fundamental step in all speech synthesizers is the concatenation of units to actually form the output speech signal. Whether the units consist of waveforms or spectral patterns, the way they are joined is usually evaluated by some similarity or distance measure, in an attempt to minimize discontinuities at the unit boundaries. In natural human speech, the acoustic signal flows smoothly from phone to phone and from word to word, because one speaker produces the entire signal, guided by the objective of producing a coherent utterance to facilitate human communication. In speech synthesis, when successive units are drawn from diverse sources (even from a single, supposedly uniform database spoken by one speaker), the units have various degrees of mismatch at the boundaries. Appropriate distance measures can assist both in choosing the units to join and in their adjustment for smoother transitions between the concatenated units. Most such measures are spectral in nature, e.g., Klatt's Weighted Spectral Slope Measure [Quac 88] or more recent auditory-based measures [Hans 98].

14.6 Case Study: Automatic Unit Selection for Waveform Speech Synthesis

To better appreciate some of the more popular trends in speech synthesis research, we now examine in more detail how some recent speech synthesis systems establish their unit databases. The major objective of such methods, as compared to other speech synthesis techniques, is to improve the naturalness of synthetic speech. Most of these newer systems specify two costs to minimize: that of choosing the proper speech unit to use (for a given phonetic context) and that of joining successive units together. While these costs are related, they may be examined separately, as each of their impacts on speech quality is quite different. One can argue that the first task (i.e., having the right units in memory and then selecting the right one each time) is more critical, because a smooth joining algorithm (the second task) is useless if the units being concatenated are poor. At the same time, if one has poor transitions between good units, the synthetic speech would sound excessively discontinuous, which is also a rather bad consequence.

The first task is the greater problem because it involves corpus design (which sentences for the training speaker(s) to utter) and database design (which and how many units to store, where to segment the training speech, what parameterization to use, etc.). At run-time (when the synthetic speech is needed), the speech synthesis algorithm tries different sequences of eligible units from its database, calculating a cost for each proposed unit, in terms of the degree of mismatch from the desired characteristics for that unit, and a cost to make the transition with the adjacent proposed units. Since a typical database will have many candidates for each position in the synthetic utterance, there will in general be many combinations to test. This is somewhat like the many possible paths to test in speech recognition, and leads similarly to increased computation time for speech synthesis in searching for the best sequence of units. The search space is considerably smaller than for recognition, because we know the text to synthesize, and durations are controlled directly by the synthesis algorithm. (In recognition, on the other hand, speaking rate variability leads to an explosion of possible paths to consider.)

One major challenge for speech synthesis design has been to determine an automatic method to populate the speech synthesis database with appropriate units [Iwa 95]. Such a well-designed database can minimize search time and maximize the quality of the eventual synthetic speech. Earlier speech-unit databases were relatively small; e.g., having only single units for each allophone or diphone. Coarticulation effects beyond a very limited range (e.g., for diphone units, 50% of each phone duration) were typically ignored, and obligatory adjustments (both spectral and prosodic) either were complex or caused significant quality loss. Increasing the size of the database, to include large numbers of units, classified by wider phonetic contexts (e.g., beyond diphones) and having different durations and pitch, can increase synthetic speech quality, but only if the expansion and selection are properly done.

"Intra-segmental distortion" is the name typically given to the cost of selecting units (the result of the combined process of populating the database, and then choosing the proper unit at run-time). It can be estimated through an acoustic distance measure, comparing the chosen unit with that of the actual unit observed in the training speaker's speech. That is, assume a single speaker is used to create the database, and separately utters a test database. We then compare the individual speech units segmented (perhaps manually) from the test speech against the units automatically chosen from the trained database for synthesis. The distance measure can involve both a spectral and prosodic match.

"Inter-segmental distortion" refers to acoustic mismatch caused by the smoothing of units in the time frames adjacent to their concatenation boundaries. Again, reference can be made to the transitions found in the test speech, in evaluating this distortion. Additionally or separately, one can hypothesize other criteria to judge how good a join is (i.e., without referring to the test speech), since the "smoothness" of a concatenation can be measured on perceptual criteria, apart from how well the transition actually matches the test speech.

While the general principle of minimizing the costs (i.e., acoustic distances) is obvious as an evaluation criterion, it is less clear how to integrate the individual costs (intra- and inter-segmental) over the many frames in a synthetic speech utterance. Do we minimize the sum? Do we employ a minimax method, choosing the units and smoothing that lead to minimizing the largest individual frame cost over all frames? This reflects issues discussed elsewhere about SNR versus segmental SNR, or the use of error energy as a measure to minimize in LPC analysis. A steady continual distortion may be unnoticeable at a low level, whereas a few individual frames with large distortions would be locally annoying.

As part of the recent trend in speech synthesis toward automatic selection of database speech units for storage, a technique from ASR, that of hidden Markov models (HMMs), has been applied to speech synthesis unit selection. Assuming knowledge of the text corresponding to speech signals used in training, HMMs for different pronunciations of the text attempt recognition of the speech data, resulting in the determination of estimated phone (or other unit) boundaries, in a process similar to automatic update training (parameter re-estimation) of ASR HMMs. Such automatic segmentation appears to approach human performance on such a task (e.g., 80% of boundaries within 11.5 ms of actual placement, versus 8 ms for humans) [Ljo 93].

Decision tree clustering is an automatic method to optimize unit selection, in which phonetic contexts for proposed units are chosen for their similar effects on the acoustic parameters of individual phonemes [Don 99]. In choosing phone-sized speech units automatically based on their coarticulation effects, decision trees are constructed without human intervention to maximize acoustic similarity within the selected classes. Then, when concatenation is done at synthesis time, there is a minimum of discontinuities at the merged boundaries.

For example, in [Don 99], the speech synthesis speech-unit database was created from an hour of read speech by a single speaker (reading a popular novel). Three-state (no-skip) context-independent HMMs were established for an extended set of almost 286 "allophones" (the basic set of phonemes, extended to incorporate syllabic, stress, and other phonetic effects). An hour's worth of speech yields about half a million frames of speech (using a 6-ms update rate). With 286 allophones, a context-dependent (triphone) set of HMMs would number 23 million. Given substantial repetition related to redundant text material typical of most languages, only 13,000 appeared in the hour of speech. Thus, for several reasons (e.g., computational resources, availability of training data), it was useful to cluster the context-dependent possibilities phonetically, using questions of the immediate triphone context, including stress, syllable and word position. The authors chose to set 12 as the minimum number of occurrences in the training data for each clustered state, which yielded about 5700 such states.

To select the synthesis parameters to correspond with each state, a single set was chosen as follows. All HMM state occurrences were discarded that either had duration or energy below 80% of average. This ensured that, at synthesis time, time expansion of units would be limited, and that energy would not be excessively scaled upwards; i.e., it is more practical to scale down duration and power, which better avoids artifacts in the synthetic speech. The actual choice for each model, among the remaining states not eliminated, was the occurrence in the training data with the highest average log-likelihood per frame in the state alignments (such a choice meant that outliers would not be chosen).

The stretching issue is important because units taken from fluent read speech are usually too short if directly used to synthesize other texts. In general, synthetic speech must be slower than normal reading rates to enhance intelligibility. In this study, state durations were extended (when synthesizing speech) by a certain percentage of their standard deviation in the training data: 10% for continuous speech and 50% when synthesizing isolated words. The policy of varying durations according to their variability in the training data reflects the fact that speaking rate variations affect different phonemes differently. Phonemes in context that showed little variability in the natural speech of the training data are likely the ones that should be altered little when speaking rate changes are needed for synthesis.

Intelligibility is always the first quality to measure in synthetic speech, since without that the speech is useless. The Modified Rhyme Test, as used in [Log 89], measures the intelligibility of individual phonemes in isolated word contexts. Natural speech appears to cause about 0.5% errors (recall that such artificial test conditions do not reflect the redundancy of continuous speech), while the best commercial system at the time (1989) had 3.25% errors. The newer prototype system here had 5.0% errors. Among the word-final consonants, nasals and voiced stops dominated the errors, with voicing errors prevalent (caused apparently mostly by poor durational modeling in the automatic clustering, not reflecting the tendency for longer vowels when preceding voiced consonants than unvoiced ones).

The other aspect of speech quality, naturalness, was evaluated more informally, and was claimed to be quite good. Nonetheless, the speech suffered from a general "certain, hesitant, character...thought to be the cumulative effect of all the tiny formant discontinuities" (p. 238 [Don 99]). The generally good quality was attributed to state-based, rather than phone-based, clustering (also good for ASR), which leads to more consistent modeling, as shorter units are used. Even the automatic segmentation has further advantages when making errors in estimating phone boundaries: such errors tend to reoccur at synthesis time with the same bias. As long as such "errors" (as objectively evaluated in terms of formal linguistic segmentation) remain consistent in training and run-time, their negative effects on speech quality can lead to much reduced degradation.

As [Beut 99] notes, calculating all possible join costs becomes very expensive as databases increase in size. Precomputing and caching such costs is unfeasible for large databases as well. However, only a small subset of the search space is used in practice; so by actually synthesizing large amounts of speech, one can gather statistics on which units and joins are most frequently examined, and cache those results for fast future access. These authors found that a cache of 0.7% of the joins saves 99% of the computation.

14.7 Intonation

Independently of the choice of basic approach for speech synthesis, deciding on an appropriate intonation is a significant task. The speech synthesis system must determine how to vary the three **prosodic** (or **suprasegmental**) parameters that contribute to intonation - pitch (or F0), duration, and intensity - as a function of each input text. In natural speech, intonation varies in very complex ways with several aspects of spoken utterances. At the lowest level – that of the segmentals or phonetics of phonemes, amplitude varies substantially with manner of articulation, somewhat with voicing, and less so with place of articulation. Vowels have more intensity than consonants, low vowels more than high vowels, etc. Many intensity variations are phoneme- dependent, and stressed syllables are more intense than unstressed ones.

Pitch (due to the vibration rate F0 of the vocal cords) is the most difficult of the intonation parameters to predict, owing to its large variation. In many languages (including English), lexically-stressed syllables of emphasized words have significant pitch changes, which cue the boundaries of syntactic phrases. Pitch often displays an overall falling pattern (a **declination line**), with excursions superimposed above and below for stress and syntactic effects. Compared to pitch, duration is more tightly linked to phonemics; e.g., vowels are longer than consonants, stressed syllables longer than unstressed ones, and consonants shorter in clusters. Vowels are longer when located prior to voiced consonants than to unvoiced ones. When speaking rate varies, vowels tend to expand or compress more than consonants.

Specifying intonation that sounds natural is difficult owing in part to a lack of reliable indicators in input text to indicate intonational phenomena such as syntactic boundaries. While sentence- and clause-final punctuation (. ? ! : ;) are good places for pauses in general, sentences often consist of many words with no internal punctuation other than commas. Unlike clause-final punctuation, commas often do not correspond well to intonation (speakers may or may not pause at a comma, depending on its specific syntactic role; e.g., 'a large, round, yellow ball' normally has no pauses). In most European languages, an intonational break often occurs after a (high-information) **content word** (e.g., noun, verb, adjective, or adverb) followed by a **function word** (these latter grammatical words are typically **closed-class** prepositions, articles, pronouns, conjunctions, etc.). Highlighting the final word in such sequences (with durational lengthening and a pitch rise) is often appropriate.

In simple voice-response synthesis, the stored units are typically large (e.g., phrases), and F0 and intensity are usually stored either explicitly along with the spectral parameters in LPC- or formant-synthesis, or implicitly as part of the signals of waveform synthesizers. However, if small units are concatenated, the synthetic voice is unnatural unless intonation is adjusted to fit the context. Since intonation varies significantly across languages and is a complex function of many factors, it remains a major problem for most speech synthesis systems.

In line with recent stochastic trends in the segmental aspects of speech synthesis, some recent research has examined statistical methods for generating F0 patterns [Ross 99]. Traditionally, intonation has been specified via an expert-system approach, based mostly on syntactic and semantic information that natural language processors could extract from the input sentences to a speech synthesis device. As found for much of automatic speech recognition, the advantage of an automatic, data-driven procedure is that much more speech data can contribute (toward the estimation of an appropriate intonation) than is normally possible when phonetic experts must interpret the data manually. In addition, nuances of intonation that often escape detection by phoneticians may be more easily incorporated into synthetic pitch contours, when such patterns derive directly from the data. The difficulty of interpreting F0 contours has led to arbitrary simplifications in modeling, e.g., the ToBI system. Such models tend to reflect the biases of the linguists who interpret the data, and are only loosely based on trends from speech production and perception experiments.

While there appear to be correlations among the acoustic correlates for intonation (F0, duration, intensity), most speech synthesis modeling of these parameters has occurred somewhat separately, due to the complex nature of their interactions. Thus we examine how speech synthesis handles each parameter individually. Certainly, the major factors to specify intonation have effects across several parameters. These factors include: stress, syntactic structure, emotion [Murr 93], speaking rate, and conversation context. Traditionally, F0 has been treated in a hierarchical way, in which layers of effects are specified, ranging from global patterns (at the sentential level or higher) to local effects (at the word level) to micro-effects (related to segmental aspects of phonemes). While such an approach is not necessarily required, the hierarchical method has the advantage of ease of interpretation.

At the highest level would normally be global conversational context. This is important because speakers adjust their style of speaking to accommodate the ease or difficulty of communicating, as a function of background noise, communication channel, the audience for the speech, and topics to be discussed. Intonation is quite different at a global level when one gives tourist instructions to foreign strangers on a noisy street than when chatting with a family member in a quiet room. In a speech synthesis situation, this would presume that the system user might have to specify a general level of emphasis to enhance (or reduce) the degree of variation to which F0 is subject globally.

Generally, as one speaks, air is slowly expelled from the lungs, and F0 tends to fall slowly from an initial high start (corresponding to full lungs). The rough correlation of F0 with lung volume has led to the term **declination** to describe the gradual downward trend of F0 during most utterances. One oversimplified model for F0 has the pitch pattern moving between two declination lines, with jumps between the levels signaling stress and syntactic phenomena. This assumes that humans have a general relaxation pattern for sentential F0 (which may relate to both lung and vocal cord muscles, which control F0), and that local F0 changes normally stay within a gradually narrowing range (for which the upper line suggests greater effort by the speaker, and the lower line corresponds to an unemphatic baseline). The commonly-used Fujisaki model [Fuji 93] employs the more generally-accepted lower declination line, with binary accent commands cueing upward jumps in F0.

14.8 Text Processing

The initial step of TTS, called front-end analysis, involves processing the textual input to obtain information about: which phonemes to pronounce (corresponding to the text), which words (and more specifically which syllables) to stress, where to place communication (juncture) breaks and pauses, and other related linguistic details. Unlike much of the rest of the speech synthesis process, text processing is necessarily language-dependent. Languages use different alphabets, and each language has its own set of phonemes, although they tend in practice to choose from a larger superset of phonemes (e.g., as characterized by Worldbet or by the standard International Phonetic Alphabet). Many speech synthesis systems are created for one specific language, but extending capacity to other languages is facilitated if the initial global phoneme set includes common phonemes outside the initial target language. There are subtle differences between various languages' versions of many phonemes (e.g., /i/ or /s/), but often such adjustments can be made relatively easily, if the system stores spectral parameters in a form directly related to speech production models. (Modern waveform concatenation systems, however, are less flexible in allowing language or speaker modifications.) Some attempts have been made recently toward multilingual speech synthesis, in which several modules are language-independent, dealing with universal aspects of speech production (e.g., a large phoneme set, coarticulation, and a general vocal-tract model), while other modules remain language-specific [Spro 98].

One major aspect of the natural language processing needed for TTS is text-to-phoneme conversion. Synthesizers that accept any text as input need a linguistic processor to convert each text into phonetic symbols, to permit access to the appropriate stored speech units. A basic task is to transform letters into phonemes. For this job, a table look-up is common: a computer dictionary with an entry for each word in the desired language, its pronunciation (as well as indications as to which syllables to stress), its syntactic category, and sometimes semantic information. Many systems also have language-dependent rules, which examine the context of each letter in a word to determine how it is pronounced; e.g., the letter [p] in English is pronounced /p/, except before the letter [h] (e.g., 'telephone', but note an exception 'haphazard').

For some languages (e.g., Korean, Spanish), this task is trivial because their written languages derive directly from the spoken versions, i.e., each written letter or symbol corresponds simply to one phoneme. Many others (e.g., Italian, Finnish, German) have a relatively small set of rules by which native speakers pronounce written words. Other languages are much more complex. For example, English needs hundreds of these rules. Obviously, a language like Mandarin with thousands of symbols representing words requires an extensive mapping from text to phonemes. Even languages with small alphabets (e.g., the 26 letters of English) can be quite complex in their pronunciation rules. Nonetheless, basic text-to- phoneme conversion seems to be a straightforward task for computational linguists. Errors in phonetic transcription with advanced natural-language-processing systems are almost always due to proper nouns (i.e., names) or foreign words. Such words are sometimes specifically demarcated (capitalized or italicized) in normal text, which could allow for their automatic identification for special processing.

Expert-system (or rule-based) approaches have been designed for several languages with very high success rates [Elov 76]. Alternative approaches (e.g., with neural networks) have been tried [Damp 99], following the argument that rules specified by linguists cannot be as efficient as automatically-determined (i.e., data-driven) rules. This appears to be a hypothetical argument, since few TTS systems rely solely on rules, and simple combinations of dictionaries and rules appear to be very efficient in determining accurate pronunciations. Neural networks may be helpful in initial design of rules for languages yet to be programmed for TTS, but the simplistic network models do not seem to yield efficient results.

Given today's inexpensive computer memories, even complex languages to pronounce (such as English) can be handled readily with a combination of rules and table look-up. Look-up has the advantage of simplicity, at the cost of increased memory, and can store relevant linguistic information beyond simple pronunciation. In such dictionaries, each textual entry (usually a word) specifies the corresponding phoneme sequence, the syllable(s) to stress, the word's syntactic category (i.e., part-of-speech, such as preposition), and sometimes aspects of its semantic information. Such data is pertinent to the TTS task of specifying intonation.

Since pronunciation rules need many exceptions for complex languages, one is tempted to forego the rules for TTS in such languages. (The compact rule set for a phonetic language such as Spanish is more tempting in its efficiency, but it is always harder to specify intonation well without a dictionary.) A better solution is to use a dictionary in a first pass, and use rules in a second pass if a word cannot be found in the dictionary. This allows handling a large majority of words via look-up, with the benefit of accessing data relevant for intonation specification, while rarer words (including mistyped and foreign words) would be pronounced according to the more general language rules. Such rules will always be needed, unless one can guarantee that all input words will be present in the dictionary. Since new words in all active languages are constantly being invented (e.g., slang, jargon), it is safer to use a hybrid approach.

There is the related problem of pronouncing proper names and foreign words. In the future, one might conceive of some universal dictionary of all the world's languages (perhaps including all possible spellings, as well as typographical errors), but even in such a case pronunciation varies from language to language for the same word. Usually, a TTS system should pronounce names and foreign words as native speakers in the target language would, not as native speakers of the foreign language (e.g., an English synthesizer would not normally be expected to use a nasalized vowel in 'bon voyage'). Name pronunciation is especially difficult, to the point where native speakers may frequently disagree. One recent approach derived name pronunciation based on analogy with other words in the dictionary [Bag 98].

A hybrid approach, combining rules with a dictionary, may provide optimal performance. Memory is reduced by eliminating dictionary words that can be predicted from context or via the rules. For example, a morphophonemic analyzer can be used to decompose many lengthy input words into their component parts (prefixes, roots, and suffixes). Rather than populate a dictionary with many words deriving from each root (e.g., for the root word 'walk', we have 'walks, walked, walking'), the dictionary could store all roots and affixes, and find pronunciations indirectly. In languages which allow very long words to be assembled from smaller words and parts of words (e.g., Finnish, German), this decomposition approach is very useful. Many languages with relatively short words can benefit as well (e.g., in French, most verbs in standard dictionaries just list the infinitival form, but TTS needs to handle all of the 39 differently-spelled versions resulting from different conjugations and tenses). In addition, lexical stress and part-of-speech are often determined by affixes; so a decomposition approach can thus supply additional information for intonation.

For simple voice-response synthesis of sentences, phrases, or individual words, there is no problem extending our discussions here to other languages, because such systems just play back the previously-stored speech units as is (i.e., no natural-language processing is needed). More generally, speech synthesis in each different language requires major modifications to several synthesizer components. The **front end** of speech synthesis systems especially, which handles letter-to-phoneme rules, relationships between text and intonation, and different sets of phonemes, is certainly language-dependent [Pavl 97]. The **back end**, corresponding to simulation of the vocal tract using digital filters, is relatively invariant for different languages. Even languages with sounds (e.g., clicks, snorts) other than the typical pulmonic egressives of most languages require only simple modifications.

14.9 Evaluation of Speech Synthesis Output

It is often difficult to judge which aspect of the multi-step speech synthesis procedure is most responsible for speech quality. While mistakes in text-to-phoneme conversion can be easily identified, 'errors' in choosing units or specifying intonation are often difficult to detect. One recent evaluation [Don 01] suggests that the proper choice of units may be the biggest remaining factor in achieving excellent naturalness in speech synthesis and that proper intonation modification is also quite important. As in speech coding applications, synthetic speech often requires subjective human evaluation.

Commercial synthesizers are widely available, but only for some of the world's major languages. They often combine software, memory, and processing chips to provide output speech quality ranging from speech that sounds close to natural to intelligible yet flawed voices (as in inexpensive personal computer programs). Basic DSP chips are widely used for speech synthesis. Current microprocessors can easily handle the computation speeds needed for synthesis, and many synthesizers exist entirely in software. However, memory requirements can still be a concern, particularly for the newer waveform concatenation systems.

Typical spectra-based synthesis (e.g., using formant targets for each phoneme, or diphones of LPC patterns) imposes relatively simple computational requirements. However, the recent trend toward time-domain synthesis, with its large memory requirements for unlimited-text synthesis (and a corresponding potentially large search time), can strain computer resources. Thus, recent work has tried to shown that real-time high-quality synthesis is still feasible with some time-domain techniques [She 02]. Resource issues can be pertinent for portable, lightweight synthesis devices.

14.10 Summary

This final chapter provided the third selected area of speech technology, speech synthesis, which forms a core module in larger text-to-speech systems. After a brief introduction to speech synthesis, where major steps in a typical system were outlined, basic appoaches and methods for speech synthesis were presented. We covered three main methods — articulatory method, spectral method, and waveform method. A recent trend in speech synthesis has centered on the use of data-driven, optimization principles to automatically select speech units in waveform-based concatenation methods. We provided a case study to elaborate on such a new trend. We also covered intonation, text pre-processing, and evaluation aspects of speech synthesis and text-to-speech.

Text-to-speech (TTS) and voice response systems have been improved significantly in recent years. Speech synthesis, as the back end of a TTS system, which simulates the actual airflow in the human vocal tract, has existed in mechanical form for centuries. However, it is only in the last few decades, as practical computers have become commonplace, that much better synthetic speech has been developed. Before the advent of digital signal processing (DSP) chips in the 1970s, such speech was only possible with large computers. Now, voice response can be found using inexpensive machines.

Limited-vocabulary voice response yields high-quality speech and suffices for many applications. Increasingly inexpensive memory has led to use of large inventories of speech units, which can overcome some coarticulation and intonation problems. Such trends follow those of automatic speech recognition methods, where stochastic methods involve very simple network models, but require massive amounts of training and memory to accommodate the large amount of variability in the way speakers talk. Speech synthesis does not normally need to model speaker variability, but it must accommodate the large variability found in the many different phonetic contexts that occur in normal speech. In the future, increased understanding of how humans produce and perceive speech will yield more efficient speech synthesis, and some combination of stochastic- and knowledge-based methods will yield general synthetic speech quite similar to that of humans.

References

[Abb 76] Abbas P. J. and Sachs M. B. (1976) "Two-tones suppression in auditory-nerve fibers: Extension of a stimulus-response relationship," *J. Acoust. Soc. Am.*, 59(1), 112–122.

[Ace 93] Acero A. (1993) **Acoustic and Environmental Robustness in Automatic Speech Recognition**, Kluwer Academic, Boston.

[Ace 00] Acero A., Deng L., Kristjansson T., and Zhang J. (2000) "HMM Adaptation using vector Taylor series for noisy speech recognition," *Proceedings of the ICSLP*, 3, 869–872.

[Adl 60] Adler R., Chu L. & Fano R. (1960) **Electromagnetic Energy Transmission and Radiation**, Wiley & Sons, New York.

[Ain 96] Ainsworth W. and Greenberg S. (eds.) (1996) *Proceedings of ESCA Tutorial and Research Workshop on the Auditory Basis of Speech Recognition,* Keele University, UK.

[Ains 96b] Ainsworth W., Greenberg S., Hirahara T., Cariani P. and Delgutte B. (1996) "Representation of low-frequency vowel formants in the auditory nerve", in **Auditory Basis of Speech Perception**, 83–86.

[Ains 96a] Ainsworth W., Greenberg S., Sachs M., Le Prell G., May B. and Hienz R. (1996) "Adequacy of auditory-nerve rate representations of vowels: Comparison with behavioral measures in cat", in **Auditory Basis of Speech Perception**, 120–126.

[Ains 02] Ainsworth W. and Greenberg S. (eds.) (2002) **Listening to Speech: An Auditory Perspective**, Oxford University Press, New York.

[All 85] Allen J. (1985) "Cochlear modeling," *IEEE Signal Processing Magazine*, 2, 3–29.

[All 92] Allen J. (1992) "Overview of text-to-speech systems," in **Advances in Speech Signal Processing** (S. Furui and M. Sondhi, eds.), Marcel Dekker, New York, 741–790.

[All 94] Allen J. (1994) "How do humans process and recognize speech?" *IEEE Trans. Speech & Audio Processing*, 2, 567–577.

[Ana 96] Anastasakos T., McDonough T., Schwarts R. and Makhoul J. (1996) "A compact model for speaker-adaptive training," *Proceedings of the ICSLP*, 2, 1137-1140.

[Arch 97] Archangeli D. and Langendoen D. (eds.) (1997) **Optimality Theory: An Overview**, Blackwell, Cambridge, U.K.

[Arle 91] Arle J. E. and Kim D. O. (1991) "Simulations of cochlear nucleus neural circuitry: Excitatory-Inhibitory response-Area types I-IV," *J. Acoust. Soc. Am.*, 90(6), 3106–3121.

[Atal 71] Atal B. S. and Hanauer S. L. (1971) "Speech analysis and synthesis by linear prediction of the speech wave," *J. Acoust. Soc. Am.*, 50 (2), 637–655.

[Atal 79] Atal B. and Schroeder M. (1979) "Predictive coding of speech signals and subjective error criteria," *IEEE Trans. Acoustics, Speech & Signal Processing,* ASSP-27, 247–254.

[Bag 98] Bagshaw P. (1998) "Phonemic transcription by analogy in text- to-speech synthesis: novel word pronunciation and lexicon compression", *Computer Speech and Language*, 12, 119–142.

[Bahl 75] Bahl L. R. and Jelinek F. (1975) "Decoding for channels with insertions, deletions, and substitutions, with applications to speech recognition," *IEEE Trans. Information Theory*, 21, 404–411.

[Bahl 83] Bahl L., Jelinek F. and Mercer, R. L. (1983) "A Maximum likelihood approach to continuous speech recognition," *IEEE Transactions on Pattern Analysis and Machine Intelligence*, PAMI-5, 179–190.

[Bahl 89] Bahl L., Brown P., de Souza P., and Mercer R. (1989) "A tree-based statistical language model for natural language speech recognition," *IEEE Trans. Acoustics, Speech & Signal Processing,* ASSP-34, 1001-1008.

[Baker 75] Baker J. (1975) "The Dragon system — An overview," *IEEE Trans. Acoustics, Speech & Signal Processing,* ASSP-23, 24–29.

[Bakis 91] Bakis R. (1991) "Coarticulation modeling with continuous-state HMMs," *Proc. IEEE Workshop on Automatic Speech Recognition*, 20–21.

[Bakis 92] Bakis R. (1993) "An articulatory-like speech production model with controlled use of prior knowledge," notes from **Frontiers in Speech Processing**, CD-ROM.

[Bar 93] Bar-Shalom Y. and Li X. (1993) **Estimation and Tracking**, Artech House, Boston.

[Baum 67] Baum L. E. and Egon J. A. (1967) "An inequality with applications to statistical estimation for probabilistic functions of a Markov process and to a model for ecology," *Bull. Amer. Meteorol. Soc.* 73, 360–363.

[Baum 68] Baum L. E. and Sell G. R. (1968) "Growth functions for transformations of manifolds," *Pac. J. Math.* 27 (2), 211–227.

[Baum 72] Baum L.E. (1972) "An inequality and associated maximization technique in statistical estimation for probabilistic functions of Markov processes," *Inequalities*, 3, 1–8.

[Bek 41] von Békésy G. (1941) "Über die Messung der Schwingungsamplitude der Gehörknöchelchen mittels einer Kapizitiven Sonde," *Akustische Zeitschrift*, 6, 1–16.

[Bek 49] von Békésy G. (1949) "The vibration of the cochlear partition in anatomical preparations and in models of the inner ear," *J. Acoust. Soc. Am.*, 21, 233–245.

[Bell 81] Bell-Berti F. & Harris K. (1981) "A temporal model of speech production," Phonetica, 38, 9–20.

[Belle 82] Bellegarda J.R. and Farden d.C. (1982) "Continuously adaptive linear predictive coding of speech," *Proceedings of the ICASSP*, 347–350.

[Bell 89] Bellegarda J. and Nahamoo D. (1989) "Tied mixture continuous parameter models for large vocabulary isolated speech recognition," *IEEE Proceedings of the ICASSP*, 1, 13–16.

[Beut 99] Beutnagel M., Mohri M. and Riley M. (1999) "Rapid unit selection from a large speech corpus for concatenative speech synthesis," *Proceedings of the Eurospeech*, 607–610.

[Bhat 43] Bhattacharyya A. (1943) "On a Measure of divergence between two statistical populations defined by their probability distributions," *Bull. Calcutta Math. Soc.* 35, 99–110.

[Bird 90] Bird S. and Klein E. (1990) "Phonological events," *Journal of Linguistics*, 36, 33–57.

[Bird 95] Bird S. (1995) **Computational Phonology - A Constraint-Based Approach**, Cambridge Univ. Press, Cambridge, U.K.

[Bishop 97] Bishop C. (1997) **Neural Networks for Pattern Recognition**, Clarendon Press, Oxford.

[Blac 90] Blackburn C. C. and Sachs M. B. (1990) "The representation of the steady-state vowel sound /ε/ in the discharge patterns of cat anteroventral nucleus neurons," *Journal of Neurophysiology*, 63(5), 1191–1212.

[Blac 95] Blackburn C., and Young. S. (1995) "Towards improved speech recognition using a speech production model," *Proceedings of the Eurospeech*, , 2, 1623–1626.

[Blad 83] Bladon A. (1983) "Two formant models of vowel perception: Shortcomings and enhancements," *Speech Communication*, 2, 305–313.

[Blum 79] Blumstein S. & Stevens K. (1979) "Acoustic invariance in speech production: Evidence from measurements of the spectral characteristics of stop consonants," *J. Acoust. Soc. Am.*, 66, 1001–1017.

[Blom 84] Blomberg M., Carlson R., Elenius K. and Granstrom B. (1984) "Auditory neural feedback as a basis for speech processing," *Quarterly Progress Status Report, Speech Transmiss. Lab.*, Royal Institute of Technology (Stockholm),1–15.

[Bol 58] Bolinger D. (1958) "A theory of pitch accent in English," *Word*, 14, 109–149.

[Boll 79] Boll S. (1979) "Suppression of acoustic noise in speech using spectral subtraction," *IEEE Trans. Acoustics, Speech & Signal Processing*, ASSP-27, 113–120.

[Boll 91] Boll S. (1991) "Speech enhancement in the 1980s: Noise suppression with pattern matching," in **Advances in Speech Signal Processing**, (S. Furui and M. M. Sondhi, eds.), Ch. 10, 309–325, Marcel Deckker, New York.

[Bou 81] Bourk T. R., Mielcarz J. P. and Norris B. E. (1981) "Tonotopic organization of the anteroventral cochlear nucleus of the Cat," *Hearing Research*, 4, 215–241.

[Boyce 77] Boyce S. & Espy-Wilson C. (1997) "Coarticulatory stability in American English /r/," *J. Acoust. Soc. Am.*, 101, 3741–3753.

[Broc 82] Brockx J. & Nooteboom S. (1982) "Intonation and the perceptual separation of simultaneous voices," *Journal of Phonetics*, 10, 23–36.

[Broe 83] van der Broecke M. & van Heuven V. (1983) "Effect and artifact in the auditory discrimination of rise & decay time: Speech & nonspeech," *Perception and Psychophys.*, 33, 305–313.

[Brow 86] Browman C. and Goldstein L. (1986), "Towards an articulatory phonology," *Phonology Yearbook,* 3, 219–252.

[Brow 89] Browman C. and Goldstein L. (1989), "Articulatory gestures as phonological units," *Phonology,* 6, 201–251.

[Brow 90] Browman C. & Goldstein L. (1990) "Gestural specification using dynamically defined articulatory structures," *Journal of Phonetics*, 18, 299–320.

[Brow 92] Browman C. and Goldstein L. (1992), "Articulatory phonology: An overview," *Phonetica*, 49, 155–180.

[Brug 92] Brugge J. F., Popper A. N. and Fay R. R. (1992) "An overview of central auditory processing," **The Mammalian Auditory Pathway: Neurophysiology**, Spinger-Verlag, New York, 1–33.

[Burt 85] Burton D. K., Shore J. E. and Buck J. T. (1985) "Isolated-word speech recognition using multisection vector quantization codebooks," *IEEE Trans. Acoustics, Speech & Signal Processing,* 33(4), 837–849.

[Buzo 80] Buzo A., Gray A. H., Gray R. M., and Markel J. D. (1980) "Speech coding based on vector quantization," *IEEE Trans. Acoustics, Speech & Signal Processing,* ASSP-28, 562–574.

[Byrn 97] Byrne B., Finke M., Khudanpur S., McDonough J., Nock H., Riley M., Saraclar M., Wooters C., and Zavaliagkos G. (1997) "Pronunciation modeling for conversational speech recognition: A status report from WS97," *Proc. IEEE workshop Automatic Speech Recognition and Understanding,* 26–33.

[Camp 86] Campbell J. & Tremain T. (1986) "Voiced/unvoiced classification of speech with applications to the U.S. government LPC-10E algorithm," *IEEE Proceedings of the ICASSP*, 473–476.

[Cant 79] Cant N. B. and Morest D. K. (1979) "Organization of the neurons in the anterior division of the anteroventral cochlear nucleus of the cat," *Neuroscience*, 4, 1909–1923.

[Cant 92] Cant N. B., Webster D. B., Popper A. N. and Fay R. R. (1992) "The cochlear nucleus: Neural types and their synaptic organization," **The Mammalian Auditory Pathway: Neuroanatomy**, 66–116, Spinger-Verlag, New York.

[Cari 92] Cariani P. and Delgutte B. (1992) "The pitch of complex sounds is simply coded in interspike interval distributions of auditory nerve fibers" *Society of Neuroscience Abstracts* 18, 383.

[Cari 93] Cariani P. and Delgutte B. (1993) "Interspike interval distributions of auditory nerve fibers in response to concurrent vowels with same and different fundamental frequencies" *Association for Research in Otolaryngology (ARO) Abstracts.*

[Carl 74] Carlson R., Fant G. & Granstrom B. (1974) "Vowel perception and auditory theory," *Acustica*, 31.

[Carl 79] Carlson R., Granstrom B. & Klatt D. (1979) "Vowel perception: the relative perceptual salience of selected acoustic manipulations," *Quarterly Progress Status Report, Speech Transmiss. Lab.*, Royal Institute of Technology, Stockholm, 3, 73–83.

[Carl 96] Carlin B. P. and Louis T. A. (1996) **Bayes and Empirical Bayes Methods for Data Analysis**, Chapman & Hall, London.

[Carn 86] L. H. Carney and C. D. Geisler (1986) "A temporal analysis of auditory-nerve fiber responses to spoken stop consonant-vowel syllables," *J. Acoust. Soc. Am.*, 79(6), 1896–1914.

[Carn 88] Carney L. H. and Yin T. (1988) "Temporal coding of resonances by low-frequency auditory nerve fibers: Single-fiber responses and a population model," *Journal of Neurophysiology*, 60, 1653–1677.

[Carn 89] Carney L. H. (1989) **Temporal Encoding and Processing of Information in the Auditory Nerve and Anteroventral Cochlear Nucleus of the cat,** Ph.D. thesis, University of Wisconsin-Madison.

[Carn 90] Carney L. H. (1990) "Sensitivities of cells in anteroventral cochlear nucleus of cat to spatiotemporal discharge patterns across primary afferents," *Journal of Neurophysiology*, 64(2), 437–456.

[Carn 92] Carney L. H. (1992) "Modeling the sensitivity of cells in the anteroventral cochlear nucleus to spatiotemporal discharge patterns," *Philosophical Transactions of the Royal Society B*, 336, 403–406.

[Carn 93] Carney L. H. (1993) "A model for the responses of low-frequency auditory nerve fibers in Cat," *J. Acoust. Soc. Am.*, 93 (1), 401–417.

[Carn 94] Carney L. H. (1994) "Spatiotemporal encoding of sound level: Models for normal encoding and recruitment of loudness," *Hearing Research*, 76, 31–44..

[Carn 99] Carney, L.H., McDuffy, M.J., and Shekhter, I. (1999) "Frequency glides in the impulse responses of auditory-nerve fibers," *J. Acoust. Soc. Am.*, 105, 2384– 2391.

[Cars 01] Carson-Berndsen J. and M. Walsh. (2001) "Defining constraints for multilinear processing," *Proceedings of the Eurospeech*, 1, 2281–2284.

[Chen 97] Chen M. (1997) "Acoustic correlates of English and French nasalized vowels," *J. Acoust. Soc. Am.*, 102, 2360–2370.

[Chhik 89] Chhikara R. and Folks J. (1989) **The Inverse Gaussian Distribution-Theory, Methodology, and Applications,** Marcel Dekker, New York.

[Chis 85] Chistovich L. (1985) "Central auditory processing of peripheral vowel spectra," *J. Acoust. Soc. Am.*, 77, 789–805.

[Chom 68] Chomsky N. and Halle M. (1968) **The Sound Pattern of English**, Harper and Row, New York.

[Chou 96] Chou W., Juang B., and Lee C. (1996) "A minimum error rate pattern recognition approach to speech recognition," *International J. Pattern Recognition and Artificial Intelligence,* 8, 5–31.

[Chou 00] Chou W., Juang B., and Lee C. (2000) "Discriminant-function-based minimum recognition error rate pattern-recognition approach to speech recognition," *Proceedings of the IEEE*, 88(8), 1201–1223.

[Chris 97] Christensen L. & Humes L. (1997) "Identification of multidimensional stimuli containing speech cues and the effects of training," *J. Acoust. Soc. Am.,* 102, 2297–2310.

[Church 87] Church K. W. (1987) **Phonological Parsing in Speech Recognition**, Kluwer Academic, Boston.

[Clark 95] J. Clark and C. Yallop (1995) **An Introduction to Phonetics and Phonology**, 2nd Ed., Blackwell, Oxford, U.K.

[Clem 95] Clements N. and Hume E. (1995) "The internal organization of speech sounds," in **The Handbook of Phonological Theory,** J. Goldsmith (ed.), Blackwell, Cambridge, U.K., 206–244.

[Coh 89] Cohen J. R. (1989) "Application of an auditory model to speech recognition," *J. Acoust. Soc. Am.,* 85 (6), 2623–2629.

[Cohen 95] Cohen L. (1995) **Time-Frequency Analysis**, Prentice Hall, Englewood Cliffs, NJ.

[Cok 76] Coker C. (1976) "A model of articulatory dynamics and control," *Proceedings of the IEEE*, , 64, 452–460.

[Colb 90] Colburn L. H., Han Y-A., and Culotta C. P. (1990) "Coincidence model of MSO responses,"*Hearing Research* 49, 335–346.

[Cole 74] Cole R. & Scott B. (1974) "Toward a theory of speech perception," Psych. Rev., 81, 348–374.

[Cole 90] R. Cole, A. Rudnicky, V. Zue, & D. Reddy (1980) "Speech as patterns on paper," in **Perception and Production of Fluent Speech,** (Cole A. ed.), Lawrence Erlbaum, Hillside, N.J., 3-50.

[Cole 75] Cole R. & Cooper W. (1975) "The perception of voicing in English affricates and fricatives," *J. Acoust. Soc. Am.,* 58, 1280–1287.

[Cole 98] Coleman J. (1998) **Phonological Representation – Their names, forms, and powers**, Cambridge University Press, Cambridge, U.K.

[Cooke 93] Cooke M. (1993) **Modeling Auditory Processing and Organization**, Cambridge University Press , Cambridge, U.K.

[Cry 88] Crystal T. & House A. (1988) "Segmental durations in connected speech signals," *J. Acoust. Soc. Am.,* 83, 1553–1585.

[Cry 90] Crystal T. & House A. (1990) "Articulation rate and the duration of syllables and stress groups in connected speech," *J. Acoust. Soc. Am.*, 88, 101–112.

[Damp 99] Damper R., Marchand Y., Adamson M. and Gustafson K. (1999) "Evaluating the pronunciation component of text-to-speech systems for English: a performance comparison of different approaches," *Computer Speech and Language*, 13, 155–176.

[Dau 96] Dau T., Puschel D., and Kohlrausch A. (1996) "A quantitative model of the effective signal processing in the auditory system: I. Model structure," *J. Acoust. Soc. Am.*, 99, 3615–3622.

[Dau 96a] Dau T., Puschel D., and Kohlrausch A. (1996) "A quantitative model of the effective signal processing in the auditory system: II. Simulation and measurements," *J. Acoust. Soc. Am.*, 99, 3623–3631.

[Dav 80] Davis S. & Mermelstein P. (1980) "Comparison of parametric representations for monosyllabic word recognition in continuously spoken sentences," *IEEE Trans. Speech & Audio Processing*, 28, 357–366.

[Dav 94] Davis K. A. and Voigt H. F. (1994) "Functional interconnections between neurons in the dorsal cochlear nucleus (DCN) of unanesthetized, decerebrate gerbil (*Meriones Unguiculatus*)," *Abstracts of the 17th midwinter research meeting of ARO*, 17.

[Daw 79] Dawid A. P. (1979) "Conditional independence in statistical theory (with discussion)," *J. Royal Statistical Soc. Series B*, 41, 1–31.

[DeG 70] DeGroot M. H. (1970) **Optimal Statistical Decisions**, McGraw-Hill, New York.

[Dela 90] Delgutte B. (1990) "Two-tone rate suppression in auditory-nerve fibers: Dependence on suppressor frequency and level," *Hearing Research*, 49, 225–246.

[Dela 55] Delattre P.C., Liberman A. M., and Cooper F. S. (1955) "Acoustic loci and transitional cues for consonants," *J. Acoust. Soc. Am.*, 27, 769-773.

[Dela 70] Delattre P. (1970) "Syntax and intonation, a study in disagreement," *Modern Language J.*, 54, 3–9.

[Delg 80] Delgutte B. (1980) "Representation of speech-like sounds in the discharge patterns of auditory-nerve fibers," *J. Acoust. Soc. Am.*, 68 (3), 843–857.

[Delg 84a] Delgutte B. and Kiang N. Y. S. (1984) "Speech coding in the auditory nerve: I. Vowel-like sounds," *J. Acoust. Soc. Am.*, 75, 866–878.

[Delg 84b] Delgutte B. and Kiang N. Y. S. (1984) "Speech coding in the auditory nerve: IV. Sounds with Consonant-like dynamic Characteristics," *J. Acoust. Soc. Am.*, 75, 897–907.

[Delg 84c] Delgutte B. and Kiang N. Y. S. (1984) "Speech coding in the auditory nerve: V. Vowels in Background noise," *J. Acoust. Soc. Am.*, 75, 908–918.

[Delg 84d] Delgutte B. (1984) "Speech coding in the auditory nerve: II. Processing schemes for vowel-like sounds," *J. Acoust. Soc. Am.*, 75, 879–886.

[Delg 97] Delgutte B. (1997) "Auditory neural processing of speech," in **The Handbook of Phonetic Sciences** (W. J. Handcastle and J. Lavar eds.), Blackwell, Cambridge, 507–538.

[Dell 93] Deller J., Proakis J. and Hansen J.H.L. (1993) **Discrete Time Processing of Speech Signals**, Macmillan, New York.

[Demp 77] Dempster A.P., Laird N.M., and Rubin D.B. (1977) "Maximum likelihood from incomplete data via the EM algorithm," *J. Roy. Stat. Soc.*, 39, 1–38.

[Deng 85] Deng L. and Geisler D. (1985) "Changes in the phase of excitatory-tone responses in cat auditory nerve fibers by suppressor tones and fatigue," *J. Acoust. Soc. Am.,* 78 (5), 1633–1643.

[Deng 86] Deng L. (1986) **Experimental and Modeling Studies of Complex Sound Processing in the Peripheral Auditory System,** Ph.D. thesis, University of Wisconsin-Madison, Madison, WI.

[Deng 87a] Deng L. and Geisler D. (1987) "A composite auditory model for processing speech sounds," *J. Acoust. Soc. Am.,* 82(6), 2001–2012.

[Deng 87b] Deng L., Geisler D., and Greenberg S. (1987) "Responses of auditory-nerve fibers to multiple-tone complexes," *J. Acoust. Soc. Am.,* 82(6), 1989– 2000.

[Deng 87c] Deng L. and Geisler D. (1987) "Responses of auditory-nerve fibers to nasal consonant-vowel syllables," *J. Acoust. Soc. Am.,* 82(6), 1977–1988.

[Deng 88a] Deng L., Lennig M. and Mermelstein P. (1988) "Modeling acoustic-phonetic detail in a hidden-Markov-model-based large vocabulary speech recognizer," *IEEE Proceedings of the ICASSP*, 1, 509–512.

[Deng 88b] Deng L., Geisler D. and Greenberg S. (1988) "A composite model of the auditory periphery for the processing of speech," *Journal of Phonetics* 16, 93–108.

[Deng 89] Deng L., Kenny P., Gupta V. and Mermelstein P. (1989) "A Locus model of coarticulation in an HMM speech recognizer," *IEEE Proceedings of the ICASSP*, 1, 97–100.

[Deng 90] Deng L., Lennig M. and Mermelstein P. (1990) "Modeling microsegments of stop consonants in a Hidden Markov based word recognizer," *J. Acoust. Soc. Am.,* 87, 2738–2747.

[Deng 90a] Deng L., Lennig M., Seitz P. and Mermelstein P. (1990) "Large vocabulary word recognition using context-dependent allophonic hidden Markov models," *Computer Speech and Language*, 4, 345-357.

[Deng 91a] Deng L. and Erler K. (1991) "Microstructural speech units and their HMM representation for discrete utterance speech recognition," *IEEE Proceedings of the ICASSP*, 1, 193–196.

[Deng 91b] Deng L., Kenny P., Lennig M., Gupta V., Seitz F. and Mermelstein P. (1991) "Phonemic hidden Markov models with continuous mixture output densities for large vocabulary word recognition," *IEEE Transactions on Signal Processing*, 39, 1677–1681.

[Deng 91c] Deng L. (1991) "Speech modeling and recognition using a time series model containing trend functions with Markov modulated parameters," *Proc. IEEE Workshop on Automatic Speech Recognition*, 1991, 24-26.

[Deng 92a] Deng L. (1992-1993) "A Computational model of the phonology-phonetics interface for automatic speech recognition," Summary Report of Research in Spoken Language, Laboratory for Computer Science, Massachusetts Institute of Technology.

[Deng 92b] Deng L. (1992) "A generalized hidden Markov model with state-conditioned trend functions of time for the speech signal," *Signal Processing*, 27, 65–78.

[Deng 92c] Deng L. (1992) "Processing of acoustic signals in a cochlear model incorporating laterally coupled suppressive elements," *Neural Networks*, 5(1), 19–34.

[Deng 92d] Deng L., and Erler K. (1992) "Structural design of a hidden Markov model based speech recognizer using multi-valued phonetic features: Comparison with segmental speech units," *J. Acoust. Soc. Am.*, 92, 3058–3067.

[Deng 92e] Deng L., Kenny P., Lennig M., and Mermelstein P. (1992) "Modeling acoustic transitions in speech by state-interpolation hidden Markov models," *IEEE Transactions on Signal Processing*, 40(2), 265–272.

[Deng 93a] Deng L. and Kheirallah I. (1993) "Numerical property and efficient solution of a nonlinear transmission-line model for basilar-membrane wave motions," *Signal Processing*, 33(3), 9–286.

[Deng 93b] Deng L. and Kheirallah I. (1993) "Dynamic formant tracking of noisy speech using temporal analysis on outputs from a nonlinear cochlear model," *IEEE Trans. Biomedical Engineering*, 40(5), 456–467.

[Deng 93c] Deng L. (1993) "A stochastic model of speech incorporating hierarchical nonstationarity," *IEEE Trans. Speech & Audio Processing*, 1(4), 471–475.

[Deng 93d] Deng L. and Mark J. W. (1993) "Parameter estimation of Markov modulated Poisson processes via the EM algorithm with time discretization," *Telecommunication Systems*, 1(1), 321–338.

[Deng 94a] Deng L. and Braam D. (1994), "Context-dependent Markov model structured by locus equations: Application to phonetic classification," *J. Acoust. Soc. Am.*, 96, 2008–2025.

[Deng 94b] Deng L., Hasanein K., and Elmasry M. (1994) "Analysis of correlation structure for a neural predictive model with application to speech recognition," *Neural Networks*, 7(2), 331–339.

[Deng 94c] Deng L., Aksmanovic M., Sun D., and Wu J. (1994) "Speech recognition using hidden Markov models with polynomial regression functions as nonstationary states," *IEEE Trans. Speech & Audio Processing*, 2(4), 507–520.

[Deng 94d] Deng L., and Sameti H. (1994) "Speech recognition using dynamically defined speech units," *Proceedings of the ICSLP*, 4, 2167– 2170.

[Deng 94e] Deng L., and Sun D. (1994), "A statistical approach to automatic speech recognition using the atomic speech units constructed from overlapping articulatory features," *J. Acoust. Soc. Am.*, 95, 2702–2719.

[Deng 95a] Deng L. and Rathinavalu, C. (1995) "A Markov model containing state-conditioned second-order nonstationarity: Application to speech recognition," *Computer Speech and Language*, 9(1), 63–86.

[Deng 95b] Deng L., Wu J., and Sameti H. (1995) "Improved speech modeling and recognition using multi-dimensional articulatory states as primitive speech units," *Proceedings of the ICSLP*, 385–388.

[Deng 96] Deng L. and Sameti H. (1996) "Transitional speech units and their representation by the regressive Markov states: Applications to speech recognition," *IEEE Trans. Speech & Audio Processing*, 4(4), 301–306.

[Deng 97a] Deng L. (1997) "Autosegmental representation of phonological units of speech and its phonetic interface," *Speech Communication*, 23(3), 211–222.

[Deng 97b] Deng L. (1997) "Integrated-multilingual speech recognition using universal phonological features in a functional speech production model," *IEEE Proceedings of the ICASSP*, 2, 1007–1010.

[Deng 97c] Deng L., Ramsay G., and Sun D. (1997) "Production models as a structural basis for automatic speech recognition," *Speech Communication*, 22(2), 93–111.

[Deng 97d] Deng L. and Aksmanovic M. (1997) "Speaker-independent phonetic classification using hidden Markov models with mixtures of trend functions," *IEEE Trans. Speech & Audio Processing*, 5, 319–324.

[Deng 97e] Deng L. and Shen X. (1997) "Maximum likelihood in statistical estimation of dynamical systems: Decomposition algorithm and simulation results," *Signal Processing*, 57, 65-79.

[Deng 98] Deng L. (1998) "A dynamic, feature-based approach to the interface between phonology and phonetics for speech modeling and recognition," *Speech Communication,* 24(4), 299–323.

[Deng 99a] Deng L. and Ma J. (1999) "A statistical coarticulatory model for the hidden vocal-tract-resonance dynamics," *Proceedings of the Eurospeech*, 4, 1499–1502.

[Deng 99b] Deng L. (1999) "Computational models for auditory speech processing," in **Computational Models of Speech Pattern Processing**, (K. Ponting ed.) (NATO ASI Series), Springer, 67–77.

[Deng 99c] Deng L. (1999) "Computational models for speech production," in **Computational Models of Speech Pattern Processing**, (K. Ponting ed.) (NATO ASI Series), Springer, 199–214.

[Deng 99d] Deng L. (1999) "Articulatory features and associated production models in statistical speech recognition," in **Computational Models of Speech Pattern Processing**, (K. Ponting ed.) (NATO ASI Series), Springer, 214–224.

[Deng 00a] Deng L. (2000) "Integrated-multilingual speech recognition – Impact on Chinese spoken language processing," *Communications of Oriental Language and Information Processing*, 8(2), 191–209.

[Deng 00b] Deng L., Acero A., Plumpe M., and Huang X.D. (2000) "Large-vocabulary speech recognition under adverse acoustic environments," *Proceedings of the ICSLP*, 2, 806–809.

[Deng 00c] Deng L. and Ma Z. (2000) "Spontaneous speech recognition using a statistical coarticulatory model for the hidden vocal-tract-resonance dynamics," *J. Acoust. Soc. Am.*, 108(6), 3036–3048.

[Deng 01] Deng L., Droppo J., and Acero A. (2001) "Recursive estimation of nonstationary noise using a nonlinear model with iterative stochastic approximation," *Proceedings of Automatic Speech Recognition and Understanding,* Italy.

[Deng 01] Deng L., Acero A., Jiang L., Droppo J., and Huang X.D. (2001) "High-performance robust speech recognition using stereo training data," *IEEE Proceedings of the ICASSP*, 1, 301–304.

[Deng 02] Deng L., Droppo J., and Acero A. (2002) "A Bayesian approach to speech feature enhancement using the dynamic cepstral prior," *IEEE Proceedings of the ICASSP*, 1, 829-832.

[Diehl 89] Diehl R. and Kluender K. (1989) "On the object of speech perception," *Ecological Psychology,* 1, 1–45.

[Diga 93] Digalakis V., Rohlicek J., and Ostendorf M. (1993) "ML estimation of a stochastic linear system with the *EM* algorithm and its application to speech recognition," *IEEE Trans. Speech & Audio Processing,* 1, 431–442.

[Ding 97] Ding W. and Campbell N. (1997) "Optimizing unit selection with voice source and formants in the CHATR speech synthesis system," *Proceedings of the Eurospeech*, 537–540.

[Don 99] Donovan R. and Woodland P. (1999) "A hidden Markov model-based trainable speech synthesizer," *Computer Speech and Language*, 13, 223–241.

[Don 01] Donovan R. (2001) "A component by component listening test analysis of the IBM trainable speech synthesis system", *Proceedings of the Eurospeech*, 329–332.

[Drop 01] Droppo J., Deng L., and Acero A. (2001) "Evaluation of the SPLICE algorithm on the Aurora2 database," *Proceedings of the Eurospeech*, 217–220.

[Duif 73] Duifhuis H. (1973) "Consequences of peripheral frequency selectivity for non-simultaneous masking," *J. Acoust. Soc. Am.*, 54, 1471–1488.

[Duif 82] Duifhuis H., Willems L., & Sluyter R. (1982) "Measurement of pitch in speech: An implementation of Goldstein's theory of pitch perception," *J. Acoust. Soc. Am.*, 71, 1568–1580.

[Duif 93] Duifhuis H., Horst J. W., Van Dijk P., Van Netten, A. M. Geisler C. D., Bender A. and Liotopoulos F. K. (1993) "Time-Domain modeling of a nonlinear, active model of the cochlea," in **Biophysics of Hair Cell Sensory Systems**, World Scientific, 330–337.

[Dur 95] Durand J. and Katamba, F. (1995) **Frontiers of Phonology: Atoms, Structures, Derivations,** Longman, London.

[Durb 59] Durbin J. (1959) "Efficient estimation of parameters in moving-Average models," *Biometrika*, 46, 306–316.

[Dutoit 97] Dutoit T. (1997) **From text to speech: A concatenative approach**, Kluwer, Boston.

[Eady 86] Eady S. & Cooper W. (1986) "Speech intonation and focus location in matched statements and questions," *J. Acoust. Soc. Am.*, 80, 402–415.

[Egge 85] Eggermont J. J. (1985) "Peripheral auditory adaptation and fatigue: A model oriented review," *Hearing Research*, 18, 57–71.

[Eisn 02] Eisner J. (2002) "Comprehension and compilation in optimality theory," *Proc. 40th Annual Meeting of the ACL*, 1–8.

[Elli 95] Elliott R., Aggoun L., and Moore J. (1995) **Hidden Markov Models — Estimation and Control,** Springer, New York.

[Elov 76] Elovitz H.S., Johnson R., McHugh A. and Shore J.E. (1976) "Letter-to-Sound rules for automatic translation of English text to phonetics," *IEEE Trans. Acoustics, Speech & Signal Processing,* ASSP-24, 446–459.

[Ephr 89] Ephraim Y., Malah D., and Juang B. H. (1989) "On the application of hidden Markov models for enhancing noisy speech," *IEEE Trans. Acoustics, Speech & Signal Processing,* ASSP-37, 1846–1856.

[Ephr 90] Ephraim Y. (1990) "A Minimum mean square error approach for speech enhancement," *IEEE Proceedings of the ICASSP*, 829–832.

[Ephr 92] Ephraim Y. (1992) "Statistical-model-based speech enhancement systems," *Proceedings of the IEEE*, 80(10), 1526–1555.

[Eva 78] Evans E. F. (1978) "Place and time coding of frequency in ther peripheral auditory system: Some physiological pros and cons,"*Audiology*, 17, 369–420.

[Everitt 81] Everitt B. S. and Hand D. J. (1981) **Finite Mixture Distributions**, Chapman and Hall, London.

[Fant 67] Fant G. (1967) "Auditory patterns of speech," in **Models for the perception of speech and Visual Form** (W. Wathen-Dunn, ed.), MIT Press, Cambridge, MA, 111–125.

[Fant 68] Fant G. (1968) "Analysis and synthesis of speech processes," in **Manual of Phonetics** (B. Malmberg, ed.), North Holland, Amsterdam, 173–277.

[Flan 72] Flanagan J. (1972) **Speech Analysis, Synthesis and Perception**, 2nd Ed., Springer Verlag, New York.

[Flet 95] Fletcher H. (J. B. Allen, ed.) (1995) **Speech and Hearing in Communication** (The ASA Edition), Acoustical Soc. Am., New York.

[Forn 73] Forney G. D. (1973) "The Viterbi algorithm," *Proceedings of the IEEE*, 61, 268–278.

[Foul 69] Foulke E. & Sticht T. (1969) "Review of research on the intelligibility and comprehension of accelerated speech," Psych. Bull., 72, 50–62.

[Fowl 86] Fowler C. (1986) "An event approach to the study of speech perception from a direct-realist perspective," *Journal of Phonetics*, 14, 3–28.

[Fowl 96] Fowler C. (1996) "Listeners do hear sounds, not tongues," *J. Acoust. Soc. Am.*, 99, 1730–1741.

[Fowler 86] Fowler C. (1986) "An event approach to the study of speech perception from a direct-realist perspective," *Journal of Phonetics*, 14, 3–28.

[Free 88] Freeman G. H. (1988) "Trellis source codes designed by conjugate gradient optimization," *IEEE Trans. Communications*, 36, 1–12.

[Frey 01] Frey B., Deng L., Acero A., and Kristjansson T. (2001) "ALGONQUIN: Iterating Laplace's method to remove multiple types of acoustic distortion for robust speech recognition," *Proceedings of the Eurospeech*, 901–904.

[Frie 78] Friedman D. (1978) "Multidimensional pseudo-maximum- likelihood pitch estimation," *IEEE Trans. Speech & Audio Processing*, 26, 185–196.

[Fries 94] Fries G. (1994) "Hybrid time- and frequency-domain speech synthesis with extended glottal source generation," *IEEE Proceedings of the ICASSP*, 1, 581–584.

[Fry 55] Fry D. (1955) "Duration and intensity as physical correlates of linguistic stress," *J. Acoust. Soc. Am.*, 27, 765–768.

[Fry 58] Fry D. (1958) "Experiments in the perception of stress," *Language and Speech*, 1, 126–152.

[Fuji 77] Fujisaki H. (1977) "Functional models of articulatory and phonatory dynamics," *Proceedings of Articulatory Modeling Symposium*, 127–136.

[Fuji 78] Fujimura O., Macchi M. & Streeter L. (1978) "Perception of stop consonants with conflicting transitional cues: A cross-linguistic study," *Language and Speech*, 21, 337–346.

[Fuji 93] Fujisaki H., Ljungqvist M. & Murata H. (1993) "Analysis and modeling of word accent and sentence intonation in Swedish," *IEEE Proceedings of the ICASSP*, 1, 211–214.

[Fuk 69] Fukunaga K. and Krile T. F. (1969) "Calculation of Bayes' recognition error for two multivariate Gaussian distribution," *IEEE Trans. Computers* 18 (3), 220–229.

[Fuk 90] Fukunaga K. (1990) **Statistical Pattern Recognition,** 2nd Ed., Academic Press, San Diego, 51–110.

[Fur 86] Furui S. (1986) "Speaker-independent isolated word recognition using dynamic features of the speech spectrum," *IEEE Trans. Acoustics, Speech & Signal Processing,* ASSP-34, 52-59.

[Fur 89] Furui S. (1989) **Digital Speech Processing, Synthesis, and Recognition,** Marcel Dekker, New York.

[Gales 93] Gales M. and Young S. (1993) "Segmental HMMs for speech recognition," *Proceedings of the Eurospeech*, , 1579-1582.

[Gales 98] Gales M. (1998) "Cluster adaptive training for speech recognition," *Proceedings of the ICSLP*, 1783–1786.

[Gales 00] Gales M. (2000) "Cluster adaptive training of hidden Markov models," *IEEE Trans. Speech & Audio Processing*, 8(4), 417–428.

[Gales 02] Gales M. (2002) "Maximum likelihood multiple projection schemes for hidden Markov models," *IEEE Trans. Speech & Audio Processing*, 10(2), 37–47.

[Gao 00] Gao Y., Bakis R., Huang J., and Zhang B. (2000) "Multistage coarticulation model combining articulatory, formant and cepstral features," *Proceedings of the ICSLP*, 1, 25– 28.

[Gauv 94] Gauvain J.L. and Lee C.-H. (1994) "Maximum a posteriori estimation for multivariate Gaussian mixture observation of Markov chains", *IEEE Trans. Speech & Audio Processing*, 2(2), 291-298.

[Geis 93] Geisler C. D. (1993) "A realizable cochlear model using feedback from motile outer hair cells," *Hearing Research*, 68, 253–262.

[Gelf 79] Gelfand S. & Silman S. (1979) "Effects of small room reverberation upon the recognition of some consonant features," *J. Acoust. Soc. Am.*, 66, 22–29.

[Geo 97] George E. and Smith M. (1997) "Speech analysis/synthesis and modification using an analysis-by-synthesis/overlap-add sinusoidal model," *IEEE Trans. Speech & Audio Processing*, 5, 389–406.

[Ger 92] Gersho A. and Gray R. M. (1992) **Vector Quantization and Signal Compression**, Kluwer Academic Publishers, Boston.

[Gersch 92] Gersch W. (1992) "Smoothness priors," in **New Directions in Time Series Analysis**, (D. Brillinger et al. eds.), Springer, New York, 111–146.

[Ghah 99] Ghahramani Z. and Roweis S. (1999) "Learning nonlinear dynamic systems using an EM algorithm," *Advances in Neural Information Processing Systems*, 11, 1–7.

[Ghah 00] Ghahramani Z. and Hinton G. (2000) "Variational learning for switching state-space model," *Neural Computation*, 12, 831–864.

[Ghit 86] Ghitza O. (1986) "Auditory nerve representation as a front-end for speech recognition in a noisy environment," *Computer Speech and Language*, 1(2), 109–131.

[Ghit 88] Ghitza O. (1988) "Auditory neural feedback as a basis for speech processing," *IEEE Proceedings of the ICASSP*, 1, 91–94.

[Ghit 92] Ghitza O. (1992) "Auditory nerve representation as a basis or speech processing," in **Advances in Speech Signal Processing** (S. Furui & M. Sondhi, eds.), Marcel Dekker, New York, 453–485.

[Ghit 93a] Ghitza O. (1993) "Adequacy of auditory models to predict internal human representation of speech sounds," *J. Acoust. Soc. Am.*, 93 (4), 2160–2171.

[Ghit 93b] Ghitza O., and Sondhi M. (1993) "Hidden Markov models with templates as nonstationary states: an application to speech recognition," *Computer Speech and Language*, 7, 101–119.

[Ghit 94] Ghitza O. (1994) "Auditory models and human performance in tasks related to speech coding and speech recognition," *IEEE Trans. Speech & Audio Processing*, 2(1), 115– 132.

[Gig 94a] Giguere C. and Woodland P. (1994) "A computational model of the auditory periphery for speech hearing research: I. Ascending path," *J. Acoust. Soc. Am.*, 95, 331–342.

[Gig 94b] Giguere C. and Woodland P. (1994) "A computational model of the auditory periphery for speech hearing research: II. Descending path," *J. Acoust. Soc. Am.*, 95, 343–349.

[Gish 93] Gish H. and Ng K. (1993) "A segmental speech model with applications to word spotting," *IEEE Proceedings of the ICASSP*, 1, 447-450.

[Glas 90] Glasberg B. & Moore B. (1990) "Derivation of auditory filter shape from notched-noise data," Hearing Research47, 103–138.

[Gold 69] Gold B. andRabiner L. (1969) "Parallel processing techniques for estimating pitch periods of speech in the time domain," *J. Acoust. Soc. Am.*, 46, 442–448.

[Golds 90] Goldsmith J.A. (1990) **Autosegmental and Metrical Phonology**, Blackwell, Oxford, UK.

[Gray 80] Gray R. M., Buzo A., Gray A. H., and Matsuyama Y. (1980) "Distortion measures for speech processing," *IEEE Trans. Acoustics, Speech & Signal Processing,* ASSP-28, 367–376.

[Gray 93] Gray R. M. (1993) "Toeplitz and circulant matrices: a Review," Technical Report, Stanford University.

[Gree 61] Greenwood D. D. (1961) "Critical bandwidth and the frequency coordinates on the basilar membrane," *J. Acoust. Soc. Am.*, 33, 1344–1356.

[Gree 88] Greenberg S. (1988) *Journal of Phonetics* (Theme issue on Representation of Speech in the Auditory Periphery), Academic Press, U.K.

[Gree 01] Greenberg S. and Slaney M. (eds.) (2001) **Computational Models of Auditory Function**, (NATO Science Series), IOS Press, Amsterdam.

[Gree 88a] Greenberg S. (1988) "Acoustic transduction in the auditory periphery," *Journal of Phonetics* 16, 3–17.

[Gree 95] Greenberg S. (1995) "Auditory processing of speech," in **Principles of Experimental Phonetics** (N. Lass ed.), Mosby, London, 362–407.

[Guen 95] Guenther F.H. (1995) "A modeling framework for speech motor development and kinematic articulator control," *Proc. XIII'th Int. Conf. Phonetic Sci.*, 2, 92–99.

[Gui 67] Guinan J. J. JR. and Peake W. T. (1967) "Middle-ear characteristics of anesthetized cats," *J. Acoust. Soc. Am.*, 41 (5), 1237–1261.

[Hag 83] Hagerman B. (1983) "Clinical measurements of speech reception threshold in noise," *Technical Audiology*, 107, 13.

[Hagg 70] Haggard M., Ambler S. and Callow M. (1970) "Pitch as a voicing cue," *J. Acoust. Soc. Am.*, 47, 613–617.

[Halle 62] Halle M. and Stevens K. (1962) "Speech recognition: A model and a program for research," *IRE Trans. Information Theory*, 7, 155–159.

[Hans 97] Hanson H. (1997) "Glottal characteristics of female speakers: Acoustic correlates," *J. Acoust. Soc. Am.*, 101, 466–481.

[Hans 98] Hansen J. and Chappell D. (1998) "An auditory-based distortion measure with application to concatenative speech synthesis", *IEEE Trans. Speech & Audio Processing*, 6(5), 489– 495.

[Hart 74] Hart J. (1974) "Discriminability of the size of pitch movements in speech," IPO Annual Progress Report, 9, 56–63.

[Hart 90] Hart J., Collier R. and Cohen A. (1990) **A Perceptual Study of Intonation: An experimental-phonetic approach to speech melody**, Cambridge University Press, Cambridge, U.K.

[Hawk 96a] Hawkins H., McMullen T., Popper A., and Fay R. (eds.) (1996) **Auditory Computation**, Springer-Verlag, New York.

[Hawk 96b] Hawkins H., McMullen T., Popper A. and Fay R. (1996) "Analysis and synthesis of cochlear mechanical function using models," in **Auditory Computation**, (A. Hubbard and D. Mountain, eds.), Springer-Verlag, New York, 121–156.

[Haykin 94] Haykin S. (1994) **Neural Networks – A Comprehensive Foundation**, Maxwell Macmillan, Toronto.

[Haykin 96] Haykin S. (1996) **Adaptive Filter Theory**, Prentice Hall, New Jersey.

[Hed 93] Hedrick M. and Ohde R. (1993) "Effect of relative amplitude of frication on perception of place of articulation," *J. Acoust. Soc. Am.*, 94, 2005–2026.

[Heinz 01a] Heinz M., Colburn S., and Carney L. (2001) "Evaluating auditory performance limits: I. One-parameter discrimination using a computational model for auditory nerve," *Neural Computation*, 13, 2273–2316.

[Heinz 01b] Heinz M., Colburn S., and Carney L. (2001) "Evaluating auditory performance limits: II. One-parameter discrimination with random-level variation," *Neural Computation*, 13, 2217–2338.

[Henke 96] Henke W. (1966) **Dynamic Articulatory Model of Speech Production Using Computer Simulation,** Ph.D. thesis, MIT.

[Herm 90] Hermansky H. (1990) "Perceptual linear prediction (PLP) analysis of speech," *J. Acoust. Soc. Am.*, 87, 1738–1752.

[Herm 93] Hermes D. (1993) "Pitch analysis," in **Visual Representations of Speech Signals**, (M. Cooke et al, eds.), Wiley, W. Sussex, 3–24.

[Herm 97] Hermes D. (1997) "Timing of pitch movements and accentuation of syllables in Dutch," *J. Acoust. Soc. Am.*, 102, 2390–2402.

[Herm 94] Hermansky H. and Morgan N. (1994) "RASTA processing of speech," *IEEE Trans. Speech & Audio Processing*, 2, 578–589.

[Herm 98] Hermansky H. (1998) "Should recognizers have ears?," *Speech Communication*, 25, 3–27.

[Hess 83] Hess W. (1983) **Pitch Determination of Speech Signals**, Springer Verlag, New York.

[Hew 92] Hewitt M. J., Meddis R. and Shackleton T. M. (1992) "A Computer model of a cochlear-Nucleus stellate cell: Responses to amplitude-modulated and pure-tone stimuli," *J. Acoust. Soc. Am.*, 91(4), 2096–2109.

[Hew 94] Hewitt M. J. and Meddis R. (1994) "A Computer model of amplitude-modulation sensitivity of single units in the inferior colliculus," *J. Acoust. Soc. Am.*, 95(4), 2145–2159.

[Hiki 70] Hiki S. (1970) "Control rule of the tongue movement for dynamic analog speech synthesis," *J. Acoust. Soc. Am.*, 47, 85-86.

[Hill 84] Hillenbrand J., Ingrisano D., Smith B. and Flege J. (1984) "Perception of the voiced-voiceless contrast in syllable-final stops," *J. Acoust. Soc. Am.*, 76, 18–26.

[Hill 93] Hillenbrand J. and Gayvert R. (1993) "Identification of steady-state vowels synthesized from the Peterson and Barney measurements," *J. Acoust. Soc. Am.*, 94, 668– 674.

[Hill 95] Hillenbrand J., Getty L., Clark M. and Wheeler K. (1995) "Acoustic characteristics of American English vowels," *J. Acoust. Soc. Am.*, 97, 3099–3111.

[Hira 90] Hirahara and Iwamida (1990) "Auditory spectrograms in HMM phoneme recognition," *Proceedings of the ICSLP*, 1, 381–384.

[Holm 86] Holmes J. (1986) "Normalization in vowel perception," in **Invariance and Variability in Speech Processes** (J. Perkell & D. Klatt, eds.), Erlbaum, Hillsdale, NJ, 346–357.

[Holm 95] Holmes W. and Russell M. (1995) "Speech recognition using a linear dynamic segmental HMM," *Proceedings of the Eurospeech*, 2, 1611-1641.

[Holm 96] Holmes, W. and Russell, M. (1996) "Modeling speech variability with segmental HMMs," *Proceedings of the Eurospeech*, 1, 447–450.

[Holm 97] Holmes J., Holmes W., and Garner P. (1997) "Using formant frequencies in speech recognition," *Proceedings of the Eurospeech*, 3, 2083–2086.

[Holm 99] Holmes W. and Russell M. (1999) "Probabilistic-trajectory segmental HMMs," *Computer Speech and Language*, 13, 3–27.

[Hon 98] Hon H., Acero A., Huang X., Liu J. andPlumpe M. (1998) "Automatic generation of synthesis units for trainable text-to-speech systems," *IEEE Proceedings of the ICASSP*, 1, 273–276.

[Hon 00] Hon H. and Wang K. (2000) "Unified frame and segment based models for automatic speech recognition," *IEEE Proceedings of the ICASSP*, 2, 1017–1020.

[Honda 94] Honda K. (1994) "Organization of tongue articulation for vowels," *Journal of Phonetics*, 24, 39–52.

[Honda 95] Honda M., and Kaburagi T. (1995) "A dynamical articulatory model using potential task representation," *Proceedings of the ICSLP*, 1, 179–182.

[Howe 83] Howell P. and Rosen S. (1983) "Production and perception of rise time in the voiceless affricate/fricative distinction," *J. Acoust. Soc. Am.*, 73, 976– 984.

[Huang 90] Huang X.D. and Jack M. (1988) "Semi-continuous hidden Markov models with maximum likelihood vector quantization," *Proc. IEEE Workshop on Speech Recognition.*

[Huang 90] Huang X.D. and Jack M. (1990) "Semi-continuous hidden Markov models for speech signals," *Computer Speech and Language*, 3(3).

[Huang 01] Huang X., Acero A. and Hon S. (2001) **Spoken Language Processing,** Prentice Hall, New York.

[Hub 93] Hubbard A. (1993) "A Traveling-wave amplifier model of the cochlea," *Science*, 259, 68–71.

[Hub 96] Hubbard A. and Mountain D. (1996) "Analysis and synthesis of cochlear mechanical functions using models," in **Auditory Computation** (Hawkins H., McMullen T., Popper A. and Fay R. eds.), Springer-Verlag, New York.

[Hugg 72a] Huggins A. (1972) "Just noticeable differences for segment duration in natural speech," *J. Acoust. Soc. Am.*, 51, 1270–1278.

[Hugg 72b] Huggins A. (1972) "On the perception of temporal phenomena in speech," *J. Acoust. Soc. Am.*, 51, 1279–1290.

[Hunt 96] Hunt A. and Black W. (1996) "Unit selection in a concatenative speech synthesis system using a large speech database," *IEEE Proceedings of the ICASSP*, 373–376.

[Huo 97] Huo Q. and Lee C.-H. (1997) "On-line adaptive learning of the continuous density hidden Markov model based on approximate recursive Bayes estimate," *IEEE Trans. Speech & Audio Processing*, 5(2), 161–172.

[Huo 99] Huo Q. and Ma B. (1999) "Irrelevant variability normalization in learning structure from data: a case study on decision-tree based HMM state tying," *IEEE Proceedings of the ICASSP*, 1, 577–580.

[Irv 92] Irvine D. R. F., Popper A. N. and Fay R. R. (1992) "Auditory brainstem processing," **The Mammalian Auditory Pathway: Neurophysiology**, 153–231, Springer-Verlag, New York.

[Iwa 95] Iwahashi N. and Sagisaka Y. (1995) "Speech segment network approach for optimization of synthesis unit set," *Computer Speech and Language*, 9, 335–352.

[Jaco 52] Jakobson R, Fant G. and Halle M. (1952) **Preliminaries to Speech Analysis**, MIT Press, Cambridge, MA.

[Jak 61] Jakobson R., Fant G. and Halle M. (1961) **Preliminaries to Speech Analysis: The Distinctive Features and Their Correlates**, MIT Press, Cambridge, MA.

[Jel 76] Jelinek F. (1976) "Continuous speech recognition by statistical methods," *Proceedings of the IEEE*, 64 (4), 532–556.

[Jen 91] Jenison R., Greenberg S., Kluender K. and Rhode W. (1991) "A composite model of the auditory periphery for the processing of speech based on the filter response functions of single auditory-nerve fibers," *J. Acoust. Soc. Am.*, 90(2), 773–786.

[Jenk 94] Jenkins J., W. Strange W. and Miranda S. (1994) "Vowel identification in mixed-speaker silent-center syllables," *J. Acoust. Soc. Am.*, 95, 1030–1041.

[Jens 93] Jensen J. T. (1993) **English Phonology**, John Benjamins Publishing Company, Amsterdam.

[Jia 98] Jiang H., Hirose K. and Huo Q. (1998) "A minimax search algorithm for CDHMM based robust continuous speech recognition," *Proceedings of the ICSLP*, 1, 389–392.

[Jia 99] Jiang H., Hirose K. and Huo Q. (1999) "Robust speech recognition based on Bayesian prediction approach," *IEEE Trans. Speech & Audio Processing*, 7(4), 426–440.

[Jia 01] Jiang H. and Deng L. (2001) "A Bayesian approach to speaker verification," *IEEE Trans. Speech & Audio Processing*, 9(8), 874–884.

[Jia 02] Jiang H. and Deng L. (2002) "A robust compensation strategy against extraneous acoustic variations in spontaneous speech recognition," *IEEE Trans. Speech & Audio Processing*, 10(1), 9–17.

[John 93] Johnson K., Ladefoged P. and Lindau M. (1993) "Individual differences in vowel production," *J. Acoust. Soc. Am.*, 94, 701–714.

[Juang 84] Juang B. H. (1984) "On the hidden Markov model and dynamic time warping for speech recognition - A unified view," *Bell Laboratories Technical Journal*, 63(7), 1213–1243.

[Juang 85] Juang B. and Rabiner L. (1985) "Mixture autoregressive hidden Markov models for speech signals," *IEEE Trans. Acoustics, Speech & Signal Processing,* ASSP-33, 1404–1413.

[Juang 86] Juang B., Levinson S. and Sondhi M. (1986) "Maximum likelihood estimation for multivariate mixture observations of Markov chain," *IEEE Trans. Information Theory*, IT-32, 307–309.

[Juang 88] Juang B. and Rabiner L. R. (1988) "Mixture autoregressive hidden Markov models for speech signals," *IEEE Trans. Acoustics, Speech & Signal Processing,* 33(6), 1404–1413.

[Juang 92] Juang B. and Katagiri S. (1992) "Discriminative learning for minimum error classification," *IEEE Trans. Acoustics, Speech & Signal Processing,* ASSP-40(12), 3043–3054.

[Juang 00] Juang F. and Furui S. (eds.) (2000) *Proceedings of the IEEE*, (special issue), Vol. 88.

[Kab 96] Kaburagi T. and Honda, M. (1996) "A model of articulator trajectory formation based on the motor tasks of vocal-tract shapes," *J. Acoust. Soc. Am.*, 99, 3154–3170.

[Kata 98] Katagiri S., Juang B., Lee, C. (1998) "Pattern recognition using a family of design algorithms based upon the generalized probabilistic descent method," *Proceedings of the IEEE*, 86(11), 2345–2373.

[Katam 89] Katamba F. (1989) **An Introduction to Phonology**, Longman, London.

[Kay 88] Kay S. M. (1988) **Modern Spectral Estimation**, Prentice Hall, NJ.

[Kay 93] Kay S. (1993) **Fundamentals of Statistical Signal Processing: Estimation Theory**, Prentice Hall, Upper Saddle River, NJ.

[Kay 98] Kay S. (1998) **Fundamentals of Statistical Signal Processing: Detection Theory**, Prentice Hall, Upper Saddle River, NJ.

[Keat 90] Keating P. (1990) "The window model of coarticulation: articulatory evidence," in **Papers in Laboratory Phonology I**, Chapter 26 (J. Kingston and M. Beckman eds.), Cambridge University Press, Cambridge, UK, 451–470.

[Keid 75] Keidel W. and Neff W. (1975) **Handbook of Sensory Physiology of Auditory System,** Springer-Verlag, Berlin.

[Keid 76] Keidel W., Neff W., and Geisler C. D. (1976) "Mathematical models of the mechanics of the inner ear," in **Handbook of Sensory Physiology of Auditory System**, Springer-Verlag, New York, 391–415.

[Kel 84] Kelso J., Tuller B., Vatikiotis-Bateson E., and Fowler C. (1984) "Functionally specific articulatory cooperation following jaw perturbations during speech: evidence for coordinative structures," *J. Exp. Psych. Human Percep. & Perf.* 10, 812–832.

[Kel 86] Kelso J., Saltzman E., and Tuller B. (1986) "The dynamical perspectives on speech production: data and theory," *Journal of Phonetics*, 14, 29–59.

[Ken 90] Kenny P., Lennig M. and Mermelstein P. (1990) "A linear predictive HMM for vector-valued observations with applications to speech recognition," *IEEE Trans. Acoustics, Speech & Signal Processing,* ASSP-38(2), 220–225.

[Kent 82] Kent R. and Murray A. (1982) "Acoustic features of infant vocalic utterances at 3, 6, and 9 months," *J. Acoust. Soc. Am.*, 72, 353–365.

[Kent 95] Kent R., Adams S. and Turner G. (1995) "Models of speech production," in **Principles of Experimental Phonetics**, (Lass N. ed.), Mosby, London, 3–45.

[Kerk 93] Kerkhoff J. and Boves L. (1993) "Designing control rules for a serial pole-zero vocal tract model," *Proceedings of the Eurospeech*, 893–896.

[Kew 82] Kewley-Port D. (1982) "Measurement of formant transitions in naturally produced stop consonant-vowel syllables," *J. Acoust. Soc. Am.*, 70, 379–389.

[Kew 98] Kewley-Port D. (1998) "Auditory models of formant frequency discrimination for isolated vowels," *J. Acoust. Soc. Am.*, 103, 1654–1666.

[Key 94] Keyser J. and Stevens, K. (1994) "Feature geometry and the vocal tract," *Phonology*, 11(2), 207–236.

[Khei 91] Kheirallah I. (1991) **Experiments on a Nonlinear Transmission-Line Cochlear Model with Applications to Speech Processing,** Master thesis, University of Waterloo, Canada.

[Kim 91] Kim D. O., Parham K., Sirianni J. G. and Chang S. O. (1991) "Spatial response profiles of posteroventral cochlear nucleus neurons and auditory-nerve fibers in unanesthetized decerebrate cats: Response to pure tones," *J. Acoust. Soc. Am.*, 89(6), 2804–2817.

[Kim 94] Kim C.-J. (1994) "Dynamic linear models with Markov switching," *Journal of Econometrics*, 60, 1–22.

[Kim 98] Kim N.S. (1998) "Nonstationary environment compensation based on sequential estimation," *IEEE Signal Processing Letters*, 5, 57–60.

[King 97] Kingsbury B., Morgan N. and Greenberg S. (1998) "Robust speech recognition using the modulation spectrogram," *Speech Communication*, 25, 117–132.

[King 98] King S., Stephenson T., Isard S., Taylor P., and Strachan A. (1998) "Speech recognition via phonetically featured syllables," *Proceedings of the ICSLP*, 1, 1031–1034.

[Kirc 96] Kirchhoff K. (1996) "Syllable-level desynchronization of phonetic features for speech recognition," *Proceedings of the ICSLP*, 2274–2276.

[Kirc 98] Kirchhoff K. (1998) "Robust speech recognition using articulatory information," *ICSI Technical Report*, TR-98-037.

[Kitag 96] Kitagawa G. and W. Gersch W. (1996) **Smoothness Priors Analysis of Time Series**, Springer, New York.

[Klatt 73] Klatt D. (1973) "Discrimination of fundamental frequency contours in synthetic speech: Implications for models of pitch perception," *J. Acoust. Soc. Am.*, 53, 8- 16.

[Klatt 75a] Klatt D. and Cooper W. (1975) "Perception of segment duration in sentence contexts," in **Structure and Process in Speech Perception** (A. Cohen & S. Nooteboom, eds.), Springer-Verlag, New York, 69–89.

[Klatt 75b] Klatt D. (1975) "Vowel lengthening is syntactically determined in a connected discourse," *Journal of Phonetics*, 3, 129–140.

[Klatt 76] Klatt D. (1976) "Structure of a phonological rule component for a synthesis-by-rule program," *IEEE Trans. Speech & Audio Processing*, 24, 391–398.

[Klatt 80] Klatt D. (1980) "Software for a cascade/parallel formant synthesizer," *J. Acoust. Soc. Am.*, 67, 971–995.

[Klatt 87a] Klatt D. (1987) "Review of text-to-speech conversion for English," *J. Acoust. Soc. Am.*, 82, 737–793.

[Klatt 87b] Allen A., Hunnicutt S., and Klatt D. (1987) **From Text to Speech: The MITalk System**, Cambridge University Press, Cambridge, UK.

[Klatt 89] Klatt D. (1989) "Review of selected models of speech perception," in **Lexical Representation and Process**, (W. Marslen-Wilson ed.), 169– 226.

[Klein 70] Klein W., Plomp R. and Pols L. (1970) "Vowel spectra, vowel spaces, and vowel identification," *J. Acoust. Soc. Am.,* 48, 999–1009.

[Kohn 88] Kohn R. and Ansley C. (1988) "Equivalence between Bayesian smoothness priors and optimal smoothing for function estimation," in **Bayesian Analysis of Time Series and Dynamic Models**, (J. Spall ed.), Marcel Dekker, New York, 393–430.

[Kooij 71] Kooij J. (1971) **Ambiguity in Natural Language**, North Holland, Amsterdam.

[Kroger 95] Kroger B., Schroder G., and Opgen C. (1995) "A gesture-based dynamic model describing articulatory movement data," *J. Acoust. Soc. Am.,* 98, 1878–1889.

[Kuhn 90] Kuhn R. and DeMori R. (1990) "A cache-based natural language model for speech recognition," *IEEE Trans. Pattern Analysis & Machine Intelligence*, PAMI-12(6) 570–583.

[Labo 95] Laboissière R., Ostry D.J., and Perrier P. (1995) "A model of human jaw and hyoid motion and its implications for speech production," *Proc. XIII'th Int. Conf. Phonetic Sci.*, 2, 60–67.

[Ladd 84] Ladd D. R. (1984) "Declination: A review and some hypotheses," *Phonology Yearbook*, 1, 53–74.

[Ladd 85] Ladd D. R., Silverman K., Tolkmitt F., Bergmann G. and Scherer K. (1985) "Evidence for the independent function of intonation contour type, voice quality, and F0 range in signaling speaker affect," *J. Acoust. Soc. Am.,* 78, 435–444.

[Lafon 68] Lafon J-C. (1968) "Auditory basis of phonetics," in **Manual of Phonetics** B. Malmberg, ed., North Holland (Amsterdam), 76–104.

[Lamel 81] Lamel L., Rabiner L. R, Rosenberg A. E. and Wilpon J. G. (1981) "An Improved endpoint detection for isolated word recognition," *IEEE Trans. Acoustics, Speech & Signal Processing,* ASSP-29(4), 777–785.

[Lane 91] Lane H. and Webster J. (1991) "Speech deterioration in postlingually deafened adults," *J. Acoust. Soc. Am.,* 89, 859–866.

[Lang 92] Langner G. (1992) "Periodicity coding in the auditory system," *Hearing Research*, 60, 115–142.

[Lee 89] Lee P. (1989) **Bayesian Statistics: An Introduction**, Oxford University Press, New York.

[Lee 91] Lee C-H., Lin C. H. and Juang B. H. (1991) "A study on speaker adaptation of the parameters of continuous density hidden Markov models", *IEEE Trans. Signal Processing*, 39(4), 806-814.

[Lee 96a] Lee C.-H. and Gauvain J.-L. (1996) "Bayesian adaptive learning and MAP estimation of HMM" in **Automatic Speech and Speaker Recognition – Advanced Topics,** C. Lee, F. Soong, and K. Paliwal (eds.), Kluwer Academic, Boston, 83-107.

[Lee 96] Lee C-H., Soong F., and Paliwal K. (eds.) (1996) **Automatic Speech and Speaker Recognition-Advanced Topics,** Kluwer Academic, Boston.

[Legg 95] Leggetter C. and Woodland P. (1995) "Maximum likelihood linear regression for speaker adaptation of continuous density hidden Markov models," *Computer Speech and Language*, 9, 171-185.

[Leh 70] Lehiste I. (1970) **Suprasegmentals**, MIT Press, Cambridge, MA.

[Leh 76] Lehiste I., Olive J. and Streeter L. (1976) "The role of duration in disambiguating syntactically ambiguous sentences," *J. Acoust. Soc. Am.*, 60, 119–1202.

[Leh 77] Lehiste I. (1977) "Isochrony reconsidered," *Journal of Phonetics*, 5, 253–263.

[Leon 72] Leon P. and Martin P. (1972) "Machines and measurements," in **Intonation** (D. Bolinger, ed.), Penguin, Harmondsworth, UK, 30–47.

[Lev 83] Levinson S., Rabiner L., and Sondhi M. (1983) "An introduction to the application of the theory of probabilistic functions of a Markov process to automatic speech recognition," *Bell System Technical J.*, 62, 1035–1074.

[Levi 85] Levinson S. E. (1985) "Structural methods in automatic speech recognition," *Proceedings of the IEEE*, 73 (11), 1625–1650.

[Levi 90] Levinson S. E. and Roe D. B. (1990) "A Prospective on speech recognition," *IEEE Communications Magazine*, 28 (1), 28–34.

[Lib 56] Liberman A., Delattre P., Gerstman L. and Cooper F. (1956) "Tempo of frequency change as a cue for distinguishing classes of speech sounds," *J. Exp. Psychol.*, 52, 127–137.

[Lib 58] Liberman A., Delattre P. and Cooper F. (1958) "Some cues for the distinction between voiced and voiceless stops in initial position," *Language and Speech*, 1, 153–167.

[Lib 67] Liberman A., Cooper F., Shankweiler D. and Studdert- Kennedy M. (1967) "Perception of the speech code," *Psychology Review*, 74, 431–461.

[Lib 82] Liberman M. C. (1982)"The cochlear frequency map for the cat: labeling auditory nerve fibers of known characteristic frequency," *J. Acoust. Soc. Am.*, 72, 1441–1449.

[Lib 85] Liberman A. and Mattingly I. (1985) "The motor theory of speech perception revised," *Cognition*, 21, 1–36.

[Lien 73] Lien M. D. and Cox J. R. (1973) "A mathematical model for the mechanics of the cochlea," Technical Report, Sever Institute of Washington University, St. Louis.

[Lim 78a] Lim J.S. and Oppenheim A. (1978) "All-pole modeling of degraded speech," *IEEE Trans. Speech & Audio Processing*, 26, 197–210.

[Lim 78b] Lim. J (1978) "Evaluation of a correlation subtraction method for enhancing speech degraded by additive white noise," *IEEE Trans. Acoustics, Speech & Signal Processing,* ASSP-26, 471–472.

[Lim 88] Lim J. and Oppenheim A. (1988) **Advanced Topics in Signal Processing**, Prentice Hall, Englewood Cliffs, NJ.

[Lind 80] Lindblom B., Oehman S., Schroeder M. R., Atal B. S. and Hall J. L. (1980) "Objective measures of certain speech signal degradations based on masking properties of human auditory perception", in **Frontiers of Speech Communication Research**, Academic Press, New York, 217–229.

[Lind 63] Lindblom B. (1963) "On vowel reduction," Technical Report No. 29, The Royal Institute of Technology, Speech Transmission Laboratory, Stockholm, Sweden.

[Lind 90] Lindblom B. (1990) "Explaining phonetic variation: A sketch of the H&H theory," in *Proc. NATO Workshop on Speech Production and Speech Modeling*, (W. Hardcastle and A. Marchal eds.), 403–439.

[Lind 96] Lindblom B. (1996) "Role of articulation in speech perception: Clues from production," *J. Acoust. Soc. Am.,* 99(3), 1683–1692.

[Lipo 82] Liporace L. A. (1982) "Maximum likelihood estimation for multivariate observations of Markov sources," *IEEE Trans. Information Theory*, IT-28(5), 729–734.

[Lisk 78] Lisker L., Liberman A., Erickson D., Dechovitz D. and Mandler R. (1978) "On pushing the voicing-onset-time (VOT) boundary" *Language and Speech*, 20, 209–216.

[Ljo 93] Ljolie A. and Riley M. (1993) "Automatic segmentation of speech for TTS," *Proceedings of the Eurospeech*, 1445–1448.

[Log 89] Logan J., Greene B. and Pisoni D. (1989) "Segmental intelligibility of synthetic speech produced by rule," *J. Acoust. Soc. Am.,* 86, 566–581.

[Lyon 96] Lyon R. and Shamma S. (1996) "Auditory representation of timbre and pitch," in **Auditory Computation**, (H. Hawkins et al., eds.), Springer, New York, 221–270.

[Ma 99] Ma J. and L. Deng L. (1999) "Optimization of dynamic regiems in a statistical hidden dynamic model for spontaneous speech recognition," *Proceedings of the Eurospeech*, 2, 1339–1342.

[Ma 00] Ma J. and Deng L. (2000) "A path-stack algorithm for optimizing dynamic regimes in a statistical hidden dynamic model of speech," *Computer Speech and Language*, 14, 101–104.

[Ma 02] Ma J. and Deng L. (2002) "A mixture linear model with target-directed dynamics for spontaneous speech recognition," *IEEE Proceedings of the ICASSP*, 1, 961–964.

[Macd 76] Macdonald N. (1976) "Duration as a syntactic boundary cue in ambiguous sentences," *IEEE Proceedings of the ICASSP*, 569–572.

[MacG 87] MacGregor R. J. (1987) **Neural and Brain Modeling**, Academic Press, New York.

[Mack 83] Mack M. and Blumstein S. (1983) "Further evidence of acoustic invariance in speech production: The stop-glide contrast," *J. Acoust. Soc. Am.*, 73, 1739–1750.

[Madd 94] Madden J. (1994) "The role of frequency resolution and temporal resolution in the detection frequency modulation," *J. Acoust. Soc. Am.*, 95, 454–462.

[Makh 75] Makhoul J. (1975) "Linear prediction: A tutorial review," *Proceedings of the IEEE*, 63(4), 561–580.

[Manis 83] Manis P. B. and Brownell W. E. (1983) "Synaptic organization of eighth nerve afferents to cat dorsal cochlear nucleus," *Journal of Neurophysiology*, 50, 1156–1181.

[Mari 89] Mariani J. (1989) "Recent advances in speech processing," *IEEE Proceedings of the ICASSP*, 1, 429–440.

[Markel 76] Markel J. D. and Gray A. H. (1976) **Linear Prediction of Speech**, Springer-Verlag, Berlin.

[Mars 85] Marslen-Wilson W. (1985) "Speech shadowing and speech comprehension," *Speech Communication*, 4, 55–73.

[Mass 80] Massaro D. and Oden G. (1980) "Evaluation and integration of acoustic features in speech perception," *J. Acoust. Soc. Am.*, 67, 996–1013.

[Math 99] Mathews J. H. and Fink K. D. (1999) **Numerical Methods — Using MATLAB**, 3rd editon, Prentice Hall, New Jersey.

[Matt 70] Mattingly I. (1970) "Speech synthesis for phonetic and phonological models," Haskins Labs. Stat. Rep. Speech, Vol. SR–23, 117–149; also in **Current Trends in Linguistics**, Vol. XII (T. Sebeok, ed.), Mouton, The Hague.

[McA 98] McAllister D., Gillick L., Scattone F., and Newman M. (1998) "Fabricating conversational speech data with acoustic models: A program to examine model-data mismatch," *Proceedings of the ICSLP*, 3, 1847–1850.

[McC 74] McCandless S. (1974) "An algorithm for automatic formant extraction using linear prediction spectra," *Proceedings of the IEEE*, 22, 135–141.

[McG 77] McGonegal C., Rabiner L., and Rosenberg A. (1977) "A subjective evaluation of pitch detection methods using LPC synthesized speech," *IEEE Trans. Speech & Audio Processing*, 25, 221–229.

[McGow 94] McGowan R. (1994) "Recovering articulatory movement from formant frequency trajectories using task dynamics and a genetic algorithm: Preliminary model tests," *Speech Communication*, 14, 19–48.

[McGow 96] McGowan R. and Lee M. (1996) "Task dynamic and articulatory recovery of lip and velar approximations under model mismatch conditions," *J. Acoust. Soc. Am.*, 99, 595–608.

[McGow 97] McGowan R. and Faber A. (1997) "Speech production parameters for automatic speech recognition," *J. Acoust. Soc. Am.*, 101(1), 28.

[McL 97] McLachlan G. and Krishnan T. (1997) **The EM Algorithm and Extensions,** Wiley, New York.

[Medd 91] Meddis R. and Hewitt M. (1991) "Virtual pitch and phase sensitivity of a computer model of the auditory periphery", *J. Acoust. Soc. Am.,* 89, 2866–2882.

[Medd 97] Meddis R. and O'Mard L. (1997) "A unitary model of pitch perception," *J. Acoust. Soc. Am.,* 102, 1811–1820.

[Mend 95] Mendel J. M. (1995) **Lessons in Estimation Theory for Signal Processing, Communications, and Control**, Prentice Hall, New Jersey.

[Meng 91] Meng H. (1991) **The Use of Distinctive Features for Automatic Speech Recognition,** Master thesis, MIT.

[Merm 73] Mermelstein P. (1973) "Articulatory model for the study of speech production," *J. Acoust. Soc. Am.,* 53, 1070–1082.

[Merm 78] Mermelstein P. (1978) "Difference limens for formant frequencies of steady-state and consonant-bound vowels," *J. Acoust. Soc. Am.,* 63, 572–580.

[Mik 84] Mikami N. and Ohba R. (1984) "Pole-zero analysis of voiced speech using group delay characteristics," *IEEE Trans. Speech & Audio Processing*, 32, 1095–1097.

[Mill 55] Miller G. and Nicely P. (1955) "An analysis of perceptual confusions among some English consonants," *J. Acoust. Soc. Am.,* 27, 338–352.

[Mill 77] Miller J. (1977) "Nonindependence of feature processing in initial consonants," *Journal of Speech & Hearing Research*, 20, 510–518.

[Mill 81] Miller J. (1981) "Effects of speaking rate on segmental distinctions," in **Perspectives on the Study of Speech**, (P. Eimas & J. Miller, eds.), Erlbaum, Hillsdale, NJ, 39–74.

[Mill 83] Miller M. I. and Sachs M. B. (1983) "Representation of stop consonants in the discharge pattern of auditory-nerve fibers," *J. Acoust. Soc. Am.,* 74, 502–517.

[Mill 84] Miller J. and Shermer T. (1984) "A distinction between the effects of sentential speaking rate and semantic congruity on word identification," *Perception and Psychophys.*, 36, 329–337.

[Mitra 98] Mitra S. (1998) **Digital Signal Processing: A Computer- Based Approach**, McGraw-Hill, NY.

[Mol 56] Mol H. and Uhlenbeck E. (1956) "The linguistic relevance of intensity in stress," *Lingua*, 5, 205–213.

[Moll 83] Møller A. R. (1983) **Auditory Physiology**, Academic Press, Toronto.

[Mon 83] Monsen R. and Engebretson A.M. (1983) "The accuracy of formant frequency measurements: A comparison of spectrographic analysis & linear prediction," *J. Speech & Hearing Research*, 26, 89–97.

[Moon 94] Moon S-J. and Lindblom B. (1994) "Interaction between duration, context, and speaking style in English stressed vowels," *J. Acoust. Soc. Am.*, 96, 40–55.

[Moore 99] Moore R. (1999) "Using natural-language knowledge sources in speech recognition," in **Computational Models of Speech Pattern Processing** (NATO ASI Series) (K. Ponting Ed.), Springer, 304-327.

[Moo 84] Moore B., Glasberg B., and Shailer M. (1984) "Frequency and intensity difference limens for harmonics within complex tones," *J. Acoust. Soc. Am.*, 75, 550–561.

[More 96a] Moreno P. (1996) **Speech Recognition in Noisy Environments**, Ph.D. thesis, CMU.

[More 96b] Moreno P., Raj B., and Stern R. (1996) "A vector Taylor series approach for environment-independent speech recognition," *IEEE Proceedings of the ICASSP*, 1, 733–736.

[Morg 94] Morgan N. and Bourlard H. (1994) **Connectionist Speech Recognition - A hybrid approach**, Kluwer Academic Publishers, Boston.

[Mort 65] Morton J. and Jassem W. (1965) "Acoustic correlates of stress," *Language and Speech*, 8, 159–181.

[Murr 93] Murray I. and Arnott J. (1993) "Toward the simulation of emotion in synthetic speech: A review of the literature on human vocal emotion," *J. Acoust. Soc. Am.*, 93, 1097–1108.

[Nab 97] Nábalek A. and Ovchinnikov A. (1997) "Perception of nonlinear and linear formant trajectories," *J. Acoust. Soc. Am.*, 101, 488–497.

[Near 87] Nearey T. and Shammass S. (1987) "Formant transitions as partly distinctive invariant properties in the identification of voiced stops," *Canadian Acoustics*, 15, 17–24.

[Neely 81] Neely S. T. (1981) "Finite-difference solution of a two-dimensional mathematical model of the cochlea," *J. Acoust. Soc. Am.*, 69, 1386–1393.

[Ney 00] Ney H. and Ortmanns S. (2000) "Progress in dynamic programming search for LVCSR", *Proceedings of the IEEE*, , 88(8), 1224–1240.

[Nils 94] Nilsson M., Soli S. and Sullivan J. (1994) "Development of the Hearing in Noise Test for the measurement of speech reception thresholds in quiet and in noise," *J. Acoust. Soc. Am.*, 95, 1085–1099.

[Niy 91] Niyogi P. (1991) **Modeling Speaker Variability and Imposing Speaker Constraints in Phonetic Classification,** Master thesis, MIT.

[Nock 00] Nock H. and Young S. (2000) "Loosely coupled HMMs for ASR: A preliminary study," Technical Report TR386, Cambridge University.

[Nut 81] Nuttall A. (1981) "Some windows with very good sidelobe behavior," *IEEE Trans. Speech & Audio Processing*, 29, 84–91.

[O'Br 01] O'Brien D. and Monaghan A. I. C. (2001) "Concatenative synthesis based on a harmonic model", *IEEE Trans. Speech & Audio Processing*, 9 (1), 11–20.

[O'Sh 74] O'Shaughnessy D. (1974) "Consonant durations in clusters," *IEEE Trans. Acoustics, Speech & Signal Processing*, ASSP-22, 282–295.

[O'Sh 83] O'Shaughnessy D. and Allen J. (1983) "Linguistic modality effects on fundamental frequency in speech," *J. Acoust. Soc. Am.*, 74, 1155–1171.

[O'Sh 00] O'Shaughnessy D. (2000) **Speech Communications: Human and Machine**, IEEE Press, Piscataway, NJ.

[Oba 91] Obara K. and Hirahara T. (1991) "Auditory front-end in DTW word recognition under noisy, reverberant, and multi-Speaker conditions," *J. Acoust. Soc. Am.*, 90(4) (Pt.2), S2274.

[Oha 83] Ohala J. (1983) "The origin of sound patterns in vocal tract constraints," in **The Production of Speech**, (P. MacNeilage, ed.), Springer-Verlag, NY,189–216.

[Ohm 66] Ohman S. (1966) "Co-articulation in VCV utterances: Spectrographic measurements," *J. Acoust. Soc. Am.*, 39, 151–168.

[Opp 81] Oppenheim A. and Lim J. (1981) "The importance of phase in signals," *Proceedings of the IEEE*, 69, 529–541.

[Opp 89] Oppenheim A. and Schafer R. (1989) **Discrete-Time Signal Processing**, Prentice-Hall, Englewood Cliffs, NJ.

[Opp 99] Oppenheim A., Schafer R., and Buck J. (1999) **Discrete-Time Signal Processing**, 2nd Ed., Prentice Hall, Upper Saddle River, NJ.

[Osen 81] Osen K. K. and Mugnaini E. (1981) "Neuronal circuits in the dorsal cochlear nucleus," **Neuronal Mechanisms of Hearing**, (J. Syka and L. Aitkin eds.), Plenum Press, New York, 119–125.

[Osen 88] Osen K. K. (1988) "Anatomy of the mammalian cochlear nuclei: a Review," in **Auditory Pathway: Structure and Function**, (J. Syka and R. B. Masterton eds.), Plenum Press, New York, 65–75.

[Ost 96] Ostendorf, M., Digalakis, V. and Kimball, O. (1996) "From HMMs to segment models: A unified view of stochastic modeling for speech recognition," *IEEE Trans. Speech & Audio Processing*, 4(5), 360–378.

[Ost 00] Ostendorf M. (2000) "Moving beyond the beads-on-a-string model of speech," *Proc. IEEE Automatic Speech Recognition and Understanding Workshop.*

[Paez 72] Paez M. and Glisson T. (1972) "Minimum mean squared error quantization in speech," *IEEE Trans. Communications*, 20, 225–350.

[Pal 84] Paliwal K. and Aarskog A. (1984) "A comparative performance evaluation of pitch estimation methods for TDHS/Sub-band coding of speech," *Speech Communication*, 3, 253–259.

[Park 84] Parker E. and Diehl R. (1984) "Identifying vowels in CVC syllables: Effects of inserting silence and noise," *Perception and Psychophys.*, 36, 369–380.

[Pars 76] Parsons T.W. (1976) "Separation of speech from interfering of speech by means of harmonic selection," *J. Acoust. Soc. Am.*, 60, 911–918.

[Past 82] Pastore R., Harris L. and Kaplan J. (1982) "Temporal order identification: Some parameter dependencies," *J. Acoust. Soc. Am.*, 71, 430–436.

[Patt 78] Patterson R. and Green D. (1978) "Auditory masking," in **Handbook of Perception**, IV (E. Carterette & M. Friedman, eds.), Academic, New York, 337-361.

[Patt 91] Patterson R. and Holdsworth J. (1991) "A functional model of neural activity patterns and auditory images", in **Advances in Speech, Hearing, and Language Processing**, Vol. 3, (W. Ainsworth, ed.), JAI Press, London.

[Patt 95] Patterson R., Allerhand M. and Giguere C. (1995) "Time-domain modeling of peripheral auditory processing: A modular architecture and a software platform," *J. Acoust. Soc. Am.*, 98, 1890-1894.

[Pavl 97] Pavlova E., Pavlov Y., Sproat R., Shih C. and van Santen P. (1997) "Bell Laboratories Russian text-to-speech system," *Computer Speech and Language*, 6, 37-76.

[Pavl 84] Pavlovic C. and Studebaker G. (1984) "An evaluation of some assumptions underlying the articulation index," *J. Acoust. Soc. Am.*, 75, 1606-1612.

[Pavl 99] Pavlovic V. (1999) **Dynamic Bayesian Networks for Information Fusion with Applications to Human-Computer Interfaces**, Ph.D. thesis, University of Illinois at Urbana-Champaigne.

[Payt 88] Payton K. L. (1988) "Vowel processing by a model of the auditory periphery: A comparison to Eighth-nerve responses," *J. Acoust. Soc. Am.*, 83, 145-162.

[Payt 94] Payton K., Uchanski R. and Braida L. (1994) "Intelligibility of conversational and clear speecc in noise and reverberation for listeners with normal and impaired hearing," *J. Acoust. Soc. Am.*, 95, 1581-1592.

[Pea 01] Pearce D. (Ed.) (2001) *ESE2 Special sessions on noise robust recognition, Proceedings of the Eurospeech*, 1, 421-442.

[Perk 67] Perkel D. J., Gerstein G. L. and Moore G. P. (1967) "Neuronal spike trains and stochastic point processes II. Simultaneous spike trains," *Journal of Biophysics*, 7, 419-440.

[Perk 80] Perkell J.S. (1980) "Phonetic features and the physiology of speech production," in **Language Production**, (B. Butterworth ed.), Academic Press, London.

[Perk 86] Perkell J. and Klatt D. (1986) **Invariance and Variability in Speech Processes**, Erlbaum, Hillsdale, New York.

[Perk 95] Perkell J.S., Matthies M.L., Svirsky M.A., and Jordan M.I. (1995) "Goal-based speech motor control: a theoretical framework and some preliminary data," *Journal of Phonetics*, 23, 23-35.

[Perk 97] Perkell J.S. (1997) "Articulatory processes," in **The Handbook of Phonetic Sciences**, (W. J. Handcastle and J. Lavar eds.), Blackwell, Cambridge, U.K., 333-370.

[Perr 96a] Perrier P., Ostry D., and Laboissière R. (1996) "The equilibrium point hypothesis and its application to speech motor control," *J. Speech & Hearing Research*, 39, 365-378.

[Perr 96b] Perrier P. et al. (eds.) (1996) *Proceedings of the first ESCA Tutorial & Research Workshop on Speech Production Modeling – From Control Strategies to Acoustics*, Autrans, France.

[Pet 52] Peterson G. and Barney H. (1952) "Control methods used in a study of vowels," *J. Acoust. Soc. Am.*, 24, 175–184.

[Pet 60] Peterson G. and Lehiste I. (1960) "Duration of syllable nuclei in English," *J. Acoust. Soc. Am.*, 32, 693–703.

[Pic 99] Picone J., Pike S., Reagan R., Kamm T., Bridle J., Deng L., Ma Z., Richards H., Schuster M. (1999) "Initial evaluation of hidden dynamic models on conversational speech," *IEEE Proceedings of the ICASSP*, 1, 109–112.

[Pick 82] Pickles J. (1982) **An Introduction to the Physiology of Hearing**, Academic, London.

[Pick 88] Pickles J. (1988) **An Introduction to the Physiology of Hearing**, 2nd Ed., Academic Press, U.K., 256–296.

[Pier 79] Pierrehumbert J. (1979) "The perception of fundamental frequency declination," *J. Acoust. Soc. Am.*, 66, 363–369.

[Piso 97] Pisoni D. (1997) "Some thoughts on 'normalization' in speech perception," in **Talker Variability in Speech Processing**, (K. Johnson & J. Mullennix eds.), Academic Press, San Diego, CA, 9–32.

[Poll 63] Pollack I. and Pickett J. (1963) "Intelligibility of excerpts from conversational speech," *Language and Speech*, 6, 165–171.

[Pont 89] Pont M.J. and Damper R.I. (1989) "Software for a computational model of afferent neural activity from the cochlea to dorsal acoustic stria," VSSP Research Group Technical Report, University of Southampton.

[Pont 91] Pont M. and Damper R. (1991) "A computational model of afferent neural activity from the cochlea to the dorsal acoustic stria," *J. Acoust. Soc. Am.*, 89, 1213–1228.

[Popp 92] Popper A., Fay R., Greenberg S. and Rhode W. (1992) "Physiology of the cochlear nuclei," in **The Mammalian Auditory Pathway: Neurophysiology**, Springer-Verlag, New York, 94–152.

[Porat 97] Porat B. (1997) **A Course in Digital Signal Processing**, John Wiley & Sons, New York.

[Pori 82] Poritz A. B. (1982) "Linear predictive hidden Markov models and the speech signal," *IEEE Proceedings of the ICASSP*, 1291–1294.

[Pori 88] Poritz A. B. (1988) "Hidden Markov models: A guided tour," *IEEE Proceedings of the ICASSP*, 1, 7–13.

[Pow 96] Power M. and Braida L. (1996) "Consistency among speech parameter vectors: Application to predicting speech intelligibility," *J. Acoust. Soc. Am.*, 100, 3882–3898.

[Price 91] Price P., Ostendorf M., Shattuck-Hufnagel S. and Fong C. (1991) "The use of prosody in syntactic disambiguation," *J. Acoust. Soc. Am.*, 90, 2956–2970.

[Quac 88] Quackenbush S., Barnwell T. and Clements M. (1988) **Objective Measures for Speech Quality**, Prentice-Hall, Englewood Cliffs, NJ.

[Quat 92] Quatieri T. and McAulay R. (1992) "Shape invariant time- scale and pitch modification of speech," *IEEE Trans. Acoustics, Speech & Signal Processing,* ASSP-40, 497–510.

[Rab 68] Rabiner L. (1968) "Speech synthesis by rule: an acoustic domain approach," *Bell Systems Technical Journal*, 47, 17–37.

[Rab 75] Rabiner L. and Gold B. (1975) **Theory and Application of Digital Signal Processing**, Prentice-Hall, Englewood Cliffs, NJ.

[Rab 76] Rabiner L., Cheng M., Rosenberg A., and McGonegal C. (1976) "A comparative performance study of several pitch detection algorithms," *IEEE Trans. Acoustics, Speech & Signal Processing,* ASSP-24, 300–418.

[Rab 79] Rabiner L. and Schafer R. (1979) **Digital Processing of Speech Signals**, Prentice-Hall, Englewood Cliffs, NJ.

[Rab 83] Rabiner L. R., Levinson S. E. and Sondhi M. M. (1983) "On the application of vector quantization and hidden Markov models to speaker-independent, isolated word recognition ," *Bell Systems Technical Journal*, 62(4), 1075–1105.

[Rab 86] Rabiner L. R. and Juang B. H. (1986) "An introduction to hidden Markov models," *IEEE Trans. Acoustics, Speech & Signal Processing,* ASSP-33(3), 561–573.

[Rab 89] Rabiner L. R. (1989) "A tutorial on hidden Markov models and selected applications in speech recognition," *Proceedings of the IEEE*, 77(2), 257–285.

[Rab 93] Rabiner L. R. and Juang B. H. (1993) **Fundamentals of Speech Recognition**, Prentice-Hall, Englewood Cliffs, NJ.

[Rao 84] Rao S. (1984) **Optimization: Theory and Applications**, Wiley Eastern Ltd., New Delhi.

[Rasm 99] Rasmussen, C. E. (1999) "The infinite Gaussian mixture model," *Advances in Neural Information Processing Systems*, 1, MIT Press, Cambridge, MA.

[Rathi 97] Rathinavelu C. and Deng L. (1997) "HMM-based speech recognition using state-dependent, discriminatively derived transforms on Mel-warped DFT features," *IEEE Trans. Speech & Audio Processing*, 5, 243–256.

[Rathi 98] Rathinavelu C. and Deng L. (1998) "Speech trajectory discrimination using the minimum classification error learning," *IEEE Trans. Speech & Audio Processing*, 6, 505–515.

[Rathi 01] Rathinavelu C. and Deng L. (2001) "A maximum a posteriori approach to speaker adaptation using the trended hidden Markov model," *IEEE Trans. Speech & Audio Processing*, 9(5), 549–557.

[Ratt 98] Rattay F., Gebeshuber I. and Gitter A. (1998) "The mammalian auditory hair cell: A simple electric circuit model," *J. Acoust. Soc. Am.,* 103, 1558–1565.

[Reca 83] Recasens D. (1983) "Place cues for nasal consonants with special reference to Catalan," *J. Acoust. Soc. Am.,* 73, 1346–1353.

[Reca 97] Recasens D. (1997) "A model of lingual coarticulation based on articulatory constraints," *J. Acoust. Soc. Am.*, 102, 544–561.

[Red 84] Reddy R. (1976) "Speech recognition by machine: A review," *Proceedings of the IEEE*, 64, 501-531.

[Red 84] Reddy N. S. and Swamy M. (1984) "High-resolution formant extraction from linear prediction phase spectra," *IEEE Trans. Speech & Audio Processing*, 32, 1136–1144.

[Repp 79] Repp B. (1979) "Relative amplitude of aspiration noise as a voicing cue for syllable-initial stop consonants," *Language and Speech*, 27, 173–189.

[Repp 83] Repp B. (1983) "Categorical perception: Issues, methods, findings," in **Speech and Language: Advances in Basic Research and Practice** 10, (N. Lass, ed.), Academic Press, New York.

[Rhod 85] Rhode W. S. (1985) "The Use of intracellular techniques in the study of the cochlear nucleus," *J. Acoust. Soc. Am.,* 78(1), 320–327.

[Rhod 86] Rhode W. S. and Smith P. H. (1986) "Physiological studies on neurons in the dorsal cochlear nucleus of cat," *Journal of Neurophysiology*, 56(2), 287–307.

[Rhod 92] Rhode W. S., Greenberg S., Popper A. N. and Fay R. R. (1992) "Physiology of the cochlear nuclei," in **The Mammalian Auditory Pathway: Neurophysiology**, Spinger-Verlag, New York, 94–152.

[Rich 99] Richards H.B. and Bridle J.S. (1999) "The HDM: A segmental hidden dynamic model of coarticulation," *IEEE Proceedings of the ICASSP*, 1, 357–360.

[Rob 99] Robert A. and Eriksson J. (1999) "A composite model of the auditory periphery for simulating responses to complex sounds," *J. Acoust. Soc. Am.,* 106, 1852–1864.

[Rose 96] Rose R., Schroeter J., and Sondhi M. (1996) "The potential role of speech production models in automatic speech recognition," *J. Acoust. Soc. Am.,* 99(3), 1699–1709.

[Rosen 71] Rosenberg A., Schafer R. and Rabiner L. (1971) "Effects of smoothing and quantizing the parameters of formant-coded voiced speech," *J. Acoust. Soc. Am.,* 50, 1532–1538.

[Rosen 96] Rosenfeld R. (1999) "A maximum entropy approach to adaptive statistical language modeling." *Computer Speech and Language*, 10, 187–228.

[Ross 99] Ross K. and Ostendorf M. (1999) "A dynamical system model for generating fundamental frequency for speech synthesis", *IEEE Trans. Speech & Audio Processing*, 7(3), 295–309.

[Rub 81] Rubin P., Baer T. and Mermelstein P. (1981) "An articulatory synthesizer for perceptual research," *J. Acoust. Soc. Am.,* 70, 321–328.

[Rub 96] Rubin P., Saltzman E., Goldstein L., McGowan R., Tiede M, and Browman C. (1996) "CASY and extensions to the task-dynamic model," *Proc. 4th European Speech Production Workshop*, 125–128.

[Rus 93] Russell M. (1993) "A segmental HMM for speech pattern matching," *IEEE Proceedings of the ICASSP*, 1, 499-502.

[Rust 94] Rustagi J. S. (1994) **Optimization Techniques in Statistics**, Academic Press, Boston.

[Sac 79] Sachs M. B. and Young E. D. (1979) "Encoding of steady-state vowels in the auditory nerve: Representation in terms of discharge rate," *J. Acoust. Soc. Am.*, 66(2), 470–479.

[Sac 86] Sachs M. B., Young E. D., Winslow R. L., Shofner W. P., Moore B. C. J. and Patterson R. D. (1986) "Some aspects of rate coding in the auditory nerve," in **Auditory Frequency Selectivity**, Plenum, New York, 121–128.

[Sac 88] Sachs M. B., Blackburn C. and Young E. D. (1988) "Rate-Place and temporal-Place representations of vowels in the auditory nerve and anteroventral cochlear nucleus," *Journal of Phonetics*, 16, 37–53.

[Sag 86] Sagey E. (1986) "The representation of features and relations in nonlinear phonology," PH.D. dissertation, M.I.T., Cambridge, MA.

[Sait 68] Saito S. and Itakura F. (1968) "The theoretical consideration of statistically optimal methods for speech spectral density," *Report of Electrical Communication Lab.*, NTT (Tokyo).

[Sait 85] Saito S. and Nakata K. (1985) **Fundamentals of Speech Signal Processing** Academic Press (Tokyo).

[Salt 89] Saltzman E. and Munhall K. (1989) "A dynamical approach to gestural patterning in speech production," *Ecological Psychology*, 1, 333– 382.

[Sam 98] Sameti H., Sheikhzadeh H., Deng L., and Brennan R. (1998) "HMM-based strategies for enhancement of speech embedded in nonstationary noise," *IEEE Trans. Speech & Audio Processing*, 6(5), 445–455.

[Sam 02] Sameti H. and Deng L. (2002) "Nonstationary-state HMM representation of speech signals for speech enhancement," *Signal Processing*, 82, 205-227.

[Sav 95] Savariaux C., Perrier P. and Orliaguet J-P. (1995) "Compensation strategies for the perturbation of the rounded vowel [u] using a lip tube," *J. Acoust. Soc. Am.*, 98, 2528–2442.

[Sche 84] Scheffers M. (1984) "Discrimination of fundamental frequency of synthesized vowel sounds in a noise background," *J. Acoust. Soc. Am.*, 76, 428–434.

[Scho 80] Schouten M. (1980) "The case against a speech mode of perception," *Acta Psychologica*, 44, 71–98.

[Scho 92] Schouten M. and van Hessen A. (1992) "Modeling phoneme perception. I: Categorical perception," *J. Acoust. Soc. Am.*, 92, 1841–1855.

[Schr 73] Schroeder M. R. and Hall J. L. (1973) "A model for mechanical-to-Neural transduction in the auditory receptor," *J. Acoust. Soc. Am.*, 54, 263.

[Schr 75] Schroeder M. (1975) "Models of hearing," *Proceedings of the IEEE*, 63, 1332–1350.

[Schr 94] Schroeter J. and Sondhi M. (1994) "Techniques for estimating vocal-tract shapes from the speech signal," *IEEE Trans. Speech & Audio Processing*, 2(1), 133–150.

[Sear 79] Searle C. L., Jacobson J. and Rayment S. G. (1979) "Stop consonant discrimination based on human audition," *J. Acoust. Soc. Am.*, 65, 799–809.

[Seck 90] Secker-Walker H. E. and Searle C. L. (1990) "Time-Domain analysis of auditory-nerve-fiber firing rates," *J. Acoust. Soc. Am.*, 88(3), 1427–1436.

[Seg 01] Segura J., Torre A., Benitez M., and Peinado A. (2001) "Model-based compensation of the additive noise for continuous speech recognition: Experiments using the AURORA2 database and tasks," *Proceedings of the Eurospeech*, 1, 221–224.

[Sham 85a] Shamma S. A. (1985) "Speech processing in the auditory system I: The representation of speech sounds in the responses of the auditory nerve," *J. Acoust. Soc. Am.*, 78(5), 1612–1621.

[Sham 85b] Shamma S. A. (1985) "Speech processing in the auditory system II: Lateral inhibition and the central processing of speech evoked activity in the auditory nerve," *J. Acoust. Soc. Am.*, 78(5), 1622–1632.

[Sham 88] Shamma S. (1988) "The acoustic features of speech sounds in a model of auditory processing: Vowels and voiceless fricatives," *Journal of Phonetics* 16, 77–91.

[Sham 89] Shamma S. (1989) "Spatial and temporal processing in central auditory networks," in **Methods in Neuronal Modeling**, C. Koch and I. Segev, MIT Press, Cambridge, 247–283.

[Sharf 81] Sharf D. and Ohde R. (1981) "Physiologic, acoustic, & perceptual aspects of coarticulation: Implications for the remediation of articulatory disorders," *Sound and Language* 5, 153–245.

[She 94a] Sheikhazed H. and Deng L. (1994) "Waveform-based speech recognition using hidden filter models: Parameter selection and sensitivity to power normalization," *IEEE Trans. Speech & Audio Processing*, 2(1), 80–91.

[She 94b] Sheikhzadeh H., Sameti H., Deng L. and Brennan R. L. (1994) "Comparative performance of spectral subtraction and HMM-based speech enhancement strategies with application to hearing aid design," *IEEE Proceedings of the ICASSP*, 1, 13–16.

[She 94c] Sheikhzadeh H. (1994) **Temporal Signal Processing Techniques with Applications to Automatic Speech Recognition,** Ph.D. thesis, University of Waterloo, Canada.

[She 95] Sheikhzadeh H. and Deng L. (1995) "Interval statistics generated from a cochlear model in response to speech sounds," *J. Acoust. Soc. Am.*, 95(5, pt. 2), 2842.

[She 98] Sheikhzadeh H. and Deng L. (1998) "Speech analysis and recognition using interval statistics generated from a composite auditory model," *IEEE Trans. Speech & Audio Processing*, 6(1), 50–54.

[She 99] Sheikhzadeh H. and Deng L. (1999) "A layered neural network interfaced with a cochlear model for the study of speech encoding in the auditory system," *Computer Speech and Language*, 13, 39–64.

[She 02] Sheikhzadeh H., Cornu E., Brennan R., and Schneider T. (2002) "Real-time speech synthesis on an ultra-low resource, programmable DSP system," *IEEE Proceedings of the ICASSP*, 1, 433–436.

[Shir 76] Shirai K. and Honda M. (1976) "Estimation of articulatory motion," in **Dynamic Aspects of Speech Production**, University of Tokyo Press, Tokyo, 279–304.

[Shor 83] Shore J. E. and Burton D. K. (1983) "Discrete utterance speech recognition without time alignment," *IEEE Trans. Information Theory*, 29(4), 473–491.

[Shou80] Shoup J. (1980) "Phonological aspects of speech recognition," in *Trends in Speech Recognition* (W. Lea, ed.), Prentice-Hall, NJ, 125–138.

[Shum 91] Shumway R. and Stoffer D. (1991) "Dynamic linear models with switching," *J. American Statistical Assoc.*, 86, 763–769.

[Sine 83] Sinex D. and Geisler D. (1983) "Responses of auditory-nerve fibers to consonant-vowel syllables," *J. Acoust. Soc. Am.*, 73, 602–615.

[Siou 96] Siouris G. (1996) **An Engineering Approach to Optimal Control and Estimation Theory**, Wiley & Sons, New York.

[Siv 96] Sivia D. S. (1996) **Data Analysis – A Bayesian Tutorial**, Clarendon Press, Oxford.

[Slan 92] Slaney M. and Lyon R. (1992) "On the importance of time — A temporal representation of sound," in **Visual Representations of Speech Signals**, (M. Cook et al, eds.), Wiley, Chichester, UK.

[Smit 75] Smith R. L. and Zwislocki J. J. (1975) "Short-term adaptation and incremental responses of single auditory-nerve fibers," *Biological Cybernetics*, 17, 169–182.

[Smit 82] Smith R. L. and Brachman M. L. (1982) "Adaptation in auditory-nerve fibers: A revised model," *Biological Cybernetics*, 44, 107–120.

[Smits 96] Smits R., Bosch L. and Collier R. (1996) "Evaluation of various sets of acoustic cues for the perception of prevocalic stop consonants," *J. Acoust. Soc. Am.*, 100, 3852–3881.

[Som 97] Sommers M. and Kewley-Port D. (1997) "Modeling formant frequency discrimination of female vowels," *J. Acoust. Soc. Am.*, 99, 3770–3781.

[Sond 68] Sondhi M. (1968) "New methods of pitch extraction," *IEEE Trans. Audio and Electroac.* AU-16, 262–266.

[Sond 83] Sondhi M. and Resnick J. (1983) "The inverse problem for the vocal tract: Numerical methods, acoustical experiments & speech synthesis," *J. Acoust. Soc. Am.*, 73, 985–1002.

[Sond 86] Sondhi M. (1986) "Resonances of a bent vocal tract," *J. Acoust. Soc. Am.*, 79, 1113–1116.

[Spir 91] Spirou G. A. and Young E. D. (1991) "Organization of dorsal cochlear nucleus Type IV unit response and their relationship to activation by bandlimited noise," *Journal of Neurophysiology*, 66(5), 1750–1768.

[Spra 72] Spragins J. (1972) "A note on the iterative application of Bayes's rule," *IEEE Trans. Information Theory*, 3(1), 1–8.

[Spro 98] Sproat R. (1998) **Multi-Lingual Text-To-Speech Synthesis: The Bell Labs Approach**, Kluwer Academic, Boston.

[Srin 78] Srinath M. (1978) **An Introduction to Statistical Signal Processing with Applications**, John Wiley & Sons, New York.

[Stev 71] Stevens K. (1971) "Airflow and turbulence noise for fricative and stop consonants: Static considerations," *J. Acoust. Soc. Am.*, 50, 1180–1192.

[Stev 81] Stevens K. and Blumstein S. (1981) "The search for invariant acoustic correlates of phonetic features," in **Perspectives on the Study of Speech**, (P. Eimas and J. Miller, eds.), 1–38.

[Stev 89] Stevens K. (1989) "On the quantal nature of speech," *Journal of Phonetics*, 17, 1989, 3–45.

[Stev 92] Stevens K., et al. (1992) "Implementation of a model for lexical access based on features," *Proceedings of the ICSLP*, 1, 499–502.

[Stev 93] Stevens K. (1993) Course notes, "Speech Synthesis with a Formant Synthesizer," MIT.

[Stev 97] Stevens K. (1997) "Articulatory-acoustic-auditory relations," in **The Handbook of Phonetic Sciences**, (W. J. Handcastle and J. Lavar eds.), Blackwell, Cambridge, U.K., 462–506.

[Stev 98] Stevens K. (1998) **Acoustic Phonetics**, The MIT Press, Cambridge, MA.

[Stew 81] Stewart L. C. (1981) "Trellis data compression ," Technical Report, Stanford University.

[Stre 83] Streeter L., Macdonald N., Apple W., Krauss R. and Galotti K. (1983) "Acoustic and perceptual indicators of emotional stress," *J. Acoust. Soc. Am.*, 73, 1354–1360.

[Stud 70] Studdert-Kennedy M., Liberman A., Harris K. and Cooper F. (1970) "The motor theory of speech perception: A reply to Lane's critical review," *Psych. Rev.* 77, 234–249.

[Styl 01] Stylianou Y. (2001) "Removing linear phase mismatches in concatenative speech synthesis", *IEEE Trans. Speech & Audio Processing*, 9(3), 232–239.

[Summ 77] Summerfield Q. and Haggard M. (1977) "On the dissociation of spectral and temporal cues to the voicing distinction in initial stop consonants," *J. Acoust. Soc. Am.*, 62, 435–448.

[Sun 00] Sun J., Deng L., and Jing X. (2000) "Data-driven model construction for continuous speech recognition using overlapping articulatory features," *Proceedings of the ICSLP*, 1, 437–440.

[Sun 01] Sun J. and Deng L. (2001) "An overlapping-feature based phonological model incorporating linguistic constraints: Applications to speech recognition," *J. Acoust. Soc. Am.*, 111(2), 1086-1101.

[Sus 91] Sussman H., McCaffrey H., and Matthews S. (1991) "An investigation of locus equations as a source of relational invariance for stop place categorization," *J. Acoust. Soc. Am.*, 90, 1309–1325.

[Sus 97] Sussman H., Bessell N., Dalston E. and Majors T. (1997) "An investigation of stop place of articulation as a function of syllable position: A locus equation perspective," *J. Acoust. Soc. Am.*, 101, 2826–2838.

[Syka 81] Syka J., Aitkin L. and Evans E. F. (1981) "The Dynamic range problem: Place and time coding at the level of cochlear nerve and nucleus," in **Neural Mechanisms of Hearing**, Plenum, New York, 69–85.

[Syrd 98] Syrdal A., Stylianou Y., Conkie A. and Schroeter J. (1998) "TD-PSOLA versus harmonic plus noise model in diphone based speech synthesis," *IEEE Proceedings of the ICASSP*, 1, 273–276.

[Tani 96] Tanizaki H. (1996) **Nonlinear Filters — Estimation and Applications**, 2nd Ed., Springer, Berlin.

[Tart 94] Tartter V. and Brown D. (1994) "Hearing smiles and frowns in normal and whisper registers," *J. Acoust. Soc. Am.*, 96, 2101–2107.

[Tish 91] Tishby N. Z. (1991) "On the application of mixture AR hidden Markov models to text independent speaker recognition," *IEEE Trans. Acoustics, Speech & Signal Processing*, ASSP-39(3), 563–570.

[vB 94] van Bergem D. (1994) "A model of coarticulatory effects on the schwa," *Speech Communication*, 14, 143–162.

[Vier 80] Viergever M. A. (1980) "Mechanics of the inner ear - A mathematical approach," Delft University Press, Delft, Netherlands.

[Vii 01] O. Viikki (Ed.) (2001) *Speech Communication* (Special issue on Noise Robust ASR), Vol. 34.

[Voi 88] Voigt H. F. and Young E. D. (1988) "Evidence of inhibitory interactions between neurons in dorsal cochlear nucleus," *Journal of Neurophysiology*, 44, 76–96.

[Voi 90] Voigt H. F. and Young E. D. (1990) "Cross-Correlation analysis of inhibitory interactions in dorsal cochlear nucleus," *Journal of Neurophysiology*, 64(5), 1590–1610.

[Wang 94] Wang K. and Shamma S. (1994) "Self-normalization and noise-robustness in early auditory representations," *IEEE Trans. Speech & Audio Processing*, 2(3), 412–435.

[Watk 96] Watkins A. and Makin S. (1996) "Effects of spectral contrast on perceptual compensation for spectral-envelope distortion," *J. Acoust. Soc. Am.*, 99, 3749–3757.

[Way 94] Wayland S., Miller J. and Volaitis L. (1994) "Influence of sentential speaking rate on the internal structure of phoentic categories," *J. Acoust. Soc. Am.*, 95, 2694–2701.

[Wei 85] Weiss T. F. and Leong R. (1985) "A model for signal transmission in an ear having hair cells with free-standing cilia. IV. Mechanoelectrical transduction stage,"*Hearing Research*, 20, 175–195.

[West 88] Westerman L. A. and Smith R. L. (1988) "A diffusion model of the transient response of the cochlear inner hair cell synapse," *J. Acoust. Soc. Am.*, 83(6), 2266–2276.

[Wilh 86] Wilhelms R., Meyer P. and Strube H. (1986) "Estimation of articulatory trajectories by Kalman filtering," in **Signal Processing III: Theories and Applications**, (I. Young et. al. eds.), 477–480.

[Will 72] Williams C. and Stevens K. (1972) "Emotions and speech: Some acoustical correlates," *J. Acoust. Soc. Am.*, 52, 1238–1250.

[Wils 84] Wilson R., Arcos J. and Jones H. (1984) "Word recognition with segmented-alternated CVC words: A preliminary report on listeners with normal hearing," *Journal of Speech & Hearing Research*, 27, 378–386.

[Wins 86] Winslow R. L., Parta P. E., Sachs M. B., Yost W. A. and Watson C. S. (1986) "Rate coding in the auditory nerve," in **Auditory Processing of Complex Sounds**, Lawrence Erlbaum Associates, Hillsdale, NJ, 212–224.

[Wint 90] Winter I. M. and Palmer A. R. (1990) "Temporal responses of primary-like anteroventral cochlear nucleus units to the steady-state vowel /i/," *J. Acoust. Soc. Am.*, 88(3), 1437–1441.

[Wood 79] Wood S. (1979) "A radiographic analysis of constriction locations for vowels," *Journal of Phonetics*, 7, 25–43.

[Wout 01] Wouters J. and Macon M. (2001) "Control of spectral dynamics in concatenative speech synthesis", *IEEE Trans. Speech & Audio Processing*, 9(3), 30–38.

[Wu 83] Wu C.F.J. (1983) "On the convergence properties of the EM algorithm," *Annals of Statistics*, 11, 95–103.

[Xu 94] Xu Y. (1994) "Production and perception of coarticulated tones," *J. Acoust. Soc. Am.*, 95, 2240–2253.

[Young 76] Young E. D. and Brownell W. E. (1976) "Response to tones andoise of single cells in dorsal cochlear nucleus of unanesthetized cats," *Journal of Neurophysiology*, 39, 282–300.

[Young 79] Young E. D. and Sachs M. D. (1979) "Representation of steady-state vowels in the temporal aspects of the discharge patterns of populations of auditory-nerve fibers," *J. Acoust. Soc. Am.*, 66, 1381–1403.

[Young 81] Young E. D. and Voigt H. F. (1981) "The internal organization of the dorsal cochlear nucleus," **Neuronal Mechanisms of Hearing** (J. Syka and L. Aitkin eds.), Plenum Press, New York, 127–133.

[Young 82] Young E. D. and Voigt H. D. (1982) "Response properties of Type II and Type III units in dorsal cochlear nucleus,"*Hearing Research*, 6, 153–169.

[Young 93] Young E. D. and Nelken I. (1993) "Nonlinearity of spectra processing in the dorsal cochlear nucleus (DCN)," *J. Acoust. Soc. Am.*, 94(Pt.2), 2294.

[Young 94] Young S., Odell J. and Woodland P. (1994) "Tree-based state tying for high accuracy acoustic modeling," *Proc. ARPA Human Language Technology Workshop*, 307–312.

[Young 95] Young S. (1995) "Large vocabulary continuous speech recognition: A review," *Proc. IEEE Workshop on Automatic Speech Recognition*, 3– 28.

[Young 98] Young S., Odell J., Ollason D., Valtchev V., and Woodland P. (1998) **The HTK Book**, Cambridge University.

[Zem 81] Zemlin W. R. (1981) **Speech and Hearing Science: Anatomy and Physiology**, Prentice-Hall, New Jersey.

[Zha 01] Zhang X., Heins M., Bruce I., and Carney L. (2001) "A phenomenological model for the responses of auditory-nerve fibers: I. Nonlinear tuning with compression and suppression," *J. Acoust. Soc. Am.*, 109, 648–670.

[Zue 76] Zue V. (1976) **Acoustic Characteristics of Stop Consonants: A Controlled Study,** Ph.D. thesis, MIT.

[Zue 85] Zue V. W. (1985) "The use of speech knowledge in automatic speech recognition," *Proceedings of the IEEE*, 73, 1616–1624.

[Zue 91] Zue V. W. (1991) "Notes on Speech Spectrogram Reading," MIT, Cambridge MA.

[Zwi 74] Zwicker E. and Terhardt E. (1974) **Facts and Models in Hearing**, Springer-Verlag, Berlin.

[Zwi 79] Zwicker E., Terhardt E. and Poulus E. (1979) "Automatic speech recognition using psychoacoustic models," *J. Acoust. Soc. Am.*, 65, 487–498.

[Zwi 84] Zwicker E. (1984) "Dependence of post-masking on masker duration and its relation to temporal effects in loudness," *J. Acoust. Soc. Am.*, 75, 219–223.

Index